TABLE III
Values of t_α

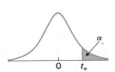

df	$t_{0.10}$	$t_{0.05}$	$t_{0.025}$	$t_{0.01}$	$t_{0.005}$	df
1	3.078	6.314	12.706	31.821	63.657	1
2	1.886	2.920	4.303	6.965	9.925	2
3	1.638	2.353	3.182	4.541	5.841	3
4	1.533	2.132	2.776	3.747	4.604	4
5	1.476	2.015	2.571	3.365	4.032	5
6	1.440	1.943	2.447	3.143	3.707	6
7	1.415	1.895	2.365	2.998	3.499	7
8	1.397	1.860	2.306	2.896	3.355	8
9	1.383	1.833	2.262	2.821	3.250	9
10	1.372	1.812	2.228	2.764	3.169	10
11	1.363	1.796	2.201	2.718	3.106	11
12	1.356	1.782	2.179	2.681	3.055	12
13	1.350	1.771	2.160	2.650	3.012	13
14	1.345	1.761	2.145	2.624	2.977	14
15	1.341	1.753	2.131	2.602	2.947	15
16	1.337	1.746	2.120	2.583	2.921	16
17	1.333	1.740	2.110	2.567	2.898	17
18	1.330	1.734	2.101	2.552	2.878	18
19	1.328	1.729	2.093	2.539	2.861	19
20	1.325	1.725	2.086	2.528	2.845	20
21	1.323	1.721	2.080	2.518	2.831	21
22	1.321	1.717	2.074	2.508	2.819	22
23	1.319	1.714	2.069	2.500	2.807	23
24	1.318	1.711	2.064	2.492	2.797	24
25	1.316	1.708	2.060	2.485	2.787	25
26	1.315	1.706	2.056	2.479	2.779	26
27	1.314	1.703	2.052	2.473	2.771	27
28	1.313	1.701	2.048	2.467	2.763	28
29	1.311	1.699	2.045	2.462	2.756	29
∞	1.282	1.645	1.960	2.326	2.576	∞

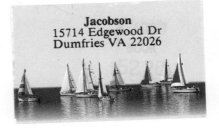

INTRODUCTORY STATISTICS

INTRODUCTORY STATISTICS

THIRD EDITION

Neil A. Weiss
Arizona State University

Matthew J. Hassett
Arizona State University

 ADDISON-WESLEY PUBLISHING COMPANY

Reading, Massachusetts . Menlo Park, California . New York
Don Mills, Ontario . Wokingham, England . Amsterdam . Bonn
Sydney . Singapore . Tokyo . Madrid . San Juan

On the cover: Among its countless uses, statistics is applied regularly in manufacturing and production. For example, a company that produces snack foods uses a packaging machine to package 454-gram bags of pretzels. Although the actual net weights deviate somewhat from 454 grams and vary from one bag to another, it is nevertheless important to the company that the mean net weight of the bags be kept at 454 grams. A sample of 50 bags of pretzels is found to have a mean net weight of 451.2 grams with a standard deviation of 7.9 grams. Can we conclude that the packaging machine is not working properly?

Sponsoring Editor: Charles B. Glaser
Production Supervisor: Marion E. Howe
Electronic Production Supervisor: Mona Zeftel
Technical Art Consultant: Dick Morton
Compositor: Carol and Neil Weiss
Illustrator: Scientific Illustrators
Designer: Catherine L. Johnson
Cover Designer: Marshall Henrichs
Manufacturing Manager: Roy Logan

Library of Congress Cataloging-in-Publication Data

Weiss, N. A. (Neil A.)
    Introductory statistics / Neil A. Weiss, Matthew J. Hassett. —
3rd ed.
        p.   cm.
      ISBN 0–201–17833–8
      1. Statistics.   I. Hassett, Matthew J.   II. Title.
QA276.12.W45   1991
519.5—dc20   89-17663
                                                         CIP

2 3 4 5 6 7 8 9 10 HA 9594939291

PREFACE

Statistics is not only an indispensible tool, used by government, business, and virtually every academic discipline, but it is also a subject with which some familiarity is essential in order to comprehend the world around us. The purpose of this book is to provide the reader with a clear understanding of basic statistical techniques and to present well-organized procedures for applying those techniques. Introductory high school algebra is a sufficient prerequisite.

We have designed the text so that the book can be used for a one-quarter, one-semester, two-quarter, two-semester, or three-quarter course. The amount of time that is to be devoted to the book can be varied by both choice of topics and depth of coverage.

FEATURES

The following features have been included in the text to assist the reader in learning introductory statistics:

Emphasis on application. We have concentrated on the application of statistical techniques to the analysis of data. Although the statistical theory has been kept to a minimum, we have made every effort to give a thorough development of the rationale for using each statistical procedure.

Detailed and careful explanations. We have attempted to include every step of explanation that a typical reader might need. Our guiding principle is to avoid "cognitive jumps" and thereby make the learning process smooth and enjoyable. We feel that detailed and careful explanations will result in a better understanding.

Data sets. In most examples and exercises, we have provided the raw data sets instead of just the summary statistics. There are two advantages to this. First, it gives the reader a more concrete and real picture of statistics (supplying only the summary statistics places the reader in a position of being once removed from the statistical analysis). Second, having the raw data sets provides the opportunity to solve the problems using a computer, if so desired. We should point out here that a floppy disk containing the data sets for the odd-numbered basic exercises has been prepared and is available from Addison-Wesley Publishing Company.

Procedure boxes. To help the reader with the application of statistical procedures, we have presented easy-to-follow, step-by-step methods for carrying out those procedures. For easy access, each procedure is displayed with a color background. A unique feature of this book is that when a procedure is illustrated by an example, each step in the procedure is presented explicitly within the example. This serves a two-fold purpose: It shows the reader how the procedure is applied and helps the reader master the steps in the procedure.

Procedure index. With the vast number of statistical procedures, it is sometimes difficult to locate a specific procedure, especially when the book is being used for reference purposes. Thus, we have included a *procedure index,* which can be found on the back inside cover of the book. This provides a quick and easy way to find the required procedure for performing any particular statistical analysis.

Computer usage. Nowadays, virtually all professional applications of statistics are done by computer. It is therefore important, we feel, that every student of statistics have some familiarity with statistical computer packages. We have chosen Minitab® to illustrate the basic ideas of statistical computer packages. At the end of most sections, we have included optional material that explains how Minitab can be used to solve problems that were solved by hand earlier in the section. Each solution consists of introducing the required commands, displaying the computer output, and interpreting the results.

Exercises dealing with the computer material are explicitly identified. For students who do not actually have access to Minitab, we have provided problems of two types; those that ask which Minitab command and subcommands (if any) should be used to accomplish a specific task and those that ask for an interpretation of a Minitab printout. Additionally, for those students who do have access to Minitab, we have prepared problems that ask for a particular statistical analysis to be carried out using Minitab.

The instructor has the following three options regarding the coverage of computer usage: (1) no coverage, (2) moderate integration using the optional material in the text, or (3) complete integration using the *Minitab Supplement.* For those instructors selecting the second option, we should note that Section 9.4, on P-values, is essential for practically all subsequent computer sections.

Chapter introductions and chapter outlines. At the beginning of each chapter, we have presented a description of the chapter and an explanation of how the chapter relates to the text as a whole. As a further aid, a chapter outline follows the chapter introduction. The chapter outline lists the sections in the chapter and provides a short summary of the content of each section.

Definitions, formulas, key facts. As an aid to learning and for reference, we have set off all definitions, formulas, and key facts. These items are printed in color to make them easy to locate.

Minitab is a registered trademark of Minitab, Inc., 3081 Enterprise Drive, State College, PA 16801. Phone: (814) 238-3280. Fax: (814) 238-4383. Telex: 881612. We would like to thank Minitab, Inc., for their assistance.

Real examples. Since we believe that the majority of students learn by example, every concept discussed in the book is illustrated by at least one detailed example. The examples are, for the most part, based on real-life situations and have been chosen for their interest as well as for their illustrative value.

Extensive and diverse exercise sets. We have presented both extensive and diverse exercise sets. Most of the exercises are based on information found in newspapers, magazines, statistical abstracts, and journal articles. The exercises are designed not only to help the reader learn the material but also to show that statistics is a lively and relevant discipline.

Since students in introductory statistics courses often have different mathematical backgrounds, we have included three levels of exercises: basic, intermediate, and advanced. Each exercise set contains several basic exercises. The *basic exercises* provide routine applications of material presented in the text and every reader should master these. We have organized the basic exercises so that each concept is covered by at least two problems. For each odd-numbered basic exercise that involves a particular concept, there is also an even-numbered basic exercise that involves that same concept. An icon consisting of a single horizontal rule (—) preceding an exercise number identifies that exercise as a basic exercise. The answers to the odd-numbered basic exercises are given in Appendix B and the answers to the even-numbered basic exercises can be found in the *Instructor's Manual.*

In addition to the basic exercises, most of the exercise sets include intermediate and advanced problems. The *intermediate exercises* contain supplementary material that is not necessarily covered in the text, but that may be of interest to some of the more highly motivated students. An icon consisting of two horizontal rules (=) preceding an exercise number identifies that exercise as an intermediate exercise.

The *advanced exercises* cover abstract concepts, theory, and algebraic derivations. These exercises are intended for the students with special mathematical background and aptitude. An icon consisting of three horizontal rules (≡) preceding an exercise number identifies that exercise as an advanced exercise. The solutions to all the intermediate and advanced exercises can be found in the *Instructor's Manual.*

Chapter reviews. Frequently, students in introductory statistics courses feel a certain amount of anxiety and confusion about how to study and review. To help the student, we have written a chapter-review section at the end of every chapter. The chapter reviews include: (1) a list of key terms with page references, (2) formulas, (3) chapter objectives, and (4) a review test. These pedagogical aids provide the student with an organized method for reviewing and studying. The answers to the review tests are presented in Appendix B.

ORGANIZATION

We have attempted to write a text that offers a great deal of flexibility in the choice of material to be covered. Chapters 1–3 introduce the nature of statistics and the fundamentals of descriptive statistics. In Chapters 4–6, we present probability, discrete random variables, and the normal distribution. Chapter 7 discusses sampling and the sampling distribution of the mean. Following that, in Chapters 8 and 9, we examine confidence intervals and hypothesis tests for one mean or proportion. We consider Chapters 1–9 the core of an introductory statistics course.

Chapter 10 presents inferences for two means or proportions and is recommended strongly for inclusion in the course. In Chapter 11, we discuss chi-square procedures (goodness-of-fit test, independence test, and inferences for a population standard deviation).

We have divided the traditional material on regression and correlation into two chapters. Chapter 12 examines descriptive methods in regression and correlation and can be covered at any time after Chapter 3. Chapter 13 presents inferential methods in regression and correlation and can be covered once Chapters 9 and 12 have been completed.

Analysis of variance (ANOVA) is discussed in Chapter 14. Topics covered include one-way analysis of variance for a completely randomized design and two-way analysis of variance for a randomized block design. Chapter 15 introduces nonparametric statistics and presents some of the most widely used nonparametric procedures. Finally, in Chapter 16, we examine the key aspects of planning and conducting a complete statistical study. Although this chapter refers to some of the material discussed in Chapter 14, it can, with some minor modifications, be covered after Chapter 9.

The flowchart at the top of the next page summarizes the preceding discussion and shows the interdependence among chapters. In the flowchart, the prerequisites for a given chapter consist of all chapters that have a path leading to that chapter. Thus, the flowchart provides an aid to the instructor in selecting the topics that he or she wishes to cover.

CHANGES IN THE THIRD EDITION

We have made several significant changes in the third edition of *Introductory Statistics*. However, we would like to emphasize that the basic pedagogical philosophy has been retained. All chapters have been rewritten for the purpose of updating and expanding the material, fine-tuning the organization, and adding new sections where appropriate. Here are some of the changes to be found in the third edition:

- There has been a substantial increase in the number of exercises: Approximately 1500 exercises (not counting parts) now appear in the book, about 400 exercises more than in the second edition.
- Chapter 1 (The Nature of Statistics) now includes an optional section, Section 1.4, that explains how to input, print, and name data sets when using Minitab. This prepares the student to work with the optional computer material in the remainder of the book.
- The computer sections, which were located at the end of chapters in the second edition, have been integrated into each chapter to occur immediately following the particular statistical concept under consideration. In addition, computer exercises have now been incorporated into the exercise sets. All computer material is optional.

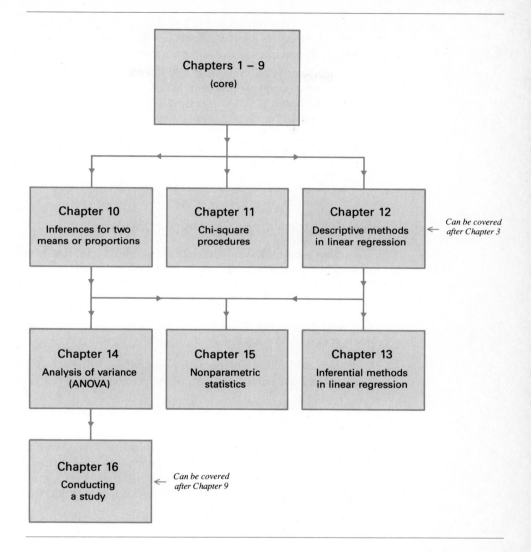

- z-scores are now introduced in Chapter 3 (Descriptive Measures), with further treatment in Chapter 6 (The Normal Distribution).
- Chapter 4 (Probability Concepts) now contains an optional section on Bayes' rule and an optional section on counting rules.
- The method for determining the z-values corresponding to a specified area under the standard normal curve is now presented in Chapter 6, instead of waiting until Chapter 8.
- A new optional section on P-values has been included in Chapter 9 (Hypothesis Tests for One Mean or Proportion). This offers further integration of the computer into the course, if so desired.

- Chapter 10 (Inferences for Two Means or Proportions) now discusses confidence intervals for the difference between two means or proportions in the text, instead of only in the exercises.
- The material on regression and correlation has been split into two chapters, one for descriptive methods, Chapter 12, and the other for inferential methods, Chapter 13. This makes it possible to cover descriptive methods in regression and correlation once Chapter 3 has been completed.
- Optional sections on multiple regression have been added to both Chapters 12 and 13.
- Chapter 14 (Analysis of Variance) now includes a section on two-way analysis of variance for a randomized block design.

SUPPLEMENTS AND OTHER SUPPORT

The following supplements have been prepared specifically to accompany the third edition of *Introductory Statistics:*

Minitab Supplement. This supplement, by Professor Peter W. Zehna, provides in-depth coverage of Minitab. It is designed to be used in conjunction with *Introductory Statistics* and is keyed to the book. No prerequisite knowledge of computers or statistical software is presumed.

Instructor's Manual. This manual, prepared by Professor Bernard J. Morzuch, contains worked out solutions to all of the exercises in *Introductory Statistics.*

Student's Solutions Manual. This manual, also by Professor Morzuch, presents detailed solutions to every fourth exercise in *Introductory Statistics.*

OmniTest Computerized Testing. This unique testing software offers a virtually endless supply of quizzes, tests, and final examinations. Features include multiple-exam versions, customized question editing, on-screen preview and edit functions, and pull-down menus. *OmniTest* is available for use with IBM or compatibles.

Printed Test Bank. This supplement provides several printed examinations for each chapter of *Introductory Statistics.*

DataDisk. This is a floppy disk containing files for the data sets appearing in the odd-numbered basic exercises in *Introductory Statistics.* DataDisk makes it possible to store those data sets in Minitab without having to enter them manually.

Additional support is available with the following items:

The Student Edition of Minitab. This version of Minitab, distributed by Addison-Wesley, is for use with IBM or compatibles and includes three floppy disks, a quick-reference card, and *The Student Edition of Minitab* manual.

Adventures in Statistics. This software is a unique, interactive, tutorial package produced and distributed by Quant Systems. The program focuses on three approaches: Technical Skill Building, Concept Development, and Thinking Statistically. All three help develop important skills and demonstrate the relevance of statistics. *Adventures in Statistics* is available for use with IBM or compatibles. For a demonstration package, call Quant Systems at (803) 571-2825 or consult your local Addison-Wesley sales representative.

Against All Odds. These videotapes, produced by the Consortium for Mathematics and its Applications (COMAP), consist of 26 half-hour programs on statistics. For information, phone Intelemation at 1-800-LEARNER.

ACKNOWLEDGMENTS

It is our pleasure to thank the following reviewers whose comments and suggestions were invaluable in writing the third edition of *Introductory Statistics:*

George Anderson
Central Piedmont Community College

Gwen Applebaugh
University of Wisconsin, Eau Claire

Gary Beus
Brigham Young University

William Beyer
University of Akron

Jerry Bloomberg
Essex Community College

Toni Carroll
Siena Heights College

Constance Cutchins
University of Central Florida

Elizabeth Eltinge
Texas A&M University

Eugene Enneking
Portland State University

Larry Griffey
Florida Community College, Jacksonville

Linda Malone
University of Central Florida

Jacinta Mann
Seton Hill College

Joseph McWilliams
Stephen F. Austin State University

Bernard J. Morzuch
University of Massachusetts, Amherst

Charles B. Peters
University of Houston

James Schott
University of Central Florida

Rana P. Singh
Virginia State University

Joseph Walker
Georgia State University

We also wish to thank Professor Michael Driscoll, Professor Dennis Young, Dr. Mikel Aicken, Dr. Terry Woodfield, Lt. Colonel Jon Epperson, Lt. Colonel Ray Mitchell, and Major Sam Thompson for their assistance. Our special thanks go to Professors Ronald Jacobowitz and Jerry Bloomberg for their many suggestions and comments. To Professor Larry Griffey, we would like to express our appreciation not only for his help with this book but also for his continuing aid on many projects.

We are grateful to Professor Peter Zehna for preparing the *Minitab Supplement* and to Professor Bernard Morzuch for preparing the *Instructor's Manual* and *Student's Solutions Manual.* Our appreciation also goes to Mr. Howard Blaut and Mr. Rick Hanna for providing data on real estate; to Mr. Jeffrey Jirele for supplying data on automobile insurance; and to Ms. Mary M. Neary for furnishing data on SAT scores and grade-point averages.

To Professor Toni Carroll and Mr. Gregory Weiss, we give our thanks for their scrupulous proofreading of the text. Ms. Carolyn Morrow did a marvelous job coordinating the flow of material. Our thanks go as well to Mr. Dick Morton for his expertise as technical art consultant. Ms. Mona Zeftel, our electronic production supervisor, ensured that the complex task of electronic production proceeded accurately and efficiently. We thank her for that and for her astuteness in handling several other parts of the production process.

As usual, it was a pleasure working with our production supervisor, Ms. Marion Howe. She not only took care of all her responsibilities with admirable precision, but was also a constant source of help for the seemingly countless problems that arise in the production of a book. To Mr. Charles Glaser, our editor, we extend our deep appreciation. He most certainly played a fundamental role—from pre-revision, through planning and development, through reviewing, through production. We had literally hundreds of conversations regarding all facets of the book. In short, whatever merits the book may have are in no small part due to him.

Our thanks to our designer, Ms. Catherine Johnson, for what we think is a superb design. We are also grateful to Professor Donald E. Knuth and his associates for developing the TEX typesetting system which has made it possible to produce technical manuscripts with incredible typographical precision. Finally, we would like to thank Ms. Carol Weiss for her help, support, and encouragement. It is fair to say that, apart from writing the book, she was involved in every single aspect of development and production that the authors were. Moreover, Carol took on the arduous task of typesetter. She did an outstanding job processing the manuscript using TEX.

Tempe, Arizona   *N.A.W.*
                                                                           M.J.H.

CONTENTS

CHAPTER 4
PROBABILITY CONCEPTS   141

CHAPTER 5
DISCRETE RANDOM VARIABLES   231

CHAPTER 6
THE NORMAL DISTRIBUTION   283

CHAPTER 1

THE NATURE OF STATISTICS

What does the word "statistics" bring to mind? Most people immediately think of numerical facts or data, such as unemployment figures, farm prices, or the number of marriages and divorces. *Webster's New World Dictionary* gives two definitions of the word "statistics":

> 1. facts or data of a numerical kind, assembled, classified, and tabulated so as to present significant information about a given subject. 2. [construed as sing.], the science of assembling, classifying, and tabulating such facts or data.

Technically, however, statistics means more than this. Not only do statisticians assemble, classify, and tabulate data, but they also analyze data in order to make generalizations and decisions. For example, a political analyst can use data from a portion of the voting population to predict the political preferences of the entire voting population. In this chapter, we introduce some basic terminology so that the various meanings of the word "statistics" will become clear.

CHAPTER OUTLINE

1.1 Two kinds of statistics   Discusses the two major types of statistics, descriptive statistics and inferential statistics.

1.2 Classification of statistical studies   Presents several examples that illustrate how to classify a statistical study as either descriptive or inferential.

1.3 The development of inferential statistics   Gives a brief sketch of the historical development of inferential statistics.

1.4 Using the computer (Optional)   Discusses the role of the computer in contemporary statistics and introduces the statistical computer package Minitab.

1.1   Two kinds of statistics

You probably feel that you already know something about statistics. If you read newspapers, watch the news on television, or follow sports, then you see and hear the word "statistics" frequently. In this section we will use familiar examples such as baseball statistics and voter polls to introduce the two major types of statistics, **descriptive statistics** and **inferential statistics.**

Each spring in the late 1940s, the major-league baseball season was officially opened when President Harry S. Truman threw out the "first ball" of the season at the opening game of the Washington Senators. Both President Truman and the Washington Senators had reason to be interested in statistics. Consider, for instance, the year 1948.

EXAMPLE 1.1   *Illustrates descriptive statistics*

In 1948 the Washington Senators played 153 games, winning 56 and losing 97. They finished seventh in the American League and were led in hitting by B. Stewart whose batting average was .279. These and many other statistics were compiled by baseball statisticians who took the complete records for each game of the season and organized that large mass of information effectively and efficiently.

Although baseball fans take these statistics for granted, a great deal of time and effort is required to gather and organize them. Moreover, without such statistics, baseball would be much harder to understand. For instance, picture yourself trying to select the best hitter in the American League with only the official score sheets for each game. [More than 600 games were played in 1948; the best hitter was Ted Williams who led the league with a batting average of .369.]   ■

The work of baseball statisticians provides us with a good illustration of descriptive statistics. Below we give a formal definition of descriptive statistics.

DEFINITION 1.1   Descriptive statistics

Descriptive statistics consists of methods for organizing and summarizing information in a clear and effective way.

Among other things, descriptive statistics includes the construction of graphs, charts, and tables, and the calculation of various descriptive measures, such as averages and percentiles. We will discuss descriptive statistics in detail in Chapters 2 and 3. Descriptive statistics is one of the two major types of statistics. The other major type, inferential statistics, is illustrated in the next example.

EXAMPLE 1.2   *Illustrates inferential statistics*

In the fall of 1948, President Truman was also quite concerned about statistics. The Gallup Poll just before the election predicted that he would win only 44.5% of the vote and be defeated by the Republican nominee, Thomas E. Dewey. This time, however, the statisticians had predicted incorrectly. Truman won more than 49% of the vote and with it the presidency. The Gallup Organization modified some of its procedures and has correctly predicted the winner since. ■

Political polling provides us with an example of inferential statistics. It would be tremendously expensive to interview all Americans on their voting preferences. Statisticians who wish to gauge the sentiment of the entire **population** of American voters can afford to interview only a carefully chosen group of a few thousand voters. This group is referred to as a **sample** of the population. Statisticians analyze the information obtained from the sample to make inferences (draw conclusions) about the preferences of the entire voting population. Inferential statistics provides methods for making such inferences.

The terminology introduced above in the context of political polling is used in general in statistics. Specifically, we have the following definitions:

DEFINITION 1.2   Population and sample

Population: The collection of all individuals, items, or data under consideration in a statistical study.
Sample: That part of the population from which information is collected.

Figure 1.1 provides a graphical display of the relationship between a population and a sample from the population.

FIGURE 1.1
Population and sample

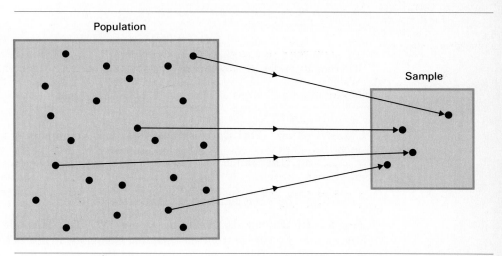

With Definition 1.2 in mind, we now present the following definition of inferential statistics:

DEFINITION 1.3   Inferential statistics

Inferential statistics consists of methods of drawing conclusions about a population based on information obtained from a sample of the population.

Exercises 1.1

__ 1.1 Define the following terms:
a) population   b) sample

__ 1.2   What are the two major types of statistics? Describe them in detail.

1.2 Classification of statistical studies

In later chapters we will conduct a thorough examination of both descriptive and inferential statistics. At this point, however, you need only be able to classify statistical studies as either descriptive or inferential. The examples below are intended to give you some practice. In each example, we have presented the result of a statistical study and have classified the study as descriptive or inferential. You should attempt to classify each study yourself before reading our explanation.

EXAMPLE 1.3   *Illustrates the classification of statistical studies*

The study   Table 1.1 gives the voting results for the 1948 presidential election.

TABLE 1.1
Final results
of the 1948
presidential election

Ticket	Votes	Percent
Truman-Barkley (Democrat)	24,179,345	49.7
Dewey-Warren (Republican)	21,991,291	45.2
Thurmond-Wright (States Rights)	1,176,125	2.4
Wallace-Taylor (Progressive)	1,157,326	2.4
Thomas-Smith (Socialist)	139,572	0.3

Classification   This study is descriptive. It is a summary of the votes cast by the entire population of voters. No inferences were made.   ■

EXAMPLE 1.4   *Illustrates the classification of statistical studies*

The study   For the 101 years preceding 1977, the baseballs used by the major leagues were purchased from the Spalding Company. In 1977 that company stopped manufacturing major-league baseballs, and the major leagues arranged to buy their baseballs from the Rawlings Company.

Early in the 1977 season, pitchers began to complain that the Rawlings ball was "livelier" than previous balls. They claimed it was harder, bounced farther and faster, and gave an unfair advantage to hitters. There was some evidence for this. For instance, in the first 616 games of 1977, there were 1033 home runs, compared to only 762 home runs in the first 616 games of the previous year.

Sports Illustrated magazine sponsored a careful study of the liveliness question, and the results appeared in the June 13, 1977 issue. In this study, an independent testing company randomly selected 85 baseballs from the current (1977) supplies of various major league teams. The bounce, weight, and hardness of the baseballs chosen were carefully measured. These measurements were then compared with measurements obtained previously from similar tests on baseballs used in the years 1952, 1953, 1961, 1963, 1970, and 1973. The conclusion, given on page 24 of the *Sports Illustrated* article, was as follows: "...the 1977 Rawlings ball is livelier than the 1976 Spalding, but not as lively as it could be under big league rules, or as the ball has been in the past."

Classification   This is an inferential study. The independent testing company used a sample of 85 baseballs from the 1977 supplies of major league teams to make an inference about the population of all such baseballs. [It has been estimated that approximately 360,000 baseballs were used by the major leagues in 1977.]   ■

The *Sports Illustrated* study also provides an excellent illustration of a situation in which it is not feasible to obtain data for the entire population. Indeed, after the bounce and hardness tests, all of the baseballs sampled were taken to a butcher in Plainfield, New Jersey, to be sliced in half so that researchers could look inside them. Clearly, it would not have been practical to test every baseball in this way.

EXAMPLE 1.5   *Illustrates the classification of statistical studies*

The study   In the late 1940s and early 1950s there was great public concern over epidemics of polio. In an attempt to alleviate this serious problem, Jonas Salk of the University of Pittsburgh developed a vaccine for polio. Various preliminary experiments indicated that the vaccine was safe and potentially effective. Nonetheless, it was deemed necessary to conduct a large-scale study to determine whether the vaccine would truly work.

A test was devised involving a sample of nearly two million grade-school children. All of the children were inoculated, but only half received the Salk vaccine. The other half were given a placebo, a harmless solution. An evaluation center kept records of who received the Salk vaccine and who did not. The center found that the incidence of polio was far less among the children inoculated with the Salk vaccine. From that information it was concluded that the Salk vaccine would be effective in preventing polio for all American schoolchildren. The Salk vaccine was then made available for general use.

Classification   This is an inferential study. The group of children inoculated may have been quite large, but it was still just a sample; and the information obtained from that sample was used to make an inference about the effectiveness of the Salk vaccine among the population of all American schoolchildren.   ■

Finally, we should emphasize that it is possible to perform a descriptive study on a sample, as well as on a population. It is only when an inference is made about the population from information obtained from the sample that the study becomes inferential.

Exercises 1.2

In Exercises 1.3–1.10, classify each of the studies as either descriptive or inferential.

___ 1.3   The A. C. Nielsen Company collects and publishes information on the television-viewing habits of Americans. Data collected in 1986 from a sample of Americans gave the following estimates of average TV viewing time per week for all Americans. The times are given in hours and minutes. [SOURCE: Nielsen Media Research, *Nielsen Report on Television*.]

Group (by age)		Time
Average all persons		*30:20*
Women	Total 18+	34:56
	18–24	26:05
	55+	43:58
Men	Total 18+	29:21
	18–24	20:20
	55+	39:14
Teens	Female	20:33
	Male	22:38
Children	2–5	28:06
	6–11	23:31

___ 1.4   In 1936, the voters of North Carolina cast their presidential votes as follows:

Candidate	Number of votes
Roosevelt, Democratic	616,414
Landon, Republican	223,283
Thomas, Socialist	21
Browder, Communist	11
Lemke, Union	2

___ 1.5   The U.S. National Center for Health Statistics published the following rate estimates in *Vital Statistics of the United States* for the leading causes of death in 1987. The estimates are based on a 10% sampling of all 1987 United States death certificates. Rates are per 100,000 population.

Cause	Rate
Major cardiovascular diseases	397.0
Malignant neoplasms (cancers)	196.1
Accidents	39.0
Chronic obstructive pulmonary diseases	32.2
Influenza and pneumonia	28.8

___ 1.6   The U.S. National Institute on Drug Abuse published the following table on drug use, by type of drug, in *National Household Survey on Drug Abuse*. The percentages given are estimates obtained from national samples.

Type of drug	Percent of young adults			
	Ever used		*Current user*	
	1974	1985	1974	1985
Marijuana	52.7	60.5	25.2	21.9
Inhalants	9.2	12.8	(z)	1.0
Hallucinogens	16.6	11.5	2.5	1.6
Cocaine	12.7	25.2	3.1	7.7
Heroin	4.5	1.2	(z)	(z)
Analgesics	(NA)	11.4	(NA)	2.1
Stimulants[1]	17.0	17.3	3.7	4.0
Sedatives[1]	15.0	11.0	1.6	1.7
Tranquilizers[1]	10.0	12.2	1.2	1.7
Alcohol	81.6	92.8	69.3	71.5
Cigarettes	68.8	76.0	48.8	37.2

NA = Not available.   z = Less than .5 percent.
1 = Prescription drugs.

___ 1.7   The following table displays the 1986 attendance figures for selected spectator sports. Data are given in thousands, rounded to the nearest thousand.

Sport	Attendance (thousands)
Baseball, major leagues	48,452
Basketball	
college	31,645
pro	11,491
Football	
college	36,388
pro	19,684
Horse racing	70,580
Greyhound racing	25,759

Year	High price ($)	Low price ($)
1900	47,500	37,500
1920	115,000	85,000
1929	625,000	550,000
1935	140,000	65,000
1940	60,000	33,000
1960	162,000	135,000
1970	320,000	130,000
1987	1,150,000	575,000

1.8 Newspapers publish weather data for cities all over the world. Below are low and high temperatures and sky-condition readings for some selected cities in Mexico and Canada on March 29, 1989.

MEXICO	**Lo**	**Hi**	**Sky**
Acapulco	64	86	Pcy
Guadalajara	44	82	Pcy
Mazatlan	57	84	F
Merida	68	100	Cdy
Mexico City	46	82	F
Monterrey	57	93	F
Veracruz	73	87	F

CANADA	**Lo**	**Hi**	**Sky**
Calgary	24	41	Clr
Edmonton	11	44	Clr
Halifax	31	53	Pcy
Montreal	32	39	Clr
Ottawa	31	41	Pcy
Regina	18	24	Sno
St. John's	16	43	Sno
Toronto	31	38	Clr
Vancouver	37	53	Clr
Winnipeg	26	30	Sno

1.9 The New York Stock Exchange keeps records of the selling prices for seats on the Exchange. The following table provides the high and low prices for some years in this century:

1.10 A 1984 study by the Gallup Organization concluded that an estimated 83% of American households are involved in at least one form of indoor or outdoor gardening. [SOURCE: National Gardening Association, *National Gardening Survey*.]

1.11 For each situation below, decide whether the indicated study would be descriptive or inferential. Give a reason for each of your answers.
a) A tire manufacturer wants to estimate the average life of a new type of steel-belted radial.
b) A sports writer plans to list the winning times for all swimming events in the 1992 Olympics.
c) A politician obtains the exact number of votes that were cast for her opponent in 1990.
d) A medical researcher tests an anticancer drug that may have harmful side effects.
e) A candidate for governor estimates the percentage of voters that will vote for him in the upcoming gubernatorial election.

1.12 The chairperson of a mathematics department at a large state university wants to estimate the average final exam score for the 2476 students in basic algebra. She randomly selects 50 exams from the 2476 and finds that the average score on the 50 exams chosen is 78.3%. From this she estimates that the average score for all 2476 students is about 78.3%.
a) What kind of study has the chairperson done?
b) What kind of study would she have done had she averaged all 2476 exam scores?

1.3 The development of inferential statistics

As we have discovered, the science of statistics includes both descriptive statistics and inferential statistics. Descriptive statistics appeared first. As a matter of fact, censuses were taken as long ago as Roman times. Over the years, records of such things as births, deaths, and taxes have led naturally to the development of descriptive statistics.

Inferential statistics is a newer arrival. Major developments began to occur with the research of Karl Pearson (1857–1936) and Ronald Fisher (1890–1962), who published their results in the early years of this century. Since the work of Pearson and Fisher, inferential statistics has evolved rapidly and is now applied in a wide range of subject areas. In fact, an understanding of the fundamental concepts of inferential statistics has become mandatory for virtually every professional.

Familiarity with statistics will also help you make more sense out of many things you read in newspapers and magazines. For instance, in the description of the *Sports Illustrated* baseball test in Example 1.4, it may have struck you as unreasonable that a sample of only 85 baseballs could be used to draw a conclusion about a population of roughly 360,000 baseballs. By the time you have completed Chapter 9, you will understand why such inferences are not unreasonable.

The primary objective of this text is to present the fundamentals of inferential statistics. However, since almost any inferential study involves aspects of descriptive statistics, we will first investigate descriptive statistics.

1.4 Using the computer (Optional)

Computers are ideal for performing statistical calculations and analyses. Rarely is it necessary to write your own computer programs, since programs already exist for almost all aspects of statistics. The most commonly used programs for statistical work are taken from **computer packages.** Computer packages are collections of statistical computer programs written by some organization or individual. Nowadays, computer packages are available for use on mainframes, minicomputers, and microcomputers.

There are many high-quality computer packages on the market. We have chosen Minitab[†] to illustrate the basic ideas of computer packages. At the end of most sections, we will discuss briefly how Minitab can be used to solve problems that were solved by hand earlier in the section. Each solution will consist of introducing the appropriate commands, displaying the computer output, and interpreting the results. For a more detailed treatment of Minitab, we refer the reader to the *Minitab Supplement* by Peter W. Zehna (Reading, Massachusetts: Addison-Wesley Publishing Company, 1991).

To use Minitab, we must first gain access to it on the computer. The method for doing that depends on the type of computer system being employed. Consequently, the precise method for accessing Minitab should be obtained from the instructor or an appropriate member of the staff at the computer center. Once Minitab has been accessed, the prompt MTB > is displayed on the screen. This indicates that a Minitab command may now be typed.

[†] Minitab is a registered trademark of Minitab, Inc.

In this section, we will explain how to input data sets into the computer, how to print data sets, and how to name data sets. We commence with a discussion of how to input data sets.

SET AND READ

There are two Minitab commands that can be used to input data. The **SET** command is employed to input one data set at a time and the **READ** command is employed to simultaneously input two or more data sets of the same size. We first consider the SET command.

EXAMPLE 1.6   *Illustrates the SET command*

Table 1.2 gives the number of days to maturity for 40 short-term investments. The data are from *Barron's National Business and Financial Weekly.*

TABLE 1.2
Days to maturity for
40 short-term
investments

70	64	99	55	64	89	87	65
62	38	67	70	60	69	78	39
75	56	71	51	99	68	95	86
57	53	47	50	55	81	80	98
51	36	63	66	85	79	83	70

Input this data set into the computer.

SOLUTION   To begin, we need to understand that, with Minitab, a data set is stored in a column. A column is designated by a "C" followed by a number. For example, C1 stands for Column 1, C2 stands for Column 2, and so forth.

To store the days-to-maturity data from Table 1.2, we first type the command SET followed by the column in which the data are to be stored. So, to store the data in, say, C1, we first type SET C1. Note that we have underlined the command SET C1. This means that the command is to be typed by the person using the computer.

Next we press the RETURN key to signify that, for the moment, we are done typing. On some computers that key is replaced by an ENTER key, on others by an ENDLINE key, and on still others by a different key. After the RETURN key is pressed, the prompt DATA> is displayed on the screen to indicate that the data may now be entered.

Now we type the numbers in the data set, separating the numbers by spaces. Any number of spaces will do as long as the numbers are separated. The days-to-maturity data in Table 1.2 will not all fit on one line and so they must be entered on several lines, say, five lines of eight numbers each. (It is not necessary for each line to contain the same number of pieces of data.) Remember that the RETURN key must be pressed after each line!

When we have finished entering the data, we type the command END (and press the RETURN key). The entire procedure that we have just discussed is depicted in Printout 1.1 at the top of the following page.

PRINTOUT 1.1
Storing the
days-to-maturity
data using the
SET command

```
MTB > SET C1
DATA> 70 64 99 55 64 89 87 65
DATA> 62 38 67 70 60 69 78 39
DATA> 75 56 71 51 99 68 95 86
DATA> 57 53 47 50 55 81 80 98
DATA> 51 36 63 66 85 79 83 70
DATA> END
```

The days-to-maturity data from Table 1.2 are now stored in the column C1. Those data will be available for analysis until we type **SET C1** again or sign off the computer. Finally, we should emphasize that although the days-to-maturity data have been entered using five lines of eight numbers each, all 40 numbers are stored in C1 in one long column. ∎

Note: If an error is made while entering a data set, there are several ways that the error can be corrected. One way is to go immediately to the END command, retype the SET command, and start all over again. The old data will be erased and the new data inserted in its place. Other (and more efficient) ways for correcting an error are discussed in the *Minitab Supplement*.

As we know, the SET command is used to input one data set at a time. To input two or more data sets simultaneously, we can use the READ command, provided that the data sets are of the same size. The next example shows how the READ command is applied.

EXAMPLE 1.7   *Illustrates the READ command*

Table 1.3 displays data on age and price for a sample of 11 Nissan Zs. The data were obtained from the January 4, 1990, *Asian Import* edition of the *Auto Trader* magazine. Note that the age data are in years and the price data are in hundreds of dollars, rounded to the nearest hundred dollars.

TABLE 1.3
Age and price
data for Nissan Zs

Car	Age (yrs) x	Price ($100s) y
1	5	85
2	4	103
3	6	70
4	5	82
5	5	89
6	5	98
7	6	66
8	6	95
9	2	169
10	7	70
11	7	48

Input both the age data and the price data into the computer simultaneously.

SOLUTION To simultaneously store the age and price data from Table 1.3, we first type the command READ followed by the columns in which the data are to be stored. So, to store the data in, say, C2 and C3, we first type READ C2 C3 (and press the RETURN key). Then we enter the data row by row. That is, we enter 5 85 and press RETURN, enter 4 103 and press RETURN, and so on. See Printout 1.2.

PRINTOUT 1.2
Storing the age
and price data
simultaneously using
the READ command

```
MTB > READ C2 C3
DATA> 5 85
DATA> 4 103
DATA> 6 70
DATA> 5 82
DATA> 5 89
DATA> 5 98
DATA> 6 66
DATA> 6 95
DATA> 2 169
DATA> 7 70
DATA> 7 48
DATA> END
     11 ROWS READ
```

The age data from the second column of Table 1.3 are now stored in C2 and the price data from the third column of Table 1.3 are now stored in C3. ∎

PRINT

To display data on the screen, we use the **PRINT** command followed by the column or columns to be printed. Example 1.8 illustrates the use of the PRINT command.

EXAMPLE 1.8 *Illustrates the PRINT command*

Employ the PRINT command to
a) print the days-to-maturity data found in Table 1.2 on page 9.
b) simultaneously print both the age data and the price data for Nissan Zs, given in Table 1.3 on page 10.

SOLUTION a) Recall that the days-to-maturity data are stored in C1. Thus, to print that data, we simply type PRINT C1. This command and its results are displayed in Printout 1.3 at the top of page 12.
b) Since the age and price data for Nissan Zs are stored in C2 and C3, respectively, we simultaneously print those data sets by typing PRINT C2 C3. Printout 1.4 on page 12 shows that command and its results. ∎

PRINTOUT 1.3
Printing the
days-to-maturity
data using the
PRINT command

```
MTB > PRINT C1
C1
    70   64   99   55   64   89   87   65   62   38   67   70   60
    69   78   39   75   56   71   51   99   68   95   86   57   53
    47   50   55   81   80   98   51   36   63   66   85   79   83
    70
```

PRINTOUT 1.4
Printing the age
and price data for
Nissan Zs using the
PRINT command

```
MTB > PRINT C2 C3
 ROW    C2      C3

  1      5      85
  2      4     103
  3      6      70
  4      5      82
  5      5      89
  6      5      98
  7      6      66
  8      6      95
  9      2     169
 10      7      70
 11      7      48
```

NAME

It is often useful to give a column a name. This allows us to reference the column by its name instead of trying to remember the number of the column. Once a column has been named, any output involving that column is labeled with the name.

A column name can consist of between one and eight characters. Any characters can be used to form the name with the following exceptions: A name cannot begin or end with a blank and neither an apostrophe (') nor a number sign (#) can be used in a name. To name a column, we use the **NAME** command followed first by the column to be named and then by the name.

To illustrate, recall that we stored data on the number of days to maturity for 40 short-term investments in C1. We might want to name that column "Maturity." To do so, we type **NAME C1 'MATURITY'**. Note that the name is enclosed by *apostrophes*. We emphasize that apostrophes must be used to enclose the name. Otherwise, the command will not work and an error message will result. The apostrophes must be used both when naming and referencing the column.

After a column has been named, we can use either its number or its name in any succeeding command. Thus, for example, we can print the days-to-maturity data by typing either **PRINT C1** or **PRINT 'MATURITY'**. As we noted, however, once a column has been named, the output will always use the name, regardless of whether the column number or column name is used in a command. For instance, whether we type **PRINT C1** or **PRINT 'MATURITY'** the output will be labeled **MATURITY**.

To change the name of a column, we simply use the NAME command again. For example, to change the name of C1 from "Maturity" to "Days," we would type `NAME C1 'DAYS'`.

Finally, we can name more than one column at a time. For instance, consider again the age and price data for Nissan Zs given in Table 1.3 on page 10. We stored those data sets in C2 and C3, respectively. To name C2 "Age" and C3 "Price," we type `NAME C2 'AGE' C3 'PRICE'`.

Exercises 1.4

___ 1.13 (**Computer exercise**) The U.S. Energy Information Administration collects data on residential energy consumption and expenditures. Results are published in the document *Residential Energy Consumption Survey: Consumption and Expenditures.* The following table gives last year's energy consumptions for a sample of 50 households in the South. Data are in millions of BTU.

130	55	45	64	155	66	60	80	102	62
58	101	75	111	151	139	81	55	66	90
97	77	51	67	125	50	136	55	83	91
54	86	100	78	93	113	111	104	96	113
96	87	129	109	69	94	99	97	83	97

Explain how Minitab can be used to
a) store the data set in C1.
b) name C1 "Energy."
c) print the data set.
If you have access to Minitab, use it to carry out parts (a)–(c).

___ 1.14 (**Computer exercise**) The Bureau of Economic Analysis gathers information on the length of stay in Europe and the Mediterranean by U.S. travelers. Data are published in *Survey of Current Business.* A sample of 36 U.S. residents who traveled to Europe and the Mediterranean this year yielded the following data, in days, on length of stay:

41	16	6	21	1	21
5	31	20	27	17	10
3	32	2	48	8	12
21	44	1	56	5	12
3	13	15	10	18	3
1	11	14	12	64	10

Explain how Minitab can be used to

a) store the data set in C2.
b) name C2 "Stays."
c) print the data set.
If you have access to Minitab, use it to carry out parts (a)–(c).

___ 1.15 (**Computer exercise**) The U.S. National Center for Health Statistics publishes data on heights and weights by age and sex in *Vital and Health Statistics.* A random sample of 11 males, aged 18–24 years, yielded the following data:

Height (inches) x	Weight (lb) y
65	175
67	133
71	185
71	163
66	126
75	198
67	153
70	163
71	159
69	151
69	155

Explain how Minitab can be used to
a) store the height data and weight data simultaneously in C3 and C4, respectively.
b) name C3 "Height" and C4 "Weight" at once.
c) print both data sets at the same time.
If you have access to Minitab, use it to carry out parts (a)–(c).

___ 1.16 (**Computer exercise**) Hanna Properties specializes in custom-home resales in the Equestrian Estates, an exclusive subdivision in Phoenix, Arizona.

A random sample of nine custom homes, currently listed for sale, provided the information on size and price shown at the right. The size data are in hundreds of square feet, rounded to the nearest hundred square feet. The price data are in thousands of dollars, rounded to the nearest thousand dollars. Explain how Minitab can be used to

a) store the size data and price data simultaneously in C5 and C6, respectively.
b) simultaneously name C5 "Size" and C6 "Price."
c) print both data sets at the same time.

If you have access to Minitab, use it to carry out parts (a)–(c).

Size (100 sq ft) x	Price ($1000s) y
26	235
27	249
33	267
29	269
29	295
34	345
30	415
40	475
22	195

Chapter review

KEY TERMS

computer packages,* 8
descriptive statistics, 2
END,* 9
inferential statistics, 4
NAME,* 12

population, 3
PRINT,* 11
READ,* 9
sample, 3
SET,* 9

YOU SHOULD
BE ABLE TO

1. classify statistical studies as either descriptive or inferential.
2. identify the population and the sample in an inferential study.
3. use the Minitab commands covered in this chapter.*

REVIEW TEST

1. Give an example of
   a) a descriptive study.
   b) an inferential study.

2. At the end of Section 1.3, we stated that almost any inferential study involves aspects of descriptive statistics. Why do you think that is true?

In Problems 3–7 classify each of the studies as either descriptive or inferential.

3. A metropolitan newspaper displayed the following college football scores on the front page of the Sunday paper:

COLLEGE FOOTBALL

Michigan St.	12	Arizona	12
ASU	3	Washington St.	7
BYU	31	Rutgers	28
Washington	3	Florida	28
Michigan	20	UCLA	26
Notre Dame	12	Tennessee	26

4. A National Institute of Mental Health survey concluded that "about 20% of adult Americans suffer

---

* An asterisk indicates material that is optionally covered.

from at least one psychiatric disorder." This and other estimates were obtained from results of interviews with thousands of Americans in St. Louis, Baltimore, and New Haven, Connecticut.

5. On February 14, 1985, the Bureau of the Census released a survey indicating that "fewer Americans have health insurance coverage than previously thought." The survey, which was based on a sample of 20,000 households, concluded that about 85% of the population is covered by health insurance—a far cry from the 97.3% figure found in a 1978 survey by the Department of Health and Human Services.

6. A newspaper reporter who was doing library research for an article on civil aviation obtained the following data on fatalities. [SOURCE: U.S. Federal Aviation Administration.]

Year	Fatalities
1979	1731
1980	1392
1981	1414
1982	1480
1983	1135
1984	1218
1985	1267
1986	1029

7. The Bureau of Justice Statistics (BJS) conducts monthly surveys of about 60,000 households in the United States. These monthly surveys are then used to determine annual estimates on criminal victimization. For example, the BJS reported the following victimization rate estimates for crimes against households between 1979 and 1985 in the publication *Criminal Victimization in the United States*. Rates are per 1000 households.

Year	Burglary	Larceny	Motor vehicle theft
1979	84	134	18
1980	84	127	17
1981	88	121	17
1982	78	114	16
1983	70	105	15
1984	64	99	15
1985	63	97	14

*8. (**Computer problem**) A study was conducted by a research physician on the ages of persons with diabetes. The following data were obtained for the ages of a sample of 35 diabetics:

48	41	57	83	41	55	59
61	38	48	79	75	77	7
54	23	47	56	79	68	61
64	45	53	82	68	38	70
10	60	83	76	21	65	47

Explain how Minitab can be used to
a) store the data set in C1.
b) name C1 "Ages."
c) print the data set.
If you have access to Minitab, use it to carry out parts (a)–(c).

*9. (**Computer problem**) Colleges and universities often require prospective students to take college entrance examinations. The theory is that scores on those examinations are positively correlated with success in college. In the following table are the mathematics percentile scores on a college entrance examination and the freshman GPAs for a random sample of 10 students attending a southeastern university.

Math percentile score x	Freshman GPA y
85	3.23
77	2.95
98	3.70
86	3.85
66	2.37
55	2.75
75	2.73
64	2.16
71	2.59
82	2.78

Explain how Minitab can be used to
a) store the data on math percentile scores and the data on freshman GPAs simultaneously in C2 and C3, respectively.
b) simultaneously name C2 "Mathpsc" and C3 "Freshgpa."
c) print both data sets at the same time.
If you have access to Minitab, use it to carry out parts (a)–(c).

CHAPTER 2

ORGANIZING DATA

As we discovered in Chapter 1, *descriptive statistics* consists of methods for organizing and summarizing information in a clear and effective way. In this chapter, we will begin our study of descriptive statistics. Specifically, we will learn how to classify data by type, organize data into tables, and summarize data with graphical displays. We will also see that care must be taken in interpreting graphical displays since they can often be misleading.

CHAPTER OUTLINE

2.1 Data   Introduces the different types of data that are found in statistics.

2.2 Grouping data   Describes several procedures for grouping data into classes and constructing tables.

2.3 Graphs and charts   Presents some useful ways to portray data graphically.

2.4 Stem-and-leaf diagrams   Examines a recently invented descriptive procedure that combines both grouping and graphing techniques.

2.5 Misleading graphs   Discusses several ways in which graphs and charts can be misleading and emphasizes the importance of reading and constructing graphs carefully.

2.1 Data

The information collected, organized, and analyzed by statisticians is called **data.** There are several different types of data, and a statistician's choice of methodology is partly determined by the type of data being considered. In this section we will discuss some of the more important types of data.

EXAMPLE 2.1   *Introduces three major types of data*

At noon on April 17, 1989, more than 5200 men and women set out to run from Hopkinton Center to the John Hancock Building in Boston. Their run, covering 26 miles and 385 yards, would be watched by thousands of people lining metropolitan Boston streets and by millions more on television news reports. It was the ninety-third running of the Boston Marathon.

A great deal of information was accumulated that afternoon and was recorded by the Boston Athletic Association. The men's competition was won by Abebe Mekonnen of Ethiopia with a time of two hours, nine minutes, and six seconds. The winner of the women's competition was Ingrid Kristiansen of Norway; her time was two hours, 24 minutes, and 33 seconds. There were 4239 men and 864 women who finished before the official cutoff time of four hours and 30 minutes. [After four hours and 30 minutes no official times are recorded, although hundreds of runners are usually still on the course.]

The Boston Marathon provides us with examples of three major types of data. The simplest type is illustrated by the information that classifies each entrant as either male or female. Such data, which give qualitative information about an individual or item, are called **qualitative data.** For instance, the information that Ingrid Kristiansen is a female is qualitative data. *(Nominal)*

Most racing fans are interested in the places of the finishers. Information on place is an example of **ordinal data,** data about order or rank. The information that Abebe Mekonnen and Juma Ikangaa finished first and second among the men while Mark Horwitz and Ken Travis finished 500th and 501st is ordinal data.

Ordinal data give information about place but do not measure the difference in performance between places. For instance, Abebe Mekonnen finished 50 seconds ahead of Juma Ikangaa, but Mark Horwitz beat Ken Travis by only one second. More can be learned about what happened in a race by looking at the times of the finishers. Differences between times indicate exactly how far apart two runners finished, whereas differences in places do not. Information on time is an example of **metric data,** data obtained from measurement. The information that Abebe Mekonnen ran his race in 2:09:06 is metric data. *(Note:* The term *metric* refers to *measurement* and not to the metric system. Metric data can be given in units other than those in the metric system, e.g., inches.*)* ■

We now summarize the definitions of the three types of data that were introduced in Example 2.1.

Nominal (handwritten)

DEFINITION 2.1 Qualitative, ordinal, and metric data

Qualitative data: Data that give <u>non-numerical</u> information such as gender, eye color, and blood type.

Ordinal data: Data about <u>order or rank on a scale</u> such as 1, 2, 3, ... or A, B, C, ...

Ratio (handwritten)

Metric data: Data obtained from the <u>measurement of quantities such as time</u>, <u>height, and weight.</u> *Has a "zero".* (handwritten)

Interval (handwritten)

Another important type of data is **frequency data** (or **count data**). Counting the number of individuals or items that fall into categories such as "male" and "female" yields frequency data. For instance, the information that 4239 men and 864 women finished the Boston Marathon in under four hours and 30 minutes is frequency data.

DEFINITION 2.2 Frequency data *(Does not have "absolute zero"* (handwritten)

Frequency data: Data on the <u>number</u> of individuals or items falling in various categories.

The next three examples provide additional illustrations of qualitative, ordinal, metric, and frequency data.

EXAMPLE 2.2 *Illustrates types of data*

Humans are classified as having one of the four blood types A, B, AB, and O.
a) What kind of data do you receive when you are told your blood type?
b) Geneticists and anthropologists record the number of individuals of each blood type. What kind of data are they collecting?

SOLUTION a) Your blood type is qualitative data. It places you in one of four non-numerical categories—A, B, AB, or O.
b) Recording the number of individuals in each of the four blood-type categories gives frequency data. ∎

EXAMPLE 2.3 *Illustrates types of data*

At most colleges and universities, students completing a course receive a grade of A, B, C, D, or F.
a) What type of data is the information that Carol Scott received a grade of A in Professor H's statistics class?
b) What type of data is provided by the information that the final grades in Professor H's statistics class were 17 As, 16 Bs, and 8 Cs, with no Ds or Fs?

SOLUTION a) The information that Carol Scott received a grade of A is ordinal data since the grades, A, B, C, D, and F, rank students' performance.

b) The information on the number of As, Bs, Cs, Ds, and Fs gives frequency data, obtained by counting the number of students falling into each of the five grade categories.   ∎

Note that Professor H's grade distribution, given in Example 2.3, is somewhat higher than the grades usually observed in statistics classes. This illustrates that not all ranking schemes are the same. Professor H's grade of A may have been easier to get than Professor W's. Thus, we see that ordinal data may have different meanings when the ranking is done by different people.

EXAMPLE 2.4   *Illustrates types of data*

The *Information Please Almanac* lists the world's highest waterfalls.
a) The list shows that Angel Falls in Venezuela is 3281 feet high, more than twice as high as Ribbon Falls at Yosemite, California, which is 1612 feet high. What kind of data are these heights?
b) From the list, we also find that of the world's 40 highest waterfalls, four are more than 1700 feet high, five are between 1000 and 1700 feet high, and 31 are less than 1000 feet high. What type of data do these counts provide?

SOLUTION   a) The waterfall heights are metric data, determined by taking measurements.
b) The counts of the number of waterfalls in the three height categories (more than 1700 ft, 1000–1700 ft, less than 1000 ft) are frequency data.   ∎

DISCRETE AND CONTINUOUS DATA

Qualitative and ordinal data are referred to as **discrete,** because they sort individuals or items into separate, or discrete, classes. For instance, track and field authorities classify entrants for competition as either male or female (qualitative data) and assign the first two finishers places 1 and 2 (ordinal data). There is no way to be classified between male and female or to be assigned place 1.357.
      On the other hand, most metric data are called **continuous,** since they involve measurement on a continuous scale. For example, the time that it takes a runner to complete a marathon can conceptually be *any* positive number.

CLASSIFICATION AND THE CHOICE OF A STATISTICAL METHOD

Statisticians make other distinctions in classifying data, but the types just presented are sufficient for most applications. Data classification is sometimes difficult— statisticians themselves will often disagree over data type. For example, some classify data involving amounts of money as metric data, while others classify it as frequency data. In most cases, however, the classification of data is fairly clear and serves as an aid in the choice of the correct statistical method.

Exercises 2.1

__ **2.1** Give a reason why the classification of data is important. *aids in choice of correct stats methods*

For each part of Exercises 2.2–2.10, classify the data as qualitative, ordinal, metric, or frequency.

__ **2.2** According to Sidney S. Culbert of the University of Washington, the principal languages of the world in 1988 were as follows:

Rank	Language	Speakers (millions)
1	Mandarin	825
2	English	431
3	Hindustani	325
4	Spanish	320
5	Russian	289
6	Arabic	187
7	Bengali	178

a) What type of data is given in the first column of the table?
b) What type of data is provided by the information that Neil Armstrong speaks English?
c) What type of data is provided by the information in the last column of the table?

__ **2.3** Tobacco production in the United States for the years 1982–1987 is displayed in the following table. [SOURCE: U.S. Department of Agriculture, Economic Research Service, *Agricultural Statistics*.]

Year	Pounds (millions)
1982	1994
1983	1428
1984	1728
1985	1512
1986	1164
1987	1226

metric

What type of data is given in the second column of this table?

__ **2.4** On May 4, 1961, Commander Malcolm Ross, USNR, ascended 113,739.9 feet in a free balloon. What kind of data is the height given here?

__ **2.5** The following table is reprinted by permission from the October issue of *Science 85*. Copyright © 1985 by the American Association for the Advancement of Science.

DO YOUR WORRIES MATCH THOSE OF THE EXPERTS?

Experts and lay people were asked to rank the risk of dying in any year from various activities and technologies. The experts' ranking closely matches known fatality statistics.

PUBLIC		EXPERTS
1	Nuclear power	20
2	Motor vehicles	1
3	Handguns	4
4	Smoking	2
5	Motorcycles	6
6	Alcoholic beverages	3
7	General (private) aviation	12
8	Police work	17
9	Pesticides	8
10	Surgery	5
11	Fire fighting	18
12	Large construction	13
13	Hunting	23
14	Spray cans	26
15	Mountain climbing	29
16	Bicycles	15
17	Commercial aviation	16
18	Electric power (nonnuclear)	9
19	Swimming	10
20	Contraceptives	11
21	Skiing	30
22	X rays	7
23	High school and college football	27
24	Railroads	19
25	Food preservatives	14
26	Food coloring	21
27	Power mowers	28
28	Prescription antibiotics	24
29	Home appliances	22
30	Vaccinations	25

ordinal

What type of data are provided by the numbers?

__ **2.6** The table at the top of the first column on the next page provides figures on U.S. industrial employment for 1988. [SOURCE: U.S. Bureau of Labor Statistics, *Employment and Earnings*.]

Industry	Employees (millions)
Agriculture	3.1
Mining	0.7
Construction	5.3
Manufacturing	19.5
Transportation	5.6
Trade	25.3
Services	25.4

What kind of data are given by the employee numbers?

__ **2.7**   The continental statistics in the following table provide approximate land areas and 1988 population estimates. [SOURCE: Population Reference Bureau, Inc.]

Continent	Land area (1000 sq mi)	Population (millions)
Asia	10,644	2,995
Africa	11,707	623
North America	9,360	416
South America	6,883	286
Europe	1,905	497
Oceana	3,284	26

a) What type of data is provided by the area figures?
b) What type of data is contained in the statement that "Africa is largest in area and second largest in population"?
c) What type of data is provided by the population figures?
d) What type of data do we obtain from the fact that Madame Curié was born in Europe?

__ **2.8**   The table at the top of the next column gives the rank by population and the population of the 10 largest cities in the U.S. in 1980 and 1986. [SOURCE: U.S. Bureau of the Census, *Census of Population*.]

a) What type of data is provided by the statement "in 1980 Houston was the fifth largest city in the United States"?
b) What type of data is given in the last column of the table?
c) What type of data is provided by the information that Theodore Roosevelt was born in New York?

City	1980 Rank	1980 Population	1986 Rank	1986 Population
New York	1	7,071,639	1	7,262,700
Los Angeles	3	2,966,850	2	3,259,340
Chicago	2	3,005,072	3	3,009,530
Houston	5	1,595,138	4	1,728,910
Philadelphia	4	1,688,210	5	1,642,900
Detroit	6	1,203,339	6	1,086,220
San Diego	8	875,538	7	1,015,190
Dallas	7	904,078	8	1,003,520
San Antonio	11	785,880	9	914,350
Phoenix	9	789,704	10	894,070

__ **2.9**   The five largest U.S. commercial banks (by deposits as of December 31, 1987) are listed below. [SOURCE: American Banker.]

Bank	Rank	Deposits ($millions)
Citibank NA, New York	1	104,970
Bank of America NT&SA San Francisco	2	70,221
Chase Manhattan Bank NA, New York	3	63,741
Morgan Guaranty Trust Co., New York	4	45,756
Manufacturers Hanover Trust Co., New York	5	45,667

a) What kind of data is given in the second column of the table?
b) What kind of data is given in the third column of the table?

__ **2.10**   What kinds of data would be collected in each of the following situations?
a) A quality-control engineer measures the lifetimes of electric light bulbs.
b) A businessperson wants to know the number of families with preteen children in Pueblo, CO.
c) A sporting-goods manufacturer classifies each major league baseball player as right-handed or left-handed and counts the number in each category.
d) A sociologist needs to estimate the average annual income of residents of Ossining, New York.

e) A pollster plans to classify each individual in a sample of voters as Democrat or Republican and count the total number in each group.

f) An administrator at a community college needs to know how many men and women participated in varsity sports during the spring semester and how much money was spent on men's sports and on women's sports.

___ 2.11 Of the three data types qualitative, metric, and frequency, only one involves non-numerical data. Which one is that?

___ 2.12 Of the data types qualitative, ordinal, metric, and frequency, only one involves continuous data. Which one is that?

___ 2.13 Below are several items pertaining to individuals.
a) height
b) weight
c) age
d) sex
e) number of siblings
f) religion
g) place of birth
h) high school class rank
Which involve continuous data?

2.2 Grouping data

The data collected in a real-world situation can sometimes be overwhelming. For example, a list of American colleges and universities with information on enrollment, number of teachers, highest degree offered, and governing official can be found in *The World Almanac*. These data occupy 28 pages of small type!

By suitably organizing data, it is often possible to make a rather complicated set of data easier to understand. In this section we will discuss **grouping,** one of the most common methods for organizing data.

EXAMPLE 2.5 **Illustrates grouping data**

Table 2.1 gives the number of days to maturity for 40 short-term investments. The data are from *Barron's National Business and Financial Weekly*.

TABLE 2.1
Days to maturity for 40 short-term investments

70	64	99	55	64	89	87	65
62	38	67	70	60	69	78	39
75	56	71	51	99	68	95	86
57	53	47	50	55	81	80	98
51	36	63	66	85	79	83	70

It is somewhat difficult to get a clear picture of the data in Table 2.1. By grouping the data into categories, or **classes,** we can make it much simpler to comprehend.

The first step is to decide on the classes. One convenient way to group these data is by tens. Since the smallest piece of data is 36 and the largest is 99, grouping by tens results in the classes 30–39, 40–49, and so on up to 90–99. These classes are given in the first column of Table 2.2 at the top of the following page.

The second (and final) step in grouping the data is to determine how many investments are in each class. We do this by going through the data in Table 2.1 and making a tally mark in the appropriate line of Table 2.2 for each investment. For instance, the first investment in Table 2.1 has a 70-day maturity period. This calls for a tally mark on the line for the class 70–79. The results of the tallying

TABLE 2.2

Days to maturity	Tally	Number of investments
30–39	|||	3
40–49	|	1
50–59	⅃Ⅱ |||	8
60–69	⅃Ⅱ ⅃Ⅱ	10
70–79	⅃Ⅱ ||	7
80–89	⅃Ⅱ ||	7
90–99	||||	4
		40

procedure are shown in the second column of Table 2.2. Now we count the tallies for each class and record the totals in the third column of Table 2.2.

By simply glancing at Table 2.2 we can obtain various pieces of useful information. For instance, we see that there are more investments in the 60–69 days-to-maturity range than in any other. Comparing Tables 2.1 and 2.2, we see that grouping the data makes it much easier to read and understand.   ■

In Example 2.5 we used a common-sense approach to grouping data into classes. Some of that common sense can be written down as guidelines for grouping. Three of the most important guidelines are the following:

1. *The number of classes should be between five and twenty.*

In Example 2.5, seven classes are used. Generally, the number of classes should be small enough to provide an effective summary, but large enough to display the relevant characteristics of the data.

2. *Each piece of data must belong to one, and only one, class.*

Careless planning in Example 2.5 could have led to classes like 30–40, 40–50, 50–60, and so on. Then, for instance, to which class would the investment with a 50-day maturity period belong? The classes in Table 2.2 do not cause such confusion, they cover all maturity periods and do not overlap.

3. *Whenever feasible, all classes should have the same width.*

The classes in Table 2.2 all have a width of 10 days. Among other things, choosing classes of equal width facilitates the graphical display of the data.

The list of guidelines could go on, but that would be artificial. The purpose of grouping is to organize the data into a reasonable number of classes in order to make it more accessible and understandable.

FREQUENCY AND RELATIVE-FREQUENCY DISTRIBUTIONS

The number of pieces of data that fall into a particular class is called the **frequency** of that class. For example, as we see from Table 2.2, the frequency of the class 50–59 is eight, since there are eight investments in the 50–59 days-to-maturity range. A table listing all classes and their frequencies is called a **frequency distribution.** The first and third columns of Table 2.2 on page 24 constitute a frequency distribution for the days-to-maturity data.

In addition to the frequency of a class, we are often interested in the **percentage** of a class. We find the percentage by dividing the frequency of the class by the total number of pieces of data. Referring again to Table 2.2, we see that the percentage of investments in the class 50–59 is

Frequency of
class 50–59

$$\frac{8}{40} = 0.20 \text{ or } 20\%$$

Total number of
pieces of data

In other words, 20% of the investments have a maturity period of between 50 and 59 days, inclusive.

The percentage of a class, expressed as a decimal, is called the **relative frequency** of the class. For the class 50–59, the relative frequency is 0.20. A table listing all classes and their relative frequencies is called a **relative-frequency distribution.** Table 2.3 displays a relative-frequency distribution for the days-to-maturity data.

TABLE 2.3
Relative-frequency
distribution for the
days-to-maturity
data in Table 2.1

Days to maturity	Relative frequency		
30–39	0.075	←	3/40
40–49	0.025	←	1/40
50–59	0.200	←	8/40
60–69	0.250	←	10/40
70–79	0.175	←	7/40
80–89	0.175	←	7/40
90–99	0.100	←	4/40
	1.000		

Note that the relative frequencies must always add up to 1 (100%). Why?

TERMINOLOGY

There is considerable terminology associated with grouping. We have already discussed several of the terms used in grouping. To introduce some additional ones, let us return to the days-to-maturity data.

Consider, for example, the class 50–59. The smallest maturity period that can go in this class is 50. This is called the **lower class limit** of the class. The largest maturity period that can go in this class is 59. This is called the **upper class limit** of the class.

The midpoint of the class 50–59 is $(50 + 59)/2 = 54.5$, and this is called the **class mark.** Finally, the width of this class, obtained by subtracting its lower class limit from the lower class limit of the next higher class, is $60 - 50 = 10$. This is called the **class width** of the class. Definition 2.3 summarizes the terminology of grouping.

DEFINITION 2.3   Terms used in grouping

Classes: Categories for grouping data.
Frequency: The number of data values in a class.
Relative frequency: The ratio of the frequency of a class to the total number of pieces of data.
Frequency distribution: A listing of classes and their frequencies.
Relative-frequency distribution: A listing of classes and their relative frequencies.
Lower class limit: The smallest value that can go into a class.
Upper class limit: The largest value that can go into a class.
Class mark: The midpoint of a class.
Class width: The difference between the lower class limit of the given class and the lower class limit of the next higher class.

A table giving the classes, frequencies, relative-frequencies, and class marks for a data set is called a **grouped-data table.** A grouped-data table for the days-to-maturity data is presented in Table 2.4.

TABLE 2.4
Grouped-data table for days-to-maturity data

Days to maturity	Frequency	Relative frequency	Class mark
30–39	3	0.075	34.5
40–49	1	0.025	44.5
50–59	8	0.200	54.5
60–69	10	0.250	64.5
70–79	7	0.175	74.5
80–89	7	0.175	84.5
90–99	4	0.100	94.5
	40	1.000	

In the next example, we will find the grouped-data table for another data set.

EXAMPLE 2.6   *Illustrates grouped-data tables*

A pediatrician began testing cholesterol levels of young patients and was alarmed to find that a large number had levels over 200 milligrams per 100 milliliters. The readings of 20 patients with high levels are presented in Table 2.5 at the top of page 27. Construct a grouped-data table. Use a class width of five and start at 195.

TABLE 2.5 Cholesterol levels for 20 high-level patients			
210	209	212	208
217	207	210	203
208	210	210	199
215	221	213	218
202	218	200	214

$$\frac{220}{195}{5} = \text{class width}$$

SOLUTION Since we are to use a class width of five and start at 195, the first class will be 195–199. From Table 2.5 we see that the highest cholesterol level is 221. Thus, we choose the classes displayed in the first column of Table 2.6.

TABLE 2.6

Cholesterol level	Tally	Frequency
195–199	I	1
200–204	III	3
205–209	IIII	4
210–214	IIII II	7
215–219	IIII	4
220–224	I	1
		20

The results of tallying the data in Table 2.5 are shown in the second and third columns of Table 2.6. From Table 2.6 we obtain the grouped-data table given in Table 2.7.

TABLE 2.7
Grouped-data table for cholesterol-level data

Cholesterol level	Frequency	Relative frequency	Class mark
195–199	1	0.05	197
200–204	3	0.15	202
205–209	4	0.20	207
210–214	7	0.35	212
215–219	4	0.20	217
220–224	1	0.05	222
	20	1.00	

To illustrate some typical computations for the third and fourth columns, consider the class 210–214. We have

$$\text{Relative frequency} = \frac{7}{20} = 0.35$$

and

MTB

(midpoint)

$$\text{Class mark} = \frac{210 + 214}{2} = 212$$

SINGLE-VALUE GROUPING

Up to this point, each class we have used for grouping data represents several possible numerical values. For instance, as we see from Table 2.7, each cholesterol-level class represents five possible cholesterol levels. The first class, 195–199, is for levels of 195, 196, 197, 198, or 199; the second class, 200–204, is for levels of 200, 201, 202, 203, or 204; and so on.

In some cases, however, it is more appropriate to use classes that each represent a single possible numerical value. For instance, it is often preferable to group *discrete data* using such classes. Here is an example.

EXAMPLE 2.7   *Illustrates single-value grouping*

A planner is collecting data on the number of school-age children per family in a small town. Thirty families are selected at random. Table 2.8 displays the number of school-age children in each of the 30 families chosen.

TABLE 2.8
Number of
school-age children
in each of 30 families

0	3	0	0	3	0
2	2	0	1	2	1
0	0	1	2	4	0
4	2	1	0	1	0
0	2	0	1	3	2

a) Group these data using classes that each represent a single numerical value.
b) Identify the class limits and the class marks.
c) Construct a grouped-data table.

SOLUTION   a) We first note that since each class is to represent a single numerical value, the classes must be 0, 1, 2, 3, and 4. These are displayed in the first column of Table 2.9.

TABLE 2.9
Frequency and
relative-frequency
distributions
for number of
school-age children

Number of school-age children	Frequency	Relative frequency
0	12	0.400
1	6	0.200
2	7	0.233
3	3	0.100
4	2	0.067
	30	1.000

Applying the tallying procedure to the data in Table 2.8, we obtain the frequencies in the second column of Table 2.9. Dividing each frequency by the total number of pieces of data, 30, gives the relative frequencies shown in the third column of Table 2.9. The table indicates, for example, that seven of the 30 families, or 23.3%, have two school-age children.

b) For this part we are to identify the class limits and class marks. Consider, for instance, the class "3" (i.e., three school-age children). We have

Lower class limit = 3 (the smallest value that can go into the class)

Upper class limit = 3 (the largest value that can go into the class)

and

$$\text{Class mark} = \frac{3+3}{2} = 3 \text{ (the midpoint of the class)}$$

Thus, for the class "3," the lower class limit, the upper class limit, and the class mark are all equal to 3. A similar statement is true for the other classes.

c) Finally, we are to construct a grouped-data table. This requires us to append a class-mark column to Table 2.9. However, from part (b) we know that the class mark for each class is the same as the class itself. Thus, it is unnecessary to add a class-mark column to Table 2.9, since such a column would be identical to the first column. In other words, Table 2.9 can serve as a grouped-data table. ∎

MTB

We now summarize the important points made about single-value grouping in Example 2.7: When a class for grouping data represents a single numerical value, then that value is the lower class limit, the upper class limit, and the class mark for the class. If every class for grouping a data set is based on a single value, then the class-mark column may be omitted from the grouped-data table because the first column gives the class marks as well as the classes.

ANOTHER WAY OF GROUPING DATA

It is often convenient to group data using classes that consist of values from one number up to, but not including, another number. This is particularly true when dealing with *continuous data* or *decimal data*. Consider Example 2.8.

EXAMPLE 2.8   *Illustrates another way to group data*

The U.S. National Center for Health Statistics publishes data on weights and heights by age and sex in *Vital and Health Statistics*. The following weights, given to the nearest tenth of a pound, were obtained from a sample of 18–24-year-old males:

TABLE 2.10
Weights of 37 males, aged 18–24 years

129.2	185.3	218.1	182.5	142.8	155.2	170.0	151.3
187.5	145.6	167.3	161.0	178.7	165.0	172.5	191.1
150.7	187.0	173.7	178.2	161.7	170.1	165.8	
214.6	136.7	278.8	175.6	188.7	132.1	158.5	
146.4	209.1	175.4	182.0	173.6	149.9	158.6	

We will group these weights using the classes "120–under 140," "140–under 160," and so on. The class "120–under 140" is for weights of *at least* 120 lb, but *less than* 140 lb; the class "140–under 160" is for weights of *at least* 140 lb, but *less than* 160 lb; and so forth. These classes are displayed in the first column of Table 2.11.

TABLE 2.11
Grouped-data table
for the weights
of 37 males,
aged 18–24 years

Weight (lb)	Frequency	Relative frequency	Class mark
120–under 140	3	0.081	130
140–under 160	9	0.243	150
160–under 180	14	0.378	170
180–under 200	7	0.189	190
200–under 220	3	0.081	210
220–under 240	0	0.000	230
240–under 260	0	0.000	250
260–under 280	1	0.027	270
	37	0.999	

Applying the tallying procedure to the data in Table 2.10, we obtain the frequencies in the second column of Table 2.11. The relative frequencies, given in the third column of Table 2.11, are then found in the usual manner.

Finally, let us discuss the class marks. For a class of the form "*a*–under *b*," we define the class mark to be $(a + b)/2$. For instance, the class mark for the class "120–under 140" is $(120 + 140)/2 = 130$. Similar calculations give the remaining class marks shown in the fourth column of Table 2.11. ■

Note: As we know, the sum of the relative frequencies must always be exactly equal to 1 in any relative-frequency distribution. However, the sum of the relative frequencies in the third column of Table 2.11 is given as 0.999. The reason for this discrepancy is that each relative frequency is displayed rounded to three decimal places; and when we add these rounded relative frequencies, the resulting sum differs from 1 by a little.

FREQUENCY AND RELATIVE-FREQUENCY DISTRIBUTIONS FOR QUALITATIVE DATA

The concepts of class limits and class marks are applicable to metric data and certain kinds of ordinal data. However, they are not appropriate for qualitative data. For instance, with data that categorize runners as male or female, the classes are "male" and "female." For qualitative-data classes such as these, it makes no sense to look for class limits or class marks. We can, of course, still compute frequencies and relative frequencies for qualitative data. Consider Example 2.9.

EXAMPLE 2.9   *Illustrates frequency and relative-frequency distributions for qualitative data*

Professor W asked his introductory statistics students to state their political party affiliations as Democratic, Republican, or Other. The responses are depicted in Table 2.12 at the top of page 31. [D = Democratic, R = Republican, O = Other.] Determine the frequency and relative-frequency distributions for the data.

TABLE 2.12
Political party
affiliations
of students in
introductory statistics

D	R	O	R	R	R	R	R
D	O	R	D	O	O	R	D
D	R	O	D	R	R	O	R
D	O	D	D	D	R	O	D
O	R	D	R	R	R	R	D

SOLUTION   The classes for grouping the data are "Democratic," "Republican," and "Other." Tallying the data in Table 2.12, we obtain the frequency distribution displayed in the first two columns of Table 2.13.

TABLE 2.13
Frequency and
relative-frequency
distributions
for political
party affiliations

Party	Frequency	Relative frequency
Democratic	13	0.325
Republican	18	0.450
Other	9	0.225
	40	1.000

Dividing each frequency in the second column by the total number of students (40), we get the relative frequencies in the third column of Table 2.13. The first and third columns of the table constitute the relative-frequency distribution for the data. ■

Exercises 2.2

___ 2.14 What is one of the main reasons for grouping data?

___ 2.15 Do class limits and class marks make sense for qualitative-data classes? Explain your answer.

___ 2.16 State the three most important guidelines in choosing the classes for grouping a data set.

___ 2.17 Explain the difference between each of the following pairs of terms:
a) frequency and relative frequency
b) percentage and relative frequency

___ 2.18 A research physician conducted a study on the ages of persons with diabetes. The following data were obtained for the ages of a sample of 35 diabetics:

48	41	57	83	41	55	59
61	38	48	79	75	77	7
54	23	47	56	79	68	61
64	45	53	82	68	38	70
10	60	83	76	21	65	47

Construct a grouped-data table for these ages. Use classes of equal width, beginning with the class 0–9.

___ 2.19 A soft-drink bottler sells "one-liter" bottles of soda. A consumer group is concerned that the bottler may be shortchanging the customers. Thirty bottles of soda are randomly selected. The contents, in milliliters, of the bottles chosen are shown below.

1025	977	1018	975	977
990	959	957	1031	964
986	914	1010	988	1028
989	1001	984	974	1017
1060	1030	991	999	997
996	1014	946	995	987

Construct a grouped-data table for these soft-drink data. Use classes of equal width, starting with the class 910–929.

___ 2.20 The Bureau of Economic Analysis gathers information on the length of stay in Europe and the Mediterranean by U.S. travelers. Data are published in *Survey of Current Business*. A sample of 36 U.S. residents who traveled to Europe and the Mediterranean this year yielded the following data, in days, on length of stay:

41	16	6	21	1	21
5	31	20	27	17	10
3	32	2	48	8	12
21	44	1	56	5	12
3	13	15	10	18	3
1	11	14	12	64	10

Construct a grouped-data table using classes of equal width, beginning with the class 1–7.

__ 2.21   The U.S. Energy Information Administration collects data on residential energy consumption and expenditures. Results are published in the document *Residential Energy Consumption Survey: Consumption and Expenditures*. The table below gives last year's energy consumptions for a sample of 50 households in the South. Data are in millions of BTU.

130	55	45	64	155	66	60	80	102	62
58	101	75	111	151	139	81	55	66	90
97	77	51	67	125	50	136	55	83	91
54	86	100	78	93	113	111	104	96	113
96	87	129	109	69	94	99	97	83	97

Construct a grouped-data table for these energy consumption figures. Use classes of equal width, starting with the class 40–49.

__ 2.22   The Bureau of the Census conducts nation-wide surveys on characteristics of American households. Below are data on the number of persons per household for a sample of 40 households.

2	5	2	1	1	2	3	4
1	4	4	2	1	4	3	3
7	1	2	2	3	4	2	2
6	5	2	5	1	3	2	5
2	1	3	3	2	2	3	3

Construct a grouped-data table for these household sizes using classes based on a single value.

__ 2.23   A car salesperson keeps track of the number of cars she sells per week. The number of cars she sold per week last year are as follows:

1	0	3	3	1	0	2	1	4	0	4	1	2
3	6	4	3	0	2	2	1	1	2	2	2	3
5	1	0	2	5	3	1	3	1	1	1	1	2
2	3	0	4	4	1	0	1	1	3	2	5	2

Construct a grouped-data table for the number of sales per week. Use classes based on a single value.

__ 2.24   A fifth-grade class consisting of 25 students was given an eight-question quiz on fractions. The following data provide the number of incorrect answers for each student:

3	4	2	3	1
1	1	6	1	2
0	5	1	2	1
0	3	4	0	3
2	0	3	1	1

Construct an appropriate grouped-data table.

__ 2.25   Cudahey Masonry, Inc., employs 80 brick-layers. Records are kept on the number of days missed by each of the employees. The absentee records for the past year are as follows:

2	3	6	2	6	5	5	2	4	7	5	3	6	4	4	4
2	2	4	5	4	0	2	1	6	3	5	3	6	6	4	7
5	2	5	0	5	6	5	2	4	2	6	2	4	3	5	4
2	4	4	3	3	4	0	5	6	3	5	5	2	4	4	2
0	7	5	5	7	6	1	5	3	3	4	7	7	2	5	5

Construct an appropriate grouped-data table for these absentee data.

__ 2.26   The Food and Nutrition Board of the National Academy of Sciences states that the recommended daily allowance (RDA) of iron is 18 mg for adult females under the age of 51. The amounts of iron intake during a 24-hour period for a sample of 45 such females are as follows. Data are in milligrams.

15.0	18.1	14.4	14.6	10.9	18.1	18.2	18.3	15.0
16.0	12.6	16.6	20.7	19.8	11.6	12.8	15.6	11.0
15.3	9.4	19.5	18.3	14.5	16.6	11.5	16.4	12.5
14.6	11.9	12.5	18.6	13.1	12.1	10.7	17.3	12.4
17.0	6.3	16.8	12.5	16.3	14.7	12.7	16.3	11.5

Construct a grouped-data table for these iron intakes. Use classes of equal width, beginning with the class 6–under 8.

__ 2.27   The National Education Association does surveys on starting salaries for college graduates. A sample of 35 liberal arts graduates yielded the following starting annual salaries. Data are in thousands of dollars, rounded to the nearest hundred dollars.

20.0	16.8	21.3	20.6	21.0	18.7	23.8
18.3	17.7	18.0	19.1	21.1	19.6	19.0
18.7	20.8	20.4	17.1	19.5	19.9	19.2
19.1	17.2	18.3	22.7	20.0	19.2	20.9
19.1	20.8	20.5	21.4	16.3	16.3	20.5

Using classes of equal width and starting with the class 16–under 17, construct a grouped-data table for these starting annual salaries.

The following Associated Press article appeared in a major metropolitan newspaper. We will use the article in Exercises 2.28 and 2.29.

DOLLAR LEADERS

NEW YORK (AP) — The following is a list of the most active NYSE stocks based on the dollar volume. The total is based on the median price of the stock traded multiplied by the shares traded.

	Tot. ($1000)	Sales (hds.)	Last
Reyl pfC	$100,9720	80536	127
IBM	$705,516	55011	127½
RchVck	$523,317	10755	48¾
SCM	$358,257	49245	72⅝
Revlon	$271,385	63113	43
WstgE	$261,517	68148	38⅜
CocaCl	$259,958	37607	68⅜
GnFds	$251,132	29545	83¼
Digital	$218,483	20636	105⅜
RckCtr n	$217,851	09611	19⅞
CessAir	$206,341	73366	29½
GMot	$193,443	28500	68¼
PhilMr	$173,890	22258	75⅝
AmExp	$158,911	37949	42
McDnld	$148,393	22273	65
MCA	$143,096	20190	68
GenEl	$142,523	23509	60
Exxon	$140,513	27087	51⅛
AtlRich	$138,506	23181	59
MartM s	$137,368	38832	33⅝
Morgn s	$126,692	26672	46⅝
Burrgh	$126,061	19469	63⅛
BeafCo	$125,221	36695	34⅜
CBS	$120,761	10234	117⅛

___ **2.28**   The column headed "Tot." in the article shows the total dollar volumes for the most active New York Stock Exchange (NYSE) stocks. The totals are given in thousands of dollars.

a) Group these totals into a grouped-data table with classes as in the first column of the following table. Note that the classes use values in millions of dol-

lars (e.g., the class "100–under 150" is for volumes of at least $100 million but less than $150 million).

b) Why is there no class mark for the last class?

Dollar volume ($millions)	Frequency	Relative frequency	Class mark
100–under 150			
150–under 200			
200–under 250			
250–under 300			
300–under 350			
350–under 400			
400 and over			—

___ **2.29**   The second column of the Associated Press article displays the numbers of shares traded for the most active New York Stock Exchange (NYSE) stocks. The sales are given in hundreds. Group these sales data into a grouped-data table using classes of equal width. Begin with the class "0–under 1," where the values are in millions of sales.

___ **2.30**   The following table gives the all-time top television programs by number of viewers as of March, 1988. [SOURCE: Nielsen Media Research, *Nielsen Report on Television.*]

Program	Date	Network	Households
M*A*S*H Special	2/28/83	CBS	50,150,000
Super Bowl XX	1/26/86	NBC	41,490,000
Dallas	11/21/80	CBS	41,470,000
Super Bowl XIX	1/20/85	ABC	40,900,000
Super Bowl XVII	1/30/83	NBC	40,480,000
Super Bowl XXI	1/25/87	CBS	40,030,000
Super Bowl XVI	1/24/82	CBS	40,020,000
Super Bowl XVIII	1/22/84	CBS	38,800,000
The Day After	11/20/83	ABC	38,550,000
Super Bowl XXII	1/31/88	ABC	37,120,000
Roots Pt. VIII	1/30/77	ABC	36,380,000
Thorn Birds Pt. III	3/29/83	ABC	35,990,000
Thorn Birds Pt. IV	3/30/83	ABC	35,900,000
Thorn Birds Pt. II	3/28/83	ABC	35,400,000
Super Bowl XIV	1/20/80	CBS	35,330,000
Super Bowl XIII	1/21/79	NBC	35,090,000
CBS NFC Championship Game	1/10/82	CBS	34,960,000
Super Bowl XV	1/25/81	NBC	34,540,000
Super Bowl XII	1/15/78	CBS	34,410,000
Gone With The Wind Pt. 1	11/7/76	NBC	33,960,000

Construct a frequency and relative-frequency distribution for the network data. *(Hint:* The classes are "ABC," "CBS," and "NBC.")

___ 2.31   The National Collegiate Athletic Association wrestling champions for the years 1963–1988 are as follows. [SOURCE: *The World Almanac, 1989.*]

Year	Champion	Year	Champion
1963	Oklahoma	1976	Iowa
1964	Oklahoma State	1977	Iowa State
1965	Iowa State	1978	Iowa
1966	Oklahoma State	1979	Iowa
1967	Michigan State	1980	Iowa
1968	Oklahoma State	1981	Iowa
1969	Iowa State	1982	Iowa
1970	Iowa State	1983	Iowa
1971	Oklahoma State	1984	Iowa
1972	Iowa State	1985	Iowa
1973	Iowa State	1986	Iowa
1974	Oklahoma	1987	Iowa State
1975	Iowa	1988	Arizona State

Construct a frequency and relative-frequency distribution for the champions.

___ 2.32   The exam scores for the students in an introductory statistics class are as follows:

88	82	89	70	85
63	100	86	67	39
90	96	76	34	81
64	75	84	89	96

a) Group these exam scores using the classes 30–39, 40–49, 50–59, 60–69, 70–79, 80–89, 90–100.
b) What are the widths of the classes?
c) If you wanted all the classes to have equal widths, what classes would you use?

Contingency tables: The methods that we have examined in this section apply to grouping data that involve *one* characteristic of the members of a population or sample. Such data are called **univariate.** For instance, in Example 2.8 we considered data on the single characteristic, weight, for a sample of 18–24-year-old males. Thus, those data are univariate. We could have considered not only the weights of the males, but also their heights. Then we would have data on two characteristics, height and weight. Data that involve *two* characteristics of the members of a population or sample are called **bivariate.** Bivariate data can be grouped using tables called **contingency ta-**

bles. Exercises 2.33 and 2.34 deal with the grouping of bivariate data using contingency tables.

___ 2.33   The following bivariate data on age and sex were obtained from the students in a freshman calculus course. The data show, for example, that the first student on the list is 21 years old and is a male.

Age	Sex	Age	Sex	Age	Sex	Age	Sex	Age	Sex
21	M	29	F	22	M	23	F	21	F
20	M	20	M	23	M	44	M	28	F
42	F	18	F	19	F	19	M	21	F
21	M	21	M	21	M	21	F	21	F
19	F	26	M	21	F	19	M	24	F
21	F	24	F	21	F	25	M	24	F
19	F	19	M	20	F	21	M	24	F
19	M	25	M	20	F	19	M	23	M
23	M	19	F	20	F	18	F	20	F
20	F	23	M	22	F	18	F	19	M

We will discuss the grouping of these data into the following contingency table:

Age (yrs)

	Under 21	21–25	Over 25	Total
Male		I		
Female				
Total				

Sex

a) To tally the data for the first student, we place a tally mark in the box labeled by the "21–25"-column and the "Male"-row, as indicated. Tally the data for the remaining 49 students.
b) Construct a table like the one in part (a) but with the frequencies replacing the tally marks. Add the frequencies in each row and column of your table and record the sums in the proper "Total" boxes.
c) What do the row and column totals represent?
d) Add the row totals and the column totals. Why are these two sums equal, and what does their common value represent?
e) Construct a table giving the relative frequencies for the data. [Divide each frequency by the grand total of 50 students.]
f) Interpret the entries in your table from part (e) in terms of percentages.

___ 2.34   The heights and weights of the students in Exercise 2.33 are as follows:

Height	Weight	Height	Weight	Height	Weight
68	140	67	155	74	215
67	140	67	130	67	129
72	145	68	160	72	275
69	145	64	127	68	135
66	115	74	170	60	95
72	185	73	180	75	175
64	130	63	142	61	120
65	145	69	170	73	180
62	127	62	103	64	125
69	135	68	160	66	130
66	110	75	185	63	105
69	178	64	122	69	155
63	130	64	130	70	170
72	185	70	215	64	105
67	120	63	105	65	132
68	135	76	200	65	115
64	130	71	169		

Repeat parts (a)–(f) of Exercise 2.33 for the bivariate data on heights and weights. Use the following contingency table:

Height (in)

Weight (lb)	60–65	66–71	72–77	Total
90–129				
130–169				
170–209				
210–249				
250–289				
Total				

2.3 Graphs and charts

Besides grouping, another method for organizing and summarizing data is to draw a picture of some kind. The old saying "a picture is worth a thousand words" has particular relevance in statistics—a graph or chart of a data set often provides the simplest and most efficient display. In this section, we will examine various techniques for organizing and summarizing data using graphs and charts. We begin by discussing *histograms*.

EXAMPLE 2.10 **Introduces histograms**

Table 2.4 gives a grouped-data table for the number of days to maturity for 40 short-term investments. The first three columns of that table are repeated in Table 2.14.

TABLE 2.14
Frequency and
relative-frequency
distributions for
days-to-maturity data

Days to maturity	Frequency (no. of investments)	Relative frequency
30–39	3	0.075
40–49	1	0.025
50–59	8	0.200
60–69	10	0.250
70–79	7	0.175
80–89	7	0.175
90–99	4	0.100
	40	1.000

One way to display the data pictorially is to construct a graph with the classes depicted on the horizontal axis and the frequencies depicted on the vertical axis. This can be done using a **frequency histogram** as shown in Figure 2.1.

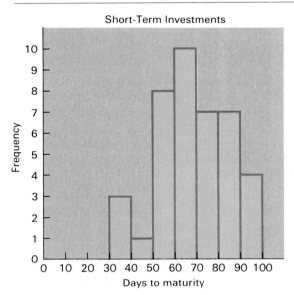

The height of each bar is equal to the frequency of the class it represents. Note that each bar extends from the lower class limit of a class to the lower class limit of the next higher class. [This is only one of several methods for depicting the classes. Additional methods are discussed in Exercises 2.64 and 2.65.]

Some other important observations about Figure 2.1 are that each axis of the frequency histogram has a label and the frequency histogram as a whole has a title. These are common elements that all such graphical displays should possess.

A frequency histogram displays the frequencies of the classes. To display the relative frequencies (or percentages), we can use a **relative-frequency histogram,** which is quite similar to a frequency histogram. In fact, the only difference is that the height of each bar in a relative-frequency histogram is equal to the relative frequency of the class instead of the frequency of the class. The relative-frequency histogram for the days-to-maturity data is shown in Figure 2.2 on page 37.

Note that the relative-frequency histogram in Figure 2.2 and the frequency histogram in Figure 2.1 have identical shapes. This is because the frequencies and relative frequencies are in the same proportion.

Finally, we present a word of caution. We have used the lower class limits 30, 40, 50, ... to label the horizontal axes of the histograms in Figures 2.1 and 2.2, since it seems natural to do so. However, care must be taken when reading the histograms. For instance, the bar for 80 to 90 represents only maturity periods in the 80s. A maturity period of 90 days is included in the next bar to the right. ▪

MTB

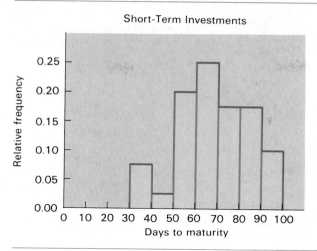

Below we present the definition of a frequency histogram and the definition of a relative-frequency histogram.

DEFINITION 2.4   Frequency and relative-frequency histograms

Frequency histogram: A graph that displays the classes on the horizontal axis and the frequencies of the classes on the vertical axis. The frequency of each class is represented by a vertical bar whose height is equal to the frequency of the class.

Relative-frequency histogram: A graph that displays the classes on the horizontal axis and the relative frequencies of the classes on the vertical axis. The relative frequency of each class is represented by a vertical bar whose height is equal to the relative frequency of the class.

HISTOGRAMS FOR SINGLE-VALUE GROUPING

For the days-to-maturity data, each class represents 10 possible days to maturity and the histogram bar for each class extends over those 10 possible days (see Figures 2.1 and 2.2). When data are grouped using classes based on a single value, we proceed somewhat differently. In that case, each bar is placed directly over the only possible numerical value in the class, as illustrated in the next example.

EXAMPLE 2.11   *Illustrates histograms for single-value grouped data*

In Example 2.7 on pages 28–29 we considered data on the number of school-age children in each of 30 families. We grouped the data using classes based on a single value. The frequency and relative-frequency distributions are given in Table 2.9 and are repeated here in Table 2.15 at the top of the following page. Construct a frequency and relative-frequency histogram for this grouped data.

TABLE 2.15
Frequency and
relative-frequency
distributions
for number of
school-age children

Number of school-age children	Frequency	Relative frequency
0	12	0.400
1	6	0.200
2	7	0.233
3	3	0.100
4	2	0.067
	30	1.000

SOLUTION   As we just mentioned, for single-value grouping such as in Table 2.15, we place the middle of each histogram bar directly over the single numerical value represented by the class. Hence, the frequency and relative-frequency histograms for the grouped data in Table 2.15 are as pictured in Figure 2.3 below and Figure 2.4 at the top of the following page.

FIGURE 2.3
Frequency histogram
for number of
school-age children
in each of 30 families

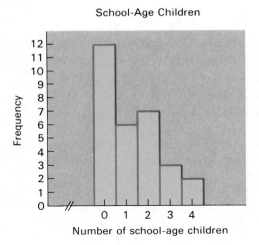

Note that in Figures 2.3 and 2.4 the symbol "//" has been placed on the horizontal axis. This indicates that the zero point (0) on that axis is not in its usual position at the intersection of the horizontal and vertical axes. Whenever any such modification is made, whether on the horizontal or vertical axis, the symbol "//," or some similar symbol, should be used to indicate that fact. ■

MTB

DOTPLOTS

Another kind of graphical display of numerical data is the **dotplot.** Dotplots are particularly useful for showing the relative positions of the data in a data set or for comparing two or more data sets. We introduce dotplots in Example 2.12.

FIGURE 2.4
Relative-frequency
histogram for
number of school-age
children in each
of 30 families

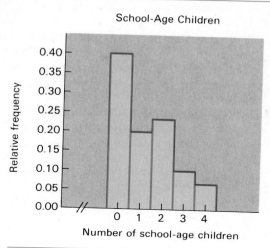

FIGURE 2.4
Relative-frequency
histogram for
number of school-age
children in each
of 30 families

EXAMPLE 2.12   Illustrates dotplots

A farmer is interested in a newly developed fertilizer that supposedly will increase his yield of oats. He uses the fertilizer on a sample of 15 one-acre plots. The yields, in bushels, are depicted in Table 2.16. Construct a dotplot for the data.

TABLE 2.16
Oat yields

67	65	55	57	58
61	61	61	64	62
62	60	62	60	67

SOLUTION   To construct a dotplot for the data in Table 2.16, we begin by drawing a horizontal axis that displays the possible oat yields. Then we go through the data and record each yield by placing a dot over the appropriate value on the horizontal axis. For instance, the first yield is 67 bushels. This calls for a dot over the "67" on the horizontal axis. The dotplot for the data in Table 2.16 is pictured in Figure 2.5. ■

MTB

FIGURE 2.5
Dotplot for oat yields

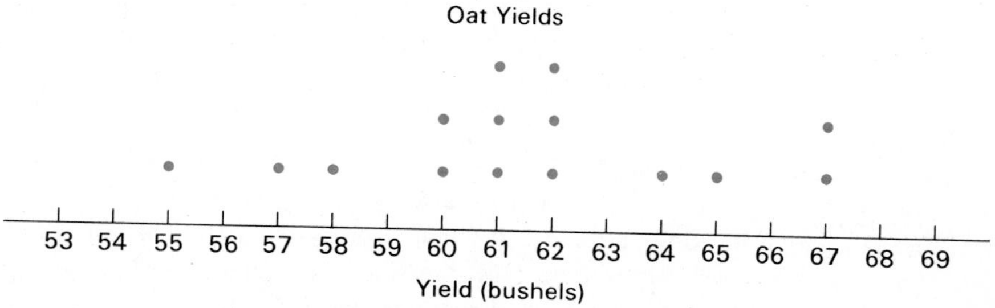

Oat Yields

53 54 55 56 57 58 59 60 61 62 63 64 65 66 67 68 69

Yield (bushels)

As you can see, dotplots are quite similar to histograms. In fact, when data are grouped in classes based on a single value, a dotplot and frequency histogram are essentially identical. However, dotplots are generally more convenient to use than histograms when dealing with single-value grouped data that involve decimals.

GRAPHICAL DISPLAYS FOR QUALITATIVE DATA

Histograms and dotplots are designed for use with numerical data. Qualitative data are portrayed using different techniques. Two common methods for portraying qualitative data graphically are *pie charts* and *bar graphs*.

EXAMPLE 2.13   *Illustrates pie charts and bar graphs*

In Example 2.9 (pages 30–31), we obtained frequency and relative-frequency distributions for the political party affiliations of the students in Professor W's introductory statistics class. We repeat those distributions in Table 2.17.

TABLE 2.17
Frequency and
relative-frequency
distributions
for political
party affiliations

Party	Frequency	Relative frequency
Democratic	13	0.325
Republican	18	0.450
Other	9	0.225
	40	1.000

The table shows, for instance, that 32.5% of the students in Professor W's introductory statistics class are Democrats.

Note that we are dealing here with qualitative-data classes; namely, "Democratic," "Republican," and "Other." We can use a pie chart to display the relative-frequency distribution given in the first and third columns of Table 2.17. A **pie chart** is a disk divided into pie-shaped pieces proportional to the relative frequencies. In this case, we need to divide a disk into three pie-shaped pieces comprising 32.5%, 45.0%, and 22.5% of the disk. We can do this by using a protractor and the fact that there are 360° in a circle. Thus, for instance, the first piece of the disk is obtained by marking off 117° (32.5% of 360°). The pie chart for the relative-frequency distribution in Table 2.17 is shown in Figure 2.6 at the top of page 41.

We can also use a bar graph to portray the same information. A **bar graph** is like a histogram except that its bars are separated. The bar graph for the relative-frequency distribution in Table 2.17 is pictured in Figure 2.7 on page 41.   ■

Pie charts are generally preferable to bar graphs for qualitative data since people are accustomed to having the horizontal axis of a graph show order. For example, someone might infer from Figure 2.7 that "Republican" is less than "Other" because "Republican" is shown to the left of "Other" on the horizontal axis. Pie charts do not lead to such inferences.

FIGURE 2.6
Pie chart for
relative-frequency
distribution of
political party
affiliations

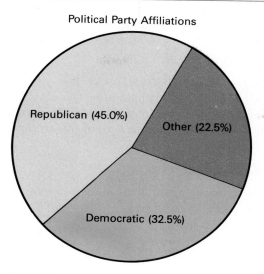

FIGURE 2.7
Bar graph for
relative-frequency
distribution of
political party
affiliations

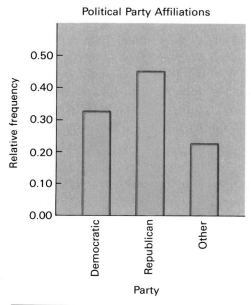

OTHER GRAPHICAL DISPLAYS

Histograms, dotplots, pie charts, and bar graphs are only a few of the countless ways that data can be portrayed pictorially. Fortunately, most graphical displays are based on common sense and are easy to understand.

We will consider some additional graphical displays in the exercises. These include relative-frequency polygons and ogives. See Exercises 2.66–2.69.

USING THE COMPUTER (OPTIONAL)

Minitab has a program called **HISTOGRAM** that will simultaneously group a data set and display a frequency histogram for the data. We illustrate the use of HISTOGRAM by applying it to the days-to-maturity data that we considered in Example 2.5 of Section 2.2 and Example 2.10 of this section.

EXAMPLE 2.14   *Illustrates the HISTOGRAM command*

The number of days to maturity for 40 short-term investments, given in Table 2.1, are repeated here in Table 2.18.

TABLE 2.18
Days to maturity for 40 short-term investments

70	64	99	55	64	89	87	65
62	38	67	70	60	69	78	39
75	56	71	51	99	68	95	86
57	53	47	50	55	81	80	98
51	36	63	66	85	79	83	70

Use Minitab to construct a frequency distribution and frequency histogram for these data based on the classes 30–39, 40–49, ..., 90–99.

SOLUTION   To begin, we enter the data from Table 2.18 into C1 using the SET command and name C1 "Maturity" using the NAME command. (Refer to Section 1.4 for details.) Next we type the command HISTOGRAM followed by the storage location of the data. That is, we type HISTOGRAM 'MATURITY'; and press RETURN. Note that there is a semicolon after 'MATURITY'. This tells Minitab that there will be a *subcommand* on the next line.

To specify that the class width is to be 10, we employ the **INCREMENT** subcommand; that is, we type INCREMENT=10; and press RETURN. Note that there is again a semicolon at the end of the subcommand. This is because we will use another subcommand, **START,** to specify that the first class mark is to be 34.5. Thus, we type START=34.5. and press RETURN. This time we put a period after the subcommand to tell Minitab that there will be no more subcommands.[†] Printout 2.1, on page 43, shows all of the above commands and the resulting output.

The first column of the output, headed **Midpoint,** gives the class marks corresponding to the classes 30–39, 40–49, ..., 90–99. The second column, headed **Count,** displays the frequencies of those classes. Finally, by turning the output 90° counterclockwise, we obtain, from the asterisks, a frequency histogram. You should compare Minitab's frequency histogram to the one we drew by hand in Figure 2.1 on page 36.   ∎

---

† The two subcommands, INCREMENT and START, are optional. We have used them to obtain a frequency distribution and histogram that are based on the classes 30–39, 40–49, ..., 90–99. If the subcommands are not used, Minitab will select its own classes.

```
MTB > SET C1
DATA> 70 64 99 55 64 89 87 65
DATA> 62 38 67 70 60 69 78 39
DATA> 75 56 71 51 99 68 95 86
DATA> 57 53 47 50 55 81 80 98
DATA> 51 36 63 66 85 79 83 70
DATA> END
MTB > NAME C1 'MATURITY'
MTB > HISTOGRAM 'MATURITY';
SUBC> INCREMENT=10;
SUBC> START=34.5.

Histogram of MATURITY    N = 40

Midpoint    Count
    34.5       3   ***
    44.5       1   *
    54.5       8   ********
    64.5      10   **********
    74.5       7   *******
    84.5       7   *******
    94.5       4   ****
```

One final note about the HISTOGRAM command. If we want our frequency distribution and histogram to be based on single-value classes, then we use the INCREMENT subcommand to ensure that each class represents a single numerical value. For integer data, the subcommand INCREMENT=1 will do.

On pages 38–40 we discussed dotplots. Recall that a dotplot for a data set is constructed by displaying the possible numerical values on a horizontal axis and by recording each piece of data with a dot over the appropriate value on the horizontal axis. The Minitab program for obtaining dotplots is called **DOTPLOT**. We will apply DOTPLOT to obtain a dotplot for the data on oat yields considered previously in Example 2.12 on page 39.

EXAMPLE 2.15   *Illustrates the DOTPLOT command*

A farmer is interested in a new fertilizer that supposedly will increase his yield of oats. He uses the fertilizer on a sample of 15 one-acre plots. The yields, in bushels, are depicted in Table 2.19. Use Minitab to construct a dotplot of the data.

TABLE 2.19
Oat yields

67	65	55	57	58
61	61	61	64	62
62	60	62	60	67

SOLUTION   We first enter the data into C2 using the SET command and name C2 "Oats" using the NAME command. Then we type the command DOTPLOT followed by the storage location of the data; that is, we type <u>DOTPLOT 'OATS'</u>. See Printout 2.2.

PRINTOUT 2.2
Minitab output
for DOTPLOT

```
MTB > SET C2
DATA> 67 65 55 57 58 61 61 61
DATA> 64 62 62 60 62 60 67
DATA> END
MTB > NAME C2 'OATS'
MTB > DOTPLOT 'OATS'
```

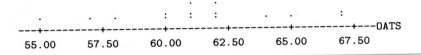

Note that Minitab uses 10 dashes (–) from one tick (+) mark to the next. Since, in this case, there are 2.5 units between tick marks, we see that each dash represents 0.25 (= 2.5/10) units. You should compare the dotplot in Printout 2.2 to the one that we obtained by hand in Figure 2.5 on page 39. ■

We should point out that the DOTPLOT command has two optional subcommands, **INCREMENT** and **START**. The INCREMENT subcommand allows us to specify the distance between the tick (+) marks. The START subcommand allows us to specify the first, and optionally the last, tick mark on the horizontal axis.

Exercises 2.3

__ 2.35   What is the difference between a frequency histogram and a relative-frequency histogram?

In Exercises 2.36–2.45, we have given the frequency and relative-frequency distributions obtained in Exercises 2.18–2.27. For each exercise,
a) construct a frequency histogram.
b) construct a relative-frequency histogram.

__ 2.36   The ages of a sample of 35 diabetics:

Age (years)	Frequency	Relative frequency
0–9	1	0.029
10–19	1	0.029
20–29	2	0.057
30–39	2	0.057
40–49	7	0.200
50–59	6	0.171
60–69	7	0.200
70–79	6	0.171
80–89	3	0.086

__ 2.37   The contents, in milliliters, of a sample of 30 "one-liter" bottles of soda obtained by a consumer group from a soft-drink bottler:

Content (ml)	Frequency	Relative frequency
910–929	1	0.033
930–949	1	0.033
950–969	3	0.100
970–989	9	0.300
990–1009	7	0.233
1010–1029	6	0.200
1030–1049	2	0.067
1050–1069	1	0.033

__ 2.38   The lengths of stay in Europe and the Mediterranean obtained from a sample of 36 United States residents who travelled there this year (the lengths of stay are given in days):

Length of stay (days)	Frequency	Relative frequency
1–7	10	0.278
8–14	10	0.278
15–21	8	0.222
22–28	1	0.028
29–35	2	0.056
36–42	1	0.028
43–49	2	0.056
50–56	1	0.028
57–63	0	0.000
64–70	1	0.028

___ 2.39 Last year's energy consumptions for a sample of 50 households in the South (the data on energy consumptions are given in millions of BTU):

Energy consumption (millions of BTU)	Frequency	Relative frequency
40–49	1	0.02
50–59	7	0.14
60–69	7	0.14
70–79	3	0.06
80–89	6	0.12
90–99	10	0.20
100–109	5	0.10
110–119	4	0.08
120–129	2	0.04
130–139	3	0.06
140–149	0	0.00
150–159	2	0.04

___ 2.40 The number of persons per household for a sample of 40 American households:

Number of persons	Frequency	Relative frequency
1	7	0.175
2	13	0.325
3	9	0.225
4	5	0.125
5	4	0.100
6	1	0.025
7	1	0.025

___ 2.41 The number of cars sold per week last year by a car salesperson:

Number of cars sold	Frequency	Relative frequency
0	7	0.135
1	15	0.288
2	12	0.231
3	9	0.173
4	5	0.096
5	3	0.058
6	1	0.019

___ 2.42 The number of questions answered incorrectly on an eight-question fraction quiz by each of the 25 students in a fifth-grade class:

Number incorrect	Frequency	Relative frequency
0	4	0.16
1	8	0.32
2	4	0.16
3	5	0.20
4	2	0.08
5	1	0.04
6	1	0.04

___ 2.43 The number of days missed last year by each of the 80 brick layers employed at Cudahey Masonry:

Number of days missed	Frequency	Relative frequency
0	4	0.050
1	2	0.025
2	14	0.175
3	10	0.125
4	16	0.200
5	18	0.225
6	10	0.125
7	6	0.075

___ 2.44 The 24-hour iron intakes, in milligrams, for a sample of 45 women under the age of 51:

Iron intake (mg)	Frequency	Relative frequency
6–under 8	1	0.022
8–under 10	1	0.022
10–under 12	7	0.156
12–under 14	9	0.200
14–under 16	9	0.200
16–under 18	9	0.200
18–under 20	8	0.178
20–under 22	1	0.022

___ **2.45**   The starting annual salaries of a sample of 35 liberal arts graduates:

Starting salary ($thousands)	Frequency	Relative frequency
16–under 17	3	0.086
17–under 18	3	0.086
18–under 19	5	0.143
19–under 20	9	0.257
20–under 21	9	0.257
21–under 22	4	0.114
22–under 23	1	0.029
23–under 24	1	0.029

In Exercises 2.46–2.49, construct a dotplot for each of the data sets given.

___ **2.46**   The exam scores for the students in an introductory statistics class are as follows:

88	82	89	70	85
63	100	86	67	39
90	96	76	34	81
64	75	84	89	96

___ **2.47**   A paint manufacturer claims that the average drying time for his new latex paint is two hours. To test this claim, the drying times are obtained for a sample of 20 cans of paint. The results are displayed below in minutes.

123	109	115	121	130
127	106	120	116	136
131	128	139	110	133
122	133	119	135	109

___ **2.48**   A consumer advocacy group suspects that Wheat Flakes brand cereal contains, on the average, less than the advertised weight of 15 ounces per box. A sample of 40 boxes of Wheat Flakes gives the following weights:

15.8	15.1	15.2	15.4	14.8	15.6	15.7	14.5
14.8	15.4	15.3	15.5	15.2	14.6	15.4	15.4
15.5	14.7	14.7	15.1	14.7	15.3	15.3	15.5
14.0	14.2	14.6	15.0	15.1	14.9	14.9	15.8
15.0	14.4	15.4	14.3	15.4	15.9	15.2	15.6

___ **2.49**   The Motor Vehicle Manufacturers Association of the United States publishes information on the ages of cars and trucks in use in *Motor Vehicle Facts and Figures.* A sample of 37 trucks yields the following ages, in years:

8	12	14	16	15	5	11	13
4	12	12	15	12	3	10	9
11	3	18	4	9	11	17	
7	4	12	12	8	9	10	
9	9	1	7	6	9	7	

In Exercises 2.50 and 2.51 we have given the frequency and relative-frequency distributions obtained in Exercises 2.30 and 2.31, respectively. For each exercise,
a) draw a pie chart for the relative-frequencies.
b) construct a bar chart for the relative-frequencies.

___ **2.50**   The network data for the all-time top television programs by number of viewers as of March, 1988:

Network	Frequency	Relative frequency
ABC	7	0.35
CBS	8	0.40
NBC	5	0.25

___ **2.51**   The winners of the NCAA wrestling championships for the years 1963–1988:

Champion	Frequency	Relative frequency
Oklahoma	2	0.077
Oklahoma State	4	0.154
Iowa State	7	0.269
Michigan State	1	0.038
Iowa	11	0.423
Arizona State	1	0.038

___ **2.52** The IRS groups income tax returns by adjusted gross income. Below is a relative-frequency histogram for one year's federal individual income tax returns showing an adjusted gross income less than $50,000. [SOURCE: U.S. Internal Revenue Service, *Statistics of Income, Individual Income Tax Returns*.]

Income Tax Returns

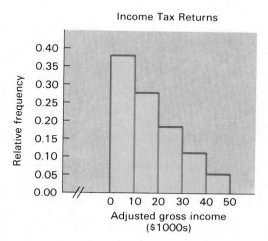

a) About what percentage of the returns had an adjusted gross income between $10,000 and $19,999, inclusive?
b) About what percentage had an adjusted gross income less than $30,000?
c) Given that there were 89,928,000 returns with an adjusted gross income less than $50,000, roughly how many had an adjusted gross income between $30,000 and $49,999, inclusive?

___ **2.53** Below is a relative-frequency histogram for the cholesterol-level data discussed in Example 2.6.

Cholesterol

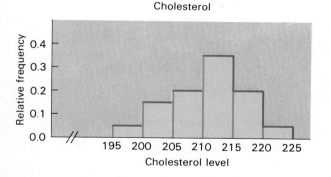

Answer the following questions using only the graph:
a) What percentage of the patients have cholesterol levels between 205 and 209, inclusive?
b) What percentage have levels of 215 or higher?
c) Given that the number of patients is 20, how many have levels between 210 and 214, inclusive?

Exercises 2.54–2.63 are computer exercises.

___ **2.54 (Computer exercise)** A study was conducted by a research physician on the ages of persons with diabetes. The following data were obtained for the ages of a sample of 35 diabetics:

48	41	57	83	41	55	59
61	38	48	79	75	77	7
54	23	47	56	79	68	61
64	45	53	82	68	38	70
10	60	83	76	21	65	47

Suppose the data are stored in a column named AGES.
a) Which Minitab command and subcommands (if any) should be employed to obtain a frequency distribution and frequency histogram for the age data using classes of equal width beginning with the class 0–9?
b) If you have access to Minitab, use it to carry out part (a).

___ **2.55 (Computer exercise)** A soft-drink bottler sells "one-liter" bottles of soda. A consumer group is concerned that the bottler may be shortchanging the customers. Thirty bottles of soda are randomly selected. The contents, in milliliters, of the bottles chosen are shown below.

1025	977	1018	975	977
990	959	957	1031	964
986	914	1010	988	1028
989	1001	984	974	1017
1060	1030	991	999	997
996	1014	946	995	987

Suppose that these data on contents are stored in a column named CONTENTS.
a) Which Minitab command and subcommands (if any) should be employed to obtain a frequency distribution and frequency histogram for these data using classes of equal width starting with the class 910–929?
b) If you have access to Minitab, use it to carry out part (a).

___ 2.56 (**Computer exercise**) The U.S. Bureau of the Census conducts nationwide monthly surveys to obtain data on characteristics of American households. Below are data on the number of persons per household for a sample of 40 households.

2	5	2	1	1	2	3	4
1	4	4	2	1	4	3	3
7	1	2	2	3	4	2	2
6	5	2	5	1	3	2	5
2	1	3	3	2	2	3	3

Suppose that these data are stored in a column named PERSONS.
a) Which Minitab command and subcommands (if any) should be employed to obtain a frequency distribution and frequency histogram for these data using classes based on a single value?
b) If you have access to Minitab, use it to carry out part (a).

___ 2.57 (**Computer exercise**) A new-car salesperson keeps track of the number of cars she sells per week. The number of cars she sold per week last year are as follows:

1	0	3	3	1	0	2	1	4	0	4	1	2
3	6	4	3	0	2	2	1	1	2	2	2	3
5	1	0	2	5	3	1	3	1	1	1	1	2
2	3	0	4	4	1	0	1	1	3	2	5	2

Suppose that these data are stored in a column named CARSALES.
a) Which Minitab command and subcommands (if any) should be employed to obtain a frequency distribution and frequency histogram for these data using classes based on a single value?
b) If you have access to Minitab, use it to carry out part (a).

___ 2.58 (**Computer exercise**) The Bureau of Economic Analysis gathers information on the length of stay in Europe and the Mediterranean by U.S. travelers. Data are published in *Survey of Current Business*. We used Minitab's HISTOGRAM command to obtain a frequency distribution and frequency histogram for the lengths of stay, in days, of a sample of U.S. residents who traveled to Europe and the Mediterranean this year. The Minitab output is shown below.

```
Histogram of STAYS    N = 36

Midpoint    Count
   4.00       10    **********
  11.00       10    **********
  18.00        8    ********
  25.00        1    *
  32.00        2    **
  39.00        1    *
  46.00        2    **
  53.00        1    *
  60.00        0
  67.00        1    *
```

a) How many U.S. residents were sampled here?
b) What is the class mark for the sixth class?
c) What common class width was used to construct the frequency distribution?
d) How many stayed between 15 and 21 days?

___ 2.59 (**Computer exercise**) The U.S. Energy Information Administration collects data on residential energy consumption and expenditures. Results are published in *Residential Energy Consumption Survey: Consumption and Expenditures*. Data, in millions of BTU, on last year's energy consumptions were collected for a sample of households in the South. We employed Minitab's HISTOGRAM command to obtain a frequency distribution and frequency histogram of the data. Here is the computer output:

```
Histogram of ENERGY    N = 50

Midpoint    Count
   44.5       1     *
   54.5       7     *******
   64.5       7     *******
   74.5       3     ***
   84.5       6     ******
   94.5      10     **********
  104.5       5     *****
  114.5       4     ****
  124.5       2     **
  134.5       3     ***
  144.5       0
  154.5       2     **
```

a) How many households were sampled here?
b) What is the class mark for the sixth class?
c) What common class width was used to construct the frequency distribution?

d) How many of the households sampled had an energy consumption of between 100 and 109 million BTU, inclusive?

___ 2.60 **(Computer exercise)** Suppose the data in Exercise 2.46 are stored in a column named SCORES.
a) Which Minitab command should be used to obtain a dotplot for the data?
b) If you have access to Minitab, use it to carry out part (a).

___ 2.61 **(Computer exercise)** Suppose the data in Exercise 2.47 are stored in a column named TIMES.
a) Which Minitab command should be used to obtain a dotplot for the data?
b) If you have access to Minitab, use it to carry out part (a).

___ 2.62 **(Computer exercise)** We used Minitab to obtain the following dotplot of the weights of the sampled boxes of Wheat Flakes considered in Exercise 2.48 (weights are in ounces):

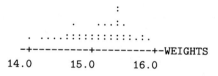

a) How many boxes of Wheat Flakes were sampled?
b) How many weighed between 15 and 15.5 ounces, inclusive?
c) In obtaining the dotplot, we used the INCREMENT subcommand. What increment did we use?
d) We also used the START subcommand to specify the first and last tick marks on the horizontal axis. What values did we specify?

___ 2.63 **(Computer exercise)** The dotplot below was obtained for the ages of the trucks from Exercise 2.49 by employing Minitab's DOTPLOT command (ages are in years):

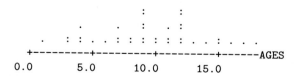

a) How many trucks were sampled?
b) How many trucks were between 5 and 8 years old, inclusive?
c) In obtaining the dotplot, we used the INCREMENT subcommand. What increment did we use?

d) We also used the START subcommand to specify the first tick mark on the horizontal axis. What value did we specify?

Class boundaries: Another method of depicting the classes on the horizontal axis of a histogram is to use *class boundaries*. The **lower class boundary** of a class is defined to be the number halfway between the lower class limit of the class and the upper class limit of the next lower class. Similarly, the **upper class boundary** of a class is defined to be the number halfway between the upper class limit of the class and the lower class limit of the next higher class. For instance, consider the class 50–59 of the days-to-maturity data (see Table 2.14 on page 35). We have

$$\text{Lower class boundary} = \frac{49 + 50}{2} = 49.5$$

$$\text{Upper class boundary} = \frac{59 + 60}{2} = 59.5$$

=== 2.64 Refer to Exercise 2.36.
a) Add a lower-class-boundary column and an upper-class-boundary column to the table given in Exercise 2.36. Then fill in these columns by computing the lower and upper class boundaries for all the classes.
b) Construct a frequency histogram for the data using class boundaries on the horizontal axis instead of class limits.
c) What are the advantages of using class boundaries instead of class limits? What are the disadvantages of using class boundaries?

=== 2.65 Refer to Exercise 2.37.
a) Add a lower-class-boundary column and an upper-class-boundary column to the table given in Exercise 2.37. Then fill in these columns by computing the lower and upper class boundaries for all the classes.
b) Construct a frequency histogram for the data using class boundaries on the horizontal axis instead of class limits.
c) What are the advantages of using class boundaries instead of class limits? What are the disadvantages of using class boundaries?

Relative-frequency polygons: Another commonly used graphical display is the *relative-frequency polygon*. In a **relative-frequency polygon**, a point is plotted above each *class mark* at a height equal to the relative frequency of the class. Then the points are joined with connecting lines. For instance, the relative-frequency polygon for the days-to-maturity data is as shown below:

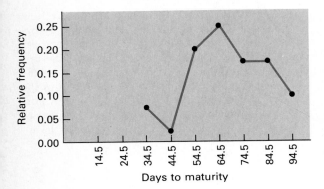

Short-Term Investments

= **2.66** Construct a relative-frequency polygon for the length-of-stay data given in Exercise 2.38.

= **2.67** Construct a relative-frequency polygon for the energy-consumption data given in Exercise 2.39.

Ogives: Cumulative information can be portrayed by using a graphical display called an *ogive* (ō′jīv). To construct an ogive, we first make a table showing cumulative frequencies and cumulative relative frequencies. Table 2.20 provides such a table for the days-to-maturity data shown in Table 2.1 on page 23.

TABLE 2.20

Less than	Cumulative frequency	Cumulative relative frequency
30	0	0.000
40	3	0.075
50	4	0.100
60	12	0.300
70	22	0.550
80	29	0.725
90	36	0.900
100	40	1.000

The first column of Table 2.20 gives the class limits and the second column of Table 2.20 gives the *cumulative frequencies*. A **cumulative frequency** is obtained by summing the frequencies of all classes representing values *less than* the specified class limit. For

instance, by referring to Table 2.14 on page 35, we can find the cumulative frequency of investments with a maturity period of less than 50 days. We see that:

$$\text{Cumulative frequency} = 3 + 1 = 4$$

This means that four of the investments have a maturity period of less than 50 days.

The third column of Table 2.20 gives the *cumulative relative frequencies*. A **cumulative relative frequency** is found by dividing the corresponding cumulative frequency by the total number of pieces of data. For instance, the cumulative relative frequency of investments with a maturity period of less than 50 days is obtained as follows:

$$\text{Cumulative relative frequency} = \frac{4}{40} = 0.100$$

This means that 10% of the investments have a maturity period of less than 50 days.

Using Table 2.20 we can now construct an ogive for the days-to-maturity data. In an **ogive,** a point is plotted above each *class limit* at a height equal to the cumulative relative frequency. Then the points are joined with connecting lines. Consequently, the ogive for the days-to-maturity data is as displayed in the following figure:

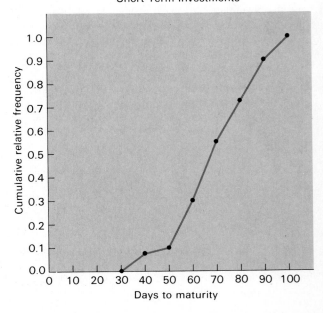

Short-Term Investments

═ 2.68   Refer to Exercise 2.38.
a) Construct a table similar to Table 2.20 for the length-of-stay data given in Exercise 2.38. Interpret the values in your table.
b) Draw an ogive for the data.

═ 2.69   Refer to Exercise 2.39.
a) Construct a table similar to Table 2.20 for the energy-consumption data given in Exercise 2.39. Interpret the values in your table.
b) Draw an ogive for the data.

2.4   Stem-and-leaf diagrams

New ways of displaying data are constantly being invented. One recently developed method is called a *stem-and-leaf diagram*. This ingenious diagram, invented in the late 1960s by Professor John Tukey of Princeton University, is often easier to construct than either a frequency distribution or a histogram, and generally displays more information. To illustrate stem-and-leaf diagrams, we return once more to the days-to-maturity data.

EXAMPLE 2.16   **Illustrates stem-and-leaf diagrams**

The data on the number of days to maturity for 40 short-term investments, given in Table 2.1, are repeated here in Table 2.21.

TABLE 2.21
Days to maturity for 40 short term investments

70	64	99	55	64	89	87	65
62	38	67	70	60	69	78	39
75	56	71	51	99	68	95	86
57	53	47	50	55	81	80	98
51	36	63	66	85	79	83	70

We grouped these data using the classes 30–39, 40–49, ..., 90–99 (Table 2.2, page 24), and we portrayed the data graphically with a frequency histogram (Figure 2.1, page 36). Now we will construct a stem-and-leaf diagram for the data. As you will see, with a stem-and-leaf diagram we can simultaneously group the data and obtain a graphical display similar to a histogram.

A stem-and-leaf diagram for the days-to-maturity data can be constructed as follows: First, select the leading digits from the data in Table 2.21. This gives the numbers 3, 4, ..., 9. Next, list those leading digits on the left-hand side of a page (see the colored numbers in Figure 2.8 at the top of the next page).

Finally, go through the data in Table 2.21 and write the final digit of each number to the right of the appropriate leading digit: The first investment has a maturity period of 70 days. This calls for a "0" to the right of the "7" in colored type in Figure 2.8. Reading down the first column of Table 2.21, the second investment has a maturity period of 62 days, so this calls for a "2" to the right of the "6" in colored type. Continuing in this manner, we obtain the diagram displayed in Figure 2.8. As indicated in the figure, the leading digits are called **stems,** the final digits are called **leaves,** and the entire diagram is called a **stem-and-leaf diagram.**

FIGURE 2.8
Stem-and-leaf
diagram for
days-to-maturity data

Stems   Leaves
        ↓   ↓

3 | 8 6 9
4 | 7
5 | 7 1 6 3 5 1 0 5
6 | 2 4 7 3 6 4 0 9 8 5
7 | 0 5 1 0 9 8 0
8 | 5 9 1 7 0 3 6
9 | 9 9 5 8

The stem-and-leaf diagram for the days-to-maturity data is similar to the frequency histogram for that data. This is because the length of the row of leaves for a class equals the frequency of the class. [Turn the stem-and-leaf diagram 90° counterclockwise and compare it to the frequency histogram in Figure 2.1 on page 36.] By shading each row of leaves, as in Figure 2.9, we get a diagram that looks even more like the frequency histogram of the data.

FIGURE 2.9
Shaded stem-and-leaf
diagram for
days-to-maturity data

The diagram in Figure 2.9 is called a **shaded stem-and-leaf diagram.** Because the numbers in the diagram are not shaded over completely, the shaded stem-and-leaf diagram exhibits the original, or raw, data as well as providing a graphical display of the frequency distribution. On the other hand, although a frequency histogram gives a graphical display of the frequency distribution, it is generally not possible to recover the raw data from the frequency histogram. (Why?)

Another form of stem-and-leaf diagram is called an **ordered stem-and-leaf diagram.** For this type of stem-and-leaf diagram, the leaves in each row are ordered from smallest to largest. This makes it easier to comprehend the data and also facilitates the computation of descriptive measures such as the median (to be discussed in Chapter 3). The ordered stem-and-leaf diagram for the days-to-maturity data is presented in Figure 2.10 at the top of the next page. ∎

FIGURE 2.10
Ordered
stem-and-leaf
diagram for
days-to-maturity data

```
3 | 6 8 9
4 | 7
5 | 0 1 1 3 5 5 6 7
6 | 0 2 3 4 4 5 6 7 8 9
7 | 0 0 0 1 5 8 9
8 | 0 1 3 5 6 7 9
9 | 5 8 9 9
```

EXAMPLE 2.17   **Illustrates stem-and-leaf diagrams**

The cholesterol-level data from Example 2.6 are repeated in Table 2.22. Data are in milligrams per 100 milliliters.

TABLE 2.22
Cholesterol levels
of 20 patients

210	209	212	208
217	207	210	203
208	210	210	199
215	221	213	218
202	218	200	214

Since these data are three-digit numbers, we use the first *two* digits as the stems and the third digits as the leaves. A stem-and-leaf diagram for the cholesterol levels is displayed in Figure 2.11.

FIGURE 2.11
Stem-and-leaf
diagram for
cholesterol-level data

```
19 | 9
20 | 8 2 9 7 0 8 3
21 | 0 7 5 0 8 2 0 0 3 8 4
22 | 1
```

The stem-and-leaf diagram in Figure 2.11 is only moderately helpful because there are so few stems. We can construct a better stem-and-leaf diagram by using two lines for each stem, with the first line for the leaf digits 0–4 and the second line for the leaf digits 5–9. This stem-and-leaf diagram is shown in Figure 2.12 at the top of the following page.   ■

MTB

As we have seen, stem-and-leaf diagrams have several advantages over the more classical techniques for grouping and graphing. However, stem-and-leaf diagrams do have some drawbacks. For instance, they are sometimes awkward to use with decimal data. Nonetheless, stem-and-leaf diagrams offer a viable alternative for organizing and summarizing data.

```
19 |
19 | 9
20 | 2 0 3
20 | 8 9 7 8
21 | 0 0 2 0 0 3 4
21 | 7 5 8 8
22 | 1
22 |
```

USING THE COMPUTER (OPTIONAL)

Minitab has a program, called **STEM-AND-LEAF,** that will construct a stem-and-leaf diagram for a set of data. In Example 2.17 we showed how to construct stem-and-leaf diagrams for some cholesterol-level data. In the next example, we will explain how Minitab's STEM-AND-LEAF command can be used to obtain those same stem-and-leaf diagrams.

EXAMPLE 2.18   *Illustrates the STEM-AND-LEAF command*

Consider again the data on cholesterol levels given in Table 2.22 on page 53. Use Minitab to obtain a stem-and-leaf diagram for these data using one line per stem; using two lines per stem.

SOLUTION   We begin by storing the cholesterol-level data in, say, C3 by applying the SET command. Then we use the NAME command to name C3 "Chollevs."

To obtain a stem-and-leaf diagram for the data using one line per stem, we first type the command STEM-AND-LEAF followed by the storage location of the data; that is, we type STEM-AND-LEAF 'CHOLLEVS';. Since, in this case, we are to use one line per stem, we also employ the (optional) **INCREMENT** subcommand. Specifically, we type INCREMENT=10.. This instructs Minitab to use a distance of 10 between the smallest possible number on one line and the smallest possible number on the next line. See Printout 2.3 at the top of the next page.

On the first line of the output we find a verbal description of what follows (Stem-and-leaf of CHOLLEVS) and the number of pieces of data (N = 20). The second line of the output (Leaf Unit = 1.0) indicates where the decimal point goes, in this case directly after each leaf digit.

Let us now discuss in detail the stem-and-leaf diagram in Printout 2.3. The second column of numbers gives the stems, and to the right of the stems are the leaves. Since the leaves on each line are ordered, we see that Printout 2.3 actually provides us with an *ordered* stem-and-leaf diagram.

The numbers in the first column of the stem-and-leaf diagram in Printout 2.3, called **depths,** are used to display cumulative frequencies. Starting from the top,

PRINTOUT 2.3
Minitab output for
STEM-AND-LEAF
using one
line per stem

```
MTB > SET C3
DATA> 210 209 212 208
DATA> 217 207 210 203
DATA> 208 210 210 199
DATA> 215 221 213 218
DATA> 202 218 200 214
DATA> END
MTB > NAME C3 'CHOLLEVS'
MTB > STEM-AND-LEAF 'CHOLLEVS';
SUBC> INCREMENT=10.


Stem-and-leaf of CHOLLEVS   N  = 20
Leaf Unit = 1.0


    1     19 9
    8     20 0237889
  (11)    21 00002345788
    1     22 1
```

the depths indicate the number of leaves (pieces of data) that lie in the given row or before. For instance, the 8 in the second row shows that there are eight leaves (pieces of data) in the first two rows.

When the row that contains the middle observation(s) is reached, the cumulative frequency is replaced by the number of leaves in that row enclosed by parentheses. Thus, in this case, we see that the middle observations lie in the 210s and that there are 11 data values in the 210s.

The depths following the row that contains the middle observation(s) indicate the number of leaves that lie in the given row or after. In this case, there is only one row following the row in which the middle observations lie, namely the fourth row. The 1 in the fourth row shows that there is one leaf (piece of data) in the last row. (We will be able to give a better illustration of depths shortly.) You should compare the stem-and-leaf diagram in Printout 2.3 to the one in Figure 2.11 on page 53.

Next let us obtain a stem-and-leaf diagram for the cholesterol-level data using two lines per stem. We proceed as before, except this time we type INCREMENT=5. instead of INCREMENT=10.. The command INCREMENT=5 instructs Minitab to use a distance of five between the smallest possible number on one line and the smallest possible number on the next line. See Printout 2.4 at the top of the following page.

Let us use this printout to again discuss depths. The 8 in the third row of the stem-and-leaf diagram indicates that there are eight leaves (pieces of data) in the first three rows; and the 5 in the fifth row indicates that there are five leaves (pieces of data) in the last two rows. You should compare the stem-and-leaf diagram in Printout 2.4 to the one obtained by hand in Figure 2.12 on page 54. ∎

PRINTOUT 2.4

Minitab output for
STEM-AND-LEAF
using two
lines per stem

```
MTB > STEM-AND-LEAF 'CHOLLEVS';
SUBC> INCREMENT=5.

Stem-and-leaf of CHOLLEVS   N  = 20
Leaf Unit = 1.0


       1     19 9
       4     20 023
       8     20 7889
      (7)    21 0000234
       5     21 5788
       1     22 1
```

Exercises 2.4

In each of Exercises 2.70–2.73,
a) construct a stem-and-leaf diagram.
b) construct an ordered stem-and-leaf diagram.

___ **2.70** A research physician conducted a study on the ages of persons with diabetes. The following data were obtained for the ages of a sample of 35 diabetics:

48	41	57	83	41	55	59
61	38	48	79	75	77	7
54	23	47	56	79	68	61
64	45	53	82	68	38	70
10	60	83	76	21	65	47

___ **2.71** A soft-drink bottler sells "one-liter" bottles of soda. A consumer group is concerned that the bottler may be shortchanging the customers. Thirty bottles of soda are randomly selected. The contents, in milliliters, of the bottles chosen are shown below.

1025	977	1018	975	977
990	959	957	1031	964
986	914	1010	988	1028
989	1001	984	974	1017
1060	1030	991	999	997
996	1014	946	995	987

[Use the stems 91, 92, . . . , 106.]

___ **2.72** The Bureau of Economic Analysis gathers information on the length of stay in Europe and the Mediterranean by U.S. travelers. Data are published in

Survey of Current Business. A sample of 36 U.S. residents who traveled to Europe and the Mediterranean this year yielded the following data, in days, on length of stay:

41	16	6	21	1	21
5	31	20	27	17	10
3	32	2	48	8	12
21	44	1	56	5	12
3	13	15	10	18	3
1	11	14	12	64	10

___ **2.73** The U.S. Energy Information Administration collects data on residential energy consumption and expenditures. Results are published in the document *Residential Energy Consumption Survey: Consumption and Expenditures.* The following table provides the data on last year's energy consumptions for a sample of 50 households in the South. Data are given in millions of BTU.

130	55	45	64	155	66	60	80	102	62
58	101	75	111	151	139	81	55	66	90
97	77	51	67	125	50	136	55	83	91
54	86	100	78	93	113	111	104	96	113
96	87	129	109	69	94	99	97	83	97

___ **2.74** As reported by the U.S. Bureau of the Census in *Current Population Reports,* the percentage of the adult population in each state that has completed high school is as follows:

State	Percent	State	Percent	State	Percent
AL	57	LA	58	OH	67
AK	83	ME	69	OK	66
AZ	72	MD	67	OR	76
AR	56	MA	72	PA	65
CA	74	MI	68	RI	61
CO	79	MN	73	SC	54
CT	70	MS	55	SD	68
DE	69	MO	64	TN	56
FL	67	MT	74	TX	63
GA	56	NE	73	UT	80
HI	74	NV	76	VT	71
ID	74	NH	72	VA	62
IL	67	NJ	67	WA	78
IN	66	NM	69	WV	56
IA	72	NY	66	WI	70
KS	73	NC	55	WY	78
KY	53	ND	66		

Construct a stem-and-leaf diagram for the percentages
a) using one line per stem.
b) using two lines per stem.

___ 2.75   The U.S. Federal Bureau of Investigation (FBI) published the following annual crime rates, by state, in *Crime in the United States*. Rates are per 1000 population.

State	Rate	State	Rate	State	Rate
AL	43	LA	61	OH	44
AK	62	ME	35	OK	60
AZ	73	MD	56	OR	71
AR	39	MA	47	PA	31
CA	68	MI	65	RI	49
CO	70	MN	44	SC	51
CT	48	MS	33	SD	27
DE	48	MO	47	TN	45
FL	82	MT	45	TX	74
GA	55	NE	39	UT	55
HI	57	NV	63	VT	40
ID	42	NH	33	VA	39
IL	55	NJ	52	WA	69
IN	39	NM	66	WV	23
IA	42	NY	58	WI	41
KS	48	NC	43	WY	44
KY	31	ND	26		

Construct a stem-and-leaf diagram for the crime rates
a) using one line per stem.
b) using two lines per stem.

___ 2.76   The annual average maximum temperatures for selected cities in the United States are given below. [SOURCE: U.S. National Oceanic and Atmospheric Administration, *Climatography of the United States*.]

City	Annual average max. temp.	City	Annual average max. temp.
Mobile, AL	77	Reno, NV	67
Juneau, AK	47	Concord, NH	57
Phoenix, AZ	85	Atlantic City, NJ	63
Little Rock, AR	73	Albuquerque, NM	70
Los Angeles, CA	70	Albany, NY	58
Sacramento, CA	73	Buffalo, NY	56
San Francisco, CA	65	New York, NY	62
Denver, CO	64	Charlotte, NC	71
Hartford, CT	60	Raleigh, NC	70
Wilmington, DE	64	Bismarck, ND	54
Washington, DC	67	Cincinnati, OH	64
Jacksonville, FL	79	Cleveland, OH	59
Miami, FL	83	Columbus, OH	62
Atlanta, GA	71	Oklahoma City, OK	71
Honolulu, HI	84	Portland, OR	62
Boise, ID	63	Philadelphia, PA	63
Chicago, IL	59	Pittsburgh, PA	60
Peoria, IL	60	Providence, RI	59
Indianapolis, IN	62	Columbia, SC	75
Des Moines, IA	59	Sioux Falls, SD	57
Wichita, KS	68	Memphis, TN	72
Louisville, KY	66	Nashville, TN	70
New Orleans, LA	78	Dallas-Ft. Worth, TX	77
Portland, ME	55	El Paso, TX	78
Baltimore, MD	65	Houston, TX	79
Boston, MA	59	Salt Lake City, UT	64
Detroit, MI	58	Burlington, VT	54
Sault Ste. Marie, MI	49	Norfolk, VA	68
Duluth, MN	48	Richmond, VA	69
Mnpls-St. Paul, MN	54	Seattle-Tacoma, WA	59
Jackson, MS	76	Spokane, WA	57
Kansas City, MO	64	Charleston, WV	66
St. Louis, MO	66	Milwaukee, WI	55
Great Falls, MT	56	Cheyenne, WY	58
Omaha, NE	62	San Juan, PR	86

Construct a stem-and-leaf diagram for these data
a) using two lines per stem.
b) using five lines per stem.

___ 2.77   The annual average minimum temperatures for selected cities in the United States are given below. [SOURCE: U.S. National Oceanic and Atmospheric Administration, *Climatography of the United States*.]

City	Annual average min. temp.	City	Annual average min. temp.
Mobile, AL	58	Reno, NV	32
Juneau, AK	33	Concord, NH	33
Phoenix, AZ	57	Atlantic City, NJ	43
Little Rock, AR	51	Albuquerque, NM	42
Los Angeles, CA	55	Albany, NY	37
Sacramento, CA	48	Buffalo, NY	39
San Francisco, CA	48	New York, NY	47
Denver, CO	36	Charlotte, NC	49
Hartford, CT	40	Raleigh, NC	48
Wilmington, DE	45	Bismarck, ND	29
Washington, DC	49	Cincinnati, OH	45
Jacksonville, FL	57	Cleveland, OH	41
Miami, FL	69	Columbus, OH	42
Atlanta, GA	51	Oklahoma City, OK	49
Honolulu, HI	70	Portland, OR	44
Boise, ID	39	Philadelphia, PA	45
Chicago, IL	40	Pittsburgh, PA	41
Peoria, IL	41	Providence, RI	41
Indianapolis, IN	42	Columbia, SC	51
Des Moines, IA	40	Sioux Falls, SD	34
Wichita, KS	45	Memphis, TN	52
Louisville, KY	46	Nashville, TN	49
New Orleans, LA	59	Dallas-Ft. Worth, TX	55
Portland, ME	35	El Paso, TX	49
Baltimore, MD	45	Houston, TX	57
Boston, MA	44	Salt Lake City, UT	39
Detroit, MI	39	Burlington, VT	35
Sault Ste. Marie, MI	31	Norfolk, VA	51
Duluth, MN	29	Richmond, VA	47
Mnpls-St. Paul, MN	35	Seattle-Tacoma, WA	44
Jackson, MS	53	Spokane, WA	37
Kansas City, MO	44	Charleston, WV	44
St. Louis, MO	45	Milwaukee, WI	38
Great Falls, MT	33	Cheyenne, WY	33
Omaha, NE	40	San Juan, PR	73

Construct a stem-and-leaf diagram for these data
a) using two lines per stem.
b) using five lines per stem.

Exercises 2.78–2.81 are computer exercises.

___ **2.78** (**Computer exercise**) Suppose the data in Exercise 2.70 are stored in a column named AGES.
a) Which Minitab command and subcommands (if any) should be employed to construct a stem-and-leaf diagram for these data
  (i) using one line per stem?
  (ii) using two lines per stem?
  (iii) using five lines per stem?
b) If you have access to Minitab, use it to obtain the three stem-and-leaf diagrams in part (a).

___ **2.79** (**Computer exercise**) Suppose that the data in Exercise 2.71 are stored in a column named CONTENTS.
a) Which Minitab command and subcommands (if any) should be employed to construct a stem-and-leaf diagram for these data
  (i) using one line per stem?
  (ii) using two lines per stem?
  (iii) using five lines per stem?
b) If you have access to Minitab, use it to obtain the three stem-and-leaf diagrams in part (a).

___ **2.80** (**Computer exercise**) The U.S. Bureau of the Census reports the percentage of the adult population in each state that has completed high school in the publication *Current Population Reports*. We applied Minitab's STEM-AND-LEAF command and the INCREMENT subcommand to obtain a stem-and-leaf diagram of those data. Here is the Minitab output:

```
Stem-and-leaf of HSCOMP    N = 50
Leaf Unit = 1.0

    2      5  34
   10      5  55666678
   14      6  1234
  (15)     6  566667777788999
   21      7  00122223334444
    7      7  66889
    2      8  03
```

a) How many pieces of data are there?
b) What increment was used for the INCREMENT subcommand?
c) How many lines per stem are used in this stem-and-leaf diagram?
d) Use the depths to determine how many of the percentages are at least 75%, i.e., 75% or greater.
e) Use the depths to determine how many of the percentages are at most 64%, i.e., 64% or less.
f) What is the largest percentage?
g) List the percentages that are in the 50s.

___ **2.81** (**Computer exercise**) A hardware manufacturer produces 10-millimeter diameter bolts. The manufacturer knows that the diameters of the bolts produced vary somewhat from 10 mm and also from each other. To get an idea of the variation in diameters, the manufacturer takes a sample of bolts and obtains the following stem-and-leaf diagram for the diameters:

```
Stem-and-leaf of BOLTDIAM  N  = 20
Leaf Unit = 0.010

    1    98 9
    1    99
    1    99
    3    99 45
    5    99 67
    9    99 8899
  (2)   100 01
    9    100 2333
    5    100 55
    3    100
    3    100 88
    1    101 0
```

a) How many bolts were sampled?
b) What is the smallest diameter, in millimeters, of the bolts sampled? *(Hint:* Note that the leaf unit is 0.010.)
c) How many lines per stem are used in this stem-and-leaf diagram?
d) Use the depths to determine how many of the bolts sampled have diameters of at least 10.04 mm, i.e., 10.04 mm or greater.
e) Use the depths to determine how many of the bolts sampled have diameters of at most 9.97 mm, i.e., 9.97 mm or less.
f) List the diameters that are less than 10 mm.

2.5 Misleading graphs

Graphs and charts are frequently constructed in a manner that causes them to be misleading. Sometimes this is intentional and sometimes it is not. Regardless of the intent, it is important to read and interpret graphs and charts with a great deal of care. In this section we will examine some misleading graphs and charts. We begin with the following example.

EXAMPLE 2.19   *Illustrates truncated graphs*

Figure 2.13 shows a bar graph from an article in a major metropolitan newspaper. The graph displays the unemployment rates in the United States from September of one year to March of the next year.

FIGURE 2.13
Unemployment rates

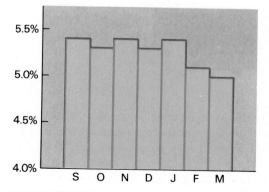

A quick look at Figure 2.13 might lead you to the conclusion that the unemployment rate dropped by roughly one-fourth between January and March, since the bar for March is about one-fourth smaller than the bar for January. In reality, however, the unemployment rate dropped less than one-thirteenth, from 5.4% to 5.0%. Consequently, we see that a more careful analysis of the graph is required than first meets the eye.

The unemployment-rate graph in Figure 2.13 is an example of a **truncated graph** because the vertical axis, which should start at 0%, starts at 4% instead. Thus, the part of the graph from 0% to 4% has been cut off or *truncated*. This truncation causes the bars to be out of correct proportion and hence creates a misleading impression, as we discovered earlier. The graph would be even more misleading if it started at 4.5%. To see this, slide a piece of paper over the bottom of Figure 2.13 so that the bars begin at 4.5%. By how much does it now appear that the unemployment rate dropped between January and March?

As we have observed, the truncated graph in Figure 2.13 is potentially misleading. However, it is probably safe to say that the truncation was done to give a picture of the "ups" and "downs" in the unemployment-rate pattern, rather than to intentionally mislead the reader.

A version of Figure 2.13 that is not truncated is shown in Figure 2.14. Although Figure 2.14 provides a correct graphical display of the unemployment-rate data, the "ups" and "downs" are not as easy to spot as they are in the truncated graph in Figure 2.13. ■

FIGURE 2.14
"Untruncated"
version of
Figure 2.13

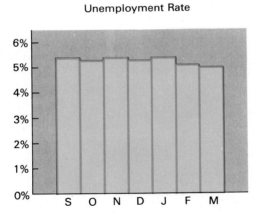

Truncated graphs have been a target of statisticians for a long time. Many statistics books discuss them and warn against their use. Nonetheless, as we have seen in Example 2.19, truncated graphs are still used today, even in reputable publications.

On the other hand, Example 2.19 also suggests that it may be desirable to cut off part of the vertical axis of a graph in order to more easily convey certain relevant information (such as the "ups" and "downs" of the monthly unemployment rates). In these cases, a truncated graph should *not* be used. Instead, a special symbol such as "//" should be employed to signify that the vertical axis has been modified.

The two graphs in Figure 2.15 provide an excellent illustration.[†] Both graphs portray the number of new single-family homes sold per month over several months. The first graph, Figure 2.15(a), is a truncated graph, truncated most likely in an attempt to present the reader with a clear visual display of the variation in sales. The second graph, Figure 2.15(b), accomplishes the same result, but is less subject to misinterpretation. Indeed, the reader is aptly warned by the slashes that part of the vertical axis between 0 and 500 has been removed.

FIGURE 2.15
New single-family
home sales

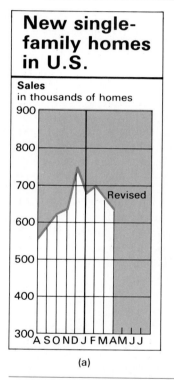

(a)

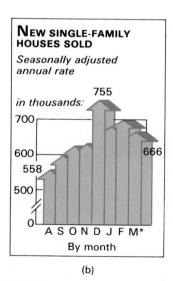

(b)

IMPROPER SCALING

Misleading graphs and charts can also result because of **improper scaling.** The following example shows how this can happen.

† Figure 2.15(a) reprinted by permission Tribune Media Services. Figure 2.15(b) data from U.S. Department of Commerce and U.S. Department of Housing and Urban Development.

EXAMPLE 2.20   **Illustrates improper scaling**

A developer is preparing a brochure to attract investors for a new shopping center that is to be built in an area of Denver, Colorado. The area is growing rapidly; this year twice as many homes will be built there as last year. To illustrate that fact, the developer draws a pictogram as in Figure 2.16.

FIGURE 2.16

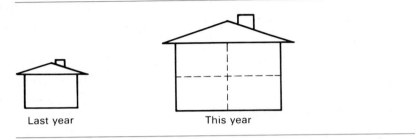

The house on the left represents the number of homes built last year. Since the number of homes that will be built this year is double the number built last year, the developer makes the house on the right twice as tall and twice as wide as the house on the left. However, this scaling is improper because it gives the visual impression that *four times* as many homes will be built this year as last. So, the developer's brochure may mislead the unwary investor. ■

There are countless ways in which graphs and charts can be misleading, besides the two ways we have discussed. Many more can be found in the entertaining and classic book *How to Lie with Statistics* by Darrell Huff (New York: W.W. Norton & Co., 1955). The main purpose of this section has been to show that graphs and charts should be read and constructed with care.

Exercises 2.5

___ **2.82** Find two examples of graphs in a current newspaper or magazine that might be misleading. Explain why you think the graphs you obtain are potentially misleading.

___ **2.83** Each year, the director of the reading program in a school district administers a standard test of reading skills. Then the director compares the average score for his district with the national average. Figure 2.17, shown at the right, was presented to the school board in 1990.

a) Observe a truncated version of Figure 2.17 by sliding a piece of paper over the bottom of the graph so that the bars start at 16.

FIGURE 2.17   Average reading scores

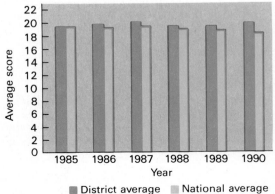

b) Observe a truncated version of Figure 2.17 by sliding a piece of paper over the bottom of the graph so that the bars start at 18.

c) What misleading impression about the 1990 scores is given by the truncated graphs that you observed in parts (a) and (b)?

___ **2.84**   The following bar graph was taken from a newspaper article entitled "Immigrants add seasoning to America's melting pot." [Used with permission from American Demographics, ©1985, Ithaca, NY.]

Race and Ethnicity in America
(in millions)

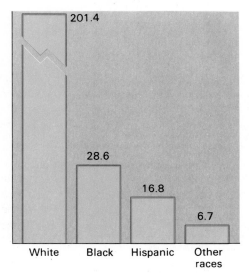

Data from Census Bureau July 1984 Estimates

a) Explain why a break is shown in the first bar on the left.

b) Why was it necessary to construct the graph with a broken bar?

c) Do you think this bar graph is potentially misleading? Why?

___ **2.85**   The following bar graph, taken from the *Arizona Republic,* provides data on the M2 money supply over several months in 1989. M2 consists of cash in circulation, deposits in checking accounts, non-bank traveler's checks, accounts such as savings deposits, and money-market mutual funds. [SOURCE: Federal Reserve System.]

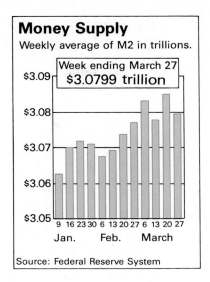

Money Supply
Weekly average of M2 in trillions.

Week ending March 27
$3.0799 trillion

Source: Federal Reserve System

a) What is wrong with the bar graph?

b) Draw a version of the bar graph with an untruncated and unmodified vertical axis.

c) Draw a version of the bar graph in which the vertical axis is modified in an acceptable manner.

=== **2.86**   Refer to Example 2.20 on page 62. Indicate a correct way in which the developer can illustrate the fact that twice as many homes will be built in the area this year as last year.

=== **2.87**   A manufacturer of golf balls has determined that a newly developed process results in a ball lasting roughly twice as long as a ball produced using the current process. To illustrate this graphically, she designs a brochure showing a "new" ball with twice the radius of the "old" ball.

Old ball   New ball

a) What is wrong with this? (*Hint:* The volume of a sphere is directly proportional to the cube of its radius.)

b) Indicate a correct way in which the manufacturer can illustrate the fact that the "new" ball lasts twice as long as the "old" ball.

Chapter review

YOU SHOULD
BE ABLE TO

1. classify data as qualitative, ordinal, metric, or frequency data.
2. distinguish between discrete data and continuous data.
3. group data into a frequency distribution and a relative-frequency distribution.
4. construct a grouped-data table.
5. draw a frequency histogram and a relative-frequency histogram.
6. construct a dotplot.
7. draw a pie chart and a bar graph.
8. construct a stem-and-leaf diagram, a shaded stem-and-leaf diagram, and an ordered stem-and-leaf diagram.
9. identify and correct misleading graphs.
10. use the Minitab commands covered in this chapter.*
11. interpret the output obtained from the application of the Minitab commands discussed in this chapter.*

REVIEW TEST

1. The world's five largest hydroelectric plants, based on ultimate capacity, are as shown in the table at the right. Capacities are in megawatts. [SOURCE: T. W. Mermel, *Intl. Waterpower & Dam Construction Handbook, 88.*]
   a) What type of data is given in the left-hand column of the table?
   b) What type of data is given in the right-hand column of the table?

Rank	Name	Country	Capacity
1	Turukhansk	USSR	20,000
2	Itaipu	Brazil/Para.	13,320
3	Grand Coulee	U.S.A.	10,830
4	Guri	Venezuela	10,300
5	Tucurui	Brazil	7,260

c) What type of data is given in the third column of the table?

d) What type of data is the information that three of the five largest hydroelectric plants in the world are in South America?

2. The ages at inauguration for the first 40 Presidents of the United States are as follows. [SOURCE: *The World Almanac, 1989.*]

George Washington....57	Chester A. Arthur.....50
John Adams.........61	Grover Cleveland......47
Thomas Jefferson.....57	Benjamin Harrison 55
James Madison........57	Grover Cleveland......55
James Monroe........58	William McKinley.....54
John Quincy Adams...57	Theodore Roosevelt ...42
Andrew Jackson.......61	William Howard Taft..51
Martin Van Buren.....54	Woodrow Wilson......56
William H. Harrison...68	Warren G. Harding....55
John Tyler............51	Calvin Coolidge.......51
James K. Polk 49	Herbert C. Hoover 54
Zachary Taylor........64	Franklin D. Roosevelt .51
Millard Fillmore.......50	Harry S. Truman......60
Franklin Pierce........48	Dwight D. Eisenhower.62
James Buchanan 65	John F. Kennedy......43
Abraham Lincoln......52	Lyndon B. Johnson....55
Andrew Johnson 56	Richard M. Nixon......56
Ulysses S. Grant 46	Gerald Rudolph Ford..61
Rutherford B. Hayes ..54	Jimmy Carter.........52
James A. Garfield.....49	Ronald Reagan........69

a) Construct a grouped-data table for these inauguration ages using equal-width classes and beginning with the class 40–44.

b) Draw a frequency histogram for the inauguration ages based on your grouping in part (a).

3. Refer to Problem 2. Construct a dotplot for the ages at inauguration of the first 40 Presidents of the United States.

4. Refer to Problem 2. Construct an *ordered* stem-and-leaf diagram for the inauguration ages of the first 40 Presidents of the United States
a) using one line per stem.
b) using two lines per stem.
c) Which of the two stem-and-leaf diagrams that you just constructed corresponds to the frequency distribution of Problem 2(a)?

*5. (**Computer problem**) Suppose the age data in Problem 2 are stored in a column named AGES.
a) Which Minitab command and subcommands (if any) should be used to obtain a frequency distribution and frequency histogram of the data similar to the ones in Problem 2?

b) Which Minitab command should be used to obtain a dotplot of the data?

c) Which Minitab command and subcommands (if any) should be used to obtain the stem-and-leaf diagram in Problem 4(b)?

*6. (**Computer problem**) A city planner working on bikeways needs information about local bicycle commuters. She designs a questionnaire. One of the questions asks how many minutes it takes the rider to pedal from home to his or her destination. The times, rounded to the nearest minute, for a sample of local bicycle commuters are stored in Minitab and the following output is obtained by applying the HISTOGRAM command:

```
Histogram of TIMES    N = 22

Midpoint    Count
  12.00       1    *
  17.00       3    ***
  22.00       6    ******
  27.00       7    *******
  32.00       3    ***
  37.00       1    *
  42.00       0
  47.00       1    *
```

a) How many times are in the sample?
b) What is the class mark for the fourth class?
c) How many times are between 30 and 34 minutes, inclusive?
d) Both the INCREMENT and START subcommands were used to get the above Minitab output. What value was specified for each?
e) What column name was chosen for the data?

*7. (**Computer problem**) Refer to Problem 6. The Minitab output below provides a stem-and-leaf diagram for the sample of bicycle-commuter times.

```
Stem-and-leaf of TIMES     N  = 22
Leaf Unit = 1.0

   1      1 2
   4      1 569
  10      2 122334
  (7)     2 6678999
   5      3 011
   2      3 7
   1      4
   1      4 8
```

a) How many lines per stem are used in this stem-and-leaf diagram?

b) Use the depths to determine how many of the times are 35 minutes or more.

c) Use the depths to determine how many of the times are less than 20 minutes.

d) What is the longest time in the sample?

e) List the times that are in the 30s.

8. The Prescott National Bank has six tellers available to serve customers. The data in the following table provide the number of busy tellers observed at 25 spot checks:

6	5	4	1	5
6	1	5	5	5
3	5	2	4	3
4	5	0	6	4
3	4	2	3	6

a) Construct a grouped-data table for these data using single-value grouping.

b) Draw a relative-frequency histogram for the data based on the grouping that you obtained in part (a).

9. The National Safety Council reports that in 1986 the number of accidental deaths by type in the United States were as follows:

Type	Frequency
Motor vehicle	48,700
Work	11,100
Home	20,300
Public	18,000

a) Draw a pie chart of the data that displays the percentage of accidental deaths in each of the four categories.

b) Draw a bar graph of the data that displays the relative frequency of accidental deaths in each of the four categories.

10. According to *The World Almanac, 1989*, the highs for the Dow Jones Industrial Averages for the years 1952 through 1987 are as shown in the table at the top of the next column.

Year	High	Year	High
1952	292.00	1970	842.00
1953	293.79	1971	950.82
1954	404.39	1972	1036.27
1955	488.40	1973	1051.70
1956	521.05	1974	891.66
1957	520.77	1975	881.81
1958	583.65	1976	1014.79
1959	679.36	1977	999.75
1960	685.47	1978	907.74
1961	734.91	1979	897.61
1962	726.01	1980	1000.17
1963	767.21	1981	1024.05
1964	891.71	1982	1071.55
1965	969.26	1983	1287.20
1966	995.15	1984	1286.64
1967	943.08	1985	1553.10
1968	985.21	1986	1955.57
1969	968.85	1987	2722.42

a) Construct a grouped-data table for the highs using equal-width classes. Start with the class "200–under 400."

b) Draw a relative-frequency histogram for the highs based on your result from part (a).

11. The following graph is from a newspaper article entitled "Hand that rocked cradle turns to work as women reshape U.S. labor force." [Used with permission from American Demographics, ©1985.]

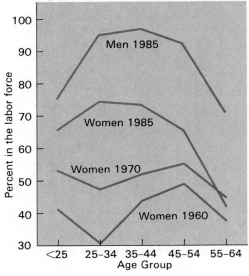

Working Men and Women by Age, 1960–1985

a) Cover up the numbers on the vertical axis of the graph with a piece of paper.

b) Look at the 1970 and 1985 graphs for women. Focus on the 25–34-year-old age group. What impression does the graph convey regarding the ratio of the percentages of women in the labor force for 1985 and 1970?

c) Now remove the piece of paper from the graph. Using the vertical scale, find the actual ratio of the percentages of 25–34-year-old women in the labor force for 1985 and 1970.

d) Why is the graph potentially misleading?

e) What can be done to make the graph less potentially misleading?

CHAPTER 3

DESCRIPTIVE MEASURES

In Chapter 2 we began our study of descriptive statistics. We learned how to organize data into tables and summarize data with graphical displays. Another method of summarizing data is to compute a number, such as an average, that describes the data set. Numbers that are used to describe data sets are called **descriptive measures.** In this chapter we will continue our study of descriptive statistics by examining some of the most important descriptive measures.

CHAPTER OUTLINE

3.1 Measures of central tendency Introduces the mean, the median, and the mode—descriptive measures that are used to indicate where the center or most typical value of a data set lies.

3.2 Summation notation; the sample mean Presents some useful shorthand notation employed in statistics.

3.3 Measures of dispersion; the sample standard deviation Discusses the range and standard deviation, descriptive measures that are used to indicate the amount of variation in a data set.

3.4 Interpretation of the standard deviation; z-scores Provides both a qualitative and quantitative interpretation of the standard deviation; and introduces z-scores for sample data.

3.5 Computing $\bar{x}$ and s for grouped data Indicates how to compute the sample mean and sample standard deviation of a data set that has been grouped into a frequency distribution.

3.6 Percentiles; box-and-whisker diagrams Examines some additional descriptive measures and introduces an associated graphical display.

3.7 Descriptive measures for populations; use of samples Defines the mean and standard deviation of a population, explains the role of samples with regard to those two descriptive measures, and introduces z-scores for population data.

3.1 Measures of central tendency

Descriptive measures that indicate where the center or most typical value of a data set lies are called **measures of central tendency,** often more simply referred to as averages. In this section we will discuss the three most important measures of central tendency—the *mean,* the *median,* and the *mode.*

THE MEAN

The most commonly used measure of central tendency is the *mean.* When people speak of taking an average, it is the mean that they are most often referring to. The definition of the mean is as follows:

DEFINITION 3.1   Mean of a data set

The *mean* of a data set is defined to be the sum of the data divided by the number of pieces of data:

$$\text{Mean} = \frac{\text{Sum of the data}}{\text{Number of pieces of data}}$$

EXAMPLE 3.1   *Illustrates Definition 3.1*

One of the authors of this text spent a summer working for a small mathematical consulting firm. The firm employed a few senior consultants who made between $700 and $950 per week, a few junior consultants who made between $300 and $350 per week, and several clerical helpers who made $200 per week. There was more work in the first half of the summer than in the second half, so there were more employees during the first half. Tables 3.1 and 3.2 give typical lists of weekly earnings for the two halves of the summer.

TABLE 3.1
Data Set I

$200	200	200	840	200	200	300
200	300	350	700	350	950	

TABLE 3.2
Data Set II

$200	200	840	350	300
300	200	200	950	200

Determine the mean of each of the two data sets.

SOLUTION   According to Definition 3.1, the mean of a data set is obtained by summing all the data and then dividing that sum by the total number of pieces of data. As we see from Table 3.1, there are 13 pieces of data in Data Set I. Adding the 13 pieces of data in that data set, we find that the sum of the data is $4990. Consequently,

$$\text{Mean of Data Set I} = \frac{\$4990}{13} = \$383.85 \text{ (to the nearest cent)}$$

Similarly,

$$\text{Mean of Data Set II} = \frac{\$3740}{10} = \$374.00$$

MTB   Thus, the mean salary of the 13 employees in Data Set I is $383.85 and that of the
10 employees in Data Set II is $374.00. ∎

THE MEDIAN

Another frequently used measure of central tendency is the *median.* Essentially, the median of a data set is the number that divides the bottom 50% of the data from the top 50%. To obtain the median of a data set, we arrange the data in increasing order and then determine the middle value in the ordered list. More precisely, we have the following definition:

DEFINITION 3.2   Median of a data set

The *median* of a data set is defined to be

1. the data value exactly in the middle of its ordered list, if the number of pieces of data is odd.
2. the mean of the two middle data values in its ordered list, if the number of pieces of data is even.

EXAMPLE 3.2   *Illustrates Definition 3.2*

Consider again the two sets of salary data given in Tables 3.1 and 3.2 on page 70. Determine the median of each of the two data sets.

SOLUTION   The number of pieces of data in Data Set I is 13, an *odd* number. Consequently, the median of that data set is the data value exactly in the middle of its ordered list. We have arranged the salaries in Data Set I in increasing order and have obtained the median, which is the seventh salary in the ordered list:

six lowest salaries   six highest salaries

200 200 200 200 200 200 300   300 350 350 700 840 950

↓

Median = 300

Thus, the median salary of the 13 employees in Data Set I is $300.
   The number of pieces of data in Data Set II is 10, an *even* number. Consequently, the median of that data set is the mean of the two middle data values in its ordered list. We have arranged the salaries in Data Set II in increasing order and have obtained the median, which is the mean of the fifth and sixth salaries in the ordered list:

five lowest salaries   five highest salaries

200   200   200   200   200   300   300   350   840   950
                        ↓   ↓

$$\text{Median} = \frac{200 + 300}{2} = 250$$

MTB   Thus, the median salary of the 10 employees in Data Set II is $250.   ■

THE MODE

The final measure of central tendency that we will discuss is the *mode*. Basically, the mode is the data value that occurs most frequently in a data set. To be exact, we make the following definition:

DEFINITION 3.3   Mode of a data set

If no data value in a data set occurs more than once, then we say that the data set has no mode. Otherwise, the *mode* of the data set is defined to be the data value or data values that occur most frequently.

Thus, a data set may have no mode, one mode, two modes, and so on. If a data set has two modes, it is said to be **bimodal;** if it has three modes, it is said to be **trimodal;** and so forth.

To obtain the mode(s) of a data set, we construct a frequency distribution for the data using classes based on a single value. The mode(s) can then be determined easily from the frequency distribution. Here is an example.

EXAMPLE 3.3   *Illustrates Definition 3.3*

Determine the mode(s) of each of the two sets of salary data given in Tables 3.1 and 3.2 on page 70.

SOLUTION   First we consider the salary data in Data Set I. From Table 3.1 we obtain a frequency distribution of the data using classes based on a single value. This is shown in Table 3.3.

TABLE 3.3
Frequency
distribution for
Data Set I using
single-value grouping

Salary	Frequency
200	6
300	2
350	2
700	1
840	1
950	1

We see from Table 3.3 that the most frequently occurring data value in Data Set I is 200 (which occurs six times). Consequently, the mode of the 13 salaries in Data Set I is $200.

Next we determine the mode for Data Set II. Referring to Table 3.2, we construct the following frequency distribution using classes based on a single value:

TABLE 3.4
Frequency
distribution for
Data Set II using
single-value grouping

Salary	Frequency
200	5
300	2
350	1
840	1
950	1

MTB

Table 3.4 shows that 200 is the data value that occurs most often in Data Set II. Therefore, the mode of the 10 salaries in Data Set II is $200. ∎

As we noted, a data set can have more than one mode if there is a tie for the most frequently-occurring data value. For instance, suppose two of the clerical helpers in Data Set I, who make $200 per week, were promoted to $300-per-week jobs. Then both the data value 200 and the data value 300 would occur most frequently (four times each). So this new data set would be bimodal and have the two modes $200 and $300.

COMPARISON OF THE MEAN, MEDIAN, AND MODE

The mean, median, and mode of a data set are frequently different. Table 3.5 summarizes the definitions of these three measures of central tendency along with their values for Data Set I and Data Set II, which we computed in Examples 3.1–3.3.

TABLE 3.5
Means, medians,
and modes of
salaries in Data Set I
and Data Set II

Measure of central tendency	Definition	Data Set I	Data Set II
Mean	$\dfrac{\text{Sum of the data}}{\text{Number of pieces of data}}$	$383.85	$374.00
Median	Middle value in ordered list	$300.00	$250.00
Mode	Most frequent data value	$200.00	$200.00

In both Data Sets I and II, the mean is larger than the median. This is because the mean is affected strongly by the few large salaries in each data set.

In general, the mean is sensitive to very large or very small data values, whereas the median is not. Consequently, when the choice for the measure of central tendency is between the mean and the median, the median is ordinarily preferred for data sets that have exceptional (very large or small) values. On the other hand, the mean takes into account the numerical value of every piece of data in a data set, whereas the median does not.

The mode for each of the two data sets differs from both the mean and the median. Whereas the mean and the median are aimed at finding the center of a data set, the mode is not—the most frequently occurring data value may not be near the center.

It should now be clear that the mean, median, and mode generally provide different information. There is no simple rule for deciding which one of these measures of central tendency to use in any given situation. You will attain skill in making such decisions through practice. However, even experts may disagree about which measure of central tendency is most suitable for a particular data set. The next example discusses three data sets and suggests which measure of central tendency is probably most appropriate for each.

EXAMPLE 3.4   *Illustrates the selection of the appropriate measure*
of central tendency

a) A student takes four exams in a biology class. His grades are 88, 75, 95, and 100. If asked for his average, which measure of central tendency is the student likely to report?   *mean*

b) The National Association of REALTORS® publishes data on resale prices of homes in the United States. Which measure of central tendency is most appropriate for such resale prices?   *median*

c) In the 1989 Boston Marathon, there were two categories of official finishers, male and female. The following table provides a frequency distribution for those data. Which measure of central tendency should be used here?

Sex	Frequency
Male	4239
Female	864

More freq. data value

SOLUTION   a) Chances are that the student would report the mean of his four exam scores, which is 89.5. The mean is probably the most reasonable measure of central tendency to use here since it takes into account the numerical value of each score and, therefore, indicates total overall performance.

b) The most appropriate measure of central tendency for resale home prices is the median. This is because the median is aimed at finding the center of the data on resale home prices and because it is not affected strongly by the relatively few homes with extremely high or low resale prices. Thus, the median provides a better indication of the "typical" resale price than either the mean or the mode.

c) The only appropriate measure of central tendency for these data is the mode, which in this case is "male." Each piece of data in this data set is either "male" or "female." There is no way to compute a mean or median for such data. In general, the mode is the only measure of central tendency that can be used for qualitative data.   ∎

POPULATION AND SAMPLE AVERAGES

We have defined the mean, the median, and the mode of a data set. Whether these averages (measures of central tendency) are considered population averages or sample averages depends on various things, such as the use to be made of them. For example, suppose the data set consists of the ages of the students in Professor W's introductory statistics class, as shown in Table 3.6.

TABLE 3.6
Ages of students in Professor W's introductory statistics class

19	24	24	24	23	20	22	21
18	20	19	19	21	19	19	23
36	22	20	35	22	23	19	26
22	17	19	20	20	21	19	21
20	20	21	19	24	21	22	21

We computed the mean, median, and mode of this data set and found that

$$\text{Mean} = 21.6 \text{ years}$$
$$\text{Median} = 21.0 \text{ years}$$
$$\text{Mode} = 19.0 \text{ years}$$

Now, the age data in Table 3.6 may constitute the entire population or it may be only a sample from a population. For a researcher interested only in the ages of the students in this particular introductory statistics class, the data set is a population; and its mean, median, and mode are a *population mean, population median,* and *population mode.* For another researcher, interested in the ages of students in all introductory statistics classes, the data set is only a sample; and its mean, median, and mode are a *sample mean, sample median,* and *sample mode.* Figure 3.1 shows the two ways in which the mean of a data set may be interpreted.

FIGURE 3.1
Possible interpretations for the mean of a data set

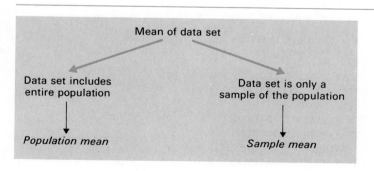

Recall that inferential statistics consists of methods of making inferences (drawing conclusions) about a population based on information obtained from a sample of the population. Consequently, the data we actually deal with in inferential statistics is sample data. Since the majority of this text is devoted to inferential statistics, we will concentrate on the calculations required for computing descriptive measures of sample data.

However, keep in mind that descriptive measures of sample data are rarely an end in themselves. Rather they are usually a means of drawing conclusions about the population from which the sample was taken. The relationships between population and sample will be discussed further in Section 3.7.

USING THE COMPUTER (OPTIONAL)

There are several ways that we can apply Minitab to obtain measures of central tendency for a data set. To determine the mean of a data set using Minitab, we can employ the **MEAN** command; and to determine the median of a data set using Minitab, we can employ the **MEDIAN** command. The following example illustrates the use of these two commands.

EXAMPLE 3.5   *Illustrates the MEAN and MEDIAN commands*

The salary data in Data Set I, displayed in Table 3.1, are repeated here in Table 3.7. Use Minitab to find the mean and median of these salaries.

TABLE 3.7
Data Set I

$200	200	200	840	200	200	300
200	300	350	700	350	950	

SOLUTION   First we use the SET command to store the salary data from Table 3.7 into C1 and use the NAME command to name C1 "DatasetI." To obtain the mean of the salary data in Data Set I, we type the command MEAN followed by the storage location of the data; that is, we type MEAN 'DATASETI'. Similarly, to obtain the median of the salary data in Data Set I, we type the command MEDIAN followed by the storage location of the data; that is, we type MEDIAN 'DATASETI'. Printout 3.1 summarizes this discussion and shows the resulting output.

PRINTOUT 3.1
Minitab output for
MEAN and MEDIAN

```
MTB > SET C1
DATA> 200 200 200 840 200 200 300
DATA> 200 300 350 700 350 950
DATA> END
MTB > NAME C1 'DATASETI'
MTB > MEAN 'DATASETI'
    MEAN    =      383.85
MTB > MEDIAN 'DATASETI'
    MEDIAN =      300.00
```

From Printout 3.1 we see that, for Data Set I, the mean salary is $383.85 and the median salary is $300.00.

Minitab does not have a MODE command. Nonetheless, we can still use Minitab to obtain the mode of a data set. The details for doing this are discussed in the *Minitab Supplement*. Also, see Exercise 3.19.

Exercises 3.1

__ **3.1** What is the purpose of a measure of central tendency?

__ **3.2** Name and describe the three most important measures of central tendency.

Determine the mean, median, and mode(s) for each of the data sets in Exercises 3.3–3.10. For the mean and the median, round each answer to one more decimal place than the original data.

__ **3.3** The U.S. National Science Foundation, Division of Science Resources Studies, collects data on the ages of recipients of science and engineering doctoral degrees. Results are published in *Survey of Earned Doctorates*. A sample of this year's recipients yields the following ages:

37	28	36	33
37	43	41	28
24	44	27	24

__ **3.4** The American Hospital Association publishes figures on the costs to community hospitals per patient per day in *Hospital Statistics*. A sample of 10 such costs in New York yields the data below.

$602	539	569	916	768
335	776	887	806	422

__ **3.5** The average retail price for oranges in 1983 was 38.5 cents per pound, as reported by the U.S. Department of Agriculture in *Food Cost Review*. Recently, a sample of 15 markets gave the following prices for oranges in cents per pound:

43.0	40.0	42.6	40.2	37.5
44.1	45.2	41.8	35.6	34.6
37.9	44.2	44.5	38.2	42.4

__ **3.6** A biologist is studying the gestation period (duration of pregnancy) of domestic dogs. Fifteen dogs are observed during pregnancy and are found to have the following gestation periods, in days:

62.0	61.4	59.8	62.2	60.3
60.4	59.4	60.2	60.4	60.8
61.8	59.2	61.1	60.4	60.9

__ **3.7** A liquid-soap manufacturer produces a bottle with an advertised content of 310 ml. A sample of 16 bottles yields the following contents:

297	318	306	300
311	303	291	298
322	307	312	300
315	296	309	311

__ **3.8** A city planner working on bikeways needs information about local bicycle commuters. She designs a questionnaire. One of the questions asks how many minutes it takes the rider to pedal from home to his or her destination. A sample of 22 local bicycle commuters yields the following times:

22	19	24	31
29	29	21	15
27	23	37	31
30	26	16	26
12	23	48	
22	29	28	

__ **3.9** The National Center for Education Statistics surveys college and university libraries to obtain information on the number of volumes held. The number of volumes, in thousands, for a sample of seven public colleges and universities are as follows:

79	516	24	265
41	15	411	

__ **3.10** As reported by the College Entrance Examination Board in *National College-Bound Senior*, the average verbal score on the Scholastic Aptitude Test in 1987 was 430 points out of a possible 800. A sample of 25 verbal scores for last year gave the data below.

346	496	352	378	315
491	360	385	500	558
381	303	434	562	496
420	485	446	479	422
494	289	436	516	615

__ 3.11   The U.S. Energy Information Administration conducts annual surveys to estimate the average number of liveable square feet for housing units. Results are published in *Residential Energy Consumption Survey: Housing Characteristics*. In 1984 it was reported that the mean was 1440 square feet and the median was 1225 square feet. Which measure of central tendency do you think is more appropriate? Explain your answer.

__ 3.12   The Bureau of the Census provides figures on the average annual income of all U.S. households in the publication *Current Population Reports*. For 1987, the mean household income was $32,144 and the median household income was $25,986. Which measure of central tendency do you think is more appropriate? Explain your answer.

__ 3.13   The National Collegiate Athletic Association wrestling champions for the years 1963–1988 are as follows. [SOURCE: *The World Almanac, 1989.*]

Year	Champion	Year	Champion
1963	Oklahoma	1976	Iowa
1964	Oklahoma State	1977	Iowa State
1965	Iowa State	1978	Iowa
1966	Oklahoma State	1979	Iowa
1967	Michigan State	1980	Iowa
1968	Oklahoma State	1981	Iowa
1969	Iowa State	1982	Iowa
1970	Iowa State	1983	Iowa
1971	Oklahoma State	1984	Iowa
1972	Iowa State	1985	Iowa
1973	Iowa State	1986	Iowa
1974	Oklahoma	1987	Iowa State
1975	Iowa	1988	Arizona State

a) Determine the mode for the champion data.
b) Would it be appropriate to use either the mean or the median here? Explain.

__ 3.14   The following table gives the all-time top television programs by number of viewers as of March, 1988. [SOURCE: Nielsen Media Research, *Nielsen Report on Television*.]

Program	Date	Network	Households
M*A*S*H Special	2/28/83	CBS	50,150,000
Super Bowl XX	1/26/86	NBC	41,490,000
Dallas	11/21/80	CBS	41,470,000
Super Bowl XIX	1/20/85	ABC	40,900,000
Super Bowl XVII	1/30/83	NBC	40,480,000
Super Bowl XXI	1/25/87	CBS	40,030,000
Super Bowl XVI	1/24/82	CBS	40,020,000
Super Bowl XVIII	1/22/84	CBS	38,800,000
The Day After	11/20/83	ABC	38,550,000
Super Bowl XXII	1/31/88	ABC	37,120,000
Roots Pt. VIII	1/30/77	ABC	36,380,000
Thorn Birds Pt. III	3/29/83	ABC	35,990,000
Thorn Birds Pt. IV	3/30/83	ABC	35,900,000
Thorn Birds Pt. II	3/28/83	ABC	35,400,000
Super Bowl XIV	1/20/80	CBS	35,330,000
Super Bowl XIII	1/21/79	NBC	35,090,000
CBS NFC Championship Game	1/10/82	CBS	34,960,000
Super Bowl XV	1/25/81	NBC	34,540,000
Super Bowl XII	1/15/78	CBS	34,410,000
Gone With The Wind Pt. 1	11/7/76	NBC	33,960,000

a) Determine the mode for the network data.
b) Would it be appropriate to use either the mean or the median here? Explain.

Most statisticians recommend using the median for indicating the center of an ordinal data set, but some researchers also use the mean. In Exercises 3.15 and 3.16 we have presented ordinal data sets. For each exercise,
a) *compute the mean of the data.*
b) *compute the median of the data.*
c) *decide which measure of central tendency provides a better descriptive summary of the data.*

__ 3.15   A distance runner entered seven marathons. His finishing places in the first six races were 4, 5, 3, 2, 7, and 4. In the seventh race he decided to go all out to win and ran in first place for 20 miles. This tired him out so badly that he ended up walking parts of the last six miles. He did finish, but only in 72nd place. The runner's places give the following ordinal data set: 4, 5, 3, 2, 7, 4, 72.

__ 3.16   Twenty-one algebra students were asked to rate the change in "test anxiety" produced by their algebra course. Negative ratings meant that they worried more about tests at the end of the course than at the beginning, whereas positive ratings meant they worried less at the end than at the beginning. Their ratings were:

0	0	0	−1	−1	1	0
−2	1	0	−2	0	−2	2
−2	1	0	−1	−3	−3	−1

Exercises 3.17–3.19 are computer exercises.

— **3.17 (Computer exercise)** Suppose the data in Exercise 3.3 are stored in a column named AGES.
a) Which Minitab command and subcommands (if any) should be used to obtain the mean of the data on ages?
b) Which Minitab command and subcommands (if any) should be used to obtain the median of the data on ages?
c) If you have access to Minitab, use it to determine the mean and the median of the age data.

— **3.18 (Computer exercise)** Suppose the data in Exercise 3.4 are stored in a column named COSTS.
a) Which Minitab command and subcommands (if any) should be used to obtain the mean of the data on costs?
b) Which Minitab command and subcommands (if any) should be used to obtain the median of the data on costs?
c) If you have access to Minitab, use it to determine the mean and the median of the cost data.

= **3.19 (Computer exercise)** One way Minitab can be used to determine the mode of a data set is as follows: (1) store the data in a column, (2) apply the HISTOGRAM command and the INCREMENT subcommand to obtain a frequency distribution in which each class is based on a single value, and (3) scan the asterisks in the output to find the mode(s).
a) The following printout was obtained by applying steps (1) and (2) to the test-anxiety data given in Exercise 3.16.

```
Histogram of ANXIETY    N = 21

Midpoint    Count
      -3       2   **
      -2       4   ****
      -1       4   ****
       0       7   *******
       1       3   ***
       2       1   *
```

Apply step (3) to obtain the mode of this data set.
b) Suppose the data on bicycle-commuter times from Exercise 3.8 are stored in a column named TIMES.

Which Minitab command and subcommands (if any) should be used to obtain the mode of the data on bicycle-commuter times?
c) If you have access to Minitab, use it to carry out part (b).

= **3.20** In some data sets there are values called **outliers** that probably should not be included in the data set for one reason or another. Suppose, for example, you are interested in the ability of high-school algebra students to compute square roots. You decide to give a square-root exam to 10 of these students. Unfortunately, one of the students had a fight with his girlfriend and cannot concentrate—he gets a 0. The 10 scores in order are as follows:

0	58	61	63	67
69	70	71	78	80

The score of 0 is an *outlier*.

Many data sets contain obvious outliers. Statisticians have a systematic method for avoiding outliers when they calculate means. They compute **trimmed means,** in which high and low values are deleted or "trimmed off" before the mean is calculated. For instance, to compute the 10% trimmed mean of the test-score data, we first delete *both* the top 10% and the bottom 10% of the ordered data. Then we calculate the mean of the remaining data. Thus, the 10% trimmed mean of the test-score data is

$$\frac{58 + 61 + 63 + 67 + 69 + 70 + 71 + 78}{8} = 67.1$$

[We deleted 0 and 80 from the original data set, since they are the bottom 10% and the top 10% of the data, respectively.]

Below is a set of algebra final-exam scores for a 40-question test.

2	15	16	16	19	21	21	25	26	27
4	15	16	17	20	21	24	25	27	28

a) Do any of the scores look like outliers?
b) Compute the usual mean for the data.
c) Compute the 5% trimmed mean for the data.
d) Compute the 10% trimmed mean for the data.
e) Compare the three means you obtained in parts (b)–(d). Which of the three means do you think provides the best measure of central tendency for the data?

3.21 Another measure of central tendency is the *midrange*. The **midrange** of a data set is defined to be the mean of the smallest and largest data values in the data set:

$$\text{Midrange} = \frac{\text{Smallest} + \text{Largest}}{2}$$

For instance, the midrange of the four exam scores, 88, 75, 95, and 100, in Example 3.4(a) on page 74 is

$$\text{Midrange} = \frac{75 + 100}{2} = 87.5$$

a) Compute the midrange of the ages in Exercise 3.3.
b) Compute the midrange of the costs in Exercise 3.4.
c) Compute the midrange of the gestation periods in Exercise 3.6.
d) What advantages does the midrange have as a measure of central tendency? What disadvantages does it have?

The modal class: Suppose that we are given a data set that is already grouped into a frequency distribution and that we do not have access to the raw data. Then, in general, it is not possible to determine the mode(s). In such cases, we can find the *modal class* instead. The **modal class** is defined to be the class(es) having the largest frequency. Note that the mode of the data set may or may not be contained in the modal class. We will discuss the concept of the modal class in Exercises 3.22–3.25.

3.22 The table at the top of the next column provides a frequency distribution for the number of days to maturity for 40 short-term investments.
a) Determine the modal class for the days-to-maturity data using the grouping given in the table.
b) Now look back at the raw data given in Table 2.1 on page 23. Determine the mode of the data.
c) Is the mode contained in the modal class?

Days to maturity	Frequency (no. of investments)
30–39	3
40–49	1
50–59	8
60–69	10
70–79	7
80–89	7
90–99	4
	40

3.23 Below is a frequency distribution for the cholesterol levels of 20 patients with high levels.

Cholesterol level	Frequency
195–199	1
200–204	3
205–209	4
210–214	7
215–219	4
220–224	1
	20

a) Determine the modal class for the cholesterol-level data using the grouping given in the above table.
b) Now look back at the raw data given in Table 2.5 on page 27. Determine the mode of the data.
c) Is the mode contained in the modal class?

3.24 There is a case when it is always possible to obtain the mode of a data set from the frequency distribution. When is that?

3.25 True or False: Suppose all the classes in a frequency distribution are based on a single value. Then the modal class and the mode are identical.

3.2   Summation notation; the sample mean

In Definition 3.1, we defined the mean of a data set with an equation written in words:

$$\text{Mean} = \frac{\text{Sum of the data}}{\text{Number of pieces of data}}$$

It is possible to write such equations more concisely by using mathematical notation. We begin by introducing the mathematical notation for "sum of the data."

EXAMPLE 3.6 *Introduces summation notation*

The exam scores for the student in Example 3.4(a) are 88, 75, 95, 100. In mathematical notation, we let the letter x_i denote the ith value in the data set. For the exam scores, we have:

$$x_1 = \text{score on Exam 1} = 88$$
$$x_2 = \text{score on Exam 2} = 75$$
$$x_3 = \text{score on Exam 3} = 95$$
$$x_4 = \text{score on Exam 4} = 100$$

More simply, we can just write $x_1 = 88$, $x_2 = 75$, $x_3 = 95$, and $x_4 = 100$. The numbers 1, 2, 3, and 4 written below the xs are called **subscripts.** Using this notation, the sum of the exam-score data can be expressed symbolically as

$$x_1 + x_2 + x_3 + x_4$$

Summation notation provides a shorthand for this last sum. The notation uses the upper-case Greek letter Σ (sigma). That letter, which corresponds to the English letter S, stands for the phrase "the sum of." Thus, in place of the lengthy expression, $x_1 + x_2 + x_3 + x_4$, we can simply use Σx, read "summation x" or "the sum of the x values:"

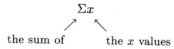

For the exam-score data,

$$\Sigma x = x_1 + x_2 + x_3 + x_4 = 88 + 75 + 95 + 100 = 358$$

MTB

In other words, the sum of the four exam scores is 358. ■

NOTATION FOR THE SAMPLE MEAN

To save us the trouble of writing the phrase "sample mean," we use the symbol $\bar{x}$, read "x bar." If we also use the letter n to denote the sample size (number of pieces of data), then the definition of the sample mean can be expressed in the following concise form:

$$\text{sample mean} \longrightarrow \bar{x} = \frac{\Sigma x}{n}$$

sum of the data

number of pieces of data

We summarize this discussion in Definition 3.4.

DEFINITION 3.4   Sample mean

The *sample mean*, $\overline{x}$, of a sample of size n is given by

$$\overline{x} = \frac{\Sigma x}{n}$$

EXAMPLE 3.7   *Illustrates Definition 3.4*

Each year, manufacturers perform mileage tests on new car models and submit the results to the Environmental Protection Agency (EPA). The EPA then tests the vehicles to determine whether the manufacturers are correct. In 1989, one company reported that a particular model, equipped with a four-speed manual transmission, averaged 29 miles per gallon (mpg) on the highway. Let us suppose that the EPA tested 15 of the cars and obtained the gas mileages given in Table 3.8.

TABLE 3.8
Gas mileages

27.3	31.2	29.4	31.6	28.6
30.9	29.7	28.5	27.8	27.3
25.9	28.8	28.9	27.8	27.6

Determine the sample mean of these gas mileages.

SOLUTION   Summing the gas mileages in Table 3.8, we obtain $\Sigma x = 431.3$. Since the number of pieces of data is $n = 15$, the sample mean gas mileage is

$$\overline{x} = \frac{\Sigma x}{n} = \frac{431.3}{15} = \boxed{28.75 \text{ mpg}}$$

MTB

OTHER IMPORTANT SUMS

We must often find sums other than the sum of the data, Σx. One such sum is the sum of the *squares* of the data, Σx^2. In the next section, we will need to obtain Σx, Σx^2, and various other sums. So that we can concentrate on the concepts to be presented there, instead of on the computations, we will discuss those sums now.

EXAMPLE 3.8   *Illustrates some other important sums*

The exam-score data from Example 3.6 are repeated in the first column of Table 3.9 at the top of the next page. The remaining columns of the table give some related quantities of importance. [Do not be concerned right now about the meaning of these other quantities. Their significance will become apparent in the next section.]

TABLE 3.9
Exam-score data and
related quantities

x	x^2	$x - \bar{x}$	$(x - \bar{x})^2$
88	7,744	−1.5	2.25
75	5,625	−14.5	210.25
95	9,025	5.5	30.25
100	10,000	10.5	110.25
358	**32,394**	**0**	**353.00**

In Example 3.6, we found that the sum of the exam-score data is $\Sigma x = 358$. This is recorded as the first entry in the last row of Table 3.9. The second column of Table 3.9 displays the squares, x^2, of the exam scores and the sum of those squares, which is $\Sigma x^2 = 32,394$.

To obtain the third column of Table 3.9, we must first compute the mean, $\bar{x}$, of the four exam scores. Since $n = 4$ and $\Sigma x = 358$,

$$\bar{x} = \frac{\Sigma x}{n} = \frac{358}{4} = 89.5$$

Subtracting $\bar{x}$ ($= 89.5$) from each of the four exam scores in the first column of the table, we get the $x - \bar{x}$ values shown in the third column. The sum of those values is $\Sigma(x - \bar{x}) = 0$. Finally, the fourth column of Table 3.9 gives the squares, $(x - \bar{x})^2$, of the $x - \bar{x}$ values. The sum of those squares is $\Sigma(x - \bar{x})^2 = 353$, as indicated. ■

Exercises 3.2

3.26   Let $x_1 = 1$, $x_2 = 7$, $x_3 = 4$, $x_4 = 5$, and $x_5 = 10$.
a) Compute Σx.
b) Find n.
c) Determine $\bar{x}$.

3.27   Let $x_1 = 12$, $x_2 = 8$, $x_3 = 9$, and $x_4 = 17$.
a) Compute Σx.
b) Find n.
c) Determine $\bar{x}$.

For each data set in Exercises 3.28–3.31,
a) *compute Σx.*
b) *find n.*
c) *determine the sample mean (round your answer to one more decimal place than the original data).*

3.28   Atlas Fishing Line, Inc., manufactures a 10-lb test line. A sample of 12 spools is subjected to tensile-strength tests. The results are:

9.8	10.2	9.8	9.4
9.7	9.7	10.1	10.1
9.8	9.6	9.1	9.7

3.29   A manufacturer of tires needs information about the life of a new steel-belted radial he is going to sell. The results of tests on a sample of these tires are given below (data are in miles).

43,725	37,732	44,473	37,396
40,652	41,868	43,097	42,200

3.30   In 1985, the average annual motor fuel expenditure per U.S. household was $1274, as reported by the Energy Information Administration in *Residential Transportation Energy Consumption Survey: Consumption Patterns of Household Vehicles.* That same year a random sample of 16 households within metropolitan areas gave the following annual motor fuel expenditures:

$1390	1459	2043	1551
415	1359	1778	1537
1167	1716	1463	904
560	1592	1710	638

___ **3.31** According to the Salt River Project (SRP), a supplier of electricity to the greater Phoenix area, the mean annual electric bill in 1984 was $852.31. An economist wants to estimate the mean for last year. He takes a sample of SRP customers and obtains the following amounts for last year's electric bills:

$1875	1478	2206	1740	1830	1516
1738	1486	1941	1608	1794	1828
1264	1999	1798	1794	1598	1568

In Exercises 3.32–3.35,
a) *compute $\bar{x}$.*
b) *compute Σx^2, $\Sigma(x-\bar{x})$, and $\Sigma(x-\bar{x})^2$ by constructing a table similar to Table 3.9 on page 83.*

___ **3.32** A sample of five families gave the following data on the number of children per family: 2, 3, 4, 4, 3.

___ **3.33** The amount of money a salesperson earned on six randomly selected days yielded the following data set: $75, 98, 130, 63, 112, 107.

___ **3.34** As reported by the R. R. Bowker Company of New York in *Library Journal,* the mean annual subscription rate to law periodicals was $39.82 in 1987. A sample of this year's law periodicals provided the following subscription rates:

$30	46	44	47
42	38	62	55
52	48	43	54

___ **3.35** A team of medical researchers has developed an exercise program to help reduce hypertension (high blood pressure). To determine whether the program is effective, the team selects a sample of 10 hypertensive individuals and places them in the exercise program for one month. Below are the diastolic blood pressures for the hypertensive individuals *before* beginning the exercise program.

106	118	118	99	109
94	109	95	97	106

In Exercises 3.36–3.40, assume that we have two data sets each with n data values. Denote the data values in one data set by $x_1, x_2, \ldots, x_n$ and the data values in the other data set by $y_1, y_2, \ldots, y_n$.

≡ **3.36** Explain the difference between $(\Sigma x)^2$ and Σx^2. Give an example to show that generally those two quantities are unequal.

≡ **3.37** Explain what the difference is between Σxy and $\Sigma x \Sigma y$. Give an example to show that generally those two quantities are unequal.

≡ **3.38** Prove the algebraic identity

$$\Sigma(x+y) = \Sigma x + \Sigma y$$

(Hint: Write $\Sigma(x+y) = (x_1 + y_1) + \cdots + (x_n + y_n)$ and rearrange the terms to obtain $\Sigma x + \Sigma y$.)

≡ **3.39** If c is a constant, prove that

$$\Sigma cx = c\Sigma x$$

≡ **3.40** If c is a constant, prove that

$$\Sigma c = nc$$

3.3 Measures of dispersion; the sample standard deviation

Up to this point, the only descriptive measures that we have discussed are measures of central tendency; namely, the mean, the median, and the mode. These descriptive measures indicate where the center or most typical value of a data set lies.

However, two data sets can have the same mean, the same median, or the same mode and yet still be quite different in other respects. For example, consider the heights of the five starting players on each of two men's college basketball teams, as shown in Figure 3.2.

FIGURE 3.2
Heights of the five starting players on each of two men's college basketball teams

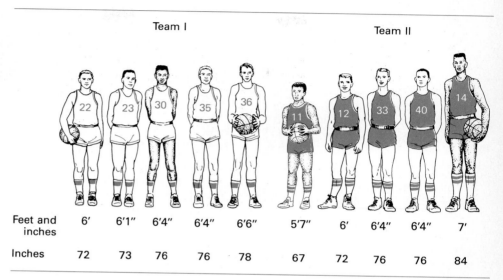

	Team I					Team II				
Feet and inches	6'	6'1"	6'4"	6'4"	6'6"	5'7"	6'	6'4"	6'4"	7'
Inches	72	73	76	76	78	67	72	76	76	84

The two teams have the same mean heights; namely, 75 inches (6' 3"). The median heights for the two teams are also identical; namely, 76 inches (6' 4"). Nonetheless, it is clear that the two data sets differ. In particular, there is much more *variation* in the heights of the players on Team II than on Team I. To describe this difference quantitatively, we use a **measure of dispersion**—a descriptive measure that indicates the amount of variation in a data set.

Just as there are several different measures of central tendency, there are also several different measures of dispersion. In this section we will discuss two of the most frequently used measures of dispersion, the *range* and the *standard deviation*.

THE RANGE

We will first discuss the *range*, since it is simpler to understand and compute. Referring to the previous example, we see that the contrast between the two teams becomes clear if we place the shortest player on each team next to the tallest. See Figure 3.3 at the top of the next page.

The **range** of a data set is obtained by computing the difference between the largest and smallest data values in the data set. Hence, as we see from Figure 3.3:

Team I: Range = 78 − 72 = 6 inches
Team II: Range = 84 − 67 = 17 inches

In general, we have the following definition:

FIGURE 3.3
Heights of the
shortest and tallest
starting players
on each of two
men's college
basketball teams

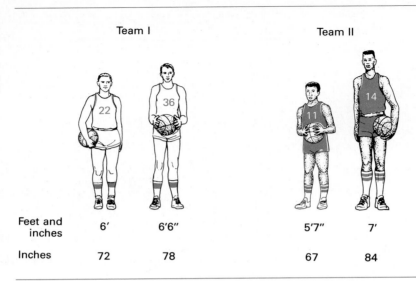

	Team I		Team II	
Feet and inches	6′	6′6″	5′7″	7′
Inches	72	78	67	84

DEFINITION 3.5   Range of a data set

The *range* of a data set is defined to be the difference between the largest and smallest data values in the data set:

$$\text{Range} = \text{Largest value} - \text{Smallest value}$$

The range of a data set is quite easy to compute. However, in using the range a great deal of information is ignored—only the largest and smallest data values are considered; the remainder of the data is disregarded.

MTB

THE SAMPLE STANDARD DEVIATION

In contrast to the range, the *standard deviation* takes into account all the data values. For this and other reasons, the standard deviation is preferable to the range as a measure of dispersion. The computations required to determine the standard deviation are more involved than those for the range. However, this is not a serious problem since computers and sophisticated calculators are available to do the necessary computations.

Roughly speaking, the standard deviation measures the variation in a data set by determining how far the data values are from the mean, on the average. If there is a large amount of variation in the data, then, on the average, the data values will be far from the mean. Hence, the standard deviation will be large. On the other hand, if there is only a small amount of variation in the data, then, on the average, the data values will be close to the mean. Consequently, the standard deviation will be small.

To actually compute the standard deviation of a data set, we need to know whether the data set constitutes the entire population or whether it is only a sample from a population. This is because the formula used to obtain the standard deviation of a sample is slightly different from the formula used to obtain the standard deviation of a population. In this section we will concentrate on the sample standard deviation. We will discuss the population standard deviation in Section 3.7.

The first step in computing the sample standard deviation is to find how far each data value is from the mean, the so-called **deviations from the mean.** We illustrate the calculation of the deviations from the mean in the next example.

EXAMPLE 3.9   *Illustrates the deviations from the mean*

The heights, in inches, of the five starting players on Team I are 72, 73, 76, 76, 78. (See Figure 3.2, page 85.) Determine the deviations from the mean.

SOLUTION   The mean height of the starting players on Team I is

$$\bar{x} = \frac{\Sigma x}{n} = \frac{72 + 73 + 76 + 76 + 78}{5} = \frac{375}{5} = 75 \text{ inches} \left(mean\right)$$

To obtain the deviation from the mean for a particular data value, we simply subtract the mean from that data value; that is, we compute $x - \bar{x}$. For instance, the deviation from the mean for the height of 72 inches is $x - \bar{x} = 72 - 75 = -3$. The deviations from the mean for all five data values are given in the second column of Table 3.10 and are displayed graphically in Figure 3.4.   ■

TABLE 3.10
Deviations
from the mean

Height x	Deviation from the mean $x - \bar{x}$
72	−3
73	−2
76	1
76	1
78	3

FIGURE 3.4
Deviations from the
mean (dots represent
data values)

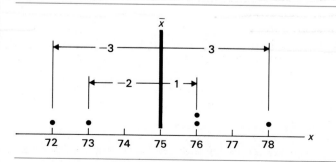

The second step in computing the sample standard deviation is to obtain a measure of the *total deviation* from the mean for all the data values. Although the quantities $x - \bar{x}$ represent deviations from the mean, adding them to get a total deviation from the mean is of no value because their sum, $\Sigma(x - \bar{x})$, *always* equals zero. [Summing the data in the second column of Table 3.10 shows that $\Sigma(x - \bar{x}) = 0$ for the height data. See Exercise 3.60 for a verification that $\Sigma(x - \bar{x}) = 0$ in general.]

In the calculation of the standard deviation, the deviations from the mean, $x - \bar{x}$, are *squared* to obtain quantities that do not add up to zero. The sum of the squared deviations from the mean, $\Sigma(x - \bar{x})^2$, is called the **sum of squared deviations** and is a measure of the total deviation from the mean of all the data.

EXAMPLE 3.10   *Illustrates the sum of squared deviations*

Compute the sum of squared deviations for the heights of the starting players on Team I.

SOLUTION   In Table 3.11 we have appended a column for $(x - \bar{x})^2$ to Table 3.10.

TABLE 3.11
Table for computing the sum of squared deviations

Height x	Deviation from mean $x - \bar{x}$	Squared deviation $(x - \bar{x})^2$
72	−3	9
73	−2	4
76	1	1
76	1	1
78	3	9
		24

From the third column of Table 3.11, we find that

$$\text{Sum of squared deviations} = \Sigma(x - \bar{x})^2 = 24 \text{ inches}^2$$

The third step in computing the sample standard deviation is to take an *average* of the squared deviations. This is accomplished by dividing the sum of squared deviations by $n - 1$ (one less than the sample size, n). The resulting quantity is called the **sample variance** and is denoted by s^2. In symbols,

$$s^2 = \frac{\Sigma(x - \bar{x})^2}{n - 1}$$

EXAMPLE 3.11   *Illustrates the sample variance*

Compute the sample variance of the heights of the starting players on Team I.

SOLUTION   From Example 3.10, the sum of squared deviations is $\Sigma(x - \bar{x})^2 = 24$ inches^2. Also, $n = 5$ since there are five pieces of data. Thus, the sample variance is

$$s^2 = \frac{\Sigma(x - \bar{x})^2}{n - 1} = \frac{24}{5 - 1} = 6 \text{ inches}^2$$

Note: If instead of dividing by $n - 1$, we divided by n, then the sample variance would be the mean of the squared deviations. Although dividing by n seems more natural, we divide by $n - 1$ for the following reason: One of the main uses of the sample variance is to estimate the population variance (to be defined in Section 3.7). Division by n tends to underestimate the population variance, whereas division by $n - 1$ does not.

It is important to realize that the sample variance is in units that are the square of the original units. This results from squaring the deviations from the mean. For instance, the sample variance of the heights of the players on Team I is $s^2 = 6$ inches^2 (Example 3.11). Since it is desirable to have descriptive measures in the original units, the final step in computing the sample standard deviation is to take the square root of the sample variance. In other words, the **sample standard deviation, s,** is

$$s = \sqrt{\frac{\Sigma(x - \bar{x})^2}{n - 1}}$$

EXAMPLE 3.12   *Illustrates the sample standard deviation*

Compute the sample standard deviation of the heights of the starting players on Team I.

SOLUTION   From Example 3.11, the sample variance is $s^2 = \Sigma(x - \bar{x})^2/(n - 1) = 6$ inches^2. Thus, the sample standard deviation is

$$s = \sqrt{\frac{\Sigma(x - \bar{x})^2}{n - 1}} = \sqrt{6} = 2.4 \text{ inches}$$

We summarize the above discussion regarding the sample standard deviation in Definition 3.6.

DEFINITION 3.6   Sample standard deviation

The *sample standard deviation, s,* of a sample of size n is defined by

$$s = \sqrt{\frac{\Sigma(x - \bar{x})^2}{n - 1}}$$

The steps required to compute the sample standard deviation were illustrated in Examples 3.9–3.12. The computations were performed in four separate examples so that we could explain the interpretation of the sample standard deviation as well as the calculations involved. However, now that we have done that, we can give a relatively simple procedure for computing the sample standard deviation, s, for any data set:

1. *Calculate the sample mean, $\bar{x}$.*
2. *Construct a table to determine the sum of squared deviations, $\Sigma(x - \bar{x})^2$.*
3. *Apply Definition 3.6 to obtain the sample standard deviation, s.*

EXAMPLE 3.13   *Illustrates Definition 3.6*

The heights, in inches, of the five starting players on Team II are 67, 72, 76, 76, 84. Compute the sample standard deviation of these heights.

SOLUTION   We apply the procedure described above. First, we calculate the sample mean, $\bar{x}$:

$$\bar{x} = \frac{\Sigma x}{n} = \frac{67 + 72 + 76 + 76 + 84}{5} = \frac{375}{5} = 75 \text{ inches}$$

Next, we construct a table, with columns for x, $x - \bar{x}$, and $(x - \bar{x})^2$, in order to determine the sum of squared deviations, $\Sigma(x - \bar{x})^2$:

TABLE 3.12

x	$x - \bar{x}$	$(x - \bar{x})^2$
67	−8	64
72	−3	9
76	1	1
76	1	1
84	9	81
		156

From the third column of Table 3.12, we see that the sum of squared deviations is

$$\Sigma(x - \bar{x})^2 = 156 \text{ inches}^2$$

Finally, we apply Definition 3.6. We have $n = 5$ and $\Sigma(x - \bar{x})^2 = 156$. Consequently, the sample standard deviation of the heights for Team II is

$$s = \sqrt{\frac{\Sigma(x - \bar{x})^2}{n - 1}} = \sqrt{\frac{156}{5 - 1}} = \sqrt{39} = 6.2 \text{ inches}$$

■

In Example 3.12, we found that the sample standard deviation of the heights of the players on Team I is $s = 2.4$ inches; and in Example 3.13, we found that the sample standard deviation of the heights of the players on Team II is $s = 6.2$ inches. Consequently, we see that Team II, which has more variation in height than Team I, also has a larger standard deviation. This is the way a measure of dispersion is supposed to work.

KEY FACT 3.1

The more variation there is in a data set, the larger its standard deviation.

A SHORTCUT FORMULA FOR S

We are going to present an alternative formula for computing the sample standard deviation, s. Consequently, it will be convenient to have a name for the original formula

$$s = \sqrt{\frac{\Sigma(x - \overline{x})^2}{n - 1}}$$

Since this is the formula used to define the sample standard deviation, we will call it the **defining formula** for s.

The alternative formula for computing the sample standard deviation is given in Formula 3.1. We will call this formula the **shortcut formula** for s.

FORMULA 3.1 Shortcut formula for the sample standard deviation

The sample standard deviation, s, of a sample of size n can be computed from the formula

$$s = \sqrt{\frac{n(\Sigma x^2) - (\Sigma x)^2}{n(n - 1)}}$$

The shortcut formula for s is equivalent to the defining formula. That is, both formulas give the same result (differences due to roundoff error are possible). However, the shortcut formula is generally faster and easier for doing calculations and reduces the chance of roundoff errors.

Before illustrating the shortcut formula for s, we should comment on the similar-looking expressions Σx^2 and $(\Sigma x)^2$ that occur in the formula. The expression Σx^2 represents the sum of the squares of the data values. It is obtained by first squaring each data value and then summing those squared values. On the other hand, the expression $(\Sigma x)^2$ represents the square of the sum of the data values; and it is obtained by first summing the data values and then squaring that sum.

EXAMPLE 3.14 *Illustrates Formula 3.1*

In Example 3.13 we computed the sample standard deviation of the heights for the five starting players on Team II using the defining formula for s. Compute that sample standard deviation using the shortcut formula for s.

SOLUTION To apply the shortcut formula, we need the sums, Σx and Σx^2. These are determined in Table 3.13 at the top of the next page.

TABLE 3.13
Table for
computation
of *s* using the
shortcut formula

x	x^2
67	4,489
72	5,184
76	5,776
76	5,776
84	7,056
375	28,281

We have $n = 5$ and, from the last row of Table 3.13, we see that $\Sigma x = 375$ and $\Sigma x^2 = 28{,}281$. Thus, by the shortcut formula,

$$s = \sqrt{\frac{n(\Sigma x^2) - (\Sigma x)^2}{n(n-1)}} = \sqrt{\frac{5(28{,}281) - (375)^2}{5(5-1)}}$$

MTB

$$= \sqrt{\frac{780}{20}} = \sqrt{39} = \text{6.2 inches}$$

∎

We have now computed the sample standard deviation of the heights of the players on Team II in two ways—in Example 3.13 we used the defining formula for *s* and in Example 3.14 we used the shortcut formula for *s*. As you can see, both formulas give the same value for the sample standard deviation (6.2 inches). For these height data, both formulas are relatively easy to apply. However, for most data sets, and especially for those in which the mean is not a whole number, the shortcut formula is better.

Some words of warning: When computing the sample standard deviation, it is important not to perform any rounding until the end of the computation. Otherwise significant roundoff error may result. This same warning applies to all calculations.

USING THE COMPUTER (OPTIONAL)

In this section, we have discussed two measures of dispersion, the range and the sample standard deviation. We will now indicate how Minitab can be used to obtain those measures of dispersion for a data set. Let us begin with the range.

Recall that the *range* of a data set is the difference between the largest and smallest data values in the data set. Minitab does not have a RANGE command, but it is easy to determine the range using two other Minitab commands—the MAX and MIN commands. The **MAX** command gives the largest value (maximum) in a data set and the **MIN** command gives the smallest value (minimum) in a data set. Consequently, to determine the range of a data set, we need only obtain the difference between the MAX and the MIN, as the following example shows.

EXAMPLE 3.15   *Illustrates how to find the range using Minitab*

The heights, in inches, of the five starting players on Team II are 67, 72, 76, 76, 84. Use Minitab to obtain the range of these height data.

SOLUTION   First we store the height data in C2 and name C2 "Team II." The range for Team II is obtained by subtracting the height of the shortest player on Team II from the height of the tallest player on Team II.

To determine the height of the tallest player on Team II, we type the command MAX followed by the storage location of the data; that is, we type `MAX 'TEAM II'`. Similarly, to determine the height of the shortest player on Team II, we type the command MIN followed by the storage location of the data; that is, we type `MIN 'TEAM II'`. Printout 3.2 summarizes the above discussion and also displays the output that results from the MAX and MIN commands.

PRINTOUT 3.2
Minitab output
for MAX and MIN

```
MTB > SET C2
DATA> 67 72 76 76 84
DATA> END
MTB > NAME C2 'TEAM II'
MTB > MAX 'TEAM II'
   MAXIMUM =        84.000
MTB > MIN 'TEAM II'
   MINIMUM =        67.000
```

The output in Printout 3.2 shows that the height of the tallest player on Team II is 84 inches and the height of the shortest player on Team II is 67 inches. Subtracting these two heights, we obtain the range: Range = 84 − 67 = 17 inches. ∎

For the height data considered in the previous example, it is actually simpler to determine the range "by hand" than by using Minitab. This is only because the data set is so small. For large data sets, where the data are not ordered, it is much easier to use Minitab.

Next we show how Minitab can be used to obtain the sample standard deviation of a data set. The appropriate command is **STDEV.** We illustrate the use of that command in Example 3.16.

EXAMPLE 3.16   *Illustrates the STDEV command*

The heights, in inches, of the five starting players on Team II are 67, 72, 76, 76, 84. Use Minitab to determine the sample standard deviation of these heights.

SOLUTION   In Example 3.15 we stored the heights of the players on Team II in C2 and named C2 "Team II." To obtain the sample standard deviation of those heights, we apply the STDEV command. Specifically, we type `STDEV 'TEAM II'`. See Printout 3.3.

```
MTB > STDEV 'TEAM II'
  ST.DEV. =      6.2450
```

Thus, the sample standard deviation of the heights of the five starting players on Team II is $s = 6.2450$ inches. ∎

Exercises 3.3

___ **3.41** What is the purpose of a measure of dispersion?

___ **3.42** Why is the standard deviation preferable to the range as a measure of dispersion?

In Exercises 3.43–3.50 we have repeated the data from Exercises 3.3–3.10. For each exercise,
a) *determine the range.*
b) *compute the sample standard deviation, s, using the defining formula.*
c) *compute the sample standard deviation, s, using the shortcut formula.*
d) *state which formula you found easier to use in computing s.*
Note: In parts (b) and (c) round your final answers to one more decimal place than the original data.

___ **3.43** The National Science Foundation, Division of Science Resources Studies, collects data on the ages of recipients of science and engineering doctoral degrees. Results are published in *Survey of Earned Doctorates*. A sample of this year's recipients yields the following ages:

37	28	36	33
37	43	41	28
24	44	27	24

___ **3.44** The American Hospital Association reports data on the costs to community hospitals per patient per day in *Hospital Statistics*. A sample of 10 such costs in New York yields the data below.

$602	539	569	916	768
335	776	887	806	422

___ **3.45** The average retail price for oranges in 1983 was 38.5 cents per pound, as reported by the U.S.

Department of Agriculture in *Food Cost Review*. Recently, a sample of 15 markets gave the following prices for oranges in cents per pound:

43.0	40.0	42.6	40.2	37.5
44.1	45.2	41.8	35.6	34.6
37.9	44.2	44.5	38.2	42.4

___ **3.46** A biologist is studying the gestation period of domestic dogs. Fifteen dogs are observed during pregnancy and are found to have the gestation periods, in days, given below.

62.0	61.4	59.8	62.2	60.3
60.4	59.4	60.2	60.4	60.8
61.8	59.2	61.1	60.4	60.9

___ **3.47** A manufacturer of liquid soap produces a bottle with an advertised content of 310 ml. A sample of 16 bottles yields the following contents:

297	318	306	300
311	303	291	298
322	307	312	300
315	296	309	311

___ **3.48** A city planner working on bikeways needs information about local bicycle commuters. She designs a questionnaire. One of the questions asks how many minutes it takes the rider to pedal from home to his or her destination. A sample of 22 local bicycle commuters yields the following times:

22	19	24	31
29	29	21	15
27	23	37	31
30	26	16	26
12	23	48	
22	29	28	

__ **3.49** The National Center for Education Statistics surveys college and university libraries to obtain information on the number of volumes held. The number of volumes, in thousands, for a sample of seven public colleges and universities are as follows:

79	516	24	265
41	15	411	

__ **3.50** As reported by the College Entrance Examination Board in *National College-Bound Senior,* the average verbal score on the Scholastic Aptitude Test in 1987 was 430 points out of a possible 800. A sample of 25 verbal scores for last year gave the data below.

346	496	352	378	315
491	360	385	500	558
381	303	434	562	496
420	485	446	479	422
494	289	436	516	615

__ **3.51** TSI, an independent testing agency, tested the lifetimes of two brands of light bulbs. Light bulb life is defined as the number of hours that a bulb will operate continuously before it burns out. The results of tests on seven bulbs of each brand are displayed in the table below (data in hundreds of hours).

Brand A	Brand B
10.5	11.3
9.1	7.0
10.0	9.7
10.3	9.6
9.4	10.5
9.6	11.8
9.7	8.7

a) Compute $\bar{x}$ for each data set.
b) Determine the median of each data set.
c) Although the two data sets have the same means and medians, they are quite different in another respect. How are they different?
d) Which data set appears to have less variation?
e) Compute s for each data set.
f) Are your answers in parts (d) and (e) consistent? Why?

__ **3.52** Consider the four data sets shown in the table at the top of the next column.

Data Set I		Data Set II		Data Set III		Data Set IV	
1	5	1	9	5	5	10	4
1	8	1	9	5	5	10	4
2	8	1	9	5	5	4	4
2	9	1	9	5	5	4	4
5	9	1	9	5	5	4	2

a) Compute $\bar{x}$ for each data set.
b) Although the four data sets have the same means, they are quite different in another respect. How are they different?
c) Which data set appears to have the least variation? the greatest variation?
d) Compute the range of each data set.
e) Compute s for each data set.
f) From your answers to parts (d) and (e), which measure of dispersion better distinguishes the variation in the four data sets, the range or the standard deviation? Explain your answer.
g) Are your answers from parts (c) and (e) consistent? Why?

__ **3.53** Below are 10 IQ scores:

110	122	132	107	101
97	115	91	125	142

Time each of the following calculations on a watch or timer.
a) Calculate the sample standard deviation of the 10 IQs using the defining formula for s, and record the time your calculation requires.
b) Calculate the sample standard deviation of the 10 IQs using the shortcut formula for s, and record the time your calculation requires.
c) Was the shortcut formula really a time-saver?

Exercises 3.54–3.57 are computer exercises.

__ **3.54 (Computer exercise)** Suppose the data in Exercise 3.44 are stored in a column named COSTS.
a) Which Minitab commands should be used to obtain the range of the data?
b) If you have access to Minitab, use it to determine the range of the cost data.

__ **3.55 (Computer exercise)** Suppose the data in Exercise 3.45 are stored in a column named PRICES.
a) Which Minitab commands should be used to obtain the range of the data?
b) If you have access to Minitab, use it to determine the range of the price data.

__ **3.56 (Computer exercise)** Suppose the data in Exercise 3.44 are stored in a column named COSTS.

a) Which Minitab command and subcommands (if any) should be used to obtain the sample standard deviation of the data?

b) If you have access to Minitab, use it to determine the sample standard deviation of the cost data.

__ **3.57 (Computer exercise)** Suppose the data in Exercise 3.45 are stored in a column named PRICES.

a) Which Minitab command and subcommands (if any) should be used to obtain the sample standard deviation of the data?

b) If you have access to Minitab, use it to determine the sample standard deviation of the price data.

= **3.58** Another measure of dispersion is the **mean absolute deviation (MAD)**. The mean absolute deviation of a data set is defined to be the mean of the absolute values of the deviations from the mean. That is, the mean absolute deviation is defined by

$$\text{MAD} = \frac{\Sigma |x - \bar{x}|}{n}$$

a) Compute the mean absolute deviation of the ages in Exercise 3.43.

b) Compute the mean absolute deviation of the costs in Exercise 3.44.

= **3.59** In Exercise 3.20 (page 79), we discussed *outliers*, extreme values in a data set. At the top of the next column are two data sets. Data Set II is obtained by removing the outliers from Data Set I.

Data Set I					Data Set II			
0	12	14	15	23	10	14	15	17
0	14	15	16	24	12	14	15	
10	14	15	17		14	15	16	

a) Compute the sample standard deviation for both data sets.

b) Compute the range for both data sets.

c) What effect do outliers have on the variation in a data set? Explain.

≡ **3.60** On page 88, we pointed out that the sum of the deviations from the mean is always equal to zero; that is, $\Sigma(x - \bar{x}) = 0$ for any data set. Prove that this is true. (*Hint:* Use the results of Exercises 3.38–3.40 on page 84.)

≡ **3.61** This exercise shows that the shortcut formula for s is equivalent to its defining formula.

a) Verify that

$$\Sigma(x - \bar{x})^2 = \Sigma x^2 - \frac{(\Sigma x)^2}{n}$$

(*Hint:* Expand the square on the left and then apply the results of Exercises 3.38–3.40 on page 84.)

b) Conclude from part (a) that

$$\frac{\Sigma(x - \bar{x})^2}{n - 1} = \frac{n(\Sigma x^2) - (\Sigma x)^2}{n(n - 1)}$$

c) Deduce from part (b) that the defining formula and shortcut formula for s are equivalent.

3.4   Interpretation of the standard deviation; z-scores

As you know, the standard deviation is a measure of dispersion—it is a descriptive measure used to indicate the amount of variation in a data set. Key Fact 3.1 on page 91 states that the standard deviation has an important property that any measure of dispersion must have; namely, *the more variation there is in a data set, the larger its standard deviation.*

Up to this point, we have been concentrating more on the calculation of the standard deviation than on its meaning. In this section, we will present material that should give you a better idea of how to interpret the standard deviation.

Table 3.14 displays two data sets, each of which has 10 data values. A brief inspection of the table reveals that there is more variation in Data Set II than in Data Set I.

TABLE 3.14

Data Set I	Data Set II
51	37
44	61
41	49
58	20
48	70
47	53
53	48
47	48
45	50
66	64

Using the formulas

$$\bar{x} = \frac{\Sigma x}{n} \quad \text{and} \quad s = \sqrt{\frac{n(\Sigma x^2) - (\Sigma x)^2}{n(n-1)}}$$

we computed the sample mean and sample standard deviation of each data set. The results are shown in Table 3.15.

TABLE 3.15

Data Set I	Data Set II
$\bar{x} = 50.0$	$\bar{x} = 50.0$
$s = 7.4$	$s = 14.2$

So, as we would expect, the standard deviation of Data Set II is larger than that of Data Set I.

To enable you to visually compare the variations in the two data sets, we have drawn the graphs in Figures 3.5 and 3.6, shown at the top of the next page. On each graph we have marked the data values with dots. In addition, we have located the sample mean, $\bar{x} = 50$, and have measured off intervals equal in length to the standard deviation, 7.4 for Data Set I and 14.2 for Data Set II.

Let us examine Figure 3.5. Note, for example, that the horizontal position labeled $\bar{x} + 2s$ represents the number that is two standard deviations to the right of the mean, which in this case is $\bar{x} + 2s = 50.0 + 2 \cdot 7.4 = 64.8$. Likewise, the horizontal position labeled $\bar{x} - 3s$ represents the number that is three standard deviations to the left of the mean, which in this case is $\bar{x} - 3s = 50.0 - 3 \cdot 7.4 = 27.8$. Similar remarks apply to Figure 3.6.

The graphs in Figures 3.5 and 3.6 vividly illustrate that there is more variation in Data Set II than in Data Set I. The graphs also show that, for each data set, all the data values lie within a few standard deviations to either side of the mean. This is no accident—*almost all of the data in any data set will lie within three standard deviations to either side of the mean.*

A data set with a great deal of variation will have a large standard deviation and, consequently, three standard deviations to either side of its mean will be rather

FIGURE 3.5   Data Set I. $\bar{x} = 50$, $s = 7.4$

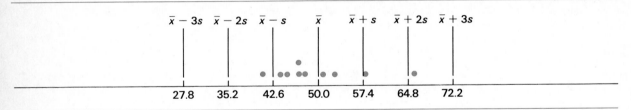

FIGURE 3.6   Data Set II. $\bar{x} = 50$, $s = 14.2$

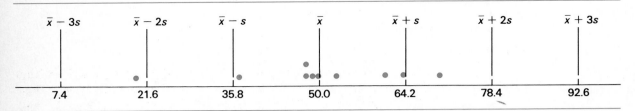

extensive, as in Figure 3.6. On the other hand, a data set with little variation will have a small standard deviation and, hence, three standard deviations to either side of its mean will be rather narrow, as in Figure 3.5.

CHEBYCHEV'S RULE

As we said, almost all of the data in any data set will lie within three standard deviations to either side of the mean. The Russian mathematician P. L. Chebychev (1821–1894) developed a rule that generalizes this last statement and also makes it more precise.

KEY FACT 3.2   Chebychev's rule

For *any* data set:

PROPERTY 1 At least 75% of the data lies within *two* standard deviations to either side of the mean, that is, between $\bar{x} - 2s$ and $\bar{x} + 2s$.

PROPERTY 2 At least 89% of the data lies within *three* standard deviations to either side of the mean, that is, between $\bar{x} - 3s$ and $\bar{x} + 3s$.

PROPERTY 3 In general, for any number $k > 1$, at least $1 - 1/k^2$ of the data lies within k standard deviations to either side of the mean, that is, between $\bar{x} - k \cdot s$ and $\bar{x} + k \cdot s$.

Figure 3.7, at the top of the following page, provides a graphical display of the first two properties in Chebychev's rule.

Note that the first two properties in the statement of Chebychev's rule are special cases of the third property. For instance, let us apply Property 3 with $k = 2$.

FIGURE 3.7
Graphical illustration
of Chebychev's rule

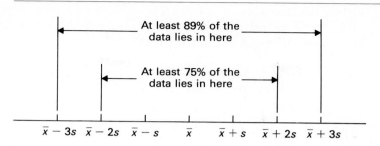

Then

$$1 - 1/k^2 = 1 - 1/2^2 = 1 - 1/4 = 0.75$$

or 75%. Thus, by Property 3, for any data set, at least 75% of the data lies within two standard deviations to either side of the mean. This last statement is the one given in Property 1. Similarly, Property 2 can be obtained from Property 3 by setting $k = 3$.

We should also point out that Property 3 of Chebychev's rule can be applied using any value of k that is greater than 1. For instance, let us take $k = 2.5$. Then,

$$1 - 1/k^2 = 1 - 1/2.5^2 = 1 - 1/6.25 = 0.84$$

or 84%. Thus, by Property 3, for any data set, at least 84% of the data lies within 2.5 standard deviations to either side of the mean.

It must be emphasized that the value of Chebychev's rule is not in its precision but in its generality. For example, consider the data set portrayed in Figure 3.5 on page 98. Chebychev's rule says that *at least* 75% of the data will lie within two standard deviations of the mean. However, as we can see from Figure 3.5, 90% of the data actually lies within two standard deviations of the mean. Similarly, Chebychev's rule says that *at least* 89% of the data will lie within three standard deviations of the mean, whereas, in fact, 100% of the data actually lies within three standard deviations of the mean.

Thus, we see that Chebychev's rule does not necessarily give precise estimates for the percentage of data that lies within a given number of standard deviations to either side of the mean. Its power is its generality—Chebychev's rule holds for *any* data set. Moreover, Chebychev's rule permits us to make pertinent statements about a data set when we know only its mean and standard deviation, and frequently that is all we do know. Consider the following example.

EXAMPLE 3.17   *Illustrates Chebychev's rule*

The R. R. Bowker Company of New York collects data on annual subscription rates to periodicals. Results are published in *Library Journal*. A sample of 40 sociology periodicals taken this year has a mean subscription rate of $\overline{x} = \$63.75$ with a standard deviation of $s = \$10.42$. Using this information only, apply Properties 1 and 2 of Chebychev's rule to make some observations about the subscription rates of the 40 sociology periodicals in the sample.

SOLUTION   Based on the fact that $\bar{x} = \$63.75$ and $s = 10.42$, we can construct Figure 3.8. [Note, for instance, that $\bar{x} - 2s = 63.75 - 2 \cdot 10.42 = 42.91$.]

FIGURE 3.8

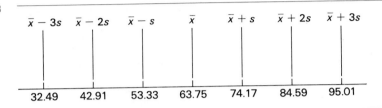

By Property 1 of Chebychev's rule, at least 75% of the 40 sociology periodicals have subscription rates within two standard deviations to either side of the mean. Since 75% of 40 is 30, we can conclude, in view of Figure 3.8, that at least 30 of the 40 sociology periodicals have subscription rates between $42.91 and $84.59.

Applying Property 2 of Chebychev's rule, we can say that at least 89% of the 40 sociology periodicals have subscription rates within three standard deviations to either side of the mean. Now, 89% of 40 is 35.6. So, in view of Figure 3.8, we can conclude that at least 36 of the 40 sociology periodicals have subscription rates between $32.49 and $95.01.   ∎

A final remark concerning Chebychev's rule: Many data sets that occur in practice have histograms that are bell-shaped. Although Chebychev's rule will apply to this type of data set (as well as to any other type of data set), we can naturally provide more precise estimates given this additional information. These more precise estimates, which apply only to bell-shaped distributions, are stated in the so-called **empirical rule.** We will examine that rule when we study bell-shaped distributions in Chapter 6.

Z-SCORES

A quantity that is frequently used in statistical analysis is the *z-score*. The **z-score** (or **standard score**) for a data value is the number of standard deviations that the data value is away from the mean of the data set. We introduce *z*-scores in Example 3.18.

EXAMPLE 3.18   *Introduces z-scores*

A researcher in nutrition is studying daily protein intake. She takes a sample of 500 daily protein intakes. The mean intake turns out to be 77 grams with a standard deviation of 8 grams. Determine the *z*-score for an intake of 93 grams.

SOLUTION   The z-score for the intake of 93 grams is the number of standard deviations that intake is away from the mean. We know that the mean intake is $\bar{x} = 77$ grams. So the intake of 93 grams is 16 grams $(93 - 77)$ away from the mean. Since the standard deviation is $s = 8$ grams, 16 grams away from the mean is

$$\frac{16}{8} = 2 \text{ standard deviations}$$

away from the mean. Consequently, the z-score for the intake of 93 grams is $z = 2$. See Figure 3.9. Note that, in Figure 3.9, the numbers in the row labeled x represent protein intakes in grams and the numbers in the row labeled z represent z-scores (i.e., number of standard deviations away from the mean).   ∎

FIGURE 3.9

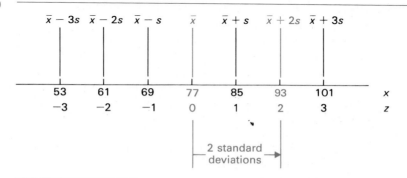

If we study the solution in the previous example, we can determine the general formula for computing z-scores. First we subtracted the mean of $\bar{x} = 77$ grams from the specified data value of 93 grams in order to determine how far the data value is away from the mean. Then we divided that difference by the standard deviation of $s = 8$ grams to find how many *standard deviations* the specified data value of 93 grams is away from the mean. In summary, we computed the z-score for the intake of 93 grams as follows:

$$z = \frac{93 - \bar{x}}{s} = \frac{93 - 77}{8} = \frac{16}{8} = 2$$

More generally, if we use x to denote a data value, then we have the following definition:

DEFINITION 3.7   Sample z-score

The *sample z-score* for a data value, x, is defined to be the number of standard deviations that x is away from the mean. The sample z-score is computed by using the formula

$$z = \frac{x - \bar{x}}{s}$$

where $\bar{x}$ and s are the mean and standard deviation of the sample data. A negative z-score indicates that the data value is smaller than the mean, whereas a positive z-score indicates that the data value is larger than the mean.

Note: When computing a z-score, round the answer to two decimal places.

EXAMPLE 3.19   *Illustrates Definition 3.7*

Consider again the protein-intake data from Example 3.18. Determine the z-score for an intake of
a) 99 grams.
b) 65 grams.

SOLUTION   Recall that for the protein-intake data, the sample mean is $\bar{x} = 77$ grams and the sample standard deviation is $s = 8$ grams. So, the z-score for a protein intake, x, in the sample is

$$z = \frac{x - \bar{x}}{s} = \frac{x - 77}{8}$$

a) For an intake of 99 grams, we have $x = 99$ and so the z-score is

$$z = \frac{99 - 77}{8} = \boxed{2.75}$$

Thus, a protein intake of 99 grams is 2.75 standard deviations above the mean.
b) For an intake of 65 grams, the z-score is

$$z = \frac{65 - 77}{8} = \boxed{-1.50}$$

So, a protein intake of 65 grams is 1.50 standard deviations below the mean.   ∎

THE Z-SCORE AS A MEASURE OF RELATIVE STANDING

We often want to describe the position of a particular data value in a data set relative to the rest of the data in the data set. Descriptive measures that indicate the relative position of a data value within a data set are called **measures of relative standing.**

The z-score can be used as a measure of relative standing. If a data value, x, has a large positive z-score, then x is larger than most of the other data values in the data set; if a data value, x, has a large negative z-score, then x is smaller than most of the other data values in the data set; and if a data value, x, has a z-score near 0, then x is located near the mean of the data set.

We can make the statements in the previous paragraph somewhat more precise by applying Chebychev's rule to the z-score of a data value. Here is an example.

EXAMPLE 3.20   *Illustrates the z-score as a measure of relative standing*

Consider once more the data on protein intakes. Use the z-score and Chebychev's rule to estimate the relative standing of a protein intake of 51 grams.

SOLUTION   For the protein-intake data, $\bar{x} = 77$ grams and $s = 8$ grams. Thus, the z-score for an intake of 51 grams is

$$z = \frac{51 - 77}{8} = -3.25$$

Thus, a protein intake of 51 grams is 3.25 standard deviations below the mean.

By Property 2 of Chebychev's rule, at least 89% of the protein intakes in the sample are within three standard deviations to either side of the mean. Consequently, a protein intake of 51 grams, which is 3.25 standard deviations below the mean, is smaller than at least 89% of all the protein intakes in the sample.

We can use Property 3 of Chebychev's rule to get an even better idea of the relative standing of the protein intake of 51 grams. Taking $k = 3.25$ (the absolute value of the z-score for 51 grams), we see that

$$1 - 1/k^2 = 1 - 1/3.25^2 = 1 - 1/10.5625 = 0.905$$

or 90.5%. Consequently, by Property 3 of Chebychev's rule, at least 90.5% of the protein intakes in the sample are within 3.25 standard deviations to either side of the mean. Therefore, a protein intake of 51 grams, which is 3.25 standard deviations below the mean, is smaller than at least 90.5% of all the protein intakes in the sample. ∎

Exercises 3.4

__ **3.62**   Consider the following data sets:

Data Set 1		Data Set 2	
30	20	14	9
16	24	56	32
22	19	13	8
23	13	26	3
18	9	9	16
18	28	31	23

a) In which data set does there appear to be more variation?
b) Compute $\bar{x}$ and s for each data set.
c) Draw graphs similar to Figure 3.5 on page 98.
d) Interpret your graphs from part (c).

e) Do most of the data in each data set lie within three standard deviations to either side of the mean?

__ **3.63**   Consider the following data sets:

Data Set 3		Data Set 4	
82	78	97	59
85	94	100	100
65	84	30	95
91	86	87	79
81	84	90	93

a) In which data set does there appear to be more variation?
b) Compute $\bar{x}$ and s for each data set.

c) Draw graphs similar to Figure 3.5 on page 98.

d) Interpret your graphs from part (c).

e) Do most of the data in each data set lie within three standard deviations to either side of the mean?

___ **3.64** What does Chebychev's rule say about the percentage of data in a data set that lies within

a) four standard deviations to either side of the mean?

b) 2.5 standard deviations to either side of the mean?

c) 3.75 standard deviations to either side of the mean?

___ **3.65** What does Chebychev's rule say about the percentage of data in a data set that lies within

a) 1.25 standard deviations to either side of the mean?

b) 3.5 standard deviations to either side of the mean?

c) five standard deviations to either side of the mean?

___ **3.66** Refer to Exercise 3.62.

a) What percentage of the data does Chebychev's rule say will lie within two standard deviations to either side of the mean? Within three standard deviations to either side of the mean?

b) Using your graph from Exercise 3.62(c), determine the percentage of the data in Data Set 1 that actually lies within two standard deviations to either side of the mean; within three standard deviations to either side of the mean.

c) Repeat part (b) for Data Set 2.

d) What do parts (b) and (c) illustrate about Chebychev's rule?

___ **3.67** Refer to Exercise 3.63.

a) What percentage of the data does Chebychev's rule say will lie within two standard deviations to either side of the mean? Within three standard deviations to either side of the mean?

b) Using your graph from Exercise 3.63(c), determine the percentage of the data in Data Set 3 that actually lies within two standard deviations to either side of the mean; within three standard deviations to either side of the mean.

c) Repeat part (b) for Data Set 4.

d) What do parts (b) and (c) illustrate about Chebychev's rule?

___ **3.68** We stated on page 97 that almost all of the data in any data set will lie within three standard deviations to either side of the mean. But in most examples and exercises so far, *all* of the data were within three standard deviations of the mean. Verify that this is not the case for the following data set:

100	28	69	85	85	98
97	100	87	97	96	94
89	97	93	92	95	94
79	83	97	90	96	57
64	74	80	87	89	58

(*Note:* $\Sigma x = 2550$ and $\Sigma x^2 = 224{,}272$.)

In Exercises 3.69–3.74 we have given the sample mean, sample standard deviation, and sample size for a data set. For each exercise,

a) *construct a graph similar to Figure 3.8 on page 100.*

b) *apply Property 1 of Chebychev's rule to make some observations about the data.*

c) *apply Property 2 of Chebychev's rule to make some observations about the data.*

___ **3.69** The Philadelphia CPA firm Laventhol and Horwath conducts annual surveys to obtain characteristics of full-service and economy lodging establishments. Suppose that the room rates (for a double) for 300 lodging establishments have a mean of $\bar{x} = \$40.97$ with a standard deviation of $s = \$8.66$.

___ **3.70** The U.S. National Center for Health Statistics collects data on cigarette smokers by sex and age and publishes the results in *Vital and Health Statistics*. Suppose that a sample of 2760 females who presently smoke has a mean age of $\bar{x} = 38.7$ years with a standard deviation of $s = 12.6$ years.

___ **3.71** The Gallup Organization conducts annual surveys on home gardening for the National Association for Gardening. Data appear in *National Gardening Survey*. Suppose that a sample of 250 households with vegetable gardens yields a mean garden size of $\bar{x} = 643$ square feet with a standard deviation of $s = 247$ square feet.

___ **3.72** As reported by the Health Insurance Association of America in *Survey of Hospital Semi-Private Room Charges*, the average daily charge for a semi-private room in U.S. hospitals was $253 in 1988. In that same year, a sample of 30 Massachusetts hospitals yielded a mean semi-private room charge of $\bar{x} = \$260.68$ with a standard deviation of $s = \$12.77$.

___ **3.73** According to the publication, *Annual Report of the Commissioner and Chief Counsel of the Internal Revenue Service*, the average income tax per taxable return was $4471 in 1986. A sample of 30 taxable returns from last year has a mean income tax of $\bar{x} = \$4239.8$ with a standard deviation of $s = \$2215.6$.

___ **3.74** *The World Almanac, 1985* reports that in 1980 the average travel time to work for residents of South Dakota was 13 minutes. For this year, a sample of 35 travel times for South Dakota residents gave a mean of $\bar{x} = 14.0$ minutes with a standard deviation of $s = 13.3$ minutes.

___ **3.75** Refer to Exercise 3.69.
a) Find the sample z-scores for room rates in the sample of $67, $20, and $37. Interpret your results.
b) Use your results from part (a) to estimate the relative standing in the sample of each of the room rates in part (a).

___ **3.76** Refer to Exercise 3.70.
a) Determine the sample z-scores for female smokers in the sample whose ages are 58 years, 29 years, and 73 years. Interpret your results in words.
b) Use your results from part (a) to estimate the relative standing in the sample of each of the ages in part (a).

___ **3.77** Refer to Exercise 3.71.
a) Determine the sample z-scores for vegetable gardens in the sample whose sizes are 1631, 211, and 618 square feet. Interpret your results in words.
b) Use your results from part (a) to estimate the relative standing in the sample of each of the sizes in part (a).

___ **3.78** Refer to Exercise 3.72.
a) Find the sample z-scores for semi-private rooms in the sample whose daily charges are $311, $343, and $262. Interpret your results in words.
b) Use your results from part (a) to estimate the relative standing in the sample of each of the daily charges in part (a).

___ **3.79** Suppose you are thinking of buying a resale home in a large tract. The owner is asking $95,500. Your realtor obtains the sale prices of 25 comparable homes in the area that have sold recently. The sample mean of the 25 sale prices is $110,258 with a sample standard deviation of $5,237. Does it appear that the home you are contemplating buying is a bargain? Explain by using the z-score and Chebychev's rule.

___ **3.80** Suppose you take an exam with 400 possible points. The instructor tells you that the mean score is 280 and that the standard deviation is 20. He also tells you that you got 350. Did you do well on the exam? Explain by using the z-score and Chebychev's rule.

= **3.81** Given a data set, how many standard deviations to either side of the mean must we go to be assured that
a) at least 95% of the data lies within?
b) at least 99% of the data lies within?

≡ **3.82** Consider a data set with n data values, say $x_1, x_2, \ldots, x_n$.
a) Show that if all the data values are equal, then the standard deviation is zero.
b) Show that if the standard deviation of the data set is zero, then all the data values must be equal.

≡ **3.83** Suppose a data set consists of $2m^2 - 1$ zeros, one $-m$, and one m, where m is a positive integer.
a) Compute $\bar{x}$ and s for this data set.
b) How many standard deviations from the mean is the data value m?
c) Assuming that $m \geq 4$, what percentage of the data lies within three standard deviations to either side of the mean? What does this percentage approach as m increases without bound?

3.5 Computing $\bar{x}$ and s for grouped data

We have learned how to compute the sample mean, $\bar{x}$, and the sample standard deviation, s, of a data set. The formulas that we have developed apply only to raw (ungrouped) data sets. Frequently, however, the data we encounter are grouped into a frequency distribution. In this section we will present methods for computing $\bar{x}$ and s when the data are in grouped form.

COMPUTING $\bar{X}$ FOR GROUPED DATA

In Example 3.1 on page 70, we considered two sets of salary data representing weekly earnings during two halves of the summer for the employees of a small mathematical consulting firm. The salary data for the second data set, Data Set II, are repeated here in Table 3.16.

$200	200	840	350	300
300	200	200	950	200

A frequency distribution for Data Set II, using classes based on a single value, is presented in Table 3.17.

TABLE 3.17
Frequency
distribution for
Data Set II using
single-value grouping

Salary ($)	Frequency
200	5
300	2
350	1
840	1
950	1
	10

We computed the mean of Data Set II in Example 3.1. In doing so, we had to determine the sum of the salaries displayed in Table 3.16. That sum can be found more quickly using the frequency distribution given in Table 3.17. To see how, we first express the sum of the salaries in Data Set II as follows:

$$\text{Sum of salaries} = 200 + 200 + 840 + 350 + 300 + 300 + 200 + 200 + 950 + 200$$

$$= \overbrace{200 + 200 + 200 + 200 + 200}^{5} + \overbrace{300 + 300}^{2} + \overbrace{350}^{1} + \overbrace{840}^{1} + \overbrace{950}^{1}$$

In the second sum we combined like salaries and placed the frequency of each of the different salaries above each group. Consequently, the sum of the salaries can be rewritten as shown in the following diagram:

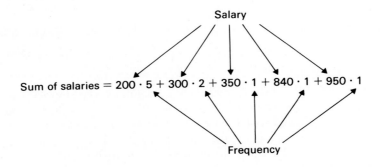

In other words, we can determine the sum of the salaries by multiplying each of the *different* salaries by its corresponding frequency and then adding the results. This can be done most efficiently by appending a column to the frequency distribution in Table 3.17 in order to perform the necessary computations. See Table 3.18.

TABLE 3.18

Salary x	Frequency f	Salary · Frequency xf
200	5	1000
300	2	600
350	1	350
840	1	840
950	1	950
	10	3740	⟵ *Sum of salaries*

We have also introduced some mathematical notation in Table 3.18: x represents salary-class value (not each individual salary, as before); f represents class frequency; and xf represents salary-class value times class frequency. Note that for a given class, xf equals the sum of the salaries in that class.

We will now show how Table 3.18 can be used to compute the mean of the salary data. The sum of all the salaries is Σxf, the sum of the third column in Table 3.18. The number of salaries is Σf, the sum of the second column in Table 3.18. Consequently, the mean of the salaries in Data Set II is

$$\bar{x} = \frac{\text{Sum of salaries}}{\text{Number of salaries}} = \frac{\Sigma xf}{\Sigma f} = \frac{3740}{10} = \$374$$

Note: Previously we have used the letter n to denote the sample size (number of pieces of data). As the above illustration shows, when the data are in grouped form, the sample size is equal to the sum of the class frequencies; that is, $n = \Sigma f$. To minimize the amount of notation, we will generally use n instead of Σf to denote the sample size.

The preceding discussion shows how we can compute the mean of a data set when the data are grouped in a frequency distribution. We summarize that discussion in Formula 3.2.

FORMULA 3.2 Computing the sample mean for grouped data

The sample mean, $\bar{x}$, of a data set that is grouped in a frequency distribution can be computed using the formula

$$\bar{x} = \frac{\Sigma xf}{n}$$

where f denotes class frequency and n ($= \Sigma f$) denotes the sample size.

EXAMPLE 3.21 *Illustrates Formula 3.2*

Table 3.19, at the top of the next page, provides a frequency distribution, using classes based on a single value, for the salary data in Data Set I (Table 3.1, page 70).

TABLE 3.19
Frequency
distribution for
Data Set I using
single-value grouping

Salary ($) x	Frequency f
200	6
300	2
350	2
700	1
840	1
950	1

Compute the sample mean of these salaries.

SOLUTION   Since the data are in grouped form, we apply Formula 3.2. Appending an xf-column to Table 3.19 we obtain Table 3.20.

TABLE 3.20
Table for
calculation of $\bar{x}$

x	f	xf
200	6	1200
300	2	600
350	2	700
700	1	700
840	1	840
950	1	950
	13	4990

Using the final row of Table 3.20, we can compute $\bar{x}$:

$$\bar{x} = \frac{\Sigma xf}{n} = \frac{4990}{13} = \$383.85$$

MTB   Thus, the mean of the salaries in Data Set I is $383.85. ∎

COMPUTING S FOR GROUPED DATA

We have just seen how to compute the sample mean of a data set when the data are grouped into a frequency distribution. Using similar reasoning, we can also obtain a formula for computing the sample standard deviation of a data set when the data are grouped into a frequency distribution. That formula and its shortcut version are presented in Formula 3.3.

FORMULA 3.3   Computing the sample standard deviation for grouped data

The sample standard deviation, s, of a data set that is grouped in a frequency distribution can be computed using the formula

$$s = \sqrt{\frac{\Sigma(x - \bar{x})^2 f}{n - 1}}$$

or its shortcut version

$$s = \sqrt{\frac{n(\Sigma x^2 f) - (\Sigma x f)^2}{n(n-1)}}$$

where f denotes class frequency and $n\ (= \Sigma f)$ denotes the sample size.

EXAMPLE 3.22   *Illustrates Formula 3.3*

A frequency distribution for the salary data in Data Set I is given in Table 3.19 and is repeated below in the first two columns of Table 3.21. Compute the sample standard deviation of these salaries.

SOLUTION   We will apply the shortcut formula presented in Formula 3.3. As you can see from the formula, we will need a table with columns for x, f, xf, x^2, and $x^2 f$. We will also need to sum the second, third, and fifth columns of that table to obtain n, $\Sigma x f$, and $\Sigma x^2 f$. See Table 3.21.

TABLE 3.21
Table for calculation of s using the shortcut formula

x	f	xf	x^2	$x^2 f$
200	6	1200	40,000	240,000
300	2	600	90,000	180,000
350	2	700	122,500	245,000
700	1	700	490,000	490,000
840	1	840	705,600	705,600
950	1	950	902,500	902,500
	13	4990		2,763,100

From the final row of Table 3.21 we find that

$$s = \sqrt{\frac{n(\Sigma x^2 f) - (\Sigma x f)^2}{n(n-1)}} = \sqrt{\frac{13(2,763,100) - (4990)^2}{13(12)}}$$

$$= \sqrt{\frac{35,920,300 - 24,900,100}{156}} = \sqrt{\frac{11,020,200}{156}} = 265.79$$

MTB

Thus, the sample standard deviation of the salaries in Data Set I is $265.79. ∎

COMPUTING $\bar{X}$ AND S WHEN THE CLASSES ARE NOT BASED ON A SINGLE VALUE

Up to this point, we have considered the computations of $\bar{x}$ and s for grouped data in which each class is based on a single value. The appropriate formulas are Formulas 3.2 and 3.3. In case each class represents a number of different values, such as the grouped data on page 24, Formulas 3.2 and 3.3 cannot be used to compute the exact values of $\bar{x}$ and s. However, they can be used to compute the approximate values of $\bar{x}$ and s. This is accomplished by substituting the *class marks* for the x-values.

EXAMPLE 3.23   *Illustrates Formulas 3.2 and 3.3*

A grouped-data table for the number of days to maturity for 40 short-term investments is given in Table 2.4 on page 26. The first, second, and fourth columns of that table are repeated here in Table 3.22.

TABLE 3.22

Days to maturity	Frequency	Class mark
30–39	3	34.5
40–49	1	44.5
50–59	8	54.5
60–69	10	64.5
70–79	7	74.5
80–89	7	84.5
90–99	4	94.5
	40	

Use Table 3.22 to compute the approximate values of $\bar{x}$ and s.

SOLUTION   As we said, the approximate values of $\bar{x}$ and s can be obtained by using the class marks for the x-values in Formulas 3.2 and 3.3. The required sums are obtained in Table 3.23.

TABLE 3.23

x	f	xf	x^2	x^2f
34.5	3	103.5	1190.25	3,570.75
44.5	1	44.5	1980.25	1,980.25
54.5	8	436.0	2970.25	23,762.00
64.5	10	645.0	4160.25	41,602.50
74.5	7	521.5	5550.25	38,851.75
84.5	7	591.5	7140.25	49,981.75
94.5	4	378.0	8930.25	35,721.00
	40	2720.0		195,470.00

From the second, third, and fifth columns of Table 3.23, we find that

$$\bar{x} \approx \frac{\Sigma xf}{n} = \frac{2720}{40} = 68.0$$

and

$$s \approx \sqrt{\frac{n(\Sigma x^2 f) - (\Sigma x f)^2}{n(n-1)}} = \sqrt{\frac{40(195{,}470) - (2720)^2}{40(39)}} = \sqrt{\frac{420{,}400}{1560}} = 16.4$$

MTB

Thus, $\bar{x} \approx 68.0$ days and $s \approx 16.4$ days.

Since we have the raw (ungrouped) data for the days-to-maturity data in Table 2.1 on page 23, we can compute the exact values of $\bar{x}$ and s using Definition 3.4 on page 82 and Formula 3.1 on page 91. We did this and obtained

$$\bar{x} = 68.3 \text{ days} \qquad \text{and} \qquad s = 16.7 \text{ days}$$

Comparing these values to the ones found in Example 3.23, we see that the grouped-data formulas really do give only the approximate values of $\bar{x}$ and s when the classes are not based on a single value. This is because the class mark of each class provides only a typical value for the data values in the class and not the data values themselves.

For the days-to-maturity data we could actually compute the exact values of $\bar{x}$ and s because the raw data were available. In practice, when data are given in a grouped form and the classes are not based on a single value, the raw data are usually not accessible. For such cases, we must be content with approximate values of $\bar{x}$ and s.

Exercises 3.5

In each of Exercises 3.84–3.87, round your answers to one more decimal place than the original data.

— 3.84   An English professor wants to estimate the average study time per week for students in basic English courses at her school. To accomplish this, she selects a sample of 25 students and records their weekly study times. The times, to the nearest hour, are:

Study time	Frequency
2	2
3	1
4	3
5	1
6	5
7	5
8	5
9	2
11	1

a) compute the sample mean, $\bar{x}$, using Formula 3.2 on page 107.

b) compute the sample standard deviation, s, using one of the formulas in Formula 3.3 on page 108.

— 3.85   The U.S. Bureau of the Census collects data on family size and publishes the results in *Current Population Reports*. A sample of 500 American families yields the sizes indicated here:

Size of family	Frequency
2	198
3	118
4	101
5	59
6	12
7	3
8	8
9	1

a) Compute the sample mean, $\bar{x}$, using Formula 3.2 on page 107.
b) Compute the sample standard deviation, s, using one of the formulas in Formula 3.3 on page 108.

__ 3.86   The heights to the nearest inch of the basketball players on the 1988–89 Phoenix Suns roster are given in the frequency distribution below. [SOURCE: *Phoenix Suns 1988–89 Media Guide*]

Height	Frequency
74	2
75	2
76	1
77	0
78	2
79	2
80	1
81	2
82	3
83	1

a) Compute the sample mean, $\bar{x}$, using Formula 3.2.
b) Compute the sample standard deviation, s, using one of the formulas in Formula 3.3.

__ 3.87   The ages of the students in Professor W's introductory statistics class are given in Table 3.6 on page 75. A frequency distribution for those ages is displayed below.

Age	Frequency
17	1
18	1
19	9
20	7
21	7
22	5
23	3
24	4
26	1
35	1
36	1

a) Compute the sample mean, $\bar{x}$, using Formula 3.2.
b) Compute the sample standard deviation, s, using one of the formulas in Formula 3.3.

In Exercises 3.88 and 3.89 you will be given raw (ungrouped) data. For each data set,
a) *compute $\bar{x}$ and s by applying Definition 3.4 on page 82 and Formula 3.1 on page 91.*

b) *group the data into a frequency distribution in which each class is based on a single value.*
c) *compute $\bar{x}$ and s using your frequency distribution from part (b) and the formulas discussed in this section.*
d) *compare your answers from parts (a) and (c).*

__ 3.88   The Prescott National Bank has six tellers available to serve customers. The data below give the number of busy tellers observed at 25 spot checks.

6	5	4	1	5
6	1	5	5	5
3	5	2	4	3
4	5	0	6	4
3	4	2	3	6

__ 3.89   A wheat farmer tests a new fertilizer on a sample of 30 one-acre plots. The yields of wheat, in bushels, are as follows:

39	41	38	39	38
41	40	38	37	39
39	41	43	38	40
39	39	43	39	37
38	37	40	43	40
42	40	39	36	43

Compute the approximate values of $\bar{x}$ and s for the grouped data in Exercises 3.90–3.93.

__ 3.90   The National Center for Education Statistics publishes cost estimates for various aspects of higher education in *Digest of Education Statistics*. One such cost is the annual tuition and fees in private, four-year universities. The annual tuitions for a sample of 35 private, four-year universities yield the following frequency distribution:

Annual tuition	Frequency
$2000–$2999	2
$3000–$3999	2
$4000–$4999	6
$5000–$5999	7
$6000–$6999	7
$7000–$7999	6
$8000–$8999	4
$9000–$9999	1

__ 3.91   A study by Lewis M. Terman, published in his book *The Intelligence of School Children* (Boston:

Houghton Mifflin, 1946), gave the following frequency distribution for the IQs of 112 children who were attending kindergarten in San Jose and San Mateo, CA:

IQ	Frequency
60–69	1
70–79	5
80–89	13
90–99	22
100–109	28
110–119	23
120–129	14
130–139	3
140–149	2
150–159	1

___ **3.92**   According to the Food and Nutrition Board of the National Academy of Sciences, the recommended daily allowance (RDA) of calcium for adults is 800 mg (milligrams). A nutritionist thinks that people with incomes below the poverty level average less than the RDA of 800 mg. To test her suspicion, the daily intakes of calcium are determined for a sample of 50 people with incomes below the poverty level. The results are displayed in the following frequency distribution:

Intake (mg)	Frequency
Under 200	1
200–under 400	1
400–under 600	12
600–under 800	16
800–under 1000	12
1000–under 1200	7
1200–under 1400	1

___ **3.93**   *Runner's World* magazine conducts studies on the finishing times of runners in the New York City 10-km run. Suppose a sample of 40 finishing times yields the following frequency distribution:

Finishing time (minutes)	Frequency
30–under 40	1
40–under 50	2
50–under 60	14
60–under 70	13
70–under 80	8
80–under 90	2

In each of Exercises 3.94 and 3.95 you will be given a set of raw data and a frequency distribution for that data. For each data set,

a) compute $\overline{x}$ and s by applying Definition 3.4 on page 82 and Formula 3.1 on page 91 to the raw data.
b) compute the approximate values of $\overline{x}$ and s using the frequency distribution and the formulas discussed in this section.
c) compare your answers from parts (a) and (b).

___ **3.94**   The Bureau of Economic Analysis gathers information on the length of stay in Europe and the Mediterranean by U.S. travelers. Data are published in *Survey of Current Business*. A sample of 36 U.S. residents who traveled to Europe and the Mediterranean this year yielded the following data on length of stay. The data are in days.

Raw data

41	16	6	21	10	21
5	31	20	27	17	10
3	32	2	48	8	12
21	44	1	56	5	12
3	13	15	10	18	3
1	11	14	12	64	10

$[\Sigma x = 643, \ \Sigma x^2 = 20{,}185]$

Grouped data

Length of stay (days)	Frequency
1–7	10
8–14	10
15–21	8
22–28	1
29–35	2
36–42	1
43–49	2
50–56	1
57–63	0
64–70	1

___ **3.95**   The U.S. Energy Information Administration collects data on residential energy consumption and expenditures. Results are published in the document *Residential Energy Consumption Survey: Consumption and Expenditures.* The tables below give last year's energy consumptions for a sample of 50 households in the South. Data are in millions of BTU.

Raw data

130	55	45	64	155
66	60	80	102	62
58	101	75	111	151
139	81	55	66	90
97	77	51	67	125
50	136	55	83	91
54	86	100	78	93
113	111	104	96	113
96	87	129	109	69
94	99	97	83	97

$[\Sigma x = 4486, \ \Sigma x^2 = 438{,}942]$

Grouped data

Energy consumption (million BTU)	Frequency
40–49	1
50–59	7
60–69	7
70–79	3
80–89	6
90–99	10
100–109	5
110–119	4
120–129	2
130–139	3
140–149	0
150–159	2

___ **3.96** If data are grouped into a frequency distribution using classes based on a single value, is it possible to compute $\bar{x}$ and s without using the grouped-data formulas of this section? Explain your answer.

___ **3.97** If data are grouped into a frequency distribution using classes that are not based on a single value, why do the grouped data formulas of this section give only the approximate values of $\bar{x}$ and s?

Computing $\bar{x}$ for data grouped into a relative-frequency distribution: Formula 3.2 for computing the mean of grouped data is $\bar{x} = (\Sigma x f)/n$. Using algebra, we can rewrite this formula as

$$(1) \qquad \bar{x} = \Sigma x \cdot \frac{f}{n}$$

Since f/n represents the relative frequency of a class, we see that the sample mean of data that are grouped into a relative-frequency distribution can be obtained as follows: For each class, multiply the class mark by the relative frequency of the class, and then sum all of the resulting products. We will apply Formula (1) in Exercises 3.98 and 3.99.

═ **3.98** Refer to Exercise 3.84.
a) Construct a relative-frequency distribution.
b) Use Formula (1) to compute $\bar{x}$, and compare your answer to the one obtained in Exercise 3.84.

═ **3.99** Refer to Exercise 3.85.
a) Construct a relative-frequency distribution.
b) Use Formula (1) to compute $\bar{x}$, and compare your answer to the one obtained in Exercise 3.85.

≡ **3.100** Below is a set of 20 scores on a 100-point, true-false psychology quiz in which each question was worth 10 points.

70	70	80	80	80
70	70	80	80	80
70	70	80	80	90
70	70	80	80	90

a) Compute the sample mean, $\bar{x}$, of the exam scores from the raw data.
b) Construct a frequency distribution for the data using the classes 70–79, 80–89, and 90–99.
c) Use your frequency distribution from part (b) to compute the approximate value of $\bar{x}$.
d) Compare your answers from parts (a) and (c). Why is the approximation in part (c) so poor?

3.6 Percentiles; box-and-whisker diagrams

As we learned in Section 3.1, the *median* of a data set divides the data into two equal parts, the bottom 50% and the top 50%. Frequently it is useful, for both descriptive and inferential purposes, to divide a data set into a larger number of parts. **Quartiles** divide the data into quarters, or four equal parts; **deciles** divide the data into tenths, or 10 equal parts; and **percentiles** divide the data into hundredths, or 100 equal parts.

QUARTILES

A data set has *three* quartiles, which we denote by Q_1, Q_2, and Q_3. Roughly speaking, the first quartile, Q_1, is the number that divides the bottom 25% of the data from the top 75%; the second quartile, Q_2, is the median which, as we know, is the number that divides the bottom 50% of the data from the top 50%; and, finally, the third quartile, Q_3, is the number that divides the bottom 75% of the data from the top 25%. Example 3.24 illustrates the computation of quartiles.

EXAMPLE 3.24   *Illustrates quartiles*

The A. C. Nielsen Company publishes data on the TV viewing habits of Americans by various characteristics in *Nielsen Report on Television*. A sample of 20 people yields the weekly viewing times, in hours, displayed in Table 3.24.

TABLE 3.24
Weekly TV
viewing times

25	41	27	32	43
66	35	31	15	5
34	26	32	38	16
30	38	30	20	21

Determine and interpret the quartiles for these data.

SOLUTION   To determine the quartiles, we use the following procedure:

1. *Arrange the data in increasing order.*
2. *Divide the data into quarters.*
3. *Find the numbers dividing the quarters.*

The results of applying these three steps are displayed in the diagram below.

First quarter	Second quarter	Third quarter	Fourth quarter
5 15 16 20 21	25 26 27 30 30	31 32 32 34 35	38 38 41 43 66

First quartile
$$= \frac{21 + 25}{2}$$
$$= \mathbf{23.0}$$

Second quartile
$$= \frac{30 + 31}{2}$$
$$= \mathbf{30.5}$$

Third quartile
$$= \frac{35 + 38}{2}$$
$$= \mathbf{36.5}$$

Consequently, we see that $Q_1 = 23.0$, $Q_2 = 30.5$, and $Q_3 = 36.5$. Thus, 25% of the weekly viewing times are less than 23.0 hours, 25% are between 23.0 and 30.5 hours, 25% are between 30.5 and 36.5 hours, and 25% are greater than 36.5 hours. ■

MTB

It was easy to find the quartiles for the TV-viewing-times data because the number of viewing times (20) is a multiple of four. When the number of pieces of data is not a multiple of four, the determination of quartiles is somewhat more complicated. A general method for obtaining quartiles will be explored in the exercises.

DECILES AND PERCENTILES

We will now briefly discuss deciles and percentiles. As we said, deciles divide a data set into tenths, or 10 equal parts. A data set has *nine* deciles, which we denote by D_1, D_2, ..., D_9. Basically, the first decile, D_1, is the number that divides the bottom 10% of the data from the top 90%; the second decile, D_2, is the number that divides the bottom 20% of the data from the top 80%; and so on. To obtain the deciles of a data set, we divide the ordered data into tenths and then determine the numbers dividing the tenths.

The percentiles of a data set divide it into hundredths, or 100 equal parts. A data set has 99 percentiles, denoted by P_1, P_2, ..., P_{99}. The first percentile, P_1, is the number that divides the bottom 1% of the data from the top 99%; the second percentile, P_2, is the number that divides the bottom 2% of the data from the top 98%; and so forth. To obtain the percentiles of a data set, we divide the ordered data into hundredths and then determine the numbers dividing the hundredths.

The second quartile, fifth decile, and fiftieth percentile of a data set are all the same and all equal the median. In symbols,

$$\text{Median} = Q_2 = D_5 = P_{50}$$

Similarly, we have equalities such as $Q_1 = P_{25}$, $D_1 = P_{10}$, and $Q_3 = P_{75}$.

In practice, quartiles, deciles, and percentiles are most useful with large data sets, and for such data sets the required calculations are almost always done by computer. Our purpose here has been simply to present the basic concepts so that you will be familiar with the terminology and understand the principles behind the computations. We will discuss percentiles again when we study the normal distribution in Chapter 6.

BOX-AND-WHISKER DIAGRAMS

Box-and-whisker diagrams, or **boxplots,** are used to provide a graphical display of the variation in a data set. These diagrams, like stem-and-leaf diagrams, were invented by Professor John Tukey. The method for constructing a box-and-whisker diagram is presented as Procedure 3.1.

PROCEDURE 3.1
To construct a box-and-whisker diagram.

STEP 1 *Determine the quartiles for the data.*
STEP 2 *Find the smallest and largest data values.*

> **STEP 3** *Draw a horizontal axis on which the values obtained in Steps 1 and 2 can be located. Above this axis, mark the quartiles and the smallest and largest data values with vertical lines.*
>
> **STEP 4** *Connect the quartiles to each other to make a box, and then connect the box to the largest and smallest data values by lines.*

Note: The lines connecting the box to the largest and smallest data values are called **whiskers.** Also, in the context of box-and-whisker diagrams, the first and third quartiles are called **hinges.**[†]

EXAMPLE 3.25   **Illustrates Procedure 3.1**

The weekly TV viewing times for a sample of 20 people, given in Table 3.24, are repeated here in Table 3.25. Construct a box-and-whisker diagram for these data. Interpret the results.

TABLE 3.25
Weekly TV
viewing times

25	41	27	32	43
66	35	31	15	5
34	26	32	38	16
30	38	30	20	21

SOLUTION   To obtain a box-and-whisker diagram for the TV viewing times, we apply the step-by-step method given in Procedure 3.1.

STEP 1 *Determine the quartiles for the data.*

The quartiles for the TV viewing times were obtained in Example 3.24 (see the diagram on page 115). We found that $Q_1 = 23.0$, $Q_2 = 30.5$, and $Q_3 = 36.5$.

STEP 2 *Find the smallest and largest data values.*

From Table 3.25 or, more simply, by referring to the diagram on page 115, we see that the smallest data value is 5 and the largest data value is 66.

STEP 3 *Draw a horizontal axis on which the values obtained in Steps 1 and 2 can be located. Above this axis, mark the quartiles and the smallest and largest data values with vertical lines.*

This is done in Figure 3.10 at the top of the next page.

STEP 4 *Connect the quartiles to each other to make a box, and then connect the box to the largest and smallest data values by lines.*

This is also done in Figure 3.10. (Remember that the lines connecting the box to the largest and smallest data values are called *whiskers.*)

[†] We should point out that not all statisticians define quartiles in the same way. The formulas that we use to define the first and third quartiles (discussed in the exercises) are actually the defining formulas for the hinges. If some other formulas are used to define the first and third quartiles, then the hinges may not be the same as the first and third quartiles. In such cases, use the hinges and not the first and third quartiles in the box-and-whisker diagram.

FIGURE 3.10
Box-and-whisker
diagram for TV
viewing times

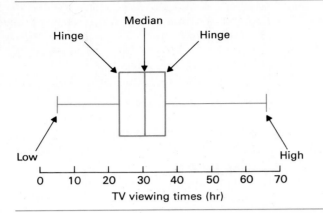

Figure 3.10 is the box-and-whisker diagram for the TV-viewing-times data. Observe that the two boxes in the box-and-whisker diagram indicate the spread of the second and third quarters of the data. Also observe that the two whiskers indicate the spread of the first and fourth quarters. Thus, we see that there is less variation in the middle two quarters of the TV-viewing-times data than in the first and fourth quarters, and that the fourth quarter has the greatest variation of all.   ■

MTB

USING THE COMPUTER (OPTIONAL)

We can use Minitab to obtain a box-and-whisker diagram by employing the **BOX-PLOT** command. We will illustrate the use of that command by applying it to the TV-viewing-times data considered in Examples 3.24 and 3.25.

EXAMPLE 3.26   *Illustrates the BOXPLOT command*

Use Minitab to obtain a box-and-whisker diagram for the TV viewing times displayed in Table 3.25 on page 117.

SOLUTION   First we use the SET command to store the TV viewing times in C3. Next we apply the NAME command to name C3 "Times." Now we can obtain the box-and-whisker diagram for the TV viewing times by typing the command BOXPLOT followed by the storage location of the sample data; that is, by typing BOXPLOT 'TIMES'. See Printout 3.4 at the top of the following page.

Note that in this box-and-whisker diagram there are 12 units between tick (+) marks on the horizontal axis (the user can specify the number of units between tick marks by applying the INCREMENT subcommand). Also note that there are 10 dashes (-) between tick marks. Hence each dash represents 1.2 units.

The Minitab output for BOXPLOT uses Is to mark the hinges (first and third quartiles) and a + to mark the median. Also, the largest value is marked here by an asterisk (*), indicating that it is a possible outlier (outliers are discussed in

PRINTOUT 3.4
Minitab output
for BOXPLOT

```
MTB > SET C3
DATA> 25 41 27 32 43
DATA> 66 35 31 15  5
DATA> 34 26 32 38 16
DATA> 30 38 30 20 21
DATA> END
MTB > NAME C3 'TIMES'
MTB > BOXPLOT 'TIMES'
```

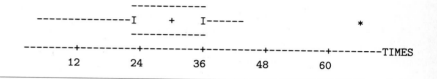

Exercise 3.20 on page 79). You should compare the box-and-whisker diagram in Printout 3.4 to the one we obtained by hand in Figure 3.10 on page 118. ∎

We can use a box-and-whisker diagram generated by Minitab's BOXPLOT command to estimate the smallest data value, the largest data value, and the quartiles. For instance, consider the box-and-whisker diagram for the TV viewing times, shown in Printout 3.4. Recall that each dash (-) in that diagram represents 1.2 units. Reading from left to right, we see that the smallest data value is about 4.8, the first quartile is about 22.8, the median is about 30.0, the third quartile is about 36.0, and the largest data value is about 66.0.

But, we don't have to be content with estimating these values—Minitab has a command that will determine the exact values of the largest and smallest data values and the quartiles. That command is called **LVALS**, which is short for *letter-values*.[†] We applied the LVALS command to the TV-viewing-times data and obtained the output shown in Printout 3.5.

PRINTOUT 3.5
Minitab output
for LVALS

	DEPTH	LOWER	UPPER	MID	SPREAD
MTB > LVALS 'TIMES'					
N=	20				
M	10.5		30.500		30.500	
H	5.5	23.000	36.500	29.750	13.500
E	3.0	16.000	41.000	28.500	25.000
D	2.0	15.000	43.000	29.000	28.000
	1	5.000	66.000	35.500	61.000

The row of the output that reads N= 20 indicates that there are 20 data values (TV viewing times) in the data set. The row beneath that begins with the letter M,

---

† The LVALS command is not available on all versions of Minitab.

which stands for *median*. To the right of the letter M, under the column labeled DEPTH, is the number 10.5. This indicates that the median of the data set is halfway between the tenth and eleventh data values when the data are arranged in increasing order. In other words, the median is equal to the mean of the tenth and eleventh data values in an ordered list. The numerical value of the median is the number to the right of the depth of 10.5, namely, 30.5.

In the next row we can find the information about the hinges (first and third quartiles). As you might suspect, the letter H stands for *hinges*. The depth of 5.5 that appears next indicates that the first quartile is halfway between the fifth and sixth data values in an ordered list, counting in from the left; and that the third quartile is halfway between the fifth and sixth data values in an ordered list, counting in from the right. Following the depth of 5.5 are the values of the first and third quartiles, respectively. So, we see that the first quartile equals 23.0 and the third quartile equals 36.5.

To obtain the smallest and largest data values in the data set, we use the last row of Printout 3.5. The smallest data value is the entry in that row under the column labeled LOWER and the largest data value is the entry in that row under the column labeled UPPER. Thus, the smallest and largest data values in the data set are 5 and 66, respectively.

As you can see, there is a lot of other information contained in the output of the LVALS command. For a detailed discussion of that information, we refer the reader to the *Minitab Supplement*.

Exercises 3.6

Determine and interpret the quartiles for each of the data sets in Exercises 3.101–3.104.

___ **3.101** The exam scores for the students in an introductory statistics course are displayed in the following table:

88	67	64	76	86
85	82	39	75	34
90	63	89	90	84
81	96	100	70	96

___ **3.102** A water-heater manufacturer guarantees the electric heating element for a period of five years. The lifetimes, in months, for a sample of 20 such elements are:

49.3	79.3	86.4	68.4	62.6
64.1	53.2	30.0	66.6	66.9
65.1	67.8	95.9	40.1	79.3
92.1	92.0	93.0	48.9	63.7

___ **3.103** The U.S. Energy Information Administration publishes figures on residential energy consumption and expenditures in *Residential Energy Consumption Survey: Consumption and Expenditures*. A sample of 36 households using electricity as their primary energy source yields the following data on last year's energy expenditures:

$1376	1452	1235	1480	1185	1327
1059	1400	1227	1102	1168	1070
1180	1221	1351	1014	1461	1102
976	1394	1379	987	1002	1532
1450	1177	1150	1352	1266	1109
949	1351	1259	1179	1393	1456

___ **3.104** The following data provide the hourly temperature readings in Colorado Springs, Colorado, for Tuesday, August 22, 1978. The data are in degrees Fahrenheit.

69	63	61	74	88	87	74	68
66	61	64	79	87	85	71	65
65	60	70	84	85	73	70	62

___ **3.105**   Obtain the deciles for the 20 exam scores given in Exercise 3.101. Interpret your results.

___ **3.106**   Refer to Exercise 3.102. Find the deciles for the lifetimes of the sample of 20 electric heating elements. Interpret your results.

In each of Exercises 3.107–3.110, construct a box-and-whisker diagram for the specified data set and use the diagram to discuss the variation in the data set.

___ **3.107**   The exam scores for the 20 students in an introductory statistics class, given in Exercise 3.101.

___ **3.108**   The lifetimes, in months, shown in Exercise 3.102, for a sample of 20 electric heating elements.

___ **3.109**   The energy expenditures for a sample of 36 households using electricity as their primary energy source, from Exercise 3.103.

___ **3.110**   The hourly-temperature data for Colorado Springs, Colorado, on Tuesday, August 22, 1978, given in Exercise 3.104.

Exercises 3.111–3.114 are computer exercises.

___ **3.111** (**Computer exercise**)   Suppose that the data in Exercise 3.101 are stored in a column named SCORES.
a) Which Minitab command and subcommands (if any) should be used to obtain a box-and-whisker diagram for the data?
b) Which Minitab command and subcommands (if any) should be used to determine the smallest and largest data values and the quartiles of the data?
c) If you have access to Minitab, use it to carry out parts (a) and (b).

___ **3.112** (**Computer exercise**)   Suppose the data in Exercise 3.102 are stored in a column named LIFE.
a) Which Minitab command and subcommands (if any) should be used to obtain a box-and-whisker diagram for the data?
b) Which Minitab command and subcommands (if any) should be used to determine the smallest and largest data values and the quartiles of the data?
c) If you have access to Minitab, use it to carry out parts (a) and (b).

___ **3.113** (**Computer exercise**)   The U.S. Department of Agriculture collects information on annual per capita beef consumption and publishes the data in *Food Consumption, Prices, and Expenditures*. Each person in a sample is asked to estimate his or her beef consumption, in pounds, for last year.
a) Printout 3.6 at the top of the next page shows a box-and-whisker diagram for the beef-consumption data generated by Minitab's BOXPLOT command. Use the diagram to discuss the variation in the data set. Also, use the diagram to determine approximately the smallest and largest data values and the quartiles for the data.
b) In Printout 3.7, shown on the next page, we have given the Minitab output obtained by applying the LVALS command to the beef-consumption data. Use the output to determine exactly the smallest and largest data values and the quartiles for the data. Compare your answers to the approximate values that you obtained in part (a).

___ **3.114** (**Computer exercise**)   A prospective investor in a limited-partnership firm wants information on the monthly rental charges for three-bedroom homes in the area. She obtains the monthly rental charges for a sample of three-bedroom homes.
a) Printout 3.8, on the next page, gives the output obtained by applying Minitab's BOXPLOT command to the sample of monthly rental charges. Use the diagram to discuss the variation in the data. Also, employ the diagram to find approximately the smallest and largest data values and the quartiles for the data.
b) In Printout 3.9, shown on the following page, we have displayed the output generated by applying Minitab's LVALS command to the monthly-rental-charge data. Use the output to determine exactly the smallest and largest data values and the quartiles for the data. Compare your answers to the approximate values obtained in part (a).

A general method for obtaining quartiles: We will now describe a general procedure for determining the quartiles of a data set. This procedure is to be used in Exercises 3.115–3.118.

1. Arrange the data in increasing order.
2. The median of the data set is at position $(n+1)/2$, where n is the number of pieces of data.
3. Let m denote the whole number obtained by rounding the position of the median down to the nearest integer. Then the first quartile is at position $(m+1)/2$, counting in from the left; and the third quartile is at position $(m+1)/2$, counting in from the right.

PRINTOUT 3.6   Minitab output for Exercise 3.113(a)

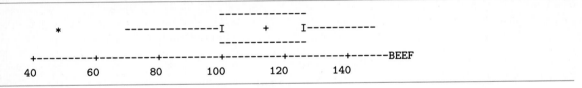

PRINTOUT 3.7   Minitab output for Exercise 3.113(b)

	DEPTH	LOWER	UPPER	MID	SPREAD
N=	50				
M	25.5		113.500	113.500	
H	13.0	100.000	126.000	113.000	26.000
E	7.0	91.000	132.000	111.500	41.000
D	4.0	81.000	135.000	108.000	54.000
C	2.5	74.000	140.000	107.000	66.000
B	1.5	58.500	145.000	101.750	86.500
	1	48.000	148.000	98.000	100.000

PRINTOUT 3.8   Minitab output for Exercise 3.114(a)

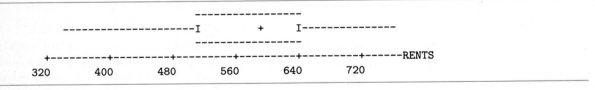

PRINTOUT 3.9   Minitab output for Exercise 3.114(b)

	DEPTH	LOWER	UPPER	MID	SPREAD
N=	35				
M	18.0		594.000	594.000	
H	9.5	512.500	638.000	575.250	125.500
E	5.0	493.000	721.000	607.000	228.000
D	3.0	453.000	740.000	596.500	287.000
C	2.0	381.000	756.000	568.500	375.000
	1	341.000	757.000	549.000	416.000

For example, suppose a data set has 18 pieces of data. Then $n = 18$ and so the median of the data set is at position $(n+1)/2 = (18+1)/2$, or 9.5, in the ordered list. In other words, the median of the data set is the mean of the ninth and tenth data values in the ordered list. To obtain the positions of the first and third quartiles, we first round the position of the median, 9.5 in this case, down to the nearest whole number. This gives the number 9, so $m = 9$. Now we calculate $(m + 1)/2$, which is $(9 + 1)/2$, or 5. Therefore, the first quartile is at position 5, counting in from the left in the ordered list; and the third quartile is at position 5, counting in from the right in the ordered list.

3.115   Consider once more the data on energy expenditures given in Exercise 3.103. Apply the three-step procedure just presented to obtain the quartiles of the data. Compare your answers with the ones obtained in Exercise 3.103.

━━ **3.116** A social psychologist is studying the relationship between alcohol-consumption patterns of drinkers and IQ. The IQs of a sample of nine drinkers provided the following data:

121	122	112
108	94	98
120	109	105

Determine the quartiles for this data set.

━━ **3.117** The U.S. Federal Highway Administration conducts studies on motor vehicle travel by type of vehicle. Results are published annually in *Highway Statistics*. Suppose that a sample of 15 cars yields the following data on number of thousands of miles driven for last year:

10.2	10.3	8.9	12.7	8.3
9.2	13.7	7.7	3.3	10.6
11.8	6.6	8.6	5.7	12.0

Obtain the quartiles for these data.

━━ **3.118** In January 1984, the U.S. Department of Agriculture reported in *Family Economic Review* that a typical American family of four with an intermediate budget would spend about $92 per week for food. A consumer researcher in Kansas suspected that the median weekly cost was less in her state. She took a sample of 10 Kansas families of four, each with an intermediate budget, and obtained the following weekly food costs:

$78	104	84	70	96
73	87	85	76	94

Find the quartiles for this data set.

3.7 Descriptive measures for populations; use of samples

As you know, in inferential studies we use data obtained from a sample of a population to draw conclusions about the entire population. Thus, the data we analyze in such studies are sample data. For this reason, we have concentrated on the calculation and interpretation of descriptive measures for sample data. But, it is important to keep in mind that the ultimate objective is to use descriptive measures obtained from the sample to make inferences about the population.

In this section, we will discuss some descriptive measures for populations. Although, in reality, we usually don't have access to the data for an entire population, it is nonetheless helpful to see the notation and formulas used for descriptive measures of populations.

THE POPULATION MEAN

Recall that for a sample of size n, the sample mean, $\bar{x}$, is defined by

$$\bar{x} = \frac{\Sigma x}{n}$$

The mean of a finite population is computed in the same way—sum the data and then divide by the total number of pieces of data. However, to distinguish a population mean from a sample mean, we use the lower-case Greek letter μ (pronounced "mew") to denote the mean of a population. Furthermore, we use the upper-case English letter N to represent the size of a population. Table 3.26, at the top of the next page, summarizes the notations used for both a sample and a population.

	Size	Mean
Sample	n	$\bar{x}$
Population	N	μ

Using the notation displayed in the second row of Table 3.26, we now define the mean of a finite population.

DEFINITION 3.8   Population mean

The *population mean*, μ, **of a finite population of size N is defined by**

$$\mu = \frac{\Sigma x}{N}$$

In inferential studies, the size of the population is ordinarily quite large and the mean of the population is usually difficult or impossible to determine exactly. However, for the sake of illustration, we will sometimes consider small populations for which the mean, μ, can easily be computed.

EXAMPLE 3.27   *Illustrates Definition 3.8*

Table 3.27 presents some data for the starting offense of the 1988–1989 Los Angeles Rams football team.

TABLE 3.27
Los Angeles Rams
Offense, 1988–89

Name	Position	Height	Weight (lb)
Jim Everett	QB	6'5"	212
Doug Smith	C	6'3"	272
Jackie Slater	RT	6'4"	285
Tom Newberry	RG	6'2"	285
Duval Love	LG	6'3"	287
Irv Pankey	LT	6'5"	295
Damone Johnson	TE	6'4"	250
Henry Ellard	WR	5'11"	182
Aaron Cox	WR	5'10"	178
Greg Bell	RB	5'11"	210
Buford McGee	RB	6'0"	210

Compute the population mean weight of these 11 players.

SOLUTION   The sum of the weights in Table 3.27 is $\Sigma x = 2666$ lb. Since there are 11 players, $N = 11$. Thus, the population mean weight of the players is

$$\mu = \frac{\Sigma x}{N} = \frac{2666}{11} = \boxed{242.4 \text{ lb}}$$

MTB

If we were interested in the mean weight of *all* starting-offense players in professional football, then the 11 weights of the starting-offense players on the Rams would be considered a sample instead of a population. The mean of those 11 weights would then be a sample mean instead of a population mean, and we would write $\bar{x} = 242.4$ lb instead of $\mu = 242.4$ lb. Consequently, we see that whether a data set is considered population data or sample data often depends on the interests of the researcher conducting the study.

USING A SAMPLE MEAN TO ESTIMATE A POPULATION MEAN

As we have mentioned, although we usually deal with sample data in inferential studies, the objective is to describe the entire population. The reason for resorting to a sample is that it is generally more practical. We illustrate this last point in Example 3.28.

EXAMPLE 3.28   *Illustrates a use of the sample mean*

The U.S. Bureau of the Census annually reports figures on the mean income of American households in the publication *Current Population Reports*. To obtain the complete population data—the incomes of *all* American households—would be extremely expensive and time-consuming. It is also unnecessary because accurate estimates for the mean income of all American households can be obtained from the mean income of a sample of such households.

In reality, the Census Bureau samples 72,000 households out of a total of more than 83 million American households. The population of interest here is comprised of the incomes of all American households, and the mean of those incomes is the population mean, μ. The sample consists of the incomes of the 72,000 households sampled by the Census Bureau, and the mean of those incomes is the sample mean, $\bar{x}$. We summarize the situation in Figure 3.11.

FIGURE 3.11
Population and
sample for incomes
of American
households

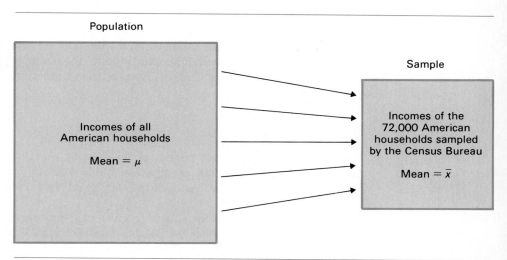

Once the sample is taken, the Census Bureau can compute the sample mean income, $\bar{x}$, of the 72,000 households obtained. Using the value of $\bar{x}$, the Census Bureau can then estimate the population mean income, μ, of all American households. We will study these kinds of inferences in Chapter 8.　■

THE POPULATION STANDARD DEVIATION

Recall that the standard deviation of a sample is denoted by the letter s and is defined by

$$s = \sqrt{\frac{\Sigma(x - \bar{x})^2}{n - 1}}$$

The standard deviation of a finite population is computed in a similar, but slightly different, way. It is denoted by the lower-case Greek letter σ (pronounced "sigma") and is defined as follows:

DEFINITION 3.9   Population standard deviation

The *population standard deviation*, σ, of a finite population of size N is defined by

$$\sigma = \sqrt{\frac{\Sigma(x - \mu)^2}{N}}$$

and can also be computed using the *shortcut formula*

$$\sigma = \sqrt{\frac{\Sigma x^2}{N} - \mu^2}$$

Just as s^2 is called the sample variance, σ^2 is called the **population variance.** Note that in the defining formula for the population standard deviation, σ, we divide by N, the size of the population. On the other hand, in the defining formula for the sample standard deviation, s, we divide by $n - 1$, one less than the size of the sample. This is because in inferential studies, the main use of s is as an estimate of σ. Dividing by $n - 1$, instead of by n, in the formula for s results in a better estimate of σ, on the average.

EXAMPLE 3.29   *Illustrates Definition 3.9*

The weights, to the nearest pound, of the starting offense of the 1988–1989 Los Angeles Rams football team are given in Table 3.27 on page 124. Compute the population standard deviation, σ, of the weights of these 11 players.

SOLUTION   We will apply the shortcut formula for σ given in Definition 3.9. In the first column of Table 3.28, we have copied the weights from Table 3.27. The second column of Table 3.28 gives the squares of those weights.

TABLE 3.28

x	x^2
212	44,944
272	73,984
285	81,225
285	81,225
287	82,369
295	87,025
250	62,500
182	33,124
178	31,684
210	44,100
210	44,100
2666	666,280

Recalling that $\mu = \Sigma x/N$, we see from the last row of Table 3.28 that

$$\sigma = \sqrt{\frac{\Sigma x^2}{N} - \mu^2} = \sqrt{\frac{666{,}280}{11} - \left(\frac{2666}{11}\right)^2} = 42.8$$

MTB

The population standard deviation of the weights of the 11 players is 42.8 lb. ∎

Note: In the previous example, we computed the population standard deviation, σ, of the weights of 11 football players. The formula for calculating σ involves the population mean, μ, of the weights. In Example 3.27, we found that $\mu = 242.4$ lb. But when we computed σ in the previous example, we didn't use that value of μ. Instead, we employed the formula $\mu = \Sigma x/N$ all over again. The reason is that the value 242.4 for μ is a rounded value; and as we said earlier, no rounding should be done until the computation is complete.

USING A SAMPLE STANDARD DEVIATION TO ESTIMATE A POPULATION STANDARD DEVIATION

In the next example, we will illustrate how a sample standard deviation can be used to estimate a population standard deviation.

EXAMPLE 3.30   *Illustrates a use of the sample standard deviation*

A hardware manufacturer produces 10-millimeter bolts. The manufacturer knows that the diameters of the bolts produced vary somewhat from 10 mm and also from each other. But even if he is willing to accept some variation in bolt diameters, he cannot tolerate too much variation. For if the variation is too large, then too many of the bolts produced will be unusable.

Therefore, the manufacturer must make sure that the standard deviation, σ, of the bolt diameters is not unduly large. Since, in this case, it is not possible to determine σ exactly (why?), inferential statistics must be employed.

Suppose the manufacturer decides to estimate σ by taking a sample of 20 bolts. The population of interest here consists of the diameters of all bolts that have been or ever will be produced by the manufacturer, and the standard deviation of those diameters is the population standard deviation, σ. The sample consists of the diameters of the 20 bolts sampled by the manufacturer, and the standard deviation of those diameters is the sample standard deviation, s. Figure 3.12 summarizes the above discussion.

FIGURE 3.12
Population and sample for bolt diameters

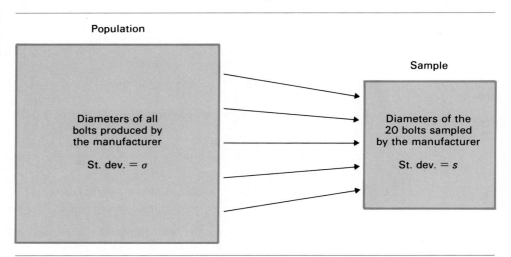

After the sample is taken, the manufacturer can compute the sample standard deviation, s, of the diameters of the 20 bolts obtained. Using the value of s, he can then estimate the population standard deviation, σ, of the diameters of *all* bolts being produced. We will examine these types of inferences in Chapter 11. ∎

PARAMETER AND STATISTIC

There is statistical terminology that helps us to distinguish between descriptive measures for populations and descriptive measures for samples.

DEFINITION 3.10   Parameter and statistic

Parameter: A descriptive measure for a population.
Statistic: A descriptive measure for a sample.

Thus, for example, μ and σ are *parameters,* whereas $\bar{x}$ and s are *statistics.*

FORMULAS FOR GROUPED POPULATION DATA

There are formulas for computing the mean and standard deviation of a population when the data are grouped in a frequency distribution. These formulas are derived in the same way as the grouped-data formulas for the sample mean and sample standard deviation (Section 3.5).

FORMULA 3.4 Computing μ for grouped data

The mean of a finite population whose data are grouped in a frequency distribution can be computed using the formula

$$\mu = \frac{\Sigma x f}{N}$$

where f denotes class frequency and N $(= \Sigma f)$ denotes the population size.

FORMULA 3.5 Computing σ for grouped data

The standard deviation of a finite population whose data are grouped in a frequency distribution can be computed using the formula

$$\sigma = \sqrt{\frac{\Sigma(x - \mu)^2 f}{N}}$$

or its shortcut version

$$\sigma = \sqrt{\frac{\Sigma x^2 f}{N} - \mu^2}$$

where f denotes class frequency and N $(= \Sigma f)$ denotes the population size.

EXAMPLE 3.31 *Illustrates Formulas 3.4 and 3.5*

A real-estate development consists of 385 homes. In Table 3.29 we have presented a frequency distribution for the number of bedrooms in each of the 385 homes.

TABLE 3.29
Frequency distribution for number of bedrooms

No. bedrooms x	Frequency f
2	38
3	177
4	128
5	42

The table shows, for instance, that 38 of the homes have two bedrooms. Regarding this development as a population of interest, compute the mean and standard deviation of the number of bedrooms per home.

SOLUTION Since the population data are in grouped form, we apply Formulas 3.4 and 3.5 (we will use the shortcut formula to compute σ). As the formulas indicate, we will need a table with columns for x, f, xf, x^2, and $x^2 f$. See Table 3.30.

TABLE 3.30

x	f	xf	x^2	x^2f
2	38	76	4	152
3	177	531	9	1593
4	128	512	16	2048
5	42	210	25	1050
	385	1329		4843

From the second and third columns of Table 3.30, we see that

$$\mu = \frac{\Sigma xf}{N} = \frac{1329}{385} = 3.5$$

Now using the fifth column also, we find that

$$\sigma = \sqrt{\frac{\Sigma x^2 f}{N} - \mu^2} = \sqrt{\frac{4843}{385} - \left(\frac{1329}{385}\right)^2} = 0.8$$

MTB

Thus, the (population) mean number of bedrooms per home in the development is 3.5, with a standard deviation of 0.8.   ∎

The frequency distribution in Table 3.29 uses classes based on a single value. When population data are grouped into a frequency distribution in which each class represents several values, we use the class marks for the x-values in Formulas 3.4 and 3.5. However, we must keep in mind that the values for μ and σ so obtained are only approximately equal to the actual population mean and standard deviation.

INTERPRETATION OF THE POPULATION STANDARD DEVIATION; Z-SCORES

The interpretation of the standard deviation of a population is essentially identical to that of the standard deviation of a sample. In particular, the more variation there is in a population, the larger its standard deviation. Chebychev's rule also holds for populations. Specifically, we have the following key fact:

KEY FACT 3.3   Chebychev's rule

For *any* population:

PROPERTY 1   At least 75% of the population values lie within *two* standard deviations to either side of the mean, that is, between $\mu - 2\sigma$ and $\mu + 2\sigma$.

PROPERTY 2   At least 89% of the population values lie within *three* standard deviations to either side of the mean, that is, between $\mu - 3\sigma$ and $\mu + 3\sigma$.

PROPERTY 3   In general, for any number $k > 1$, at least $1 - 1/k^2$ of the population values lie within k standard deviations to either side of the mean, that is, between $\mu - k \cdot \sigma$ and $\mu + k \cdot \sigma$.

As we pointed out in Section 3.4, the power of Chebychev's rule is its generality; it holds for any population. On the other hand, if we know something about the distribution of the population, then we can obtain better estimates than those given by Chebychev's rule. In particular, if the population has a bell-shaped distribution, then the **empirical rule** gives precise estimates for the percentage of population values that lie within a specified number of standard deviations to either side of the mean. We will discuss the empirical rule in Chapter 6.

Finally, let us discuss population z-scores. In Section 3.4 we introduced the concept of z-scores. Recall that the z-score for a data value is the number of standard deviations that the data value is away from the mean of the data set. The sample z-score is computed by first subtracting the sample mean, $\bar{x}$, from the data value, x, and then dividing the result by the sample standard deviation, s:

$$z = \frac{x - \bar{x}}{s}$$

The population z-score is defined in the same way except, of course, we use μ in place of $\bar{x}$ and σ in place of s.

DEFINITION 3.11   Population z-score

The z-score for a population value, x, is defined to be the number of standard deviations that x is away from the mean. The population z-score is computed by using the formula

$$z = \frac{x - \mu}{\sigma}$$

where μ and σ are the mean and standard deviation of the population. A negative z-score indicates that the population value is smaller than the mean, while a positive z-score indicates that the population value is larger than the mean.

The z-score can be used as a measure of relative standing. If a population value, x, has a large positive z-score, then x is larger than most of the other members of the population; if a population value, x, has a large negative z-score, then x is smaller than most of the other members of the population; and if a population value, x, has a z-score near 0, then x is located near the mean of the population.

EXAMPLE 3.32   *Illustrates Chebychev's rule and z-scores*

It is known that a coffee machine dispenses an average of 6 fluid ounces of coffee with a standard deviation of 0.2 fluid ounces.

a) Fill in the blanks: At least 75% of the cups of coffee dispensed by the machine contain between _____ and _____ fluid ounces.
b) A cup of coffee dispensed from the machine is found to contain 7.1 fluid ounces of coffee. Determine and interpret the z-score for this cup of coffee. Does this cup of coffee contain an unusually large amount?

SOLUTION First of all, by assumption, we know that the population mean amount dispensed is $\mu = 6$ fluid ounces and that the standard deviation of the amounts dispensed is $\sigma = 0.2$ fluid ounces.

a) By Property 1 of Chebychev's rule, at least 75% of the cups of coffee dispensed by the machine have contents within two standard deviations to either side of the mean. Two standard deviations to the left of the mean is $\mu - 2\sigma = 6 - 2 \cdot 0.2 = 5.6$ and two standard deviations to the right of the mean is $\mu + 2\sigma = 6 + 2 \cdot 0.2 = 6.4$. Thus, at least 75% of the cups of coffee dispensed by the machine have contents between **5.6** and **6.4** fluid ounces.

b) The z-score for a content of 7.1 fluid ounces is

$$z = \frac{x - 6}{0.2} = \frac{7.1 - 6}{0.2} = \boxed{5.5}$$

In other words, the content of 7.1 fluid ounces is 5.5 standard deviations above the mean.

By Property 2 of Chebychev's rule, at least 89% of the cups of coffee dispensed by the machine have contents within three standard deviations to either side of the mean. Consequently, the cup of coffee containing 7.1 fluid ounces, whose content is 5.5 standard deviations above the mean, contains more coffee than at least 89% of all cups dispensed.

Using Property 3 of Chebychev's rule, we can get even a better idea of the relative standing of this particular cup of coffee. Taking $k = 5.5$ (the z-score for 7.1 fluid ounces), we see that

$$1 - 1/k^2 = 1 - 1/5.5^2 = 1 - 1/30.25 = 0.967$$

or 96.7%. Consequently, by Property 3 of Chebychev's rule, at least 96.7% of the cups of coffee dispensed by the machine have contents within 5.5 standard deviations to either side of the mean. Therefore, the cup of coffee containing 7.1 fluid ounces, whose content is 5.5 standard deviations above the mean, contains more coffee than at least 96.7% of all cups dispensed. It does seem that this cup of coffee contains an unusually large amount. ∎

THE POPULATION MEAN AND POPULATION STANDARD DEVIATION IN INFERENTIAL STUDIES

We would like to emphasize again that in inferential studies the population mean, μ, and population standard deviation, σ, are rarely known. In fact, two major topics in inferential statistics are:

1. Using the mean, $\bar{x}$, of a sample to make inferences about the population mean, μ.

2. Using the standard deviation, s, of a sample to make inferences about the population standard deviation, σ.

Topic 1 will be examined in Chapters 8 and 9 and Topic 2 in Chapter 11.

Exercises 3.7

___ **3.119** We generally deal with sample data in inferential studies, but what is the ultimate objective?

___ **3.120** In Section 3.3 we considered the heights of the starting five players on each of two men's basketball teams. The heights, in inches, for the players on Team I are 72, 73, 76, 76, and 78. Considering the heights as a sample of the heights of all male starting college basketball players,
a) compute the sample mean height, $\bar{x}$.
b) compute the sample standard deviation, s.
Considering the heights now as a population,
c) compute the population mean height, μ.
d) compute the population standard deviation, σ.
Comparing your answers from parts (a) and (c) and from parts (b) and (d),
e) why are the values for the sample mean, $\bar{x}$, and population mean, μ, equal?
f) why are the values for the sample standard deviation, s, and the population standard deviation, σ, different?

___ **3.121** In Section 3.3 we considered the heights of the starting five players on each of two men's basketball teams. The heights, in inches, for the players on Team II are 67, 72, 76, 76, and 84. Considering the heights as a sample of the heights of all male starting college basketball players,
a) compute the sample mean height, $\bar{x}$.
b) compute the sample standard deviation, s.
Considering the heights now as a population,
c) compute the population mean height, μ.
d) compute the population standard deviation, σ.
Comparing your answers from parts (a) and (c) and from parts (b) and (d),
e) why are the values for the sample mean, $\bar{x}$, and population mean, μ, equal?
f) why are the values for the sample standard deviation, s, and the population standard deviation, σ, different?

In each of Exercises 3.122–3.125 consider the given data set a population of interest. For each exercise,
a) *compute the population mean, μ.*
b) *compute the population standard deviation, σ.*

___ **3.122** Last year's monthly electric bills for a family living in the southeast are as follows. The data are in dollars, rounded to the nearest dollar.

$87	54	125	89
92	68	142	63
65	95	140	84

___ **3.123** Last year's monthly phone bills for the family in Exercise 3.122 are presented below. The data are in dollars, rounded to the nearest dollar.

$81	35	75	48
55	46	61	41
24	25	33	32

___ **3.124** As reported by the U.S. Department of Agriculture in *Crop Production*, the acreages of tobacco harvested in 1987 by each of the tobacco producing states are as follows:

State	Acreage (thousands)	State	Acreage (thousands)
SC	42	MD	15
TN	49	FL	6
NC	225	VA	39
GA	32	OH	7
PA	11	KY	148

___ **3.125** The U.S. Agency for International Development compiles information on U.S. foreign aid commitments for economic assistance. Data are published in *U.S. Overseas Loans and Grants and Assistance from International Organizations*. In 1987 the commitments to Latin American countries were as follows:

Country	Aid ($millions)	Country	Aid ($millions)
Bolivia	29	Haiti	75
Costa Rica	161	Honduras	175
Domin. Rep.	20	Jamaica	44
Ecuador	37	Panama	12
El Salvador	414	Peru	21
Guatemala	154		

___ **3.126** In Example 3.27 we found that the population mean weight of the starting offensive players on the 1988–1989 Los Angeles Rams football team is

$\mu = 242.4$ lb. In this context, is the number 242.4 a parameter or a statistic? Explain your answer.

___ 3.127   In Example 3.29 we determined that the population standard deviation of the weights of the starting offense on the 1988–1989 Los Angeles Rams football team is $\sigma = 42.8$ lb. In this context, is the number 42.8 a parameter or a statistic? Explain.

___ 3.128   The ages of the students in Professor W's introductory statistics class are given in Table 3.6 on page 75. A frequency distribution for those ages is:

Age	Frequency
17	1
18	1
19	9
20	7
21	7
22	5
23	3
24	4
26	1
35	1
36	1

Regarding the ages of these students as a population,
a) compute the population mean age, μ.
b) compute the population standard deviation, σ.

___ 3.129   An eight-question quiz on fractions is given to a fifth-grade class. A frequency distribution for the number of incorrect answers is:

Number incorrect	Frequency
0	4
1	8
2	4
3	5
4	2
5	1
6	1

a) Compute the population mean number of questions missed.
b) Compute σ.

___ 3.130   The U.S. Bureau of the Census collects and publishes annual figures for state expenditures on highways. The following is a frequency distribution of the 1987 highway expenditures by the 50 states:

Expenditures ($millions)	Number of states
0–under 400	12
400–under 800	12
800–under 1200	15
1200–under 1600	3
1600–under 2000	3
2000–under 2400	1
2400–under 2800	1
2800–under 3200	0
3200–under 3600	2
3600–under 4000	1

a) Compute the approximate value of the mean 1987 highway expenditure, μ, per state.
b) Compute the approximate value of the standard deviation, σ, of these highway expenditures.
c) Why are your values for μ and σ only approximately correct?

___ 3.131   The following table gives the 1986 age distribution for Americans between 3 and 34 years old who are enrolled in school. [SOURCE: U.S. Bureau of the Census, *Current Population Reports*.]

Age (yrs)	Frequency (thousands)
3–4	2,813
5–6	6,917
7–13	22,987
14–15	7,007
16–17	6,861
18–19	3,872
20–21	2,430
22–24	2,154
25–29	1,882
30–34	1,230

a) Compute the approximate value of the mean age, μ, for Americans between 3 and 34 years old who are enrolled in school.
b) Compute the approximate value of the standard deviation, σ, of the ages.
c) Why are your values for μ and σ only approximately correct?

___ 3.132   As reported in *News* by the U.S. Department of Agriculture, the mean weekly food cost for a couple with two children 6–11 years old is $\mu = \$95.40$.

Assume that the standard deviation of the weekly food costs is $\sigma = \$17.20$.

a) Fill in the blanks: At least 75% of such couples have weekly food costs between $\$\_\_\_\_\_$ and $\$\_\_\_\_\_$.

b) Fill in the blanks: At least 89% of such couples have weekly food costs between $\$\_\_\_\_\_$ and $\$\_\_\_\_\_$.

c) Find the population z-scores for such couples with weekly food costs of $149.59, $96.60, and $56.70. Interpret your results in words.

d) Use your results from part (c) to estimate the relative standing of each of the weekly food costs in part (c).

__ 3.133 According to a 1973 study by R. R. Paul, the mean gestation period of the Morgan horse is $\mu = 339.6$ days with a standard deviation of $\sigma = 13.3$ days.

a) Fill in the blanks: At least 75% of such gestation periods are between _____ and _____ days.

b) Fill in the blanks: At least 89% of such gestation periods are between _____ and _____ days.

c) Find the population z-scores for gestation periods of 293 days, 368 days, and 338 days.

d) Use your results from part (c) to estimate the relative standing of each of the gestation periods in part (c).

__ 3.134 A company produces cans of stewed tomatoes with an advertized weight of 14 oz. The standard deviation of the weights is known to be 0.4 oz. The quality-control engineer selects a can of stewed tomatoes at random and finds its net weight to be 17.28 oz.

a) Estimate the relative standing of that can of stewed tomatoes, assuming the true mean weight is 14 oz.

b) Should the quality-control engineer be concerned that the true mean weight of all cans of stewed tomatoes being produced is not 14 oz? Explain.

__ 3.135 Suppose you buy a new car whose advertized mileage is 25 mpg (miles per gallon). After driving the car for several months, you find that its mileage is 21.4 mpg. You telephone the manufacturer and learn that the standard deviation of gas mileages for all cars of the model you bought is 1.15 mpg.

a) Find the z-score for the gas mileage of your car.

b) Does it appear that your car is getting unusually low gas mileage? Explain.

Computing μ for data grouped into a relative-frequency distribution: Formula 3.4 for computing the population mean of grouped data is $\mu = (\Sigma x f)/N$. Using algebra, we can rewrite this formula as

(2) $$\mu = \Sigma x \cdot \frac{f}{N}$$

Since f/N represents the relative frequency of a class, we see that the population mean of data that are grouped into a relative-frequency distribution can be obtained as follows: For each class, multiply the class mark by the relative frequency of the class, and then sum all of the resulting products. We will apply Formula (2) in Exercises 3.136 and 3.137.

$=$ 3.136 Refer to Exercise 3.128.

a) Construct a relative-frequency distribution for the population of ages.

b) Use Formula (2) to compute μ, and compare your answer to the one obtained in Exercise 3.128(a).

$=$ 3.137 Refer to Exercise 3.129.

a) Construct a relative-frequency distribution for the number of incorrect answers.

b) Use Formula (2) to compute μ, and compare your answer to the one obtained in Exercise 3.129(a).

$=$ 3.138 Derive two formulas that can be used to compute the standard deviation, σ, of a population when it is grouped in a relative-frequency distribution.

$=$ 3.139 Use one of the formulas that you derived in Exercise 3.138 to compute σ from the relative-frequency distribution that you constructed in Exercise 3.137(a). Compare your answer to the one that you obtained in Exercise 3.129(b).

$=$ 3.140 Use one of the formulas that you derived in Exercise 3.138 to compute σ from the relative-frequency distribution that you constructed in Exercise 3.136(a). Compare your answer to the one that you obtained in Exercise 3.128(b).

$=$ 3.141 Consider the three data sets below:

Data Set 1	Data Set 2	Data Set 3
2 4	7 5 5 3	4 7 8 9 7
7 3	9 8 6	4 5 3 4 5

a) Assuming that these data sets are each a sample of a population, compute their standard deviations. [Round your final answers to two decimal places.]

b) Assuming these data sets are each a population, compute their standard deviations. [Round your final answers to two decimal places.]

c) Using your results from parts (a) and (b), make an educated guess about the answer to the following question: Suppose that both s and σ are computed for the same data set. Will they tend to be closer together if the data set is large or if it is small?

≡ 3.142   Consider a data set with m data values. If the data set is a sample from a population, then we compute the sample standard deviation, s. If the data set is a population, then we compute the population standard deviation, σ.

a) Derive a formula that gives the precise relationship between the values of s and σ calculated from the same data set.

b) Refer to the three data sets displayed in Exercise 3.141. Verify that your formula in part (a) works for each of the three data sets.

c) Suppose a data set consists of 15 data values. You compute the population standard deviation of the data and obtain $\sigma = 38.6$. Then you realize that the data set is actually a sample of a population and that you should have computed the sample standard deviation, s, instead. Use your result from part (a) to obtain s.

Chapter review

KEY TERMS

bimodal, 72
box-and-whisker diagram, 116
BOXPLOT,* 118
Chebychev's rule, 98, 130
deciles, 115
descriptive measures, 69
deviation from the mean, 87
hinges, 117
LVALS,* 119
MAX,* 92
mean, 70
MEAN,* 76
measures of central tendency, 70
measures of dispersion, 85
measures of relative standing, 102
median, 71
MEDIAN,* 76
MIN,* 92
mode, 72
parameter, 128

percentiles, 115
population mean (μ), 124
population standard deviation (σ), 126
population variance (σ^2), 126
population z-score, 131
quartiles, 115
range, 86
sample mean ($\overline{x}$), 82
sample standard deviation (s), 89
sample variance (s^2), 88
sample z-score, 101
standard score, 100
statistic, 128
STDEV,* 93
subscripts, 81
sum of squared deviations, 88
summation notation, 81
trimodal, 72
whiskers, 117

FORMULAS

In the formulas below,

f = class frequency
μ = population mean
n = sample size
N = population size
x = data value

$\overline{x}$ = sample mean
s = sample standard deviation
σ = population standard deviation
z = z-score

Sample mean, 82

$$\overline{x} = \frac{\Sigma x}{n}$$

Range, 86

Range = Largest value − Smallest value

Sample standard deviation, 89, 91

$$s = \sqrt{\frac{\Sigma(x - \bar{x})^2}{n - 1}} \qquad \text{or} \qquad s = \sqrt{\frac{n(\Sigma x^2) - (\Sigma x)^2}{n(n - 1)}}$$

Sample z-score, 101

$$z = \frac{x - \bar{x}}{s}$$

Sample mean (grouped form), 107

$$\bar{x} = \frac{\Sigma x f}{n}$$

(x = class mark)

Sample standard deviation (grouped form), 108, 109

$$s = \sqrt{\frac{\Sigma(x - \bar{x})^2 f}{n - 1}} \qquad \text{or} \qquad s = \sqrt{\frac{n(\Sigma x^2 f) - (\Sigma x f)^2}{n(n - 1)}}$$

(x = class mark)

Population mean, 124

$$\mu = \frac{\Sigma x}{N}$$

Population standard deviation, 126

$$\sigma = \sqrt{\frac{\Sigma(x - \mu)^2}{N}} \qquad \text{or} \qquad \sigma = \sqrt{\frac{\Sigma x^2}{N} - \mu^2}$$

Population mean (grouped form), 129

$$\mu = \frac{\Sigma x f}{N}$$

(x = class mark)

Population standard deviation (grouped form), 129

$$\sigma = \sqrt{\frac{\Sigma(x - \mu)^2 f}{N}} \qquad \text{or} \qquad \sigma = \sqrt{\frac{\Sigma x^2 f}{N} - \mu^2}$$

(x = class mark)

Population z-score, 131

$$z = \frac{x - \mu}{\sigma}$$

YOU SHOULD
BE ABLE TO

1. use and understand each of the preceding formulas.
2. explain the purpose of a measure of central tendency.
3. compute and interpret the mean, the median, and the mode of a data set.
4. choose an appropriate measure of central tendency for a data set.
5. use and understand summation notation.
6. explain the purpose of a measure of dispersion.
7. compute and interpret the range and the standard deviation of a data set.
8. state and apply Chebychev's rule.
9. compute and interpret sample z-scores.
10. determine and interpret the quartiles, deciles, and percentiles of a data set.
11. construct and interpret a box-and-whisker diagram.
12. distinguish between parameters and statistics.
13. compute and interpret population z-scores.
14. use the Minitab commands covered in this chapter.*
15. interpret the output obtained from the application of the Minitab commands discussed in this chapter.*

REVIEW TEST

1. Euromonitor Publications Limited, London, England, compiles information on per capita food consumption of major food commodities in various countries. Data are published in *European Marketing Data and Statistics*. Samples of 10 Germans and 15 Russians yield the following fish consumptions for last year. Data are in kilograms.

Germans		Russians		
17	17	16	21	12
1	9	11	5	23
15	6	19	19	22
10	13	16	23	12
14	11	18	7	17

a) Compute the mean of each data set.
b) Compute the median of each data set.
c) Compute the mode of each data set.

2. The National Center for Health Statistics publishes information on the duration of marriages in *Vital Statistics of the United States*. Which measure of central tendency is more appropriate for data on the duration of marriages, the mean or the median?

3. Death certificates provide data on the causes of death. Which measure of central tendency is appropriate for such data?

4. Telephone companies conduct surveys to obtain information on the lengths of telephone conversations. Suppose a sample of 12 phone calls yields the following durations to the nearest minute:

4	2	1	2	2	8
6	3	1	3	1	15

a) Compute the sample mean duration, $\bar{x}$, of the 12 telephone calls.
b) Compute the range of the durations of the 12 telephone calls.
c) Compute the sample standard deviation, s, of the durations of the 12 telephone calls.

5. Refer to Problem 4.
a) Draw a graph similar to Figure 3.5 on page 98.
b) Fill in the blank: By Chebychev's rule, at least _____% of the data lies within two standard deviations to either side of the mean.
c) What percentage of the telephone data actually lies within two standard deviations to either side of the mean?
d) From parts (b) and (c) we see that Chebychev's rule does not necessarily give precise estimates. What then makes it so valuable?

6. Dr. Thomas Stanley of Georgia State University has collected information on millionaires, including their ages, since 1973. Suppose that the mean

age of a sample of 36 millionaires is $\bar{x} = 58.5$ years with a standard deviation of $s = 13.4$ years.

a) Complete the graph below.

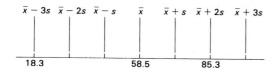

b) Apply Chebychev's rule to fill in the blanks: At least 27 of the 36 millionaires are between _____ and _____ years old.

c) Apply Chebychev's rule to fill in the blank: At least _____% of the 36 millionaires are between 18.3 and 98.7 years old.

7. Refer to Problem 6.
   a) Determine and interpret the sample z-scores for millionaires in the sample whose ages are 59 years, 31 years, and 79 years.
   b) Use your results from part (a) to estimate the relative standing in the sample of each of the ages in part (a).

8. Referring to Problem 6, the actual ages of the 36 millionaires sampled are as follows:

31	64	39	66	68	48	69	71	52
68	45	60	54	66	79	38	48	77
53	52	79	75	67	71	42	39	57
47	74	59	64	42	55	61	79	48

a) Determine the quartiles for the data.
b) Construct a box-and-whisker diagram.

*9. **(Computer problem)** Suppose the age data in Problem 8 are stored in a column named AGES.
   a) Which Minitab command and subcommands (if any) should be used to obtain the mean of the data?
   b) Which Minitab command and subcommands (if any) should be used to obtain the median of the data?
   c) Which Minitab commands should be used to obtain the range of the data?
   d) Which Minitab command and subcommands (if any) should be used to obtain the sample standard deviation of the data?
   e) If you have access to Minitab, use it to determine the mean, median, range, and sample standard deviation of the age data.

*10. **(Computer problem)** Suppose the age data in Problem 8 are stored in a column named AGES.
   a) Which Minitab command and subcommands (if any) should be used to obtain a box-and-whisker diagram for the data?
   b) Which Minitab command and subcommands (if any) should be used to determine the smallest and largest data values and the quartiles of the data?
   c) If you have access to Minitab, use it to carry out parts (a) and (b).

*11. **(Computer problem)** Refer to Problem 8.
   a) Printout 3.10 at the top of the next page shows a box-and-whisker diagram for the age data generated by Minitab's BOXPLOT command. Use the diagram to discuss the variation in the data set. Also, use the diagram to determine approximately the smallest and largest data values and the quartiles for the data.
   b) In Printout 3.11 on the following page we have given the Minitab output obtained by applying the LVALS command to the age data. Use the output to determine exactly the smallest and largest data values and the quartiles for the data. Compare your answers to the approximate values that you obtained in part (a).

12. An exercise physiologist measured the heart rates of 30 people who had been placed on a long distance running program. A frequency distribution for these rates is displayed below.

Rate x	Frequency f	Rate x	Frequency f
67	2	72	3
68	2	73	4
69	2	74	4
70	3	75	3
71	6	77	1

a) Compute the mean heart rate, $\bar{x}$.
b) Compute the sample standard deviation, s, of the heart rates.

13. The table shown at the top of the first column on the following page displays the 1987–1988 enrollment figures for the University of California campuses. Data are in thousands, rounded to the nearest hundred. [SOURCE: *Peterson's Guides*.]

PRINTOUT 3.10   Minitab output for Problem 11(a)

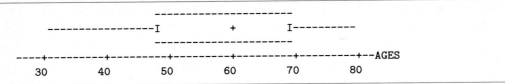

PRINTOUT 3.11   Minitab output for Problem 11(b)

	DEPTH	LOWER	UPPER	MID	SPREAD
N=	36				
M	18.5	59.500		59.500	
H	9.5	48.000	68.500	58.250	20.500
E	5.0	42.000	75.000	58.500	33.000
D	3.0	39.000	79.000	59.000	40.000
C	2.0	38.000	79.000	58.500	41.000
	1	31.000	79.000	55.000	48.000

Campus	Enrollment (thousands)
Berkeley	32.1
Davis	20.9
Irvine	15.1
Los Angeles	35.4
Riverside	6.6
San Diego	16.6
San Francisco	3.7
Santa Barbara	17.9
Santa Cruz	8.6

a) Compute the population mean enrollment, μ, per UC campus.
b) Compute σ.
c) Compute and interpret the population z-scores for the enrollments at the Los Angeles and San Diego campuses.

14. The table at the top of the next column provides a frequency distribution for the 1988 annual salaries of the state governors in the United States. [SOURCE: *The World Almanac, 1989.*]
   a) Find the approximate value of μ for this population of salaries.
   b) Find the approximate value of σ.
   c) Why are the values that you computed for μ and σ in parts (a) and (b) only approximate?

Salary ($thousands)	Frequency
30–under 40	1
40–under 50	0
50–under 60	4
60–under 70	11
70–under 80	14
80–under 90	13
90–under 100	5
100–under 110	1
110–under 120	0
120–under 130	0
130–under 140	1

15. The Energy Information Administration reports figures on retail gasoline prices in the publication *Monthly Energy Review*. Data are obtained by sampling 10,000 gasoline service stations from a total of more than 185,000. For the 10,000 stations sampled, suppose the mean price per gallon for unleaded regular gasoline is $1.03.
   a) Is the mean price given here a sample mean or a population mean? Why?
   b) What letter would you use to designate the mean of $1.03?
   c) Is the mean price given here a statistic or a parameter? Explain.

Won't be tested on this ch.

CHAPTER 4

PROBABILITY CONCEPTS

Up to this point we have been concentrating on *descriptive statistics,* methods for organizing and summarizing data. However, the main purpose of this text is to present the fundamentals of *inferential statistics,* methods of drawing conclusions about a population based on information obtained from a sample of the population.

Since inferential statistics involves using information obtained from part of the population (the sample) to draw conclusions about the entire population, we can never be certain that our conclusions are correct—we are dealing with uncertainty. Consequently, before we can understand, develop, and apply the methods of inferential statistics, we need to become familiar with uncertainty.

The science of uncertainty is called **probability.** Probability enables us to evaluate and control the likelihood that a statistical inference is correct. The next few chapters are devoted to the study of probability.

CHAPTER OUTLINE

4.1 Introduction; classical probability

Most applications of probability to statistical inference involve large populations. However, the fundamental concepts of probability are most easily illustrated and explained using relatively small populations and games of chance. So, keep in mind that many of the examples in this chapter are designed expressly to present the principles of probability in a lucid manner.

CLASSICAL PROBABILITY

We will begin our study of probability by discussing **classical probability**. Classical probability applies when the possible outcomes of an experiment are equally likely to occur.

EXAMPLE 4.1   *Introduces classical probability*

The ages of the students in Professor W's introductory statistics class are listed in Table 3.6 on page 75. A grouped-data table for those ages is given in Table 4.1.

TABLE 4.1
Grouped-data table for ages of students in Professor W's introductory statistics class

Age (yrs) x	Frequency f	Relative frequency
17	1	0.025
18	1	0.025
19	9	0.225
20	7	0.175
21	7	0.175
22	5	0.125
23	3	0.075
24	4	0.100
26	1	0.025
35	1	0.025
36	1	0.025
	40	1.000

Suppose that one of the students is selected **at random,** meaning that each student is equally likely to be the one selected. What is the probability that the student selected is 20 years old?

SOLUTION   As we see from the second column of Table 4.1, seven of the 40 students in the class are 20 years old. Thus, the chances are seven out of 40 for selecting a student who is 20 years old. The probability is therefore

$$\frac{\text{Number of 20-year-olds}}{\text{Total number of students}} = \frac{7}{40}$$

Note that the probability, 7/40, of selecting a student aged 20 is exactly the same as the relative frequency, 0.175, of the students aged 20. ∎

DEFINITION 4.1 Classical probability

Suppose that there are N *equally-likely* possible outcomes for an experiment. Then the probability that a specified event occurs equals the number of ways, f, that the event can occur, divided by the total number, N, of possible outcomes. That is, the probability equals

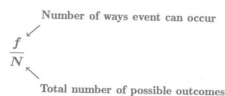

For convenience, we will usually refer to Definition 4.1 as the **f/N-rule.**

In Example 4.1, the experiment consists of selecting a student at random from Professor W's introductory statistics class. The possible outcomes are equally likely since, by assumption, the student is selected at random. Also, $N = 40$ since there are 40 students in the class. If we are considering the event that the student selected is 20 years old, then $f = 7$ since seven of the students in the class are 20 years old. Therefore, the probability that a student selected at random is 20 years old equals

$$\frac{f}{N} = \frac{7}{40} = 0.175$$

as we noted in Example 4.1.

EXAMPLE 4.2 **Illustrates Definition 4.1**

For the situation in Example 4.1, find the probability that a student selected at random is younger than 21.

SOLUTION From Table 4.1, we see that there are 18 $(1 + 1 + 9 + 7)$ students in the class younger than 21. So, $f = 18$ and the probability is

$$\frac{f}{N} = \frac{18}{40} = 0.450$$

In terms of percentages, this means that 45.0% of the students in the class are under 21 years of age. ∎

EXAMPLE 4.3 **Illustrates Definition 4.1**

A frequency distribution for the incomes of the families in a small city is displayed in Table 4.2 at the top of the next page. Find the probability that a randomly selected family makes between $15,000 and $49,999 (inclusive).

Income	Frequency
Under $5000	285
$5,000–$9,999	474
$10,000–$14,999	584
$15,000–$24,999	1213
$25,000–$34,999	1135
$35,000–$49,999	1310
$50,000–$74,999	980
$75,000 & over	506
	6487

SOLUTION   The second column of Table 4.2 shows that there are a total of 6487 families in the city; thus, $N = 6487$. The event in question here is that the randomly selected family makes between $15,000 and $49,999. To find the probability of that event, we first need to determine the number of ways, f, that it can occur; that is, the number of families that make between $15,000 and $49,999. We see from Table 4.2 that the number of such families is $1213 + 1135 + 1310 = 3658$, so $f = 3658$. Therefore, the probability that a randomly selected family makes between $15,000 and $49,999 is

$$\frac{f}{N} = \frac{3658}{6487} = 0.564$$

MTB   (to three decimal places). In other words, approximately 56.4% of the families in the city make between $15,000 and $49,999. ∎

EXAMPLE 4.4   Illustrates Definition 4.1

When a pair of *balanced* dice is rolled, there are a total of 36 possible equally-likely outcomes. We have depicted the possibilities in Figure 4.1 at the top of the next page. Find the probability that
a) the sum of the dice is 11.
b) doubles are rolled (that is, both dice come up the same number).

SOLUTION   For this experiment, $N = 36$.
a) The sum of the dice can be 11 in $f = 2$ ways, as is readily seen from Figure 4.1. Thus, the probability is

$$\frac{f}{N} = \frac{2}{36} = 0.056$$

b) There are $f = 6$ ways that doubles can be rolled, and so the probability of rolling doubles is

$$\frac{f}{N} = \frac{6}{36} = 0.167$$

∎

FIGURE 4.1
Possible outcomes for
rolling a pair of dice

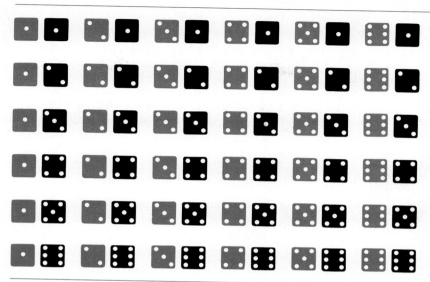

PROBABILITIES AND PERCENTAGES

We pointed out in Example 4.1 that the probability of randomly selecting a student aged 20 equals the relative frequency (percentage) of the students aged 20. More generally, we have the following fact:

KEY FACT 4.1 Probabilities and percentages

Suppose that an experiment consists of selecting one member at random from a finite population. Then the probability that a specified event occurs is equal to the percentage (relative frequency) of members of the population that satisfy the conditions prescribed by the specified event.

Thus, for example, the fact that 12.4% of the United States population is black also means that the probability is 0.124 that a randomly selected United States resident will be black.

You should note that the probabilities in Examples 4.2 and 4.3 arise from the random selection of a member from a finite population, whereas those in Example 4.4 do not.

BASIC PROPERTIES OF PROBABILITIES

It is important to note some simple but basic properties of probability.

KEY FACT 4.2 Basic properties of probabilities

PROPERTY 1 The probability of an event is always between 0 and 1.

PROPERTY 2   The probability of an event that cannot occur is 0. [An event that cannot occur is said to be *impossible.*]

PROPERTY 3   The probability of an event that must occur is 1. [An event that must occur is called *certain.*]

Property 1 indicates that numbers such as 5 or -0.23 could not possibly be probabilities. Thus, if you calculate a probability and get an answer like 5 or -0.23, then you made an error. The next example illustrates Properties 2 and 3.

EXAMPLE 4.5   *Illustrates Properties 2 and 3 of Key Fact 4.2*

Consider again the age data from Example 4.1. [Refer to Table 4.1, page 142.]
a) What is the probability that a student selected at random is 25 years old?
b) What is the probability that a student selected at random is younger than 50?

SOLUTION   a) Since *none* of the students in the class are 25 years old, we have $f = 0$. Thus, the probability that a student selected at random is 25 years old equals

$$\frac{f}{N} = \frac{0}{40} = \boxed{0}$$

This illustrates Property 2 of Key Fact 4.2.

b) We next determine the probability that a student selected at random is younger than 50. Because *all* of the 40 students in the class are younger than 50, we have $f = 40$. The probability is therefore

$$\frac{f}{N} = \frac{40}{40} = \boxed{1}$$

This illustrates Property 3 of Key Fact 4.2.   ■

We need to discuss two points regarding the method for computing probabilities presented in this section. First, in all of the examples considered thus far, either we have been given both the number of ways, f, that an event can occur and the total number, N, of possible outcomes or we have been able to obtain those numbers by a simple computation or direct listing. Often, however, neither f nor N will be given and a direct listing will be impractical because the number of possible outcomes is so large. In those situations, we must employ alternative techniques, such as the ones that will be described in Section 4.8.

Second, the procedure that we have learned for computing probabilities applies only to experiments in which each outcome is equally likely to occur. When that is not the case, it is necessary to resort to other methods in order to determine probabilities. Some of those methods will be examined later.

Exercises 4.1

In the exercises below, express each probability as a decimal rounded to three places.

___ **4.1**   The absentee records over the past year for the employees of Cudahey Masonry, Inc., are presented below in grouped form.

Days missed x	Number of employees f
0	4
1	2
2	14
3	10
4	16
5	18
6	10
7	6
	80

If an employee is selected at random, find the probability that, over the past year, the employee missed
a) three days of work.
b) at most two days of work (that is, two or fewer).
c) between one and five days of work, inclusive.
d) eight days of work.
e) at most eight days of work.

___ **4.2**   A frequency distribution of the number of cars owned by each of the 6487 families in Example 4.3 is as follows:

Cars owned	Number of families
0	27
1	1422
2	2865
3	1796
4	324
5	53
	6487

What is the probability that a randomly selected family owns
a) two cars?
b) more than three cars?
c) between one and three cars, inclusive?
d) seven cars?
e) at most seven cars? (that is, seven or fewer)

___ **4.3**   As reported by the Bureau of Labor Statistics in the publication *Employment and Earnings,* the age distribution of employed persons 16 years old and over is as shown in the following table. The frequencies are given in thousands.

Age (yrs) x	Number of persons f
16–19	5,899
20–24	11,748
25–34	28,429
35–44	23,597
45–54	15,216
55–64	10,163
65 & over	2,737
	97,789

If an employed person is selected at random, find the probability that the person is
a) between 25 and 34 years old.
b) at least 45 years old (that is, 45 years old or older).
c) between 20 and 44 years old.
d) under 20 or over 54.

___ **4.4**   According to the report *U.S. Scientists and Engineers,* published by the National Science Foundation, the distribution of American scientists by field is as given in the following table. Frequencies are in thousands.

Type	Frequency
Life scientists	452
Physical scientists	313
Computer scientists	576
Social scientists	282
Psychologists	274
Mathematical scientists	140
Environmental scientists	123
	2160

What is the probability that a randomly selected scientist is
a) a psychologist?
b) a life or social scientist?
c) not a computer scientist?

___ **4.5**   Suppose two balanced dice are rolled. Calculate the probabilities of the following events. [Refer to Figure 4.1 on page 145.]
a) The sum of the dice is 6.
b) The sum of the dice is even.
c) The sum of the dice is 7 or 11.
d) The sum of the dice is either 2, 3, or 12.

___ **4.6**   A balanced dime is tossed three times. The possible outcomes can be represented as

HHH	HTH	THH	TTH
HHT	HTT	THT	TTT

where, for example, HHT means the first two tosses came up heads and the third tails. Since the dime is balanced, each of the eight outcomes above is equally likely. What is the probability that
a) exactly two of the three tosses yield heads?
b) the last two tosses come up tails?
c) all three tosses come up the same?
d) the second toss is a head?

___ **4.7**   The U.S. Senate consists of 100 senators, two from each state. If a senator is selected at random to serve as the chair of a subcommittee, what is the probability that the senator selected is from Georgia?

___ **4.8**   According to the Census Bureau publication, *Current Housing Reports,* housing in the U.S. is occupied as follows. Frequencies are given in thousands.

Owner-occupied	Renter-occupied	Vacant year-round
56,145	32,280	8,324

What is the probability that a housing unit selected at random is
a) owner-occupied?
b) not renter-occupied?
c) vacant year-round?

___ **4.9**   The Internal Revenue Service (IRS) reports in *Statistics of Income* that the number of U.S. businesses in various categories are as follows. Frequencies are in thousands.

Proprietorships	Partnerships	Corporations
11,929	1,714	3,277

What is the probability that a business selected at random is

a) a proprietorship?
b) not a partnership?
c) a corporation?

___ **4.10**   Which of the following numbers could not possibly be probabilities?
a) 0.462   b) −0.201   c) 1
d) 5/6   e) 3.5   f) 0
Justify your answer.

___ **4.11**   Which of the following numbers could not possibly be probabilities?
a) 0.732   b) 4.75   c) 0.0
d) −3/4   e) 1   f) 4/5
Justify your answer.

___ **4.12**   Refer to Exercise 4.2. Which, if any, of the events in parts (a)–(e) are certain? impossible?

___ **4.13**   Refer to Exercise 4.1. Which, if any, of the events in parts (a)–(e) are certain? impossible?

═ **4.14**   Explain what is wrong with the following argument: When two balanced dice are rolled, the sum of the dice can be 2, 3, 4, 5, 6, 7, 8, 9, 10, 11, or 12. This gives 11 possibilities. Therefore, the probability that the sum is 12 equals 1/11.

═ **4.15**   Explain what is wrong with the following argument: When a balanced coin is tossed twice, the total number of heads obtained can be either 0, 1, or 2. This gives three possibilities. Therefore, the probability of getting two heads is 1/3.

═ **4.16**   A study of the methods that Americans use to get to work revealed the following data, in hundreds of thousands of workers. [SOURCE: U.S. Bureau of the Census, *1980 Census of Population.*]

	Type of worker		
Method		Urban	Rural	Total
Automobile		450	150	600
Public trans.		65	5	70
Total		515	155	670

If a worker is selected at random, what is the probability that he or she
a) is an urban worker?
b) drives an automobile to work?
c) is an urban worker who drives an automobile to work?

d) is an urban worker or drives an automobile to work (or both)?

e) is a rural worker who uses public transportation to get to work?

= 4.17 The table shown in the column at the right gives a frequency distribution of institutions of higher education in the United States by region and type. [SOURCE: U.S. National Center for Education Statistics, *Digest of Education Statistics*.] If an institution of higher education is selected at random, what is the probability that

a) it is a public school?

b) it is in the Midwest?

c) it is a public school and in the Midwest?

d) it is a public school or in the Midwest?

e) it is a Northeastern private school?

		Type		
		Public	Private	Total
Region	Northeast	266	555	821
	Midwest	359	504	863
	South	533	502	1035
	West	313	242	555
	Total	1471	1803	3274

≡ 4.18 Consider an experiment for which there are N equally-likely possible outcomes. Use Definition 4.1 on page 143 to establish that Properties 1–3 of Key Fact 4.2 (page 145) hold for such an experiment.

4.2   Events

Before we can continue our study of probability, we need to discuss *events* in greater detail. In the previous section we used the word "event" intuitively. To be more precise, an **event** consists of a collection of outcomes. Consider Example 4.6.

EXAMPLE 4.6   *Illustrates events*

A deck of playing cards consists of 52 cards, as displayed in Figure 4.2 at the top of the next page. When we perform the experiment of randomly selecting a card from the deck, there are $N = 52$ possible outcomes; namely, the 52 cards pictured in Figure 4.2. This collection of all possible outcomes is called the **sample space** for the experiment.

There are many events associated with this experiment. Let us consider the following four events:

1. The event that the card selected is the king of hearts.
2. The event that the card selected is a king.
3. The event that the card selected is a heart.
4. The event that the card selected is a face card.

List the outcomes comprising each of these four events.

SOLUTION   The first event above consists of the single outcome "king of hearts" and is pictured in Figure 4.3 on the following page.

FIGURE 4.2
Deck of cards

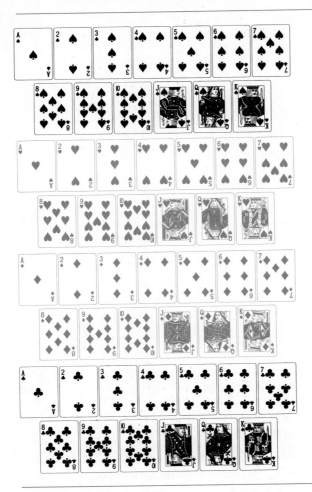

FIGURE 4.3
The event the king
of hearts is selected

The second event, the event that the card selected is a king, consists of the four outcomes "king of spades," "king of hearts," "king of diamonds," and "king of clubs." This event is depicted in Figure 4.4.

FIGURE 4.4
The event a
king is selected

There are 13 outcomes comprising the third event, the event that the card selected is a heart; namely, the outcomes "ace of hearts," "two of hearts," ..., "king of hearts." See Figure 4.5.

FIGURE 4.5
The event a
heart is selected

Finally, the fourth event, the event that the card selected is a face card, consists of 12 outcomes; namely, the 12 face cards shown in Figure 4.6.

FIGURE 4.6
The event a face
card is selected

Note that, when the experiment of selecting a card from the deck is performed, an event **occurs** if it includes the card selected. For instance, if the card selected from the deck turns out to be the "king of spades," then the second and fourth events above occur, while the first and third events do not occur. ∎

We now summarize the terminology discussed so far in this section:

DEFINITION 4.2   Sample space and events

Sample space: The collection of all possible outcomes for an experiment.
Event: Any collection of outcomes for an experiment; that is, any subset of the sample space.

NOTATION AND GRAPHICAL DISPLAYS FOR EVENTS

It is convenient and less cumbersome to employ letters such as A, B, C, D, ... to represent events. For instance, in the card-selection experiment of Example 4.6, we might let

A = event the card selected is the king of hearts
B = event the card selected is a king
C = event the card selected is a heart
D = event the card selected is a face card

Graphical displays of events are also quite useful for explaining and understanding probability. This can be accomplished by the use of **Venn diagrams,** named after the English logician John Venn, 1834–1923. The sample space is portrayed as a rectangle with the various events drawn as disks inside the rectangle. In the simplest case, only one event is displayed, as in Figure 4.7. The shaded portion of the figure represents the event E.

FIGURE 4.7
Venn diagram
for event E

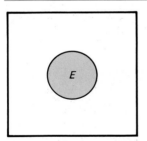

RELATIONSHIPS AMONG EVENTS

To each event E there corresponds another event defined by the condition that "E does not occur." This is called the **complement of E** and is denoted by **(not E).** The event (not E) consists of all outcomes not in E. A Venn diagram makes this idea clearer. See Figure 4.8 at the top of the next page.

With any two events, A and B, we can associate two new events. The first new event is defined by the condition that "*both* event A *and* event B occur" and is denoted by **(A & B).** This event consists of all outcomes common to both event A and event B. A Venn diagram for the event (A & B) is given in Figure 4.9 on the next page.

FIGURE 4.8
Venn diagram
for (not *E*)

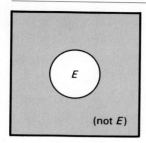

FIGURE 4.9
Venn diagram
for (*A* & *B*)

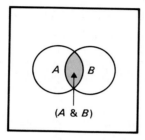

The second new event associated with two events, A and B, is defined by the condition that "*either* event A *or* event B *or* both occur;" or, equivalently, that "*at least one* of the events A and B occurs." This event is denoted by (***A* or *B***) and consists of all outcomes in either event A or event B or both. A Venn diagram for the event (A or B) is displayed in Figure 4.10.

FIGURE 4.10
Venn diagram
for (*A* or *B*)

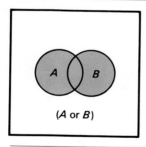

Definition 4.3 summarizes the terms describing relationships among events.

DEFINITION 4.3 Relationships among events

Suppose that E, A, and B denote events. Then the events (not E), (A & B), and (A or B) are defined as follows:

(not E): The event that "E does not occur."
(A & B): The event that "both event A and event B occur."
(A or B): The event that "either event A or event B or both occur."

Note: It is important to realize that the event $(A \& B)$ is the same as the event $(B \& A)$. This is because the event that "both event A and event B occur" is the same as the event that "both event B and event A occur." Similarly, the event $(A \text{ or } B)$ is the same as the event $(B \text{ or } A)$.

EXAMPLE 4.7   *Illustrates Definition 4.3*

For the experiment of randomly selecting a card from a deck of 52, let

A = event the card selected is the king of hearts
B = event the card selected is a king
C = event the card selected is a heart
D = event the card selected is a face card

The outcomes making up each of these four events are shown, respectively, in Figures 4.3–4.6 on pages 150–151. Determine the following events:
a) $(\text{not } D)$ b) $(B \& C)$ c) $(B \text{ or } C)$ d) $(C \& D)$

SOLUTION   a) $(\text{not } D)$ is the event that "D does not occur"—the event that a face card is *not* selected. Event $(\text{not } D)$ consists of the 40 cards in the deck that are not face cards. See Figure 4.11.

FIGURE 4.11
The event $(\text{not } D)$

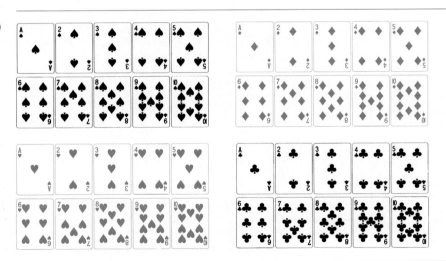

b) $(B \& C)$ is the event that "both event B and event C occur"—the event that the card selected is *both* a king *and* a heart. This can only happen if the card selected is the king of hearts. In other words, $(B \& C)$ is the event that the card selected is the king of hearts. It consists of the single outcome shown in Figure 4.12 at the top of the next page. [Note that the event $(B \& C)$ is the same as event A so we can write $A = (B \& C)$.]

FIGURE 4.12
The event (*B* & *C*)

c) (*B* or *C*) is the event that "either event *B* or event *C* or both occur"—the event that the card selected is *either* a king *or* a heart *or* both. Event (*B* or *C*) consists of 16 outcomes; namely, the four kings and the 12 non-king hearts. See Figure 4.13. [Note that there are 16 (not 17) ways that the event (*B* or *C*) can occur since the outcome "king of hearts" is common to both event *B* and event *C*.]

FIGURE 4.13
The event (*B* or *C*)

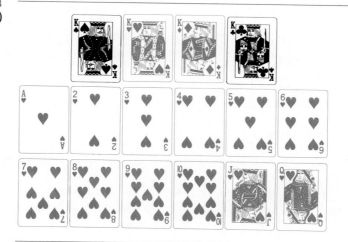

d) (*C* & *D*) is the event that "both event *C* and event *D* occur"—the event that the card selected is *both* a heart *and* a face card. For this to occur, the card selected must be either the king, queen, or jack of hearts. Thus, event (*C* & *D*) consists of the three outcomes displayed in Figure 4.14. These are precisely the outcomes common to both events *C* and *D*. ∎

FIGURE 4.14
The event (*C* & *D*)

In the previous example, we listed the outcomes comprising the events in question. Sometimes this is unnecessary, inconvenient, or undesirable. In cases such as the following, we can describe events verbally instead of listing their outcomes.

EXAMPLE 4.8   *Illustrates Definition 4.3*

A frequency distribution for the ages of the students in Professor W's introductory statistics class is presented in Table 4.3.

TABLE 4.3
Frequency distribution for ages of students in Professor W's introductory statistics class

Age (yrs) x	Frequency f
17	1
18	1
19	9
20	7
21	7
22	5
23	3
24	4
26	1
35	1
36	1
	40

Suppose a student is selected at random. Consider the following events:

$$A = \text{event the student selected is under 21}$$
$$B = \text{event the student selected is over 30}$$
$$C = \text{event the student selected is in his or her 20s}$$
$$D = \text{event the student selected is over 18}$$

Determine each of the following events:
a) (not D)   b) ($A \& D$)   c) (A or D)   d) (B or C)

SOLUTION   a) (not D) is the event that "D does not occur"—the event that the student selected is *not* over 18. In other words, (not D) is the event that the student selected is 18 or under. This event is comprised of the two students in the class who are 18 or under (see Table 4.3).

b) ($A \& D$) is the event that "both event A and event D occur"—the event that the student selected is *both* under 21 *and* over 18. Consequently, ($A \& D$) is the event that the student selected is either 19 or 20. This event is comprised of the 16 students in the class who are 19 or 20.

c) (A or D) is the event that "either event A or event D or both occur"—the event that the student selected is *either* under 21 *or* over 18 (*or* both). But every student is either under 21 or over 18. Consequently, $(A$ or $D)$ contains all students and hence is certain to occur.

d) $(B$ or $C)$ is the event that "either event B or event C or both occur"—the event that the student selected is *either* over 30 *or* in his or her 20s. A glance at Table 4.3 shows that $(B$ or $C)$ is comprised of the 29 students in the class who are 20 or over. ∎

MUTUALLY EXCLUSIVE EVENTS

Another important concept in probability is that of mutually exclusive events. Two events are said to be **mutually exclusive** if at most one of them can occur when the experiment is performed. This requires that the two events have no outcomes in common.

DEFINITION 4.4   Mutually exclusive events

Two events are said to be *mutually exclusive* if they cannot both occur when the experiment is performed; that is, if at most one of the events can occur.

The Venn diagrams in Figures 4.15 and 4.16 portray the difference between events that are mutually exclusive and events that are not mutually exclusive.

FIGURE 4.15
Mutually exclusive events: the events have no outcomes in common

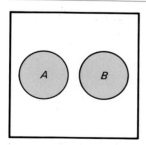

FIGURE 4.16
Events that are not mutually exclusive: the events have outcomes in common

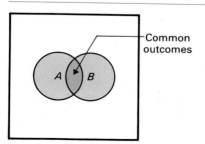

EXAMPLE 4.9   *Illustrates Definition 4.4*

For the experiment of randomly selecting a card from a deck of 52 playing cards, let

$$C = \text{event the card selected is a heart}$$
$$D = \text{event the card selected is a face card}$$
$$E = \text{event the card selected is an ace}$$

Which of the following pairs of events are mutually exclusive?
a) C, D b) C, E c) D, E

SOLUTION   a) Event C and event D are *not* mutually exclusive because they have the common outcomes "king of hearts," "queen of hearts," and "jack of hearts." In other words, event C and event D can both occur if the card selected is either the king, queen, or jack of hearts.

b) Event C and event E are *not* mutually exclusive since they have the common outcome "ace of hearts." Both events can occur if the card selected is the ace of hearts.

c) Event D and event E *are* mutually exclusive since they have no common outcomes. They cannot both occur when the experiment is performed because it is impossible to select a card that is both a face card and an ace.   ■

EXAMPLE 4.10   *Illustrates Definition 4.4*

For the experiment of selecting a student at random from Professor W's introductory statistics course, let

$$A = \text{event the student selected is under 21}$$
$$B = \text{event the student selected is over 30}$$
$$C = \text{event the student selected is in his or her 20s}$$

Which of the following pairs of events are mutually exclusive?
a) A, B b) B, C c) A, C

SOLUTION   a) Event A and event B *are* mutually exclusive since it is not possible to select a student who is both under 21 and over 30.

b) Event B and event C *are* mutually exclusive since it is not possible to select a student who is both over 30 and in his or her 20s.

c) Event A and event C are *not* mutually exclusive. They can both occur if the student selected is 20 years old.   ■

The concept of mutually exclusive events can be extended to more than two events. Consider, for instance, three events, A, B, and C. These events are said to be *mutually exclusive* if at most one of them can occur when the experiment is performed. This requires that no two of them have any outcomes in common. The Venn diagram in Figure 4.17 portrays three mutually exclusive events.

FIGURE 4.17
Three mutually
exclusive events

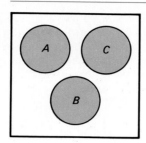

In general, we have the following definition:

DEFINITION 4.5 Mutually exclusive events

Two or more events are said to be *mutually exclusive* if no two of them can occur simultaneously; that is, if at most one of the events can occur when the experiment is performed.

EXAMPLE 4.11 *Illustrates Definition 4.5*

In the card selection experiment, let

$$D = \text{event the card selected is a face card}$$
$$E = \text{event the card selected is an ace}$$
$$F = \text{event the card selected is an 8}$$
$$G = \text{event the card selected is a 10 or a jack}$$

a) Are the three events, D, E, and F, mutually exclusive?
b) Are the four events, D, E, F, and G, mutually exclusive?

SOLUTION a) Events D, E, and F *are* mutually exclusive since no two of them can occur simultaneously.

b) Events D, E, F, and G are *not* mutually exclusive since event D and event G can both occur if the card selected is a jack.   ∎

Exercises 4.2

__ **4.19** When a single die is rolled, there are six possible outcomes:

List the outcomes comprising each of the events below.

$A = $ event the die comes up even
$B = $ event the die comes up four or more
$C = $ event the die comes up at most two
$D = $ event the die comes up three

___ 4.20   In a horse race, the odds are as follows:

Horse	Odds (against winning)
#1	8
#2	15
#3	2
#4	3
#5	30
#6	5
#7	10
#8	5

Let

     $A =$ event that one of the top two favorites wins
           (The top two favorites are the two horses
           with the lowest odds, Horses #3 and #4.)
     $B =$ event that the winning horse has a number
           above five
     $C =$ event that the winning horse has a number
           of at most three
     $D =$ event that one of the two long shots wins
           (The two long shots are the two horses
           with the highest odds.)

List the outcomes comprising each of the four events given above.

___ 4.21   When a dime is tossed four times, there are 16 possible outcomes:

HHHH	HTHH	THHH	TTHH
HHHT	HTHT	THHT	TTHT
HHTH	HTTH	THTH	TTTH
HHTT	HTTT	THTT	TTTT

Here, for example, HTTH represents the outcome that the first toss is heads, the next two tosses are tails, and the fourth toss is heads. Let

     $A =$ event exactly two heads are tossed
     $B =$ event the first two tosses are tails
     $C =$ event the first toss is a head
     $D =$ event all four tosses come up the same

List the outcomes that make up each of the four events given above.

___ 4.22   A committee is formed consisting of five executives, three women and two men. Their first names are Maria (M), John (J), Susan (S), Bill (B), and

Carol (C). The committee needs to select a chairperson and a secretary. It decides to make the selection randomly by drawing straws. The person getting the longest straw will be appointed chairperson and the one getting the shortest straw will be appointed secretary. We can represent the possible outcomes in the following manner:

MS	SM	CM	JM	BM
MC	SC	CS	JS	BS
MJ	SJ	CJ	JC	BC
MB	SB	CB	JB	BJ

Here, for example, MS represents the outcome that Maria is appointed chairperson and Susan is appointed secretary. Let

     $A =$ event a male is appointed chairperson
     $B =$ event Carol is appointed chairperson
     $C =$ event Bill is appointed secretary
     $D =$ event only females are appointed

List the outcomes comprising each of the four events given above.

___ 4.23   Refer to Exercise 4.19. For each of the following events, list the outcomes that make up the event and describe the event in words.
a) (not A)   b) (A & B)   c) (B or C)

___ 4.24   Refer to Exercise 4.20. For each of the following events, list the outcomes that make up the event and describe the event in words.
a) (not C)   b) (C & D)   c) (A or C)

___ 4.25   Refer to Exercise 4.21. For each of the following events, list the outcomes that make up the event and describe the event in words.
a) (not B)   b) (A & B)   c) (C or D)

___ 4.26   Refer to Exercise 4.22. For each of the following events, list the outcomes that make up the event and describe the event in words.
a) (not A)   b) (B & D)   c) (B or C)

___ 4.27   The absentee records over the past year for the employees of Cudahey Masonry, Incorporated, are presented in the frequency distribution at the top of the first column on the next page.

Days missed	Number of employees
0	4
1	2
2	14
3	10
4	16
5	18
6	10
7	6
	80

For an employee selected at random, let

A = event that the employee missed at most three days

B = event that the employee missed at least one day

C = event that the employee missed between four and six days, inclusive

D = event that the employee missed more than six days

Describe each of the following events in words and determine the number of outcomes (employees) that comprise each event.

a) (not A)   b) (A & B)   c) (C or D)

___ **4.28** A frequency distribution for the number of cars owned by each of the families in a small city is:

Cars owned	Number of families
0	27
1	1422
2	2865
3	1796
4	324
5	53
	6487

For a family selected at random, let

A = event the family owns at most three cars

B = event the family owns at least one car

C = event the family owns between two and four cars, inclusive

D = event the family owns at least three cars

Describe each of the following events in words and determine the number of outcomes (families) that comprise each event.

a) (not B)   b) (C & D)   c) (A or D)

___ **4.29** As reported by the Bureau of Labor Statistics in the publication *Employment and Earnings,* the age distribution of employed persons 16 years old and over is as shown in the following table. The frequencies are given in thousands.

Age (yrs) x	Number of persons f
16–19	5,899
20–24	11,748
25–34	28,429
35–44	23,597
45–54	15,216
55–64	10,163
65 & over	2,737
	97,789

An employed person is selected at random. Let

A = event the person is under 20

B = event the person is between 20 and 54

C = event the person is under 45

D = event the person is at least 55

Describe each of the following events in words and determine the number of outcomes (persons) that comprise each event.

a) (not C)   b) (not B)   c) (B & C)   d) (A or D)

___ **4.30** Each part of this exercise lists events from Exercise 4.20. Indicate whether the events are mutually exclusive:

a) Events A and B

b) Events B and C

c) Events A, B, and C

d) Events A, B, and D

e) Events A, B, C, and D

___ **4.31** Refer to Exercise 4.19.

a) Are the events A and B mutually exclusive?

b) Are the events B and C mutually exclusive?

c) Are the events A, C, and D mutually exclusive?

d) Are there three mutually exclusive events among A, B, C, and D? How about four?

___ **4.32** For the following groups of events from Exercise 4.28, determine which are mutually exclusive.

a) Events A and B

b) Events (not B) and C

c) Events A and D

d) Events (not B), C, and D

___ **4.33**   For the following groups of events from Exercise 4.29, determine which are mutually exclusive.
a) Events C and D
b) Events B and C
c) Events A, B, and D
d) Events A, B, and C
e) Events A, B, C, and D

___ **4.34**   Draw a Venn diagram portraying four mutually exclusive events.

___ **4.35**   Construct a Venn diagram portraying four events, A, B, C, and D, with the following property: Events A, B, and C are mutually exclusive; events A, B, and D are mutually exclusive; but no other three of the four events are mutually exclusive.

___ **4.36**   Suppose that A, B, and C are three events with the property that they cannot all occur simultaneously. Does this necessarily imply that A, B, and C are mutually exclusive? Justify your answer and illustrate with a Venn diagram.

4.3   Some rules of probability

In this section, we will discuss several rules of probability. In preparation, let's look at some additional notation used in probability.

EXAMPLE 4.12   *Introduces some additional probability notation*

When a balanced die is rolled, there are $N = 6$ equally-likely possible outcomes. These are pictured in Figure 4.18.

FIGURE 4.18

Consider the event that the die comes up even. This event can occur in $f = 3$ ways; namely, if "2," "4," or "6" comes up. Since

$$\frac{f}{N} = \frac{3}{6} = 0.5$$

we see that *the probability that the die comes up even is 0.5.*

Employing probability notation enables us to express the italicized phrase much more concisely. Let

$$A = \text{event the die comes up even}$$

We use the notation $P(A)$ to stand for the probability that event A occurs. Thus, the italicized statement above can be written simply as

$$P(A) = 0.5$$

and is read "the probability of A is 0.5." We should emphasize that the notation A refers to the *event* that the die comes up even, while the notation $P(A)$ refers to the *probability* of that event. ∎

DEFINITION 4.6 Probability notation

If E is an event, then the notation $P(E)$ stands for the probability that the event E occurs, and is read "the probability of E."

THE SPECIAL ADDITION RULE

The first rule of probability that we will study is called the **special addition rule**. It states that if two events are mutually exclusive, then the probability that either occurs is equal to the sum of their probabilities.

EXAMPLE 4.13   *Introduces the special addition rule*

The ages of the 40 students in Professor W's introductory statistics class are displayed in a frequency distribution in Table 4.4.

TABLE 4.4
Frequency distribution for ages of students in Professor W's introductory statistics class

Age (yrs) x	Frequency f
17	1
18	1
19	9
20	7
21	7
22	5
23	3
24	4
26	1
35	1
36	1
	40

Suppose that a student is selected at random. Let

$$E = \text{event the student selected is 19}$$
$$F = \text{event the student selected is 20}$$

The probability that the student selected is 19 is computed as usual:

$$P(E) = \frac{f}{N} = \frac{9}{40} = 0.225$$

Similarly, the probability that the student selected is 20 equals

$$P(F) = \frac{f}{N} = \frac{7}{40} = 0.175$$

Next, let us compute the probability that the student selected is either 19 or 20; that is, $P(E \text{ or } F)$. As we see from Table 4.4, a total of $9 + 7 = 16$ students in the class are either 19 or 20. Thus,

$$P(E \text{ or } F) = \frac{f}{N} = \frac{9+7}{40} = \frac{16}{40} = 0.400$$

Now here is the important point of this example. In computing $P(E \text{ or } F)$, we can take a slightly different route:

$$P(E \text{ or } F) = \frac{f}{N} = \frac{9+7}{40} = \frac{9}{40} + \frac{7}{40} = 0.225 + 0.175 = 0.400$$

Note that the two shaded numbers are, respectively, $P(E)$ and $P(F)$. Consequently, in this case we have

$$P(E \text{ or } F) = P(E) + P(F)$$

This formula is the *special addition rule*. It holds not only for the events E and F given here, but for any two events that are mutually exclusive.   ■

FORMULA 4.1   The special addition rule

If event A and event B are *mutually exclusive*, then

$$P(A \text{ or } B) = P(A) + P(B)$$

More generally, if events A, B, C, ... are *mutually exclusive*, then

$$P(A \text{ or } B \text{ or } C \text{ or } \cdots) = P(A) + P(B) + P(C) + \cdots$$

EXAMPLE 4.14   *Illustrates Formula 4.1*

The U.S. Bureau of the Census compiles information about American farms and publishes the results in *Census of Agriculture*. According to that publication, a relative-frequency distribution for the size of farms in the U.S. is as shown in the first two columns of Table 4.5.

TABLE 4.5
Size of farms
in the U.S.

Size (acres)	Relative frequency	Event
Under 10	0.087	A
10–49	0.192	B
50–99	0.156	C
100–179	0.173	D
180–259	0.098	E
260–499	0.143	F
500–999	0.085	G
1000–1999	0.040	H
2000 & over	0.026	I

Table 4.5 shows, for instance, that 17.3% (0.173) of the farms in the U.S. have between 100 and 179 acres.

In the third column of Table 4.5 we have introduced events that correspond to each size class. For example, if a farm is selected at random, then D is the event that the farm obtained has between 100 and 179 acres. The probabilities for the events in the third column are simply the relative frequencies given in the second column. Consequently, the probability that a randomly selected farm has between 100 and 179 acres is $P(D) = 0.173$.

Use Table 4.5 and the special addition rule to determine the probability that a randomly selected farm has

a) under 50 acres.

b) between 100 and 499 acres inclusive.

SOLUTION To find the required probabilities, we express each event in (a) and (b) in terms of the *mutually exclusive* events A through I in Table 4.5 and then apply the special addition rule.

a) The event that a randomly selected farm has under 50 acres can be expressed as $(A$ or $B)$, as we see from Table 4.5. Since event A and event B are mutually exclusive (why?), the special addition rule implies that

$$P(A \text{ or } B) = P(A) + P(B) = 0.087 + 0.192 = \boxed{0.279}$$

In terms of percentages, this means that 27.9% of the farms in the U.S. have under 50 acres.

b) The event that a randomly selected farm has between 100 and 499 acres can be expressed as $(D$ or E or $F)$. The events D, E, and F are mutually exclusive. Thus by the special addition rule

$$P(D \text{ or } E \text{ or } F) = P(D) + P(E) + P(F)$$
$$= 0.173 + 0.098 + 0.143$$
$$= \boxed{0.414}$$

In other words, 41.4% of U.S. farms have between 100 and 499 acres. ∎

THE COMPLEMENTATION RULE

The special addition rule has several consequences. One such consequence is called the **complementation rule.** It states that the probability that an event occurs is equal to one minus the probability that the event doesn't occur.

EXAMPLE 4.15 *Introduces the complementation rule*

The first two columns of Table 4.6, shown at the top of the following page, provide a relative-frequency distribution for the number of cars owned by each of the families in a small city. Find the probability that a randomly selected family owns at least one car.

TABLE 4.6

Cars owned	Relative frequency	Event
0	0.004	A
1	0.219	B
2	0.442	C
3	0.277	D
4	0.050	E
5	0.008	F
	1.000	

SOLUTION   Let

$$G = \text{event the family selected owns at least one car}$$

We can express G in terms of the events B through F given in the third column of Table 4.6:

$$G = (B \text{ or } C \text{ or } D \text{ or } E \text{ or } F)$$

Since the events B through F are mutually exclusive, the special addition rule implies that

$$
\begin{aligned}
P(G) &= P(B \text{ or } C \text{ or } D \text{ or } E \text{ or } F) \\
&= P(B) + P(C) + P(D) + P(E) + P(F) \\
&= 0.219 + 0.442 + 0.277 + 0.050 + 0.008 \\
&= \boxed{0.996}
\end{aligned}
$$

There is, however, an easier way to find $P(G)$. Consider the event that "G does not occur"—(not G). Since G is the event that the family selected owns at least one car, (not G) is the event that the family selected does not own at least one car; that is, the family owns 0 cars. From Table 4.6 we find that

$$P(\text{not } G) = 0.004$$

Now, here is the important point. Either the event G occurs or it doesn't. In other words, either event G occurs or event (not G) occurs. This means that the event (G or (not G)) is certain. Consequently,

$$P(G \text{ or } (\text{not } G)) = 1$$

Moreover, event G and event (not G) are mutually exclusive, so that the special addition rule implies

$$P(G \text{ or } (\text{not } G)) = P(G) + P(\text{not } G)$$

This and the previous equation yield the equation

$$P(G) + P(\text{not } G) = 1$$

Rearranging terms, we get the *complementation rule:*

$$P(G) = 1 - P(\text{not } G)$$

Since, as we have seen, $P(\text{not } G) = 0.004$, we can quickly find $P(G)$ using the complementation rule:

$$P(G) = 1 - P(\text{not } G) = 1 - 0.004 = \boxed{0.996}$$

This, of course, is the same value for $P(G)$ that we obtained earlier. However, the second method for determining $P(G)$ is quicker and easier. ∎

FORMULA 4.2   The complementation rule

For any event E,

$$P(E) = 1 - P(\text{not } E)$$

In words, the probability that an event E occurs equals 1 minus the probability that its complement, (not E), occurs.

EXAMPLE 4.16   *Illustrates Formula 4.2*

In Example 4.14 we considered the distribution for the size of farms in the U.S. The relative-frequency distribution from Table 4.5 is repeated here as Table 4.7.

TABLE 4.7
Size of farms
in the U.S.

Size (acres)	Relative frequency	Event
Under 10	0.087	A
10–49	0.192	B
50–99	0.156	C
100–179	0.173	D
180–259	0.098	E
260–499	0.143	F
500–999	0.085	G
1000–1999	0.040	H
2000 & over	0.026	I

Suppose that a farm is selected at random. Determine the probability that the farm selected has
a) less than 2000 acres.
b) at least 50 acres.

SOLUTION   a) Let

$$J = \text{event the farm selected has less than 2000 acres}$$

To find $P(J)$ we will use the complementation rule, since it is easier to compute $P(\text{not } J)$. Note that $(\text{not } J)$ is the event that the farm selected has 2000 or more acres, which is event I in Table 4.7. So,

$$P(\text{not } J) = P(I) = 0.026$$

Applying the complementation rule we find that

$$P(J) = 1 - P(\text{not } J) = 1 - 0.026 = \boxed{0.974}$$

b) Let

$$K = \text{event the farm selected has at least 50 acres}$$

We will apply the complementation rule to find $P(K)$. Now, $(\text{not } K)$ is the event that the farm selected has less than 50 acres. From Table 4.7 we see that $(\text{not } K)$ is the same as the event $(A \text{ or } B)$. Since event A and event B are mutually exclusive, the special addition rule implies that

$$P(\text{not } K) = P(A \text{ or } B) = P(A) + P(B)$$
$$= 0.087 + 0.192 = 0.279$$

Using this and the complementation rule we conclude that

$$P(K) = 1 - P(\text{not } K) = 1 - 0.279 = \boxed{0.721}$$

The probability is 0.721 that a randomly selected farm has at least 50 acres.   ■

THE GENERAL ADDITION RULE

Recall that the *special addition rule* is a formula for obtaining $P(A \text{ or } B)$ from $P(A)$ and $P(B)$; namely,

$$P(A \text{ or } B) = P(A) + P(B)$$

This formula holds provided event A and event B are mutually exclusive. What happens if event A and event B are not mutually exclusive? The answer is supplied by the **general addition rule,** which we introduce in Example 4.17.

EXAMPLE 4.17   *Introduces the general addition rule*

Consider again the experiment of randomly selecting a card from a deck of 52 playing cards. Let

$$A = \text{event a king is selected}$$
$$B = \text{event a heart is selected}$$

The event A can occur in four ways, as shown in Figure 4.19.

FIGURE 4.19
The event A

The event B can occur in 13 ways, as shown in Figure 4.20.

FIGURE 4.20
The event B

Consequently,

$$P(A) = \frac{f}{N} = \frac{4}{52} = \boxed{0.077}$$

and

$$P(B) = \frac{f}{N} = \frac{13}{52} = \boxed{0.250}$$

Now let's look at the event $(A \text{ or } B)$:

$$(A \text{ or } B) = \text{event a king or a heart is selected}$$

This event can occur in 16 (not 17) ways, as we can see by studying Figure 4.21 shown at the top of the next page. Thus,

$$P(A \text{ or } B) = \frac{f}{N} = \frac{16}{52} = \boxed{0.308}$$

Note that

$$P(A \text{ or } B) \neq P(A) + P(B)$$

since $0.308 \neq 0.077 + 0.250$. The reason that $P(A \text{ or } B) \neq P(A) + P(B)$ is because event A and event B are not mutually exclusive. They have the common outcome

FIGURE 4.21
The event (*A* or *B*)

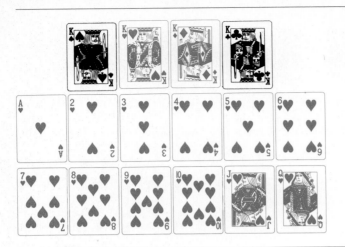

"king of hearts." So when we add $P(A)$ to $P(B)$ we count the king of hearts twice, instead of once. The probability of that common outcome is

$$P(A \,\&\, B) = \frac{f}{N} = \frac{1}{52} = \boxed{0.019}$$

and when we subtract this probability from $P(A) + P(B)$ we do get $P(A \text{ or } B)$:

$$P(A \text{ or } B) = P(A) + P(B) - P(A \,\&\, B)$$

since $0.308 = 0.077 + 0.250 - 0.019$. ∎

FORMULA 4.3   The general addition rule

If A and B are any two events, then

$$P(A \text{ or } B) = P(A) + P(B) - P(A \,\&\, B)$$

In words, the probability that either event A or event B occurs equals the probability that event A occurs plus the probability that event B occurs minus the probability that both occur.

We often have a choice for computing $P(A \text{ or } B)$ either directly or by using the general addition rule. The next example illustrates such a case.

EXAMPLE 4.18   *Illustrates Formula 4.3*

In the card selection experiment, determine the probability that the card selected is either a spade or a face card
a) without using the general addition rule.
b) with the aid of the general addition rule.

SOLUTION   a) Let

E = event the card selected is either a spade or a face card

The event E consists of 22 cards; namely, the 13 spades plus the other nine face cards that are not spades. See Figure 4.22.

FIGURE 4.22
The event E

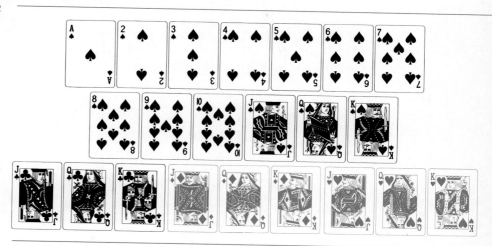

Consequently,

$$P(E) = \frac{f}{N} = \frac{22}{52} = \boxed{0.423}$$

b) To find $P(E)$ using the general addition rule, we first note that we can write $E = (C \text{ or } D)$, where

C = event the card selected is a spade
D = event the card selected is a face card

Event C consists of the 13 spades and event D consists of the 12 face cards. Also, event $(C \& D)$ consists of the three spades that are face cards—the king, queen, and jack of spades. Applying the general addition rule we get

$$P(E) = P(C \text{ or } D) = P(C) + P(D) - P(C \& D)$$

$$= \frac{13}{52} + \frac{12}{52} - \frac{3}{52} = 0.250 + 0.231 - 0.058 = \boxed{0.423}$$

which agrees with the result obtained in part (a). ∎

We just computed the probability of selecting either a spade or a face card in two ways, with and without the aid of the general addition rule. In this case, computing the probability is easier without using the general addition rule. There are many cases, however, in which the use of the general addition rule is the easier or even the only way to compute a probability. Consider the following example.

EXAMPLE 4.19   *Illustrates Formula 4.3*

The U.S. Federal Bureau of Investigation (FBI) provides annual reports on persons arrested by sex, age, and race. Reports are published in *Crime in the United States*. For the year 1987, 82.3% of persons arrested were male, 16.5% were under 18, and 12.8% were males under 18. Suppose a person arrested in 1987 is selected at random. Find the probability that the person selected is either a male or under 18.

SOLUTION   Let

$$M = \text{event the person selected is a male}$$
$$E = \text{event the person selected is under 18}$$

From the percent data given above, we have

$$P(M) = 0.823 \qquad P(E) = 0.165 \qquad P(M \,\&\, E) = 0.128$$

The probability that the person selected at random is either a male or under 18 is $P(M \text{ or } E)$. Employing the general addition rule we conclude that

$$P(M \text{ or } E) = P(M) + P(E) - P(M \,\&\, E)$$
$$= 0.823 + 0.165 - 0.128$$
$$= \boxed{0.860}$$

In terms of percentages this means that 86.0% of those arrested in 1987 were either male or under 18.   ∎

It is important to realize that the general addition rule is consistent with the special addition rule. That is, if two events are mutually exclusive, then the general addition rule gives the same result as the special addition rule. To see this, suppose event A and event B are mutually exclusive. Then they cannot both occur when the experiment is performed. This means that the event $(A \,\&\, B)$ is impossible and hence has a probability of 0. Applying the general addition rule, we get

$$P(A \text{ or } B) = P(A) + P(B) - P(A \,\&\, B)$$
$$= P(A) + P(B) - 0$$
$$= P(A) + P(B)$$

which is the same result given by the special addition rule. Thus, if you are not sure whether to use the special addition rule or the general addition rule, it is always safe to use the general addition rule.

Finally, the general addition rule we have given here is for two events only. We will examine how it can be extended to apply to three or more events in Exercises 4.52 through 4.54.

Exercises 4.3

4.37 A bowl contains 10 marbles of which three are red, two are white, and five are blue. If a marble is selected at random from the bowl, let

E = event the marble selected is white

Find the probability that the marble selected is white, and use probability notation to express your answer.

4.38 Suppose that you hold 20 out of a total of 500 tickets sold for a lottery. The grand-prize winner is determined by the random selection of one of the 500 tickets. Let

G = event you win the grand prize

Find the probability that you win the grand prize, and express your answer using probability notation.

4.39 An age distribution for U.S. Senators in the 98th Congress is given in the table below. [SOURCE: U.S. Congress, Joint Committee on Printing, *Congressional Directory.*]

Age (yrs)	Number of senators
Under 40	7
40–49	28
50–59	39
60–69	20
70–79	3
80 and over	3
	100

Suppose a senator is selected at random. Let

A = event the senator is under 40
B = event the senator is in his or her 40s
C = event the senator is in his or her 50s
S = event the senator is under 60

a) Use the table and the f/N-rule to find $P(S)$.
b) Express the event S in terms of the events A, B, and C.
c) Determine $P(A)$, $P(B)$, and $P(C)$.
d) Compute $P(S)$ using the special addition rule and your results from parts (b) and (c). Compare your answer with the one found in part (a).

4.40 As reported by Dun & Bradstreet in the publication *Business Failure Record*, the number of commercial failures for the year 1987 by type of industry are as follows:

Industry	Failures
Mining	622
Wholesale trade	4,321
Retail trade	12,242
Construction	6,735
Transportation, public utilities	2,249
	26,169

Suppose a failed business is selected at random. Let

A = event it was in wholesale trade
B = event it was in retail trade
T = event it was in either wholesale trade or retail trade

a) Use the table and the f/N-rule to find $P(T)$.
b) Express the event T in terms of the events A and B.
c) Determine $P(A)$ and $P(B)$.
d) Compute $P(T)$ using the special addition rule and your results from parts (b) and (c). Compare your answer with the one found in part (a).

4.41 According to the report *Statistics of Income*, compiled by the Internal Revenue Service, a relative-frequency distribution for U.S. businesses by receipts received from sales and services is as shown below.

Receipts	Relative frequency	Event
Under $25,000	0.613	A
$25,000–$49,999	0.108	B
$50,000–$99,999	0.095	C
$100,000–$499,999	0.132	D
$500,000–$999,999	0.025	E
$1,000,000 or more	0.029	F

The table shows, for example, that 61.3% of U.S. businesses had receipts of under $25,000. Now, suppose a business is selected at random. Let

A = event the business selected had receipts under $25,000
B = event the business selected had receipts between $25,000 and $49,000

and so on (see the third column of the table). Determine the probability that a randomly selected business had receipts of
a) under $100,000.
b) at least $500,000.
c) between $25,000 and $499,999.
d) Interpret each of your results in parts (a)–(c) in terms of percentages.

___ 4.42 The following is a percentage distribution for the number of years of school completed by U.S. adults 25 years old and over. [SOURCE: U.S. Bureau of the Census, *Current Population Reports*.]

	Years completed	Percent	Event
Elementary school	0–4	3.1	A
	5–7	5.6	B
	8	7.1	C
High school	9–11	13.3	D
	12	37.9	E
College	13–15	15.3	F
	16 or more	17.7	G

The table shows, for instance, that 37.9% of adults, 25 years old and over, have completed exactly 12 years of school. Suppose an adult, 25 years old or over, is randomly selected from the population. Let

A = event the person selected has completed between zero and four years of school
B = event the person selected has completed between five and seven years of school

and so forth (see the third column of the table). Determine the probability that a randomly selected U.S. adult 25 years old or over
a) has at most an elementary-school education.
b) has at most a high-school education.
c) has completed at least one year of college.
d) Interpret each of your results in parts (a)–(c) in terms of percentages.

In Exercises 4.43–4.46, find the designated probabilities with the aid of the complementation rule.

___ 4.43 Refer to Exercise 4.39. Compute the probability that a randomly selected senator in the 98th Congress is
a) at least 40 years old.
b) under 70 years old.

___ 4.44 Refer to Exercise 4.40. Determine the probability that a failed business selected at random was not in construction.

___ 4.45 In Exercise 4.41, find the probability that a randomly selected business had receipts of
a) under $1,000,000.
b) at least $50,000.

___ 4.46 In Exercise 4.42, calculate the probability that a randomly selected person has completed
a) less than four years of college.
b) at least five years of school.

___ 4.47 In the game of craps, two balanced dice are rolled. There are 36 equally likely outcomes possible, as shown in Figure 4.1 on page 145. Let

A = event the sum of the dice is 7
B = event the sum of the dice is 11
C = event the sum of the dice is 2
D = event the sum of the dice is 3
E = event the sum of the dice is 12
F = event the sum of the dice is 8
G = event doubles are rolled

a) Compute the probability of each of the seven events listed above. [See Figure 4.1.]
b) The player wins on the first roll if the sum of the dice is 7 or 11. Find the probability of this event using the special addition rule and your results from part (a).
c) The player loses on the first roll if the sum of the dice is 2, 3, or 12. Determine the probability of that event using the special addition rule and your results from part (a).
d) Compute the probability that either the sum of the dice is 8 or doubles are rolled
   (i) without using the general addition rule.
   (ii) with the aid of the general addition rule.

___ 4.48 As reported by the U.S. Bureau of Justice Statistics in *Profile of Jail Inmates*, about 56.5% of jail inmates are white, 94.0% are male, and 53.5% are white males. Suppose that a jail inmate is selected at random. Let

W = event the inmate selected is white
M = event the inmate selected is male

a) Find $P(W)$, $P(M)$, and $P(W \& M)$.
b) Determine $P(W \text{ or } M)$ and interpret your results in terms of percentages.

c) What is the probability that a randomly selected inmate is female?

— **4.49** According to the Census Bureau's *Current Population Reports,* of U.S. adults, 52.6% are female, 7.0% are divorced, and 4.2% are divorced females. For a U.S. adult selected at random, let

> F = event the person selected is female
> D = event the person selected is divorced

a) Find $P(F)$, $P(D)$, and $P(F \& D)$.
b) Determine $P(F \text{ or } D)$ and interpret your results in terms of percentages.
c) What is the probability that a randomly selected adult is male?

— **4.50** Suppose A and B are events with $P(A) = 1/4$, $P(B) = 1/3$, and $P(A \text{ or } B) = 1/2$.
a) Are A and B mutually exclusive?
b) Determine $P(A \& B)$.

— **4.51** Suppose $P(A) = 1/3$, $P(A \text{ or } B) = 1/2$, and $P(A \& B) = 1/10$. Find $P(B)$.

≡ **4.52** The general addition rule for three events is

$$P(A \text{ or } B \text{ or } C) = P(A) + P(B) + P(C)$$
$$- P(A \& B) - P(A \& C)$$
$$- P(B \& C) + P(A \& B \& C)$$

Prove the above rule. *Hint:* Let $D = (B \text{ or } C)$ and apply the general addition rule to $(A \text{ or } D)$. Also, note that $(A \& D) = ((A \& B) \text{ or } (A \& C))$.

≡ **4.53** When a balanced dime is tossed three times, there are eight equally likely outcomes possible:

HHH	HTH	THH	TTH
HHT	HTT	THT	TTT

Let

> A = event the first toss is a head
> B = event the second toss is a tail
> C = event the third toss is a head

a) List the outcomes comprising each of the three events above.
b) Find $P(A)$, $P(B)$, and $P(C)$.
c) Describe each of the following events in words and list the outcomes comprising each one: $(A \& B)$, $(A \& C)$, $(B \& C)$, $(A \& B \& C)$.
d) Find the probability for each event in part (c).
e) Describe the event $(A \text{ or } B \text{ or } C)$ in words and list the outcomes that make it up.
f) Find $P(A \text{ or } B \text{ or } C)$ using your result from part (e) and the f/N-rule.
g) Determine $P(A \text{ or } B \text{ or } C)$ using your results from parts (b) and (d) and the general addition rule for three events given in Exercise 4.52.

≡ **4.54** On page 170, we stated the general addition rule for two events and, in Exercise 4.52, we stated the general addition rule for three events.
a) Guess what the general addition rule is for four events.
b) Prove your answer in part (a).

4.4 Contingency tables; joint and marginal probabilities

An important application of probability and statistics involves the analysis of data obtained by cross classifying the members of a population or sample according to two characteristics. Some examples of cross classifications are annual income versus educational level, automobile accident frequency versus age, and political affiliation versus religion. Frequencies for cross-classified data are most easily displayed using *contingency tables,* which we introduce in the following example.

EXAMPLE 4.20 **Introduces contingency tables**

The data displayed in Table 4.8 are adapted from the *Arizona State University Statistical Summary.* The table provides a frequency distribution obtained by cross classifying the faculty according to the two characteristics, age and rank.

TABLE 4.8
Contingency table
for age vs. rank of
faculty members

Rank

Age	Full Professor R_1	Associate Professor R_2	Assistant Professor R_3	Instructor R_4	Total
Under 30 A_1	2	3	57	6	68
30–39 A_2	52	170	163	17	402
40–49 A_3	156	125	61	6	348
50–59 A_4	145	68	36	4	253
60 & over A_5	75	15	3	0	93
Total	430	381	320	33	1164

The number 2 in the upper left-hand corner of Table 4.8 tells us that two faculty members are both under 30 and at the rank of full professor. The number 170 diagonally below the upper left-hand corner shows that 170 faculty members are associate professors in their 30s.

The row total in the first row of the table indicates that 68 $(2 + 3 + 57 + 6)$ of the faculty members are under 30. Similarly, the column total in the third column shows that 320 of the faculty members are assistant professors. The number 1164 in the lower right-hand corner of the table gives the total number of faculty. That total can be found by summing either the row totals or the column totals. It can also be found by summing all the numbers inside the box formed by the heavy lines.

A table such as Table 4.8 that gives a frequency distribution for cross-classified data is called a **contingency table.** The table includes all the different possibilities in the cross classification—it accounts for all contingencies. The boxes inside the heavy lines give the frequencies for the various contingencies. These boxes are called the **cells** of the contingency table. There are 20 cells in Table 4.8.   ■

MTB

JOINT AND MARGINAL PROBABILITIES

We will now use the age and rank data from Example 4.20 to introduce the concepts of *joint and marginal probabilities.*

EXAMPLE 4.21   *Introduces joint and marginal probabilities*

Refer to Table 4.8. Suppose a faculty member is selected at random. If you study Table 4.8, you will note that the rows and columns are labeled with letters. The first row is labeled A_1, which represents the event that the faculty member selected is under 30:

$$A_1 = \text{event the faculty member selected is under 30}$$

Similarly,

$$R_2 = \text{event the faculty member selected is an associate professor}$$

and so forth. Note that the events A_1 through A_5 are mutually exclusive, as are the events R_1 through R_4 (why?).

In addition to considering the events A_1 through A_5 and R_1 through R_4 separately, we can also consider them jointly. For example, the event that the faculty member selected is under 30 (event A_1) *and* is also an associate professor (event R_2) can be expressed as $(A_1 \,\&\, R_2)$:

$$(A_1 \,\&\, R_2) = \text{event the faculty member selected is}$$
$$\text{an associate professor under 30}$$

The event $(A_1 \,\&\, R_2)$ is represented by the cell in the first row and second column of Table 4.8.

There are 20 different joint events of this type, one for each cell of the table. In fact, it is useful to think of the table as a sort of Venn diagram. This is pictured in Figure 4.23. The Venn diagram makes it clear that the 20 joint events $(A_1 \,\&\, R_1)$, $(A_1 \,\&\, R_2)$, ..., $(A_5 \,\&\, R_4)$ are mutually exclusive.

FIGURE 4.23
Venn diagram to
accompany Table 4.8

	R_1	R_2	R_3	R_4
A_1	$(A_1 \,\&\, R_1)$	$(A_1 \,\&\, R_2)$	$(A_1 \,\&\, R_3)$	$(A_1 \,\&\, R_4)$
A_2	$(A_2 \,\&\, R_1)$	$(A_2 \,\&\, R_2)$	$(A_2 \,\&\, R_3)$	$(A_2 \,\&\, R_4)$
A_3	$(A_3 \,\&\, R_1)$	$(A_3 \,\&\, R_2)$	$(A_3 \,\&\, R_3)$	$(A_3 \,\&\, R_4)$
A_4	$(A_4 \,\&\, R_1)$	$(A_4 \,\&\, R_2)$	$(A_4 \,\&\, R_3)$	$(A_4 \,\&\, R_4)$
A_5	$(A_5 \,\&\, R_1)$	$(A_5 \,\&\, R_2)$	$(A_5 \,\&\, R_3)$	$(A_5 \,\&\, R_4)$

Let us now move on to an examination of probabilities. Here we have $N = 1164$, since there are 1164 faculty members altogether. To find, for example, the probability that a randomly selected faculty member is an associate professor (event R_2), we note from Table 4.8 that $f = 381$, and so

$$P(R_2) = \frac{f}{N} = \frac{381}{1164} = \boxed{0.327}$$

Similarly, the probability that a randomly selected faculty member is under 30 equals

$$P(A_1) = \frac{f}{N} = \frac{68}{1164} = 0.058$$

[In terms of percentages, these last two probabilities indicate that 32.7% of the faculty are associate professors and that 5.8% of the faculty are under 30.]

We can also find probabilities for joint events, so-called **joint probabilities.** For instance, the probability that a randomly selected faculty member is an associate professor under 30 equals

$$P(A_1 \& R_2) = \frac{f}{N} = \frac{3}{1164} = 0.003$$

In other words, 0.3% of the faculty are associate professors under 30.

In Table 4.9 we have replaced the joint frequency distribution in Table 4.8 with a **joint probability distribution.** The probabilities in the table are determined in the same way as the three probabilities we just computed.

TABLE 4.9
Joint probability
distribution
corresponding
to Table 4.8

Rank

Age		Full Professor R_1	Associate Professor R_2	Assistant Professor R_3	Instructor R_4	$P(A_i)$
Under 30	A_1	0.002	0.003	0.049	0.005	0.058
30–39	A_2	0.045	0.146	0.140	0.015	0.345
40–49	A_3	0.134	0.107	0.052	0.005	0.299
50–59	A_4	0.125	0.058	0.031	0.003	0.217
60 & over	A_5	0.064	0.013	0.003	0.000	0.080
$P(R_j)$		0.369	0.327	0.275	0.028	1.000

Note that the joint probabilities are given inside the heavy lines of the table. Also observe that the row labeled "Total" and the column labeled "Total" in Table 4.8 have been relabeled as $P(R_j)$ and $P(A_i)$, respectively. This is because the last row of Table 4.9 gives the probabilities of the events R_1 through R_4, and the last column gives the probabilities of the events A_1 through A_5. These probabilities are often called **marginal probabilities** since they are "in the margin" of the joint probability distribution table.

Finally, we should point out that the sum of the joint probabilities in a given row or column equals the marginal probability in that row or column. (Any observed discrepancy is due to roundoff error.) For example, consider the row A_4 of Table 4.9. The sum of the joint probabilities in that row equals

$$P(A_4 \& R_1) + P(A_4 \& R_2) + P(A_4 \& R_3) + P(A_4 \& R_4)$$
$$= 0.125 + 0.058 + 0.031 + 0.003$$
$$= 0.217$$

which is precisely the marginal probability at the end of the row A_4; that is, $P(A_4)$.

The fact that the sum of a row or column of joint probabilities equals the marginal probability in that row or column follows from the special addition rule. See Exercise 4.68.

MTB

Exercises 4.4

4.55 The following contingency table cross classifies institutions of higher education in the United States by region and type. [SOURCE: U.S. National Center for Education Statistics, *Digest of Education Statistics.*]

Type

Region	Public T_1	Private T_2	Total
Northeast R_1	266	555	821
Midwest R_2	359	504	863
South R_3	533	502	1035
West R_4	313	242	555
Total	1471	1803	3274

a) How many cells does this contingency table have?
b) What is the total number of institutions of higher education in the United States?
c) How many institutions of higher education are in the Midwest?

d) How many are public?
e) How many institutions of higher education are private schools in the South?

4.56 As reported by the Motor Vehicle Manufacturers Association of the United States in *Motor Vehicle Facts and Figures*, the number of cars and trucks in use by age are as shown in the following contingency table. Frequencies are in millions.

Type

Age (yrs)	Car V_1	Truck V_2	Total
Under 3 A_1	21.5	3.2	24.7
3–5 A_2	29.9	2.5	32.4
6–8 A_3	22.2	2.0	24.2
9–11 A_4	17.9	1.8	19.7
12 & over A_5	15.4	3.6	19.0
Total	106.9	13.1	120.0

a) How many cells does this contingency table have?
b) What is the total number of cars and trucks in use?
c) How many vehicles are trucks?
d) How many vehicles are between three and five years old, inclusive?
e) How many vehicles are trucks that are between nine and 11 years old, inclusive?

___ **4.57**   The following contingency table gives a joint frequency distribution for the civilian labor force in the U.S. based on a cross classification of employment status and educational level. Frequencies are given in thousands. [SOURCE: U.S. Bureau of Labor Statistics, unpublished data.]

Employment status

Years of school completed	Employed E_1	Unemployed E_2	Total
Less than 8 S_1	3,535.5	612.2	4,147.7
8 S_2	3,240.9	517.1	3,758.0
9–11 S_3	12,767.0	2,807.4	15,574.4
12 S_4	40,068.9	4,601.5	44,670.4
13–15 S_5	18,266.7	1,340.4	19,607.1
16 or more S_6	20,230.8	686.0	20,916.8
Total	98,109.8	10,564.6	108,674.4

a) How many cells does this contingency table have?
b) What is the size of the civilian labor force?
c) How many people in the civilian labor force have completed exactly 12 years of school?
d) How many are unemployed?
e) How many with 16 or more years of school are unemployed?

___ **4.58**   The contingency table at the top of the next column results from cross classifying U.S. hospitals by type and number of beds. [SOURCE: American Hospital Association, *Hospital Statistics*.]

Number of beds

Type	6–24 B_1	25–74 B_2	75+ B_3	Total
General H_1	299	1894	3945	6138
Psychiatric H_2	17	121	378	516
Chronic H_3	0	7	40	47
Tuberculosis H_4	0	1	10	11
Other H_5	22	131	162	315
Total	338	2154	4535	7027

a) How many hospitals have at least 75 beds?
b) How many psychiatric hospitals are there?
c) How many general hospitals are there with between 25 and 74 beds?

___ **4.59**   According to the document *Census of Agriculture*, published by the U.S. Bureau of the Census, a contingency table for the number of farms, in thousands, by acreage and tenure of operator is as follows:

Tenure of operator

Acreage	Full owner T_1	Part owner T_2	Tenant T_3	Total
Under 50 A_1	532	74	84	690
50–179 A_2	563		94	814
180–499 A_3	262		87	596
500–999 A_4	57	128		215
1000+ A_5	36	107	18	161
Total	1450	713	313	2476

a) Fill in the three empty cells.
b) How many cells does this contingency table have?
c) How many farms have under 50 acres?
d) How many farms are tenant operated?
e) How many farms are operated by part owners and have between 500 and 999 acres?
f) How many farms are not full-owner operated?

___ 4.60   The U.S. Bureau of the Census compiles information on the annual income of families by type of family. Data are published in *Current Population Reports*. Below is a contingency table that provides a joint frequency distribution for that data. Frequencies are in thousands of families.

Type of family

Income level	Married couple F_1	Husband only F_2	Wife only F_3	Total
Under $10,000 I_1	5,852	392	4,308	10,552
$10,000–$19,999 I_2	12,326		2,992	15,925
$20,000–$49,999 I_3	26,240			29,109
$50,000 and over I_4	5,213	107	111	5,431
Total	49,631	1,984	9,402	61,017

a) Fill in the three empty cells.
b) How many cells does this contingency table have?
c) Determine the number of families that make between $20,000 and $49,999.
d) How many families have only the husband present?
e) How many families have only the wife present and make under $10,000?
f) How many families make at least $20,000?
g) Find the number of married couples that make less than $20,000.

___ 4.61   Refer to Exercise 4.55.
a) For a randomly selected institution of higher education, describe each of the following events in

words: T_2, R_3, and $(T_1 \, \& \, R_4)$.
b) Compute the probability of each event in part (a) and interpret your results in terms of percentages.
c) Construct a joint probability distribution similar to Table 4.9 on page 178.
d) Verify that the sum of each row and column of joint probabilities equals the marginal probability in that row or column. (Rounding may cause slight deviations here).

___ 4.62   Refer to Exercise 4.56. Suppose a vehicle (car or truck) in use is selected at random.
a) Describe each of the following events in words: A_3, V_1, and $(A_3 \, \& \, V_1)$.
b) Determine the probability of each event in part (a) and interpret your results in terms of percentages.
c) Construct a joint probability distribution similar to Table 4.9 on page 178.
d) Verify that the sum of each row and column of joint probabilities equals (up to rounding error) the corresponding marginal probability.

___ 4.63   The contingency table in Exercise 4.57 cross classifies the civilian labor force according to employment status and educational level. Suppose a person in the civilian labor force is selected at random.
a) Describe each of the following events in words: E_1, S_5, and $(E_1 \, \& \, S_5)$.
b) Compute the probability of each event in part (a).
c) Compute $P(E_1 \text{ or } S_5)$ by
   (i) using the contingency table and the f/N-rule.
   (ii) using the general addition rule and your results from part (b).
d) Construct a joint probability distribution.

___ 4.64   Refer to Exercise 4.58. Suppose a hospital is selected at random.
a) Describe each of the following events in words: H_2, B_2, $(H_2 \, \& \, B_2)$, and $(H_4 \, \& \, B_1)$.
b) Compute the probability of each event in part (a).
c) Compute $P(H_2 \text{ or } B_2)$ by
   (i) using the contingency table and the f/N-rule.
   (ii) using the general addition rule and your results from part (b).
d) Construct a joint probability distribution.

___ 4.65   The contingency table in Exercise 4.59 cross classifies U.S. farms by acreage and tenure of operator. Suppose a U.S. farm is selected at random.
a) Use the letters in the margins of the contingency table to represent each of the following events:
   (i) The farm selected has between 180 and 499 acres, inclusive.

(ii) The farm selected is part-owner operated.
(iii) The farm selected is full-owner operated with at least 1000 acres.
b) Compute the probability of each event in part (a).
c) Construct a **joint percentage distribution.** [A joint percentage distribution is like a joint probability distribution but with percentages replacing probabilities.]

___ **4.66**   Refer to Exercise 4.60. Suppose a family is selected at random.
a) Use the letters in the margins of the contingency table to represent each of the following events:
   (i) The family selected has only the wife present.
   (ii) The family selected makes at least $50,000.
   (iii) The family selected makes between $10,000 and $19,999 and has only the husband present.
b) Determine the probability of each event in part (a).
c) Construct a joint percentage distribution.

≣ **4.67**   Explain why the joint events in a contingency table are mutually exclusive.

≣ **4.68**   This exercise supplies a proof of the fact that the sum of the joint probabilities in a given row or column equals the marginal probability in that row or column. Consider a joint probability distribution in the following form:

	C_1	$\cdots$	C_n	$P(R_i)$
R_1	$P(R_1 \ \& \ C_1)$	$\cdots$	$P(R_1 \ \& \ C_n)$	$P(R_1)$
.	.	$\cdots$	.	.
.	.	$\cdots$	.	.
.	.	$\cdots$	.	.
R_m	$P(R_m \ \& \ C_1)$	$\cdots$	$P(R_m \ \& \ C_n)$	$P(R_m)$
$P(C_j)$	$P(C_1)$	$\cdots$	$P(C_n)$	1

a) Explain why we can write

$$R_1 = [(R_1 \ \& \ C_1) \text{ or } \cdots \text{ or } (R_1 \ \& \ C_n)]$$

b) Why are the events $(R_1 \ \& \ C_1), \ldots, (R_1 \ \& \ C_n)$ mutually exclusive?
c) Why can we conclude from parts (a) and (b) that

$$P(R_1) = P(R_1 \ \& \ C_1) + \cdots + P(R_1 \ \& \ C_n)$$

This equation shows that the first row of joint probabilities sums to the marginal probability at the end of that row. A similar argument applies to any other row or column.

4.5 Conditional probability

In this section, we will introduce the concept of *conditional probability*. The **conditional probability** of an event is the probability of the event, given that another event has occurred. Consider the following example.

___

EXAMPLE 4.22   *Introduces conditional probability*

When a balanced die is rolled, there are $N = 6$ equally likely outcomes. These are displayed in Figure 4.24.

FIGURE 4.24   ___

___

If, for instance, we let

$$F = \text{event a 4 is rolled}$$

then the probability that event F occurs is:

$$P(F) = \frac{f}{N} = \frac{1}{6} = 0.167$$

To introduce conditional probability, let us suppose that the die is rolled and, although we cannot see the die, we are told that the result is an even number. Now what is the probability that event F occurs? That is, if we are given that the die comes up even, what is the probability that it is a 4?

The answer is quite simple. If we are told that the die comes up even, there are no longer six equally likely outcomes possible, there are only three, as shown in Figure 4.25.

FIGURE 4.25

So, the probability that event F (a 4) occurs is no longer $1/6 = 0.167$. It is now

$$\frac{f}{N} = \frac{1}{3} = 0.333$$

Thus, the information that the die comes up even affects the probability that event F occurs. The new probability, 0.333, is called a *conditional probability* since it is computed under the *condition* that the die comes up even.

It is useful to develop a notation for conditional probabilities, just as it is for unconditional probabilities. To this end, recall that in this example,

$$F = \text{event a 4 is rolled}$$

and let

$$E = \text{event the die comes up even}$$

Then we use the notation

$$P(F \mid E)$$

to represent the conditional probability that event F (a 4) occurs, *given* that event E (even) occurs. The symbol $P(F \mid E)$ is read "the probability of F given E." Just remember that the vertical bar "$\mid$" stands for "given." In summary, we have

$$P(F) = \frac{1}{6} = 0.167, \qquad P(F \mid E) = \frac{1}{3} = 0.333$$

To further illustrate conditional probability using this example, let

$$C = \text{event the die comes up odd}$$
$$D = \text{event the die comes up three or less}$$

Let us compute the conditional probability that event C occurs, given that event D occurs; that is, the probability the die comes up odd, given that it comes up three or less. Given that event D occurs, the possible outcomes are the three shown in Figure 4.26 at the top of the next page.

FIGURE 4.26

And, under these circumstances, event C (odd) can occur in $f = 2$ ways. Thus,

$$P(C \mid D) = \frac{f}{N} = \frac{2}{3} = \boxed{0.667}$$

Note that the unconditional probability that event C occurs is

$$P(C) = \frac{f}{N} = \frac{3}{6} = \boxed{0.500}$$

since the die can come up odd in three of the six possibilities. So again, additional information concerning the experiment has affected the probability. ∎

We now summarize the concepts and notation introduced in Example 4.22.

DEFINITION 4.7   Conditional probability

Suppose that A and B are events. Then the probability that event B occurs given that event A has occurred is called a *conditional probability*. It is denoted by the symbol $P(B \mid A)$, which is read "the probability of B given A."

Conditional probability plays a crucial role in the analysis of cross-classified data. In Section 4.4 we introduced contingency tables as a method for tabulating such data. We now illustrate the procedure for obtaining conditional probabilities for cross-classified data directly from a contingency table.

EXAMPLE 4.23   *Illustrates Definition 4.7*

In Example 4.20 we presented a contingency table resulting from cross classifying the faculty at Arizona State University according to age and rank. We repeat that table at the top of the next page as Table 4.10. Suppose a faculty member is selected at random.
a) Find the (unconditional) probability that the faculty member selected is in his or her 50s.
b) Find the (conditional) probability that the faculty member selected is in his or her 50s, given that an assistant professor is selected.
c) Interpret the probabilities obtained in parts (a) and (b) in terms of percentages.

SOLUTION   a) The probability that the faculty member selected is in his or her 50s is $P(A_4)$. From Table 4.10 we see that $N = 1164$ since there are 1164 faculty members altogether. Also, $f = 253$ since 253 of the faculty are in their 50s. Therefore,

$$P(A_4) = \frac{f}{N} = \frac{253}{1164} = \boxed{0.217}$$

TABLE 4.10
Contingency table
for age vs. rank of
faculty members

Rank

	Full Professor R_1	Associate Professor R_2	Assistant Professor R_3	Instructor R_4	Total
Under 30 A_1	2	3	57	6	68
30–39 A_2	52	170	163	17	402
40–49 A_3	156	125	61	6	348
50–59 A_4	145	68	36	4	253
60 & over A_5	75	15	3	0	93
Total	430	381	320	33	1164

Age

b) Here we are to find the probability that the faculty member selected is in his or her 50s (event A_4), given that an assistant professor is selected, that is, given that event R_3 occurs. In other words, we want to determine $P(A_4 \mid R_3)$. To find this probability, we simply restrict our attention to the "assistant professor" column of the contingency table. We have $N = 320$ since there are a total of 320 assistant professors. Also, $f = 36$ since 36 of the assistant professors are in their 50s. Thus,

$$P(A_4 \mid R_3) = \frac{f}{N} = \frac{36}{320} = 0.113$$

c) $P(A_4) = 0.217$ indicates that 21.7% of the faculty are in their 50s. $P(A_4 \mid R_3) = 0.113$ indicates that 11.3% of the assistant professors are in their 50s. ∎

THE CONDITIONAL PROBABILITY RULE

In the previous examples, we computed conditional probabilities *directly*. That is, we first obtained the new sample space determined by the *given* event, and then we calculated probabilities from there in the usual manner. For instance, in Example 4.22 on page 182, we computed the conditional probability of rolling a 4, given that the die comes up even. To do that, we first obtained the new sample space (in this case, 2, 4, 6) and then went on from there.

Sometimes, however, we cannot determine conditional probabilities directly but must instead compute them in terms of unconditional probabilities. To see how this can be done, we return to the situation of Example 4.23.

EXAMPLE 4.24   *Introduces the conditional probability rule*

In Example 4.23 we computed the conditional probability that a randomly selected faculty member is in his or her 50s (event A_4), given that an assistant professor is selected (event R_3). To accomplish this, we restricted our attention to the R_3-column of Table 4.10 (page 185) and obtained

$$P(A_4 \mid R_3) = \frac{36}{320} = 0.113$$

This is a *direct* computation of the conditional probability $P(A_4 \mid R_3)$.

To find $P(A_4 \mid R_3)$ using unconditional probabilities, proceed as follows. First note that the number 36 in the numerator of the above fraction is the number of assistant professors in their 50s, and that this corresponds to the event $(R_3 \& A_4)$—"assistant professor" and "in 50s." Next observe that the 320 in the denominator of the above fraction is the total number of assistant professors, and that this corresponds to the event R_3.

Thus, the numbers 36 and 320 are precisely those used to compute the unconditional probabilities of the events $(R_3 \& A_4)$ and R_3, respectively:

$$P(R_3 \& A_4) = \frac{36}{1164} = 0.031, \qquad P(R_3) = \frac{320}{1164} = 0.275$$

So, by arithmetic and the previous three probabilities, we see that

$$P(A_4 \mid R_3) = \frac{36}{320} = \frac{36/1164}{320/1164} = \frac{P(R_3 \& A_4)}{P(R_3)}$$

Consequently, the conditional probability, $P(A_4 \mid R_3)$, can be computed from the unconditional probabilities, $P(R_3 \& A_4)$ and $P(R_3)$, by using the formula

$$P(A_4 \mid R_3) = \frac{P(R_3 \& A_4)}{P(R_3)}$$

This formula holds in general and is called the *conditional probability rule*.   ■

FORMULA 4.4   **The conditional probability rule**

If A and B are any two events, then

$$P(B \mid A) = \frac{P(A \& B)}{P(A)}$$

In words, the conditional probability of event B given event A is equal to the joint probability of event A and event B divided by the probability of event A.

Note: When using the conditional probability rule to compute a conditional probability, we divide by the probability of the *given* event (the event on the right of the "|").

We should emphasize that for the faculty-member example, conditional probabilities can be computed either directly or by using the conditional probability rule. However, as the next two examples illustrate, the conditional probability rule is sometimes the only way that conditional probabilities can be determined.

EXAMPLE 4.25   *Illustrates Formula 4.4*

The U.S. Bureau of the Census compiles data on the marital status of American adults and publishes the results in *Current Population Reports*. Table 4.11 provides a joint probability distribution for the marital status of American adults by sex.

TABLE 4.11
Joint probability distribution for marital status vs. sex

	Single M_1	Married M_2	Widowed M_3	Divorced M_4	$P(S_i)$
Male S_1	0.116	0.319	0.012	0.028	0.475
Female S_2	0.093	0.325	0.066	0.041	0.525
$P(M_j)$	0.209	0.644	0.078	0.069	1.000

Marital status — Sex

Suppose that an American adult is selected at random.
a) Determine the probability that the adult selected is divorced, given that the adult selected is a male.
b) Determine the probability that the adult selected is a male, given that the adult selected is divorced.

SOLUTION   Note that, unlike our previous illustrations with contingency tables, we do not have the frequency data here, but only the probability (relative-frequency) data. Because of this, we cannot compute conditional probabilities directly; we must use the conditional probability rule.

a) For this part, we want to find the conditional probability that the adult selected is divorced, given that the adult selected is a male: $P(M_4 \mid S_1)$. Using the conditional probability rule and Table 4.11, we obtain

$$P(M_4 \mid S_1) = \frac{P(S_1 \,\&\, M_4)}{P(S_1)} = \frac{0.028}{0.475} = \boxed{0.059}$$

In terms of percentages, this means that 5.9% of adult males are divorced.

b) Here we want to determine the conditional probability that the adult selected is a male, given that the adult selected is divorced: $P(S_1 \mid M_4)$. Referring to Table 4.11 and applying the conditional probability rule, we see that

$$P(S_1 \mid M_4) = \frac{P(M_4 \,\&\, S_1)}{P(M_4)} = \frac{0.028}{0.069} = \boxed{0.406}$$

In other words, 40.6% of divorced adults are males.   ∎

EXAMPLE 4.26   *Illustrates Formula 4.4*

As reported by the Federal Bureau of Investigation in the publication *Crime in the United States*, 4.9% of property crimes are committed in rural areas and 1.9% of property crimes are burglaries committed in rural areas. What percentage of property crimes committed in rural areas are burglaries?

SOLUTION   To solve this problem, we first let

$$R = \text{event that a property crime is committed in a rural area}$$
$$B = \text{event that a property crime is a burglary}$$

Then we need to compute $P(B \mid R)$.
From the given percentage data, we know $P(R) = 0.049$ and $P(R \& B) = 0.019$. Consequently, by the conditional probability rule,

$$P(B \mid R) = \frac{P(R \& B)}{P(R)} = \frac{0.019}{0.049} = 0.388$$

Thus, 38.8% of the property crimes committed in rural areas are burglaries.   ∎

Exercises 4.5

For Exercises 4.69–4.74, compute conditional probabilities directly; that is, do not use the conditional probability rule.

___ **4.69** Suppose a card is selected at random from an ordinary deck of 52 playing cards. Let

A = event a face card is selected
B = event a king is selected
C = event a heart is selected

Determine the following probabilities and express your results in words.
a) $P(B)$ b) $P(B \mid A)$ c) $P(B \mid C)$
d) $P(B \mid (\text{not } A))$ e) $P(A)$ f) $P(A \mid B)$
g) $P(A \mid C)$ h) $P(A \mid (\text{not } B))$

___ **4.70** A balanced dime is tossed twice. The four equally likely outcomes are HH, HT, TH, TT. Let

A = event the first toss is a head
B = event the second toss is a head
C = event at least one toss is heads

Determine the following probabilities and express your results in words.
a) $P(B)$ b) $P(B \mid A)$ c) $P(B \mid C)$
d) $P(C)$ e) $P(C \mid A)$ f) $P(C \mid (\text{not } B))$

___ **4.71** The absentee records over the past year for the employees of Cudahey Masonry, Inc., are presented below in a frequency distribution.

Days missed	Number of employees
0	4
1	2
2	14
3	10
4	16
5	18
6	10
7	6
	80

a) Find the probability that a randomly selected employee missed exactly three days of work.

b) Find the conditional probability that a randomly selected employee missed exactly three days, given that the employee missed at least one day.
c) Find the conditional probability that a randomly selected employee missed at most three days, given that the employee missed at least one day.
d) Interpret your results in parts (a)–(c) in terms of percentages.

— **4.72** As reported by the U.S. Bureau of the Census in the publication *Census of Population,* a frequency distribution for the population of the states in the U.S. and Washington, D.C., is as shown in the following table:

Population size (millions)	Number of states
Under 1	7
1–under 2	7
2–under 3	13
3–under 5	6
5–under 10	6
10 or over	12
	51

Suppose that a state (including Washington, D.C., as a state) is selected at random. Determine the probability that the population of the state selected
a) is between two and three million.
b) is between two and three million, given that it is at least one million.
c) is less than five million, given that it is at least one million.
d) Interpret your results in parts (a)–(c) in terms of percentages.

— **4.73** The contingency table at the top of the next column cross classifies institutions of higher education in the United States by region and type. [SOURCE: U.S. National Center for Education Statistics, *Digest of Education Statistics.*] Suppose that an institution of higher education is selected at random. Determine the probability that the institution selected
a) is in the Northeast.
b) is in the Northeast, given that it is a private school.
c) is a private school, given that it is in the Northeast.
d) Interpret your results in parts (a)–(c) in terms of percentages.

		Type		
		Public T_1	Private T_2	Total
Region	Northeast R_1	266	555	821
	Midwest R_2	359	504	863
	South R_3	533	502	1035
	West R_4	313	242	555
	Total	1471	1803	3274

— **4.74** As reported by the Motor Vehicle Manufacturers Association of the United States in *Motor Vehicle Facts and Figures,* the number of cars and trucks in use by age are as shown in the following contingency table. Frequencies are in millions.

		Type		
		Car V_1	Truck V_2	Total
Age (yrs)	Under 3 A_1	21.5	3.2	24.7
	3–5 A_2	29.9	2.5	32.4
	6–8 A_3	22.2	2.0	24.2
	9–11 A_4	17.9	1.8	19.7
	12 & over A_5	15.4	3.6	19.0
	Total	106.9	13.1	120.0

Suppose a vehicle is selected at random. What is the probability that the vehicle selected
a) is under three years old?
b) is under three years old, given that it is a car?
c) is a car?
d) is a car, given that it is under three years old?

e) Interpret each of your results in parts (a)–(d) in terms of percentages.

___ **4.75** According to the document *Census of Agriculture*, published by the U.S. Bureau of the Census, a contingency table for the number of farms by acreage and tenure of operator is as shown below. Frequencies are in thousands of farms.

Tenure of operator

Acreage	Full owner T_1	Part owner T_2	Tenant T_3	Total
Under 50 A_1	532	74	84	690
50–179 A_2	563	157	94	814
180–499 A_3	262	247	87	596
500–999 A_4	57	128	30	215
1000+ A_5	36	107	18	161
Total	1450	713	313	2476

Suppose a U.S. farm is selected at random.
a) Find $P(T_3)$.
b) Find $P(T_3 \& A_3)$.
c) Compute $P(A_3 \mid T_3)$ directly from the table.
d) Compute $P(A_3 \mid T_3)$ using the conditional probability rule and your results from parts (a) and (b).
e) State your results in parts (a)–(c) in words.

___ **4.76** The contingency table shown at the top of the next column was obtained by cross classifying American hospitals according to type and number of beds. [SOURCE: American Hospital Association, *Hospital Statistics*.] Suppose that an American hospital is selected at random.
a) Find $P(H_1)$.
b) Find $P(H_1 \& B_3)$.
c) Compute $P(B_3 \mid H_1)$ directly from the table.
d) Compute $P(B_3 \mid H_1)$ using the conditional probability rule and your results from parts (a) and (b).
e) State your results in parts (a)–(c) in words.

Number of beds

Type	6–24 B_1	25–74 B_2	75+ B_3	Total
General H_1	299	1894	3945	6138
Psychiatric H_2	17	121	378	516
Chronic H_3	0	7	40	47
Tuberculosis H_4	0	1	10	11
Other H_5	22	131	162	315
Total	338	2154	4535	7027

___ **4.77** The following table provides a joint probability (relative-frequency) distribution for the members of the 98th Congress by political party. [SOURCE: U.S. Congress, Joint Committee on Printing, *Congressional Directory*.]

Type

Party	Reps C_1	Senators C_2	$P(P_i)$
Democrats P_1	0.500	0.084	0.584
Republicans P_2	0.313	0.103	0.416
$P(C_j)$	0.813	0.187	1.000

If a member of the 98th Congress is selected at random, what is the probability that the member selected
a) is a senator?
b) is a Republican senator?
c) is a Republican, given that he or she is a senator?
d) is a senator, given that he or she is a Republican?
e) Interpret each of your results in parts (a)–(d) in terms of percentages.

___ **4.78** The National Center for Education Statistics publishes information on American engineers and scientists in *Digest of Education Statistics*. Below is a table that gives a joint probability (relative-frequency) distribution for engineers and scientists by highest degree obtained.

Type

	Engineers T_1	Scientists T_2	$P(D_i)$
Bachelors D_1	0.343	0.289	0.632
Masters D_2	0.098	0.146	0.244
Doctorate D_3	0.017	0.091	0.108
Other D_4	0.013	0.003	0.016
$P(T_j)$	0.471	0.529	1.000

(Highest degree)

A person is selected at random from among the engineers and scientists. Determine the probability that the person selected
a) is an engineer.
b) has a doctorate.
c) is an engineer with a doctorate.
d) is an engineer, given the person has a doctorate.
e) has a doctorate, given the person is an engineer.
f) Interpret each of your results in parts (a)–(e) in terms of percentages.

___ **4.79** According to the Census Bureau's *Current Population Reports,* 10.6% of American families are black and 3.5% are black and have incomes below the poverty level. What percentage of black families have incomes below the poverty level?

___ **4.80** The National Center for Education Statistics publishes information about educational institutions in *Digest of Education Statistics.* According to that publication, 13.4% of all students at all levels attend private institutions and 4.5% of all students attend private colleges. What percentage of students in private schools are in college?

= **4.81** Refer to Exercise 4.74.
a) Construct a joint probability distribution.
b) Find the probability distribution of age for cars in use; that is, construct a table giving the conditional probabilities that a car in use is under 3 years old, 3–5 years old, 6–8 years old, etc.
c) Find the probability distribution of type for vehicles 3–5 years old.
d) The probability distributions in parts (b) and (c) are examples of **conditional probability distributions.** Find two other conditional probability distributions for the type-versus-age data of motor vehicles in use.

= **4.82** Do you think that the conditional probability of an event B given another event A must always be different than the unconditional probability of event B? That is, do you think it is always true that $P(B\,|\,A) \neq P(B)$? (*Hint:* Try to construct an example in which $P(B\,|\,A) = P(B)$.)

4.6   The multiplication rule; independence

The *conditional probability rule,* given on page 186, is used for computing conditional probabilities in terms of unconditional probabilities. The rule is

$$P(B\,|\,A) = \frac{P(A \,\&\, B)}{P(A)}$$

By multiplying both sides of the previous equation by $P(A)$, we can also obtain a formula for computing joint probabilities in terms of marginal and conditional probabilities. That formula is called the **general multiplication rule** and is presented as Formula 4.5.

FORMULA 4.5   The general multiplication rule

If A and B are any two events, then

$$P(A \& B) = P(A)P(B \mid A)$$

In words, the probability that both event A and event B occur equals the probability that event A occurs times the conditional probability that event B occurs given that event A occurs.

Note that the conditional probability rule and the general multiplication rule are simply variations of each other. When the joint probabilities and marginal probabilities are known or easily determined directly, we can use the conditional probability rule to obtain conditional probabilities. On the other hand, when the marginal and conditional probabilities are known or easily determined directly, we can use the general multiplication rule to obtain joint probabilities.

EXAMPLE 4.27   *Illustrates Formula 4.5*

The U.S. Bureau of Labor Statistics collects data on American workers and publishes the information in *Employment and Earnings*. According to that publication, 30.9% of all employed black people are white-collar workers and 9.2% of all employed workers are black. Suppose that an employed worker is selected at random. What is the probability that the worker selected is a black white-collar worker?

SOLUTION   To determine the required probability, we first let

$B =$ event the worker selected is black

$W =$ event the worker selected is a white-collar worker

Then the probability that the worker selected is a black white-collar worker can be expressed as $P(B \& W)$.

According to the given data, $P(B) = 0.092$ since 9.2% of all employed workers are black. Also $P(W \mid B) = 0.309$ since 30.9% of all employed black people are white-collar workers. Applying the general multiplication rule, we get

$$P(B \& W) = P(B)P(W \mid B) = 0.092 \cdot 0.309 = \boxed{0.028}$$

Thus, the probability is 0.028 that a randomly selected employed worker is a black white-collar worker. Expressed in percentages, 2.8% of all employed workers are black white-collar workers.   ∎

Another application of the general multiplication rule arises in situations where two or more members are selected from a population. Consider Example 4.28.

EXAMPLE 4.28   *Illustrates Formula 4.5*

In Professor W's introductory statistics class, the numbers of males and females are as given in the following frequency distribution.

TABLE 4.12
Frequency
distribution of
males and females
in Professor W's
introductory
statistics class

Sex	Frequency
Male	17
Female	23
	40

Suppose two students are selected at random from the class. Assume that the first student selected is *not* returned to the class before the second student is selected. What is the probability that the first student selected is a female and the second is a male?

SOLUTION   Let us employ the following notation:

$$F1 = \text{event the first student selected is female}$$
$$M2 = \text{event the second student selected is male}$$

The problem is to determine the probability that the first student selected is female and the second student selected is male, that is, $P(F1 \,\&\, M2)$. By the general multiplication rule, we can write

$$P(F1 \,\&\, M2) = P(F1)P(M2 \,|\, F1)$$

We now proceed to obtain the two probabilities on the right-hand side of the previous equation. As we see from Table 4.12, 23 out of the 40 students are female. Therefore,

$$P(F1) = \frac{f}{N} = \frac{23}{40}$$

Also, given that the first student selected is female (that is, given that event $F1$ occurs), there are 39 students remaining in the class, of which 17 are male and 22 are female. Thus, the conditional probability that the second student selected is male, given that the first student selected is female equals

$$P(M2 \,|\, F1) = \frac{f}{N} = \frac{17}{39}$$

Applying the general multiplication rule, we can now conclude that

$$P(F1 \,\&\, M2) = P(F1)P(M2 \,|\, F1) = \frac{23}{40} \cdot \frac{17}{39} = \boxed{0.251}$$

Consequently, the probability is 0.251 that the first student selected is a female and the second is a male.

We can also use a device called a **tree diagram** to calculate probabilities when employing the general multiplication rule. A tree diagram for the present example is displayed in Figure 4.27 at the top of the next page.

FIGURE 4.27
Tree diagram for
student-selection
problem

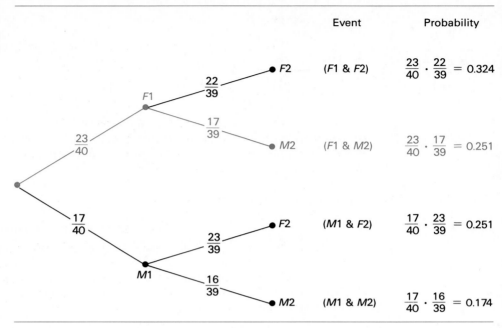

The part of the tree diagram representing the determination of the probability $P(F1 \,\&\, M2)$, which we just computed, is shown in color. ∎

Note: When two or more members of a population are selected and those members selected are not returned to the population for possible reselection, then the selection process is called **sampling without replacement.**

The general multiplication rule can be extended to more than two events. This will be explored in Exercise 4.104.

INDEPENDENCE

We next discuss the concept of *statistical independence.* Two events are **statistically independent** if the occurrence (or nonoccurrence) of one of the events does not affect the probability that the other event occurs. More formally, we have the following definition.

DEFINITION 4.8   Statistical independence

Event B is said to be *statistically independent* of event A, if the occurrence of event A does not affect the probability that event B occurs. In symbols:

$$P(B \mid A) = P(B)$$

For brevity, we will usually omit the adjective "statistically" when discussing independence. Thus, for example, we will use the term "independent" instead of "statistically independent."

EXAMPLE 4.29   *Illustrates Definition 4.8*

For the experiment of randomly selecting a card from an ordinary deck of 52 playing cards, let

$$A = \text{event a face card is selected}$$
$$B = \text{event a king is selected}$$
$$C = \text{event a heart is selected}$$

Use Definition 4.8 to
a) determine whether event B is independent of event A.
b) determine whether event B is independent of event C.

SOLUTION   First note that the unconditional probability of event B is

$$P(B) = \frac{f}{N} = \frac{4}{52} = \frac{1}{13} = \boxed{0.077}$$

a) To determine whether event B is independent of event A, we need to compute $P(B\,|\,A)$ and compare it to $P(B)$. If those two probabilities are equal, then event B is independent of event A; otherwise, event B is not independent of event A. Now, given that event A occurs, there are 12 possible outcomes (four kings, four queens, and four jacks), and event B can occur in four ways out of those 12 possibilities. So,

$$P(B\,|\,A) = \frac{f}{N} = \frac{4}{12} = \boxed{0.333}$$

This is not equal to $P(B)$. Thus, the fact that event A occurs affects the probability that event B occurs. In other words, event B is *not* independent of event A. It is useful to observe that this lack of independence stems from the following fact: The percentage of kings among the face cards (33.3%) is not the same as the percentage of kings among all the cards (7.7%).

b) Here we need to compute $P(B\,|\,C)$ and compare it to $P(B)$. Given that event C occurs, there are 13 possible outcomes (the 13 hearts), and event B can occur in one way out of those 13 possibilities. Therefore,

$$P(B\,|\,C) = \frac{f}{N} = \frac{1}{13} = \boxed{0.077}$$

This is equal to $P(B)$. Thus, the fact that event C occurs does not affect the probability that event B occurs. In other words, event B *is* independent of event C. It is useful to observe that this independence stems from the following fact: The percentage of kings among the hearts is the same as the percentage of kings among all the cards; namely, 7.7%. ∎

EXAMPLE 4.30   *Illustrates Definition 4.8*

The American Medical Association compiles information on American physicians in the publication *Physician Characteristics and Distribution in the U.S.* Table 4.13 provides a contingency table for the number of American surgeons cross classified by specialty and base of practice. Frequencies are in thousands.

TABLE 4.13
Surgeons by specialty
and base of practice

Specialty	Base			
	Office B_1	Hospital B_2	Other B_3	Total
General surgery S_1	23.6	12.0	1.6	37.2
Obstetrics/Gynecology S_2	23.6	6.6	1.1	31.3
Orthopedic S_3	13.1	4.2	0.4	17.7
Ophthalmology S_4	12.1	2.6	0.5	15.2
Total	72.4	25.4	3.6	101.4

Suppose a surgeon is selected at random. Is the event that the surgeon selected is office based independent of the event that the surgeon selected is an orthopedist? That is, is event B_1 independent of event S_3?

SOLUTION   To solve this problem, we need to compute $P(B_1 \mid S_3)$ and $P(B_1)$. If those two probabilities are equal, then event B_1 is independent of event S_3; otherwise, event B_1 is not independent of event S_3. From the table, we find that

$$P(B_1 \mid S_3) = \frac{13.1}{17.7} = \boxed{0.740}$$

and

$$P(B_1) = \frac{72.4}{101.4} = \boxed{0.714}$$

Thus, $P(B_1 \mid S_3) \neq P(B_1)$, and so event B_1 is *not* independent of event S_3. That is, the event that the surgeon selected is office based is *not* independent of the event that the surgeon selected is an orthopedist. This lack of independence results from the fact that the percentage of orthopedists who are office based (74.0%) is not the same as the percentage of all surgeons who are office based (71.4%).   ■

It can be shown that if event B is independent of event A, then it is also true that event A is independent of event B (see Exercise 4.105). So, in such cases, we often say that event A and event B are **independent** or that A and B are **independent events**. Finally, if two events are not independent, then they are said to be **dependent**. Hence, the events S_3 and B_1 in Example 4.30 are dependent.

THE SPECIAL MULTIPLICATION RULE

Recall that the *general multiplication rule* states that for any two events, A and B,

$$P(A \& B) = P(A)P(B \mid A)$$

If A and B are *independent* events, then $P(B \mid A) = P(B)$. Thus, for the special case of independent events, we can replace the term $P(B \mid A)$, occurring in the general multiplication rule, by the term $P(B)$. This yields the following rule:

FORMULA 4.6 **The special multiplication rule**

If event A and event B are *independent*, then

$$P(A \ \& \ B) = P(A)P(B)$$

and, conversely, if $P(A \ \& \ B) = P(A)P(B)$, then event A and event B are independent. In words, if two events are independent, then their joint probability equals the product of their marginal probabilities; and vice-versa.

In Examples 4.29 and 4.30, we used the definition of independence, Definition 4.8, to decide whether two specified events are independent. If the two events are, say, A and B, this means determining whether $P(B \mid A) = P(B)$. Alternatively, we can decide whether event A and event B are independent with the aid of the special multiplication rule; that is, by determining whether $P(A \ \& \ B) = P(A)P(B)$.

The special multiplication rule is also used to compute joint probabilities when we know or can reasonably assume that two events are independent. Such a situation is illustrated in the next example.

EXAMPLE 4.31   *Illustrates Formula 4.6*

A roulette wheel contains 38 numbers of which 18 are red, 18 are black, and two are green. When the roulette ball is spun it is equally likely to land on any of the 38 numbers. In two plays at the wheel, what is the probability that
a) the ball lands on red both times?
b) the ball lands on green the first time and on black the second time?

SOLUTION   First of all we note that it is reasonable to assume that outcomes on successive plays at the wheel are independent (why?).
a) For this part, we are to determine the probability that, in two plays at the wheel, the ball lands on red both times. Let

$$R1 = \text{event the ball lands on red the first time}$$
$$R2 = \text{event the ball lands on red the second time}$$

Then we need to find $P(R1 \ \& \ R2)$.

Since outcomes on successive plays at the wheel are independent, event $R1$ and event $R2$ are independent events. So, by the special multiplication rule,

$$P(R1 \ \& \ R2) = P(R1)P(R2)$$

But,

$$P(R1) = \frac{f}{N} = \frac{18}{38}$$

and the same is true for $P(R2)$. Thus,

$$P(R1 \,\&\, R2) = P(R1)P(R2) = \frac{18}{38} \cdot \frac{18}{38} = \boxed{0.224}$$

There is a 22.5% chance of the ball landing on red twice in a row.

b) Here we want to compute the probability that the ball lands on green the first time and on black the second time. Let

$$G1 = \text{event the ball lands on green the first time}$$
$$B2 = \text{event the ball lands on black the second time}$$

Then, since $G1$ and $B2$ are independent events, the special multiplication rule implies that

$$P(G1 \,\&\, B2) = P(G1)P(B2) = \frac{2}{38} \cdot \frac{18}{38} = \boxed{0.025}$$

There is a 2.5% chance that the ball will land on green the first time and on black the second time.   ■

The definition of independence for three or more events is more complicated than that for two events. Nevertheless, the special multiplication rule still holds. For example, if A, B, C, and D are independent events, then

$$P(A \,\&\, B \,\&\, C \,\&\, D) = P(A)P(B)P(C)P(D)$$

FORMULA 4.7   The special multiplication rule

If the events A, B, C, ... are *independent*, then

$$P(A \,\&\, B \,\&\, C \,\&\, \cdots) = P(A)P(B)P(C)\cdots$$

EXAMPLE 4.32   *Illustrates Formula 4.7*

What is the probability that, in five plays at the roulette wheel, the ball lands on red all five times?

SOLUTION   Let

$$R1 = \text{event the ball lands on red the first time}$$
$$R2 = \text{event the ball lands on red the second time}$$

and so forth. We need to compute $P(R1 \,\&\, R2 \,\&\, R3 \,\&\, R4 \,\&\, R5)$. Since we may assume that outcomes on successive plays at the wheel are independent, the special multiplication rule applies to give

$$P(R1 \,\&\, R2 \,\&\, R3 \,\&\, R4 \,\&\, R5) = P(R1)P(R2)P(R3)P(R4)P(R5)$$

$$= \frac{18}{38} \cdot \frac{18}{38} \cdot \frac{18}{38} \cdot \frac{18}{38} \cdot \frac{18}{38} = \boxed{0.024}$$

There is a 2.4% chance that the ball will land on red five times in a row.   ■

MUTUALLY EXCLUSIVE VERSUS INDEPENDENCE

It is not uncommon for students to confuse the concept of *mutually exclusive events* with that of *independent events*. These terms do not mean the same thing. The concept of "mutually exclusive" involves whether or not two events can occur simultaneously. On the other hand, the concept of "independence" involves whether or not the occurrence of one event affects the probability of the occurrence of another.

We will make some general statements concerning the relationship between these two ideas in Exercise 4.109. The important point, however, is that the concepts of mutually exclusive events and independent events are quite different.

Exercises 4.6

___ **4.83** According to the Census Bureau's *Current Population Reports*, 29.2% of all farm families make at least $25,000 per year and 2.6% of all families are farm families. Use the general multiplication rule to find the probability that a randomly selected family is a farm family making at least $25,000 per year. Interpret your result in terms of percentages.

___ **4.84** The National Center for Education Statistics states in *Digest of Education Statistics* that 43.9% of all public elementary schools have between 250 and 499 students. Moreover, 51.4% of all public schools are elementary schools. Use the general multiplication rule to determine the probability that a randomly selected public school is an elementary school with between 250 and 499 students. Interpret your result in terms of percentages.

___ **4.85** Cards numbered 1, 2, 3, ..., 10 are placed in a box. The box is shaken and a blindfolded person selects two successive cards without replacement.
a) What is the probability that the first card selected is numbered 6?
b) Given that the first card is numbered 6, what is the probability that the second is numbered 9?
c) What is the probability of selecting first a 6 and then a 9?
d) What is the probability that both cards selected are numbered over 5?

___ **4.86** A person has agreed to participate in an ESP experiment. He is asked to randomly pick two numbers between 1 and 6. The second number must be different from the first. Let

H = event the first number picked is a 3
K = event the second number picked exceeds 4

Determine
a) $P(H)$ b) $P(K \mid H)$ c) $P(H \& K)$
Find the probability that
d) both numbers picked are less than 3.
e) both numbers picked are greater than 3.

___ **4.87** The following table gives a frequency distribution of the political-party affiliations of U.S. governors. [SOURCE: *The World Almanac, 1989.*]

Party	Frequency
Democratic	27
Republican	23
	50

Suppose two governors are selected at random without replacement. What is the probability that
a) the first selected is a Republican and the second is a Democrat?
b) both governors selected are Republicans?
c) Draw a tree diagram for this problem similar to Figure 4.27 on page 194.
d) What is the probability that both governors selected belong to the same party?

___ **4.88** A frequency distribution for the class of students in a midwest high school is as follows:

Class	Frequency
Freshman	89
Sophomore	127
Junior	118
Senior	93
	427

Suppose two students are randomly selected without replacement. Determine the probability that
a) the first student selected is a junior and the second is a senior.
b) both students selected are sophomores.
c) Draw a tree diagram for this problem similar to Figure 4.27 on page 194.
d) What is the probability that one of the students selected is a freshman and the other student selected is a sophomore?

___ 4.89   The U.S. National Center for Health Statistics compiles data on injuries and publishes the information in *Vital and Health Statistics*. A contingency table for injuries in the U.S. by circumstance and sex is as follows. Frequencies are in millions.

Circumstance

	Work C_1	Home C_2	Other C_3	Total
Male S_1	8.0	9.8	17.8	35.6
Female S_2	1.3	11.6	12.9	25.8
Total	9.3	21.4	30.7	61.4

Sex (row label)

a) Find $P(C_1)$.
b) Find $P(C_1 \mid S_2)$.
c) Are the events C_1 and S_2 independent? Why?
d) Is the event that an injured person is male independent of the event that an injured person was hurt at home? Justify your answer.

___ 4.90   A study published in *1980 Census of Population* by the U.S. Bureau of the Census revealed the following data on the methods that Americans use to get to work by residence. Frequencies are in millions of workers.

Residence

	Urban R_1	Rural R_2	Total
Automobile M_1	45.0	15.0	60.0
Public trans. M_2	6.5	0.5	7.0
Total	51.5	15.5	67.0

Method (row label)

a) Find $P(M_1)$.
b) Find $P(M_1 \mid R_2)$.
c) Are the events M_1 and R_2 independent? Why?
d) Is the event that a worker resides in an urban area independent of the event that the worker uses an automobile to get to work? Justify your answer.

___ 4.91   When a balanced dime is tossed three times, there are eight equally likely outcomes:

HHH	HTH	THH	TTH
HHT	HTT	THT	TTT

Let

A = event the first toss is heads
B = event the third toss is tails
C = event the total number of heads is one

a) Compute $P(A)$, $P(B)$, and $P(C)$.
b) Compute $P(B \mid A)$.
c) Are the events A and B independent? Why?
d) Compute $P(C \mid A)$.
e) Are the events A and C independent? Why?

___ 4.92   When two balanced dice are rolled, there are 36 equally likely outcomes possible, as shown in Figure 4.1 on page 145. Let

A = event the orange die comes up even
B = event the black die comes up odd
C = event the sum of the dice is 10
D = event the sum of the dice is even

a) Compute $P(A)$, $P(B)$, $P(C)$, and $P(D)$.
b) Compute $P(B \mid A)$.
c) Are the events A and B independent? Why?
d) Compute $P(C \mid A)$.
e) Are the events A and C independent? Why?
f) Compute $P(D \mid A)$.
g) Are the events A and D independent? Why?

__ **4.93** The following table provides a joint probability (relative-frequency) distribution for the members of the 98th Congress by political party. [SOURCE: U.S. Congress, Joint Committee on Printing, *Congressional Directory*.]

Type

Party	Reps C_1	Senators C_2	$P(P_i)$
Democrats P_1	0.500	0.084	0.584
Republicans P_2	0.313	0.103	0.416
$P(C_j)$	0.813	0.187	1.000

a) Determine $P(P_1)$, $P(C_2)$, and $P(P_1 \ \& \ C_2)$.

b) Are the events P_1 and C_2 independent? [Use the special multiplication rule to answer this question.]

__ **4.94** The National Center for Education Statistics publishes information on American engineers and scientists in *Digest of Education Statistics*. Below is a table that gives a joint probability (relative-frequency) distribution for engineers and scientists by highest degree obtained.

Type

Highest degree	Engineers T_1	Scientists T_2	$P(D_i)$
Bachelors D_1	0.343	0.289	0.632
Masters D_2	0.098	0.146	0.244
Doctorate D_3	0.017	0.091	0.108
Other D_4	0.013	0.003	0.016
$P(T_j)$	0.471	0.529	1.000

a) Determine $P(T_2)$, $P(D_3)$, and $P(T_2 \ \& \ D_3)$.

b) Are the events T_2 and D_3 independent? Why?

__ **4.95** Two cards are drawn at random from an ordinary deck of 52 cards. Determine the probability that both cards are aces, if

a) the first card is replaced before the second card is drawn.

b) the first card is not replaced before the second card is drawn.

__ **4.96** In the game of *Yahtzee*, five balanced dice are rolled.

a) What is the probability of rolling all 2s? (*Hint:* Use the fact that the outcomes of different dice are independent and apply the special multiplication rule.)

b) What is the probability that all the dice come up the same number? (*Hint:* You will need to use both the special addition rule and the special multiplication rule.)

__ **4.97** Suppose that event E and event F are independent, and that $P(E) = 1/3$ and $P(F) = 1/4$. Determine

a) $P(E \ \& \ F)$. b) $P(E \text{ or } F)$.

__ **4.98** A family has two portable computers that run on batteries. There is a 70% chance that a given computer will run for over 60 operating hours without a change of batteries. Determine the probability that

a) both computers run for over 60 hours without a change of batteries.

b) one or the other or both will run for over 60 hours without a change of batteries.

__ **4.99** In a letter to the editor appearing in the February 23, 1987 issue of *U.S. News and World Report*, a reader discussed the issue of shuttle safety. Each "criticality 1" item must have a 99.99% reliability, according to NASA standards. This means that the probability of failure for a "criticality 1" item is only 0.0001. For Mission 25, there were 748 "criticality 1" items. Determine the probability that

a) none of the "criticality 1" items would fail.

b) at least one "criticality 1" item would fail.

c) Interpret your result in part (b) in words.

__ **4.100** A hardware manufacturer produces nuts and bolts of a particular size. Each bolt produced is attached to a nut to make a single unit. It is known that 2% of the nuts produced and 3% of the bolts produced have some sort of defect. A nut-bolt unit is considered defective if either the nut or the bolt has a defect. What percentage of the nut-bolt units are not defective?

__ **4.101** As reported by the Chicago Title Insurance Company in the publication *The Guarantor*, the probability is 0.768 that a home-buyer will purchase a resale home. In the next four home purchases, find the probability that

a) the first three will be resales and the fourth will be a new home.
b) the first will be a resale, the second will be a new home, and the last two will be resales.
c) the first will be a resale, the next two will be new homes, and the last will be a resale.
d) exactly three of the four will be resales. (*Hint:* There are four possible ways in which exactly three of the four home purchases will be resales; two of these four ways are discussed in parts (a) and (b).)

___ 4.102   The FBI compiles information on violent crimes by type and publishes the results in *Crime in the United States*. Here is a table giving the probabilities for the various types of violent crimes.

Violent crime	Probability
Murder	0.016
Forcible rape	0.064
Robbery	0.403
Aggravated assault	0.517

The table shows, for instance, that the probability is 0.016 that a violent crime will be a murder. Out of three violent crimes, find the probability that
a) the first two will be robberies and the last will be a forcible rape.
b) the first will be a murder, the second will be a robbery, and the third will be an aggravated assault.

=== 4.103   The National Center for Health Statistics collects data on activity limitations. Results are published in *Vital and Health Statistics*. The data show that 13.6% of males have an activity limitation and that 14.4% of females have an activity limitation. Are gender and activity limitation statistically independent? Explain your answer.

=== 4.104   For three events, **the general multiplication rule** is

$$P(A \& B \& C) = P(A)P(B \mid A)P(C \mid (A \& B))$$

a) Suppose three cards are randomly selected without replacement from an ordinary deck of 52. Find the probability that all three cards are hearts; that the first two cards are hearts and the third is a spade.
b) There is a general multiplication rule for any number of events. State the general multiplication rule for four events.

=== 4.105   Prove that if $P(B \mid A) = P(B)$, then it is also true that $P(A \mid B) = P(A)$. This shows that if

event B is independent of event A, then it is also true that event A is independent of event B.

=== 4.106   Three events, A, B, and C, are said to be **independent** if

$$P(A \& B) = P(A)P(B)$$
$$P(A \& C) = P(A)P(C)$$
$$P(B \& C) = P(B)P(C)$$

and

$$P(A \& B \& C) = P(A)P(B)P(C)$$

What do you think is required for four events to be independent?

=== 4.107   Consider the experiment of rolling two balanced dice. Let

A = event the orange die comes up even
B = event the black die comes up even
C = event the sum of the dice is even
D = event the orange die comes up 1, 2, or 3
E = event the orange die comes up 3, 4, or 5
F = event the sum of the dice is 5

Use the definition of independence for three events, given in Ex. 4.106, to solve the following problems.
a) Are A, B, and C independent?
b) Show that $P(D \& E \& F) = P(D)P(E)P(F)$ but that D, E, and F are not independent.

=== 4.108   Suppose that a balanced coin is tossed four times. There are 16 equally likely outcomes:

HHHH	THHH	THHT	THTT
HHHT	HHTT	THTH	TTHT
HHTH	HTHT	TTHH	TTTH
HTHH	HTTH	HTTT	TTTT

Let

A = event the first toss is heads
B = event the second toss is tails
C = event the last two tosses are heads

Use the definition of independence for three events, given in Exercise 4.106, to show that A, B, and C are independent.

=== 4.109   This exercise explores the relationship between the concepts of mutually exclusive events and independent events. Consider any two events, A and B, neither of which is impossible (i.e., assume $P(A) > 0$ and $P(B) > 0$).

a) Show that if A and B are independent events, then they cannot be mutually exclusive. (*Hint:* Use the special multiplication rule to show $P(A \& B) > 0$.)
b) Show that if A and B are mutually exclusive events, then they cannot be independent. (*Hint:* Argue that $P(B \mid A) = 0$ so that $P(B \mid A) \neq P(B)$.)
c) Find an example of two events that are neither mutually exclusive nor independent.

4.7 Bayes' rule (Optional)

In this section, we will discuss an important rule of probability developed by Thomas Bayes, an eighteenth century clergyman. This rule is aptly called **Bayes' rule.**

One of the primary uses of Bayes' rule is to revise probabilities in accordance with newly acquired information. Actually, such revised probabilities are really just conditional probabilities and so, in some sense, we have already examined much of the material in this section. However, as we will see, there are a couple of new techniques involved in Bayes' rule.

THE RULE OF TOTAL PROBABILITY

In preparation for Bayes' rule, we need to study another rule of probability that is important in its own right. This rule is called the **rule of total probability** and is introduced in Example 4.33.

EXAMPLE 4.33 *Introduces the rule of total probability*

The U.S. Bureau of the Census collects data on the resident population by region of residence and age. Results are published in *Current Population Reports.*

In the first two columns of Table 4.14, we have provided a percentage distribution for region of residence. The third column displays the percentage of seniors (age 65 or over) in each region. The table shows, for instance, that 20.6% of U.S. residents live in the Northeast region. It also shows that 13.5% of residents living in the Northeast are seniors.

TABLE 4.14

Region	Percentage of U.S. population	Percent seniors
Northeast	20.6	13.5
Midwest	24.5	12.6
South	34.5	12.1
West	20.4	10.8
	100.0	

What percentage of U.S. residents are seniors?

SOLUTION   To solve this problem, we first translate the information into the language of probability. Suppose a U.S. resident is selected at random. Let

$$S = \text{event the resident selected is a senior}$$

and

$$R_1 = \text{event the resident selected lives in the Northeast}$$
$$R_2 = \text{event the resident selected lives in the Midwest}$$
$$R_3 = \text{event the resident selected lives in the South}$$
$$R_4 = \text{event the resident selected lives in the West}$$

Then the second and third columns of Table 4.14 yield the probabilities shown in Table 4.15.

TABLE 4.15

$P(R_1) = 0.206$	$P(S \mid R_1) = 0.135$
$P(R_2) = 0.245$	$P(S \mid R_2) = 0.126$
$P(R_3) = 0.345$	$P(S \mid R_3) = 0.121$
$P(R_4) = 0.204$	$P(S \mid R_4) = 0.108$

The problem is to determine the percentage of U.S. residents who are seniors; that is, $P(S)$. We must obtain this probability using the available information—the probabilities in Table 4.15. Here is how we do that.

Event S occurs if the resident selected is a senior. But for the resident selected to be a senior, the resident selected must live in the Northeast and be a senior, live in the Midwest and be a senior, live in the South and be a senior, or live in the West and be a senior. (A senior must live in exactly one of the four regions.) In terms of our notation, this means that

(1) $S = [(R_1 \ \& \ S) \text{ or } (R_2 \ \& \ S) \text{ or } (R_3 \ \& \ S) \text{ or } (R_4 \ \& \ S)]$

A Venn diagram for Equation (1) is shown in Figure 4.28. The entire shaded area of the diagram represents all residents who are seniors; that is, event S. Now look, for instance, at the second subrectangle in the diagram. This represents all residents who live in the Midwest; that is, event R_2. The shaded portion of that subrectangle represents all seniors who live in the Midwest; that is, event $(R_2 \ \& \ S)$. Similar statements hold for the other three subrectangles. The fact that the entire shaded area is composed of the shaded portions of the four subrectangles provides us with a graphical verification of Equation (1).

FIGURE 4.28

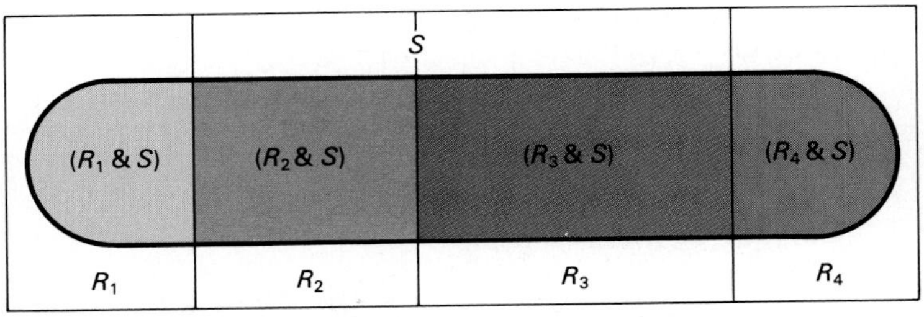

Since the events R_1, R_2, R_3, and R_4 are mutually exclusive (why?), so are the events $(R_1 \,\&\, S)$, $(R_2 \,\&\, S)$, $(R_3 \,\&\, S)$, and $(R_4 \,\&\, S)$. [Refer to Figure 4.28.] Therefore, by Equation (1) and the special addition rule, Formula 4.1 on page 164,

$$P(S) = P(R_1 \,\&\, S) + P(R_2 \,\&\, S) + P(R_3 \,\&\, S) + P(R_4 \,\&\, S)$$

Applying the general multiplication rule, Formula 4.5 on page 192, to each term on the right-hand side of the previous equation, we now obtain the *rule of total probability*:

$$(2) \quad P(S) = P(R_1)P(S\,|\,R_1) + P(R_2)P(S\,|\,R_2) + P(R_3)P(S\,|\,R_3) + P(R_4)P(S\,|\,R_4)$$

All of the probabilities on the right-hand side of Equation (2) can be found in Table 4.15. Substituting in those probabilities, we get

$$P(S) = 0.206 \cdot 0.135 + 0.245 \cdot 0.126 + 0.345 \cdot 0.121 + 0.204 \cdot 0.108$$
$$= \boxed{0.122}$$

A tree diagram for this calculation is shown in Figure 4.29. (In the figure, J represents the event that the resident selected is not a senior.) We obtain $P(S)$ from the tree diagram as follows: First we multiply the two probabilities on each branch of the tree that ends with S. Then we add up all of those products.

FIGURE 4.29

Tree diagram for calculating $P(S)$ using the rule of total probability

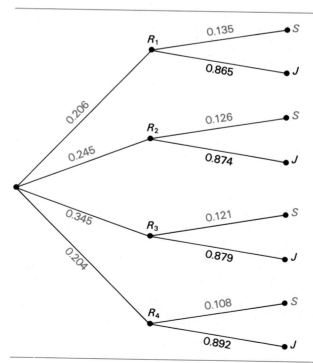

In any case, we see that the probability is 0.122 that a randomly selected U.S. resident will be a senior. In other words, 12.2% of U.S. residents are seniors. ∎

Note that, in the previous example, at least one of the events, R_1, R_2, R_3, R_4, must occur. This is because a U.S. resident must live in one of the four regions. So, the event $(R_1$ or R_2 or R_3 or $R_4)$ is certain. On account of this, we say that the events R_1, R_2, R_3, R_4 are **exhaustive.**

DEFINITION 4.9   Exhaustive events

Events A_1, A_2, ..., A_k are said to be *exhaustive* if at least one of the events must occur when the experiment is performed; that is, if the event

$$(A_1 \text{ or } A_2 \text{ or } \cdots \text{ or } A_k)$$

is certain.

The events, R_1, R_2, R_3, R_4, in Example 4.33 are not only exhaustive but, as we noted earlier, they are also mutually exclusive. If a collection of events is both exhaustive and mutually exclusive, then *exactly one* of the events must occur when the experiment is performed. Indeed, for such a collection of events, at least one of the events must occur when the experiment is performed since the events are exhaustive; and at most one of the events can occur when the experiment is performed since the events are mutually exclusive.

We are now in a position to state the rule of total probability. As we mentioned in Example 4.33, Equation (2) is the rule of total probability for the particular application considered in that example. More generally, we have the following.

FORMULA 4.8   The rule of total probability

Suppose events A_1, A_2, ..., A_k are mutually exclusive and exhaustive; that is, exactly one of those events must occur when the experiment is performed. Then for any event B,

$$P(B) = P(A_1)P(B \mid A_1) + P(A_2)P(B \mid A_2) + \cdots + P(A_k)P(B \mid A_k)$$

BAYES' RULE

We are now ready to discuss Bayes' rule. To introduce Bayes' rule, we return to the situation of Example 4.33.

EXAMPLE 4.34   *Introduces Bayes' rule*

The first two columns of Table 4.14 on page 203 display a percentage distribution for the region of residence of U.S. residents. In the third column of that table are the percentages of seniors in each region. For ease in reference, we repeat Table 4.14 here as Table 4.16.

TABLE 4.16

Region	Percentage of U.S. population	Percent seniors
Northeast	20.6	13.5
Midwest	24.5	12.6
South	34.5	12.1
West	20.4	10.8
	100.0	

What percentage of all U.S. seniors live in the Northeast?

SOLUTION We begin by converting the problem into a probability problem. Suppose a U.S. resident is selected at random. As in Example 4.33, let

$$S = \text{event the resident selected is a senior}$$

and

$$R_1 = \text{event the resident selected lives in the Northeast}$$
$$R_2 = \text{event the resident selected lives in the Midwest}$$
$$R_3 = \text{event the resident selected lives in the South}$$
$$R_4 = \text{event the resident selected lives in the West}$$

Then the second and third columns of Table 4.16 yield the probabilities shown in Table 4.17.

TABLE 4.17

$P(R_1) = 0.206$	$P(S \mid R_1) = 0.135$
$P(R_2) = 0.245$	$P(S \mid R_2) = 0.126$
$P(R_3) = 0.345$	$P(S \mid R_3) = 0.121$
$P(R_4) = 0.204$	$P(S \mid R_4) = 0.108$

The problem is to determine the percentage of all U.S. seniors that live in the Northeast; that is, $P(R_1 \mid S)$. To obtain that conditional probability from the information given in Table 4.17, we proceed as follows. First, we apply the conditional probability rule, Formula 4.4 on page 186, to write

(3)
$$P(R_1 \mid S) = \frac{P(S \& R_1)}{P(S)}$$

To compute the numerator of the term on the right-hand side of Equation (3), we use the general multiplication rule and the probability data in the first row of Table 4.17:

$$P(S \& R_1) = P(R_1 \& S) = P(R_1)P(S \mid R_1) = 0.206 \cdot 0.135 = 0.028$$

To compute the denominator of the term on the right-hand side of Equation (3), we use the rule of total probability and all the probability data in Table 4.17. We did this earlier in Example 4.33 and obtained

$$P(S) = P(R_1)P(S \mid R_1) + P(R_2)P(S \mid R_2) + P(R_3)P(S \mid R_3) + P(R_4)P(S \mid R_4)$$

$$= 0.206 \cdot 0.135 + 0.245 \cdot 0.126 + 0.345 \cdot 0.121 + 0.204 \cdot 0.108$$

$$= 0.122$$

Substituting the two probabilities we just computed into Equation (3), we find that

$$P(R_1 \mid S) = \frac{P(S \,\&\, R_1)}{P(S)} = \frac{0.028}{0.122} = \boxed{0.230}$$

In other words, 23.0% of all U.S. seniors live in the Northeast.   ∎

Note: In the previous example, we rounded the probabilities $P(S \,\&\, R_1)$ and $P(S)$ before computing $P(R_1 \mid S)$. This was done for illustrative reasons only. Generally, no rounding should be done until the final computation is complete.

By studying the method that we used to solve the previous example, we obtain the following rule, which is known as *Bayes' rule*.

FORMULA 4.9   Bayes' rule

Suppose events $A_1, A_2, \ldots, A_k$ are mutually exclusive and exhaustive; that is, exactly one of those events must occur when the experiment is performed. Then for any event B,

$$P(A_i \mid B) = \frac{P(A_i)P(B \mid A_i)}{P(A_1)P(B \mid A_1) + P(A_2)P(B \mid A_2) + \cdots + P(A_k)P(B \mid A_k)}$$

where A_i can be any one of the events $A_1, A_2, \ldots, A_k$.

Note: It is important to realize that Bayes' rule applies when the probabilities, $P(A_1), P(A_2), \ldots, P(A_k)$, and the conditional probabilities, $P(B \mid A_1), P(B \mid A_2), \ldots, P(B \mid A_k)$, are known. Also, observe that the denominator on the right-hand side of Bayes' rule is simply $P(B)$ expressed in terms of those known probabilities by applying the rule of total probability.

EXAMPLE 4.35   *Illustrates Formula 4.9*

According to the Arizona Chapter of the American Lung Association, 7.0% of the population has lung disease. Of those people having lung disease, 90.0% are smokers; and of those not having lung disease, 25.3% are smokers. Suppose a person is selected at random from the population. Determine the probability that the person selected has lung disease, given that the person selected is a smoker.

SOLUTION   Let

$$S = \text{event the person selected is a smoker}$$

and

$$L_1 = \text{event the person selected has no lung disease}$$
$$L_2 = \text{event the person selected has lung disease}$$

Note that events L_1 and L_2 are mutually exclusive and exhaustive.

The data provided in the statement of the problem imply that $P(L_2) = 0.070$, $P(S \mid L_2) = 0.900$, and $P(S \mid L_1) = 0.253$. Also, since $L_1 = (\text{not } L_2)$, we have $P(L_1) = P(\text{not } L_2) = 1 - P(L_2) = 1 - 0.070 = 0.930$. We summarize this information in Table 4.18.

TABLE 4.18

$P(L_1) = 0.930$	$P(S \mid L_1) = 0.253$
$P(L_2) = 0.070$	$P(S \mid L_2) = 0.900$

The problem is to determine the probability that the person selected has lung disease, given that the person selected is a smoker; that is, $P(L_2 \mid S)$. Applying Bayes' rule to the probability data in Table 4.18, we obtain

$$P(L_2 \mid S) = \frac{P(L_2)P(S \mid L_2)}{P(L_1)P(S \mid L_1) + P(L_2)P(S \mid L_2)} = \frac{0.070 \cdot 0.900}{0.930 \cdot 0.253 + 0.070 \cdot 0.900}$$
$$= \boxed{0.211}$$

Thus, the probability is 0.211 that a person selected at random will have lung disease, given that the person selected is a smoker. In other words, 21.1% of smokers have lung disease. ∎

We conclude this section by introducing some terminology that is frequently used in conjunction with Bayes' rule. To that end, let us return to the previous example. From the information provided, we know that the probability is 0.070 that a randomly selected person has lung disease: $P(L_2) = 0.070$. This probability does not take into account whether the person selected is a smoker. It is therefore called a **prior probability** since it represents the probability that the person selected has lung disease *before* knowing whether the person selected is a smoker.

Now suppose that the person selected is found to be a smoker. On the basis of this additional information, we can revise the probability that the person selected has lung disease. This can be done by determining the conditional probability that the person selected has lung disease, given that the person selected is a smoker: $P(L_2 \mid S)$. We found, in Example 4.35, that $P(L_2 \mid S) = 0.211$. This revised (conditional) probability is called a **posterior probability** since it represents the probability that the person selected has lung disease *after* knowing that the person selected is a smoker.

Exercises 4.7

__ **4.110** Kress, Inc., markets two types of evaporative coolers, metal and fiberglass. The company has four sales districts. In the first two columns of the following table, we give the percentage distribution of total sales by district. The third column shows the percentage of sales in each district that are metal coolers.

District	Percentage of total sales	Percent metal
I	45	35
II	26	30
III	18	60
IV	11	22
	100	

We see, for instance, that 45% of all sales occur in District I and that 35% of the coolers sold in District I are metal coolers.

a) What percentage of sales are metal coolers?
b) What percentage of metal-cooler sales occur in District II?

__ **4.111** The first two columns of the following table provide a percentage distribution for the religious affiliation of the voters in a large city. In the third column, we show the percentage of Democrats in each religious group of voters.

Religion	Percentage of voters	Percent Democrats
Catholic	28	53
Jewish	2	61
Protestant	57	42
Other	4	58
None	9	67
	100	

The table shows, for instance, that 28% of the voters in the city are Catholic and that 53% of the Catholic voters in the city are Democrats.

a) What percentage of the voters are Democrats?
b) What percentage of Democrats are Protestant?

__ **4.112** Textbook editors must estimate the sales of new books. The records of one major publishing company indicate the following: 10% of all books sell more than projected, 30% sell at (close to) projected, and 60% sell less than projected. Of those books that sell more than projected, 70% are revised for a second edition, as are 50% of those that sell close to projected and 20% of those that sell less than projected.

a) What percentage of books published by this publishing company go to a second edition?
b) What percentage of books published by this publishing company that go to a second edition sold less than projected in their first edition?

__ **4.113** EDA Products, Inc., a manufacturer of customized metal fabrication items, produces forged tools for a major retailer. EDA currently uses three 50-ton forge presses to manufacture a particular model of slip joint pliers. While each of the presses accounts for one third of production, the presses produce defective units with varying percentages. In fact, recent quality assurance tests indicate that Press 1 produces 1.1% defective units, Press 2 produces 0.8% defective units, and Press 3 produces 1.6% defective units.

a) What percentage of slip joint pliers produced by EDA are defective?
b) Of those slip joint pliers that are defective, what percentage are produced by Press 2?

__ **4.114** The National Center for Health Statistics provides information on suicides by sex and method used. Data are published in *Vital Statistics of the United States*. In 1986, there were 30,904 suicides in the U.S. of which 24,226 were males and 6,678 were females. The following table gives the relative-frequency distributions for method used by males and females who committed suicide in 1986.

Method used	Relative frequency for males	Relative frequency for females
Poisoning	0.145	0.377
Hanging/strang.	0.155	0.127
Firearms	0.641	0.395
Other	0.059	0.101

Suppose a 1986 suicide report is selected at random.

a) Determine the probability that a firearm was used for the suicide.
b) Find the prior probability that the person who committed suicide was a female.

c) Find the posterior probability that the person who committed suicide was a female, given that a firearm was used.

d) Interpret your results in parts (a)–(c) in terms of percentages.

— **4.115** A 1989 Gallup poll asked 1005 adults and 500 teen-agers the question "What is the nation's top problem?". The pie charts shown below were used to summarize the results of the survey.

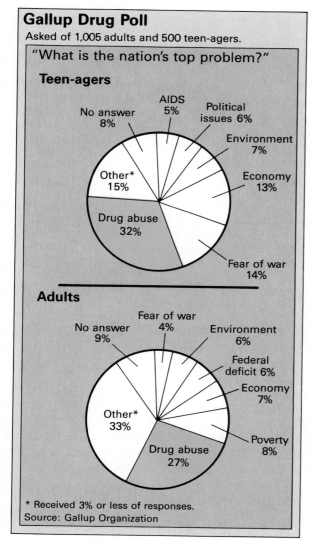

Gallup Drug Poll

Asked of 1,005 adults and 500 teen-agers.

"What is the nation's top problem?"

Teen-agers

AIDS 5%
Political issues 6%
No answer 8%
Environment 7%
Other* 15%
Economy 13%
Drug abuse 32%
Fear of war 14%

Adults

Fear of war 4%
No answer 9%
Environment 6%
Federal deficit 6%
Economy 7%
Other* 33%
Poverty 8%
Drug abuse 27%

* Received 3% or less of responses.
Source: Gallup Organization

Suppose a person who participated in the survey is selected at random.

a) Determine the probability that the person selected said that drug abuse is the nation's top problem.

b) Find the prior probability that the person selected is a teen-ager.

c) Find the posterior probability that the person selected is a teen-ager, given that the person selected said that drug abuse is the nation's top problem.

d) Interpret your results in parts (a)–(c) in terms of percentages.

— **4.116** At a grocery store, eggs come in cartons of a dozen eggs per carton. Experience indicates that 78.5% of the cartons have no broken eggs, 19.2% have one broken egg, 2.2% have two broken eggs, and 0.1% have three broken eggs (the percentage of cartons with four or more broken eggs is negligible). One egg is selected at random from a carton and is found to be broken. What is the probability that this egg is the only broken one in the carton?

— **4.117** Medical tests are frequently used to decide whether a person has a particular disease. The **sensitivity** of a test is defined to be the probability that the test is positive, given that the person has the disease; and the **specificity** of a test is defined to be the probability that the test is negative, given that the person does not have the disease. A test for a certain disease has been used for many years. Experience with the test indicates that its sensitivity is 0.934 and that its specificity is 0.968. Furthermore, it is known that about one in every 500 people has the disease.

a) Interpret the sensitivity and specificity of this test in terms of percentages.

b) Suppose a person is selected at random from the population and is given the test. Further suppose that the test is positive. What is the probability that the person actually has the disease?

c) Interpret your result from part (b) in terms of percentages.

≡ **4.118** Suppose that events $A_1, A_2, \ldots, A_k$ are mutually exclusive and exhaustive. Further suppose that those events are equally likely to occur. Show that for any event B,

$$P(B) = \frac{1}{k}[P(B \mid A_1) + P(B \mid A_2) + \cdots + P(B \mid A_k)]$$

(Hint: Use the rule of total probability.)

≡ **4.119** Suppose that events $A_1, A_2, \ldots, A_k$ are mutually exclusive and exhaustive. Further suppose that those events are equally likely to occur. Show that for any event B,

$$P(A_i \mid B) = \frac{P(B \mid A_i)}{P(B \mid A_1) + P(B \mid A_2) + \cdots + P(B \mid A_k)}$$

for $i = 1, 2, \ldots, k$. (*Hint:* Use Bayes' rule and the result of Exercise 4.118.)

≡ **4.120**   Provide a detailed proof for the rule of total probability, Formula 4.8 on page 206. (*Hint:* Refer to the argument used in Example 4.33.)

≡ **4.121**   Provide a detailed proof for Bayes' rule, Formula 4.9 on page 208.

4.8 Counting rules (Optional)

We often find it necessary to determine the number of ways that something can happen; for example, the number of possible outcomes for an experiment or the number of ways that a certain task can be performed. Sometimes we can simply list the possibilities and then count them. However, in most cases the number of possibilities is so large that a direct listing of the possibilities is impractical.

Thus, we need to develop techniques for determining the number of possibilities that do not rely on a direct listing. Such techniques are usually referred to as **counting rules.** In this section, we will discuss some of the most important counting rules.

THE FUNDAMENTAL COUNTING RULE

There is one counting rule that is basic to all the counting techniques that we will discuss. We will call this counting rule the **fundamental counting rule.**† To introduce the fundamental counting rule, we consider Example 4.36.

EXAMPLE 4.36   *Introduces the fundamental counting rule*

A woman looking for an apartment in a large apartment complex finds herself with the following choices: Level one, level two, or level three; and northern, southern, eastern, or western exposure. Each exposure is available on each level. How many possibilities are there for the selection of both level and exposure?

SOLUTION   We will first use a tree diagram to systematically obtain a direct listing of the possibilities. See Figure 4.30 at the top of the next page. Each branch of the tree corresponds to one possibility for the selection of a level and exposure. For instance, the first branch of the tree corresponds to selecting a northern exposure on level one. The number of possibilities can be obtained by counting the number of branches at the end of the tree. Thus we see that the woman has a total of 12 level-exposure possibilities.

Although the tree-diagram approach for determining the number of possibilities is a direct listing, it provides us with a clue for obtaining the number of possibilities without resorting to a direct listing. Specifically, there are three possibilities

---

† Several other names are often used for this rule; e.g., the **basic principle of counting** and the **multiplication rule.**

FIGURE 4.30
Tree diagram for
level-exposure
choices

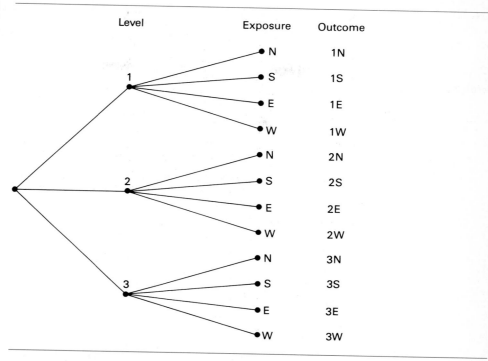

for level, indicated by the three branches emanating from the starting point of the tree; and to each possibility for level, there correspond four possibilities for exposure, indicated by the four branches emanating from the end of each level branch. Consequently, there are

$$\underbrace{4 + 4 + 4}_{\text{3 times}} = \mathbf{3 \cdot 4} = 12$$

possibilities altogether. So we see that the total number of possibilities can be obtained by *multiplying* the number of possibilities for the level by the number of possibilities for the exposure. ∎

KEY FACT 4.3　The fundamental counting rule

Suppose that two actions (choices, experiments) are to be performed in a definite order. Further suppose that there are m_1 possibilities for the first action and that corresponding to each of these possibilities there are m_2 possibilities for the second action. Then there are

$$m_1 \cdot m_2$$

possibilities altogether for the two actions.

More generally, suppose that r actions (choices, experiments) are to be performed in a definite order. Further suppose that there are m_1 possibilities for the first action; and that corresponding to each of these possibilities, there are m_2 possibilities for the second action; and that corresponding to each of these possibilities, there are m_3 possibilities for the third action; and so on. Then there are

$$m_1 \cdot m_2 \cdots m_r$$

possibilities altogether for the r actions.

In Example 4.36, there are $r = 2$ actions, selecting a floor level and selecting an exposure. Since there are three possibilities for the level, $m_1 = 3$; and since corresponding to each level there are four possibilities for the exposure, $m_2 = 4$. Therefore, by the fundamental counting rule, the total number of possibilities for the selection of both level and exposure is

$$m_1 \cdot m_2 = 3 \cdot 4 = 12$$

as we discovered in Example 4.36.

For the level-exposure problem it is not too difficult to determine the number of possibilities by a direct listing (see the tree diagram in Figure 4.30). This is because the number of possibilities is quite small. Nonetheless, it is still much easier to determine the number of possibilities by applying the fundamental counting rule. Moreover, when dealing with problems where the number of possibilities is large, a direct listing will be prohibitive and the fundamental counting rule will be the only reasonable way to determine the number of possibilities.

EXAMPLE 4.37   **Illustrates Key Fact 4.3**

The license plates of a state consist of three letters followed by three digits (whole numbers between 0 and 9).
a) How many different license plates are possible?
b) How many possibilities are there for license plates in which no letter or digit is repeated?

SOLUTION   For both parts (a) and (b), we will apply the fundamental counting rule with six actions; that is, $r = 6$.
a) There are 26 possibilities for the first letter, 26 for the second letter, 26 for the third letter; and there are 10 possibilities for the first digit, 10 for the second digit, 10 for the third digit. Hence, by the fundamental counting rule, there are

$$m_1 \cdot m_2 \cdot m_3 \cdot m_4 \cdot m_5 \cdot m_6 = 26 \cdot 26 \cdot 26 \cdot 10 \cdot 10 \cdot 10 = \boxed{17{,}576{,}000}$$

possibilities altogether for different license plates. Note that it would not be practical to obtain the number of possibilities by a direct listing—the tree diagram would have 17,576,000 branches.

b) For this part, there are again 26 possibilities for the first letter. But to each possibility for the first letter, there correspond only 25 possibilities for the second letter (since the second letter cannot be the same as the first). And to each possibility for the first two letters, there correspond 24 possibilities for the third letter (since the third letter cannot be the same as either the first or the second). Similarly, there are 10 possibilities for the first digit, nine for the second digit, and eight for the third digit. So, by the fundamental counting rule, there are

$$m_1 \cdot m_2 \cdot m_3 \cdot m_4 \cdot m_5 \cdot m_6 = 26 \cdot 25 \cdot 24 \cdot 10 \cdot 9 \cdot 8 = \boxed{11{,}232{,}000}$$

possibilities for license plates in which no letter or digit is repeated. ∎

FACTORIALS

Before we continue our presentation of counting rules, we need to discuss **factorials.** Factorials are used extensively in mathematics and its applications.

DEFINITION 4.10 Factorials

Let k be a positive integer (whole number). Then the product of the first k positive integers is called k *factorial* and is denoted by the symbol $k!$. Thus,

$$k! = k(k-1) \cdots 2 \cdot 1$$

We also define $0! = 1$.

EXAMPLE 4.38 *Illustrates Definition 4.10*

Determine $3!$, $4!$, and $5!$.

SOLUTION Applying Definition 4.10, we get

$$3! = 3 \cdot 2 \cdot 1 = \boxed{6}$$
$$4! = 4 \cdot 3 \cdot 2 \cdot 1 = \boxed{24}$$
$$5! = 5 \cdot 4 \cdot 3 \cdot 2 \cdot 1 = \boxed{120}$$

∎

Note, for example, that $6! = 6 \cdot 5!$, $6! = 6 \cdot 5 \cdot 4!$, $6! = 6 \cdot 5 \cdot 4 \cdot 3!$, and so on. In general, if $j \leq k$, then $k! = k(k-1) \cdots (k-j+1)(k-j)!$.

PERMUTATIONS

We now introduce the concept of permutations. A **permutation** of r objects from a collection of m objects is any *ordered* arrangement of r of the m objects. The total number of permutations of r objects that can be formed from a collection of m objects is denoted by $(m)_r$ or $_mP_r$. Let's look at a simple example.

EXAMPLE 4.39   *Introduces permutations*

Consider the collection of objects consisting of the first five letters of the English alphabet: *a, b, c, d, e.*

a) List all possible permutations of three letters from this collection of five letters.

b) Use part (a) to determine the total number of permutations of three letters that can be formed from the collection of five letters; that is, determine $(5)_3$.

c) Use the fundamental counting rule to determine the total number of permutations of three letters that can be formed from the collection of five letters; that is, determine $(5)_3$ using the fundamental counting rule.

SOLUTION   a) For this part we need to list all ordered arrangements of three letters from the first five letters of the English alphabet. This is done in Table 4.19.

TABLE 4.19
Possible
permutations

abc	*abd*	*abe*	*acd*	*ace*	*ade*	*bcd*	*bce*	*bde*	*cde*
acb	*adb*	*aeb*	*adc*	*aec*	*aed*	*bdc*	*bec*	*bed*	*ced*
bac	*bad*	*bae*	*cad*	*cae*	*dae*	*cbd*	*cbe*	*dbe*	*dce*
bca	*bda*	*bea*	*cda*	*cea*	*dea*	*cdb*	*ceb*	*deb*	*dec*
cab	*dab*	*eab*	*dac*	*eac*	*ead*	*dbc*	*ebc*	*ebd*	*ecd*
cba	*dba*	*eba*	*dca*	*eca*	*eda*	*dcb*	*ecb*	*edb*	*edc*

b) From Table 4.19 we see that there are 60 possible permutations of three letters from the collection of five letters. In other words, $(5)_3 = 60$.

c) Here we want to use the fundamental counting rule to determine the total number of permutations of three letters from the collection of five letters. We proceed as follows. There are five possibilities for the first letter, then four possibilities for the second letter, and then three possibilities for the third letter. Hence, by the fundamental counting rule, there are

$$m_1 \cdot m_2 \cdot m_3 = 5 \cdot 4 \cdot 3 = 60$$

possibilities altogether. So, again, we see that $(5)_3 = 60$.   ∎

We can make two relevant observations from the previous example. First, it will be tedious or impractical to list all possible permutations under consideration. Second, it is not necessary to list the possible permutations in order to determine how many there are—we can use the fundamental counting rule to count the number of possible permutations.

By studying part (c) of Example 4.39, we can see how to use the fundamental counting rule to obtain a general formula for $(m)_r$, the number of possible permutations of r objects from a collection of m objects. Indeed, there are m possibilities for the first object, $m - 1$ for the second object, $m - 2$ for the third object, and so on. Thus, we see that $(m)_r = m(m-1) \cdots (m-r+1)$. Multiplying numerator and denominator of this last expression by $(m-r)!$, we get the equivalent expression $(m)_r = m!/(m-r)!$. We summarize this discussion in Formula 4.10.

FORMULA 4.10 Permutations rule

The number of possible permutations of r objects from a collection of m objects is given by the formula

$$(m)_r = \frac{m!}{(m-r)!}$$

EXAMPLE 4.40 *Illustrates Formula 4.10*

In an *exacta* wager at the race track, the bettor picks the two horses that he or she thinks will finish first and second, respectively. Suppose that a race has 12 entrants. Determine the number of different exacta wagers that are possible.

SOLUTION Selecting two horses from the 12 horses for an exacta wager is equivalent to specifying a permutation of two objects from a collection of 12 objects; the first object is the horse selected to come in first place and the second object is the horse selected to come in second place. Thus, the number of different exacta wagers is $(12)_2$—the number of possible permutations of two objects from a collection of 12 objects. Applying the permutations rule (Formula 4.10) with $m = 12$ and $r = 2$, we obtain

$$(12)_2 = \frac{12!}{(12-2)!} = \frac{12!}{10!} = \frac{12 \cdot 11 \cdot \cancel{10!}}{\cancel{10!}} = 12 \cdot 11 = \boxed{132}$$

Hence, there are 132 different exacta wagers possible in a 12 horse race. ∎

EXAMPLE 4.41 *Illustrates Formula 4.10*

A student has 10 books to arrange on a shelf of a bookcase. How many ways is it possible to arrange the 10 books?

SOLUTION Any particular arrangement of the 10 books on the shelf is a permutation of 10 objects from a collection of 10 objects. Thus, for this problem we need to determine the number of possible permutations of 10 objects from a collection of 10 objects, more commonly expressed as the number of possible permutations of 10 objects among themselves. Applying the permutations rule, Formula 4.10, we get

$$(10)_{10} = \frac{10!}{(10-10)!} = \frac{10!}{0!} = \frac{10!}{1} = 10! = \boxed{3{,}628{,}800}$$

Thus, there are 3,628,800 ways to arrange the 10 books on the shelf. (It doesn't seem possible that there are this many possibilities, but there are!) ∎

Generalizing the previous example, let us find the number of possible permutations of m objects among themselves; that is, the number of possible permutations of m objects from a collection of m objects. Using Formula 4.10, we conclude that

$$(m)_m = \frac{m!}{(m-m)!} = \frac{m!}{0!} = \frac{m!}{1} = m!$$

Consequently, we have the following formula as a special case of Formula 4.10.

FORMULA 4.11 Special permutations rule

The number of possible permutations of m objects among themselves is

$$(m)_m = m!$$

COMBINATIONS

Next we discuss combinations. A **combination** of r objects from a collection of m objects is any *unordered* arrangement of r of the m objects; in other words, any subset of r objects from the collection of m objects. Note that in combinations order does not matter, whereas in permutations order does matter.

The total number of combinations or r objects that can be formed from a collection of m objects is denoted by $\binom{m}{r}$ or $_mC_r$. Let us return to the situation of Example 4.39.

EXAMPLE 4.42   *Introduces combinations*

Consider the collection of objects consisting of the first five letters of the English alphabet: *a, b, c, d, e*.
a) List all possible combinations of three letters from this collection of five letters.
b) Use part (a) to determine the total number of combinations of three letters that can be formed from the collection of five letters; that is, determine $\binom{5}{3}$.

SOLUTION   a) For this part we need to list all unordered arrangements (subsets) of three letters from the first five letters in the English alphabet. This is done in Table 4.20.

TABLE 4.20
Possible
combinations

$\{a,b,c\}$ $\{a,b,d\}$ $\{a,b,e\}$ $\{a,c,d\}$ $\{a,c,e\}$ $\{a,d,e\}$ $\{b,c,d\}$ $\{b,c,e\}$ $\{b,d,e\}$ $\{c,d,e\}$

b) From Table 4.20 we see that there are 10 possible combinations of three letters from the collection of five letters. In other words, $\binom{5}{3} = 10$. ∎

In the previous example, we obtained the number of possible combinations by first listing the possible combinations and then counting the number on the list. In general, it will not be practical to obtain the number of possible combinations by resorting to a direct listing. Fortunately, we can derive a simple formula for determining the number of possible combinations. To see how this is done, let us return once more to the English-letters illustration.

Look at the first combination, $\{a,b,c\}$, in Table 4.20. By the special permutations rule, Formula 4.11, there are $3! = 6$ different permutations of these three letters among themselves. They are: *abc, acb, bac, bca, cab,* and *cba.* Referring now to Table 4.19 on page 216, we see that these are the six permutations displayed in the first column of that table.

Similarly, there are $3! = 6$ different permutations of the three letters in the second combination, $\{a, b, d\}$, in Table 4.20; and these six permutations are the ones displayed in the second column of Table 4.19. The same comments apply to the other eight combinations in Table 4.20.

Therefore, we see that to each combination of three letters from the collection of five letters, there correspond $3! = 6$ permutations of three letters from the collection of five letters. Moreover, any such permutation is accounted for in this way. Consequently, there must be $3!$ times as many permutations as combinations. So, the number of possible combinations of three letters from the collection of five letters must equal $1/3!$ times the number of possible permutations:

$$\binom{5}{3} = \frac{1}{3!} \cdot (5)_3 = \frac{1}{3!} \cdot \frac{5!}{(5-3)!} = \frac{5!}{3!\,(5-3)!} = \frac{5 \cdot 4 \cdot 3!}{3!\,2!} = \frac{5 \cdot 4}{2} = 10$$

This is what we obtained in Example 4.42 by a direct listing.

The same type of argument that we just gave holds in general: Consider a collection of m objects. Then to each combination of r objects from the collection of m objects, there correspond $r!$ permutations (the $r!$ permutations of the r objects among themselves). This implies that there must be $r!$ times as many permutations of r objects from the m objects as there are combinations of r objects from the m objects. Therefore,

$$\binom{m}{r} = \frac{1}{r!} \cdot (m)_r = \frac{1}{r!} \cdot \frac{m!}{(m-r)!} = \frac{m!}{r!\,(m-r)!}$$

We summarize this result in Formula 4.12.

FORMULA 4.12   Combinations rule

The number of possible combinations of r objects from a collection of m objects is given by the formula

$$\binom{m}{r} = \frac{m!}{r!\,(m-r)!}$$

EXAMPLE 4.43   *Illustrates Formula 4.12*

In order to recruit new members, a compact-disc club advertises a special introductory offer. A new member agrees to buy one compact disc at regular club prices and then receives free any four compact discs of his or her choice from a collection of 69 compact discs. How many possibilities does a new member have for the selection of the four free compact discs?

SOLUTION   Any particular selection of four compact discs from 69 compact disks is just a combination of four objects from a collection of 69 objects. Therefore, by the combinations rule, Formula 4.12, the number of possible selections equals

$$\binom{69}{4} = \frac{69!}{4!\,(69-4)!} = \frac{69!}{4!\,65!} = \frac{69 \cdot 68 \cdot 67 \cdot 66 \cdot 65!}{4!\,65!} = \frac{69 \cdot 68 \cdot 67 \cdot 66}{24} = 864{,}501$$

There are 864,501 ways in which a new member can choose four compact disks from the collection of 69 compact disks. ■

EXAMPLE 4.44   *Illustrates Formula 4.12*

An economics professor is teaching a junior-level course using a new teaching method. The course has 42 students. The professor wants to conduct in-depth interviews with the students to get feedback on the new teaching method. There are too many students in the class to interview all of them, so she decides to interview a sample of five of the students. How many different samples are there?

SOLUTION   A sample of five students from the class of 42 students is really just a combination of five objects from a collection of 42 objects. Consequently, by the combinations rule, Formula 4.12, the number of different samples is

$$\binom{42}{5} = \frac{42!}{5!\,(42-5)!} = \frac{42!}{5!\,37!} = \boxed{850{,}668}$$

There are 850,668 different samples of five students that can be obtained from the population of 42 students in the class. ■

The preceding example shows how to determine the number of possible samples of a specified size from a finite population. This result is so important that we record it as the following formula.

FORMULA 4.13   **Number of possible samples**

The number of possible samples of size n from a population of size N is

$$\binom{N}{n} = \frac{N!}{n!\,(N-n)!}$$

SOME APPLICATIONS TO PROBABILITY

Suppose that an experiment has N *equally-likely* possible outcomes. Then, according to the f/N-rule, the probability that a specified event occurs equals the number of ways, f, that the event can occur, divided by the total number, N, of possible outcomes.

In the probability problems we have done up to this point, it was always easy to determine f and N. However, in many applications, that is not the case. Frequently we must use counting rules to obtain the number of possible outcomes, N, of the experiment and the number of ways, f, that the specified event can occur. The following two examples provide some typical ways in which counting rules are applied to solve probability problems.

EXAMPLE 4.45 *Illustrates the application of counting rules to probability*

A drug is known to have a 50% effectiveness rate in curing a certain disease. Suppose seven people with the disease are given the drug. What is the probability that exactly four of the seven people are cured?

SOLUTION Since the drug has a 50% effectiveness rate, the possible outcomes of the experiment (cure-noncure results) are equally likely. Thus, we can apply the f/N-rule to obtain the probability in question.

First let's determine the number of possible outcomes, N. There are two possibilities for the first person (cured or not cured), two possibilities for the second person, and so on. So, by the fundamental counting rule, there are

$$\underbrace{2 \cdot 2 \cdots 2}_{\text{7 times}} = 2^7 = 128$$

possibilities altogether. Thus, $N = 128$.

Next we need to determine the number of ways, f, that the specified event can occur; that is, we must determine how many outcomes there are in which exactly four of the seven people are cured. But this is just the number of ways that we can select four people (the ones who will be cured) from seven people—the number of possible combinations of four objects from seven objects. Hence, by the combinations rule, Formula 4.12 on page 219, the number of possibilities equals

$$\binom{7}{4} = \frac{7!}{4! \, (7-4)!} = \frac{7!}{4! \, 3!} = 35$$

Thus, $f = 35$.

Therefore, by the f/N-rule, the probability that exactly four of the seven people are cured is

$$\frac{f}{N} = \frac{35}{128} = \boxed{0.273}$$

∎

EXAMPLE 4.46 *Illustrates the application of counting rules to probability*

The quality-assurance engineer of a television company inspects TVs in lots of 100. He selects five of the 100 TVs at random and inspects them thoroughly. Assuming that six of the 100 TVs in the current lot are actually defective, find the probability that exactly two of the five TVs selected by the engineer are defective.

SOLUTION Since the engineer makes his selection at random, each of the possible outcomes is equally likely. So, we can apply the f/N-rule to obtain the required probability.

First we determine the number of possible outcomes, N, for the experiment. This is the number of ways that five TVs can be selected from the 100 TVs—the number of possible combinations of five objects from a collection of 100 objects. Applying the combinations rule, we obtain

$$\binom{100}{5} = \frac{100!}{5!\,(100-5)!} = \frac{100!}{5!\,95!} = 75{,}287{,}520$$

Thus, $N = 75{,}287{,}520$.

Next we determine the number of ways, f, that the specified event can occur; that is, the number of outcomes in which exactly two of the five TVs selected are defective. To accomplish this, it is helpful to think of the 100 TVs as partitioned into two groups; namely, the defective TVs and the non-defective TVs. See the first part of Figure 4.31.

FIGURE 4.31
Calculating the
number of outcomes
in which exactly
two of the five TVs
selected are defective

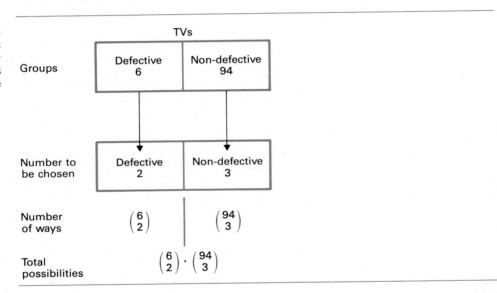

There are six TVs in the first group of which two are to be selected. This can be done in

$$\binom{6}{2} = \frac{6!}{2!\,(6-2)!} = \frac{6!}{2!\,4!} = 15$$

ways. There are 94 TVs in the second group of which three are to be selected. This can be done in

$$\binom{94}{3} = \frac{94!}{3!\,(94-3)!} = \frac{94!}{3!\,91!} = 134{,}044$$

ways. Consequently, by the fundamental counting rule, there are a total of

$$\binom{6}{2} \cdot \binom{94}{3} = 15 \cdot 134{,}044 = 2{,}010{,}660$$

outcomes in which exactly two of the five TVs selected are defective. Thus, $f = 2{,}010{,}660$. Figure 4.31 summarizes the calculations done in this paragraph.

Therefore, by the f/N-rule, the probability that exactly two of the five TVs selected are defective equals

$$\frac{f}{N} = \frac{2{,}010{,}660}{75{,}287{,}520} = \boxed{0.027}$$

There is a 2.7% chance that exactly two of the five TVs selected by the engineer will be defective.   ∎

Exercises 4.8

___ **4.122**   A zip code consists of five digits.
a) How many possible zip codes are there?
b) How many possible zip codes are there in which no digit appears more than once?

___ **4.123**   Telephone numbers in the United States consist of a three-digit area code followed by a seven-digit local number. Suppose that neither the first digit of an area code nor the first digit of a local number can be a zero but that all other choices are acceptable.
a) How many different area codes are possible?
b) For a given area code, how many local telephone numbers are possible?
c) How many telephone numbers are possible?

___ **4.124**   Nowadays, computerized testing systems are used extensively by professors. A statistics professor needs to construct a five-question quiz, one question for each of five topics. The computerized testing system that she uses provides eight choices for the question on the first topic, 10 choices for the question on the second topic, seven choices for the question on the third topic, eight choices for the question on the fourth topic, and six choices for the question on the fifth topic. How many possibilities are there for the five-question quiz?

___ **4.125**   *Scientific Computing & Automation* magazine offers free subscriptions to the scientific community. The magazine does ask, however, that a person answer six questions: primary title, type of facility, area of work, brand of computer used, type of operating system in use, and type of instruments in use. There are six choices given for the first question, eight for the second, five for the third, 19 for the fourth, 16

for the fifth, and 14 for the sixth. How many possibilities are there for answering all six questions?

___ **4.126**   Determine the value of each of the following quantities:
a) $(7)_3$ b) $(5)_2$ c) $(8)_4$ d) $(6)_0$ e) $(9)_9$

___ **4.127**   Determine the value of each of the following quantities:
a) $(4)_3$ b) $(15)_4$ c) $(6)_2$ d) $(10)_0$ e) $(8)_8$

___ **4.128**   At a movie festival, a team of judges is to pick the first, second, and third place winners from the 18 films entered. How many possibilities are there?

___ **4.129**   Investment firms usually have a large selection of mutual funds from which an investor can choose. One such firm has 30 mutual funds. Suppose that you plan to invest in four of these mutual funds, one during each quarter of next year. In how many different ways can you make these four investments?

___ **4.130**   The regional sales manager of a clothing company needs to assign seven salespeople to seven different territories. How many possibilities are there for the assignments?

___ **4.131**   A psychologist plans to conduct an ESP (extrasensory perception) experiment. For part of the experiment, the psychologist takes 10 cards, numbered 1–10, and shuffles them. Then she looks at the cards one at a time. While she looks at each card, the subject writes down the number he thinks is on the card.
a) How many possibilities are there for the order in which the subject writes down the numbers?
b) If the subject has no ESP and is just guessing each time, what is the probability that he writes down the numbers in the correct order (i.e., in the order that the cards are actually arranged)?

___ **4.132** Determine the value of each of the following quantities:

a) $\binom{7}{3}$ b) $\binom{5}{2}$ c) $\binom{8}{4}$ d) $\binom{6}{0}$ e) $\binom{9}{9}$

___ **4.133** Determine the value of each of the following quantities:

a) $\binom{4}{3}$ b) $\binom{15}{4}$ c) $\binom{6}{2}$ d) $\binom{10}{0}$ e) $\binom{8}{8}$

___ **4.134** A poker hand consists of five cards dealt from an ordinary deck of 52 playing cards.
a) How many possible poker hands are there?
b) How many different hands are there consisting of three kings and two queens?
c) The hand in part (b) is an example of a full house, three cards of one denomination and two of another. How many different full houses are there?
d) Obtain the probability of being dealt a full house.

___ **4.135** The U.S. Senate consists of 100 senators, two from each state. A committee consisting of five senators is to be formed.
a) How many different committees are possible?
b) How many are possible if no state may have more than one senator on the committee?
c) If the committee is selected at random from all 100 senators, what is the probability that no state will have both of its senators on the committee?

___ **4.136** How many samples of size $n = 5$ are possible from a population of size $N = 70$?

___ **4.137** How many samples of size six are possible from a population with 45 members?

___ **4.138** Suppose that you have a key ring with eight keys on it, one of which is your house key. Further suppose that you get home after dark and can't see the keys on the key ring. You randomly try one key at a time, being careful not to mix the keys that you have already tried with the ones that you haven't. What is the probability that you get the right key
a) on the first try?
b) on the last try?
c) on or before the fifth try?

___ **4.139** Refer to Example 4.46 on page 221. Determine the probability that
a) exactly one of the TVs selected is defective.
b) at most one of the TVs selected is defective.
c) at least one of the TVs selected is defective.

___ **4.140** *The Birthday Problem.* A biology class has 38 students. Find the probability that at least two students in the class have the same birthday. For simplicity, assume that there are always 365 days in a year and that birth rates are constant throughout the year. (*Hint:* First determine the probability that no two students have the same birthday and then apply the complementation rule.)

___ **4.141** The state lottery in Arizona is called *The Pick*. A pick ticket consists of six numbers from the numbers 1–42. The six winning numbers are selected at random from the numbers 1–42. To win a prize, a pick ticket must contain four or more of the winning numbers. Suppose that you buy one pick ticket. Determine the probability that
a) you win the jackpot; that is, your six numbers are the same as the six winning numbers.
b) your ticket contains exactly four winning numbers.
c) you don't win a prize.

___ **4.142** A student takes a true-false test having 15 questions. Assuming that the student guesses at each question, find the probability that
a) the student gets at least one question correct.
b) the student gets a 60% or better on the exam.

= **4.143** According to the Center for Political Studies at the University of Michigan, Ann Arbor, approximately 50% of U.S. adults are Democrats. Suppose that 10 U.S. adults are selected at random. Determine the approximate probability that
a) exactly five are Democrats.
b) at least eight are Democrats.

= **4.144** Refer to the *Birthday Problem* considered in Exercise 4.140, but now assume that the class consists of N students.
a) Determine the probability that at least two of the students have the same birthday.
b) This part presumes that you have access to a computer or a programmable calculator. Use part (a) to construct a table giving the probability that at least two of the students in the class have the same birthday, for $N = 2, 3, \ldots, 70$.

= **4.145** Suppose that a random sample of size n is to be taken without replacement from a population of size N. (By a *random sample*, we mean that each sample of size n is equally likely to be the one selected.)
a) Determine the probability that any particular sample of size n is the one selected.
b) Determine the probability that any specified member of the population is included in the sample.

Chapter review

FORMULAS

Classical probability (f/N-rule for equally likely outcomes), 143

$$P(E) = \frac{f}{N}$$

(f = number of ways E can occur, N = total number of possible outcomes)

Special addition rule, 164

$$P(A \text{ or } B \text{ or } C \text{ or } \cdots) = P(A) + P(B) + P(C) + \cdots$$

(A, B, C, ... mutually exclusive)

Complementation rule, 167

$$P(E) = 1 - P(\text{not } E)$$

General addition rule, 170

$$P(A \text{ or } B) = P(A) + P(B) - P(A \& B)$$

Conditional probability rule, 186

$$P(B \mid A) = \frac{P(A \& B)}{P(A)}$$

General multiplication rule, 192

$$P(A \,\&\, B) = P(A)P(B\,|\,A)$$

Special multiplication rule, 198

$$P(A \,\&\, B \,\&\, C \,\&\, \cdots) = P(A)P(B)P(C)\cdots$$

(A, B, C, ... independent)

Rule of total probability,* 206

$$P(B) = P(A_1)P(B\,|\,A_1) + P(A_2)P(B\,|\,A_2) + \cdots + P(A_k)P(B\,|\,A_k)$$

(A_1, A_2, ..., A_k mutually exclusive and exhaustive)

Bayes' rule,* 208

$$P(A_i\,|\,B) = \frac{P(A_i)P(B\,|\,A_i)}{P(A_1)P(B\,|\,A_1) + P(A_2)P(B\,|\,A_2) + \cdots + P(A_k)P(B\,|\,A_k)}$$

(A_1, A_2, ..., A_k mutually exclusive and exhaustive)

Factorial,* 215

$$k! = k(k-1)\cdots 2 \cdot 1$$

(k a positive integer)

Permutations rule,* 217

$$(m)_r = \frac{m!}{(m-r)!}$$

Special permutations rule,* 218

$$(m)_m = m!$$

Combinations rule,* 219

$$\binom{m}{r} = \frac{m!}{r!\,(m-r)!}$$

Number of possible samples,* 220

$$\binom{N}{n} = \frac{N!}{n!\,(N-n)!}$$

(N = population size, n = sample size)

YOU SHOULD
BE ABLE TO

1. use and understand the preceding formulas.
2. find and describe (not E), ($A \,\&\, B$), and (A or B).
3. determine whether two or more events are mutually exclusive.
4. read and interpret contingency tables.

5. construct a joint probability distribution from a contingency table.
6. compute conditional probabilities both directly and by using the conditional probability rule.
7. determine whether two events are independent.
8. determine whether two or more events are exhaustive.*
9. state and apply Bayes' rule.*
10. state and apply the fundamental counting rule.*
11. apply counting rules to solve probability problems when appropriate.*

REVIEW TEST

1. The first two columns of Table 4.21 give a frequency distribution of the adjusted gross incomes from 1982 federal individual income tax returns. Frequencies are in thousands of returns. [SOURCE: U.S. Internal Revenue Service, *Statistics of Income, Individual Income Tax Returns.*]

TABLE 4.21

Adjusted gross income	Number of returns	Event	Probability
Under $10,000	34,081	A	
$10,000–$19,999	24,842	B	
$20,000–$29,999	16,425	C	
$30,000–$39,999	9,863	D	
$40,000–$49,999	4,717	E	
$50,000–$99,999	3,759	F	
$100,000 & over	740	G	
	94,427		

Suppose that a 1982 federal individual income tax return is selected at random.
a) Find $P(A)$, the probability that the return selected shows an adjusted gross income under $10,000.
b) Determine the probability that the return selected shows an adjusted gross income between $30,000 and $99,999.
c) Compute the probability of each of the seven events in the third column of Table 4.21 and record those probabilities in the fourth column.

2. Refer to Problem 1. Suppose that a 1982 federal individual income tax return is selected at random. Let

H = event that the return shows an adjusted gross income between $20,000 and $99,999

I = event that the return shows an adjusted gross income of at most $49,999

J = event that the return shows an adjusted gross income of at most $99,999

K = event that the return shows an adjusted gross income of at least $50,000

Describe each of the following events in words and determine the number of outcomes (returns) that comprise each event.
a) (not J) b) (H & I)
c) (H or K) d) (H & K)

3. For the following groups of events from Problem 2, determine which are mutually exclusive.
a) The events H and I.
b) The events I and K.
c) The events H and (not J).
d) The events H, (not J), and K.

4. Refer to Problems 1 and 2.
a) Use the second column of Table 4.21 and the f/N-rule to compute the probability of each of the four events H, I, J, and K.
b) Express each of the events H, I, J, and K in terms of the mutually exclusive events A through G in the third column of Table 4.21.
c) Compute the probability of each of the four events H, I, J, and K using your results from part (b), the special addition rule, and the fourth column of Table 4.21, which you completed in Problem 1(c).

5. Consider the events (not J), (H & I), (H or K), and (H & K) discussed in Problem 2.
a) Determine the probability of each of these four events using the f/N-rule. [Refer to the second column of Table 4.21.]
b) Compute $P(J)$ using the complementation rule and your result for P(not J) from part (a).

c) In Problem 4 you found that $P(H) = 0.368$ and $P(K) = 0.048$; and in part (a) of this problem you found that $P(H \;\&\; K) = 0.040$. Using these probabilities and the general addition rule, determine $P(H \text{ or } K)$. Compare your result with the value for $P(H \text{ or } K)$ that you obtained in part (a).

6. The U.S. National Center for Education Statistics compiles information about school enrollment and publishes the data in *Digest of Education Statistics*. Table 4.22 provides a contingency table for enrollment in public and private schools by level. Frequencies are in thousands of students.

TABLE 4.22

Level	Type: Public T_1	Private T_2	Total
Elementary L_1	26,951	3,600	30,551
High school L_2	12,215	1,400	13,615
College L_3	9,612	2,562	12,174
Total	48,778	7,562	56,340

a) How many cells does this contingency table have?
b) How many students are in high school?
c) How many students attend public schools?
d) How many students attend private colleges?

7. Refer to Problem 6. Suppose a student is selected at random.
   a) Describe in words each of the following events: (i) L_3 (ii) T_1 (iii) $(T_1 \;\&\; L_3)$
   b) Find the probability of each event in part (a). Interpret your results in terms of percentages.
   c) Construct a joint probability distribution for Table 4.22.
   d) Compute $P(T_1 \text{ or } L_3)$
      (i) using Table 4.22 and the f/N-rule.
      (ii) using the general addition rule and your results from part (b).

8. Refer to Problem 6. Suppose a student is selected at random.
   a) Find $P(L_3 \mid T_1)$ directly by using Table 4.22 and the f/N-rule. Interpret the probability you obtain in terms of percentages.
   b) Find $P(L_3 \mid T_1)$ using the conditional probability rule and your results from Problem 7(b).

9. Refer to Problem 6. Suppose a student is selected at random.
   a) Using Table 4.22, find $P(T_2)$ and $P(T_2 \mid L_2)$.
   b) Are the events L_2 and T_2 independent? Why? Explain your results in terms of percentages.
   c) Are the events L_2 and T_2 mutually exclusive? Why?
   d) Is the event that a student is in elementary school independent of the event that a student attends public school? Explain.

10. During one year, the College of Public Programs at Arizona State University awarded the following number of masters degrees:

Type of degree	Frequency
Master of Arts	3
Master of Public Administration	28
Master of Science	19
	50

Suppose two students who received such masters degrees are selected at random without replacement. What is the probability that
   a) the first student selected received a master of arts and the second student selected received a master of science?
   b) both students selected received a master of public administration?
   c) Draw a tree diagram for this problem similar to Figure 4.27 on page 194.
   d) What is the probability that both students selected received the same degree?

11. A manufacturer of electric water heaters knows that 25% of the water heaters made by his company last more than 10 years. Suppose four water heaters made by this company are selected at random. Determine the probability that
   a) all four last more than 10 years.
   b) the first three selected last more than 10 years and the fourth does not.
   c) exactly three of the four selected last more than 10 years.

*12. A poll was taken in June of 1989 by *The New York Times* to gauge the sentiment of Americans on the changes in the role of women over the last 20 years. One question asked was: "Many women have better jobs and more opportunities than they did 20 years ago. Do you think women have had to give up too much in the process, or not?". The relevant data are shown in the chart at the right. Suppose a person who participated in the survey is selected at random. Find the probability that

   a) the person answered 'no' to the question, given that the person selected is a woman.

   b) the person answered 'no' to the question.

   c) the person is a woman.

   d) the person is a woman, given that the person selected answered 'no' to the question.

   e) Interpret your results in parts (a)–(d) in terms of percentages.

   f) Of the four probabilities in parts (a)–(d), which are prior and which are posterior?

*13. In Example 4.40 we considered *exacta* wagering in horse racing. Two other commonly made wagers are the *quinella* and the *trifecta*. In a quinella wager, the bettor picks the two horses that he or she believes will finish first and second, but not in a specified order. In a trifecta wager, the bettor picks the three horses that he or she thinks will finish first, second, and third in a specified order. For a 12-horse race,

   a) how many different quinella wagers are there?

   b) how many different trifecta wagers are there?

   c) Repeat parts (a) and (b) for an eight-horse race.

*14. A *bridge* hand consists of 13 cards dealt at random from an ordinary deck of 52 playing cards.

   a) How many possible bridge hands are there?

   b) Find the probability of being dealt a bridge hand that contains exactly two of the four aces.

   c) Find the probability of being dealt an 8-4-1 distribution; that is, eight cards of one suit, four of another, and one of another.

   d) Find the probability of being dealt a 5-5-2-1 distribution.

Second Thoughts About Success

Based on interviews with 1,025 women and 472 men conducted by telephone June 20-25.

Many women have better jobs and more opportunities than they did 20 years ago. Do you think women have had to give up too much in the process, or not?

Women

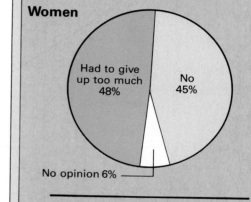

Had to give up too much 48%

No 45%

No opinion 6%

Men

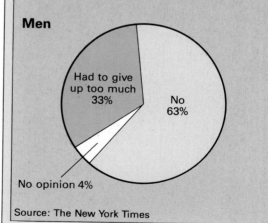

Had to give up too much 33%

No 63%

No opinion 4%

Source: The New York Times

CHAPTER 5

DISCRETE RANDOM VARIABLES

In Chapter 4, we began our investigation of probability. We continue that investigation in this chapter by studying *random variables*. A **random variable** is a numerical quantity whose value depends on chance. For instance, suppose a person is to be selected at random. Then the number of radios owned by the person chosen is an example of a random variable; as is the height of the person chosen. Both of these quantities are numerical and their values depend on chance; namely, on which person is selected.

There are two main types of random variables, discrete and continuous. A **discrete random variable** is a random variable whose possible values form a discrete data set, usually some collection of whole numbers. Thus, the first example of a random variable given in the previous paragraph is an example of a discrete random variable. A **continuous random variable** is a random variable whose possible values form a continuous data set, usually some interval of real numbers. So, the second example of a random variable given in the previous paragraph is an example of a continuous random variable.

As we will see, random variables play a key role in the design and implementation of statistical-inference procedures. In this chapter, we will study discrete random variables. Continuous random variables will be studied later.

CHAPTER OUTLINE

5.1 Discrete random variables; probability distributions Introduces discrete random variables, probability distributions, and some associated notation.

5.2 The mean and standard deviation of a discrete random variable Examines two important descriptive measures of a random variable.

5.3 Binomial coefficients; Bernoulli trials Discusses two concepts that are prerequisite to the study of the most widely used discrete probability distribution, the binomial distribution.

5.4 The binomial distribution Presents the binomial distribution and its probability formula.

5.5 The mean and standard deviation of a binomial random variable Introduces special formulas that can be used to compute the mean and standard deviation of a binomial random variable.

5.1   Discrete random variables; probability distributions

In this section we will introduce discrete random variables and their probability distributions. We begin with the following illustration.

EXAMPLE 5.1   *Introduces random variables*

Table 5.1 presents a grouped-data table for the number of cars owned by each of the families in a small city. The table shows, for instance, that 1796 out of the 6487 families, or 27.7%, own three cars.

TABLE 5.1
Grouped-data
table for number
of cars owned

Cars owned x	Frequency f	Relative frequency
0	27	0.004
1	1422	0.219
2	2865	0.442
3	1796	0.277
4	324	0.050
5	53	0.008
	6487	1.000

Since the "number of cars owned" varies from family to family, it is called a **variable.** Suppose now that a family is to be selected at random. Then the "number of cars owned" by the family selected is called a **random variable** because its value depends on chance; namely, on which family is selected.   ∎

DEFINITION 5.1   Random variable

A *random variable* is a numerical quantity whose value depends on chance.

The random variable in Example 5.1 is the "number of cars owned by a randomly selected family." The possible values of that random variable form a discrete data set; namely, the numbers 0, 1, 2, 3, 4, and 5 (see Table 5.1). Consequently, it is called a discrete random variable.

DEFINITION 5.2   Discrete random variable

A *discrete random variable* is a random variable whose possible values form a discrete data set.[†]

A discrete random variable usually involves a *count* of something, like the number of cars owned by a randomly selected family, the number of people waiting

---

[†] More precisely, a **discrete random variable** is a random variable whose possible values form a countable set (a *countable set* is a set that is finite or can be put in one-to-one correspondence with the positive integers).

for a haircut in a barber shop, or the number of households in a sample that own a color television set.

RANDOM-VARIABLE NOTATION

It is customary to denote random variables by letters such as x, y, and z. In doing so, we can develop useful notation for dealing with random variables. Let us return to the situation of Example 5.1.

In that example, we considered the random variable "number of cars owned by a randomly selected family." Suppose we let x denote that random variable. Then, for instance, we can represent the event that the family selected owns three cars by $\{x = 3\}$, read "x equals three."

The probability that the family selected owns three cars can then be expressed symbolically as $P(x = 3)$, read "the probability that x equals three." When there is no question about which random variable is under consideration, we will often write $P(3)$ instead of $P(x = 3)$.

We should emphasize a point that is frequently confused. The notation $\{x = 3\}$ refers to the *event* that the family selected owns three cars, while the notation $P(x = 3)$ refers to the *probability* of that event.

PROBABILITY DISTRIBUTION OF A RANDOM VARIABLE

We next discuss the *probability distribution of a random variable*. Let us return once more to Example 5.1 in order to introduce that concept. As before, let x denote the number of cars owned by a randomly selected family.

We want to determine the probability of each of the possible values of the random variable x. To find, for instance, $P(x = 3)$, the probability that the family selected owns three cars, we apply the f/N-rule. Referring to Table 5.1, we see that

$$P(x = 3) = \frac{f}{N} = \frac{1796}{6487} = 0.277$$

The remaining probabilities for the random variable x are obtained in the same way. Table 5.2 displays these probabilities.

TABLE 5.2
Probability distribution of the random variable x

Cars owned x	Probability $P(x)$
0	0.004
1	0.219
2	0.442
3	0.277
4	0.050
5	0.008
	1.000

Just as a table of frequencies is called a frequency distribution, a table of probabilities is called a **probability distribution.** Thus, Table 5.2 gives us the probability distribution of the random variable x; it tells us the distribution of probabilities for the various values of the random variable x.

Note that the probabilities for x, given in the second column of Table 5.2, add up to 1. This is always true (why?).

KEY FACT 5.1

The sum of the probabilities of any discrete random variable is always equal to 1. That is, if x is a discrete random variable, then

$$\Sigma P(x) = 1$$

Finally, observe that the probabilities for the random variable x, given in the second column of Table 5.2, are identical to the relative frequencies shown in the third column of Table 5.1. This is always true when dealing with finite populations.

KEY FACT 5.2   Probabilities and percentages

When dealing with a finite population, probabilities for a random variable are the same as percentages (relative frequencies).

EXAMPLE 5.2   *Illustrates random variable notation and probability distributions*

The U.S. National Center for Education Statistics compiles enrollment data on American public schools and publishes the results in *Digest of Education Statistics*. Table 5.3 displays a frequency distribution for the enrollment by grade in public elementary schools (0 = kindergarten, 1 = first grade, and so on). Frequencies are in thousands of students.

TABLE 5.3
Frequency
distribution for
enrollment by grade
in U.S. public
elementary schools

Grade y	Frequency f
0	2,700
1	2,951
2	2,786
3	2,812
4	2,926
5	3,131
6	3,180
7	3,184
8	3,062
	26,732

Suppose a student in elementary school is to be selected at random. Let y denote the grade of the student selected. Then y is a random variable whose possible values are 0, 1, 2, ..., 8.

a) Use random-variable notation to represent the event that the student selected is in the fifth grade.
b) Determine $P(y = 5)$ and express your result in terms of percentages.
c) Find $P(7)$.
d) Determine the probability distribution of the random variable y.

SOLUTION   a) The event that the student selected is in the fifth grade can be represented as $\{y = 5\}$.

b) $P(y = 5)$ is the probability that the student selected is in the fifth grade. Using Table 5.3 and the f/N-rule, we can obtain that probability:

$$P(y = 5) = \frac{f}{N} = \frac{3{,}131}{26{,}732} = 0.117$$

In terms of percentages, this means that 11.7% of the students in elementary school are in the fifth grade.

c) $P(7)$ is the probability that the student selected is in the seventh grade. So, again, using Table 5.3 and the f/N-rule we get

$$P(7) = \frac{f}{N} = \frac{3{,}184}{26{,}732} = 0.119$$

d) The probability distribution of y is determined by computing $P(y)$ for $y = 0, 1, 2, \ldots, 8$. We have already done this for $y = 5$ and $y = 7$. The other probabilities are computed similarly and are displayed in the second column of Table 5.4. ■

MTB

TABLE 5.4
Probability
distribution of the
random variable y

Grade y	Probability $P(y)$	
0	0.101	
1	0.110	
2	0.104	
3	0.105	
4	0.109	
5	0.117	
6	0.119	
7	0.119	
8	0.115	
	0.999	(rounding error)

Note: In Table 5.4 the sum of the probabilities is given as 0.999. As we know from Key Fact 5.1, the sum of the probabilities must be exactly 1. However, our computation is off by a little since we rounded the probabilities for y to three decimal places.

Once we have the probability distribution of a discrete random variable, it is easy to determine any probability involving that random variable. The basic tool for accomplishing this is the special addition rule, Formula 4.1 on page 164. Consider the following example.

EXAMPLE 5.3   *Illustrates random variable notation and probability distributions*

When a balanced dime is tossed three times, there are eight equally likely outcomes possible. See Table 5.5.

TABLE 5.5

HHH	HTH	THH	TTH
HHT	HTT	THT	TTT

Here, for example, HHT means that the first two tosses are heads and the third is tails. Let x denote the total number of heads obtained in the three tosses. Then x is a random variable whose possible values are 0, 1, 2, and 3.

a) Use random-variable notation to represent the event that exactly two heads are tossed.
b) Determine $P(x = 2)$.
c) Find the probability distribution of the random variable x.
d) Use random-variable notation to represent the event that at most two heads are tossed.
e) Find $P(x \leq 2)$.

SOLUTION

a) The event that exactly two heads are tossed can be represented as $\{x = 2\}$.
b) $P(x = 2)$ is the probability that exactly two heads are tossed. We see from Table 5.5 that

$$P(x = 2) = \frac{f}{N} = \frac{3}{8} = 0.375$$

This is because there are three ways to get a total of two heads and there are eight possible outcomes altogether.
c) The remaining probabilities for x are computed as in part (b). The probability distribution of the random variable x is shown in Table 5.6.

TABLE 5.6
Probability
distribution for
the number of
heads, x, obtained
in three tosses of
a balanced dime

No. of heads x	Probability $P(x)$
0	0.125
1	0.375
2	0.375
3	0.125
	1.000

d) The event that at most two heads are tossed can be represented as $\{x \leq 2\}$, read "x is less than or equal to two."

e) $P(x \leq 2)$ is the probability that at most two heads are tossed. The event that at most two heads are tossed can be expressed as

$$\{x \leq 2\} = (\{x = 0\} \text{ or } \{x = 1\} \text{ or } \{x = 2\})$$

Since the three events on the right are mutually exclusive (why?), we can use the special addition rule and Table 5.6 to conclude that

$$P(x \leq 2) = P(x = 0) + P(x = 1) + P(x = 2)$$
$$= 0.125 + 0.375 + 0.375 = 0.875$$

Thus, the probability is 0.875 that at most two heads are tossed.   ∎

Exercises 5.1

— **5.1**   According to the Census Bureau publication *Current Population Reports,* a frequency distribution for the number of persons per household in the United States is as follows. Frequencies are in millions.

Number of persons x	Number of households f
1	19.4
2	26.5
3	14.6
4	12.9
5	6.1
6	2.5
7†	1.6
	83.6

† Actually this is "7 or more" but we will consider it to be "7" for illustrative purposes.

Let x denote the number of persons in a household selected at random.

a) What are the possible values of the random variable x?

b) Employ random-variable notation to represent the event that the household selected has exactly five persons.

c) Determine $P(x = 5)$ and interpret your result in terms of percentages.

d) Find $P(3)$.

e) Obtain the probability distribution of x.

— **5.2**   The National Center for Education Statistics compiles enrollment data on American public schools and reports the information in the publication *Digest of Education Statistics.* Below is a table that displays a frequency distribution for the enrollment by grade in public secondary schools. Frequencies are in thousands of students.

Grade x	Enrollment f
9	3,290
10	3,223
11	3,041
12	2,908
	12,462

Suppose a student in public secondary school is to be selected at random. Let x denote the grade level of the student chosen.

a) What are the possible values of the random variable x?

b) Employ random-variable notation to represent the event that the student chosen is in the tenth grade.

c) Determine $P(x = 10)$ and interpret your result in terms of percentages.

d) Find $P(12)$.

e) Obtain the probability distribution of the random variable x.

__ **5.3** When two balanced dice are rolled, there are 36 equally likely outcomes possible. See Figure 4.1 on page 145. Let y denote the sum of the dice.
a) What are the possible values of the random variable y?
b) Employ random-variable notation to represent the event that the sum of the dice is 7.
c) Find $P(y = 7)$.
d) Determine $P(11)$.
e) Find the probability distribution of y. Leave your probabilities in fraction form.

__ **5.4** When two balanced dice are rolled, there are 36 equally likely outcomes possible. See Figure 4.1 on page 145. Let x denote the larger number showing on the two dice. For example, if the outcome is

then $x = 5$. [If both dice come up the same, then x equals that common value.]
a) What are the possible values of the random variable x?
b) Employ random-variable notation to represent the event that the larger number is four.
c) Find $P(x = 4)$.
d) Determine $P(2)$.
e) Find the probability distribution of x. Leave your probabilities in fraction form.

__ **5.5** Prescott National Bank has six tellers available to serve customers. The number of busy tellers varies from time to time, so it is a random variable, which we shall call x. From past records, the probability distribution of x is known. It is given in the following table:

Number busy x	Probability $P(x)$
0	0.03
1	0.05
2	0.08
3	0.15
4	0.21
5	0.26
6	0.22

The table indicates, for example, that the probability is 0.26 that exactly five of the tellers are busy with

customers; that is, about 26% of the time exactly five tellers are busy with customers. Use random variable notation to represent each of the following events:
a) Exactly four tellers are busy.
b) At least two tellers are busy.
c) Less than five tellers are busy.
d) At least two but less than five tellers are busy.
Use the special addition rule and the probability distribution to determine
e) $P(x = 4)$.
f) $P(x \geq 2)$.
g) $P(x < 5)$.
h) $P(2 \leq x < 5)$.

__ **5.6** On the basis of past experience, a car salesperson knows that the number of cars she sells per week is a random variable, y, with a probability distribution as shown here.

Cars sold y	Probability $P(y)$
0	0.135
1	0.271
2	0.271
3	0.180
4	0.090
5	0.036
6	0.012
7	0.003
8	0.002

Suppose a week is selected at random. Use random variable notation to represent each of the following events: The salesperson sells
a) exactly three cars.
b) at least three cars.
c) less than seven cars.
d) at least three but less than seven cars.
Use the special addition rule and the probability distribution to find
e) $P(y = 3)$.
f) $P(y \geq 3)$.
g) $P(y < 7)$.
h) $P(3 \leq y < 7)$.

__ **5.7** Benny's Barber Shop in Cleveland has five chairs for waiting customers. Previous records indicate that the probability distribution for the number of customers waiting, y, is as shown in the table at the top of the first column on the next page.

Customers waiting y	Probability $P(y)$
0	0.424
1	0.161
2	0.134
3	0.111
4	0.093
5	0.077

Represent each of the events below using random variable notation.

a) Exactly two customers are waiting.
b) At most four customers are waiting.
c) More than one customer is waiting.
d) Between two and four customers are waiting.

Apply the special addition rule and the table to find

e) $P(y = 2)$.
f) $P(y \leq 4)$.
g) $P(y > 1)$.
h) $P(2 \leq y \leq 4)$.
i) $P(1 < y \leq 4)$.

— 5.8   A small company in Tulsa, Oklahoma, has a total of 10 employees. Six are male and four are female. Three of these 10 employees are to be selected to represent the company at a small business conference sponsored by the SBA (Small Business Administration). If the selection is done randomly, then the number of women chosen is a random variable, x, whose probability distribution is as follows:

Women chosen x	Probability $P(x)$
0	0.167
1	0.500
2	0.300
3	0.033

Use random variable notation to express each of the following events: The number of women chosen is
a) exactly two.
b) at most two.
c) at least one.
d) either one or two.

Apply the special addition rule and the probability distribution to determine

e) $P(x = 2)$.
f) $P(x \leq 2)$.
g) $P(x \geq 1)$.
h) $P(x = 1 \text{ or } 2)$.

— 5.9   Suppose that z is a random variable and that $P(z > 1.96) = 0.025$. Find $P(z \leq 1.96)$. (*Hint:* Use the complementation rule, Formula 4.2 on page 167.)

≡ 5.10   Suppose t is a random variable and that $P(t > 2.02) = 0.05$ and $P(t < -2.02) = 0.05$. Determine $P(-2.02 \leq t \leq 2.02)$.

≡ 5.11   Suppose that z is a random variable and that $P(-1.64 \leq z \leq 1.64) = 0.90$. Also suppose that $P(z > 1.64) = P(z < -1.64)$. Use this information to determine $P(z > 1.64)$.

≡ 5.12   Assume that x is a random variable and that $P(x > c) = \alpha$, where c and α are numbers and $0 \leq \alpha \leq 1$. Determine $P(x \leq c)$ in terms of α.

≡ 5.13   Assume that y is a random variable and that $P(y > c) = \alpha/2$, where c and α are numbers and $0 \leq \alpha \leq 1$. Also assume that $P(y < -c) = P(y > c)$. Find $P(-c \leq y \leq c)$ in terms of α.

≡ 5.14   Suppose that t is a random variable and that $P(-c \leq t \leq c) = 1 - \alpha$, where $0 \leq \alpha \leq 1$. Further suppose that $P(t < -c) = P(t > c)$. Find $P(t > c)$ in terms of α.

5.2 The mean and standard deviation of a discrete random variable

In this section, we will study the mean and standard deviation of a discrete random variable. As you will see, the mean and standard deviation of a random variable are analogous to the mean and standard deviation of a population.

MEAN OF A DISCRETE RANDOM VARIABLE

We first introduce the concept of the mean of a discrete random variable. This is done with the aid of the following example.

EXAMPLE 5.4   *Introduces the mean of a discrete random variable*

CJ^2 Business Services is a small company specializing in word processing. The company employs eight people. A grouped-data table for the weekly salaries of these eight employees is presented in Table 5.7.

TABLE 5.7
Grouped-data table
for weekly salaries
of CJ^2 employees

Salary ($) x	Frequency f	Relative frequency
240	3	0.375
320	2	0.250
450	1	0.125
600	2	0.250
	8	1.000

a) Obtain the population mean weekly salary of CJ^2 employees.
b) Suppose a CJ^2 employee is to be selected at random. Let x denote the weekly salary of the employee chosen. Express the formula for the mean weekly salary in terms of the probability distribution of the random variable x.

SOLUTION   a) The population mean weekly salary can be obtained by employing Formula 3.4 on page 129:

$$\mu = \frac{\Sigma x f}{N}$$

This is the formula that is used to compute the mean of a population when the data are given in grouped form, as they are here. To apply the formula, we append an xf-column to the frequency distribution given in the first two columns of Table 5.7 and proceed as usual. See Table 5.8.

TABLE 5.8

x	f	xf
240	3	720
320	2	640
450	1	450
600	2	1200
	8	3010

Thus, the mean weekly salary of the employees of CJ^2 is

$$\mu = \frac{\Sigma x f}{N} = \frac{3010}{8} = \$376.25$$

b) For this part, we are to express the formula for the mean weekly salary in terms of the probability distribution of the random variable x, where x is the weekly salary of a randomly selected CJ^2 employee. Since we are dealing here with a finite population, we know from Key Fact 5.2 on page 234 that probabilities for the random variable x are the same as relative frequencies. In other words, for each x-value the probability, $P(x)$, and the relative frequency, f/N, are equal.

Because probabilities and relative frequencies are the same here, we can express the formula for μ in terms of probabilities as follows:

$$\mu = \frac{\Sigma x f}{N} = \Sigma x \cdot \frac{f}{N} = \Sigma x P(x)$$

Since, expressed in this way, the formula for the mean, μ, involves the probability distribution of the random variable x, we make the following definition. ■

DEFINITION 5.3 Mean of a discrete random variable

The *mean* of a discrete random variable, x, is defined by

$$\mu_x = \Sigma x P(x)$$

The terms *expected value* and *expectation* are also commonly used.

Note: The μ has the subscript x to emphasize that it is the mean of the random variable x. If the random variable under consideration is called y, then we would write μ_y for its mean and express the formula in Definition 5.3 as $\mu_y = \Sigma y P(y)$.

As we have seen, the computation of descriptive measures is most easily accomplished using tables. The same is true here; it is usually easiest to compute the mean of a discrete random variable using a probability distribution table. We illustrate this in Example 5.5.

EXAMPLE 5.5 *Illustrates Definition 5.3*

Suppose a CJ^2 employee is to be selected at random. Let x denote the weekly salary of the employee chosen. Determine the mean, μ_x, of the random variable x.

SOLUTION To compute μ_x, we must first obtain the probability distribution of x. In this case, probabilities for x are equal to relative frequencies. So, we can obtain the probability distribution of x by simply referring to the third column of Table 5.7 on page 240. Hence, the probability distribution of x is as shown in Table 5.9.

TABLE 5.9
Probability distribution of the random variable x

Salary x	Probability $P(x)$
240	0.375
320	0.250
450	0.125
600	0.250

Now we can compute μ_x. We append an $xP(x)$-column to Table 5.9 and apply Definition 5.3. This is done in Table 5.10.

TABLE 5.10

x	$P(x)$	$xP(x)$
240	0.375	90.00
320	0.250	80.00
450	0.125	56.25
600	0.250	150.00
		376.25

Consequently, the mean of the random variable x is

$$\mu_x = \Sigma x P(x) = \$376.25$$

MTB

Note, as expected, that the mean, μ_x, of the random variable x equals the population mean weekly salary, μ, which we computed on page 240. ∎

The definition of the mean of a discrete random variable, Definition 5.3, applies to any discrete random variable, not just to those that are associated with finite populations. Consider the following example.

EXAMPLE 5.6   *Illustrates Definition 5.3*

Prescott National Bank has six tellers available to serve customers. Of course, the number of tellers actually busy with customers varies from time to time; it is a random variable. On the basis of past records, it is known that the probability distribution for this random variable is as shown in the first two columns of Table 5.11. The table indicates, for instance, that the probability is 0.26 that exactly five tellers are busy with customers.

TABLE 5.11

No. busy x	Probability $P(x)$	$xP(x)$
0	0.03	0.00
1	0.05	0.05
2	0.08	0.16
3	0.15	0.45
4	0.21	0.84
5	0.26	1.30
6	0.22	1.32
		4.12

Find the mean, μ_x, of the random variable x. That is, find the mean number of tellers that are busy with customers.

SOLUTION Applying Definition 5.3, we obtain the mean of the random variable x by summing the $xP(x)$-values given in the third column of Table 5.11. So, $\mu_x = \Sigma x P(x) = 4.12$. In other words, the mean number of tellers busy with customers is 4.12. ∎

INTERPRETATION OF THE MEAN OF A RANDOM VARIABLE

As you know, the mean of a finite population is the arithmetic average of the population values. A similar interpretation holds for the mean of a random variable.

For instance, in Example 5.6 the random variable x denotes the number of tellers busy with customers. The mean of that random variable is $\mu_x = 4.12$. Of course, we can never observe a time when there are 4.12 busy tellers. The mean of 4.12 simply indicates that for a large number of observations, the *average* number of busy tellers will be about 4.12. This kind of interpretation holds in all cases.

KEY FACT 5.3 Interpretation of the mean of a random variable

The mean, μ_x, of a random variable x can be given the following interpretation: For a large number of observations of the random variable x, the average value will be approximately equal to μ_x.

We used the computer to simulate the number of busy tellers, x, at 50 different times; that is, we obtained 50 observations of the random variable x. The data are displayed in Table 5.12.

TABLE 5.12
Fifty observations
of the random
variable x

3	6	0	5	4	5	4	6	3	4
5	5	5	2	3	6	4	6	3	3
5	5	3	3	6	5	3	6	2	6
2	2	1	5	4	4	3	5	2	6
6	6	4	2	6	2	3	5	5	6

The average value of these 50 observations is 4.10. This is quite close to the mean, $\mu_x = 4.12$, of the random variable x. If we made, say, 1000 observations instead of 50, then the average value of those observations would be even closer to 4.12.

STANDARD DEVIATION OF A DISCRETE RANDOM VARIABLE

We next examine the concept of the standard deviation of a discrete random variable. To begin, let us return to the salary data of Example 5.4.

EXAMPLE 5.7 *Introduces the standard deviation of a discrete random variable*

A grouped-data table for the weekly salaries of the eight employees of CJ^2 Business Services is given in Table 5.7 on page 240.

a) Obtain the population standard deviation of these weekly salaries.

b) Suppose a CJ^2 employee is to be selected at random. Let x denote the weekly salary of the employee chosen. Express the formula for the population standard deviation of the weekly salaries in terms of the probability distribution of the random variable x.

SOLUTION   a) The population standard deviation of the weekly salaries can be found by employing Formula 3.5 on page 129:

$$\sigma = \sqrt{\frac{\Sigma(x-\mu)^2 f}{N}}$$

This is the formula that is used to compute the standard deviation of a population when the data are in grouped form. To apply the formula, we first recall that the population mean weekly salary is $\mu = \$376.25$. Referring to the first two columns of Table 5.7, we now construct Table 5.13.

TABLE 5.13

x	f	$x - \mu$	$(x - \mu)^2$	$(x - \mu)^2 f$
240	3	-136.25	18,564.0625	55,692.1875
320	2	-56.25	3,164.0625	6,328.1250
450	1	73.75	5,439.0625	5,439.0625
600	2	223.75	50,064.0625	100,128.1250
	8			167,587.5000

From the second and fifth columns of Table 5.13, we get

$$\sigma = \sqrt{\frac{\Sigma(x-\mu)^2 f}{N}} = \sqrt{\frac{167{,}587.5}{8}} = \$144.74$$

Thus, the population standard deviation of the weekly salaries is $\sigma = \$144.74$.

b) For this part, we are to express the formula for the population standard deviation of the weekly salaries in terms of the probability distribution of the random variable x, where x is the weekly salary of a randomly selected CJ^2 employee. Since we are dealing here with a finite population, we know from Key Fact 5.2 on page 234 that probabilities for the random variable x are the same as relative frequencies. In other words, for each x-value, the probability, $P(x)$, and the relative frequency, f/N, are equal.

Because probabilities and relative frequencies are the same here, we can express the formula for σ in terms of probabilities as follows:

$$\sigma = \sqrt{\frac{\Sigma(x-\mu)^2 f}{N}} = \sqrt{\Sigma(x-\mu)^2 \cdot \frac{f}{N}} = \sqrt{\Sigma(x-\mu)^2 P(x)}$$

This last expression serves as the definition of the standard deviation of a discrete random variable. ∎

DEFINITION 5.4   Standard deviation of a discrete random variable

The *standard deviation* of a discrete random variable, x, is defined by

$$\sigma_x = \sqrt{\Sigma(x-\mu_x)^2 P(x)}$$

Recall that the variance of a population is σ^2, the square of the population standard deviation. Similarly, σ_x^2 is called the **variance** of the random variable x.

Although we motivated Definition 5.4 using a random variable associated with a finite population, it is important to realize that the definition applies to any discrete random variable. We now present an example to show the calculations involved in computing the standard deviation of a discrete random variable.

EXAMPLE 5.8 *Illustrates Definition 5.4*

Suppose a CJ^2 employee is to be selected at random. Let x denote the weekly salary of the employee chosen. Determine the standard deviation, σ_x, of the random variable x.

SOLUTION The probability distribution of the random variable x, displayed in Table 5.9, is repeated here in the first two columns of Table 5.14.

TABLE 5.14

Salary x	Probability $P(x)$	$x - \mu_x$	$(x - \mu_x)^2$	$(x - \mu_x)^2 P(x)$
240	0.375	−136.25	18,564.0625	6,961.5234375
320	0.250	−56.25	3,164.0625	791.0156250
450	0.125	73.75	5,439.0625	679.8828125
600	0.250	223.75	50,064.0625	12,516.0156250
				20,948.4375000

The mean of the random variable x was found in Example 5.5: $\mu_x = \$376.25$. To apply Definition 5.4, we need columns for $x - \mu_x$, $(x - \mu_x)^2$, and $(x - \mu_x)^2 P(x)$. These are given in the last three columns of Table 5.14.

The last column of Table 5.14 shows that $\Sigma(x - \mu_x)^2 P(x) = 20{,}948.4375$. Taking the square root, we get σ_x:

$$\sigma_x = \sqrt{\Sigma(x - \mu_x)^2 P(x)} = \sqrt{20{,}948.4375} = \$144.74$$

Thus, the standard deviation of the random variable x is $\sigma_x = \$144.74$. Note, as expected, that the standard deviation, σ_x, of the random variable x equals the population standard deviation, σ, of the weekly salaries, which we computed in part (a) of Example 5.7 on page 244.

MTB

■

SHORTCUT FORMULA FOR σ_X

The previous example makes it clear that the computations required to find σ_x are time-consuming and tedious, even in simple cases. To deal with this problem, we can use a computer or develop a shortcut formula. The latter approach is presented in Formula 5.1.

FORMULA 5.1   Shortcut formula for σ_x

The shortcut formula for computing the standard deviation of a discrete random variable x is

$$\sigma_x = \sqrt{\Sigma x^2 P(x) - \mu_x^2}$$

EXAMPLE 5.9   *Illustrates Formula 5.1*

Let x be the weekly salary of a randomly selected CJ[2] employee. Use Formula 5.1 to compute the standard deviation, σ_x, of the random variable x.

SOLUTION   The probability distribution of the random variable x, is repeated in the first two columns of Table 5.15. To apply the shortcut formula, we need columns for x^2 and $x^2 P(x)$. These are presented in the last two columns of Table 5.15.

TABLE 5.15

x	$P(x)$	x^2	$x^2 P(x)$
240	0.375	57,600	21,600.0
320	0.250	102,400	25,600.0
450	0.125	202,500	25,312.5
600	0.250	360,000	90,000.0
			162,512.5

From the final column of Table 5.15 we see that $\Sigma x^2 P(x) = 162{,}512.5$. Recalling now that $\mu_x = \$376.25$, we apply Formula 5.1 to get

$$\sigma_x = \sqrt{\Sigma x^2 P(x) - \mu_x^2} = \sqrt{162{,}512.5 - (376.25)^2} = \sqrt{20{,}948.4375} = \boxed{\$144.74}$$

This, of course, is the same result that we obtained in Example 5.8. However, the computations done here using the shortcut formula are significantly easier.   ∎

INTERPRETATION OF THE STANDARD DEVIATION OF A RANDOM VARIABLE

Recall that the standard deviation of a population is a measure of the dispersion of the population values. Roughly speaking, it measures how far the population values are from the mean, on the average. A similar interpretation can be given to the standard deviation of a random variable.

KEY FACT 5.4   Interpretation of the standard deviation of a random variable

The standard deviation, σ_x, of a random variable x measures the dispersion of the possible values of the random variable relative to its mean; the smaller the standard deviation, the more likely that x will be close to its mean.

Exercises 5.2

In Exercises 5.15–5.22 we have presented the probability distributions of the random variables that were
considered in Exercises 5.1–5.8, respectively, of Section 5.1. For each exercise,
a) compute and interpret the mean of the random
   variable.
b) find the standard deviation of the random variable
   using the defining formula; that is, by applying Definition 5.4 on page 244.
c) find the standard deviation of the random variable
   using the shortcut formula; that is, by applying
   Formula 5.1 on page 246.

__ 5.15   The random variable x is the number of persons in a randomly selected U.S. household. Its probability distribution is:

Number of persons x	Probability $P(x)$
1	0.232
2	0.317
3	0.175
4	0.154
5	0.073
6	0.030
7	0.019

__ 5.16   The random variable x is the grade level
of a secondary-school student selected at random. Its
probability distribution is:

Grade x	Probability $P(x)$
9	0.264
10	0.259
11	0.244
12	0.233

__ 5.17   The random variable y is the sum of the
dice when two balanced dice are rolled. Its probability
distribution is as shown in the table at the top of the
following column. Perform the required computations
using fractions. That is, do not convert to decimals
until the calculations are completed.

Sum of dice y	Probability $P(y)$
2	1/36
3	1/18
4	1/12
5	1/9
6	5/36
7	1/6
8	5/36
9	1/9
10	1/12
11	1/18
12	1/36

__ 5.18   The random variable x is the larger number
showing when two balanced dice are rolled. Its probability distribution is:

Larger number x	Probability $P(x)$
1	1/36
2	1/12
3	5/36
4	7/36
5	1/4
6	11/36

Perform the computations using fractions.

__ 5.19   The random variable x is the number of
busy tellers at Prescott National Bank. Its probability
distribution is:

Number busy x	Probability $P(x)$
0	0.03
1	0.05
2	0.08
3	0.15
4	0.21
5	0.26
6	0.22

__ 5.20   The random variable y is the number of cars
sold during a week by a car salesperson. Its probability
distribution is as follows:

Cars sold y	Probability P(y)
0	0.135
1	0.271
2	0.271
3	0.180
4	0.090
5	0.036
6	0.012
7	0.003
8	0.002

___ **5.21** The random variable y is the number of customers waiting at Benny's Barber Shop. Its probability distribution is given below.

Customers waiting y	Probability P(y)
0	0.424
1	0.161
2	0.134
3	0.111
4	0.093
5	0.077

___ **5.22** The random variable x is the number of women selected to represent a company at a small business conference. Its probability distribution is:

Women chosen x	Probability P(x)
0	0.167
1	0.500
2	0.300
3	0.033

Expected value: As mentioned earlier, the mean of a random variable is also called its *expected value*. This terminology is especially useful in the areas of gambling and decision theory, as illustrated in Exercises 5.23 and 5.24.

___ **5.23** A roulette wheel contains 38 numbers, of which 18 are red, 18 are black, and two are green. When the roulette ball is spun, it is equally likely to land on any of the 38 numbers. Suppose you bet $1 on red. Then, if the ball lands on a red number, you

win $1; otherwise, you lose your $1. Let x be the amount you win on your $1 bet. Then x is a random variable. Its probability distribution is as shown in the following table:

x	P(x)
1	0.474
−1	0.526

a) Verify that the probability distribution shown in the table is correct.
b) Find the expected value (mean) of the random variable x.
c) On the average, how much will you lose per play? (*Hint:* Refer to the interpretation of the mean of a random variable given in Key Fact 5.3, page 243.)
d) About how much would you expect to lose if you bet $1 on red 100 times? 1000 times?
e) Do you think that roulette is a profitable game for a person to play?

___ **5.24** An investor plans to invest $50,000 in one of four investments. The return on each investment depends on whether next year's economy is strong or weak. The following table summarizes the possible payoffs, in dollars, for the four investments:

Next year's economy

	Strong	Weak
Certificate of deposit	6,000	6,000
Office complex	15,000	5,000
Land speculation	33,000	−17,000
Technical school	5,500	10,000

(Investment)

Let v, w, x, and y denote, respectively, the payoffs for the certificate of deposit, office complex, land speculation, and technical school. Then v, w, x, and y are random variables. Assume that next year's economy has a 40% chance of being strong and a 60% chance of being weak.

a) Find the probability distribution of each of the four random variables v, w, x, and y.

b) Determine the expected value of each of the four random variables.

c) Which investment has the best expected payoff? Which has the worst?

d) Which investment would you select? Explain your answer in detail.

___ **5.25**   A factory manager collected data on the number of equipment breakdowns per day. From this data she derived the probability distribution shown in the table below.

No. breakdowns w	Probability $P(w)$
0	0.80
1	0.15
2	0.05

a) Determine μ_w and σ_w.

b) On the average, how many breakdowns are there per day?

c) About how many breakdowns are expected per year, assuming 250 work days per year?

Properties of the mean and standard deviation of a random variable: Exercises 5.26 and 5.27 develop some important properties of the mean and standard deviation of a random variable. In doing so, the concept of *independence* for random variables will be required. Two discrete random variables, x and y, are said to be **independent** if

$$P(\{x = j\} \,\&\, \{y = k\}) = P(x = j)P(y = k)$$

for all values of j and k; in other words, if the joint probability distribution of x and y equals the product of the (marginal) probability distribution of x and the (marginal) probability distribution of y. This is equivalent to requiring that the events, $\{x = j\}$ and $\{y = k\}$, are independent for all values of j and k. A similar definition holds for independence of more than two discrete random variables.

≡ **5.26**   Refer to Exercise 5.25. Assume the number of breakdowns on different days are *independent* of one another. Let x and y denote the number of breakdowns on each of two consecutive days.

a) Complete the following joint probability distribution table:

		y			
		0	1	2	$P(x)$
x	0				
	1				
	2				
	$P(y)$				

Hint: To obtain the joint probability in the first row and third column, we use the definition of independence for random variables and the table in Exercise 5.25:

$$P(\{x = 0\} \,\&\, \{y = 2\}) = P(x = 0)P(y = 2)$$
$$= 0.80 \cdot 0.05 = 0.04$$

b) Use your answer from part (a) to find the probability distribution of the random variable $x+y$, the total number of breakdowns in two days. That is, complete the following table:

$x+y$	$P(x+y)$
0	
1	
2	
3	
4	

c) Use your answer from part (b) to determine μ_{x+y} and σ_{x+y}.

d) Use part (c) to verify that the following equations are true for this example:

$$\mu_{x+y} = \mu_x + \mu_y, \qquad \sigma_{x+y} = \sqrt{\sigma_x^2 + \sigma_y^2}$$

(*Note:* The mean and standard deviation of x and y are the same as that of w in Exercise 5.25.)

e) The equations in part (d) hold in general. That is:

If x and y are *any* two random variables, then

$$\mu_{x+y} = \mu_x + \mu_y$$

If x and y are *independent* random variables, then

$$\sigma_{x+y} = \sqrt{\sigma_x^2 + \sigma_y^2}$$

Interpret these two equations in words.

≡ **5.27**   The factory manager in Exercise 5.25 estimates that each breakdown costs the company $100
in repairs and loss of production. If w is the number
of breakdowns in a day, then $100w$ is the cost due to
breakdowns for that day.

a) Find the probability distribution of the random
   variable $100w$. [Refer to the probability distribution in Exercise 5.25.]
b) Determine the mean daily breakdown cost, μ_{100w},
   using your answer from part (a).
c) What is the relation between μ_{100w} and μ_w? [From
   Exercise 5.25, $\mu_w = 0.25$]
d) Find σ_{100w} using your answer from part (a).

e) What is the apparent relationship between σ_{100w}
   and σ_w? [From Exercise 5.25, $\sigma_w = 0.536$]
f) The results in parts (c) and (e) are true in general.
   That is:

   If w is a random variable and c is a constant, then

   $$\mu_{cw} = c\mu_w$$

   and

   $$\sigma_{cw} = |c|\sigma_w$$

   Interpret these two equations in words.

5.3 Binomial coefficients; Bernoulli trials

Many problems in probability and statistics involve situations where an experiment with *two* possible outcomes is repeated several times. Each repetition of the
experiment is called a **trial.**

For example, consider testing the effectiveness of a drug. Several patients take
the drug (the trials) and, for each patient, the drug is either effective or not effective
(the two possible outcomes). Or, consider the weekly sales of a car salesperson.
The salesperson has several customers during the week (the trials) and, for each
customer, the salesperson either makes a sale or does not make a sale (the two
possible outcomes). Or, consider taste tests for colas. A number of people taste two
different colas (the trials) and, for each person, the preference is either for the first
cola or for the second cola (the two possible outcomes).

The analysis of repeated trials of an experiment having two possible outcomes
requires knowledge of *factorials, binomial coefficients, Bernoulli trials,* and the
binomial distribution. [The "bi" in "binomial distribution" refers to the fact that
there are *two* possible outcomes for each trial.] We will study the first three concepts
in this section and the fourth in the next section.

FACTORIALS

The definition of factorials is as follows:

DEFINITION 5.5   Factorials

Let k be a positive integer (whole number). Then the product of the first k positive integers is called k *factorial* and is denoted by the symbol $k!$. Thus,

$$k! = k(k-1)\cdots 2\cdot 1$$

We also define $0! = 1$.

EXAMPLE 5.10 *Illustrates Definition 5.5*

Determine 3!, 4!, and 5!.

SOLUTION Applying Definition 5.5, we get

$$3! = 3 \cdot 2 \cdot 1 = 6$$
$$4! = 4 \cdot 3 \cdot 2 \cdot 1 = 24$$
$$5! = 5 \cdot 4 \cdot 3 \cdot 2 \cdot 1 = 120$$

■

Note, for example, that $6! = 6 \cdot 5!$, $6! = 6 \cdot 5 \cdot 4!$, $6! = 6 \cdot 5 \cdot 4 \cdot 3!$, and so on. In general, if $j \leq k$, then $k! = k(k-1)\cdots(k-j+1)(k-j)!$.

BINOMIAL COEFFICIENTS

We now define *binomial coefficients*. You may have already come upon these in algebra when you studied the binomial expansion, that is, the expansion of $(a+b)^n$. We do not require that you remember the binomial expansion but, if you would like a quick refresher, see Exercise 5.42.

DEFINITION 5.6 Binomial coefficients[†]

If n is a positive integer and x is a nonnegative integer less than or equal to n, then we define the *binomial coefficient* $\binom{n}{x}$ as follows:

$$\binom{n}{x} = \frac{n!}{x!\,(n-x)!}$$

EXAMPLE 5.11 *Illustrates Definition 5.6*

Determine the value of each of the following binomial coefficients:

a) $\binom{6}{1}$ b) $\binom{5}{3}$ c) $\binom{7}{3}$ d) $\binom{4}{4}$

SOLUTION We apply Definition 5.6.

a) $\binom{6}{1} = \frac{6!}{1!\,(6-1)!} = \frac{6!}{1!\,5!} = \frac{6 \cdot 5!}{1!\,5!} = \frac{6}{1} = 6$

b) $\binom{5}{3} = \frac{5!}{3!\,(5-3)!} = \frac{5!}{3!\,2!} = \frac{5 \cdot 4 \cdot 3!}{3!\,2!} = \frac{5 \cdot 4}{2} = 10$

[†] A reader who has covered the optional Section 4.8 will recognize the binomial coefficient, $\binom{n}{x}$, as the number of possible combinations of x objects from a collection of n objects.

c) $\dbinom{7}{3} = \dfrac{7!}{3!\,(7-3)!} = \dfrac{7!}{3!\,4!} = \dfrac{7 \cdot 6 \cdot 5 \cdot 4!}{3!\,4!} = \dfrac{7 \cdot 6 \cdot 5}{6} = \boxed{35}$

d) $\dbinom{4}{4} = \dfrac{4!}{4!\,(4-4)!} = \dfrac{4!}{4!\,0!} = \dfrac{4!}{4!\,0!} = \dfrac{1}{1} = \boxed{1}$

■

BERNOULLI TRIALS; ASSIGNMENT OF PROBABILITIES

We can now discuss Bernoulli trials, named after the Swiss mathematician James Bernoulli (1654–1705). The definition of Bernoulli trials is given as Definition 5.7.

DEFINITION 5.7   Bernoulli trials

Repeated identical trials are called *Bernoulli trials* if:

1 There are two possible outcomes for each trial, denoted generically by s (for success) and f (for failure).
2 The trials are independent.
3 The probability of a success remains the same from trial to trial. We call that probability the *success probability* and denote it by the letter p.

EXAMPLE 5.12   *Illustrates Bernoulli trials*

A drug is known to be 80% effective in curing a certain disease. Suppose that four patients with the disease are to be given the drug and the cure-noncure results recorded.
a) Formulate this process as a sequence of four Bernoulli trials.
b) Determine the possible outcomes of the four Bernoulli trials.
c) Determine the probability of each outcome in part (b).

SOLUTION   a) Each trial consists of administering the drug to one of the patients. There are two possible outcomes for each trial: cure or noncure. The trials are independent (why?). If we let a success, s, correspond to a cure, then the success probability is $p = 0.8$ (80%).

b) For this part we are to obtain the possible outcomes of the four Bernoulli trials; that is, the possible cure-noncure results for the four patients. The possible outcomes are shown in Table 5.16.

TABLE 5.16
Possible outcomes
for drug experiment

ssss	*sfss*	*fsss*	*ffss*
sssf	*sfsf*	*fssf*	*ffsf*
ssfs	*sffs*	*fsfs*	*fffs*
ssff	*sfff*	*fsff*	*ffff*

For instance, *sssf* represents the outcome that the first three patients are cured and the fourth is not.

c) Here we are to determine the probability of each of the possible outcomes. As we see from Table 5.16, there are 16 possible outcomes. However, these 16 outcomes are not equally likely. To determine the probabilities, we proceed as follows. First of all, by part (a), the success probability equals 0.8:

$$P(s) = p = 0.8$$

Therefore, the failure probability is

$$P(f) = 1 - p = 1 - 0.8 = 0.2$$

Since the trials are independent, we can apply the special multiplication rule, Formula 4.7 on page 198, to obtain the probability of each of the 16 possible outcomes. For instance, the probability of the outcome $sssf$ is

$$P(sssf) = P(s)P(s)P(s)P(f) = 0.8 \cdot 0.8 \cdot 0.8 \cdot 0.2 = 0.1024$$

and that for $fsfs$ is

$$P(fsfs) = P(f)P(s)P(f)P(s) = 0.2 \cdot 0.8 \cdot 0.2 \cdot 0.8 = 0.0256$$

Similar computations yield the probabilities of the other 14 possible outcomes. See Table 5.17.

TABLE 5.17
Outcomes and probabilities for drug experiment

Outcome	Probability
$ssss$	$(0.8)(0.8)(0.8)(0.8) = 0.4096$
$sssf$	$(0.8)(0.8)(0.8)(0.2) = 0.1024$
$ssfs$	$(0.8)(0.8)(0.2)(0.8) = 0.1024$
$ssff$	$(0.8)(0.8)(0.2)(0.2) = 0.0256$
$sfss$	$(0.8)(0.2)(0.8)(0.8) = 0.1024$
$sfsf$	$(0.8)(0.2)(0.8)(0.2) = 0.0256$
$sffs$	$(0.8)(0.2)(0.2)(0.8) = 0.0256$
$sfff$	$(0.8)(0.2)(0.2)(0.2) = 0.0064$
$fsss$	$(0.2)(0.8)(0.8)(0.8) = 0.1024$
$fssf$	$(0.2)(0.8)(0.8)(0.2) = 0.0256$
$fsfs$	$(0.2)(0.8)(0.2)(0.8) = 0.0256$
$fsff$	$(0.2)(0.8)(0.2)(0.2) = 0.0064$
$ffss$	$(0.2)(0.2)(0.8)(0.8) = 0.0256$
$ffsf$	$(0.2)(0.2)(0.8)(0.2) = 0.0064$
$fffs$	$(0.2)(0.2)(0.2)(0.8) = 0.0064$
$ffff$	$(0.2)(0.2)(0.2)(0.2) = 0.0016$

A tree diagram is useful to organize and summarize the possible outcomes and their probabilities. This is given in Figure 5.1.

FIGURE 5.1   Tree diagram corresponding to Table 5.17

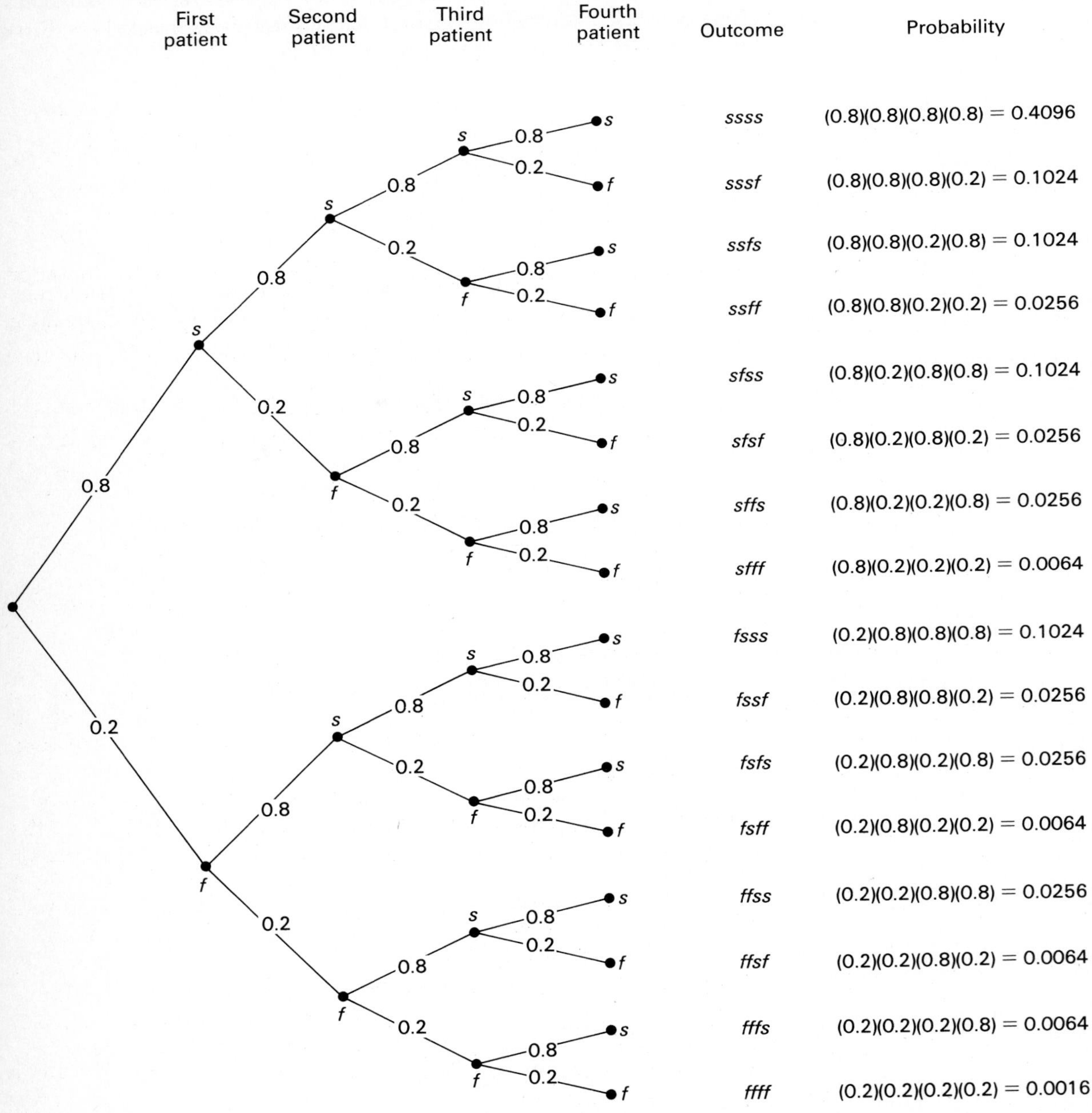

First patient	Second patient	Third patient	Fourth patient	Outcome	Probability
				ssss	(0.8)(0.8)(0.8)(0.8) = 0.4096
				sssf	(0.8)(0.8)(0.8)(0.2) = 0.1024
				ssfs	(0.8)(0.8)(0.2)(0.8) = 0.1024
				ssff	(0.8)(0.8)(0.2)(0.2) = 0.0256
				sfss	(0.8)(0.2)(0.8)(0.8) = 0.1024
				sfsf	(0.8)(0.2)(0.8)(0.2) = 0.0256
				sffs	(0.8)(0.2)(0.2)(0.8) = 0.0256
				sfff	(0.8)(0.2)(0.2)(0.2) = 0.0064
				fsss	(0.2)(0.8)(0.8)(0.8) = 0.1024
				fssf	(0.2)(0.8)(0.8)(0.2) = 0.0256
				fsfs	(0.2)(0.8)(0.2)(0.8) = 0.0256
				fsff	(0.2)(0.8)(0.2)(0.2) = 0.0064
				ffss	(0.2)(0.2)(0.8)(0.8) = 0.0256
				ffsf	(0.2)(0.2)(0.8)(0.2) = 0.0064
				fffs	(0.2)(0.2)(0.2)(0.8) = 0.0064
				ffff	(0.2)(0.2)(0.2)(0.2) = 0.0016

Note that outcomes with the same number of successes have the same probability. For instance, there are four outcomes with exactly three successes: *sssf, ssfs, sfss*, and *fsss*. All of these outcomes have the same probability, 0.1024. This is because each probability is obtained by multiplying three success probabilities of 0.8 and one failure probability of 0.2. ∎

We should point out that a success, *s*, in Bernoulli trials is often derived from a collection of outcomes. For example, a roulette wheel consists of 38 numbers of which 18 are red, 18 are black, and two are green. When the roulette ball is spun, it is equally likely to land on any of the 38 numbers. If we are interested in which number the ball lands on, then there are 38 possible outcomes (the 38 different numbers) for each play at the wheel.

Suppose, however, that we are betting on red. Then we are only interested in whether the ball lands on a red number. From this point of view, there are only two possible outcomes for each play at the wheel—either the ball lands on a red number or it doesn't. Hence, successive bets on red constitute a sequence of Bernoulli trials: Each trial consists of one play at the wheel; there are two possible outcomes for each trial, red or not red; the trials are independent because successive plays at the wheel are independent; and if we let a success, *s*, correspond to the ball landing on red, then the success probability is $p = 18/38$ since 18 of the 38 numbers are red.

BERNOULLI TRIALS AND SAMPLING FROM TWO-CATEGORY POPULATIONS

Many statistical studies are concerned with the percentage of a population that has a specified attribute. For instance, we might be interested in the percentage of U.S. households that own a color television set. Here the population consists of all U.S. households, and the specified attribute is "owns a color television set." Or we might want to know the percentage of American businesses that are minority owned. In this case, the population is comprised of all American businesses, and the specified attribute is "minority owned."

A population in which each member is classified as either having or not having a specified attribute is called a **two-category population.** Most often, the population under consideration will be large (or infinite). Thus, we usually cannot determine the exact percentage of a two-category population that has the specified attribute. We must instead estimate the percentage from sample data; that is, from the percentage of a sample from the population that has the specified attribute.

If sampling is done **with replacement** (i.e., members selected are returned to the population for possible reselection), then the sampling process constitutes Bernoulli trials. Indeed, each selection of a member from the population corresponds to a single trial. A success occurs on a given trial if the member selected in that trial has the specified attribute; otherwise, a failure occurs. The trials are independent of one another since the sampling is done with replacement. The success probability, *p*, remains the same from trial to trial—it always equals the percentage of the population that has the specified attribute.

In reality, however, sampling is ordinarily done **without replacement** (i.e., members selected are not returned to the population for possible reselection). Under these circumstances, the sampling process does not constitute Bernoulli trials because the trials are not independent and the success probability varies from trial to trial.

To illustrate our discussion, we will consider a small population; specifically, the students in Professor W's introductory statistics class. Let us take the specified attribute to be "female." From Table 4.12 on page 193, we see that the class consists of 17 males and 23 females. Thus, the percentage of the population with the specified attribute is $23/40 = 0.575$, or 57.5%.

As we said, if sampling is done with replacement, then the sampling process constitutes Bernoulli trials. Here a success occurs on a given trial if the student selected is a female; the trials are independent, since the sampling is done with replacement; and the success probability, p, always equals 0.575.

On the other hand, suppose the sampling is done without replacement. Then the sampling process does not constitute Bernoulli trials, because the trials are not independent and the success probability varies from trial to trial. For instance, the probability of a success (female) on the first trial (selection) is $23/40 = 0.575$. However, given that the first trial results in a success, the probability of a success on the second trial is $22/39 = 0.564$ (why?).

Consequently, we see that sampling without replacement from a finite two-category population does not constitute Bernoulli trials. Nonetheless, if the sample size is small relative to the size of the population, then for all practical purposes, the sampling process can be regarded as Bernoulli trials. The reason for this is as follows: If the sample size is small relative to the population size, then there is very little difference between sampling without replacement and sampling with replacement; and, in the latter case, we do indeed have Bernoulli trials. [See Exercise 5.43.] We summarize the above discussion in Key Fact 5.5.

KEY FACT 5.5

Sampling without replacement from a two-category population can be regarded as Bernoulli trials provided that the sample size is small relative to the size of the population. A rule of thumb is that Bernoulli trials may be assumed whenever the sample size is at most 5% of the population size.

Exercises 5.3

__ **5.28** Find 1!, 2!, and 6!.

__ **5.29** Compute 7!, 8!, and 9!.

__ **5.30** Evaluate the following binomial coefficients:
a) $\binom{4}{1}$ b) $\binom{6}{2}$ c) $\binom{8}{3}$ d) $\binom{9}{6}$

__ **5.31** Determine the value of each of the following binomial coefficients:
a) $\binom{5}{2}$ b) $\binom{7}{4}$ c) $\binom{10}{3}$ d) $\binom{12}{5}$

__ **5.32** Determine the value of each of the following binomial coefficients:
a) $\binom{3}{2}$ b) $\binom{6}{0}$ c) $\binom{6}{6}$ d) $\binom{7}{3}$

__ **5.33** Evaluate the following binomial coefficients:
a) $\binom{5}{3}$ b) $\binom{10}{0}$ c) $\binom{10}{10}$ d) $\binom{9}{5}$

__ **5.34** In an ESP experiment, a person in one room randomly selects one of 10 cards numbered 1 through 10, and a person in another room tries to guess the number on the card chosen. This experiment is to be repeated three times, with the card chosen being replaced before the next selection. Assuming that the person guessing lacks ESP, the probability that he guesses correctly on any given trial is $1/10 = 0.1$. Suppose we consider a success, *s*, on a given trial to be a correct guess. So, for example, *ssf* represents a correct guess on the first two trials and an incorrect guess on the third.

a) What is the success probability, *p*?

b) Construct a table similar to Table 5.17 on page 253 for the three guesses.

c) Construct a tree diagram for this problem similar to Figure 5.1 on page 254.

d) List the outcomes with exactly one success.

e) Find the probability of each outcome in part (d). Why are these probabilities all the same?

__ 5.35 Based on data from the *Statistical Abstract of the United States*, the probability that a newborn baby will be a girl is about 0.487. Suppose we consider a success, *s*, in a given birth to be "a girl."

a) What is the success probability, *p*?

b) Construct a table similar to Table 5.17 on page 253 for the next three births. Display the probabilities to three decimal places.

c) Draw a tree diagram for this problem similar to Figure 5.1 on page 254.

d) List the outcomes in which exactly two of the three babies born are girls.

e) Find the probability of each outcome in part (d). Why are these probabilities all the same?

__ 5.36 According to a survey done by the Opinion Research Corporation, 60% of all clerical workers in the U.S. "like their jobs very much." Suppose four randomly selected clerical workers are to be asked whether they like their jobs very much. Let us consider an affirmative response to be a success, *s*.

a) What is the success probability, *p*?

b) Construct a table like Table 5.17 on page 253 for the possible responses of the four clerical workers. Display the probabilities to four decimal places.

c) Draw a tree diagram for this problem similar to Figure 5.1 on page 254.

d) List the outcomes in which exactly two of the four clerical workers respond affirmatively.

e) Find the probability of each outcome in part (d). Why are these probabilities all the same?

__ 5.37 The National Institute of Mental Health reports that 20% of adult Americans suffer from at least one psychiatric disorder. Suppose that four randomly selected adult Americans are to be examined for psychiatric disorders.

a) If we let a success, *s*, correspond to an adult American having a psychiatric disorder, then what is the success probability, *p*? (*Note:* The use of the word "success" in Bernoulli trials need not conform to the ordinarily positive connotation of the word.)

b) Construct a table similar to Table 5.17 on page 253 for the four people examined. Display the probabilities to four decimal places.

c) Draw a tree diagram for this problem similar to Figure 5.1 on page 254.

d) List the outcomes in which exactly three of the four people examined have a psychiatric disorder.

e) Find the probability of each outcome in part (d). Why are these probabilities all the same?

__ 5.38 A telephone-company study in Phoenix revealed that the mean duration of residential phone calls is 3.8 minutes and that the probability is about 0.25 that a randomly selected phone call will last longer than the mean duration of 3.8 minutes. For a given phone call, suppose we consider a success, *s*, to be that the call lasts at most 3.8 minutes.

a) What is the success probability, *p*?

b) Construct a table similar to Table 5.17 on page 253 for the possible "success-failure" results of three randomly selected calls. Display the probabilities to three decimal places.

c) List the outcomes in which exactly two of the three calls last at most 3.8 minutes.

d) Find the probability of each outcome in part (c). Why are these probabilities all the same?

__ 5.39 As reported by the Chicago Title Insurance Company in *The Guarantor*, the probability is 0.768 that a home-buyer will purchase an existing (resale) home. Let a success correspond to the purchase of a new (non-resale) home.

a) What is the success probability, *p*?

b) Construct a table similar to Table 5.17 on page 253 for the possible "success-failure" results of four randomly selected home purchases. Display the probabilities to three decimal places.

c) List the outcomes in which exactly two out of the four purchases are new homes.

d) Find the probability of each outcome in part (c). Why are these probabilities all the same?

= 5.40 Show that

$$k! = k(k-1)!$$

for any positive integer *k*.

5.41 Establish the following equalities:

a) $\binom{n}{0} = 1$ b) $\binom{n}{n} = 1$ c) $\binom{n}{x} = \binom{n}{n-x}$

5.42 The binomial theorem states that

$$(a + b)^n = \binom{n}{0}a^n b^0 + \binom{n}{1}a^{n-1}b^1 + \cdots + \binom{n}{n}a^0 b^n$$

a) Verify the binomial theorem for $n = 2$ by computing $(a + b)^2 = (a + b)(a + b)$ and showing that the result is identical to

$$\binom{2}{0}a^2 b^0 + \binom{2}{1}a^1 b^1 + \binom{2}{2}a^0 b^2$$

b) Repeat part (a) for $n = 3$.

5.43 When doing this exercise, refer to the discussion of the relationship between Bernoulli trials and sampling without replacement from a finite, two-category population (pages 255–256). Consider the following frequency distribution for students in Professor W's introductory statistics class:

Sex	Frequency
Male	17
Female	23
	40

a) Suppose two students are to be selected at random. Find the probability that both students selected are male,
   (i) if the selection is done with replacement. [Apply the special multiplication rule given in Formula 4.6 on page 197.]
   (ii) if the selection is done without replacement. [Apply the general multiplication rule given in Formula 4.5 on page 192.]
b) Compare the two answers that you obtained in part (a).

Suppose Professor W's class had 10 times the students, but in the same proportions. That is, suppose the frequency distribution were:

Sex	Frequency
Male	170
Female	230
	400

c) Repeat parts (a) and (b) using this second frequency distribution.
d) In which case is there less difference between sampling without and with replacement? Can you explain why this is so?

5.4 The binomial distribution

Now that we have studied binomial coefficients and Bernoulli trials, we are in a position to examine the binomial distribution. The **binomial distribution** is the probability distribution for the *number of successes* in a sequence of Bernoulli trials. We introduce this by returning to the drug experiment of Example 5.12.

EXAMPLE 5.13   *Introduces the binomial distribution*

A drug is known to be 80% effective in curing a certain disease; that is, the probability is 0.80 that a person with the disease will be cured by the drug. If four patients with the disease are to be given the drug, find the probability that
a) exactly two are cured.
b) exactly three are cured.

SOLUTION   To begin, recall that we can regard the process of administering the drug to four patients as a sequence of four Bernoulli trials. If we let a success, s, correspond to a cure, then the success probability is $p = 0.8$ and the failure probability is $1 - p = 0.2$. The 16 possible "cure-noncure" outcomes and their probabilities are given in Table 5.17. We repeat that table here as Table 5.18.

TABLE 5.18
Outcomes and probabilities for drug experiment

Outcome	Probability
ssss	$(0.8)(0.8)(0.8)(0.8) = 0.4096$
sssf	$(0.8)(0.8)(0.8)(0.2) = 0.1024$
ssfs	$(0.8)(0.8)(0.2)(0.8) = 0.1024$
ssff	$(0.8)(0.8)(0.2)(0.2) = 0.0256$
sfss	$(0.8)(0.2)(0.8)(0.8) = 0.1024$
sfsf	$(0.8)(0.2)(0.8)(0.2) = 0.0256$
sffs	$(0.8)(0.2)(0.2)(0.8) = 0.0256$
sfff	$(0.8)(0.2)(0.2)(0.2) = 0.0064$
fsss	$(0.2)(0.8)(0.8)(0.8) = 0.1024$
fssf	$(0.2)(0.8)(0.8)(0.2) = 0.0256$
fsfs	$(0.2)(0.8)(0.2)(0.8) = 0.0256$
fsff	$(0.2)(0.8)(0.2)(0.2) = 0.0064$
ffss	$(0.2)(0.2)(0.8)(0.8) = 0.0256$
ffsf	$(0.2)(0.2)(0.8)(0.2) = 0.0064$
fffs	$(0.2)(0.2)(0.2)(0.8) = 0.0064$
ffff	$(0.2)(0.2)(0.2)(0.2) = 0.0016$

Using Table 5.18, we can solve the problems posed in parts (a) and (b).

a) The problem here is to determine the probability that exactly two of the four patients are cured. As we see from Table 5.18, the event that exactly two of the four patients are cured consists of six outcomes: *ssff, sfsf, sffs, fssf, fsfs,* and *ffss*. Moreover, these six outcomes have the same probability, 0.0256. This is because each probability is obtained by multiplying two success probabilities of 0.8 and two failure probabilities of 0.2. Consequently, by the special addition rule, Formula 4.1 on page 164, we have

$$P(\text{Exactly two are cured})$$
$$= P(ssff) + P(sfsf) + P(sffs) + P(fssf) + P(fsfs) + P(ffss)$$
$$= 0.0256 + 0.0256 + 0.0256 + 0.0256 + 0.0256 + 0.0256$$
$$= 6 \cdot 0.0256 = \boxed{0.1536}$$

Thus, the probability that exactly two are cured is 0.1536.

b) For this part we need to find the probability that exactly three of the four patients are cured. From Table 5.18, we observe that the event that exactly three of the four patients are cured consists of four outcomes: *sssf, ssfs, sfss,* and *fsss*. These outcomes all have the same probability, 0.1024. This is because each probability is obtained by multiplying three success probabilities of 0.8 and one failure probability of 0.2. So, by the special addition rule,

$$P(\text{Exactly three are cured})$$
$$= P(sssf) + P(ssfs) + P(sfss) + P(fsss)$$
$$= 0.1024 + 0.1024 + 0.1024 + 0.1024$$
$$= 4 \cdot 0.1024 = \boxed{0.4096}$$

Thus, the probability that exactly three are cured is 0.4096.   ■

As we have seen, Table 5.18 gives the 16 possible "success-failure" (cure-noncure) outcomes and their probabilities. However, in this and many other situations, it is not really the "success-failure" outcomes that are of interest. Rather it is the total *number of successes.* If we denote the total number of successes (cures) by x, then x is a random variable—its value depends on chance, namely on how many patients are cured.

Now we can use random-variable notation to represent events and probabilities concisely. For instance, the event that "exactly two of the four patients are cured" can be written simply as $\{x = 2\}$. In part (a) of Example 5.13, we determined the probability of that event to be 0.1536. We can express this as $P(x = 2) = 0.1536$ or, even more briefly, as $P(2) = 0.1536$. This is recorded in the third row of Table 5.19.

In part (b) of Example 5.13 we found that the probability is 0.4096 that exactly three of the four patients are cured. Thus, $P(3) = 0.4096$. This is recorded in the fourth row of Table 5.19. Using similar reasoning as in Exercise 5.13, we can determine the remaining probabilities for the random variable x and hence obtain its probability distribution. See Table 5.19.

TABLE 5.19
Probability distribution for the number of patients cured

Number cured x	Probability $P(x)$
0	0.0016
1	0.0256
2	0.1536
3	0.4096
4	0.4096

THE BINOMIAL PROBABILITY FORMULA

Table 5.19 displays the probability distribution of the random variable x—the number of patients, out of four, cured by a drug. Although the work done to get that probability distribution is quite involved, it is rather simple compared to most situations. This is because, in practice, the number of trials is generally much larger than four.

For instance, if 20 patients, instead of four, were given the drug, then there would be over one million possible outcomes. In such a case, the tabulation method used in Example 5.13 (Table 5.18) would certainly not be practical. Fortunately, there is a formula for binomial probabilities. The first step in developing that formula is the following key fact.

KEY FACT 5.6

Suppose that n Bernoulli trials are to be performed. Then the number of outcomes with exactly x successes is equal to the binomial coefficient $\binom{n}{x}$. That is, the event of exactly x successes in n trials consists of $\binom{n}{x}$ outcomes.

We will not stop to prove Key Fact 5.6, but let us quickly check to see that it is consistent with the results obtained in Example 5.13. In part (a) of that example, we found from Table 5.18 that there are six outcomes in which exactly two of the four patients are cured: *ssff*, *sfsf*, *sffs*, *fssf*, *fsfs*, and *ffss*. In other words, the event of exactly two successes ($x = 2$) in four trials ($n = 4$) consists of six outcomes.

Using binomial coefficients, we can determine that fact without resorting to a direct listing. Applying Key Fact 5.6, we have

$$\begin{bmatrix} \text{number of outcomes} \\ \text{comprising the event} \\ \text{of exactly two cures} \end{bmatrix} = \binom{4}{2} = \frac{4!}{2!\,(4-2)!} = \frac{4!}{2!\,2!} = \frac{4 \cdot 3 \cdot 2!}{2!\,2!} = \frac{4 \cdot 3}{2} = 6$$

The same approach holds for part (b) of Example 5.13:

$$\begin{bmatrix} \text{number of outcomes} \\ \text{comprising the event} \\ \text{of exactly three cures} \end{bmatrix} = \binom{4}{3} = \frac{4!}{3!\,(4-3)!} = \frac{4!}{3!\,1!} = \frac{4 \cdot 3!}{3!\,1!} = \frac{4}{1} = 4$$

We can now develop a probability formula for the number of successes in Bernoulli trials. We indicate briefly how that formula is derived by referring to Example 5.13. You may also wish to consult Table 5.18 on page 259.

For instance, to determine the probability of exactly three cures, $P(x = 3)$, we reason as follows. Any particular outcome with exactly three cures (e.g., *ssfs*) has probability

$$\underset{\substack{\uparrow \\ \text{probability} \\ \text{of a cure}}}{\overset{\text{three cures}}{(0.8)^3}} \quad \cdot \quad \underset{\substack{\uparrow \\ \text{probability} \\ \text{of a noncure}}}{\overset{\text{one noncure}}{(0.2)^1}} = 0.512 \cdot 0.2 = 0.1024$$

obtained by multiplying three success probabilities of 0.8 and one failure probability of 0.2. Also, by Key Fact 5.6, the number of outcomes with exactly three cures is

$$\underset{\substack{\uparrow \\ \text{number of cures}}}{\overset{\overset{\text{number of trials}}{\downarrow}}{\binom{4}{3}}} = \frac{4!}{3!\,(4-3)!} = 4$$

So, by the special addition rule, the probability of exactly three cures is

$$P(x = 3) = \binom{4}{3} \cdot (0.8)^3 (0.2)^1 = 4 \cdot 0.1024 = \boxed{0.4096}$$

Of course, this is the same result obtained in part (b) of Example 5.13. However, this time we determined the probability quickly and easily—no tabulation, no listing. More importantly, the above reasoning applies to any sequence of Bernoulli trials and leads to the following formula.

FORMULA 5.2   Binomial probability formula

Suppose that n Bernoulli trials are to be performed and that the probability of success on any given trial equals p. Let x denote the total number of successes in the n trials. Then the probability distribution of the random variable x is given by the formula

$$P(x) = \binom{n}{x} p^x (1 - p)^{n-x}$$

The random variable x is called a *binomial random variable* and is said to have the *binomial distribution with parameters n and p.*

To determine the binomial probability formula in specific problems, it is useful to have a well-organized strategy. Such a strategy is presented in Procedure 5.1.

PROCEDURE 5.1
To find a binomial probability formula.

ASSUMPTIONS

1. n identical trials are to be performed.
2. There are two possible outcomes (success and failure) for each trial.
3. The trials are independent.
4. The success probability, p, remains the same from trial to trial.

STEP 1 *Identify a success.*
STEP 2 *Determine p, the success probability.*
STEP 3 *Determine n, the number of trials.*
STEP 4 *The binomial probability formula for the number of successes, x, is*

$$P(x) = \binom{n}{x} p^x (1 - p)^{n-x}$$

Once you practice Procedure 5.1 a few times, you will find that binomial-probability problems are all quite similar. Now to some examples.

EXAMPLE 5.14   *Illustrates Procedure 5.1*

According to a telephone company study, the probability is about 0.25 that a randomly selected phone call will last longer than the mean duration of 3.8 minutes. What is the probability that, out of three randomly selected calls,
a) exactly two last longer than 3.8 minutes?
b) none last longer than 3.8 minutes?

SOLUTION   Let x denote the number of calls, out of the three selected, that last longer than 3.8 minutes. To find the required probabilities, we first apply Procedure 5.1.

STEP 1 *Identify a success.*

Here we will take a success to be a call lasting longer than 3.8 minutes.

STEP 2 *Determine p, the success probability.*

This is the probability that any particular call lasts longer than 3.8 minutes, which is 0.25. So, $p = 0.25$.

STEP 3 *Determine n, the number of trials.*

In this case, the number of trials is the number of calls selected, which is three. Thus, $n = 3$.

STEP 4 *The binomial probability formula for the number of successes, x, is*

$$P(x) = \binom{n}{x} p^x (1-p)^{n-x}$$

Since $n = 3$ and $p = 0.25$, the formula becomes

$$P(x) = \binom{3}{x} (0.25)^x (0.75)^{3-x}$$

Having completed Procedure 5.1, we can now answer the questions posed in parts (a) and (b).
a) For this part, we want the probability that exactly two of the three calls chosen last longer than 3.8 minutes; that is, the probability of exactly two successes. Applying the binomial probability formula with $x = 2$ we get

$$P(2) = \binom{3}{2} (0.25)^2 (0.75)^{3-2} = \frac{3!}{2!\,(3-2)!} (0.25)^2 (0.75)^1 = \boxed{0.141}$$

The probability is 0.141 that exactly two of the three calls chosen last longer than 3.8 minutes, a 14.1% chance.
b) The probability that none of the three calls last longer than 3.8 minutes is

$$P(0) = \binom{3}{0} (0.25)^0 (0.75)^{3-0} = \frac{3!}{0!\,(3-0)!} (0.25)^0 (0.75)^3 = \boxed{0.422}$$

There is a 42.2% chance that none of the three calls chosen last longer than 3.8 minutes. ■

EXAMPLE 5.15   *Illustrates Procedure 5.1*

A new-car salesperson knows from past experience that, on the average, she will make a sale to about 20% of her customers. What is the probability that, in five randomly selected presentations, she makes a sale to
a) exactly three customers?
b) at most one customer?
c) at least one customer?
d) Determine the probability distribution for the number of sales in five randomly selected presentations.

SOLUTION   Let x denote the number of sales in five attempts. To solve parts (a)–(d), we first apply Procedure 5.1.

STEP 1 *Identify a success.*

In this problem, a success is a sale to a customer.

STEP 2 *Determine p, the success probability.*

This is the probability that the salesperson makes a sale to any particular customer, which is given as 20%. So, $p = 0.2$.

STEP 3 *Determine n, the number of trials.*

In this case, the number of trials is just the number of customers for which a sale is attempted; namely, five. Thus, $n = 5$.

STEP 4 *The binomial probability formula for the number of successes, x, is*

$$P(x) = \binom{n}{x} p^x (1-p)^{n-x}$$

Since $n = 5$ and $p = 0.2$, the formula becomes

$$P(x) = \binom{5}{x} (0.2)^x (0.8)^{5-x}$$

Now that we have applied Procedure 5.1, it is relatively easy to solve the problems posed in parts (a) through (d).
a) Here we want the probability of exactly three sales (successes). Applying the binomial probability formula with $x = 3$ gives

$$P(3) = \binom{5}{3} (0.2)^3 (0.8)^{5-3} = \frac{5!}{3!\,(5-3)!} (0.2)^3 (0.8)^2 = \boxed{0.051}$$

The probability that the salesperson makes exactly three sales in five attempts is 0.051 or 5.1%.

b) The probability of at most one sale is

$$P(x \leq 1) = P(0) + P(1) = \binom{5}{0}(0.2)^0(0.8)^{5-0} + \binom{5}{1}(0.2)^1(0.8)^{5-1}$$

$$= 0.328 + 0.410 = 0.738$$

c) The probability of at least one sale is $P(x \geq 1)$. This can be computed by using the fact that

$$P(x \geq 1) = P(1) + P(2) + P(3) + P(4) + P(5)$$

and applying the binomial probability formula to calculate each of the five terms on the right. However, it is easier to use the complementation rule:

$$P(x \geq 1) = 1 - P(x < 1) = 1 - P(x = 0)$$

$$= 1 - P(0) = 1 - \binom{5}{0}(0.2)^0(0.8)^{5-0}$$

$$= 1 - 0.328 = 0.672$$

Consequently, we see that the salesperson has a 67.2% chance of making at least one sale in five attempts.

d) Finally, we are to determine the probability distribution of the number of sales, x, in five attempts. Thus, we need to compute $P(x)$ for $x = 0, 1, 2, 3, 4,$ and 5 using the formula

$$P(x) = \binom{5}{x}(0.2)^x(0.8)^{5-x}$$

This has already been done for $x = 0, 1,$ and 3 in parts (a) and (b). For $x = 5$ we have

$$P(5) = \binom{5}{5}(0.2)^5(0.8)^{5-5} = (0.2)^5 = 0.000$$

to three decimal places. [The exact value of $P(5)$ is 0.00032.] Similar computations yield $P(2) = 0.205$ and $P(4) = 0.006$. Therefore, the probability distribution for the number of sales, x, is as shown in Table 5.20.

MTB

TABLE 5.20
Probability distribution of the random variable x, the number of sales in five attempts

Number of sales x	Probability $P(x)$
0	0.328
1	0.410
2	0.205
3	0.051
4	0.006
5	0.000

BINOMIAL PROBABILITY TABLES

Because of the importance of the binomial distribution, tables of binomial probabilities have been extensively compiled. Such a table, Table I, can be found in the
appendix. In this table, the number of trials, n, is given in the far left column; the
number of successes, x, in the next column to the right; and the success probability,
p, along the top. The next example illustrates the use of Table I.

EXAMPLE 5.16

Illustrates the use of binomial probability tables

Use Table I to find the probabilities computed in parts (a) and (b) of Example 5.15.
That is, if a car salesperson has a 20% chance of making a sale to a customer,
determine the probability that, in five attempts, she makes a sale to
a) exactly three customers.
b) at most one customer.

SOLUTION

The number of trials is $n = 5$ and the success probability is $p = 0.2$. The binomial
distribution with those two parameters is displayed on the first page of Table I.
Using that binomial distribution, it is a simple matter to obtain the probabilities
in question.
a) The probability of exactly three sales in five attempts is given as the fourth
entry in the binomial distribution, which is 0.051.
b) To obtain the probability of at most one sale (0 or 1), we need to add the first
two entries in the binomial distribution: $0.328 + 0.410 = 0.738.$
Of course, these are the same answers that we obtained in parts (a) and (b) of
Example 5.15 where we employed the binomial probability formula. However, by
using Table I, we have reduced the computations required to almost nil.　■

As we have just observed, binomial probability tables eliminate most of the
computations necessary in dealing with the binomial distribution. But such tables
are of limited usefulness because they contain only a small number of different
values for n and p.

For example, our table, Table I, has only eleven different values for p and stops
at $n = 25$. Consequently, if we want to determine a binomial probability whose n
or p parameter is not included in the table, then we must either use the binomial
probability formula or employ a computer program. This latter method will be
discussed momentarily.

THE BINOMIAL DISTRIBUTION AND FINITE
TWO-CATEGORY POPULATIONS

Recall that a *two-category* population is one in which each member is classified as
either having or not having a specified attribute. Suppose that a random sample of
size n is to be taken from a finite two-category population in which the proportion
(percentage) of the population that has the specified attribute is equal to p. Let x
denote the number of members sampled that have the specified attribute.

If the sampling is with replacement, then the sampling process constitutes Bernoulli trials and so the random variable x has the binomial distribution with parameters n and p. However, as we mentioned in Section 5.3, sampling is usually done without replacement. If the sampling is without replacement, then the sampling process does not constitute Bernoulli trials and so the random variable x does not have a binomial distribution; rather it has what is called a **hypergeometric distribution.**

We will not present the hypergeometric probability formula here because, in practice, a hypergeometric distribution can usually be approximated by a binomial distribution. Specifically, Key Fact 5.5 on page 256 states that sampling without replacement from a finite two-category population can be regarded as Bernoulli trials provided that the size of the sample does not exceed 5% of the size of the population. This implies that, under those circumstances, the probability distribution of the random variable x, which is a hypergeometric distribution, can be approximated by a binomial distribution. Thus, we have the following fact:

KEY FACT 5.7 Binomial approximation to the hypergeometric distribution

Suppose that a random sample of size n is to be taken without replacement from a finite two-category population in which the proportion of the population that has the specified attribute equals p. Further suppose that the sample size, n, does not exceed 5% of the population size, N. Let x denote the number of members sampled that have the specified attribute. Then the random variable x has approximately a binomial distribution with parameters n and p.

For example, according to the Census Bureau publication *Current Population Reports,* 74.7% of American adults have completed high school. Suppose that eight American adults are to be selected at random (without replacement) and let x denote the number of those sampled that have completed high school. Then, since the sample size does not exceed 5% of the population size (why?), the random variable x has approximately a binomial distribution with parameters $n = 8$ and $p = 0.747$. In other words, probabilities for the random variable x are given approximately by the binomial probability formula

$$P(x) = \binom{8}{x}(0.747)^x(0.253)^{8-x}$$

In most applications, when sampling is done without replacement from a finite two-category population, the sample size will in fact not exceed 5% of the population size. Thus, practically speaking, a binomial distribution can almost always be used to approximate the probability distribution of the number of members sampled that have the specified attribute.

OTHER DISCRETE PROBABILITY DISTRIBUTIONS

The binomial distribution is the most important and most widely used discrete probability distribution. However, many other discrete probability distributions arise in practice. Here are the names of a few of the more commonly used ones: Poisson, hypergeometric, discrete uniform, geometric, negative binomial, and multinomial. We will examine the hypergeometric distribution in Exercise 5.66 and the Poisson distribution in Exercise 5.67.

USING THE COMPUTER (OPTIONAL)

Minitab has two programs for determining binomial probabilities. For one of the programs, the appropriate command is **PDF** with the **BINOMIAL** subcommand. And for the other program, the appropriate command is **CDF** with the **BINOMIAL** subcommand. We will illustrate both of these programs by applying them to the car-salesperson example.

EXAMPLE 5.17   *Illustrates PDF; BINOMIAL and CDF; BINOMIAL*

A new-car salesperson knows from past experience that, on the average, she will make a sale to about 20% of her customers. Use Minitab to obtain the probability that, in five randomly selected presentations, she makes a sale to
a) exactly three customers.
b) at most one customer.
c) Determine the probability distribution for the number of sales in five randomly selected presentations.

SOLUTION   In this problem, a success is a sale to a customer. The probability of a success is the probability that the salesperson makes a sale to a customer, which is given as 20%. So, $p = 0.2$. The number of trials is just the number of randomly selected presentations; namely, five. Thus, $n = 5$. Consequently, the appropriate binomial distribution is the one with parameters $n = 5$ and $p = 0.2$.

a) To obtain the probability of exactly three successes (sales), we apply the PDF command and the BINOMIAL subcommand. First we type the command PDF followed by the number of successes for which the probability is required; that is, we type PDF 3;. Then, on the next line, we type the BINOMIAL subcommand followed by the parameters for the binomial distribution under consideration; that is, we type BINOMIAL with n=5 and p=0.2.. These commands and the resulting output are shown in Printout 5.1.

PRINTOUT 5.1
Minitab output for
PDF 3; BINOMIAL

```
MTB > PDF 3;
SUBC> BINOMIAL with n=5 and p=0.2.
         K            P( X = K)
         3              0.0512
```

Therefore, the probability is 0.0512 that the salesperson makes exactly three sales in five attempts.

b) For this part we want the probability of at most one sale; that is, less than or equal to one sale. There are several ways to find that probability using Minitab. One way is to use the CDF command and the BINOMIAL subcommand. The CDF program calculates cumulative probabilities. A **cumulative probability** gives the probability that a random variable is *less than or equal to* a specified value. Here we need the cumulative probability for the value 1 (one success).

We first type the command CDF followed by the number of successes for which the cumulative probability is required; that is, we type CDF 1;. Then, on the next line, we type the BINOMIAL subcommand followed by the parameters for the binomial distribution under consideration; that is, we type BINOMIAL with n=5 and p=0.2.. Printout 5.2 displays these commands and the output that results.

PRINTOUT 5.2
Minitab output for
CDF 1; BINOMIAL

```
MTB > CDF 1;
SUBC> BINOMIAL with n=5 and p=0.2.
       K  P( X LESS OR = K)
       1           0.7373
```

Thus, the probability of at most one sale in five attempts is 0.7373.

c) Finally, we are to determine the probability distribution of the number of sales in five attempts. In other words, we want to obtain the binomial distribution with parameters $n = 5$ and $p = 0.2$. This is accomplished as follows.

We first type the command PDF with no other arguments on that line; that is, we type PDF;. Then, on the next line, we type the BINOMIAL subcommand followed by the parameters for the binomial distribution under consideration; that is, we type BINOMIAL with n=5 and p=0.2.. Printout 5.3 shows these commands and the resulting binomial distribution.

PRINTOUT 5.3
Minitab output for
PDF; BINOMIAL

```
MTB > PDF;
SUBC> BINOMIAL with n=5 and p=0.2.

     BINOMIAL WITH N =   5  P = 0.200000
           K          P( X = K)
           0            0.3277
           1            0.4096
           2            0.2048
           3            0.0512
           4            0.0064
           5            0.0003
```

You should compare the output in Printout 5.3 with the binomial distribution given in Table 5.20 on page 265.

Exercises 5.4

___ **5.44** This exercise makes use of the results of Exercise 5.34 on page 256. If you have not done that exercise, you should do it before proceeding. In an ESP experiment, a person in one room randomly selects one of 10 cards numbered 1 through 10, and a person in another room tries to guess the number on the card chosen. This experiment is to be repeated three times, with the card chosen being replaced before the next selection. Assuming that the person guessing lacks ESP, the probability that he guesses correctly on any given trial is $1/10 = 0.1$. Suppose we consider a success, s, on a given trial to be a correct guess.

a) Use your results from Exercise 5.34 to determine the probability of exactly one correct guess in the three tries.

b) Obtain the probability in part (a) by applying the binomial probability formula. [Use Procedure 5.1 on page 262.]

___ **5.45** This exercise makes use of the results of Exercise 5.37 on page 257. If you have not done that exercise, you should do it before proceeding. According to the National Institute of Mental Health, 20% of adult Americans suffer from a psychiatric disorder. Suppose four randomly selected adult Americans are to be examined for psychiatric disorders.

a) Use your results from Exercise 5.37 to determine the probability that exactly three of the four people examined have a psychiatric disorder.

b) Obtain the probability in part (a) by applying the binomial probability formula. [Use Procedure 5.1 on page 262.]

Use Procedure 5.1 on page 262 to solve Exercises 5.46–5.49. Express each probability as a decimal rounded to three places.

___ **5.46** According to the *Daily Racing Form*, the probability is about 0.67 that the favorite in a horse race will finish in the money (first, second, or third place). In the next five races, what is the probability that the favorite finishes in the money

a) exactly twice?

b) exactly four times?

c) at least four times?

d) between two and four times, inclusive?

e) Determine the probability distribution of the random variable x, the number of times the favorite finishes in the money in the next five races.

___ **5.47** Based on data in the *Statistical Abstract of the United States*, the probability that a newborn baby will be a girl is about 0.487. In the next three births, what is the probability that

a) exactly one is a girl?

b) at most one is a girl?

c) at least one is a girl?

d) either one or two are girls?

e) Determine the probability distribution of the random variable x, the number of newborns in the next three births that are girls.

___ **5.48** According to a survey done by the Opinion Research Corporation, 60% of all clerical workers in the U.S. "like their jobs very much." Suppose four randomly selected clerical workers are to be asked whether they like their jobs very much. Find the probability that the number of affirmative responses is

a) exactly two.

b) at most two.

c) at least one.

d) between one and three, inclusive.

e) Obtain the probability distribution of the random variable x, the number of affirmative responses.

___ **5.49** As reported by the A. C. Nielsen Company in the publication *Nielsen Report on Television*, about 85.5% of U.S. households have a color television set. If six households are randomly selected, what is the probability that the number of those households that have a color television set is

a) exactly four?

b) at least four?

c) at most five?

d) between two and five, inclusive?

e) Determine the probability distribution of the random variable x, the number of households out of six that have a color television set.

For Exercises 5.50–5.53, determine the indicated probabilities using Procedure 5.1. If possible, use Table I in the appendix to check your results.

___ **5.50** According to the Census Bureau's *Current Population Reports*, 25% of American children (under 18 years old) are not living with both parents. If 10 American children are selected at random, find the probability that the number not living with both parents is

a) exactly two.

b) at most two.

c) between three and six, inclusive.

d) either less than three or more than seven.

___ 5.51 The Energy Information Administration collects data on appliances used by American households and publishes the results in *Residential Energy Consumption Survey: Housing Characteristics*. According to that publication, 36.7% of American households use an automatic dishwasher. If 12 American households are randomly selected, find the probability that the number using an automatic dishwasher is

a) exactly three.

b) at most three.

c) at least one.

d) between two and four, inclusive.

e) either less than three or more than 10.

___ 5.52 The U.S. Bureau of the Census collects data on the educational attainment of Americans and states its findings in the publication *Current Population Reports*. That publication reveals that 74.7% of American adults have completed high school. Suppose that eight American adults are selected at random. What is the probability that

a) exactly six have completed high school?

b) at least six have completed high school?

c) at least three have completed high school?

d) between four and seven, inclusive, have completed high school?

___ 5.53 Surveys indicate that in 80% of Swedish couples both partners work. If nine Swedish couples are selected at random, what is the probability that the number of working couples is

a) exactly seven?

b) at least seven?

c) at most nine?

d) between five and eight, inclusive?

e) either less than three or more than six?

In Exercises 5.54 and 5.55, use Table I to determine the required probabilities.

___ 5.54 Studies show that 60% of American families use physical aggression to resolve conflict. Suppose that 20 families are to be selected at random. Find the probability that the number that use physical aggression to resolve conflict is

a) exactly 10.

b) between 10 and 15, inclusive.

c) over 75% of those surveyed.

d) less than eight.

___ 5.55 According to the American Bankers Association, only one in 10 people are dissatisfied with their local bank. If 15 people are randomly selected, what is the probability that the number dissatisfied with their local bank is

a) exactly two?

b) at most two?

c) at least two?

d) between one and three, inclusive?

___ 5.56 During the second round of the 1989 U.S. Open, a strange thing happened: Four golfers made holes-in-one on the sixth hole. According to the experts, the odds of a PGA golfer making a hole-in-one are 3708-1; that is, the probability is 1/3709. There were 155 golfers participating in the second round. Determine the probability that at least four of the 155 golfers would get a hole-in-one on the sixth hole.

Exercises 5.57–5.62 are computer exercises.

___ 5.57 (**Computer exercise**) For each part of Exercise 5.49,

a) state which Minitab command and subcommands (if any) should be used to determine the required probability or probabilities.

b) obtain the required probability or probabilities using Minitab, if you have access to it.

___ 5.58 (**Computer exercise**) For each part of Exercise 5.48,

a) state which Minitab command and subcommands (if any) should be used to determine the required probability or probabilities.

b) obtain the required probability or probabilities using Minitab, if you have access to it.

___ 5.59 (**Computer exercise**) The following is a Minitab printout obtained by applying the PDF command and the BINOMIAL subcommand:

```
BINOMIAL WITH N =   7  P = 0.340000
   K         P( X = K)
   0           0.0546
   1           0.1967
   2           0.3040
   3           0.2610
   4           0.1345
   5           0.0416
   6           0.0071
   7           0.0005
```

Use the printout to determine

a) the number of trials.

b) the success probability.

c) the probability of exactly three successes.

d) the probability of between two and five successes.

e) the probability of at most four successes.

f) the probability of at least four successes.

___ 5.60 **(Computer exercise)** The printout below was obtained from Minitab by employing the CDF command and the BINOMIAL subcommand.

```
BINOMIAL WITH N =   6  P = 0.590000
   K   P( X LESS OR = K)
   0          0.0048
   1          0.0458
   2          0.1933
   3          0.4764
   4          0.7819
   5          0.9578
   6          1.0000
```

Use the printout to find

a) the number of trials.

b) the success probability.

c) the probability of at most three successes.

d) the probability of at least three successes. (*Hint:* Use the complementation rule and the printout.)

e) the probability of exactly three successes. (*Hint:* Use the special addition rule to express $P(x = 3)$ as the difference of two cumulative probabilities.)

___ 5.61 **(Computer exercise)** According to an article in *Reader's Digest,* 10% of people are left-handed. Suppose 10 people are to be selected at random. Let x denote the number of people out of the 10 who are left-handed. We used Minitab to obtain the probability distribution of the random variable x. The printout is shown below. Note that Minitab stops printing the probabilities once it arrives at one that is zero to four decimal places.

```
BINOMIAL WITH N =   10  P = 0.100000
   K        P( X = K)
   0          0.3487
   1          0.3874
   2          0.1937
   3          0.0574
   4          0.0112
   5          0.0015
   6          0.0001
   7          0.0000
```

Employ the printout to obtain the probability that, out of the 10 people chosen,

a) exactly one is left-handed.

b) at least one is left-handed.

c) at most one is left-handed.

d) more than one is left-handed.

e) between one and three, inclusive, are left-handed.

___ 5.62 **(Computer exercise)** The National Center for Health Statistics reports that 27% of American adults (ages 20 to 74) have high cholesterol levels. Suppose eight American adults are to be selected at random. Let x denote the number of those chosen who have high cholesterol levels. The following Minitab printout gives the probability distribution of the random variable x.

```
BINOMIAL WITH N =   8  P = 0.270000
   K          P( X = K)
   0           0.0806
   1           0.2386
   2           0.3089
   3           0.2285
   4           0.1056
   5           0.0313
   6           0.0058
   7           0.0006
   8           0.0000
```

From the printout, determine the probability that, out of the eight adults chosen,

a) exactly three have high cholesterol levels.

b) at least three have high cholesterol levels.

c) less than two have high cholesterol levels.

d) between two and four, inclusive, have high cholesterol levels.

e) either less than two or more than five have high cholesterol levels.

═ 5.63 A sales representative for a tire manufacturer claims that the company's steel-belted radials get at least 35,000 miles. A tire dealer decides to check this claim by testing eight of the tires. If 75% or more of the eight tires he tests get at least 35,000 miles, he will purchase tires from the sales representative. If, in fact, 90% of the steel-belted radials produced by the manufacturer get at least 35,000 miles, what is the probability that the tire dealer purchases tires from the sales representative?

═ 5.64 From past experience, the owner of a restaurant knows that, on the average, 4% of the parties making reservations never show. How many reservations can the owner take and still be at least 80% sure that all parties making a reservation will show?

≡ **5.65** Suppose that x has the binomial distribution with parameters n and p. Show that

$$P(x \geq 1) = 1 - (1 - p)^n$$

≡ **5.66** In this exercise, we will discuss the **hypergeometric distribution.** When sampling without replacement from a finite two-category population, the hypergeometric distribution gives the exact probability distribution for the number of "successes" (i.e., the number of members sampled that have the specified attribute). The hypergeometric probability formula is

$$P(x) = \frac{\binom{Np}{x}\binom{N(1-p)}{n-x}}{\binom{N}{n}}$$

Here, N is the population size and n is the sample size. Also, p is the proportion of "successes" in the population; that is, the proportion of the population that has the specified attribute.

To illustrate, suppose a customer purchases four fuses from a shipment of 250 of which 94% are not defective. Let a "success" correspond to a fuse that is not defective.

a) Determine N.
b) Determine n.
c) Determine p.
d) Use the hypergeometric probability formula to find the probability distribution of the number of non-defective fuses that the customer gets.

Key Fact 5.7 on page 267 states that a hypergeometric distribution can be approximated by a binomial distribution provided the sample size does not exceed 5% of the population size. In particular, we can use the

binomial probability formula

$$P(x) = \binom{n}{x} p^x (1-p)^{n-x}$$

with $n = 4$ and $p = 0.94$ to approximate the probability distribution of the number of non-defective fuses that the customer gets.

e) Obtain the binomial distribution with parameters $n = 4$ and $p = 0.94$.
f) Compare the hypergeometric distribution that you obtained in part (d) with the binomial distribution that you obtained in part (e).

≡ **5.67** Another important discrete probability distribution is the **Poisson distribution,** named for the French mathematician and physicist Simeon Poisson (1781–1840). That probability distribution is often used to model the frequency with which a specified event occurs during a particular period of time. The Poisson probability formula is

$$P(x) = e^{-\lambda} \frac{\lambda^x}{x!}$$

where x can be any nonnegative integer and λ is a parameter equal to the mean of the distribution. The number e is the base of natural logarithms and is approximately equal to 2.7183.

As an illustration, consider the following problem: It is known that the number of telephone calls received per hour by a small CPA firm has a Poisson distribution with parameter (mean) $\lambda = 5.2$. Find the probability that during a one hour time period the firm receives
a) exactly two calls.
b) between four and six calls, inclusive.
c) at least one call.

5.5 The mean and standard deviation of a binomial random variable

In Section 5.2, we learned how to obtain the mean and standard deviation of a discrete random variable. Definition 5.3 on page 241 provides the formula for computing the mean of a discrete random variable. Definition 5.4 on page 244 and Formula 5.1 on page 246 provide formulas for computing the standard deviation of a discrete random variable. Those formulas apply to any discrete random variable and so, in particular, to a binomial random variable. Consider Example 5.18.

EXAMPLE 5.18   *Illustrates the calculation of the mean and standard deviation of a binomial random variable*

As reported by the U.S. National Center for Health Statistics in *Vital and Health Statistics,* about 60% of all eye operations are performed on females. Suppose that three eye-operation patients are to be selected at random. Let x denote the number of patients out of the three chosen that are female. Then x is a binomial random variable with parameters $n = 3$ and $p = 0.6$ (why?).
a) Find the mean of the random variable x.
b) Find the standard deviation of the random variable x.

SOLUTION   To obtain the mean and standard deviation of the random variable x, we first need to determine its probability distribution. Since x has the binomial distribution with parameters $n = 3$ and $p = 0.6$, probabilities for x can be computed using the binomial probability formula

$$P(x) = \binom{3}{x}(0.6)^x(0.4)^{3-x}$$

Employing that formula, we get the probability distribution of x. See the first two columns of Table 5.21.

TABLE 5.21

x	$P(x)$	$xP(x)$	x^2	$x^2P(x)$
0	0.064	0.000	0	0.000
1	0.288	0.288	1	0.288
2	0.432	0.864	4	1.728
3	0.216	0.648	9	1.944
		1.800		3.960

a) The mean of the random variable x can now be obtained by applying Definition 5.3 on page 241. We see from the third column of Table 5.21 that

$$\mu_x = \Sigma x P(x) = \boxed{1.8}$$

Thus, the mean of the random variable x is $\mu_x = 1.8$.

b) To determine the standard deviation of the random variable x, we will use the shortcut formula given in Formula 5.1 on page 246. From the fifth column of Table 5.21 and part (a), we conclude that

$$\sigma_x = \sqrt{\Sigma x^2 P(x) - \mu_x^2} = \sqrt{3.96 - (1.8)^2} = \sqrt{0.72} = \boxed{0.85}$$

So, the standard deviation of the random variable x is $\sigma_x = 0.85$.   ∎

In part (a) of Example 5.18 we found: If three eye-operation patients are selected at random and x denotes the number that are female, then the mean of the random variable x is $\mu_x = 1.8$. This indicates that, on the average, about 1.8 out of every three eye-operation patients are female.

   You might have guessed that result before doing any computations if you used the following reasoning: "60% of all eye-operation patients are female. Thus, out of three eye-operation patients, we would expect about 60% of them to be female. 60% of 3 is 1.8. Consequently, on the average, we would expect about 1.8 out of every three eye-operation patients to be female." This is the same result obtained for μ_x in part (a) of Example 5.18. However, here we did almost no calculations. We simply multiplied the number of trials, $n = 3$, by the success probability, $p = 0.6$.

   It can be shown that this type of reasoning always works. That is, the mean of a binomial random variable is equal to the number of trials, n, times the success probability, p.

FORMULA 5.3   Mean of a binomial random variable

Suppose that x has the binomial distribution with parameters n and p. Then the mean of the random variable x can be obtained from the formula

$$\mu_x = np$$

Formula 5.3 is derived by substituting the binomial probability formula

$$P(x) = \binom{n}{x} p^x (1 - p)^{n-x}$$

into the formula

$$\mu_x = \Sigma x P(x)$$

and simplifying. The details are somewhat tricky, but the idea is not.

   The formula that we will now present for the standard deviation of a binomial random variable is not intuitively clear like the formula for the mean. Nonetheless it is derived in basically the same way: Substitute the binomial probability formula into the formula $\sigma_x = \sqrt{\Sigma x^2 P(x) - \mu_x^2}$ and simplify. The result of doing this is given in Formula 5.4.

FORMULA 5.4   Standard deviation of a binomial random variable

Suppose that x has the binomial distribution with parameters n and p. Then the standard deviation of the random variable x can be obtained from the formula

$$\sigma_x = \sqrt{np(1 - p)}$$

   In Example 5.18 we used the general formulas[†]

$$\mu_x = \Sigma x P(x) \qquad \text{and} \qquad \sigma_x = \sqrt{\Sigma x^2 P(x) - \mu_x^2}$$

[†] The formulas are general in the sense that they can be used to find the mean and standard deviation of any discrete random variable.

to find the mean and standard deviation of a random variable x, the number of eye-operation patients out of three that are female. Alternatively, since that random variable has a binomial distribution, we can instead use the special formulas

$$\mu_x = np \quad \text{and} \quad \sigma_x = \sqrt{np(1-p)}$$

applicable only to binomial random variables. This is done in the next example.

EXAMPLE 5.19   *Illustrates Formulas 5.3 and 5.4*

According to the U.S. National Center for Health Statistics, about 60% of all eye operations are performed on females. Suppose that three eye-operation patients are to be selected at random. Let x denote the number of patients out of the three chosen that are female.
a) Find the mean of the random variable x.
b) Find the standard deviation of the random variable x.

SOLUTION   First recall that the random variable x has the binomial distribution with parameters $n = 3$ and $p = 0.6$.
a) Applying Formula 5.3, we obtain the mean of x:

$$\mu_x = np = 3 \cdot 0.6 = \boxed{1.8}$$

b) Using Formula 5.4, we get the standard deviation of x:

$$\sigma_x = \sqrt{np(1-p)} = \sqrt{3 \cdot 0.6 \cdot 0.4} = \sqrt{0.72} = \boxed{0.85}$$

The values obtained here for μ_x and σ_x are, of course, the same as those found in Example 5.18. However, the calculations required using the special formulas, Formulas 5.3 and 5.4, are significantly easier.   ■

As we have seen, the special formulas, Formulas 5.3 and 5.4, are real time-savers for computing the mean and standard deviation of a binomial random variable, even when the number of trials, n, is small. In most applications n is quite large and, in those cases, the special formulas are indispensable.

EXAMPLE 5.20   *Illustrates Formulas 5.3 and 5.4*

Based on detailed records, a small airline in the southwest has determined that the no-show rate for reservations is about 16%. In other words, the probability is 0.16 that a party making a reservation will not show up. Suppose the next flight has 42 parties with advance reservations. Find the mean and standard deviation of the number of no-shows.

SOLUTION   Let x denote the number of no-shows out of the 42 advance reservations. We need to compute the mean and standard deviation of the random variable x. The random variable x has the binomial distribution with parameters $n = 42$ and $p = 0.16$. Thus, by Formula 5.3,

$$\mu_x = np = 42 \cdot 0.16 = \boxed{6.72}$$

and, by Formula 5.4,

$$\sigma_x = \sqrt{np(1-p)} = \sqrt{42 \cdot 0.16 \cdot 0.84} = \boxed{2.38}$$

In particular, the airline can expect about 6.72 no-shows on the next flight. ∎

Exercises 5.5

__ **5.68** In an ESP experiment, a person in one room randomly selects one of 10 cards numbered 1 through 10, and a person in another room tries to guess the number on the card chosen. This experiment is to be repeated three times, with the card chosen being replaced before the next selection. Assuming the person guessing lacks ESP, the probability of a correct guess on any given trial is $1/10 = 0.1$. Let x denote the number of correct guesses out of the three tries.
a) Determine the mean and standard deviation of the random variable x by applying the general formulas

$$\mu_x = \Sigma x P(x), \qquad \sigma_x = \sqrt{\Sigma x^2 P(x) - \mu_x^2}$$

[Use the method employed in Example 5.18.]
b) Determine the mean and standard deviation of x using the special formulas

$$\mu_x = np, \qquad \sigma_x = \sqrt{np(1-p)}$$

c) Compare the amount of work required in parts (a) and (b).

__ **5.69** According to the National Institute of Mental Health, 20% of all adult Americans suffer from a psychiatric disorder. Suppose four adult Americans are to be examined for psychiatric disorders, and let x denote the number out of the four that have a psychiatric disorder.
a) Determine the mean and standard deviation of the random variable x using the general formulas

$$\mu_x = \Sigma x P(x), \qquad \sigma_x = \sqrt{\Sigma x^2 P(x) - \mu_x^2}$$

[Use the method employed in Example 5.18.]

b) Determine the mean and standard deviation of x using the special formulas

$$\mu_x = np, \qquad \sigma_x = \sqrt{np(1-p)}$$

c) Compare the amount of work required in parts (a) and (b).

In each of Exercises 5.70–5.79, use the special formulas, Formulas 5.3 and 5.4, to obtain the mean and standard deviation of the binomial random variable in question. Interpret your answer for the mean in words.

__ **5.70** According to the *Daily Racing Form*, the probability is about 0.67 that the favorite in a horse race will finish in the money (first, second, or third place). Find the mean and standard deviation of the number of favorites finishing in the money in the next five races.

__ **5.71** As reported by the A. C. Nielsen Company in the publication *Nielsen Report on Television*, about 85.5% of U.S. households have a color television set. In a random sample of six households, find the mean and standard deviation of the number that have a color television set.

__ **5.72** According to the Census Bureau's *Current Population Reports*, 25% of American children (under 18 years old) are not living with both parents. Find the mean and standard deviation of the number of children out of a sample of 10 that are not living with both parents.

__ **5.73** The Energy Information Administration collects data on appliances used by American households and publishes the results in *Residential Energy Consumption Survey: Housing Characteristics*. According to that publication, 36.7% of American households use an automatic dishwasher. If 12 households are to be selected at random, find the mean and standard deviation of the number that use an automatic dishwasher.

___ **5.74** The U.S. Bureau of the Census compiles information on the educational attainment of Americans and states its findings in the publication *Current Population Reports*. That publication reveals that 74.7% of American adults have completed high school. If eight American adults are to be selected at random, find the mean and standard deviation of the number that have completed high school.

___ **5.75** Surveys indicate that in 80% of Swedish couples both partners work. If nine Swedish couples are to be selected at random, find the mean and standard deviation of the number that both work.

___ **5.76** Studies show that 60% of American families use physical aggression to resolve conflict. If 20 families are to be selected at random, find the mean and standard deviation of the number that use physical aggression to resolve conflict.

___ **5.77** According to the American Bankers Association, only one in 10 people are dissatisfied with their local bank. If 15 people are to be selected at random, find the mean and standard deviation of the number who are dissatisfied with their local bank.

___ **5.78** An air-conditioner contractor knows from past experience that about 40% of the evaporative coolers he sells are fiberglass (as opposed to metal). Find the mean and standard deviation of the number of fiberglass units sold out of the next 20 evaporative-cooler sales.

___ **5.79** A student takes a multiple-choice test with 50 questions. Each question has four possible answers. If the student guesses at each question, find the mean and standard deviation of the number of questions answered correctly.

≡ **5.80** Suppose that n Bernoulli trials are to be performed. Let x denote the number of successes. Also, let $\hat{p}$ denote the **proportion of successes;** that is, $\hat{p} = x/n$. Use Formulas 5.3 and 5.4 and the results of part (f) of Exercise 5.27 on page 250 to show that
a) $\mu_{\hat{p}} = p$
b) $\sigma_{\hat{p}} = \sqrt{p(1-p)/n}$
[The formulas in parts (a) and (b) will be used extensively in later chapters.]

Chapter review

KEY TERMS

FORMULAS   **Mean of a discrete random variable x, 241**

$$\mu_x = \Sigma x P(x)$$

Standard deviation of a discrete random variable x, 244, 246

$$\sigma_x = \sqrt{\Sigma(x - \mu_x)^2 P(x)} \qquad \text{or} \qquad \sigma_x = \sqrt{\Sigma x^2 P(x) - \mu_x^2}$$

Factorial, 250

$$k! = k(k-1)\cdots 2 \cdot 1$$

Binomial coefficients, 251

$$\binom{n}{x} = \frac{n!}{x!\,(n-x)!}$$

Binomial probability formula, 262

$$P(x) = \binom{n}{x} p^x (1-p)^{n-x}$$

(n = number of trials, p = success probability)

Mean of a binomial random variable, 275

$$\mu_x = np$$

(n = number of trials, p = success probability)

Standard deviation of a binomial random variable, 275

$$\sigma_x = \sqrt{np(1-p)}$$

(n = number of trials, p = success probability)

YOU SHOULD
BE ABLE TO

1. use and understand each of the preceding formulas.
2. determine the probability distribution of a discrete random variable.
3. describe events using random-variable notation, when appropriate.
4. find and interpret the mean and standard deviation of a discrete random variable.
5. define and apply the concept of Bernoulli trials.
6. assign probabilities to the outcomes in a sequence of Bernoulli trials.
7. apply Procedure 5.1 on page 262 to obtain binomial probabilities.
8. use the binomial probability table (Table I in the appendix.)
9. compute the mean and standard deviation of a binomial random variable using the special formulas, Formulas 5.3 and 5.4.
10. use the Minitab commands covered in this chapter.*
11. interpret the output obtained from the application of the Minitab commands discussed in this chapter.*

REVIEW TEST

1. The table below provides a frequency distribution for the number of undergraduate students by class level at a large state university. For the class levels, the following coding is used: 1 = Freshman, 2 = Sophomore, 3 = Junior, and 4 = Senior.

Class level x	Number of students f
1	5,745
2	6,240
3	7,486
4	10,063
	29,534

Suppose an undergraduate at this university is to be selected at random. Let x denote the class level of the student chosen.

a) What are the possible values of the random variable x?

b) Use random-variable notation to represent the event that the student selected is a junior (class level 3).

c) Determine $P(x = 3)$ and interpret your results in terms of percentages.

d) Find $P(2)$.

e) Determine the probability distribution of the random variable x.

2. An accounting office has six incoming telephone lines. The probability distribution of the number of busy lines, y, is as follows:

Number busy y	Probability $P(y)$
0	0.052
1	0.154
2	0.232
3	0.240
4	0.174
5	0.105
6	0.043

Use random-variable notation to express each of the following events: The number of busy lines is
a) exactly four.
b) at least four.

c) between two and four, inclusive.
d) at least one.

Apply the special addition rule and the probability distribution to determine
e) $P(y = 4)$.
f) $P(y \geq 4)$.
g) $P(2 \leq y \leq 4)$.
h) $P(y \geq 1)$.

3. Refer to the probability distribution displayed in the table in Problem 2.
a) Find the mean of the random variable y.
b) On the average, how many lines are busy?
c) Compute the standard deviation of y using the defining formula.
d) Compute the standard deviation of y using the shortcut formula.

4. Determine $0!$, $3!$, $4!$, and $7!$.

5. Find the value of each of the following binomial coefficients:
a) $\binom{8}{3}$ b) $\binom{8}{5}$ c) $\binom{6}{6}$
d) $\binom{10}{2}$ e) $\binom{40}{4}$ f) $\binom{100}{0}$

6. The game of *craps* is played by rolling two balanced dice. A first roll of a sum of seven or 11 wins; and a first roll of a sum of two, three, or 12 loses. To win with any other first sum, that sum must be repeated before a sum of seven is thrown. It can be shown that the probability is 0.493 that a player wins a game of craps. Suppose that we consider a win by a player to be a success, *s*.
a) What is the success probability, p?
b) Construct a table giving the possible win-lose results and their probabilities for three games of craps. Express each probability as a decimal rounded to three places.
c) Draw a tree diagram.
d) List the outcomes in which the player wins exactly two out of the three times.
e) Determine the probability of each outcome in part (d). Why are these probabilities equal?

7. The nation of Surinam is located on the northern coast of South America. According to the *World Almanac*, 80% of the population is literate. Suppose that four people from Surinam are to be selected at random. Use the binomial probability formula to find the probability that the number of literate people in the sample is

a) exactly three.

b) at most three.

c) at least three.

d) Determine the probability distribution of the random variable x, the number of literate people in a sample of four.

8. Solve Problem 7 by employing Table I.

*9. **(Computer problem)** Refer to Problem 7. For each part of that problem,

   a) state which Minitab command and subcommands (if any) should be used to determine the required probability or probabilities.

   b) obtain the required probability or probabilities using Minitab, if you have access to it.

*10. **(Computer problem)** Below is a Minitab printout obtained by applying the PDF command and the BINOMIAL subcommand.

```
BINOMIAL WITH N =   5  P = 0.650000
   K        P( X = K)
   0          0.0053
   1          0.0488
   2          0.1811
   3          0.3364
   4          0.3124
   5          0.1160
```

Use the printout to determine

a) the number of trials.

b) the success probability.

c) the probability of exactly one success.

d) the probability of between one and three successes, inclusive.

e) the probability of at most one success.

f) the probability of at least one success.

*11. **(Computer problem)** The printout at the top of the next column was obtained from Minitab by employing the CDF command and the BINOMIAL subcommand. Use the printout to find

a) the number of trials.

b) the success probability.

c) the probability of at most four successes.

d) the probability of at least four successes. *(Hint: Employ the printout and the complementation rule, Formula 4.2 on page 167.)*

e) the probability of exactly four successes. *(Hint: Use the special addition rule, Formula 4.1 on page 164, to express $P(x = 4)$ as the difference of two cumulative probabilities.)*

```
BINOMIAL WITH N =   8  P = 0.570000
   K   P( X LESS OR = K)
   0          0.0012
   1          0.0136
   2          0.0711
   3          0.2235
   4          0.4762
   5          0.7440
   6          0.9216
   7          0.9889
   8          1.0000
```

*12. **(Computer problem)** In a 1987 statement, the Department of Agriculture reported that approximately four of every 10 chickens sold to consumers are contaminated by salmonella. Salmonella is a microorganism that can produce flu-like symptoms of fever, diarrhea, and vomiting within 12 to 36 hours after eating. Suppose that seven chickens are to be selected at random. Let x denote the number of those chosen that are contaminated by salmonella. The Minitab printout presented below provides the probability distribution of the random variable x.

```
BINOMIAL WITH N =   7  P = 0.400000
   K        P( X = K)
   0          0.0280
   1          0.1306
   2          0.2613
   3          0.2903
   4          0.1935
   5          0.0774
   6          0.0172
   7          0.0016
```

From the printout, determine the probability that, out of the seven chickens chosen,

a) exactly two are contaminated by salmonella.

b) at least two are contaminated by salmonella.

c) less than two are contaminated by salmonella.

d) between two and five, inclusive, are contaminated by salmonella.

13. Refer to Problem 7. Determine the mean and standard deviation of the number of literate people in a sample of four.

14. Strictly speaking, why doesn't the sampling in Problem 7 constitute a sequence of four Bernoulli trials? Why is it permissible to use the binomial distribution?

CHAPTER 6

THE NORMAL DISTRIBUTION

As we saw in Chapter 4, there are two main types of random variables, discrete and continuous. A **discrete random variable** is a random variable whose possible values form a discrete data set, usually some collection of whole numbers. Frequently, a discrete random variable involves a *count* of something, like the number of radios owned by a randomly selected person. On the other hand, a **continuous random variable** is a random variable whose possible values form a continuous data set, usually some interval of real numbers. Typically, a continuous random variable involves a *measurement* of something, like the weight of a newborn baby.

The probability distribution of a continuous random variable is called a **continuous probability distribution.** In this chapter we will discuss the most important continuous probability distribution, the **normal distribution,** or so-called "bell-shaped curve."

In our study of statistics, the normal distribution arises in two significant places. First, it has been discovered that many physical measurements involving things such as length, weight, and time have distributions that are bell-shaped. Thus, it is often appropriate to use the normal distribution as the distribution of a population or random variable. Second, the normal distribution is applied frequently in inferential statistics. For example, under certain circumstances, the normal distribution can be used to make inferences about the mean of a population.

CHAPTER OUTLINE

6.1 The standard normal curve Introduces a special normal curve and explains how to find areas under that curve.

6.2 Normal curves Discusses normal curves in general and how to determine areas under those curves.

6.3 Normally distributed populations Examines populations for which percentages are equal to areas under normal curves.

6.4 Normally distributed random variables Considers random variables for which probabilities are equal to areas under normal curves.

6.5 The normal approximation to the binomial distribution (Optional) Shows how normal curve areas can be used to approximate binomial probabilities.

6.1 The standard normal curve

We observe in the world around us a tremendous variety of populations and random variables. Although many populations and random variables are intrinsically different, a large number of them share an important characteristic. Namely, that the probabilities associated with them are equal, at least approximately, to areas under a "bell-shaped" curve; that is, a curve similar to the one shown in Figure 6.1.

FIGURE 6.1

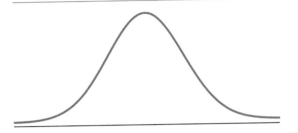

In this section and the next, we will learn how to find areas under such curves, which are called **normal curves.**

 There are many, in fact, infinitely many, normal curves. But fortunately there is a way to find areas under any normal curve once we know how to find areas under one specific normal curve—the **standard normal curve** (or **z-curve**). This curve is shown in Figure 6.2.† Note that the horizontal axis under the standard normal curve is labeled with the letter z.

FIGURE 6.2
**The standard normal
curve (z-curve)**

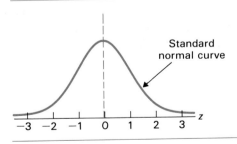

 Some of the most important properties of the standard normal curve are presented in the following key fact.

---

† The equation of the standard normal curve is

$$y = \frac{1}{\sqrt{2\pi}}\, e^{-z^2/2}$$

where $e \approx 2.718$ and $\pi \approx 3.142$.

KEY FACT 6.1   Basic properties of the standard normal curve

PROPERTY 1   The total area under the standard normal curve is equal to 1.

PROPERTY 2   The standard normal curve extends indefinitely in both directions, approaching the horizontal axis as it does so.

PROPERTY 3   The standard normal curve is *symmetric* about 0. That is, the part of the curve to the left of the dashed line in Figure 6.2 is the mirror image of the part of the curve to the right of it.

PROPERTY 4   Most of the area under the standard normal curve lies between -3 and 3.

MTB

We should point out that Property 1 in Key Fact 6.1 is not uniquely a property of the standard normal curve. In fact, the total area under any curve representing a continuous probability distribution is equal to 1.

USING THE STANDARD-NORMAL TABLE

Because of the importance of areas under the standard normal curve, tables of those areas have been constructed. We present such a table in Table II, which can be found inside the front cover of the book. A typical four-decimal-place number in the body of Table II gives the area under the standard normal curve between 0 and a specified value of z. Example 6.1 explains how to use Table II.

EXAMPLE 6.1   *Illustrates the use of the standard-normal table (Table II)*

Determine the area under the standard normal curve between 0 and 1.83.

SOLUTION   We use Table II. First we go down the left-hand column, labeled z, to "1.8." Then we go across that row until we are under the "0.03" in the top row. The number in the body of the table there, which is 0.4664, is the area under the standard normal curve between $z = 0$ and $z = 1.83$. See Figure 6.3.   ∎

FIGURE 6.3
Area under the standard normal curve between $z = 0$ and $z = 1.83$

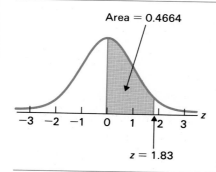

Area = 0.4664

$z = 1.83$

KEY FACT 6.2

A number in the body of Table II gives the area under the standard normal curve between 0 and a specified value of z.

To find the area under the standard normal curve between 0 and a negative value of z, we apply the fact that the standard normal curve is symmetric about 0 (Property 3 in Key Fact 6.1).

EXAMPLE 6.2   *Illustrates how to find the area between 0 and a negative z-value*

Determine the area under the standard normal curve between $z = -1.45$ and $z = 0$.

SOLUTION   We first find the area between $z = 0$ and $z = 1.45$. As in Example 6.1, we go down the column labeled z to "1.4" and then across that row until we are under the "0.05" in the top row. The number there, 0.4265, is the area under the standard normal curve between $z = 0$ and $z = 1.45$. By the symmetry of the curve, 0.4265 is also the area between $z = -1.45$ and $z = 0$. See Figure 6.4.   ■

FIGURE 6.4
Finding the area
between $z = -1.45$
and $z = 0$

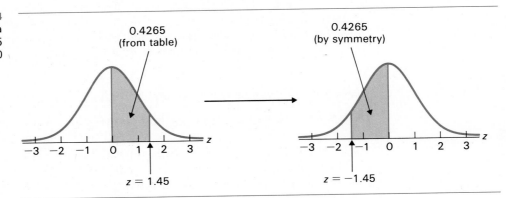

Note: Although z-values can be negative, areas must be positive. In fact, an area under any curve representing a continuous probability distribution, including the standard normal curve, must always be between 0 and 1.

Next we will learn how to obtain the area under the standard normal curve to the right of a positive value of z. To accomplish that, we use Properties 1 and 3 in Key Fact 6.1.

EXAMPLE 6.3   *Illustrates how to find the area to the right of a positive z-value*

Find the area under the standard normal curve to the right of $z = 1.25$.

SOLUTION   First of all, since the total area under the curve is 1 (Property 1) and the curve is symmetric about 0 (Property 3), it follows that the area under the standard normal curve to the right of $z = 0$ is 0.5. Next, we see from Table II that the area between $z = 0$ and $z = 1.25$ is 0.3944. Because the total area to the right of $z = 0$ is 0.5, the area to the right of $z = 1.25$ must be

$$0.5000 - 0.3944 = 0.1056$$

See Figure 6.5.

■

FIGURE 6.5
Finding the area to
the right of $z = 1.25$

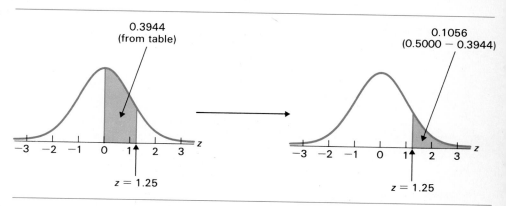

0.3944
(from table)

0.1056
(0.5000 − 0.3944)

$z = 1.25$

$z = 1.25$

Note: Using the method described in the previous example and the symmetry property, we can obtain the area to the left of a negative value of z. For instance, to find the area under the standard normal curve to the left of $z = -1.25$, we proceed as follows: First we observe that by the symmetry property (Property 3), the area to the left of $z = -1.25$ equals the area to the right of $z = 1.25$. Then we apply the method explained in Example 6.3 to determine the area to the right of $z = 1.25$, which is 0.1056. Thus, the area to the left of $z = -1.25$ is also 0.1056.

In the following example we will see how to determine the area under the standard normal curve between two positive values of z. A simple subtraction procedure does the trick.

EXAMPLE 6.4   *Illustrates how to find the area between two positive z-values*

Find the area under the standard normal curve between $z = 0.46$ and $z = 1.75$.

SOLUTION   By Table II, the area between $z = 0$ and $z = 0.46$ is 0.1772 and that between $z = 0$ and $z = 1.75$ is 0.4599. See Figure 6.6 at the top of the next page.

The area between $z = 0.46$ and $z = 1.75$ equals the area between $z = 0$ and $z = 1.75$ minus the area between $z = 0$ and $z = 0.46$. So we subtract:

$$\text{Area} = 0.4599 - 0.1772 = 0.2827$$

See Figure 6.7.

■

FIGURE 6.6

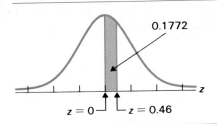

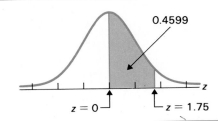

FIGURE 6.7
Area between
$z = 0.46$ and
$z = 1.75$

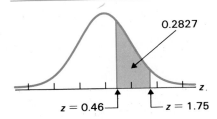

Note: To find the area under the standard normal curve between two negative z-values, we first obtain the area between the two corresponding positive z-values and then apply the symmetry property.

Our next illustration indicates the method for finding the area under the standard normal curve to the left of a positive value of z.

EXAMPLE 6.5   *Illustrates how to find the area to the left of a positive z-value*

Obtain the area under the standard normal curve to the left of $z = 1.96$.

SOLUTION   We begin by determining the area between $z = 0$ and $z = 1.96$. From Table II, that area is 0.4750. Since the area to the left of $z = 0$ is 0.5000 (why?), we conclude that the area to the left of $z = 1.96$ is

$$0.4750 + 0.5000 = 0.9750$$

This procedure is depicted in Figure 6.8.   ■

FIGURE 6.8
Area to the left of
$z = 1.96$ is the
total shaded area

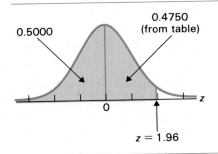

The example below shows how we can compute the area under the standard normal curve between a negative value of z and a positive value of z. This is done by adding two areas from Table II.

EXAMPLE 6.6

Illustrates how to find the area between a negative z-value and a positive z-value

Find the area under the standard normal curve between $z = -1.56$ and $z = 2.12$.

SOLUTION

To find the required area, we simply add the area between $z = -1.56$ and $z = 0$ to the area between $z = 0$ and $z = 2.12$. By symmetry, the area between $z = -1.56$ and $z = 0$ is the same as the area between $z = 0$ and $z = 1.56$, which is 0.4406. The area between $z = 0$ and $z = 2.12$ is 0.4830. Consequently, the area under the standard normal curve between $z = -1.56$ and $z = 2.12$ is

$$0.4406 + 0.4830 = \boxed{0.9236}$$

Figure 6.9 summarizes this discussion. ∎

FIGURE 6.9
Area under the standard normal curve between $z = -1.56$ and $z = 2.12$ is the total shaded area

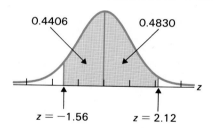

0.4406   0.4830

$z = -1.56$ $z = 2.12$

We have not covered every possible normal-curve-area problem. However, based on what you have learned in Examples 6.1 through 6.6, you should be able to use Table II and Properties 1 and 3 in Key Fact 6.1 to determine any required area under the standard normal curve.

MTB

SOME IMPORTANT AREAS UNDER THE STANDARD NORMAL CURVE

Property 4 in Key Fact 6.1 states that most of the area under the standard normal curve lies between $z = -3$ and $z = 3$. Using Table II, we can find out just how much area that is. The area between $z = 0$ and $z = 3$ is 0.4987; so is the area between $z = -3$ and $z = 0$. Thus, the area under the standard normal curve between $z = -3$ and $z = 3$ is $2 \cdot 0.4987 = 0.9974$. Since the total area under the standard normal curve is equal to 1 (Property 1 in Key Fact 6.1), we conclude that 99.74% of the area lies between $z = -3$ and $z = 3$.

By applying similar arguments, we can get the useful information given in Table 6.1. A visual display of that information is depicted in Figure 6.10.

TABLE 6.1
Some important
areas under
the standard
normal curve

z	Area under curve between −z and z	Percentage of total area
1	0.6826	68.26
2	0.9544	95.44
3	0.9974	99.74

FIGURE 6.10 Percentage of area under the standard normal curve that lies between
(a) $z = -1$ and $z = 1$, (b) $z = -2$ and $z = 2$, (c) $z = -3$ and $z = 3$

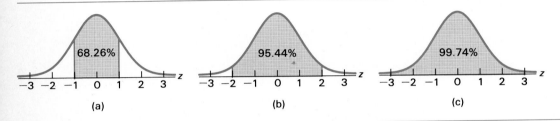

(a)   (b)   (c)

A NOTE CONCERNING TABLE II

The last value given in Table II is for $z = 3.90$. According to the table, the area between $z = 0$ and $z = 3.90$ is 0.5000. This does not mean that the area under the standard normal curve between $z = 0$ and $z = 3.90$ is exactly 0.5, but only that it is 0.5 to four decimal places (the area is 0.49995 to five decimal places). Indeed, since the area to the right of $z = 0$ is exactly 0.5 and the curve extends indefinitely to the right without ever touching the axis, the area between 0 and any value of z will be less than 0.5 no matter how far we go to the right.

FINDING THE Z-VALUE FOR A SPECIFIED AREA

Until now, we have used Table II to find areas under the standard normal curve between two specified values of z. For instance, in Example 6.1 on page 285, we used the table to find the area between $z = 0$ and $z = 1.83$, which is 0.4664. Now we will learn how to use Table II to find the z-value(s) corresponding to a specified area under the standard normal curve. Consider Example 6.7.

EXAMPLE 6.7 *Illustrates how to find the z-value for a specified area*

Find the z-value for which the area under the standard normal curve to the right of that value is 0.025. See Figure 6.11.

FIGURE 6.11

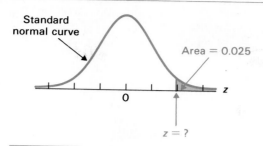

SOLUTION To find the z-value in question, first recall that the area under the standard normal curve to the right of $z = 0$ equals 0.5. Therefore, the area between $z = 0$ and the required value of z must be $0.500 - 0.025 = 0.475$, as shown in Figure 6.12.

FIGURE 6.12

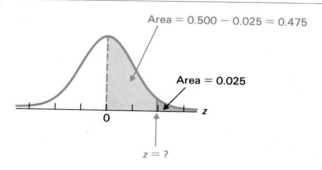

So, we need to use Table II to obtain the z-value corresponding to the area 0.475. For ease of reference we have reproduced a portion of Table II in Table 6.2 at the top of the next page.

We search the body of the table for the area 0.475 and upon finding it, note that the corresponding z-value is 1.96. Thus, the required z-value is $z = 1.96$. See Figure 6.13.

FIGURE 6.13

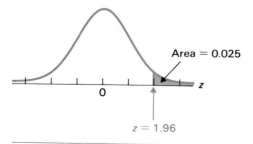

TABLE 6.2
Areas under
the standard
normal curve

z	Second decimal place in z									
	0.00	0.01	0.02	0.03	0.04	0.05	0.06	0.07	0.08	0.09
.	.	.	.	.	.	.	.	.	.	.
.	.	.	.	.	.	.	.	.	.	.
.	.	.	.	.	.	.	.	.	.	.
1.2	0.3849	0.3869	0.3888	0.3907	0.3925	0.3944	0.3962	0.3980	0.3997	0.4015
1.3	0.4032	0.4049	0.4066	0.4082	0.4099	0.4115	0.4131	0.4147	0.4162	0.4177
1.4	0.4192	0.4207	0.4222	0.4236	0.4251	0.4265	0.4279	0.4292	0.4306	0.4319
1.5	0.4332	0.4345	0.4357	0.4370	0.4382	0.4394	0.4406	0.4418	0.4429	0.4441
1.6	0.4452	0.4463	0.4474	0.4484	0.4495	0.4505	0.4515	0.4525	0.4535	0.4545
1.7	0.4554	0.4564	0.4573	0.4582	0.4591	0.4599	0.4608	0.4616	0.4625	0.4633
1.8	0.4641	0.4649	0.4656	0.4664	0.4671	0.4678	0.4686	0.4693	0.4699	0.4706
1.9	0.4713	0.4719	0.4726	0.4732	0.4738	0.4744	0.4750	0.4756	0.4761	0.4767
2.0	0.4772	0.4778	0.4783	0.4788	0.4793	0.4798	0.4803	0.4808	0.4812	0.4817
2.1	0.4821	0.4826	0.4830	0.4834	0.4838	0.4842	0.4846	0.4850	0.4854	0.4857
.	.	.	.	.	.	.	.	.	.	.
.	.	.	.	.	.	.	.	.	.	.
.	.	.	.	.	.	.	.	.	.	.

We will use the notation $z_{0.025}$ to denote the z-value with area 0.025 to its right under the standard normal curve. Consequently, as we have just seen in Example 6.7, $z_{0.025} = 1.96$. More generally, we will adopt the notation presented in the following definition.

DEFINITION 6.1

The symbol z_α is used to denote the z-value with area α (alpha) to its right under the standard normal curve. See Figure 6.14.

FIGURE 6.14

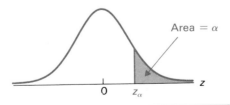

Area = α

0   z_α

As the previous example indicates, drawing a picture (or pictures) really helps to organize the solution to problems in which we must use Table II to find the z-value(s) corresponding to a specified area under the standard normal curve. The next three examples provide additional illustrations.

EXAMPLE 6.8   *Illustrates how to find the z-value for a specified area*

Use Table II to find
a) $z_{0.10}$ b) $z_{0.05}$

SOLUTION   a) $z_{0.10}$ is the z-value with area 0.10 to its right under the standard normal curve. See Figure 6.15.

FIGURE 6.15

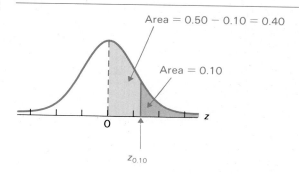

Area = 0.50 − 0.10 = 0.40

Area = 0.10

0

z

$z_{0.10}$

To find the value of $z_{0.10}$, we first note that the area under the curve between 0 and $z_{0.10}$ is $0.50 - 0.10 = 0.40$, as indicated in Figure 6.15. Thus, we search the body of Table II for the area 0.40. We find no such area in the table, and so we use the area closest to 0.40, which is 0.3997. The z-value corresponding to that area is $z = 1.28$. Consequently, $z_{0.10} = 1.28$, approximately.

b) $z_{0.05}$ is the z-value with area 0.05 to its right under the standard normal curve. See Figure 6.16.

FIGURE 6.16

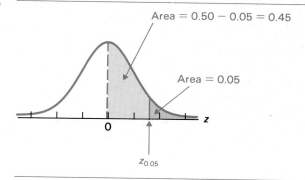

Area = 0.50 − 0.05 = 0.45

Area = 0.05

0

z

$z_{0.05}$

To find $z_{0.05}$, first observe that the area under the curve between 0 and $z_{0.05}$ is $0.50 - 0.05 = 0.45$, as shown in Figure 6.16. Consequently, we search the body of Table II for an area of 0.45. Again we find no such area in the table, and this time there are two areas closest to 0.45, namely, 0.4495 and 0.4505. Since the

z-values corresponding to these areas are $z = 1.64$ and $z = 1.65$, we take $z_{0.05}$ to be halfway between these. Thus, $z_{0.05} = 1.645$, approximately. ∎

EXAMPLE 6.9   *Illustrates how to find the z-value for a specified area*

Find the z-value for which the area under the standard normal curve to the left of that value is 0.04. See Figure 6.17.

FIGURE 6.17

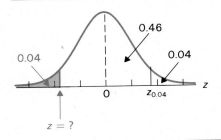

SOLUTION   We first note that by the symmetry property, the required z-value is the negative of the z-value with area 0.04 to its right; that is, the negative of $z_{0.04}$. Applying the same procedure as in the previous two examples, we find that $z_{0.04} = 1.75$, approximately. Therefore, the z-value with area 0.04 to its left is $z = -1.75$. ∎

EXAMPLE 6.10   *Illustrates how to find the z-values for a specified area*

Determine the two z-values that divide the area under the standard normal curve into a middle 0.95 area and two outside 0.025 areas. See Figure 6.18.

FIGURE 6.18

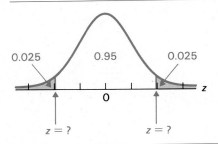

SOLUTION   As we see from Figure 6.18, the area of the shaded region on the right-hand side is 0.025. This means that the z-value on the right is $z_{0.025}$. In Example 6.7, we found that $z_{0.025} = 1.96$. Thus, the z-value on the right is 1.96. Since the standard normal curve is symmetric about 0, it follows that the z-value on the left is -1.96. Therefore we have the picture shown in Figure 6.19. ∎

MTB

FIGURE 6.19

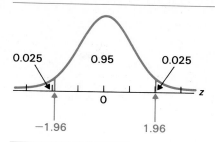

Exercises 6.1

Use Table II to find the areas under the standard normal curve specified in Exercises 6.1–6.16. Sketch a standard normal curve and shade the area of interest in each problem.

___ **6.1** Determine the area under the standard normal curve between $z = 0$ and

a) $z = 1$.   b) $z = 2$.   c) $z = 3$.
d) $z = 1.28$.   e) $z = 1.64$.   f) $z = 1.96$.

___ **6.2** Determine the area under the standard normal curve between $z = 0$ and

a) $z = 2.33$.   b) $z = 2.58$.   c) $z = 3.08$.
d) $z = 0.5$.   e) $z = 1.5$.   f) $z = 2.5$.

___ **6.3** Determine the area under the standard normal curve between $z = 0$ and

a) $z = -0.46$.   b) $z = -2.12$.   c) $z = -1.1$.
d) $z = -3.62$.   e) $z = -0.05$.   f) $z = -1.74$.

___ **6.4** Determine the area under the standard normal curve between $z = 0$ and

a) $z = -2.35$.   b) $z = -0.12$.   c) $z = -1.7$.
d) $z = -3.04$.   e) $z = -2$.   f) $z = -1.65$.

___ **6.5** Determine the area under the standard normal curve to the right of

a) $z = 1.64$.   b) $z = 1.96$.
c) $z = 2.33$.   d) $z = 2.58$.

___ **6.6** Determine the area under the standard normal curve to the right of

a) $z = 1$.   b) $z = 2$.
c) $z = 3$.   d) $z = 1.28$.

___ **6.7** Determine the area under the standard normal curve between

a) $z = 1$ and $z = 2$.   b) $z = 2$ and $z = 3$.
c) $z = -1$ and $z = -0.5$.   d) $z = -1.5$ and $z = -1$.

___ **6.8** Determine the area under the standard normal curve between

a) $z = 1.5$ and $z = 2$.   b) $z = 2$ and $z = 2.5$.
c) $z = -3$ and $z = -1$.   d) $z = -2$ and $z = -1$.

___ **6.9** Determine the area under the standard normal curve to the left of

a) $z = 1.78$.   b) $z = 0.94$.
c) $z = 2.4$.   d) $z = 3.1$.

___ **6.10** Find the area under the standard normal curve to the left of

a) $z = 2.03$.   b) $z = 0.04$.
c) $z = 1.95$.   d) $z = 1.31$.

___ **6.11** Obtain the area under the standard normal curve between

a) $z = -1$ and $z = 2$.
b) $z = -2$ and $z = 3$.
c) $z = -1.75$ and $z = 0.58$.
d) $z = -0.75$ and $z = 2.34$.

___ **6.12** Obtain the area under the standard normal curve between

a) $z = -2$ and $z = 1$.
b) $z = -3$ and $z = 1$.
c) $z = -1.83$ and $z = 2.12$.
d) $z = -0.63$ and $z = 1.77$.

___ **6.13** Find the area under the standard normal curve to the left of

a) $z = -1$.   b) $z = -2$.
c) $z = -1.28$.   d) $z = -1.64$.

___ **6.14** Find the area under the standard normal curve to the left of

a) $z = -3$.   b) $z = -1.96$.
c) $z = -2.33$.   d) $z = -2.58$.

___ 6.15 Find the area under the standard normal curve that lies
a) to the right of $z = -1.38$.
b) either to the left of $z = -2.12$ or to the right of $z = 1.67$.
c) either to the left of $z = 0.63$ or to the right of $z = 1.54$.

___ 6.16 Find the area under the standard normal curve that lies
a) to the right of $z = -1.96$.
b) either to the left of $z = -1$ or to the right of $z = 2$.
c) either to the left of $z = 1$ or to the right of $z = 2.5$.

___ 6.17 Use Table II to obtain the following shaded areas under the standard normal curve:

a)

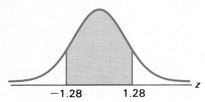

b)

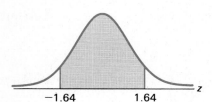

c)

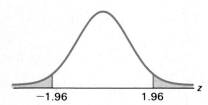

d)

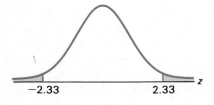

___ 6.18 Use Table II to obtain the following shaded areas under the standard normal curve:

a)

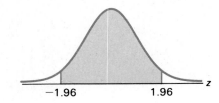

b)

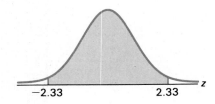

c)

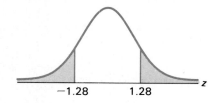

d)

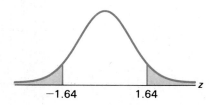

___ 6.19 Verify that the information in Table 6.1 on page 290 is correct. Draw standard normal curves for the three entries and shade the appropriate areas.

___ 6.20 Below is a standard normal curve. The total area under the curve is divided into eight regions.

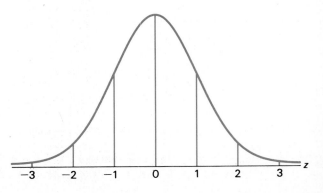

a) Determine the area of each of the eight regions.

b) Complete the following table:

Region	Area	Percentage of total area
$-\infty$ to -3	0.0013	0.13
-3 to -2		
-2 to -1		
-1 to 0		
0 to 1	0.3413	34.13
1 to 2		
2 to 3		
3 to ∞		
	1.0000	100.00

__ 6.21 Determine $z_{0.33}$; that is, find the z-value with area 0.33 to its right under the standard normal curve. Illustrate your work with a picture similar to Figure 6.16 on page 293.

__ 6.22 Determine $z_{0.015}$; that is, find the z-value with area 0.015 to its right under the standard normal curve. Illustrate your work with a picture similar to Figure 6.16 on page 293.

__ 6.23 Find the following z-values and draw graphs to illustrate your work.

a) $z_{0.03}$ b) $z_{0.005}$

__ 6.24 Obtain the z-values below and draw graphs illustrating your work.

a) $z_{0.20}$ b) $z_{0.06}$

__ 6.25 Obtain the z-value for which the area under the standard normal curve to the left of that value is equal to 0.025.

__ 6.26 Determine the z-value for which the area under the standard normal curve to the left of that value is 0.01.

__ 6.27 Find the z-value with area 0.75 to its left under the standard normal curve.

__ 6.28 Obtain the z-value with area 0.80 to its left under the standard normal curve.

__ 6.29 Determine the z-value for which the area under the standard normal curve to the right of that value is 0.95; that is, find $z_{0.95}$.

__ 6.30 Determine the z-value for which the area under the standard normal curve to the right of that value is 0.70; that is, find $z_{0.70}$.

__ 6.31 Find the two z-values that divide the area under the standard normal curve into a middle 0.90 area and two outside 0.05 areas. Illustrate your work with pictures similar to Figures 6.18 and 6.19.

__ 6.32 Find the two z-values that divide the area under the standard normal curve into a middle 0.99 area and two outside 0.005 areas. Illustrate your work with graphs similar to Figures 6.18 and 6.19.

__ 6.33 Complete the following table:

$z_{0.10}$	$z_{0.05}$	$z_{0.025}$	$z_{0.01}$	$z_{0.005}$
1.28				

== 6.34 Let $0 < \alpha < 1$. For the standard normal curve, determine

a) the z-value with area α to its right.

b) the z-value with area α to its left.

c) the two z-values that divide the area under the curve into a middle $1 - \alpha$ area and two outside areas of $\alpha/2$.

d) Draw graphs to illustrate the results that you obtained in parts (a)–(c).

6.2 Normal curves

Recall that it is important to be able to determine areas under normal curves since those areas correspond to probabilities for many populations and random variables. In the previous section we learned how to find areas under the standard normal curve. With that knowledge we can obtain areas under any normal curve, as we will discover in this section.

Each normal curve can be identified by two parameters, usually denoted by μ and σ. The parameter μ tells us where the normal curve is centered, and the parameter σ indicates its spread (shape).[†] Two normal curves that have the same μ-parameter are centered at the same place; and two normal curves that have the same σ-parameter have the same shape.

  Figure 6.20 shows three normal curves. The normal curve on the left has parameters $\mu = -2$ and $\sigma = 1$; the one in the center has parameters $\mu = 3$ and $\sigma = 1/2$; and the one on the right has parameters $\mu = 8$ and $\sigma = 2$.

FIGURE 6.20

Three normal curves

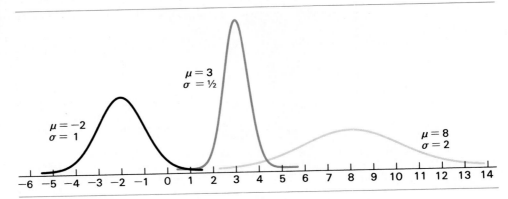

Note that each of the three normal curves displayed in Figure 6.20 is centered at its μ-parameter. Also note that the larger the σ-parameter, the more a normal curve is spread out. Some of the most important properties of normal curves are presented in Key Fact 6.3.

KEY FACT 6.3   Basic properties of normal curves

PROPERTY 1   The total area under a normal curve is equal to 1.

PROPERTY 2   A normal curve extends indefinitely in both directions, approaching the horizontal axis as it does so.

PROPERTY 3   A normal curve with parameters μ and σ is symmetric about μ. That is, the part of the curve to the left of μ is the mirror image of the part of the curve to the right of μ.

PROPERTY 4   Most of the area under the normal curve with parameters μ and σ lies between $\mu - 3\sigma$ and $\mu + 3\sigma$.

SKETCHING NORMAL CURVES

Although it is not necessary to draw normal curves exactly, it is useful to be able to sketch them. Using Key Fact 6.3, it is easy to make a quick sketch of any normal curve. We illustrate how to do this in the following example.

[†] The equation of the normal curve with parameters μ and σ is

$$y = \frac{1}{\sqrt{2\pi}\sigma} e^{-(x-\mu)^2/2\sigma^2}$$

where $e \approx 2.718$ and $\pi \approx 3.142$.

EXAMPLE 6.11   *Illustrates sketching normal curves*

Sketch the normal curve with parameters
a) $\mu = 5$ and $\sigma = 2$.
b) $\mu = 0$ and $\sigma = 1$.

SOLUTION   a) By Property 3 in Key Fact 6.3, the normal curve with parameters $\mu = 5$ and $\sigma = 2$ is symmetric about $\mu = 5$. Also, by Property 4, most of the area under that normal curve lies between

$$\mu - 3\sigma = 5 - 3 \cdot 2 = -1$$

and

$$\mu + 3\sigma = 5 + 3 \cdot 2 = 11$$

So, we sketch this curve as shown in Figure 6.21.

FIGURE 6.21
Sketch of
normal curve
with parameters
$\mu = 5$ and $\sigma = 2$

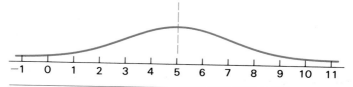

b) Next we sketch the normal curve with parameters $\mu = 0$ and $\sigma = 1$. By Property 3 in Key Fact 6.3, we know that this normal curve is symmetric about $\mu = 0$ and, from Property 4, most of the area under this normal curve lies between

$$\mu - 3\sigma = 0 - 3 \cdot 1 = -3$$

and

$$\mu + 3\sigma = 0 + 3 \cdot 1 = 3$$

MTB

Thus, we can sketch this curve as pictured in Figure 6.22.   ∎

FIGURE 6.22
Sketch of
normal curve
with parameters
$\mu = 0$ and $\sigma = 1$

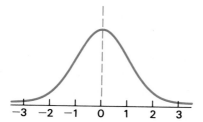

The curve in Figure 6.22 should look familiar. It is the standard normal curve. (See Figure 6.2 on page 284.) In other words:

KEY FACT 6.4

The normal curve with parameters $\mu = 0$ and $\sigma = 1$ is the standard normal curve.

FINDING AREAS UNDER NORMAL CURVES

We will now proceed with the main business of this section—to learn how to find areas under normal curves. The basic idea is to relate any given normal curve to the standard normal curve.

EXAMPLE 6.12   *Illustrates how to find areas under a normal curve*

Find the area under the normal curve with parameters $\mu = 5$ and $\sigma = 2$ that lies
a) to the right of 7.   b) between 3 and 8.

SOLUTION   a) First we sketch the normal curve with parameters $\mu = 5$ and $\sigma = 2$ and shade the area to the right of 7. See Figure 6.23. Note that we have labeled the horizontal axis in Figure 6.23 with an "x." This will help us to keep things straight.

FIGURE 6.23
Normal curve
with parameters
$\mu = 5$ and $\sigma = 2$,
with area to the
right of 7 shaded

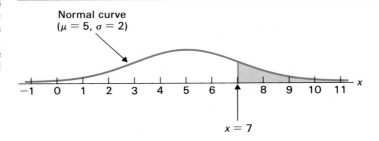

It can be shown mathematically (see Exercise 6.62) that the area under the normal curve with parameters $\mu = 5$ and $\sigma = 2$ that lies to the right of $x = 7$ is equal to the area under the standard normal curve that lies to the right of

$$z = \frac{x - \mu}{\sigma} = \frac{7 - 5}{2} = 1$$

This latter area is pictured in Figure 6.24.

FIGURE 6.24
Standard normal
curve, with area to
the right of 1 shaded

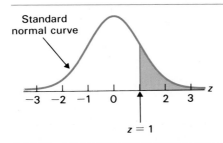

In other words, the shaded areas in Figures 6.23 and 6.24 are equal. By Table II, the area shaded in Figure 6.24 is $0.5000 - 0.3413 = 0.1587$. Thus, the area shaded in Figure 6.23 is also 0.1587. That is, the area under the normal curve with parameters $\mu = 5$ and $\sigma = 2$ that lies to the right of 7 equals 0.1587.

b) Here we need to determine the area under the normal curve with parameters $\mu = 5$ and $\sigma = 2$ that lies between 3 and 8. See Figure 6.25.

FIGURE 6.25
Normal curve with parameters $\mu = 5$ and $\sigma = 2$, with area between 3 and 8 shaded

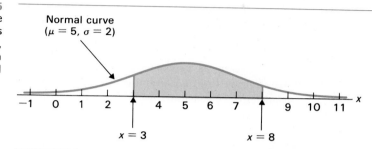

We obtain this area as follows: The area under the normal curve with parameters $\mu = 5$ and $\sigma = 2$ that lies between $x = 3$ and $x = 8$ is equal to the area under the standard normal curve that lies between

$$z = \frac{x - \mu}{\sigma} = \frac{3 - 5}{2} = -1$$

and

$$z = \frac{x - \mu}{\sigma} = \frac{8 - 5}{2} = 1.5$$

This latter area is shown in Figure 6.26.

FIGURE 6.26
Standard normal curve, with area between -1 and 1.5 shaded

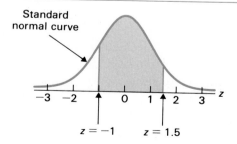

By Table II, the shaded area in Figure 6.26 equals $0.3413 + 0.4332 = 0.7745$. Consequently, the shaded area in Figure 6.25 also equals 0.7745. ∎

Z-SCORES

As indicated in Example 6.12, finding areas under normal curves entails converting x-values to z-values by first subtracting μ and then dividing by σ:

$$z = \frac{x - \mu}{\sigma}$$

That conversion process is often referred to as **standardizing.** It is precisely the same process that we use to compute population z-scores (see Definition 3.11 on page 131). In applications, the parameters μ and σ will be the mean and standard deviation of a population or random variable. Hence, we will refer to z-values obtained by standardizing as **z-scores.**

We now summarize the fundamental fact that permits us to determine areas under any normal curve by using the Table II values of areas under the standard normal curve.

KEY FACT 6.5

The area under the normal curve with parameters μ and σ that lies between $x = a$ and $x = b$ is equal to the area under the standard normal curve that lies between

$$z = \frac{a - \mu}{\sigma} \qquad \text{and} \qquad z = \frac{b - \mu}{\sigma}$$

See Figure 6.27. The area under the normal curve with parameters μ and σ that lies to the right (or left) of a given x-value is found in a similar way; namely, by first converting to the z-score and then using Table II to find the area under the standard normal curve that lies to the right (or left) of the z-score.

FIGURE 6.27

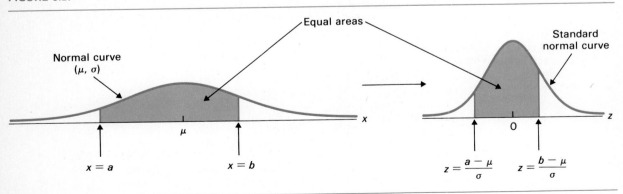

Note: The z-values in Table II are given to two decimal places. Therefore, when computing a z-score, always round the result to two decimal places.

A STREAMLINED PROCEDURE FOR FINDING NORMAL CURVE AREAS

In Example 6.12, pages 300–301, we drew two diagrams to solve each area problem—one for the normal curve with parameters $\mu = 5$ and $\sigma = 2$ and one for the standard normal curve. Generally, the two-diagram approach for finding normal curve areas is as depicted in Figure 6.27. Although that approach is helpful for explaining and understanding the necessary ideas, it is somewhat cumbersome to apply in practice. Below we will present a streamlined procedure that is faster and simpler to use.

EXAMPLE 6.13 *Introduces the streamlined procedure for finding normal curve areas*

Determine the area under the normal curve with parameters $\mu = 3$ and $\sigma = 4$ that lies between −6 and 9.

SOLUTION **STEP 1** *Sketch the normal curve with parameters μ and σ.*

Here $\mu = 3$ and $\sigma = 4$. We have sketched that normal curve in Figure 6.28. Note that the tick marks are $\sigma = 4$ units apart.

FIGURE 6.28

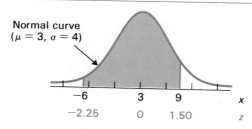

Normal curve
$(\mu = 3, \sigma = 4)$

−6	3	9	x
−2.25	0	1.50	z

z-score computations: Area between 0 and *z*:

$x = -6 \longrightarrow z = \dfrac{-6 - 3}{4} = -2.25$ 0.4878

$x = 9 \longrightarrow z = \dfrac{9 - 3}{4} = 1.50$ 0.4332

Shaded area $= 0.4878 + 0.4332 = 0.9210$

STEP 2 *Indicate on the graph the area to be determined.*

See the shaded area in Figure 6.28.

STEP 3 *Compute the required z-scores and mark them on the graph beneath the x-values.*

We need to obtain the z-scores for the x-values, $x = -6$ and $x = 9$:

$$x = -6 \longrightarrow z = \frac{-6 - 3}{4} = -2.25$$

$$x = 9 \longrightarrow z = \frac{9 - 3}{4} = 1.50$$

We have marked these z-scores beneath the x-values in Figure 6.28.

STEP 4 *Use Table II to obtain the required area.*

The area under the standard normal curve between $z = -2.25$ and $z = 0$ is 0.4878, and the area under the standard normal curve between $z = 0$ and $z = 1.50$ is 0.4332. Consequently, the required area (the area shaded in Figure 6.28) equals

$$0.4878 + 0.4332 = \boxed{0.9210}$$

MTB

This entire four-step procedure can be accomplished quickly and easily by performing all the steps in a "picture" as in Figure 6.28.   ■

The step-by-step procedure, illustrated in Example 6.13, is given below as Procedure 6.1.

PROCEDURE 6.1

To determine areas under the normal curve with parameters μ and σ.

STEP 1 *Sketch the normal curve with parameters μ and σ.*
STEP 2 *Indicate on the graph the area to be determined.*
STEP 3 *Compute the required z-scores and mark them on the graph beneath the x-values.*
STEP 4 *Use Table II to obtain the required area.*

EXAMPLE 6.14   *Illustrates Procedure 6.1*

Determine the area under the normal curve with parameters $\mu = 100$ and $\sigma = 16$ that lies to the right of 120.

SOLUTION   The results of applying Procedure 6.1 are shown in Figure 6.29. Thus, as we see from the last line of the figure, the required area is 0.1056.   ■

FIGURE 6.29

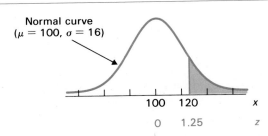

Normal curve
$(\mu = 100, \sigma = 16)$

	100	120		x
	0	1.25		z

z-score computation:

$x = 120 \longrightarrow z = \dfrac{120 - 100}{16} = 1.25$

Area between 0 and z:

0.3944

Shaded area $= 0.5000 - 0.3944 = 0.1056$

FINDING THE X-VALUE FOR A SPECIFIED AREA

The formula for converting x-values to z-scores is

$$z = \frac{x - \mu}{\sigma}$$

and, as we said, this process is called *standardizing*. If we solve the previous equation for x, we obtain a formula for converting z-scores to x-values:

$$x = \mu + z \cdot \sigma$$

We will call this latter process **destandardizing.**

Procedure 6.1 indicates how to find the area under a normal curve between two specified values of x. Frequently, however, we need to perform the reverse procedure; that is, find the x-values corresponding to a specified area under a normal curve. To accomplish this, we proceed in the following manner: First we determine, for the standard normal curve, the z-values corresponding to the specified area, as explained at the end of Section 6.1. Then we destandardize the z-values to obtain the required x-values. Example 6.15 illustrates this procedure.

EXAMPLE 6.15 *Illustrates how to find the x-value for a specified area*

For the normal curve with parameters $\mu = 100$ and $\sigma = 16$, find the x-value with area 0.04 to its left. See Figure 6.30.

FIGURE 6.30

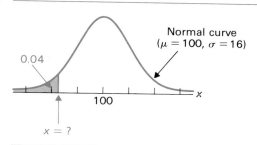

0.04

Normal curve
($\mu = 100$, $\sigma = 16$)

100

x

x = ?

SOLUTION First we determine the z-score with area 0.04 to its left; that is, the z-value with area 0.04 to its left under the standard normal curve. We did this earlier in Example 6.9 on page 294 of Section 6.1. We found that $z = -1.75$. Now we destandardize this z-score to obtain the required x-value:

$$x = \mu + z \cdot \sigma = 100 + (-1.75) \cdot 16 = 100 - 1.75 \cdot 16 = 72$$

MTB So, for the normal curve with parameters $\mu = 100$ and $\sigma = 16$, the x-value with area 0.04 to its left is $x = 72$.

Exercises 6.2

___ 6.35   Which normal curve has a wider spread, the one with parameters $\mu = 1$ and $\sigma = 2$ or the one with parameters $\mu = 2$ and $\sigma = 1$?

___ 6.36   True or False: The normal curve with parameters $\mu = -4$ and $\sigma = 3$ and the normal curve with parameters $\mu = 6$ and $\sigma = 3$ have the same shape. Explain your answer.

___ 6.37   True or False: The value of the parameter μ has no effect on the shape of a normal curve. Explain your answer.

___ 6.38   Answer true or false and give a reason for your answer: The normal curve with parameters $\mu = -4$ and $\sigma = 3$ and the normal curve with parameters $\mu = 6$ and $\sigma = 3$ are centered at the same place.

___ 6.39   Apply the procedure illustrated in Example 6.11 on page 299 to sketch the normal curve with parameters
a) $\mu = 3$ and $\sigma = 3$.
b) $\mu = 1$ and $\sigma = 3$.
c) $\mu = 3$ and $\sigma = 1$.

___ 6.40   Sketch the normal curve with parameters
a) $\mu = -2$ and $\sigma = 2$.
b) $\mu = -2$ and $\sigma = 1/2$.
c) $\mu = 0$ and $\sigma = 2$.

In Exercises 6.41–6.44, use the two-diagram procedure, portrayed in Figure 6.27 on page 302, to find the designated areas.

___ 6.41   Determine the area under the normal curve with parameters $\mu = 1$ and $\sigma = 2.5$ that lies
a) to the right of 0.
b) to the left of -1.5.
c) between -2 and 2.

___ 6.42   Find the area under the normal curve with parameters $\mu = -1.5$ and $\sigma = 1$ that lies
a) between 0 and 1.4.
b) to the left of -1.5.
c) to the right of 1.

___ 6.43   Find the area under the normal curve with parameters $\mu = 2$ and $\sigma = 1/2$ that lies
a) to the left of 2.87.
b) between 1 and 1.5.
c) to the right of 2.75.

___ 6.44   Determine the area under the normal curve with parameters $\mu = 3$ and $\sigma = 0.75$ that lies
a) to the left of 2.5.
b) between 2 and 4.
c) to the right of 1.5.

In Exercises 6.45–6.50, use Procedure 6.1 on page 304 to determine the indicated areas.

___ 6.45   Find the area under the normal curve with parameters $\mu = 74$ and $\sigma = 2$ that lies
a) between 71 and 78.
b) to the right of 76.5.

___ 6.46   Obtain the area under the normal curve with parameters $\mu = 7.3$ and $\sigma = 2$ that lies
a) between 4.3 and 11.8.
b) to the left of 9.94.

___ 6.47   Determine the area under the normal curve with parameters $\mu = 335$ and $\sigma = 10$ that lies
a) to the left of 348.5.
b) between 340 and 350.

___ 6.48   Find the area under the normal curve with parameters $\mu = 64.4$ and $\sigma = 2.4$ that lies
a) to the right of 70.
b) between 68 and 71.

___ 6.49   Find the area under the normal curve with parameters $\mu = 40.9$ and $\sigma = 7.1$ that lies
a) to the left of 35.
b) between 25 and 30.

___ 6.50   Determine the area under the normal curve with parameters $\mu = 15.6$ and $\sigma = 5.1$ that lies
a) to the left of 5.
b) between 3.42 and 10.

___ 6.51   For the normal curve with parameters $\mu = 74$ and $\sigma = 2$, determine the
a) x-value with area 0.05 to its right.
b) x-value with area 0.10 to its left.
c) two x-values that divide the area under the curve into a middle 0.95 area and two outside 0.025 areas.

___ 6.52   For the normal curve with parameters $\mu = 7.3$ and $\sigma = 2$, determine the
a) x-value with area 0.025 to its right.
b) x-value with area 0.01 to its left.
c) two x-values that divide the area under the curve into a middle 0.90 area and two outside 0.05 areas.

— **6.53**   For the normal curve with parameters $\mu = 335$ and $\sigma = 10$, obtain the x-value with
a) area 0.60 to its left.
b) area 0.60 to its right.

— **6.54**   For the normal curve with parameters $\mu = 64.4$ and $\sigma = 2.4$, find the x-value with
a) area 0.90 to its left.
b) area 0.90 to its right.

— **6.55**   For the normal curve with parameters $\mu = 40.9$ and $\sigma = 7.1$, find the three x-values that divide the area under the curve into four 0.25 areas.

— **6.56**   For the normal curve with parameters $\mu = 15.6$ and $\sigma = 5.1$, obtain the three x-values that divide the area under the curve into four 0.25 areas.

═ **6.57**   Find the indicated area under the normal curve with the given parameters.
a) $\mu = 0$, $\sigma = 1$; area between -3 and 3.
b) $\mu = 4$, $\sigma = 2$; area between -2 and 10.
c) $\mu = \mu$, $\sigma = \sigma$; area between $\mu - 3\sigma$ and $\mu + 3\sigma$.

═ **6.58**   Apply Procedure 6.1 to:
a) Complete the following table for the normal curve with parameters $\mu = 4$ and $\sigma = 3$. Draw three graphs illustrating your results.

Given x-values	Corresponding z-scores	Area between x-values
1 & 7	−1 & 1	0.6826
−2 & 10	−2 & 2	
−5 & 13		

b) Complete the following table for the normal curve with parameters μ and σ. Draw three graphs illustrating your results.

Given x-values	Corresponding z-scores	Area between x-values
$\mu - \sigma$ & $\mu + \sigma$	−1 & 1	0.6826
$\mu - 2\sigma$ & $\mu + 2\sigma$		
$\mu - 3\sigma$ & $\mu + 3\sigma$		

═ **6.59**   Employ Procedure 6.1 to solve the problems posed in parts (a) and (b).
a) Below is a sketch of the normal curve with parameters $\mu = 4$ and $\sigma = 3$. The total area under the

curve is divided into eight regions. We found the area of the region between 7 and 10 and recorded that area on the picture in color. Find and record the z-scores and areas for the remaining regions.

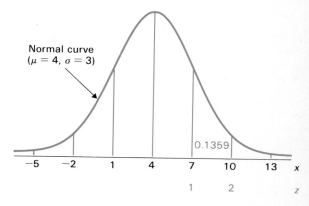

Normal curve
($\mu = 4$, $\sigma = 3$)

0.1359

b) Repeat part (a) for the normal curve with parameters μ and σ.

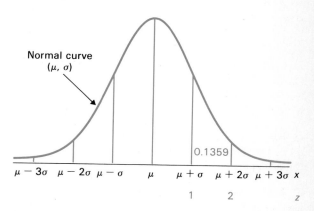

Normal curve
(μ, σ)

0.1359

═ **6.60**   Let $0 < \alpha < 1$. For the normal curve with parameters μ and σ, determine
a) the x-value with area α to its right.
b) the x-value with area α to its left.
c) the two x-values that divide the area under the curve into a middle area of $1 - \alpha$ and two outside areas of $\alpha/2$.

═ **6.61**   For the normal curve with parameters μ and σ, obtain the three x-values that divide the area under the curve into four areas of 0.25.

═ **6.62**   This exercise verifies Key Fact 6.5, which is displayed on page 302. A knowledge of elementary calculus is required to complete the exercise.

a) Show that the area under the normal curve with parameters μ and σ that lies between $x = a$ and $x = b$ equals

$$\int_{a}^{b} \frac{1}{\sqrt{2\pi}\sigma} e^{-(x-\mu)^2/2\sigma^2} \, dx$$

[See the footnote on page 298.]

b) Making the substitution $z = (x - \mu)/\sigma$, show that the integral in part (a) equals

$$\int_{(a-\mu)/\sigma}^{(b-\mu)/\sigma} \frac{1}{\sqrt{2\pi}} e^{-z^2/2} \, dz$$

c) What area does the integral in part (b) equal? [See the footnote on page 284.]

6.3 Normally distributed populations

As we mentioned in Section 6.1, many populations have distributions that can be represented by normal curves. This means that percentages for the population are equal, at least approximately, to areas under a suitable normal curve. The following example should help make this idea clear.

EXAMPLE 6.16   *Illustrates a normally distributed population*

A midwestern college has an enrollment of 3264 female students. Records show that the population mean height of these students is 64.4 inches, with a standard deviation of 2.4 inches: $\mu = 64.4$ inches and $\sigma = 2.4$ inches.

Frequency and relative-frequency distributions for this population of heights are given in Table 6.3. The table shows, for instance, that 7.35% (0.0735) of the students are between 67 and 68 inches tall.

TABLE 6.3
Frequency and relative-frequency distributions for heights

Height (inches)	Frequency f	Relative frequency
56–under 57	3	0.0009
57–under 58	6	0.0018
58–under 59	26	0.0080
59–under 60	74	0.0227
60–under 61	147	0.0450
61–under 62	247	0.0757
62–under 63	382	0.1170
63–under 64	483	0.1480
64–under 65	559	0.1713
65–under 66	514	0.1575
66–under 67	359	0.1100
67–under 68	240	0.0735
68–under 69	122	0.0374
69–under 70	65	0.0199
70–under 71	24	0.0074
71–under 72	7	0.0021
72–under 73	5	0.0015
73–under 74	1	0.0003
	3264	1.0000

A relative-frequency histogram for the heights of the 3264 female students is displayed in Figure 6.31.

FIGURE 6.31
Relative-frequency
histogram for heights

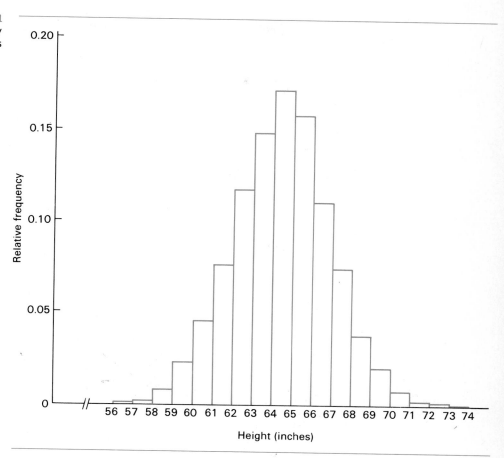

Note that the histogram in Figure 6.31 is bell-shaped. Because of this, we can approximate percentages (relative frequencies, probabilities) for the population of heights by areas under a suitable normal curve. As you might guess, the appropriate normal curve is the one with parameters μ and σ, where μ is the mean of the population and σ is its standard deviation. Here $\mu = 64.4$ and $\sigma = 2.4$. In Figure 6.32, at the top of the next page, we have superimposed the normal curve with parameters $\mu = 64.4$ and $\sigma = 2.4$ upon the histogram in Figure 6.31.

Now, let us see just how we can approximate percentages for this population of heights by areas under the normal curve with parameters $\mu = 64.4$ and $\sigma = 2.4$. To be specific, let us consider the percentage of female students who are between 67 and 68 inches tall. According to Table 6.3, the exact percentage is 7.35%, or 0.0735. Note that 0.0735 also equals the area of the cross-hatched rectangle in Figure 6.32. This is because the rectangle has height 0.0735 and width 1.

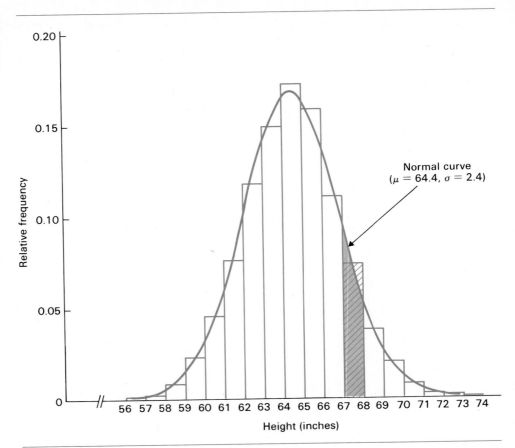

Look now at the area under the normal curve between 67 and 68, the area shaded in Figure 6.32. Observe that this area is approximately equal to the area of the cross-hatched rectangle which, as we have noted, equals the percentage of students who are between 67 and 68 inches tall. Consequently, *we can approximate the percentage of students who are between 67 and 68 inches tall by the area under the normal curve between 67 and 68.*

Although the main point of this example has now been made, it is interesting to actually compare the numerical values under consideration here. We already know that the percentage of students who are between 67 and 68 inches tall is exactly 7.35%. To determine the area under the normal curve between 67 and 68, we apply Procedure 6.1 on page 304. See Figure 6.33 at the top of the next page.

The last line of Figure 6.33 shows that the area under the normal curve between 67 and 68 is 0.0733, or 7.33%. Comparing this with the exact percentage of students who are between 67 and 68 inches tall, 7.35%, we see that the approximation by the area under the normal curve gives excellent results.   ∎

FIGURE 6.33
Determination of
the area under the
normal curve with
parameters $\mu = 64.4$
and $\sigma = 2.4$ that lies
between 67 and 68

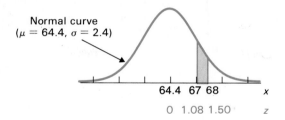

Normal curve
$(\mu = 64.4, \sigma = 2.4)$

64.4 67 68 x

0 1.08 1.50 z

z-score computations: Area between 0 and z:

$x = 67 \longrightarrow z = \dfrac{67 - 64.4}{2.4} = 1.08$ 0.3599

$x = 68 \longrightarrow z = \dfrac{68 - 64.4}{2.4} = 1.50$ 0.4332

Shaded area $= 0.4332 - 0.3599 = 0.0733$

The important point of Example 6.16 is that for certain populations, percentages are approximately equal to areas under a suitable normal curve. Not all populations have that property but, if a population does, then we say that it is **normally distributed.**

DEFINITION 6.2   Normally distributed population

A population is said to be (approximately) *normally distributed* if percentages for the population are (approximately) equal to areas under a normal curve. If such a population has mean μ and standard deviation σ, then the normal curve that is used is the one with parameters μ and σ.

Note: It is helpful to keep in mind what it means qualitatively for a population to be normally distributed; namely, that the histogram of the population is bell-shaped.

There are many instances where it is known that a population is normally distributed on the basis of past experience, previous statistical studies, or theoretical considerations. In these cases it is simple to determine any required percentage. We illustrate such a situation in the next example.

EXAMPLE 6.17   *Illustrates finding percentages for a normally distributed population*

In a 1905 study, R. Pearl determined that brain weights of Swedish men are approximately normally distributed with a mean of $\mu = 1.40$ kg (kilograms) and a standard deviation of $\sigma = 0.11$ kg.[†] Determine the percentage of Swedish men with brain weights
a) between 1.50 and 1.70 kg.
b) less than 1.20 kg.

[†] Pearl, R. 1905. Biometrical studies on man. I. Variation and correlation in brain weight. *Biometrica*, vol. 4, pp. 13–104.

SOLUTION Since the brain weights are normally distributed, percentages are equal to areas under a normal curve; namely, the normal curve whose parameters are the same as the mean and standard deviation of the population. Here $\mu = 1.40$ and $\sigma = 0.11$.

a) For this part, we want to determine the percentage of Swedish men with brain weights between 1.50 and 1.70 kg. To obtain this, we need to find the area under the normal curve with parameters $\mu = 1.40$ and $\sigma = 0.11$ that lies between 1.50 and 1.70. That area is computed in Figure 6.34.

FIGURE 6.34
Determination of the area under the normal curve with parameters $\mu = 1.40$ and $\sigma = 0.11$ that lies between 1.50 and 1.70

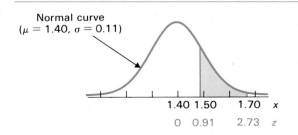

Normal curve
($\mu = 1.40$, $\sigma = 0.11$)

1.40 1.50   1.70   *x*
0   0.91   2.73   *z*

z-score computations: Area between 0 and *z*:

$x = 1.5 \longrightarrow z = \dfrac{1.50 - 1.40}{0.11} = 0.91$ 0.3186

$x = 1.7 \longrightarrow z = \dfrac{1.70 - 1.40}{0.11} = 2.73$ 0.4968

Shaded area = 0.4968 − 0.3186 = 0.1782

So, **17.82%** (0.1782) of Swedish men have brain weights between 1.50 and 1.70 kg.

b) To obtain the percentage of Swedish men with brain weights less than 1.20 kg, we need to find the area under the normal curve with parameters $\mu = 1.40$ and $\sigma = 0.11$ that lies to the left of 1.20. This is done in Figure 6.35.

FIGURE 6.35
Determination of the area under the normal curve with parameters $\mu = 1.40$ and $\sigma = 0.11$ that lies to the left of 1.20

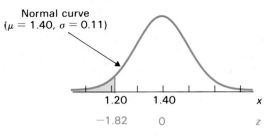

Normal curve
($\mu = 1.40$, $\sigma = 0.11$)

1.20   1.40   *x*
−1.82   0   *z*

z-score computation: Area between 0 and *z*:

$x = 1.2 \longrightarrow z = \dfrac{1.20 - 1.40}{0.11} = -1.82$ 0.4656

Shaded area = 0.5000 − 0.4656 = 0.0344

Thus, about **3.44%** of Swedish men have brain weights less than 1.20 kg. ∎

POPULATION Z-SCORES

In Section 3.7, we defined the z-score for a population value. For convenience, we repeat that definition here.

DEFINITION 6.3   Population z-score

The z-score for a population value, x, is defined to be the number of standard deviations that x is away from the mean. The population z-score is computed by using the formula

$$z = \frac{x - \mu}{\sigma}$$

where μ and σ are the mean and standard deviation of the population. A negative z-score indicates that the population value is smaller than the mean, while a positive z-score indicates that the population value is larger than the mean.

Thus, for example, a population value with a z-score of $z = 2$ is two standard deviations above the mean, a population value with a z-score of $z = -1.25$ is 1.25 standard deviations below the mean, and a population value with a z-score of $z = 0$ is equal to the mean.

Keep in mind, as we have already pointed out, that the process of computing a population z-score is identical to the process of standardizing. This implies the following fact: *For a normally distributed population, Table II gives the percentage of population values that lie between the mean and the population value that is z standard deviations above the mean.*

THE EMPIRICAL RULE

Key Fact 3.3 on page 130 states *Chebychev's rule* for populations. Recall that this rule provides estimates for the percentage of population values that lie within a specified number of standard deviations to either side of the mean. Also recall that the power of Chebychev's rule is its generality—it holds for any population. On the other hand, if we know something about the distribution of the population, then we can obtain better estimates than those provided by Chebychev's rule.

For a normally distributed population, we can use Table II to obtain the exact percentage of population values that lie within a specified number of standard deviations to either side of the mean. To illustrate, we return to the situation of Example 6.17.

EXAMPLE 6.18   *Introduces the empirical rule*

Consider again the brain weights of Swedish men. Recall that this population is normally distributed with a mean of $\mu = 1.40$ kg and a standard deviation of $\sigma = 0.11$ kg. Obtain the percentage of Swedish men that have brain weights
a) within one standard deviation to either side of the mean.
b) within two standard deviations to either side of the mean.
c) within three standard deviations to either side of the mean.

SOLUTION   Since the brain weights are normally distributed with mean $\mu = 1.40$ kg and standard deviation $\sigma = 0.11$ kg, percentages are equal to areas under the normal curve with parameters $\mu = 1.40$ and $\sigma = 0.11$. And, as we know, such areas are obtained by first converting to z-scores and then using the standard normal table, Table II.

a) The z-score for a brain weight that is one standard deviation below the mean is $z = -1$ and that for a brain weight that is one standard deviation above the mean is $z = 1$. Consequently, the percentage of Swedish men that have brain weights within one standard deviation to either side of the mean equals the area under the standard normal curve that lies between $z = -1$ and $z = 1$, which is 0.6826 or 68.26%. [To obtain the area 0.6826, we can either use Table II in the usual way or simply refer to Table 6.1 on page 290.]

b) The z-score for a brain weight that is two standard deviations below the mean is $z = -2$ and that for a brain weight that is two standard deviations above the mean is $z = 2$. Therefore, the percentage of Swedish men that have brain weights within two standard deviations to either side of the mean equals the area under the standard normal curve that lies between $z = -2$ and $z = 2$, which is 0.9544 or 95.44%.

c) The z-score for a brain weight that is three standard deviations below the mean is $z = -3$ and that for a brain weight that is three standard deviations above the mean is $z = 3$. Therefore, the percentage of Swedish men that have brain weights within three standard deviations to either side of the mean equals the area under the standard normal curve that lies between $z = -3$ and $z = 3$, which is 0.9974 or 99.74%. ∎

The arguments used in Example 6.18 apply to any normally distributed population. We never used the information that the population consists of brain weights or that $\mu = 1.40$ kg and $\sigma = 0.11$ kg. All we used was the fact that the population is normally distributed. Thus, we have the following result:

KEY FACT 6.6   Empirical rule for normally distributed populations

For any normally distributed population:

PROPERTY 1   About 68.26% of the population values lie within one standard deviation to either side of the mean; that is, between $\mu - \sigma$ and $\mu + \sigma$.

PROPERTY 2   About 95.44% of the population values lie within two standard deviations to either side of the mean; that is, between $\mu - 2\sigma$ and $\mu + 2\sigma$.

PROPERTY 3   About 99.74% of the population values lie within three standard deviations to either side of the mean; that is, between $\mu - 3\sigma$ and $\mu + 3\sigma$.

See Figure 6.36 at the top of the next page.

Using the empirical rule, we can quickly and easily obtain a good understanding of any normally distributed population. Example 6.19 provides an illustration.

FIGURE 6.36 Percentage of population values in a normally distributed population that lie within (a) one, (b) two, and (c) three standard deviations to either side of the mean

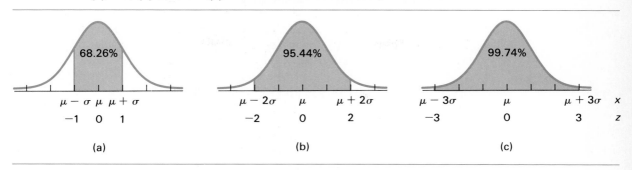

(a)

(b)

(c)

EXAMPLE 6.19 **Illustrates Key Fact 6.6**

Intelligence quotients (IQs) measured on the Stanford Revision of the *Binet-Simon Intelligence Scale* are known to be approximately normally distributed with a mean of $\mu = 100$ and a standard deviation of $\sigma = 16$. Apply the empirical rule to make some observations about IQs.

SOLUTION From Property 1 in Key Fact 6.6, about 68.26% of the population have IQs within one standard deviation to either side of the mean. Now, one standard deviation below the mean is $\mu - \sigma = 100 - 16 = 84$ and one standard deviation above the mean is $\mu + \sigma = 100 + 16 = 116$. Thus, about 68.26% of the population have IQs between 84 and 116.

From Property 2 in Key Fact 6.6, about 95.44% of the population have IQs within two standard deviations to either side of the mean. Two standard deviations below the mean is $\mu - 2\sigma = 100 - 2 \cdot 16 = 68$ and two standard deviations above the mean is $\mu + 2\sigma = 100 + 2 \cdot 16 = 132$. Therefore, about 95.44% of the population have IQs between 68 and 132.

From Property 3 in Key Fact 6.6, we see that about 99.74% of the population have IQs between 52 ($= 100 - 3 \cdot 16$) and 148 ($= 100 + 3 \cdot 16$). The results we have just obtained are summarized graphically in Figure 6.37. ∎

FIGURE 6.37 IQs

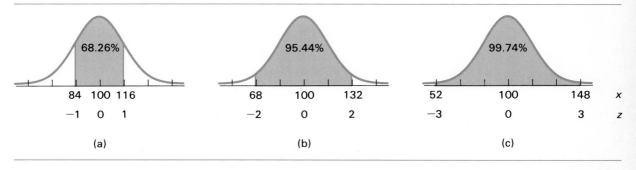

(a)

(b)

(c)

The empirical rule allows us to determine useful information about a normally distributed population quickly and easily, as illustrated in Example 6.19. We should point out, however, that similar facts are obtainable for any number of standard deviations away from the mean. For instance, we can use Table II to conclude that, for a normally distributed population, about 86.64% of the population values lie within 1.5 standard deviations to either side of the mean.

QUARTILES AND PERCENTILES FOR NORMALLY DISTRIBUTED POPULATIONS

In Chapter 3 we discussed quartiles and percentiles. Recall that **quartiles** divide a population into quarters, or four equal parts. A population has three quartiles, denoted by Q_1, Q_2, and Q_3. The first quartile, Q_1, is the number that divides the bottom 25% of the population from the top 75%; the second quartile, Q_2, is the median which, as we know, is the number that divides the bottom 50% of the population from the top 50%; and the third quartile, Q_3, is the number that divides the bottom 75% of the population from the top 25%.

For a normally distributed population, percentages are equal to areas under a normal curve. Thus, the quartiles for a normally distributed population are the three x-values that divide the area under the normal curve into four 0.25 areas. See Figure 6.38.

FIGURE 6.38
Quartiles for a
normally distributed
population

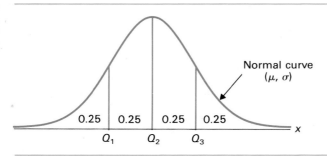

In the next example, we will obtain the quartiles for the population of IQs considered in Example 6.19.

EXAMPLE 6.20 | *Illustrates finding quartiles for a normally distributed population*

Intelligence quotients (IQs) are normally distributed with a mean of $\mu = 100$ and a standard deviation of $\sigma = 16$. Determine the quartiles for IQs.

SOLUTION | Percentages for IQs are equal to areas under the normal curve with parameters $\mu = 100$ and $\sigma = 16$. So to determine the quartiles, we need to find the three x-values that divide the area under that normal curve into four 0.25 areas. See Figure 6.39.

FIGURE 6.39

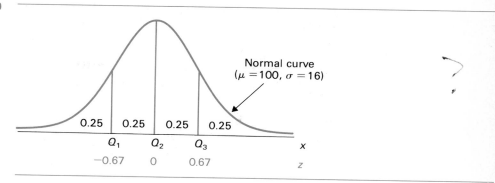

As we observe from Figure 6.39, Q_1 is the x-value with area 0.25 to its left. So the z-score corresponding to the first quartile, Q_1, is the z-value with area 0.25 to its left under the standard normal curve. Using Table II, we find that z-value to be $z = -0.67$, approximately. Similarly, we find that the z-scores corresponding to the second and third quartiles, Q_2 and Q_3, are $z = 0$ and $z = 0.67$, respectively. These three z-scores are shown in Figure 6.39.

To obtain the quartiles, we need only destandardize the three z-scores. Thus, the first quartile is

$$Q_1 = \mu + z \cdot \sigma = 100 + (-0.67) \cdot 16 = 100 - 0.67 \cdot 16$$

MTB

or $Q_1 = 89.28$. Similarly, we find that $Q_2 = 100$ and $Q_3 = 110.72$.   ∎

Finally, let us discuss percentiles. As we learned in Chapter 3, the **percentiles** of a population divide it into hundredths, or 100 equal parts. A population has 99 percentiles, denoted by $P_1, P_2, \ldots, P_{99}$. The first percentile, P_1, is the number that divides the bottom 1% of the population from the top 99%; the second percentile, P_2, is the number that divides the bottom 2% of the population from the top 98%; and so forth.

For a normally distributed population, we can obtain percentiles by using Table II, just like we did to find quartiles. Here is an example.

EXAMPLE 6.21   *Illustrates finding percentiles for a normally distributed population*

Consider once more the population of IQs, which is normally distributed with mean $\mu = 100$ and standard deviation $\sigma = 16$. Determine the 90th percentile, P_{90}.

SOLUTION   The 90th percentile, P_{90}, is the number that divides the bottom 90% of the population from the top 10%. Since percentages for IQs are equal to areas under the normal curve with parameters $\mu = 100$ and $\sigma = 16$, P_{90} is the x-value with area 0.10 to its right (or, equivalently, with area 0.90 to its left) under that normal curve. See Figure 6.40.

FIGURE 6.40

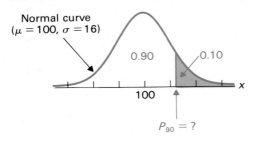

The z-score corresponding to P_{90} is, therefore, the z-value with area 0.10 to its right under the standard normal curve. From Table II, we find that z-value to be $z = 1.28$, approximately. Now we destandardize the z-score to get the 90th percentile:

$$P_{90} = \mu + z \cdot \sigma = 100 + 1.28 \cdot 16 = 120.48$$

MTB

In other words, 90% of IQs are below 120.48 and 10% are above 120.48.   ■

Exercises 6.3

A relative-frequency distribution for the heights of the 3264 female students attending a midwestern college is given in the first and third columns of Table 6.3 on page 308. The mean of that population of heights is $\mu = 64.4$ inches with a standard deviation of $\sigma = 2.4$ inches. In each part of Exercises 6.63 and 6.64, first obtain the exact percentage from Table 6.3. Then determine the corresponding area under the normal curve with parameters $\mu = 64.4$ and $\sigma = 2.4$, and compare your results.

___ **6.63** The percentage of students with heights
a) between 62 and 63 inches.
b) between 65 and 70 inches.

___ **6.64** The percentage of students with heights
a) between 71 and 72 inches.
b) between 61 and 65 inches.

___ **6.65** The annual wages, excluding board, of U.S. farm laborers in 1926 are normally distributed with a mean of $\mu = \$586$ and a standard deviation of $\sigma = \$97$. In 1926, what percentage of U.S. farm laborers had an annual wage of
a) between $500 and $700?
b) at least $400?

___ **6.66** The length of an adult yellow-bellied sapsucker is normally distributed with a mean of $\mu = 8.5$ inches and a standard deviation of $\sigma = 0.17$ inch. What percentage of yellow-bellied sapsuckers are
a) between 8.3 and 9.0 inches long?
b) less than 8.8 inches long?

___ **6.67** As reported by the U.S. National Center for Health Statistics in *Vital and Health Statistics,* males who are six feet tall and between 18 and 24 years of age have a mean weight of 175 lb. If the weights are normally distributed with a standard deviation of 14 lb, find the percentage of such males that weigh
a) between 190 and 210 lb.
b) less than 150 lb.

___ **6.68** An issue of *Scientific American* reveals that the batting averages of major league baseball players are approximately normally distributed with a mean of 0.270 and a standard deviation of 0.015. Determine the percentage of major league baseball players with batting averages
a) between 0.225 and 0.250.
b) at least 0.300.

___ **6.69** Refer to Exercise 6.65. How many standard deviations away from the mean is a 1926 U.S. farm

laborer's annual wage of
a) $780?   b) $416.25?   c) $586?

__ **6.70**   Refer to Exercise 6.66. How many standard deviations away from the mean is the length of a yellow-bellied sapsucker
a) 8.7125 inches long?
b) 8.16 inches long?
c) 8.5 inches long?

__ **6.71**   Refer to Exercise 6.67 where we considered the weights of U.S. males who are six feet tall and between 18 and 24 years of age. Assuming the weights are normally distributed, determine the percentage of such males that have weights within
a) one standard deviation to either side of the mean.
b) two standard deviations to either side of the mean.
c) three standard deviations to either side of the mean.
d) 1.5 standard deviations to either side of the mean.

__ **6.72**   Refer to Exercise 6.68. Obtain the percentage of major league baseball players who have batting averages within
a) one standard deviation to either side of the mean.
b) two standard deviations to either side of the mean.
c) three standard deviations to either side of the mean.
d) 2.5 standard deviations to either side of the mean.

__ **6.73**   The Department of Agriculture compiles information on food costs for Americans and publishes its findings in *News*. According to that document, the mean weekly food cost for a couple with two children 6–11 years old is $95.40. Presuming that these weekly food costs are normally distributed with a standard deviation of $17.20, fill in the following blanks by applying the empirical rule:
a) About 68.26% of such couples have weekly food costs between $____ and $____.
b) About 95.44% of such couples have weekly food costs between $____ and $____.
c) About 99.74% of such couples have weekly food costs between $____ and $____.
d) Draw graphs similar to the ones in Figure 6.37 on page 315 to illustrate your results.

__ **6.74**   In the publication *Nielsen Report on Television,* the A. C. Nielsen Company reveals that the mean weekly television viewing time for children ages 2–5 years is 27.15 hours. Assuming that the weekly television viewing times of these children are normally distributed with a standard deviation of 6.23 hours, fill in the following blanks by applying the empirical rule:

a) About 68.26% of all such children watch between ____ and ____ hours of TV per week.
b) About 95.44% of all such children watch between ____ and ____ hours of TV per week.
c) About 99.74% of all such children watch between ____ and ____ hours of TV per week.
d) Draw graphs similar to the ones in Figure 6.37 on page 315 to illustrate your results.

__ **6.75**   For a normally distributed population, fill in the following blanks:
a) About ____% of the population values lie within 1.96 standard deviations to either side of the mean.
b) About ____% of the population values lie within 1.64 standard deviations to either side of the mean.

__ **6.76**   For a normally distributed population, fill in the following blanks:
a) About ____% of the population values lie within 1.28 standard deviations to either side of the mean.
b) About ____% of the population values lie within 2.33 standard deviations to either side of the mean.

__ **6.77**   For a normally distributed population, fill in the following blanks:
a) About 99% of the population values lie within ____ standard deviations to either side of the mean.
b) About 80% of the population values lie within ____ standard deviations to either side of the mean.

__ **6.78**   For a normally distributed population, fill in the following blanks:
a) About 95% of the population values lie within ____ standard deviations to either side of the mean.
b) About 90% of the population values lie within ____ standard deviations to either side of the mean.

__ **6.79**   Refer to Exercise 6.65.
a) Determine the quartiles for the wages.
b) Obtain the 15th percentile.
c) Find the 98th percentile.

__ **6.80**   Refer to Exercise 6.66.
a) Determine the quartiles for the lengths.
b) Obtain the first decile (i.e., 10th percentile).
c) Find the 82nd percentile.

__ **6.81**   Refer to Exercise 6.73.
a) Determine the quartiles for the food costs.
b) Obtain the third decile (i.e., 30th percentile).
c) Find the 85th percentile.

__ **6.82**   Refer to Exercise 6.74.
a) Determine the quartiles for the viewing times.
b) Obtain the 45th percentile.
c) Find the ninth decile (i.e., 90th percentile).

≡ **6.83**   Verify the empirical rule, Key Fact 6.6 on page 314.

≡ **6.84**   Derive general formulas for the quartiles of a normally distributed population. That is, for a normally distributed population, express the quartiles, Q_1, Q_2, and Q_3, in terms of μ and σ. *(Hint: See Figure 6.38 on page 316.)*

≡ **6.85**   Let $0 < \alpha < 1$. For a normally distributed population, prove the following general form of the empirical rule: About $100 \cdot (1 - \alpha)\%$ of the population values lie within $z_{\alpha/2}$ standard deviations to either side of the mean; that is, between $\mu - z_{\alpha/2} \cdot \sigma$ and $\mu + z_{\alpha/2} \cdot \sigma$. *(Hint: Draw a picture and recall that $z_{\alpha/2}$ is the z-value with area $\alpha/2$ to its right under the standard normal curve.)*

≡ **6.86**   Derive a general formula for the kth percentile of a normally distributed population. That is, for a normally distributed population, express the kth percentile, P_k, in terms of μ, σ, and k. *(Hint: Observe that the z-score corresponding to the kth percentile is the z-value with area $1 - k/100$ to its right under the standard normal curve.)*

6.4   Normally distributed random variables

In the previous section we discussed normally distributed populations. As we know, a population is normally distributed if percentages for the population are equal to areas under a normal curve; namely, the normal curve whose parameters are the same as the mean and standard deviation of the population.

Also of great importance are *normally distributed random variables.* As you might guess, a random variable is said to be normally distributed if probabilities for the random variable are equal to areas under a normal curve.

DEFINITION 6.4   Normally distributed random variable

A random variable is said to be (approximately) *normally distributed* if probabilities for the random variable are (approximately) equal to areas under a normal curve. If such a random variable has mean μ_x and standard deviation σ_x, then the normal curve that is used is the one with parameters μ_x and σ_x.

EXAMPLE 6.22   *Illustrates finding probabilities for a normally distributed random variable*

A local bottling plant fills bottles of soda for distribution in the surrounding area. While the advertised content is 354 ml, the filling machine is actually set to a mean content of 356 ml. As a matter of fact, the amount of soda put in a bottle is normally distributed with a mean of 356 ml and a standard deviation of 1.63 ml. Determine the probability that a randomly selected bottle of soda will contain less than the advertised content of 354 ml.

SOLUTION   Let x denote the content, in ml, of a randomly selected bottle of soda. Then, by assumption, x is a normally distributed random variable with mean $\mu_x = 356$ ml and standard deviation $\sigma_x = 1.63$ ml. Therefore, probabilities for x are equal to areas under the normal curve with parameters $\mu = 356$ and $\sigma = 1.63$.

We want to determine the probability that a randomly selected bottle will contain less than 354 ml of soda, $P(x < 354)$. That probability is equal to the area under the normal curve with parameters $\mu = 356$ and $\sigma = 1.63$ that lies to the left of 354. This area is found in the usual way. See Figure 6.41.

FIGURE 6.41
Determination of
the area under the
normal curve with
parameters $\mu = 356$
and $\sigma = 1.63$ that
lies to the left of 354

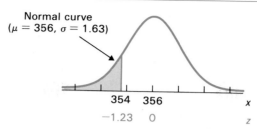

Normal curve
($\mu = 356$, $\sigma = 1.63$)

354   356

−1.23   0

x

z

z-score computation:

$x = 354 \longrightarrow z = \dfrac{354 - 356}{1.63} = -1.23$

Area between 0 and z:

0.3907

Shaded area $= 0.5000 - 0.3907 = 0.1093$

Thus, $P(x < 354) = \boxed{0.1093}$. There is roughly an 11% chance that a randomly selected bottle will contain less than the advertised content of 354 ml of soda. In other words, about 11% of the bottles will contain less than the advertised content of 354 ml.

MTB

THE EMPIRICAL RULE FOR NORMALLY DISTRIBUTED RANDOM VARIABLES

The empirical rule for normally distributed populations, Key Fact 6.6 on page 314, gives some useful properties about percentages for normally distributed populations. Here are the corresponding probability statements for normally distributed random variables.

KEY FACT 6.7   Empirical rule for normally distributed random variables

For any normally distributed random variable x:

PROPERTY 1   The probability is 0.6826 that x will be within one standard deviation to either side of its mean:

$$P(\mu_x - \sigma_x < x < \mu_x + \sigma_x) = 0.6826$$

PROPERTY 2   The probability is 0.9544 that x will be within two standard deviations to either side of its mean:

$$P(\mu_x - 2\sigma_x < x < \mu_x + 2\sigma_x) = 0.9544$$

PROPERTY 3   The probability is 0.9974 that x will be within three standard deviations to either side of its mean:

$$P(\mu_x - 3\sigma_x < x < \mu_x + 3\sigma_x) = 0.9974$$

See Figure 6.42.

FIGURE 6.42   Probability that a normally distributed random variable will be within
(a) one, (b) two, and (c) three standard deviations to either side of its mean

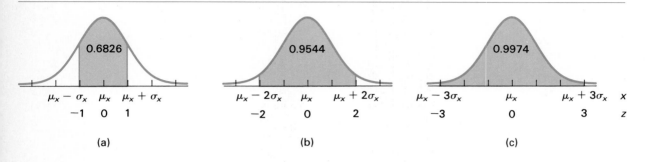

It is important to be able to interpret Properties 1–3 and similar statements in terms of percentages. For instance, Property 2 can be interpreted as follows: For a large number of observations, a normally distributed random variable will be within two standard deviations of its mean about 95.44% of the time.

EXAMPLE 6.23   *Illustrates Key Fact 6.7*

In Example 6.22, we considered the contents of soda bottles filled in a bottling plant. The amount of soda put in a bottle is normally distributed with a mean of 356 ml and a standard deviation of 1.63 ml. If we let x denote the content of a randomly selected bottle of soda, then x is a normally distributed random variable with mean $\mu_x = 356$ ml and standard deviation $\sigma_x = 1.63$ ml. Apply the empirical rule for normally distributed random variables to make some pertinent statements about the contents of the filled bottles.

SOLUTION   From Property 1 in Key Fact 6.7, the probability is 0.6826 that x will be within one standard deviation to either side of its mean. Now, one standard deviation below the mean is $\mu_x - \sigma_x = 356 - 1.63 = 354.37$ and one standard deviation above the mean is $\mu_x + \sigma_x = 356 + 1.63 = 357.63$. Thus, the probability is 0.6826 that a randomly selected bottle will contain between 354.37 and 357.63 ml of soda. That is, about 68.26% of the bottles will contain between 354.37 and 357.63 ml of soda.

From Property 2 in Key Fact 6.7, the probability is 0.9544 that x will be within two standard deviations to either side of its mean. Two standard deviations below the mean is $\mu_x - 2\sigma_x = 356 - 2 \cdot 1.63 = 352.74$ and two standard deviations above the mean is $\mu_x + 2\sigma_x = 356 + 2 \cdot 1.63 = 359.26$. So, the probability is 0.9544 that a randomly selected bottle will contain between 352.74 and 359.26 ml of soda. That is, about 95.44% of the bottles will contain between 352.74 and 359.26 ml of soda.

Applying Property 3 in Key Fact 6.7, we conclude that the probability is 0.9974 that a randomly selected bottle will contain between 351.11 ($= 356 - 3 \cdot 1.63$) and 360.89 ($= 356 + 3 \cdot 1.63$) ml of soda. In other words, about 99.74% of the bottles will contain between 351.11 and 360.89 ml of soda. ■

MTB

We should emphasize that statements similar to those in Key Fact 6.7 hold for any number of standard deviations away from the mean. For example, using Table II we find that the probability is 0.3830 that a normally distributed random variable will be within 0.5 standard deviations to either side of its mean. More generally, we have the following fact which we will need later on.

KEY FACT 6.8 **General empirical rule for normally distributed random variables**

Suppose that x is a normally distributed random variable. Then the probability is $1 - \alpha$ that x will be within $z_{\alpha/2}$ standard deviations to either side of its mean:

$$P(\mu_x - z_{\alpha/2} \cdot \sigma_x < x < \mu_x + z_{\alpha/2} \cdot \sigma_x) = 1 - \alpha$$

Here $0 < \alpha < 1$.

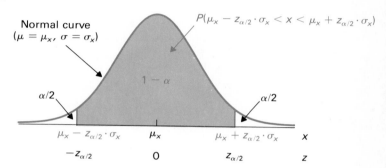

Note: Key Fact 6.8, as well as Key Fact 6.7, is simply a consequence of the fact that, for a normally distributed random variable, x, probabilities are equal to areas under the normal curve with parameters μ_x and σ_x.

To illustrate the general empirical rule, suppose we let $\alpha = 0.10$. Then $1 - \alpha = 1 - 0.10 = 0.90$ and $z_{\alpha/2} = z_{0.10/2} = z_{0.05} = 1.645$ (from Table II). So, by Key Fact 6.8, the probability is 0.90 that a normally distributed random variable will be within 1.645 standard deviations to either side of its mean. In symbols, if x is a normally distributed random variable, then

$$P(\mu_x - 1.645\sigma_x < x < \mu_x + 1.645\sigma_x) = 0.90$$

For a concrete application of the previous equation, consider again the contents of soda bottles filled in a bottling plant. Here x denotes the content of a randomly selected bottle of soda and we know that x is normally distributed. Since $\mu_x = 356$ ml and $\sigma_x = 1.63$ ml, we have $\mu_x - 1.645\sigma_x = 356 - 1.645 \cdot 1.63 = 353.32$ and $\mu_x + 1.645\sigma_x = 356 + 1.645 \cdot 1.63 = 358.68$. Thus, the probability is 0.90 that a randomly selected bottle will contain between 353.32 and 358.68 ml of soda.

AN INTERPRETATION OF THE STANDARD DEVIATION OF A NORMALLY DISTRIBUTED RANDOM VARIABLE

In Key Fact 5.4 on page 246, we gave the following interpretation to the standard deviation, σ_x, of a random variable x: It measures the dispersion of the possible values of the random variable relative to its mean; the smaller the standard deviation, the more likely that x will be close to its mean. Using the empirical rule, we can make the interpretation of σ_x even more explicit when the random variable is normally distributed.

For example, by Property 2 in Key Fact 6.7, we know that there is a 95.44% chance that a normally distributed random variable, x, will be within two standard deviations of its mean. If σ_x is small, then two standard deviations to either side of the mean will also be small, thus making it likely (a 95.44% chance) that x will be close to its mean. If σ_x is large, then two standard deviations to either side of the mean will also be large and, consequently, we cannot expect x to be close to its mean, although it may be. A graphical summary of the discussion in this paragraph is presented in Figure 6.43.

FIGURE 6.43

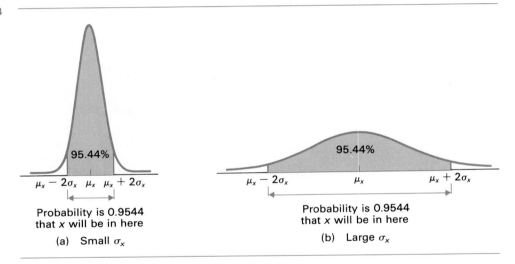

(a) Small σ_x (b) Large σ_x

STANDARDIZING A RANDOM VARIABLE

Recall that the process of standardizing involves converting x-values to z-values using the formula

$$z = \frac{x - \mu}{\sigma}$$

Standardizing is employed both to obtain population z-scores and to determine areas under a normal curve.

It is also useful to consider the process of standardizing a random variable. A random variable, x, is standardized by first subtracting its mean and then dividing by its standard deviation. The resulting random variable is called the **standardized version** of x and is denoted by the letter z. We introduce this concept in Example 6.24.

EXAMPLE 6.24   *Introduces standardizing a random variable*

A one-man barber shop has an average weekly gross income of $640, with a standard deviation of $50. Let x be the shop's gross income during a randomly selected week.
a) Determine the standardized version, z, of the random variable x.
b) If we select a week at random, then the random variable x tells us that week's gross. What does the standardized random variable z tell us?
c) If a randomly selected week has a gross of $740, then how many standard deviations away from the mean is that week's gross?

SOLUTION   We first note that x is a random variable with mean $\mu_x = 640$ and standard deviation $\sigma_x = 50$.
a) As we said, to obtain the standardized version, z, of x, we first subtract μ_x and then divide by σ_x:

$$z = \frac{x - \mu_x}{\sigma_x}$$

Here $\mu_x = 640$ and $\sigma_x = 50$. Thus, the standardized version of x is

$$z = \frac{x - 640}{50}$$

We emphasize that, since x is a random variable, so is z.
b) If we select a week at random, the standardized random variable z tells us how many standard deviations that week's gross is away from the mean.
c) If a randomly selected week has a gross of $740, then $x = 740$ and, consequently,

$$z = \frac{x - 640}{50} = \frac{740 - 640}{50} = 2$$

Hence, that week's gross is two standard deviations above the mean.   ■

The preceding discussion on standardizing a random variable is summarized in Definition 6.5.

DEFINITION 6.5   Standardized version of a random variable

Let x be a random variable. Then we define the *standardized version of x* to be the random variable

$$z = \frac{x - \mu_x}{\sigma_x}$$

The standardized random variable z gives the number of standard deviations that an observed value of x is away from the mean.

STANDARDIZING A NORMALLY DISTRIBUTED RANDOM VARIABLE

Recall that the normal curve with parameters $\mu = 0$ and $\sigma = 1$ is the standard normal curve. Suppose a normally distributed random variable has mean 0 and standard deviation 1. Then we say that it has the **standard normal distribution** because probabilities for such a random variable are equal to areas under the standard normal curve.

Now, it is important to note that we can form the standardized version of a random variable x regardless of whether x is normally distributed. If x is not normally distributed, then neither is z. But if x is normally distributed, then so is z. In fact, z then has the standard normal distribution.

KEY FACT 6.9   Standardized normally distributed random variable

Suppose that x is a normally distributed random variable. Then the standardized version of x,

$$z = \frac{x - \mu_x}{\sigma_x},$$

has the standard normal distribution. In other words, z is a normally distributed random variable with mean $\mu_z = 0$ and standard deviation $\sigma_z = 1$. Thus, probabilities for the standardized random variable z are equal to areas under the standard normal curve.

EXAMPLE 6.25   *Illustrates Key Fact 6.9*

Consider once more the contents of soda bottles filled in a bottling plant. The amount of soda put in a bottle is normally distributed with a mean of 356 ml and a standard deviation of 1.63 ml. Let x be the content of a randomly selected bottle of soda.

a) Determine the standardized version, z, of the random variable x.
b) Obtain the probability distribution of the standardized version, z, of the random variable x.

SOLUTION   a) The random variable x has mean $\mu_x = 356$ and standard deviation $\sigma_x = 1.63$. Thus, the standardized version of x is the random variable

$$z = \frac{x - 356}{1.63}$$

b) We know that the random variable x is *normally distributed*. Therefore, by Key Fact 6.9, the standardized random variable z, given in part (a), has the standard normal distribution. That is, probabilities for z are equal to areas under the standard normal curve.   ∎

Exercises 6.4

__ **6.87**   As reported by the National Education Association in *Estimates of School Statistics,* the mean salary for secondary school teachers is $28.7 thousand. Assume the salaries are normally distributed with a standard deviation of $5.2 thousand. Let x be the salary, in thousands of dollars, of a randomly selected secondary school teacher. Find
a) $P(x < 25)$.
b) $P(32 \leq x \leq 37)$.
c) Interpret your results in parts (a) and (b) in words.

__ **6.88**   *The World Almanac* reports that the mean travel time to work in New York is 29 minutes. Let x be the time, in minutes, that it takes a randomly selected worker to get to work on a randomly selected day. If the travel times are normally distributed with a standard deviation of 9.3 minutes, find
a) $P(x < 45)$.
b) $P(20 \leq x \leq 30)$.
c) Interpret your results in parts (a) and (b) in words.

__ **6.89**   The lifetime of a brand of flashlight battery is normally distributed with a mean of 30 hours and a standard deviation of 5.6 hours. Let x be the lifetime of a randomly selected flashlight battery of this brand. Determine
a) $P(x > 20)$.
b) $P(15 \leq x \leq 45)$.
c) Interpret each of your results in parts (a) and (b) in terms of percentages.

__ **6.90**   A manufacturer of timepieces claims that the weekly error, in seconds, of the watches she makes has a normal distribution with a mean of 0 and a standard deviation of 1. Let x denote the amount of time, in seconds, that one of these watches is off at the end of a randomly selected week. Find
a) $P(x < -1)$.
b) $P(x < -2 \text{ or } x > 2)$.
c) Interpret your results in parts (a) and (b) in words.

__ **6.91**   *Runner's World* magazine reports that the times of the finishers in the New York City 10-km run are normally distributed with a mean of 61 minutes and a standard deviation of nine minutes. Let x be the time of a randomly selected finisher. Find
a) $P(x > 75)$.
b) $P(x < 50 \text{ or } x > 70)$.
c) Interpret your results in parts (a) and (b) in words.

__ **6.92**   The length of the western rattlesnake is normally distributed with a mean of 42 inches and a standard deviation of 2.04 inches. Let x be the length of one of these snakes selected at random. Determine
a) $P(x > 45)$.
b) $P(35 \leq x \leq 40)$.
c) Interpret each of your results in parts (a) and (b) in terms of percentages.

__ **6.93**   A company produces bolts approximately 10 mm in diameter to fit into a circular hole 10.4 mm in diameter. In reality, the diameters of the bolts produced are normally distributed with a mean of 10 mm and a standard deviation of 0.1 mm. Let x be the diameter of a randomly selected bolt. Apply the empirical rule for normally distributed random variables, Key Fact 6.7 on page 321, to make some pertinent statements about the diameters of the bolts produced. [See Example 6.23 on page 322 for a model.]

__ **6.94**   An electronics company gives a general aptitude test to all prospective employees. The test is designed to take about one hour to complete. Experience indicates that the completion times are normally distributed with a mean of 62.4 minutes and a standard deviation of 7.2 minutes. Let x denote the time it takes a randomly selected applicant to complete the test. Apply the empirical rule for normally distributed random variables, Key Fact 6.7 on page 321, to make some pertinent statements about the completion times for the aptitude test.

__ **6.95**   Refer to Exercise 6.89. Use Key Fact 6.7 on page 321 to obtain some information regarding the lifetimes of flashlight batteries of the given brand.

__ **6.96**   Refer to Exercise 6.90. Use Key Fact 6.7 on page 321 to obtain some information about the weekly error of one of the manufacturer's watches.

__ **6.97**   According to the document *Statistical Report,* published by the U.S. Bureau of Prisons, the mean time served by prisoners released from federal institutions for the first time is 16.3 months. Assume that the standard deviation of the times served is 17.9 months. Let x be the time served by a randomly selected prisoner released for the first time from a federal institution.
a) Find the standardized version, z, of the random variable x.

b) How many standard deviations from the mean is a prison time served of 20.3 months? 64.7 months? 4.2 months?

c) Is it necessary to assume that the times served are normally distributed in order to answer parts (a) and (b)? Why?

___ 6.98   The Health Insurance Association of America compiles data on room charges in non-government, short-term, general hospitals. Results are published in *Survey of Hospital Semi-Private Room Charges*. According to that publication, the average daily charge for a semi-private hospital room is $253. Assume a standard deviation of $45. Let x be the daily semi-private room charge of a randomly selected hospital.

a) Find the standardized version, z, of the random variable x.

b) How many standard deviations from the mean is a daily room charge of $348? $107? $192?

c) Must we assume that the daily semi-private room charges are normally distributed in order to answer parts (a) and (b)? Why?

___ 6.99   Refer to Exercise 6.91.

a) Determine the standardized version, z, of the random variable x.

b) What does the random variable z represent?

c) What is the probability distribution of z?

___ 6.100   Refer to Exercise 6.92.

a) Determine the standardized version, z, of the random variable x.

b) What does the random variable z represent?

c) What is the probability distribution of the random variable z?

___ 6.101   Apply the general empirical rule for normally distributed random variables, Key Fact 6.8 on page 323, to fill in the following blanks:

a) The probability is 0.95 that a normally distributed random variable will be within _____ standard deviations to either side of its mean. (*Hint:* Here $1 - \alpha = 0.95$. Solve for α.)

b) The probability is 0.99 that a normally distributed random variable will be within _____ standard deviations to either side of its mean.

c) Express your results in parts (a) and (b) using formulas like those in Key Fact 6.7 on page 321.

___ 6.102   Apply the general empirical rule for normally distributed random variables, Key Fact 6.8 on page 323, to fill in the following blanks:

a) The probability is 0.90 that a normally distributed random variable will be within _____ standard deviations to either side of its mean.

b) The probability is 0.85 that a normally distributed random variable will be within _____ standard deviations to either side of its mean.

c) Express your results in parts (a) and (b) using formulas like those in Key Fact 6.7 on page 321.

= 6.103   This exercise concerns the filling of soda bottles discussed in Example 6.22. Recall that while the advertised content is 354 ml, the amount of soda actually put in a bottle is normally distributed with a mean of 356 ml and a standard deviation of 1.63 ml. To what setting should the mean be changed in order to insure that only 1% of the bottles will contain less than the advertised content of 354 ml?

= 6.104   Refer to Exercise 6.103. Assuming that the mean of 356 ml remains unchanged, what standard deviation must the filling process have in order to insure that only 1% of the bottles will contain less than the advertised content of 354 ml?

= 6.105   Verify the empirical rule for normally distributed random variables, Key Fact 6.7 on page 321.

= 6.106   Provide a proof of the general empirical rule for normally distributed random variables, Key Fact 6.8 on page 323. (*Hint:* Remember that probabilities for a normally distributed random variable, x, are equal to areas under the normal curve with parameters μ_x and σ_x. You will need to determine the z-scores for the x-values $\mu_x \pm z_{\alpha/2} \cdot \sigma_x$.)

6.5   The normal approximation to the binomial distribution (Optional)

In this section we will see that it is often possible to approximate binomial probabilities by areas under a normal curve. The mathematical theory for doing that is credited to Abraham DeMoivre (1667–1754) and Pierre Laplace (1749–1827).

To begin, let us briefly review the binomial distribution (see Section 5.4 for details). Suppose that n identical independent success-failure experiments (Bernoulli trials) are performed, with the probability of success on any given trial being equal to p. Let x denote the total number of successes in the n trials. Then the probability distribution of the random variable x is given by the binomial probability formula

$$P(x) = \binom{n}{x} p^x (1-p)^{n-x}$$

Moreover, we say that x has the *binomial distribution* with parameters n and p.

You might be wondering why we would use normal curve areas to approximate binomial probabilities when we can obtain them exactly by employing the binomial probability formula. The following example should help explain why.

EXAMPLE 6.26   *Illustrates the need to approximate binomial probabilities*

Insurance companies use mortality tables to determine life insurance premiums, retirement pensions, annuity payments, and other related items. Mortality tables enable actuaries to obtain the probability that a person at any given age will live a specified number of years.

According to the Department of Health and Human Services, the probability is about 0.8 (an 80% chance) that a person aged 70 will be alive at age 75. Suppose that 500 people aged 70 are selected at random. Find the probability that
a) exactly 390 of them will be alive at age 75.
b) between 375 and 425 of them, inclusive, will be alive at age 75.

SOLUTION   Let x denote the number of people out of the 500 that will be alive at age 75. Then x has the binomial distribution with parameters $n = 500$ (the 500 people) and $p = 0.8$ (the probability a person aged 70 will be alive at age 75). Thus, probabilities for x can be determined exactly by using the binomial probability formula

$$P(x) = \binom{500}{x} (0.8)^x (0.2)^{500-x}$$

Let us apply this formula to the problems posed in parts (a) and (b).
a) Here we want the probability that exactly 390 of the 500 people will still be alive at age 75, $P(x = 390)$. So, the "answer" is

$$P(390) = \binom{500}{390} (0.8)^{390} (0.2)^{110}$$

However, to actually obtain the value of the expression on the right-hand side of the previous equation is not so simple, even with a calculator. In performing the computations, we have to be quite careful to avoid such things as roundoff error and getting numbers so large or so small that they are outside the range of the calculator. Fortunately, as we will soon see, the computations can be sidestepped altogether by using normal curve areas in a very simple way.

b) In this part, we need to determine the probability that between 375 and 425 of the 500 people will be alive at age 75, $P(375 \leq x \leq 425)$. So, the "answer" is

$$P(375 \leq x \leq 425) = P(375) + P(376) + \cdots + P(425)$$

$$= \binom{500}{375}(0.8)^{375}(0.2)^{125} + \binom{500}{376}(0.8)^{376}(0.2)^{124} + \cdots + \binom{500}{425}(0.8)^{425}(0.2)^{75}$$

We see that we have the same computational difficulties as in part (a), except that here we must evaluate 51 complex expressions instead of one. [There are 51 summands in the sum.] Thus, although in theory we can use the binomial probability formula to get the answer, doing so in practice is another matter.

Surprising as it might seem, there is a way to use normal curve areas to get the (approximate) answer, and it is easy! We will return to this problem momentarily and obtain the probabilities in parts (a) and (b). ■

The previous example makes it clear why we often need to approximate binomial probabilities: Even though the binomial probability formula is available for computing binomial probabilities exactly, that formula is simply not practical to use when the number of trials, n, is large.

Under certain conditions on n and p, the histogram of a binomial distribution is bell-shaped. In such cases, we can approximate binomial probabilities by areas under a normal curve.

The following example illustrates a bell-shaped binomial distribution. For this example, it is easy to calculate binomial probabilities exactly by using the binomial probability formula. However, for purposes of illustration, we will also show how normal curve areas can be employed to approximate the binomial probabilities.

EXAMPLE 6.27   *Introduces how to approximate binomial probabilities by areas under a normal curve*

A student is taking a true-false exam with 10 questions. Suppose the student guesses at all 10 questions. Use the binomial probability formula to determine the exact probability that the student gets either seven or eight correct. Then approximate that probability by an area under a suitable normal curve.

SOLUTION   Let x be the number of correct guesses by the student. There are 10 questions, so here $n = 10$. Since the student guesses at each question, the success probability is $p = 0.5$. So x has the binomial distribution with parameters $n = 10$ and $p = 0.5$. Therefore, probabilities for x are given by the binomial probability formula

$$P(x) = \binom{10}{x}(0.5)^x(1 - 0.5)^{10-x}$$

Applying that formula, we obtain the probability distribution of x. See Table 6.4.

TABLE 6.4
Probability
distribution for
the number of
correct guesses
by the student

Number correct x	Probability $P(x)$
0	0.0010
1	0.0098
2	0.0439
3	0.1172
4	0.2051
5	0.2461
6	0.2051
7	0.1172
8	0.0439
9	0.0098
10	0.0010

The problem is to find the probability that the student gets either seven or eight questions correct, $P(x = 7 \text{ or } 8)$. From Table 6.4, the exact probability is

$$P(x = 7 \text{ or } 8) = P(7) + P(8) = 0.1172 + 0.0439 = 0.1611$$

Let us now see how we can approximate $P(x = 7 \text{ or } 8)$ by an area under a suitable normal curve. First of all, as you might expect, the normal curve that is used is the one whose parameters are the same as the mean and standard deviation of the random variable x. Since x has the binomial distribution with parameters $n = 10$ and $p = 0.5$, we can apply Formulas 5.3 and 5.4 (page 275) to obtain the mean and standard deviation of x:

$$\mu_x = np = 10 \cdot 0.5 = 5$$

and

$$\sigma_x = \sqrt{np(1 - p)} = \sqrt{10 \cdot 0.5 \cdot (1 - 0.5)} = 1.58$$

Consequently, the normal curve that is used here is the one with parameters $\mu = 5$ and $\sigma = 1.58$.

In Figure 6.44 we have drawn a histogram for the probability distribution of the random variable x. The heights of the bars are the probabilities for x given in Table 6.4. Note that the histogram is bell-shaped. Also shown in Figure 6.44 is the normal curve with parameters $\mu = 5$ and $\sigma = 1.58$.

FIGURE 6.44
Probability
histogram for x
with superimposed
normal curve

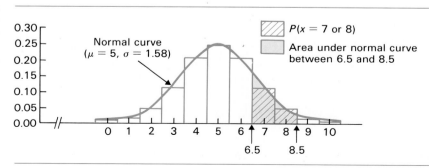

The probability, $P(x = 7 \text{ or } 8)$, is exactly equal to the combined area of the corresponding rectangles of the histogram, the cross-hatched area in Figure 6.44. By examining the figure carefully, we see that the cross-hatched area is approximately equal to the area under the normal curve between 6.5 and 8.5, the shaded area in Figure 6.44. [It should be clear from the picture why we consider the area under the normal curve between 6.5 and 8.5 instead of between 7 and 8. This is called the *correction for continuity*. Keep this in mind when we present the general procedure for approximating binomial probabilities by normal curve areas.]

Thus, we see, at least qualitatively, that the probability, $P(x = 7 \text{ or } 8)$, is approximately equal to the area under the normal curve between 6.5 and 8.5. To compare these values quantitatively, we must find the normal curve area. This is done in the usual way. See Figure 6.45.

FIGURE 6.45
Determination of
the area under the
normal curve with
parameters $\mu = 5$
and $\sigma = 1.58$
that lies between
6.5 and 8.5

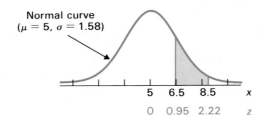

Normal curve
($\mu = 5$, $\sigma = 1.58$)

z-score computations: Area between 0 and z:

$x = 6.5 \longrightarrow z = \dfrac{6.5 - 5}{1.58} = 0.95$ 0.3289

$x = 8.5 \longrightarrow z = \dfrac{8.5 - 5}{1.58} = 2.22$ 0.4868

Shaded area $= 0.4868 - 0.3289 = 0.1579$

The last line in Figure 6.45 shows that the area under the normal curve between 6.5 and 8.5 is 0.1579. Comparing that area to the exact value of $P(x = 7 \text{ or } 8)$, which is 0.1611, we see that the normal curve area provides an excellent approximation of the exact probability.   ∎

It is not always appropriate to use normal curve areas to approximate binomial probabilities since the histograms for some binomial distributions are not sufficiently bell-shaped. The customary rule of thumb for use of the normal approximation is that *both np and $n(1 - p)$ are at least 5.*

By examining carefully what we did in Example 6.27, we can write down a general procedure for approximating binomial probabilities by areas under a normal curve. This is given as Procedure 6.2.

> **PROCEDURE 6.2**
>
> **To approximate binomial probabilities by normal curve areas.**
>
> **STEP 1** *Determine n, the number of trials, and p, the success probability.*
> **STEP 2** *Check that both np and n(1 − p) are at least 5. If they are not, the normal approximation should not be used.*
> **STEP 3** *Find μ_x and σ_x using the formulas*
>
> $$\mu_x = np \quad \text{and} \quad \sigma_x = \sqrt{np(1-p)}$$
>
> **STEP 4** *Make the correction for continuity and find the required area under the normal curve with parameters $\mu = np$ and $\sigma = \sqrt{np(1-p)}$.*

Note: Step 4 of Procedure 6.2 requires us to make the **correction for continuity.** As illustrated in Example 6.27, this means the following: When approximating the probability that a binomial random variable will be between two integers, inclusive, subtract 0.5 from the smaller integer and add 0.5 to the larger integer before finding the area under the normal curve.

We will now apply Procedure 6.2 to solve the problems posed at the beginning of this section in Example 6.26.

EXAMPLE 6.28   *Illustrates Procedure 6.2*

The probability is 0.80 that a person aged 70 will be alive at age 75. Suppose that 500 people aged 70 are selected at random. Determine the probability that
a) exactly 390 of them will be alive at age 75.
b) between 375 and 425 of them, inclusive, will be alive at age 75.

SOLUTION   We will obtain the approximate values of the probabilities in parts (a) and (b) by applying Procedure 6.2.

STEP 1 *Determine n, the number of trials, and p, the success probability.*
    We have $n = 500$ and $p = 0.8$.

STEP 2 *Check that both np and n(1 − p) are at least 5.*
    Referring to the values for n and p obtained in Step 1, we see that

$$np = 500 \cdot 0.8 = 400$$

and

$$n(1-p) = 500 \cdot 0.2 = 100$$

Thus, both np and $n(1-p)$ are at least 5.

STEP 3 *Find μ_x and σ_x using the formulas*

$$\mu_x = np \quad \text{and} \quad \sigma_x = \sqrt{np(1-p)}$$

We have

$$\mu_x = np = 500 \cdot 0.8 = 400$$

and

$$\sigma_x = \sqrt{np(1-p)} = \sqrt{500 \cdot 0.8 \cdot 0.2} = 8.94$$

STEP 4 *Make the correction for continuity and find the required area under the normal curve with parameters $\mu = 400$ and $\sigma = 8.94$.*

a) Here we want the probability that exactly 390 of the 500 people selected will be alive at age 75; that is, $P(x = 390)$. Making the correction for continuity, we need to find the area under the normal curve with parameters $\mu = 400$ and $\sigma = 8.94$ that lies between 389.5 and 390.5. [We subtracted 0.5 from 390 and added 0.5 to 390.] The required area is obtained in Figure 6.46.

FIGURE 6.46
Determination of
the area under the
normal curve with
parameters $\mu = 400$
and $\sigma = 8.94$
that lies between
389.5 and 390.5

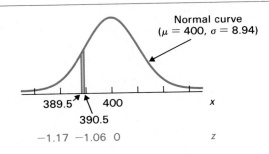

z-score computations:

$x = 389.5 \longrightarrow z = \dfrac{389.5 - 400}{8.94} = -1.17 \qquad 0.3790$

$x = 390.5 \longrightarrow z = \dfrac{390.5 - 400}{8.94} = -1.06 \qquad 0.3554$

Shaded area $= 0.3790 - 0.3554 = 0.0236$

Therefore, $P(x = 390) = 0.0236$, approximately. The probability is about 0.0236 that exactly 390 of the 500 people selected will be alive at age 75.

b) For this part we want the probability that between 375 and 425 of the 500 people selected will be alive at age 75, $P(375 \le x \le 425)$. Making the correction for continuity, we need to find the area under the normal curve with parameters $\mu = 400$ and $\sigma = 8.94$ that lies between 374.5 and 425.5. [We subtracted 0.5 from 375 and added 0.5 to 425.] See Figure 6.47.

FIGURE 6.47
Determination of
the area under the
normal curve with
parameters $\mu = 400$
and $\sigma = 8.94$
that lies between
374.5 and 425.5

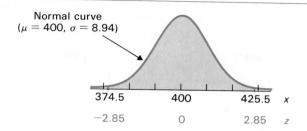

Normal curve
($\mu = 400$, $\sigma = 8.94$)

374.5   400   425.5   x

−2.85   0   2.85   z

z-score computations: Area between 0 and z:

$x = 374.5 \longrightarrow z = \dfrac{374.5 - 400}{8.94} = -2.85$ 0.4978

$x = 425.5 \longrightarrow z = \dfrac{425.5 - 400}{8.94} = 2.85$ 0.4978

Shaded area = 0.4978 + 0.4978 = 0.9956

Consequently, $P(375 \leq x \leq 425) = 0.9956$, approximately. The probability that between 375 and 425 of the 500 people selected will be alive at age 75 is approximately equal to 0.9956. ∎

Exercises 6.5

6.107 Refer to Example 6.27 on page 330.
a) Use Table 6.4 to find the exact probability that the student guesses correctly on
   (i) four or five questions, $P(x = 4 \text{ or } 5)$.
   (ii) between three and seven questions, inclusive, $P(3 \leq x \leq 7)$.
b) Apply Procedure 6.2 on page 333 to approximate the probabilities in part (a) by areas under a normal curve. Compare your answers.

6.108 Refer to Example 6.27 on page 330.
a) Use Table 6.4 to find the exact probability that the student guesses correctly on
   (i) at most five questions, $P(0 \leq x \leq 5)$.
   (ii) at least six questions, $P(6 \leq x \leq 10)$.
b) Apply Procedure 6.2 on page 333 to approximate the probabilities in part (a) by areas under a normal curve. Compare your answers.

6.109 If, in Example 6.27, the true-false exam had 30 questions instead of 10, which normal curve would you use to approximate probabilities for the number of correct guesses?

6.110 If, in Example 6.27, the true-false exam had 25 questions instead of 10, which normal curve would you use to approximate probabilities for the number of correct guesses?

In Exercises 6.111–6.118, apply Procedure 6.2 to approximate the required binomial probabilities.

6.111 According to the *Daily Racing Form*, the probability is about 0.67 that the favorite in a horse race will finish in the money (first, second, or third place). Find the probability that in the next 200 races the favorite will finish in the money
a) exactly 140 times.
b) between 120 and 130 times, inclusive.
c) at least 150 times.

6.112 A small airline in the southwest has determined that the no-show rate for reservations is about 16%. In other words, the probability is 0.16 that a party making a reservation will not show up. The next flight has 42 parties with advance reservations. What is the probability that
a) exactly five parties do not show up?

b) between nine and 12, inclusive, do not show up?

c) at least one does not show up?

d) at most two do not show up?

6.113 As reported by the U.S. National Center for Health Statistics in *Vital and Health Statistics,* about 38.3% of all injuries in the U.S. occur at home. Out of 500 randomly selected injuries, what is the probability that the number occurring at home is

a) exactly 200?

b) between 180 and 210, inclusive?

c) at most 225?

6.114 According to the *World Almanac,* the infant mortality rate in India is 139 per 1,000 live births. Determine the probability that, out of 1,000 randomly selected live births, there are

a) exactly 139 infant deaths.

b) between 120 and 150 infant deaths, inclusive.

c) at most 130 infant deaths.

6.115 A mail-order firm receives orders on about 12% of its mailings. If 750 brochures are mailed, find the probability that the number resulting in orders is

a) exactly 100.

b) either less than 80 or more than 105.

c) at least 14% of the number of brochures mailed.

6.116 Data from the *Statistical Abstract of the United States* indicate that there is about a 51.3% chance that a baby born in the U.S. will be male. What is the probability that out of the next 10,000 births at least half will be male?

6.117 A roulette wheel consists of 38 numbers of which 18 are red, 18 are black, and two are green. When the roulette ball is spun, it is equally likely to land on each of the 38 numbers. A gambler is playing roulette and bets $10 on red each time. If the ball lands on a red number, the gambler wins $10 from the house; otherwise the gambler loses $10. What is the probability that the gambler is ahead after

a) 100 bets?

b) 1000 bets?

c) 5000 bets?

(Hint: The gambler is ahead after a series of bets if and only if she has won more than half of the bets.)

6.118 A brand of flashlight battery has normally distributed lifetimes with a mean of 30 hours and a standard deviation of five hours. A supermarket purchases 500 of these batteries from the manufacturer. What is the probability that at least 80% of them last longer than 25 hours?

Chapter review

KEY TERMS

continuous probability distribution, 283
continuous random variable, 283
correction for continuity,* 333
destandardizing, 305
discrete random variable, 283
empirical rule, 314, 321
general empirical rule, 323
normal curves, 284
normal distribution, 283
normally distributed population, 311
normally distributed random
    variable, 320

percentiles, 317
population *z*-score, 313
quartiles, 316
standard normal curve, 284
standard normal distribution, 326
standardized version of a random
    variable, 325
standardizing, 302
symmetric, 285
z_α, 292
z-curve, 284
z-score, 302

FORMULAS   **z-score for an x-value, 302, 313**

$$z = \frac{x - \mu}{\sigma}$$

(μ and σ are the parameters for a normal curve or the mean and standard deviation of a population)

Standardized version of a random variable x, 325

$$z = \frac{x - \mu_x}{\sigma_x}$$

(μ_x = mean of x, σ_x = standard deviation of x)

YOU SHOULD
BE ABLE TO

1. use and understand the preceding formulas.
2. find areas under the standard normal curve using Table II.
3. use Table II to find the z-value(s) corresponding to a specified area under the standard normal curve.
4. explain the meaning of the parameters μ and σ for a normal curve.
5. sketch a normal curve.
6. find areas under any normal curve.
7. find the x-value(s) corresponding to a specified area under a normal curve.
8. determine percentages for a normally distributed population.
9. obtain and interpret a population z-score.
10. state and apply the empirical rule for normally distributed populations.
11. find quartiles and percentiles for a normally distributed population.
12. determine probabilities for a normally distributed random variable.
13. state and apply the empirical rule for normally distributed random variables.
14. state and use the general empirical rule for normally distributed random variables.
15. obtain and interpret the standardized version of a random variable.
16. approximate binomial probabilities by normal curve areas, when appropriate.*

REVIEW TEST

1. What are the two primary reasons for studying the normal distribution?

2. Determine and sketch the area under the standard normal curve that lies
   a) between $z = 0$ and $z = 2.47$.
   b) between $z = -1.85$ and $z = 0$.
   c) to the right of $z = 0.61$.
   d) to the left of $z = -3.02$.
   e) between $z = 1.11$ and $z = 2.75$.
   f) between $z = -2.06$ and $z = -0.54$.
   g) to the left of $z = 1.59$.
   h) to the right of $z = -1.34$.
   i) between $z = -2$ and $z = 1.5$.
   j) either to the left of $z = 1$ or to the right of $z = 3$.

3. For the standard normal curve, find the z-value(s)
   a) $z_{0.025}$, $z_{0.05}$, $z_{0.01}$, and $z_{0.005}$.
   b) with area 0.10 to its right.
   c) with area 0.30 to its left.
   d) with area 0.67 to its left.

   e) with area 0.85 to its right.
   f) that divide the area under the curve into a middle 0.99 area and two outside 0.005 areas.

4. Sketch the normal curve with parameters
   a) $\mu = -1$ and $\sigma = 2$.
   b) $\mu = 3$ and $\sigma = 2$.
   c) $\mu = -1$ and $\sigma = 0.5$.

5. Consider the normal curves with the following parameters: $\mu = 1.5$ and $\sigma = 3$; $\mu = 1.5$ and $\sigma = 6.2$; $\mu = -2.7$ and $\sigma = 3$; $\mu = 0$ and $\sigma = 1$.
   a) Which curve has the largest spread?
   b) Which curves are centered at the same place?
   c) Which curves have the same shape?
   d) Which curve is centered furthest to the left?
   e) Which curve is the standard normal curve?

6. Find the area under the normal curve with parameters $\mu = -1$ and $\sigma = 2.5$ that lies
   a) between 2 and 6.
   b) to the right of -5.6.
   c) to the right of 3.2.

7. For the normal curve in Problem 6, determine the x-value(s)
   a) with area 0.025 to its right.
   b) with area 0.05 to its left.
   c) with area 0.84 to its left.
   d) with area 0.975 to its right.
   e) that divide the area under the curve into a middle 0.90 area and two outside 0.05 areas.

8. Each year thousands of college seniors take the *Graduate Record Examination (GRE)*. The scores are transformed so as to have a mean of $\mu = 500$ and a standard deviation of $\sigma = 100$. Furthermore, the scores are known to be approximately normally distributed. Determine the percentage of students that score
   a) between 350 and 625.
   b) at least 375.
   c) above 750.

9. Refer to Problem 8.
   a) Obtain the quartiles for GRE scores. Interpret your results in words.
   b) Find the 99th percentile for GRE scores. Interpret your result in words.

10. Refer to Problem 8. How many standard deviations away from the mean is a score of
    a) 645?   b) 320?   c) 500?

11. Refer to Problem 8. Apply the empirical rule for normally distributed populations to fill in the following blanks:
    a) About 68.26% of students score between _____ and _____ .
    b) About 95.44% of students score between _____ and _____ .
    c) About 99.74% of students score between _____ and _____ .

12. For a normally distributed population,
    a) about ____% of the population values lie within 2.75 standard deviations of the mean.
    b) about 90% of the population values lie within _____ standard deviations of the mean.

13. According to R. R. Paul, the mean gestation period of the Morgan horse is 339.6 days (Paul, R. R. 1973. Foaling date. *The Morgan Horse* 33:40). The gestation periods are approximately normally distributed with a standard deviation of 13.3 days. Let x denote the gestation period of a randomly selected Morgan horse.

a) Determine the mean and standard deviation of the random variable x.
b) Find $P(320 \le x \le 330)$.
c) Determine $P(x > 370)$.
d) Interpret your results in parts (b) and (c).

14. Refer to Problem 13. Apply the empirical rule for normally distributed random variables to fill in the following blanks:
    a) The probability is 0.6826 that a Morgan horse will have a gestation period between _____ and _____ days.
    b) The probability is 0.9544 that a Morgan horse will have a gestation period between _____ and _____ days.
    c) The probability is 0.9974 that a Morgan horse will have a gestation period between _____ and _____ days.

15. Refer to Problem 13.
    a) Find the standardized version, z, of the random variable x.
    b) What does the random variable z represent?
    c) How many standard deviations from the mean is a gestation period of 325 days? 368 days? 339.6 days?
    d) What is the probability distribution of the random variable z?
    e) For the answer you gave in part (b), was it necessary to assume that the gestation periods are normally distributed? What about for the answer you gave in part (d)?

16. Use the general empirical rule for normally distributed random variables to fill in the following blank: The probability is 0.98 that a normally distributed random variable will be within _____ standard deviations to either side of its mean.

*17. Acute rotavirus diarrhea is the leading cause of death among children under age five, killing an estimated 4.5 million annually in developing countries. Scientists from Finland and Belgium claim that a new oral vaccine is 80% effective against rotavirus diarrhea. Assuming that the claim is correct, find the probability that, out of 1500 cases, the vaccine will be effective in
    a) exactly 1225 cases.
    b) at least 1175 cases.
    c) between 1150 and 1250 cases, inclusive.

CHAPTER 7

THE SAMPLING DISTRIBUTION OF THE MEAN

In the previous chapters we have studied descriptive statistics, probability concepts, random variables, and the normal distribution. In this chapter we will begin to see how those somewhat diverse topics can be integrated to lay the groundwork for inferential statistics.

We will first discuss the necessity for sampling and introduce some of the more common ways that are used to obtain random samples. Then we will explain why it is mandatory to incorporate probability distributions into the design of inferential studies. Following that, we will investigate the *sampling distribution of the mean.* That concept sets the stage for one of the most important statistical-inference procedures—using the mean, $\bar{x}$, of a sample from a population to draw conclusions about the mean, μ, of the entire population.

CHAPTER OUTLINE

7.1 Sampling; random samples Discusses reasons for sampling, some criteria for samples, and methods for obtaining random samples.

7.2 Sampling error; the need for sampling distributions Introduces the concept of sampling error and explains why sampling distributions are required for making statistical inferences.

7.3 The mean and standard deviation of $\bar{x}$ Shows how, for a given sample size, the mean and standard deviation of all possible sample means can be expressed in terms of the mean and standard deviation of the population.

7.4 The sampling distribution of the mean Indicates that, for a given sample size, the distribution of all possible sample means can be approximated by a normal distribution, provided certain technical conditions are met.

7.1   Sampling; random samples

Recall that inferential statistics consists of methods for drawing conclusions about a population based on information obtained from a sample of the population. We have already discussed some reasons why using a sample to acquire information about a population is often preferable to conducting a **census,** where data for the entire population are collected. Sampling is less costly than a census and it can be done more quickly than a census.

Once a researcher decides that sampling is appropriate, the next question that arises is how to select the sample. That is, what method should be employed to obtain the sample from the population? In answering this question, it is crucial to keep in mind that the sample will be used to draw conclusions about the entire population.

Clearly then, it is important for the sample to be **representative.** That is, the sample should reflect as closely as possible the relevant characteristic(s) of the population under consideration. For instance, it would not make much sense to use the mean weight of a sample of football players to make an inference about the mean weight of all U.S. adult males. Nor would it be reasonable to try to estimate the median income of California residents by sampling the incomes of residents of wealthy Beverly Hills.

To see what can happen when a sample is not representative, consider the presidential election of 1936. Before the election, the *Literary Digest* magazine conducted an opinion poll of the voting population. Its survey team asked a sample of the voting population whether they would vote for Franklin D. Roosevelt, the Democratic candidate, or for Alfred Landon, the Republican candidate. Based on the results of the survey, the magazine predicted an easy win for Landon. The actual election results, of course, were that Roosevelt won by the greatest landslide in the history of the presidency!

What happened? Some people think the error occurred because the magazine obtained its sample from among people who owned a car or had a telephone. At the time, that group included only the more well-to-do people and, historically, such people tend to vote Republican. Whatever the reason, the sample taken by the *Literary Digest* was obviously not representative.

Our goal in sampling, therefore, is to select a representative sample. For only then can we expect to make accurate inferences about a population based on information obtained from a sample of the population. Many sampling techniques have been developed to cover the diversity of circumstances in which a sample is required. However, all of these sampling techniques have one goal in common—to obtain a representative sample.

RANDOM SAMPLING

All of the statistical-inference methods that we will consider in this book are intended for use with one particular sampling procedure. That sampling procedure is called **simple random sampling,** or more briefly, **random sampling.**

DEFINITION 7.1   Random sampling and random samples

A *random-sampling* procedure is a sampling procedure for which each possible sample of a given size is equally likely to be the one selected. A sample obtained by random sampling is called a *random sample*.

The following example illustrates random sampling. For the example, we have chosen an extremely small population; namely, the population consisting of the annual salaries of the five top state officials of Oklahoma. Clearly, for such a small population, there is really no need to sample in order to acquire information.

For instance, if we were interested in the population mean salary, μ, we would undoubtedly just obtain the salaries of the five officials and compute μ exactly. However, because this population is so small, it is easy to use it to illustrate the concept of random sampling.

EXAMPLE 7.1   *Illustrates random sampling*

The population of annual salaries for the five top state officials of Oklahoma is presented in the second column of Table 7.1. Values are in thousands of dollars, rounded to the nearest thousand. [SOURCE: *The World Almanac, 1989.*]

TABLE 7.1
Top five Oklahoma
state officials and
their annual salaries

Official	Salary ($thousands)
Governor (G)	70
Lieutenant Governor (L)	40
Secretary of State (S)	37
Attorney General (A)	55
Treasurer (T)	50

a) List the possible samples of two salaries that can be obtained from the population of five salaries.

b) Describe a method for obtaining a random sample of two salaries from the population of five salaries.

c) For the sampling method described in part (b), determine the probability that any specified sample of two salaries will be the one selected.

d) Repeat parts (a)–(c) for a sample size of $n = 4$.

SOLUTION   For convenience, we will use the letters placed parenthetically after each official in Table 7.1 to represent the officials.

a) There are 10 possible samples of two salaries from the population of five salaries.[†] They are listed in the second column of Table 7.2.

[†] The number of possible samples of size n from a population of size N is given by the binomial coefficient $\binom{N}{n}$. In this case, $n = 2$ and $N = 5$ so that $\binom{N}{n} = \binom{5}{2} = 10$.

TABLE 7.2
Possible samples of
two salaries from
the population
of five salaries

Officials selected	Sample obtained
G, L	70, 40
G, S	70, 37
G, A	70, 55
G, T	70, 50
L, S	40, 37
L, A	40, 55
L, T	40, 50
S, A	37, 55
S, T	37, 50
A, T	55, 50

b) One method that we could use to obtain a random sample of size $n = 2$ is the following: Write each of the letters corresponding to the five officials, G, L, S, A, and T, on a separate piece of paper. Next place the five slips of paper into a box, shake the box and then, while blindfolded, pick two of the slips of paper.

c) The sampling procedure described in part (b) ensures that we are taking a random sample. Consequently, each of the possible samples of two salaries is equally likely to be the one selected. Since there are 10 possible samples, the probability is 1/10 that any specified sample will be the one selected.

d) For samples of size $n = 4$, there are five possibilities, as indicated in Table 7.3.

TABLE 7.3
Possible samples of
four salaries from
the population
of five salaries

Officials selected	Sample obtained
G, L, S, A	70, 40, 37, 55
G, L, S, T	70, 40, 37, 50
G, L, A, T	70, 40, 55, 50
G, S, A, T	70, 37, 55, 50
L, S, A, T	40, 37, 55, 50

MTB

In this case, a random-sampling procedure, such as picking four slips of paper out of the box, gives each of the five possible samples in the second column of Table 7.3 a probability of 1/5 of being the one selected. ∎

The method of picking slips of paper out of a box to get a random sample is not very practical, especially when the population being sampled is large. There are several standard procedures for obtaining random samples. One such procedure involves the use of a **table of random numbers.** Another employs **random-number generators,** which are available on many calculators and most computers. We will discuss those two procedures in Chapter 16.

Finally, as we mentioned earlier in this section, the statistical-inference procedures considered in this book are intended for use with (simple) random samples. Consequently, when we refer to a sample of a population, we will mean a random sample, unless we explicitly state otherwise.

Exercises 7.1

__ 7.1 Explain why the following sample is not representative: A sample of 30 dentists from Seattle is taken in order to estimate the median income of all Seattle residents.

__ 7.2 A survey of the political opinions of 150 voters in the retirement community of Sun City, Arizona is used as a sample of the political opinions of all Arizona voters. Is the sample representative? Why?

__ 7.3 The annual salaries of the five top Oklahoma state officials, displayed in Table 7.1, are repeated below. [SOURCE: *The World Almanac, 1989.*]

Official	Salary ($thousands)
Governor (G)	70
Lieutenant Governor (L)	40
Secretary of State (S)	37
Attorney General (A)	55
Treasurer (T)	50

a) List the 10 possible samples of size $n = 3$ that can be obtained from the population of five salaries. [Construct a table similar to Table 7.3 on page 342.]
b) If a random-sampling procedure is used to obtain a sample of three salaries, what is the probability that it is the first sample on your list in part (a)? the second sample? the tenth sample?

__ 7.4 According to the *World Almanac, 1989,* the members of the U.S. House of Representatives from South Carolina are Arthur Ravenel Jr. (AR), Floyd Spence (FS), Butler Derrick (BD), Liz Patterson (LP), John Spratt Jr. (JS), and Robert Tallon (RT).
a) List the 15 possible samples of two representatives that can be selected from the six. For brevity, use the initials provided.
b) Describe a procedure for taking a random sample of two representatives from the six.
c) If a random-sampling procedure is used to obtain two representatives, what is the probability of selecting AR and FS? BD and RT?

__ 7.5 Refer to Exercise 7.4.
a) List the 15 possible samples of four representatives that can be selected from the six.
b) Describe a procedure for taking a random sample of four representatives from the six.
c) If a random-sampling procedure is used to obtain four representatives, what is the probability of selecting AR, LP, JS, and RT? FS, BD, LP, and RT?

__ 7.6 Refer to Exercise 7.4.
a) List the 20 possible samples of three representatives that can be selected from the six.
b) Describe a procedure for taking a random sample of three representatives from the six.
c) If a random-sampling procedure is used to obtain three representatives, what is the probability of selecting AR, FS, and BD? BD, JS, and RT?

7.2 Sampling error; the need for sampling distributions

Now that we have seen why it is important to sample and we have discussed some methods for obtaining a random sample, we need to deal with the following problem: A sample from a population provides data for only a portion of the entire population. Consequently, it is unreasonable to expect that the sample will yield perfectly accurate information about the population. Thus, we should anticipate that a certain amount of error will result simply because we are sampling. That kind of error is called **sampling error.**

DEFINITION 7.2 Sampling error

Sampling error: The error resulting from using a sample, instead of a census, to estimate a population quantity.

EXAMPLE 7.2   *Illustrates sampling error and the need for sampling distributions*

The U.S. Bureau of the Census publishes annual figures on the mean income of American households in *Current Population Reports.* Actually, the Census Bureau reports the mean income of a sample of 72,000 households out of a total of more than 83 million households. For instance, in 1987 the mean income of American households was reported to be $32,144. This is really the sample mean income, $\bar{x}$, of the 72,000 households surveyed and not the true (population) mean income, μ, of all American households.

We certainly cannot expect the mean income, $\bar{x}$, of the 72,000 households sampled by the Census Bureau to be exactly the same as the mean income, μ, of all American households—some sampling error is to be anticipated. But how much sampling error should we expect; that is, how accurate are such estimates likely to be? Is it likely, for instance, that the sample mean household income reported by the Census Bureau will be within $1,000 of the population mean household income?

To answer these questions, we need to know the distribution of all possible sample means that could be obtained by sampling the incomes of 72,000 households. This type of distribution is called a **sampling distribution.**   ■

MTB

In the previous example we illustrated the concept of sampling error, the error resulting from the use of a sample to estimate a population quantity. That example dealt specifically with the sampling error arising from using a sample mean, $\bar{x}$, to estimate a population mean, μ. We mentioned that in order to answer questions about the accuracy of such estimates, we need to know the distribution of all possible $\bar{x}$-values (sample means) that could be obtained. That distribution is called the **sampling distribution of the mean.**

In this and the next two sections we will study the sampling distribution of the mean. Later we will examine other sampling distributions such as the sampling distribution of the standard deviation.

It is difficult to introduce the sampling distribution of the mean with an example that is both realistic and concrete. This is because, even for moderately large populations, the number of possible samples is enormous, thus prohibiting an actual listing of the possibilities. Consequently, we will use an unrealistically small population to introduce the sampling distribution of the mean. Keep in mind, however, that in real-life applications the populations are much larger.

EXAMPLE 7.3   *Introduces the sampling distribution of the mean*

Suppose the population of interest consists of the heights of the five starting players on a men's basketball team. The heights, in inches, are displayed in Table 7.4.

TABLE 7.4
Population of heights

76	78	79	81	86

a) Determine the sampling distribution of the mean for random samples of two heights from the population of five heights.

b) Make some observations about sampling error when the mean, $\bar{x}$, of a random sample of two heights is used to estimate the population mean, μ.

c) Employ the sampling distribution of the mean obtained in part (a) to find the probability that the sampling error made in estimating the population mean, μ, by the mean, $\bar{x}$, of a random sample of two heights will be at most one inch. That is, for a random sample of size $n = 2$, determine the probability that $\bar{x}$ will be within one inch of μ.

SOLUTION For future reference, we first compute the mean of the population of heights:

$$\mu = \frac{\Sigma x}{N} = \frac{76 + 78 + 79 + 81 + 86}{5} = 80$$

a) For this part we need to determine the distribution of all possible $\bar{x}$-values (sample means) for random samples of size $n = 2$. The population under consideration here is so small that we can list the possibilities directly. In fact, there are 10 possible samples of size $n = 2$. The first column of Table 7.5 displays the 10 possible samples; and the second column displays their means. As a visual aid, we have also drawn a dotplot of the $\bar{x}$-values in Figure 7.1.

TABLE 7.5
Possible samples and their means for samples of two heights from the population of five heights

Sample	$\bar{x}$
76, 78	77.0
76, 79	77.5
76, 81	78.5
76, 86	81.0
78, 79	78.5
78, 81	79.5
78, 86	82.0
79, 81	80.0
79, 86	82.5
81, 86	83.5

FIGURE 7.1
Dotplot of $\bar{x}$-values for samples of size $n = 2$

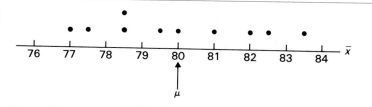

Note: From the second column of Table 7.5, we see that the value of $\bar{x}$ varies from sample to sample. For instance, the first sample has a mean of $\bar{x} = 77.0$ and the second sample has a mean of $\bar{x} = 77.5$. Thus, when sampling from a population, the sample mean, $\bar{x}$, is a *random variable* because its value depends on chance, namely, on which sample is obtained.

Since we are dealing with random samples, each of the 10 possible samples shown in the first column of Table 7.5 has the same probability of being the one selected, a probability of 1/10, or 0.1. Using that fact and the list of $\overline{x}$-values in the second column of Table 7.5, we obtain the probability distribution of the random variable $\overline{x}$ (the sampling distribution of the mean) for random samples of size $n = 2$. This is given in Table 7.6.

TABLE 7.6
Sampling distribution of the mean for random samples of size $n = 2$

Sample mean $\overline{x}$	Probability $P(\overline{x})$	
77.0	0.1	
77.5	0.1	
78.5	0.2	*(Two of the 10 samples have $\overline{x} = 78.5$)*
79.5	0.1	
80.0	0.1	
81.0	0.1	
82.0	0.1	
82.5	0.1	
83.5	0.1	

b) Using the results obtained in part (a), we can make some simple, but significant, observations about sampling error when the mean, $\overline{x}$, of a random sample of two heights is used to estimate the population mean, μ. For instance, it is unlikely that the sample mean, $\overline{x}$, of the sample selected will be identical to the population mean of $\mu = 80$. As a matter of fact, only one of the 10 samples has $\overline{x} = 80$, namely, the eighth sample in Table 7.5 on page 345. Thus, in this case, the chances are only 1/10 (10%) that $\overline{x}$ will equal μ. It is quite likely that there will be some sampling error.

c) Here we want to find the probability that, for a random sample of size two, the sampling error made in estimating μ by $\overline{x}$ will be at most one inch. Since $\mu = 80$ inches, this means we need to find $P(79 \leq \overline{x} \leq 81)$. Using the sampling distribution of the mean for samples of size $n = 2$ (Table 7.6), we see that

$$P(79 \leq \overline{x} \leq 81) = P(\overline{x} = 79.5, \ 80.0, \text{ or } 81.0)$$
$$= P(79.5) + P(80.0) + P(81.0)$$
$$= 0.1 + 0.1 + 0.1 = \mathbf{0.3}$$

If we take a random sample of two heights, there is a 30% chance that the mean of the sample selected will be within one inch of the population mean. ∎

Definition 7.3 summarizes the above discussion.

DEFINITION 7.3   Sampling distribution of the mean

When sampling from a population, the sample mean, $\overline{x}$, is a random variable because its value depends on chance, namely, on which sample is obtained. The probability distribution of the random variable $\overline{x}$ is called *the sampling distribution of the mean.*

We will now return to the situation of Example 7.3 in order to continue our discussion of the sampling distribution of the mean. Up to this point, we have considered samples of size two ($n = 2$). If we consider samples of another size, say $n = 4$, then we obtain a new and different random variable $\bar{x}$.[†]

EXAMPLE 7.4

Further illustrates the sampling distribution of the mean

The heights of the five starting players on a men's basketball team are 76, 78, 79, 81, and 86 inches.

a) Determine the sampling distribution of the mean for random samples of four heights from the population of five heights.

b) Make some observations about sampling error when the mean, $\bar{x}$, of a random sample of four heights is used to estimate the population mean, μ.

c) Employ the sampling distribution of the mean obtained in part (a) to find the probability that the sampling error made in estimating the population mean, μ, by the mean, $\bar{x}$, of a random sample of four heights will be at most one inch. That is, for a random sample of size $n = 4$, determine the probability that $\bar{x}$ will be within one inch of μ.

SOLUTION

a) There are five possible samples of size $n = 4$. These five samples and their means are listed in Table 7.7. A dotplot of the five possible $\bar{x}$-values is shown in Figure 7.2.

TABLE 7.7

Possible samples and their means for samples of four heights from the population of five heights

Sample	$\bar{x}$
76, 78, 79, 81	78.50
76, 78, 79, 86	79.75
76, 78, 81, 86	80.25
76, 79, 81, 86	80.50
78, 79, 81, 86	81.00

FIGURE 7.2

Dotplot of $\bar{x}$-values for samples of size $n = 4$

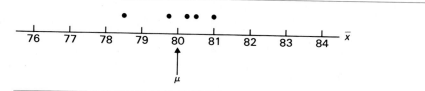

Since we are dealing with random samples, each of the five possible samples of size $n = 4$ has probability $1/5 = 0.2$ of being the one selected. Applying that fact and Table 7.7, we can now write down the probability distribution of the

---

[†] Some texts use subscripts on $\bar{x}$ to emphasize that each sample size gives rise to a different random variable $\bar{x}$. Thus, for instance, $\bar{x}_2$ and $\bar{x}_4$ are used to denote the sample means for samples of sizes two and four, respectively.

random variable $\bar{x}$ (the sampling distribution of the mean) for random samples of size $n = 4$. See Table 7.8.

TABLE 7.8
Sampling distribution
of the mean for
random samples
of size $n = 4$

Sample mean $\bar{x}$	Probability $P(\bar{x})$
78.50	0.2
79.75	0.2
80.25	0.2
80.50	0.2
81.00	0.2

b) Using the results obtained in part (a), we observe that none of the samples of four heights has a mean equal to the population mean of $\mu = 80$. Thus, when the mean, $\bar{x}$, of a random sample of four heights is used to estimate the population mean, μ, some sampling error is certain.

c) Here we want to find the probability that, for a random sample of size four, the sampling error made in estimating μ by $\bar{x}$ will be at most one inch. Since $\mu = 80$ inches, this means we need to find $P(79 \leq \bar{x} \leq 81)$. Using the sampling distribution of the mean for samples of size $n = 4$ (Table 7.8), we see that

$$P(79 \leq \bar{x} \leq 81) = P(\bar{x} = 79.75, \ 80.25, \ 80.50, \ \text{or } 81.00)$$
$$= P(79.75) + P(80.25) + P(80.50) + P(81.00)$$
$$= 0.2 + 0.2 + 0.2 + 0.2 = 0.8$$

If we take a random sample of four heights, there is an 80% chance that the mean of the sample selected will be within one inch of the population mean. ∎

We should emphasize again the following fact: To each sample size there corresponds a different random variable $\bar{x}$ and hence a different sampling distribution of the mean.

SAMPLE SIZE AND SAMPLING ERROR

In Figures 7.1 and 7.2 we drew dotplots of the possible $\bar{x}$-values for samples of size $n = 2$ and $n = 4$, respectively. Those two dotplots, as well as one for samples of size $n = 3$, are displayed together in Figure 7.3 at the top of the next page.

Figure 7.3 depicts vividly that the possible $\bar{x}$-values cluster closer around the true mean, μ, as the sample size increases. For example, when $n = 2$, three out of 10, or 30%, of the possible $\bar{x}$-values lie within one inch of μ; when $n = 3$, five out of 10, or 50%, of the possible $\bar{x}$-values lie within one inch of μ; and when $n = 4$, four out of five, or 80%, of the possible $\bar{x}$-values lie within one inch of μ.

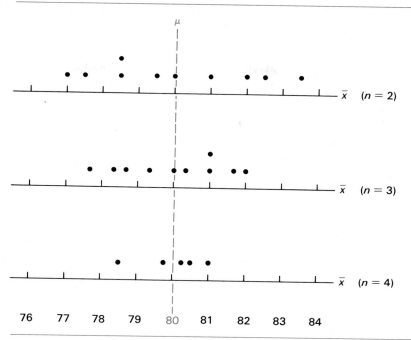

FIGURE 7.3
Dotplots of possible
$\bar{x}$-values for sample
sizes $n = 2$,
$n = 3$, and $n = 4$

More generally, we can make the following qualitative statement about the sampling distribution of the mean:

KEY FACT 7.1   Sample size and sampling error

The possible sample means cluster closer around the population mean as the sample size increases. Thus, the larger the sample size, the smaller the sampling error tends to be in estimating a population mean, μ, by a sample mean, $\bar{x}$.

CONCLUDING REMARKS

We have used a population of five heights to illustrate and explain the importance of the sampling distribution of the mean. For that small population, it is easy to obtain the sampling distribution of the mean by listing all of the possible samples.

However, as we have noted, the populations we deal with in practice are usually large. For such populations, it is not feasible to obtain the sampling distribution of the mean by a direct listing. A more serious practical problem is that, in reality, we do not even know the population values—if we did, there would be no need to sample! Realistically then, we could not list the possible samples to determine the sampling distribution of the mean even if we were willing to "put in the time."

So what can we do in the usual case of a large and unknown population? Fortunately, there exist mathematical relationships that allow us to determine, at least approximately, the sampling distribution of the mean. We will discuss those relationships in the next two sections.

Exercises 7.2

The following exercises are intended solely to provide concrete illustrations of the sampling distribution of the mean. For that reason, the populations considered are unrealistically small.

____ **7.7** As reported by the *World Almanac, 1989*, the annual salaries of the governors of the Great Lakes states are as given in the following table. The salaries are in thousands of dollars, rounded to the nearest thousand.

76	58	64	82	60

a) Compute the population mean of these five salaries.
b) List the possible samples of size $n = 2$ and their means. Construct a table similar to Table 7.5 on page 345.
c) Draw a dotplot for the possible $\bar{x}$-values like the one in Figure 7.1 on page 345.
d) Determine the probability distribution of the random variable $\bar{x}$ (the sampling distribution of the mean) for random samples of size $n = 2$. Construct a table similar to Table 7.6 on page 346.
e) For a random sample of size two, find the probability that the sample mean, $\bar{x}$, will equal the population mean, μ. That is, determine $P(\bar{x} = \mu)$.
f) For a random sample of size two, obtain the probability that the mean of the sample selected will be within four (i.e., $4,000) of the population mean. Interpret your result in terms of percentages.

____ **7.8** Repeat parts (b)–(f) of Exercise 7.7 for samples of size $n = 3$.

____ **7.9** Repeat parts (b)–(f) of Exercise 7.7 for samples of size $n = 4$.

____ **7.10** This exercise requires that you have done Exercises 7.7–7.9.
a) Draw a graph similar to Figure 7.3 on page 349 for sample sizes of $n = 2$, $n = 3$, and $n = 4$ from the population of the five governors' salaries.
b) What does your graph in part (a) illustrate about the impact that increasing sample size has on sampling error?

____ **7.11** The lengths, in centimeters, of six bullfrogs in a small pond are as follows:

19	14	15	9	16	17

Consider these six lengths a population of interest.
a) Calculate the mean length, μ, of the six bullfrogs.
b) List the possible samples of size two and their means. [There are 15 possible samples of size two.]
c) Draw a dotplot for the possible $\bar{x}$-values.
d) Determine the sampling distribution of the mean for random samples of size $n = 2$. [Leave your probabilities in fraction form.]
e) Obtain the probability that the sample mean, $\bar{x}$, of a random sample of size two will equal the population mean, μ.
f) For a random sample of size two, find the probability that the mean of the sample selected will be within 1 cm of the population mean. Interpret your result in terms of percentages.

____ **7.12** Repeat parts (b)–(f) of Exercise 7.11 for samples of size three. [There are 20 possible samples.]

____ **7.13** Repeat parts (b)–(f) of Exercise 7.11 for samples of size four. [There are 15 possible samples.]

____ **7.14** Repeat parts (b)–(f) of Exercise 7.11 for samples of size five. [There are six possible samples.]

____ **7.15** Repeat parts (b)–(f) of Exercise 7.11 for samples of size six. What is the relationship between the only possible sample here and the population?

____ **7.16** Repeat parts (b)–(f) of Exercise 7.11 for samples of size one.

____ **7.17** What do the dotplots in parts (c) of Exercises 7.11–7.16 illustrate about the impact that increasing sample size has on sampling error?

==== **7.18** Suppose a random sample is to be taken from a finite population of size N. Assume the sampling is without replacement. If the sample size is the same as the population size (that is, $n = N$), then
a) how many possible samples are there?
b) what are the possible $\bar{x}$-values?
c) what is the relationship between the only possible sample and the population?

==== **7.19** Suppose that a random sample of size $n = 1$ is to be taken from a finite population of size N.
a) How many possible samples are there?
b) What is the relationship between the possible $\bar{x}$-values and the population values?
c) What is the difference between taking a random sample of size $n = 1$ from a population and selecting a member at random from the population?

7.3 The mean and standard deviation of $\bar{x}$

In Section 7.2, we discussed the sampling distribution of the mean (the probability distribution of $\bar{x}$). That sampling distribution is used to make inferences about a population mean based on the mean of a sample from the population.

As we said earlier, it is generally not possible to know the sampling distribution of the mean exactly. Fortunately, however, we can often approximate that sampling distribution by a normal distribution. That is, under certain conditions, the random variable $\bar{x}$ is (approximately) normally distributed.

Recall that a random variable is normally distributed if probabilities for the random variable are equal to areas under a normal curve. In Chapter 6 we discovered that in order to obtain probabilities for a normally distributed random variable, we must know its mean and standard deviation. This is because the mean and standard deviation of a normally distributed random variable are also the parameters for the appropriate normal curve.

Hence, a first step in learning how to approximate the sampling distribution of the mean by a normal distribution is to study the mean and standard deviation of the random variable $\bar{x}$. That is what we will do in this section.

Before we proceed, let us review the notation used for the mean and standard deviation of a random variable. The mean of a random variable is denoted by a μ subscripted with the letter representing the random variable. So, the mean of x is written as μ_x, the mean of y as μ_y, and so forth. In particular then, the mean of $\bar{x}$ is written as $\mu_{\bar{x}}$. Similarly, the standard deviation of $\bar{x}$ is written as $\sigma_{\bar{x}}$.

THE MEAN OF $\bar{X}$

There is a very simple relationship between the mean of the random variable $\bar{x}$ and the mean of the population being sampled. Namely, the mean of $\bar{x}$ is equal to the population mean. In symbols,

$$\mu_{\bar{x}} = \mu$$

In other words, for a given sample size, the mean of all possible sample means ($\bar{x}$-values) equals the population mean. It is important to note that this equality holds regardless of the sample size under consideration. We illustrate the relationship $\mu_{\bar{x}} = \mu$ in Example 7.5.

EXAMPLE 7.5   *Illustrates that the mean of $\bar{x}$ equals the mean of the population*

The heights of the five starting players on a men's basketball team are 76, 78, 79, 81, and 86 inches. Regard these five heights as a population of interest.
a) Determine the population mean, μ.
b) Definition 5.3 on page 241 gives the defining formula for the mean of a discrete random variable. Use that formula to obtain the mean, $\mu_{\bar{x}}$, of the random variable $\bar{x}$ for samples of size two. Verify that the relation $\mu_{\bar{x}} = \mu$ holds.
c) Repeat part (b) for samples of size four.

SOLUTION   a) To determine the population mean, μ, we use Definition 3.8 (page 124):

$$\mu = \frac{\Sigma x}{N} = \frac{76 + 78 + 79 + 81 + 86}{5} = 80$$

Thus, the population mean height of the five players is $\mu = 80$ inches.

b) To obtain $\mu_{\bar{x}}$, we first need the probability distribution of $\bar{x}$ for samples of size $n = 2$. That probability distribution was determined in Example 7.3 and is repeated here in the first two columns of Table 7.9. Now we can compute $\mu_{\bar{x}}$ by proceeding in the usual way: We append an $\bar{x}P(\bar{x})$-column to the probability distribution of $\bar{x}$ and apply Definition 5.3. See Table 7.9.

TABLE 7.9
Table for computation of $\mu_{\bar{x}}$ for samples of size $n = 2$

$\bar{x}$	$P(\bar{x})$	$\bar{x}P(\bar{x})$
77.0	0.1	7.70
77.5	0.1	7.75
78.5	0.2	15.70
79.5	0.1	7.95
80.0	0.1	8.00
81.0	0.1	8.10
82.0	0.1	8.20
82.5	0.1	8.25
83.5	0.1	8.35
		80.00

From the third column of Table 7.9, we find that

$$\mu_{\bar{x}} = \Sigma \bar{x}P(\bar{x}) = 80$$

By part (a), $\mu = 80$. So, for samples of size two, we see that $\mu_{\bar{x}} = \mu$.

c) The probability distribution of $\bar{x}$ for samples of size $n = 4$, determined in Example 7.4, is repeated here in the first two columns of Table 7.10. The third column of Table 7.10 displays the $\bar{x}P(\bar{x})$-values.

TABLE 7.10
Table for computation of $\mu_{\bar{x}}$ for samples of size $n = 4$

$\bar{x}$	$P(\bar{x})$	$\bar{x}P(\bar{x})$
78.50	0.2	15.70
79.75	0.2	15.95
80.25	0.2	16.05
80.50	0.2	16.10
81.00	0.2	16.20
		80.00

Applying Definition 5.3, we find from the third column of Table 7.10 that, for samples of size four, $\mu_{\bar{x}} = 80$, which again is the same as μ. ∎

Formula 7.1 summarizes the preceding discussion. A proof of Formula 7.1 is discussed in Exercise 7.38.

FORMULA 7.1 Mean of the random variable $\bar{x}$

Suppose that a random sample of size n is to be taken from a population with mean μ. Then,

$$\mu_{\bar{x}} = \mu$$

In other words, for any sample size n, the mean of the random variable $\bar{x}$ is equal to the mean of the population.

THE STANDARD DEVIATION OF $\bar{X}$

Next we will investigate the standard deviation, $\sigma_{\bar{x}}$, of the random variable $\bar{x}$. We want to discover any apparent relationship that $\sigma_{\bar{x}}$ has with the population standard deviation, σ. Let us return once more to the population of heights.

EXAMPLE 7.6   *Investigates the standard deviation of $\bar{x}$ and its relation to the standard deviation of the population*

The heights of the five starting players on a men's basketball team are 76, 78, 79, 81, and 86 inches. We take these heights to be a population of interest.
a) Determine the population standard deviation, σ.
b) Formula 5.1 on page 246 gives a shortcut formula for the standard deviation of a discrete random variable. Use that formula to obtain the standard deviation, $\sigma_{\bar{x}}$, of the random variable $\bar{x}$ for samples of size two. Indicate any apparent relationship between $\sigma_{\bar{x}}$ and σ.
c) Repeat part (b) for samples of sizes three and four.
d) Summarize and discuss the results obtained in parts (a)–(c).

SOLUTION   a) To determine the population standard deviation, σ, we will apply the shortcut formula given in Definition 3.9 on page 126. Recalling that $\mu = 80$, we have

$$\sigma = \sqrt{\frac{\Sigma x^2}{N} - \mu^2} = \sqrt{\frac{76^2 + 78^2 + 79^2 + 81^2 + 86^2}{5} - 80^2}$$

$$= \sqrt{\frac{32{,}058}{5} - 6400} = \sqrt{11.6} = \boxed{3.41}$$

Thus, the standard deviation of the population of heights is $\sigma = 3.41$ inches.
b) The probability distribution of $\bar{x}$ for samples of size $n = 2$ is repeated in the first two columns of Table 7.11 at the top of the next page. The third column gives the $\bar{x}^2 P(\bar{x})$-values. Those values are needed to compute $\sigma_{\bar{x}}$ using the shortcut formula, Formula 5.1.

TABLE 7.11
Table for
computation of
$\sigma_{\overline{x}}$ for samples
of size $n = 2$

$\overline{x}$	$P(\overline{x})$	$\overline{x}^2 P(\overline{x})$
77.0	0.1	592.900
77.5	0.1	600.625
78.5	0.2	1232.450
79.5	0.1	632.025
80.0	0.1	640.000
81.0	0.1	656.100
82.0	0.1	672.400
82.5	0.1	680.625
83.5	0.1	697.225
		6404.350

We now apply Formula 5.1 to obtain $\sigma_{\overline{x}}$. Recalling that $\mu_{\overline{x}} = \mu = 80$, we find from the third column of Table 7.11 that

$$\sigma_{\overline{x}} = \sqrt{\Sigma \overline{x}^2 P(\overline{x}) - \mu_{\overline{x}}^2} = \sqrt{6404.350 - 80^2} = \sqrt{4.35} = 2.09$$

Thus, for samples of size $n = 2$, the standard deviation of $\overline{x}$ is $\sigma_{\overline{x}} = 2.09$. Note that this is not the same as the population standard deviation, which is $\sigma = 3.41$. Also note that $\sigma_{\overline{x}}$ is smaller than σ.

c) Using the same procedure as in part (b), we can compute $\sigma_{\overline{x}}$ for both samples of size $n = 3$ and $n = 4$. For samples of size three, $\sigma_{\overline{x}} = 1.39$. For samples of size four, $\sigma_{\overline{x}} = 0.85$.

d) We summarize our results for $\sigma_{\overline{x}}$ in Table 7.12.

TABLE 7.12
The standard
deviation of $\overline{x}$
for samples of
sizes $n = 2$,
$n = 3$, and $n = 4$

Sample size n	Standard deviation of $\overline{x}$ $\sigma_{\overline{x}}$
2	2.09
3	1.39
4	0.85

Table 7.12 suggests that the standard deviation of $\overline{x}$ gets smaller as the sample size gets larger. We could have predicted this phenomenon from the dotplots of the possible $\overline{x}$-values in Figure 7.3 on page 349 and the fact that the standard deviation of a random variable measures the spread of the possible values of the random variable. Figure 7.3 indicates graphically that the spread, and hence standard deviation, of the possible $\overline{x}$-values decreases with increasing sample size. Table 7.12 indicates that same thing numerically.   ■

The previous example provides evidence that the standard deviation of $\overline{x}$ gets smaller as the sample size increases. Our question now is whether there is a formula that relates the standard deviation of $\overline{x}$ to the sample size and to the standard deviation of the population. The answer is yes! In fact, two different formulas express the precise relationship.

One formula applies when sampling without replacement from a finite population, as in the previous example. That formula is

$$\sigma_{\bar{x}} = \sqrt{\frac{N-n}{N-1}} \cdot \frac{\sigma}{\sqrt{n}}$$

where, as usual, n denotes the sample size and N the population size.

The other formula applies when sampling with replacement from a finite population or when sampling from an infinite population. That formula is

$$\sigma_{\bar{x}} = \frac{\sigma}{\sqrt{n}}$$

If the sample size is small relative to the size of the population, then there is very little difference in the values given by the two formulas (see Exercise 7.37). Generally, in inferential statistics, the sample size *is* small relative to the size of the population. Therefore, we will use the second formula exclusively in this book since it is simpler than the first.

FORMULA 7.2   Standard deviation of the random variable $\bar{x}$

Suppose that a random sample of size n is to be taken from a population with standard deviation σ. Then,

$$\sigma_{\bar{x}} \approx \frac{\sigma}{\sqrt{n}}$$

In other words, the standard deviation of the random variable $\bar{x}$ is equal, at least approximately, to the standard deviation of the population divided by the square root of the sample size.

APPLYING THE FORMULAS

We have seen that there are simple formulas relating the mean and standard deviation of the random variable $\bar{x}$ to the mean and standard deviation of the population being sampled. Namely, $\mu_{\bar{x}} = \mu$ and $\sigma_{\bar{x}} = \sigma/\sqrt{n}$ (at least approximately). Let us apply those two formulas in an example.

EXAMPLE 7.7   *Illustrates Formulas 7.1 and 7.2*

As reported by the U.S. Bureau of the Census in *Current Housing Reports*, the mean livable square footage for single-family detached homes is 1742 square feet. The standard deviation is 568 square feet.

a) Suppose that 25 single-family detached homes are to be selected at random. Let $\bar{x}$ denote the mean livable square footage of the homes obtained. Determine the mean, $\mu_{\bar{x}}$, and the standard deviation, $\sigma_{\bar{x}}$, of the random variable $\bar{x}$.

b) Repeat part (a) for a sample of size 500.

SOLUTION   We know that $\mu = 1742$ and $\sigma = 568$.

a) Applying Formula 7.1 on page 353, we get

$$\mu_{\overline{x}} = \mu = \boxed{1742}$$

Since $n = 25$, we conclude from Formula 7.2 on page 355 that

$$\sigma_{\overline{x}} = \frac{\sigma}{\sqrt{n}} = \frac{568}{\sqrt{25}} = \boxed{113.6}$$

In other words, for random samples of size 25, the mean of all possible sample means is 1742 square feet and the standard deviation is 113.6 square feet.

b) Proceeding as in part (a), we see that

$$\mu_{\overline{x}} = \mu = \boxed{1742}$$

and, since here $n = 500$,

$$\sigma_{\overline{x}} = \frac{\sigma}{\sqrt{n}} = \frac{568}{\sqrt{500}} = \boxed{25.4}$$

That is, for random samples of size 500, the mean of all possible sample means is 1742 square feet and the standard deviation is 25.4 square feet.   ∎

SAMPLE SIZE AND SAMPLING ERROR (REVISITED)

In Key Fact 7.1 on page 349, we pointed out that the sampling error made in estimating a population mean, μ, by a sample mean, $\overline{x}$, tends to be smaller with increasing sample size. We can now make that statement somewhat more precise.

Recall from Key Fact 5.4 on page 246 that the standard deviation of a random variable measures the dispersion of the possible values of the random variable relative to its mean. So, the smaller the standard deviation, the more likely that the random variable will be close to its mean. In particular, the smaller the value of $\sigma_{\overline{x}}$, the more likely that $\overline{x}$ will be close to its mean $\mu_{\overline{x}}$.

But, as we now know, the mean of $\overline{x}$ is equal to the population mean: $\mu_{\overline{x}} = \mu$. Thus, the smaller the value of $\sigma_{\overline{x}}$, the more likely that $\overline{x}$ will be close to μ. That is, the smaller the value of $\sigma_{\overline{x}}$, the greater the tendency for small sampling error.

KEY FACT 7.2   The standard deviation of $\overline{x}$ and sampling error

The standard deviation, $\sigma_{\overline{x}}$, of $\overline{x}$ indicates the amount of sampling error to be expected when the mean, μ, of a population is estimated by the mean, $\overline{x}$, of a sample from the population. That is, the smaller the value of $\sigma_{\overline{x}}$, the smaller the sampling error tends to be. Hence $\sigma_{\overline{x}}$ is often referred to as the *standard error of the mean.*

Finally, we have seen that the standard deviation of $\bar{x}$ is equal to the standard deviation of the population divided by the square root of the sample size: $\sigma_{\bar{x}} = \sigma/\sqrt{n}$. Since n appears in the denominator, the larger the sample size, the smaller the value of $\sigma_{\bar{x}}$. Combining this last fact with Key Fact 7.2, we can make the following fundamental statement.

KEY FACT 7.3 Sample size and sampling error

When estimating a population mean by a sample mean, the larger the sample size, the greater the likelihood for small sampling error. In other words, larger sample sizes tend to produce more accurate estimates of the population mean.

Exercises 7.3

___ **7.20** Why do we need to know the mean, $\mu_{\bar{x}}$, and standard deviation, $\sigma_{\bar{x}}$, of the random variable $\bar{x}$ if we are going to use normal curve areas to approximate probabilities for $\bar{x}$?

___ **7.21** As reported by the *World Almanac, 1989*, the annual salaries of the governors of the Great Lakes states are as given in the following table. The salaries are in thousands of dollars, rounded to the nearest thousand.

76	58	64	82	60

a) Find the population mean, μ, of these five salaries.
b) Consider samples of size two without replacement. Find the mean, $\mu_{\bar{x}}$, of the random variable $\bar{x}$ by applying the formula $\mu_{\bar{x}} = \Sigma \bar{x} P(\bar{x})$. [The probability distribution of $\bar{x}$ for samples of size $n = 2$ was found in part (d) of Exercise 7.7.]
c) Find $\mu_{\bar{x}}$ using only the result of part (a).

___ **7.22** Repeat parts (b) and (c) of Exercise 7.21 for samples of size three. [The probability distribution of $\bar{x}$ for samples of size $n = 3$ was found in part (d) of Exercise 7.8.]

___ **7.23** Repeat parts (b) and (c) of Exercise 7.21 for samples of size four. [The probability distribution of $\bar{x}$ for samples of size $n = 4$ was found in part (d) of Exercise 7.9.]

In Exercises 7.24–7.28, find the mean and standard deviation of $\bar{x}$ for the indicated sample sizes by employing the formulas $\mu_{\bar{x}} = \mu$ and $\sigma_{\bar{x}} = \sigma/\sqrt{n}$.

___ **7.24** According to the Census Bureau publication *Construction Reports*, the mean price of new mo-

bile homes is $\mu = \$25,000$. The standard deviation of the prices is $\sigma = \$7200$.
a) Suppose that a random sample of 50 new mobile homes is to be selected. Let $\bar{x}$ be the mean price of the homes obtained. Find $\mu_{\bar{x}}$ and $\sigma_{\bar{x}}$.
b) Repeat part (a) for a sample size of $n = 100$.

___ **7.25** The U.S. National Center for Health Statistics compiles information on the length of stay by patients in short-stay hospitals. Data are published in *Vital and Health Statistics*. According to that publication, the average stay of female patients in short-stay hospitals is $\mu = 6.9$ days. The standard deviation is $\sigma = 4.3$ days.
a) Suppose that 75 female patients are to be selected at random. Let $\bar{x}$ denote the mean hospital stay of the 75 patients chosen. Find $\mu_{\bar{x}}$ and $\sigma_{\bar{x}}$.
b) Repeat part (a) for a random sample of size 500.

___ **7.26** The National Council of the Churches of Christ in the United States of America reports in *Yearbook of American and Canadian Churches* that the mean number of members per local church is 408. Assume the standard deviation is 116.7.
a) Suppose that 100 local churches are to be selected at random. Let $\bar{x}$ denote the mean number of members for the 100 churches chosen. Determine the mean and standard deviation of $\bar{x}$.
b) Repeat part (a) with $n = 200$.

___ **7.27** According to *Motor Vehicle Facts and Figures*, published by the Motor Vehicle Manufacturers Association of the United States, the average age of cars in use in the U.S. is 7.4 years with a standard deviation of 2.6 years.

a) Let $\bar{x}$ denote the mean age of a random sample of 50 cars. Determine the mean and standard deviation of the random variable $\bar{x}$.

b) Repeat part (a) with $n = 200$.

___ 7.28   The U.S. Bureau of Labor Statistics collects data on earnings of American workers, by industry, and publishes the results in *Employment and Earnings*. That publication indicates that the average weekly earnings of workers in the trucking industry is $421. Suppose that n such workers are to be selected at random. Let $\bar{x}$ denote the mean weekly salary of the workers chosen. Assuming a population standard deviation of $63, find the mean and standard deviation of $\bar{x}$ if

a) $n = 100$.   b) $n = 200$.   c) $n = 400$.

___ 7.29   Suppose you plan to use a sample mean, $\bar{x}$, to estimate a population mean, μ. To increase the likelihood of small sampling error, what must you do?

Finite population correction factor: For Exercises 7.30–7.37, recall the following fact from page 355: If sampling is done without replacement from a finite population of size N, then the standard deviation of $\bar{x}$ can be obtained from the formula

$$(1) \qquad \sigma_{\bar{x}} = \sqrt{\frac{N - n}{N - 1}} \cdot \frac{\sigma}{\sqrt{n}}$$

where n is the sample size. The term $\sqrt{\frac{N-n}{N-1}}$ is referred to as the **finite population correction factor.** If the sample size is small relative to the population size, then we can ignore the finite population correction factor and use the simpler formula

$$(2) \qquad \sigma_{\bar{x}} = \frac{\sigma}{\sqrt{n}}$$

A rule of thumb is that the finite population correction factor can be ignored provided that the sample size is no larger than 5% of the size of the population; that is, $n \leq 0.05N$.

___ 7.30   Refer to Example 7.6 on page 353. We computed the standard deviation of the basketball players' heights to be $\sigma = 3.41$ inches. We also applied the formula $\sigma_{\bar{x}} = \sqrt{\Sigma \bar{x}^2 P(\bar{x}) - \mu_{\bar{x}}^2}$ to determine the standard deviation of $\bar{x}$ for samples of sizes $n = 2$, $n = 3$, and $n = 4$. The results are summarized in Table 7.12 on page 354. Since the sampling is without replacement from a finite population, we can also use Formula (1) above to obtain $\sigma_{\bar{x}}$.

a) Use Formula (1) to compute $\sigma_{\bar{x}}$ for samples of size
(i) $n = 2$ (ii) $n = 3$ (iii) $n = 4$
and compare your results to those in Table 7.12.

b) Use the simpler formula, Formula (2), to compute $\sigma_{\bar{x}}$ for samples of size
(i) $n = 2$ (ii) $n = 3$ (iii) $n = 4$
and compare your results to the true values of $\sigma_{\bar{x}}$ given in Table 7.12. Why does Formula (2) yield such poor results?

c) What percentage of the population size is a sample of size
(i) $n = 2$?   (ii) $n = 3$?   (iii) $n = 4$?

=== 7.31   Refer to Exercise 7.21.
a) Determine the standard deviation, σ, of the governors' salaries.

b) Suppose that we consider samples of size two without replacement. Find the standard deviation, $\sigma_{\bar{x}}$, of the random variable $\bar{x}$ by applying the formula $\sigma_{\bar{x}} = \sqrt{\Sigma \bar{x}^2 P(\bar{x}) - \mu_{\bar{x}}^2}$.

c) Which is the appropriate relationship here, Formula (1) or Formula (2)?

d) Use Formula (1) to compute $\sigma_{\bar{x}}$ and compare your result with that from part (b).

e) Use Formula (2) to compute $\sigma_{\bar{x}}$ and compare your result with that from part (b). Why does Formula (2) yield such a poor approximation to the true value of $\sigma_{\bar{x}}$?

=== 7.32   Refer to Exercise 7.22. Repeat parts (b)–(e) of Exercise 7.31 for samples of size three. [$\sigma = 9.38$]

=== 7.33   Refer to Exercise 7.23. Repeat parts (b)–(e) of Exercise 7.31 for samples of size four. [$\sigma = 9.38$]

=== 7.34   Repeat parts (b)–(e) of Exercise 7.31 for samples of size five.

=== 7.35   Repeat parts (b)–(d) of Exercise 7.31 for samples of size one.

=== 7.36   Suppose a random sample of size n is to be taken from a population of size N.
a) Which formula, Formula (1) or Formula (2), should be used to compute $\sigma_{\bar{x}}$, if the sampling is done without replacement? with replacement?

b) Assume $n = 1$. Compute $\sigma_{\bar{x}}$ using
(i) Formula (1).   (ii) Formula (2).
Why do both formulas give the same result? Explain in words why $\sigma_{\bar{x}} = \sigma$.

c) Assume $n = N$ and that the sampling is done without replacement. Compute $\sigma_{\bar{x}}$ using Formula (1). Could you have guessed the answer without doing any computations? Explain.

≡ 7.37 Suppose that a random sample of size n is to be taken without replacement from a population of size N.

a) Prove that if $n \leq 0.05N$, then

$$0.97 \leq \sqrt{\frac{N-n}{N-1}} \leq 1$$

b) Use part (a) to explain why there is very little difference in the values given by Formulas (1) and (2) when the sample size is no larger than 5% of the population size.

≡ 7.38 Exercises 5.26(e) and 5.27(f) indicate that if x, y, and w are random variables and c is a positive constant, then

$$\mu_{x+y} = \mu_x + \mu_y$$

$$\mu_{cw} = c\mu_w$$

$$\sigma_{cw} = c\sigma_w$$

and if in addition x and y are independent, then

$$\sigma_{x+y} = \sqrt{\sigma_x^2 + \sigma_y^2}$$

These four formulas can be used to derive some of the results given in this section. Suppose a random sample of size n is to be taken from a population with mean μ and standard deviation σ. Let x_1 denote the value of the first member selected, x_2 the value of the second member selected, and so on. Then each of the random variables x_1, x_2, $\ldots$, x_n has mean μ and standard deviation σ; that is, for $i = 1, 2, \ldots, n$,

$$\mu_{x_i} = \mu, \qquad \sigma_{x_i} = \sigma$$

The sample mean is

$$\bar{x} = \frac{\Sigma x}{n} = \frac{1}{n}(x_1 + x_2 + \cdots + x_n)$$

Use the formulas given at the beginning of this exercise to prove that

a) $\mu_{\bar{x}} = \mu$

and that, if the sampling is done with replacement, then

b) $\sigma_{\bar{x}} = \sigma/\sqrt{n}$.

7.4   The sampling distribution of the mean

Recall that the probability distribution of the random variable $\bar{x}$ is called the *sampling distribution of the mean*. Also recall that we need to know that probability distribution in order to make inferences about the mean of a population based on the mean of a sample from the population.

In the previous section we took the first step for describing the sampling distribution of the mean. We discovered that it is easy to express the mean and standard deviation of the random variable $\bar{x}$ in terms of the sample size, n, and the mean, μ, and standard deviation, σ, of the population being sampled: $\mu_{\bar{x}} = \mu$ and $\sigma_{\bar{x}} = \sigma/\sqrt{n}$.

In this section, we will take the second (and final) step for describing the sampling distribution of the mean. It is helpful to distinguish between the case where the population being sampled is normally distributed and the case where it may not be so. We first consider normally distributed populations.

NORMALLY DISTRIBUTED POPULATIONS

Suppose that the population under consideration is normally distributed. Recall that qualitatively this means that the histogram of the population is bell-shaped. More precisely, a population with mean μ and standard deviation σ is normally distributed if percentages for the population are equal to areas under the normal curve with parameters μ and σ.

Although it is by no means obvious, when sampling is done from a normally distributed population, the random variable $\bar{x}$ is also normally distributed. In other words, probabilities for the random variable $\bar{x}$ are equal to areas under the normal curve with parameters $\mu_{\bar{x}}$ and $\sigma_{\bar{x}}$. Putting this together with the facts that $\mu_{\bar{x}} = \mu$ and $\sigma_{\bar{x}} = \sigma/\sqrt{n}$, we can now state the following fundamental result concerning the sampling distribution of the mean for normally distributed populations.

KEY FACT 7.4   The sampling distribution of the mean for normal populations

Suppose a random sample of size n is to be taken from a normally distributed population with mean μ and standard deviation σ. Then the random variable $\bar{x}$ is also normally distributed and has mean $\mu_{\bar{x}} = \mu$ and standard deviation $\sigma_{\bar{x}} = \sigma/\sqrt{n}$. In other words, if the population being sampled is normally distributed, then probabilities for $\bar{x}$ are equal to areas under the normal curve with parameters μ and $\sigma/\sqrt{n}$.

Note: For convenience, we will use the phrase **normal population** as an abbreviation of "normally distributed population."

An intuitive interpretation of Key Fact 7.4 can be described as follows: If for a fixed sample size n, we compute the $\bar{x}$-value for each possible sample of size n, then the histogram of all those $\bar{x}$-values will be bell-shaped.

EXAMPLE 7.8   *Illustrates Key Fact 7.4*

The U.S. Bureau of the Census compiles information on American farm operators and publishes the data in *Census of Agriculture*. According to that document, the mean age of American farm operators is $\mu = 50$ years and the standard deviation is $\sigma = 8$ years. Assuming the ages are normally distributed, the normal curve for that population of ages is displayed in Figure 7.4 at the top of the next page. Find the sampling distribution of the mean for random samples of size
a) $n = 3$.   b) $n = 10$.

SOLUTION   By assumption, the population of ages is normally distributed. Therefore, by Key Fact 7.4, the random variable $\bar{x}$ is also normally distributed and has mean $\mu_{\bar{x}} = \mu = 50$ and standard deviation $\sigma_{\bar{x}} = \sigma/\sqrt{n} = 8/\sqrt{n}$.
a) For samples of size three, the mean and standard deviation of $\bar{x}$ are $\mu_{\bar{x}} = \mu = 50$ and $\sigma_{\bar{x}} = \sigma/\sqrt{n} = 8/\sqrt{3} = 4.62$. Thus, for random samples of size $n = 3$, the sampling distribution of the mean (the probability distribution of $\bar{x}$) is a normal distribution with $\mu_{\bar{x}} = 50$ and $\sigma_{\bar{x}} = 4.62$. The normal curve for $\bar{x}$ is shown in Figure 7.5 on the next page.
b) For samples of size 10, $\mu_{\bar{x}} = \mu = 50$ and $\sigma_{\bar{x}} = \sigma/\sqrt{n} = 8/\sqrt{10} = 2.53$. So, for random samples of size 10, the sampling distribution of the mean is a normal distribution with $\mu_{\bar{x}} = 50$ and $\sigma_{\bar{x}} = 2.53$. See Figure 7.6 on the next page. ∎

FIGURE 7.4
Normal curve
for ages of
farm operators

FIGURE 7.4
Normal curve
for ages of
farm operators

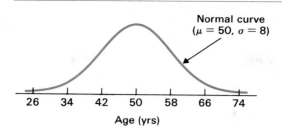

FIGURE 7.5
Sampling distribution
of the mean for
random samples
of size $n = 3$

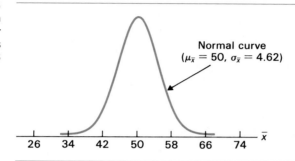

FIGURE 7.6
Sampling distribution
of the mean for
random samples
of size $n = 10$

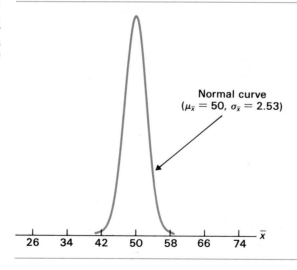

We have drawn the normal curves in Figures 7.5 and 7.6 to scale so that you can compare them visually and observe two important things that you already know. First, both curves are centered at the value of the population mean of $\mu = 50$, since $\mu_{\bar{x}} = \mu = 50$ regardless of the sample size n. Second, the spread of the second curve ($n = 10$) is less extensive than that of the first curve ($n = 3$). This is because the standard deviation, $\sigma_{\bar{x}}$, of the random variable $\bar{x}$ decreases with increasing sample size: $\sigma_{\bar{x}} = \sigma/\sqrt{n}$.

The two curves illustrate vividly that the sampling error made in estimating a population mean, μ, by a sample mean, $\bar{x}$, tends to be smaller for larger sample sizes. That is, the larger the sample size, the more likely that $\bar{x}$ will be close to μ.

Let us now illustrate how to find probabilities for the random variable $\bar{x}$ when sampling is done from a normally distributed population. Since, under those circumstances, $\bar{x}$ is normally distributed, its probabilities are equal to areas under the normal curve with parameters $\mu_{\bar{x}} = \mu$ and $\sigma_{\bar{x}} = \sigma/\sqrt{n}$.

MTB

EXAMPLE 7.9   **Illustrates Key Fact 7.4**

The ages of American farm operators are normally distributed with a mean of 50 years and a standard deviation of eight years. If 250 farm operators are to be selected at random, what is the probability that their mean age, $\bar{x}$, will be within one year of the mean age of all farm operators? That is, determine the probability that $\bar{x}$ will be between 49 and 51 years.

SOLUTION   Since the population is normally distributed, we know from Key Fact 7.4 on page 360 that the random variable $\bar{x}$ is also normally distributed. In addition, $\mu_{\bar{x}} = \mu = 50$ and $\sigma_{\bar{x}} = \sigma/\sqrt{n} = 8/\sqrt{250} = 0.51$. Therefore, probabilities for $\bar{x}$ are equal to areas under the normal curve with parameters $\mu_{\bar{x}} = 50$ and $\sigma_{\bar{x}} = 0.51$.

The problem here is to find $P(49 \leq \bar{x} \leq 51)$, the probability that the mean age of the 250 randomly selected farm operators will be between 49 and 51 years. So, we need to find the area under the normal curve with parameters 50 and 0.51 that lies between 49 and 51. That normal curve area is found in the usual way as depicted in Figure 7.7.

FIGURE 7.7
Determination of
the area under
the normal curve
with parameters
$\mu_{\bar{x}} = 50$ and
$\sigma_{\bar{x}} = 0.51$ that lies
between 49 and 51

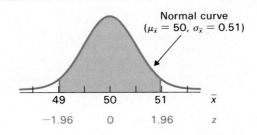

z-score computations:

Area between 0 and z:

$\bar{x} = 49 \longrightarrow z = \dfrac{49 - 50}{0.51} = -1.96$ 0.4750

$\bar{x} = 51 \longrightarrow z = \dfrac{51 - 50}{0.51} = 1.96$ 0.4750

Shaded area $= 0.4750 + 0.4750 = 0.9500$

Consequently, $P(49 \leq \bar{x} \leq 51) = 0.95$. We can interpret this result as follows: Suppose that 250 farm operators are to be selected at random. Then there is a

95% chance that the mean age, $\overline{x}$, of the 250 farm operators chosen will be within one year of the mean age, $\mu = 50$, of all farm operators. We can already begin to see the power of sampling.

∎

GENERAL POPULATIONS; THE CENTRAL LIMIT THEOREM

According to Key Fact 7.4, if the population being sampled is normally distributed, then the random variable $\overline{x}$ is also normally distributed. What if the population being sampled is not normally distributed? Remarkably, the random variable $\overline{x}$ is still approximately normally distributed, provided only that the sample size is relatively large. This extraordinary fact is called the **central limit theorem** and is one of the most important results in statistics.

KEY FACT 7.5 The central limit theorem

For a relatively large sample size, the random variable $\overline{x}$ is approximately normally distributed, regardless of how the population is distributed. The approximation becomes better and better with increasing sample size.

Generally speaking, the more "non-normal" the population, the larger the sample size must be in order for a normal distribution to provide an adequate approximation for the probability distribution of $\overline{x}$. As a rule of thumb, we will consider sample sizes of 30 or more ($n \geq 30$) large enough. Thus, we have the following statement regarding the sampling distribution of the mean.

KEY FACT 7.6 The sampling distribution of the mean for general populations

Suppose that a random sample of size $n \geq 30$ is to be taken from a population with mean μ and standard deviation σ. Then, regardless of the distribution of the population, the random variable $\overline{x}$ is approximately normally distributed and has mean $\mu_{\overline{x}} = \mu$ and standard deviation $\sigma_{\overline{x}} = \sigma/\sqrt{n}$. In other words, if the sample size, n, is 30 or more, then probabilities for $\overline{x}$ are approximately equal to areas under the normal curve with parameters μ and $\sigma/\sqrt{n}$.

EXAMPLE 7.10   *Illustrates Key Facts 7.5 and 7.6*

According to *The Guarantor,* published by the Chicago Title Insurance Company, the average monthly mortgage payment for recent home buyers is $\mu = \$732$. The standard deviation is $\sigma = \$421$. Suppose that 125 recent home buyers are to be selected at random. What is the probability that the mean monthly mortgage payment of the home buyers obtained will exceed the population mean of $732 by more than $50?

SOLUTION   The problem here is to find the probability that $\overline{x}$ will exceed $732 by more than $50; that is, $P(\overline{x} > 782)$. The random variable $\overline{x}$ has mean

$$\mu_{\overline{x}} = \mu = 732$$

and standard deviation

$$\sigma_{\overline{x}} = \frac{\sigma}{\sqrt{n}} = \frac{421}{\sqrt{125}} = 37.66$$

Moreover, since the sample size is $n = 125$, we certainly have $n \geq 30$. Consequently, by the central limit theorem, $\overline{x}$ is approximately normally distributed. This is true regardless of whether the population of monthly mortgage payments for recent home buyers is normally distributed.

Therefore, $P(\overline{x} > 782)$ is approximately equal to the area under the normal curve with parameters 732 and 37.66 that lies to the right of 782. We compute this area in Figure 7.8.

FIGURE 7.8
Determination of the area under the normal curve with parameters $\mu_{\overline{x}} = 732$ and $\sigma_{\overline{x}} = 37.66$ that lies to the right of 782

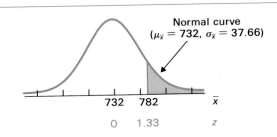

z-score computation:

$\overline{x} = 782 \longrightarrow z = \dfrac{782 - 732}{37.66} = 1.33$

Area between 0 and *z*:

0.4082

Shaded area $= 0.5000 - 0.4082 = 0.0918$

So, $P(\overline{x} > 782) = \boxed{0.0918}$. There is less than a 10% chance that the mean monthly mortgage payment of the 125 randomly selected home buyers will exceed the population mean by more than \$50.   ■

SOME VISUAL DISPLAYS

In the previous pages of this section we have examined the sampling distribution of the mean. We learned that if the population being sampled is normally distributed, then so is the random variable $\overline{x}$, regardless of the sample size. This fact is portrayed graphically in Figure 7.9(a) on the next page.

Additionally, the central limit theorem indicates that if the sample size is relatively large, then the random variable $\overline{x}$ is approximately normally distributed, regardless of the distribution of the population. Figure 7.9(b) and Figure 7.9(c) provide a visual display of the central limit theorem for two non-normal populations. In both of these cases, we see the following: For samples of size $n = 2$, the random variable $\overline{x}$ is definitely not normally distributed; for samples of size $n = 10$, the random variable $\overline{x}$ is already somewhat normally distributed; and, for samples of size $n = 30$, the random variable $\overline{x}$ is certainly very close to being normally distributed.

MTB

FIGURE 7.9 Population and sampling distributions ($n = 2$, $n = 10$, and $n = 30$) for a (a) normal population, (b) skewed population, (c) uniform population

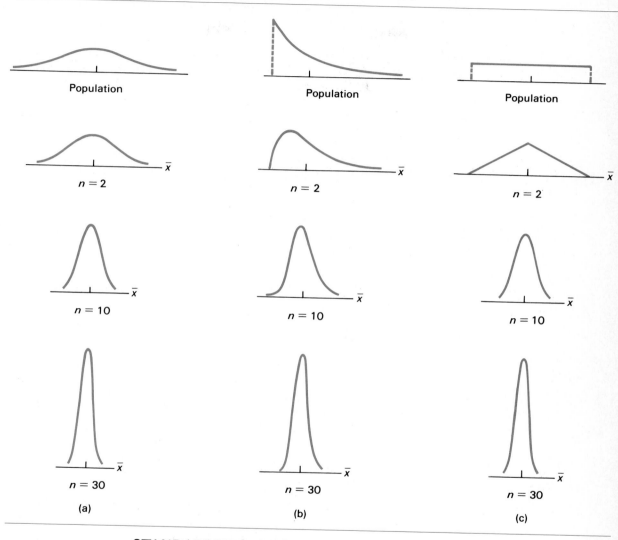

STANDARDIZED FORM

In future chapters it will be necessary for us to consider the standardized version of the random variable $\bar{x}$, that is,

$$z = \frac{\bar{x} - \mu_{\bar{x}}}{\sigma_{\bar{x}}}$$

Since $\mu_{\bar{x}} = \mu$ and $\sigma_{\bar{x}} = \sigma/\sqrt{n}$, the previous equation can be written as

$$z = \frac{\bar{x} - \mu}{\sigma/\sqrt{n}}$$

We learned in Section 6.4 (Key Fact 6.9 on page 326) that if we standardize a normally distributed random variable, then the resulting random variable has the standard normal distribution. Therefore, we can summarize the results of Section 7.3 and the present section as follows:

KEY FACT 7.7   The sampling distribution of the mean (standardized form)

Suppose that a random sample of size n is to be taken from a population with mean μ and standard deviation σ. Then the standardized random variable

$$z = \frac{\overline{x} - \mu}{\sigma/\sqrt{n}}$$

(1) has exactly the standard normal distribution if the population itself is normally distributed, regardless of sample size;
(2) has approximately the standard normal distribution if $n \geq 30$, regardless of how the population is distributed.

In other words, if either the population is normally distributed or the sample size is at least 30, then probabilities for the random variable

$$z = \frac{\overline{x} - \mu}{\sigma/\sqrt{n}}$$

are equal, at least approximately, to areas under the standard normal curve.

EXAMPLE 7.11   *Illustrates Key Fact 7.7*

As reported by the U.S. Bureau of Labor Statistics in the publication *Employment and Earnings,* the mean weekly earnings for workers in the trucking industry is $421. The standard deviation is $63. Suppose that 81 workers in the trucking industry are to be selected at random. Let $\overline{x}$ denote the mean weekly earnings of the workers obtained.
a) Determine the standardized version, z, of the random variable $\overline{x}$.
b) What is the probability distribution of the standardized version, z, of $\overline{x}$?
c) Find the probability that the standardized version, z, of $\overline{x}$ will be between -2 and 2.

SOLUTION   a) The standardized version of $\overline{x}$ is

$$z = \frac{\overline{x} - \mu_{\overline{x}}}{\sigma_{\overline{x}}} = \frac{\overline{x} - \mu}{\sigma/\sqrt{n}}$$

Since here $\mu = 421$, $\sigma = 63$, and $n = 81$, we have

$$z = \frac{\overline{x} - 421}{63/\sqrt{81}} = \frac{\overline{x} - 421}{7}$$

b) Since the sample size, $n = 81$, exceeds 30, the standardized version of $\bar{x}$,

$$z = \frac{\bar{x} - 421}{7},$$

has approximately the **standard normal distribution.** [Note that this is true regardless of how the population of weekly earnings is distributed, because $n \geq 30$. However, if that population is normally distributed, then $z = (\bar{x} - 421)/7$ has exactly the standard normal distribution.]

c) From part (b), the standardized version, z, of $\bar{x}$ has approximately the standard normal distribution. Therefore, the probability that z will be between -2 and 2 is approximately equal to the area under the standard normal curve between -2 and 2. From Table II, that area is $0.4772 + 0.4772 = 0.9544$. Thus,

$$P\left(-2 \leq z \leq 2\right) = P\left(-2 \leq \frac{\bar{x} - 421}{7} \leq 2\right) = 0.9544$$

approximately.

■

Exercises 7.4

7.39 A population has a mean of $\mu = 100$ and a standard deviation of $\sigma = 28$.

a) Determine the sampling distribution of the mean for random samples of size $n = 49$.

b) In answering part (a), what assumptions did you make about the distribution of the population?

c) Can you answer part (a) if the sample size is $n = 16$ instead of $n = 49$? Explain.

7.40 A population has a mean of $\mu = 35$ and a standard deviation of $\sigma = 42$.

a) If the population is normally distributed, obtain the sampling distribution of the mean (probability distribution of $\bar{x}$) for random samples of size nine.

b) Can you answer part (a) if the distribution of the population is unknown? Provide a detailed explanation for your answer.

c) Can you answer part (a) if the distribution of the population is unknown but the sample size is 36 instead of nine? Explain.

7.41 Suppose that a random sample of size n is to be taken from a normally distributed population with mean μ and standard deviation σ.

a) What is the probability distribution of the random variable $\bar{x}$?

b) Does your answer to part (a) depend on how large the sample size is? Explain.

c) What is the mean and standard deviation of $\bar{x}$?

7.42 The length of the western rattlesnake is normally distributed with a mean of $\mu = 42$ inches and a standard deviation of $\sigma = 2.04$ inches.

a) Sketch the normal curve for this population.

b) Find the sampling distribution of the mean for random samples of size four. Draw the normal curve for $\bar{x}$.

c) Repeat part (b) for samples of size eight.

7.43 As reported by the U.S. National Center for Health Statistics in *Vital and Health Statistics*, males who are six feet tall and between 18 and 24 years of age have a mean weight of 175 lb. Assume that the weights are normally distributed and have a standard deviation of 14 lb.

a) Sketch the normal curve for this population.

b) Find the sampling distribution of the mean for samples of size two and sketch the associated normal curve.

c) Repeat part (b) for samples of size nine.

7.44 Refer to Exercise 7.42. Suppose that a random sample of 16 snakes is to be taken. *OVER*

a) Find the probability that the mean length, $\bar{x}$, of the snakes obtained will be within one inch of the population mean of $\mu = 42$ inches; that is, between 41 and 43 inches.
b) Interpret your result in part (a) in terms of sampling error.
c) For samples of size 16, what percentage of the possible samples have means that lie within one inch of the population mean of $\mu = 42$ inches?
d) Repeat part (a) for a sample of size $n = 50$.

__ 7.45   Refer to Exercise 7.43. Suppose that nine six-foot tall males between 18 and 24 years of age are to be selected at random.
a) Find the probability that the mean weight, $\bar{x}$, of the males obtained will be within five pounds of the population mean weight of 175 lb; that is, between 170 and 180 lb.
b) Interpret your result in part (a) in terms of sampling error.
c) What percentage of all possible samples of size nine have means that lie between 170 and 180?
d) Repeat part (a) for a sample of size 30.

__ 7.46   The *Arizona Regional Multiple Listing Service (ARMLS)* publishes several books for the real estate industry. One such book deals with the region termed the "Residential Southwest," composed of Mesa, Tempe, and Chandler, Arizona. This book indicates that the mean sale price of homes in that region is $77,500. Suppose that you take a random sample of 250 homes that have recently sold.
a) Assuming a standard deviation of $57,000, find the probability that the mean sale price of the 250 homes you obtain will be within $5000 of the population mean sale price of $77,500.
b) Must you assume that the sale prices are normally distributed in order to answer part (a)? Explain.
c) Sketch the normal curve for $\bar{x}$ for random samples of size 250.
d) Repeat part (a) for a sample size of 150.

__ 7.47   According to the Census Bureau publication *Current Population Reports*, the mean annual alimony income received by women is about $4000. Assume a standard deviation of $7500. Suppose that 100 women receiving alimony are selected at random.
a) Determine the probability that the mean annual alimony received by the 100 women obtained will be within $500 of the population mean. That is, find $P(3500 \le \bar{x} \le 4500)$.
b) Must you assume that the population of annual

alimony payments is normally distributed in order to answer part (a)? Explain.
c) Sketch the normal curve for $\bar{x}$ for samples of size $n = 100$.
d) Repeat part (a) for a sample of size $n = 1000$.
e) For the alimony incomes considered here, why is it necessary to take such a large sample in order to be assured of relatively small sampling error?

__ 7.48   An air conditioning contractor is preparing to offer service contracts on the brand of compressor used in all of the units her company installs. Before she can work out the details, it is necessary for her to have an idea of how long those compressors last on the average. The contractor anticipated this need and has kept detailed records on the lifetimes of a random sample of 250 compressors. She plans to use the sample mean lifetime, $\bar{x}$, of these 250 compressors as her estimate for the population mean lifetime, μ, of all such compressors. Suppose that this brand of compressor actually has a mean lifetime of $\mu = 62$ months and a standard deviation of $\sigma = 40$ months. What is the probability that the contractor's estimate will be within five months of the true mean of 62 months?

__ 7.49   An economist employed by the Department of Agriculture needs to estimate the mean weekly food costs for couples with two children 6–11 years old. The economist plans to randomly sample 500 such families and use the mean weekly food cost, $\bar{x}$, of those families as her estimate of the true mean weekly food cost, μ. If, in fact, $\mu = \$95.40$ and $\sigma = \$17.20$, then what is the probability that the economist's estimate will be within $1.00 of the actual mean?

__ 7.50   Refer to Figure 7.9 on page 365.
a) Why are the four graphs in Figure 7.9(a) all centered at the same place?
b) Why does the spread of the graphs diminish with increasing sample size? How does this fact affect the sampling error when estimating a population mean, μ, by a sample mean, $\bar{x}$?
c) Why are the graphs in Figure 7.9(a) bell-shaped?
d) Why do the graphs in Figure 7.9(b) and the graphs in Figure 7.9(c) become bell-shaped as the sample size increases?

__ 7.51   Intelligence quotients measured on the Stanford Revision of the *Binet-Simon Intelligence Scale* are known to be approximately normally distributed with a mean of $\mu = 100$ and a standard deviation of $\sigma = 16$. Suppose nine people are to be selected at random. Let $\bar{x}$ denote the mean IQ of the people chosen.

a) Determine the standardized version, z, of the random variable $\bar{x}$.

b) What is the probability distribution of the standardized version, z, of $\bar{x}$?

c) Does your answer to part (b) depend on the fact that IQs are normally distributed? Provide a detailed explanation for your answer.

d) Find the probability that the standardized version of $\bar{x}$ will be between -1.64 and 1.64.

—— 7.52 The U.S. Energy Information Administration collects data on household vehicles and publishes the results in *Residential Transportation Energy Consumption Survey, Consumption Patterns of Household Vehicles*. This document states that the mean monthly fuel expenditure per household vehicle is $58.80. The standard deviation is $30.40. Suppose that 50 household vehicles are to be selected at random, one for each of 50 randomly selected months. Let $\bar{x}$ denote the mean monthly fuel expenditure per vehicle selected.

a) Determine the standardized version, z, of the random variable $\bar{x}$.

b) What is the probability distribution of the standardized version of $\bar{x}$?

c) Does your answer to part (b) depend on the fact that the sample size is at least 30? Explain.

d) Find the probability that the standardized version of $\bar{x}$ will be less than 1.96.

—— 7.53 A certain brand of water-softener salt comes in packages marked "net weight 40 lb." The company claims that, indeed, the bags contain an average of 40 lb of salt and that the standard deviation of the weights is 1.5 lb. Furthermore, it is known that the weights are normally distributed.

a) Obtain the probability that the weight of a randomly selected bag of water-softener salt will be 39 lb or less, if the company's claim is true.

b) Obtain the probability that the mean weight of 10 randomly selected bags of water-softener salt will be 39 lb or less, if the company's claim is true.

c) If you bought a bag of water-softener salt and it weighed 39 lb, would you consider this to be evidence that the company's claim is incorrect? Explain your answer.

d) If you bought 10 bags of water-softener salt and their mean weight was 39 lb, would you consider this to be evidence that the company's claim is incorrect? Explain your answer.

—— 7.54 According to the Salt River Project, a supplier of electricity to the greater Phoenix area, the mean annual electric bill in 1984 was $852.31. Assume that, for a given year, annual electric bills are normally distributed and that they have a standard deviation of $204. At the end of 1986, an independent consumer agency wanted to determine whether the mean annual electric bill had increased over the 1984 figure of $852.31.

a) What is the probability that the mean of a random sample of 25 annual electric bills for 1986 will be $875 or greater, if the 1986 mean annual electric bill for all customers was still $852.31?

b) If the consumer agency takes a random sample of 25 annual electric bills for 1986 and finds that $\bar{x} = \$875$, does this provide substantial evidence that the 1986 mean was more than the 1984 mean of $852.31? Why or why not? (*Hint:* Refer to your answer from part (a).)

c) Repeat parts (a) and (b) for $n = 250$.

d) To answer part (c), is it necessary to assume that the annual electric bills are normally distributed? Why?

Chapter review

Mean of the random variable $\bar{x}$, 353

$$\mu_{\bar{x}} = \mu$$

(μ = mean of the population)

Standard deviation of the random variable $\bar{x}$, 355

$$\sigma_{\bar{x}} = \frac{\sigma}{\sqrt{n}}$$

(σ = standard deviation of the population, n = sample size)

Standardized version of the random variable $\bar{x}$, 365

$$z = \frac{\bar{x} - \mu}{\sigma/\sqrt{n}}$$

YOU SHOULD
BE ABLE TO

1. use and understand the preceding formulas.
2. explain what is meant by a representative sample and by a random sample.
3. define sampling error and indicate the need for sampling distributions.
4. find the mean and standard deviation of the random variable $\bar{x}$, given the mean and standard deviation of the population and the sample size.
5. state and apply the central limit theorem.
6. determine the sampling distribution of the mean when the population being sampled is normally distributed or when the sample size is at least 30.
7. find probabilities for $\bar{x}$ when sampling from a normally distributed population.
8. find probabilities for $\bar{x}$ when the sample size is at least 30.
9. obtain the standardized version of the random variable $\bar{x}$ and identify its probability distribution when the population being sampled is normally distributed or when the sample size is at least 30.

REVIEW TEST

1. An economist wishes to estimate the mean income of the parents of college students. To accomplish that, he surveys a sample of 250 students at Yale. Is this a representative sample? Explain.

2. Classify each sampling procedure below as *random* or *not random*.
   a) A college student is hired to interview a sample of voters in her town. She stays on campus and interviews 100 students in the cafeteria.
   b) A pollster wants to interview a random sample of 20 gas-station managers in Baltimore. He pastes a list of all such managers on his wall, closes his eyes, and tosses a dart at the

list 20 times. He interviews the people whose names the dart hits.

3. In 1980, the U.S. Internal Revenue Service sampled 171,700 tax returns to obtain estimates of various parameters. Data were published in *Statistics of Income, Individual Income Tax Returns*. According to that document, the mean income tax per taxable return for the returns sampled was $\bar{x} = \$3387$.
   a) Explain the meaning of sampling error in this context.
   b) If the population mean income tax per taxable return in 1980 was actually $\mu = \$3475$, how

much sampling error was made in estimating μ by $\bar{x}$?

c) If the IRS had sampled 250,000 returns instead of 171,700, would the sampling error necessarily have been smaller? Explain.

d) In future surveys, how can the IRS increase the likelihood of a small sampling error?

4. As reported by the U.S. Department of Agriculture in *Crop Production*, the mean yield of cotton per acre for American farms is 506 pounds. The standard deviation is 237 pounds.

a) Suppose that 25 one-acre plots of cotton are to be selected at random. Let $\bar{x}$ denote the mean yield of the 25 plots obtained. Find the mean and standard deviation of $\bar{x}$.

b) Repeat part (a) for a sample size of 200.

c) Suppose 500 one-acre plots of cotton are to be selected at random. Without doing any computations, answer the following question: Will the value of $\sigma_{\bar{x}}$ here be larger, smaller, or the same as the value of $\sigma_{\bar{x}}$ in part (b)? Explain.

5. A population, which may or may not be normally distributed, has mean $\mu = 40$ and standard deviation $\sigma = 10$. Suppose that 100 members of the population are to be selected at random. Decide whether each of the following statements is true or false or whether it is not possible to tell. Give a reason for each of your answers.

a) There is about a 68.26% chance that the mean of the sample will be between 30 and 50.

b) About 68.26% of the population values lie between 30 and 50.

c) There is about a 68.26% chance that the mean of the sample will be between 39 and 41.

6. Repeat Problem 5 under the assumption that the population is normally distributed.

7. The monthly rents for studio apartments in a large city are normally distributed with a mean of $285 and a standard deviation of $45.

a) Sketch the normal curve for this population of monthly rents.

b) Find the sampling distribution of the mean for samples of size three. Draw the normal curve for $\bar{x}$.

c) Repeat part (b) for samples of size nine.

8. Refer to Problem 7. Suppose that three studio apartments are to be selected at random.

a) Find the probability that the mean monthly rent, $\bar{x}$, of the apartments chosen will be within

$10 of the population mean of $285; that is, between $275 and $295.

b) Interpret your results in part (a) in terms of sampling error.

c) What percentage of the possible samples of size three have means within $10 of the population mean?

d) Repeat part (a) for samples of size 75.

9. The following figure shows the curve for a normally distributed population (in color). Superimposed are the curves for the sampling distribution of the mean for two different sample sizes.

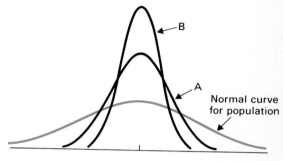

a) Why are all three curves centered at the same place?

b) Which curve corresponds to the larger sample size? Explain.

c) Why is the spread of each curve different?

d) Which of the two sampling-distribution curves corresponds to the sample size that will tend to produce less sampling error? Explain.

10. The American Council of Life Insurance reports in *Life Insurance Fact Book* that the mean life insurance in force per covered family is $\mu = \$82,800$. Assume a standard deviation of $\sigma = \$50,900$. Suppose 500 covered families are randomly selected.

a) What is the probability that the mean life insurance in force for the 500 families selected will be within $2000 of the population mean of $82,800?

b) Must you assume that the population of life-insurance amounts is normally distributed in order to answer part (a)? What if the sample size is 20 instead of 500?

c) Repeat part (a) for a sample size of 5000.

11. The U.S. Social Security Administration compiles information on benefits to retired workers and publishes the data in *Social Security Bulletin*. Ac-

cording to that publication, the mean monthly benefit to retired workers is $513. Assume a standard deviation of $93. Suppose that 125 retired workers are to be selected at random. Let $\bar{x}$ denote the mean monthly benefit received by the retired workers chosen.

a) Obtain the standardized version, z, of the random variable $\bar{x}$.

b) Find the probability distribution of the standardized version, z, of $\bar{x}$.

c) Determine the probability that the standardized version of $\bar{x}$ will take a value either greater than 2.33 or less than -2.33.

d) Could you answer part (b) if the sample size were five instead of 125? Explain.

12. A paint manufacturer located in the city of Pittsburgh claims that his paint will last for an aver-

age of five years. Assume that paint life is normally distributed and has a standard deviation of 0.5 years.

a) If you paint one house with the paint and the paint lasts 4.5 years, would you consider that to be substantial evidence against the manufacturer's claim? (*Hint:* Let x be the paint life for a randomly selected house painted with the paint. Compute $P(x \leq 4.5)$, assuming the manufacturer's claim is correct.)

b) If you paint 10 houses with the manufacturer's paint and the paint lasts an average of 4.5 years for the 10 houses, would you consider that to be substantial evidence against the manufacturer's claim?

c) Repeat part (b) if the paint lasts an average of 4.9 years for the 10 homes painted.

CHAPTER 8

CONFIDENCE INTERVALS FOR ONE MEAN OR PROPORTION

We now begin our study of inferential statistics. In the present chapter, we will examine methods for estimating the mean of a population. As you might suspect, the statistic used to estimate a population mean, μ, is the sample mean, $\bar{x}$. Because of sampling error, we cannot expect $\bar{x}$ to be exactly equal to μ. Thus, it is important to provide information about the accuracy of the estimate. This leads to the discussion of *confidence intervals*, which is the main topic of the chapter.

In addition to discovering how to estimate the mean of a population, we will also learn how to estimate the proportion (percentage) of a population that has a specified attribute. For example, we will see how it is possible to estimate the percentage of all Americans who drink alcoholic beverages from the percentage of a sample who do.

CHAPTER OUTLINE

8.1   Estimating a population mean

A common problem in statistics is to obtain information about the mean, μ, of a population. For instance, we might be interested in

1. the mean tar content, μ, of a certain brand of cigarette, or
2. the mean life, μ, of a newly developed steel-belted radial tire, or
3. the mean gas mileage, μ, of a new model car, or
4. the mean annual income, μ, of liberal arts graduates.

If the population is small, we can ordinarily obtain μ exactly by first taking a census and then computing μ from the population data. In most cases, however, the population under consideration will be large and, consequently, taking a census will be impractical, impossible, or extremely expensive. Nonetheless, we can usually get sufficiently accurate information about μ by taking a sample of the population. Let us look at an example.

EXAMPLE 8.1   *Illustrates the use of a sample mean to estimate a population mean*

The U.S. Bureau of the Census publishes annual price figures for new mobile homes in *Construction Reports*. The information is obtained from sampling, not from a census. Suppose a random sample of 40 new mobile homes yields the prices shown in Table 8.1. Data are in thousands of dollars, rounded to the nearest hundred.

TABLE 8.1
Prices ($thousands)
of 40 randomly
selected new
mobile homes

24.4	30.6	26.4	26.8	33.5	32.2	32.4	13.9
24.4	29.3	26.2	14.1	21.4	20.0	33.0	17.6
24.8	27.0	22.8	18.8	35.1	26.7	22.1	37.2
31.9	24.0	28.4	15.8	29.3	31.4	22.8	8.4
24.7	16.6	31.1	13.9	16.8	29.5	17.0	9.9

Use the sample data in Table 8.1 to estimate the population mean price, μ, of all new mobile homes.

SOLUTION   As you might expect, we will estimate the mean price, μ, of all new mobile homes by the mean price, $\bar{x}$, of the 40 new mobile homes sampled. From Table 8.1, we find

$$\bar{x} = \frac{\Sigma x}{n} = \frac{972.2}{40} = 24.31$$

Consequently, based on the sample data, we estimate that the mean price, μ, of all new mobile homes is approximately $24.31 thousand; that is, $24,310. An estimate of this type is called a **point estimate** for μ because it consists of a single number, or point.   ■

The term *point estimate* applies to the use of a statistic to estimate any parameter, not just a population mean. Thus, we have Definition 8.1.

DEFINITION 8.1 Point estimate

A *point estimate* for a parameter is the value of a statistic used to estimate the parameter.

    As we learned in Chapter 7, it is unreasonable to expect that a sample mean, $\bar{x}$, will be exactly equal to the population mean, μ; some sampling error is to be anticipated. Therefore, in addition to reporting a point estimate for μ, we need to provide information that indicates the accuracy of the estimate. This can be accomplished by giving a **confidence-interval estimate** for μ. With a confidence-interval estimate for μ, we use the mean of a sample to construct an interval of numbers and state how confident we are that μ lies in that interval. We will see shortly how to obtain such confidence intervals.
    The discussion in the previous paragraph is summarized in Definition 8.2. Again, the terminology applies to any parameter, not only to a population mean.

DEFINITION 8.2 Confidence-interval estimate; confidence level

A *confidence-interval estimate* of a parameter consists of an interval of numbers obtained from a point estimate of the parameter together with a percentage that specifies how confident we are that the parameter lies in the interval. The confidence percentage is called the *confidence level.*

    We will now present a method for obtaining a confidence interval for a population mean when the sample size is large. Before proceeding, we recommend that you review the following two facts: (1) Key Fact 6.7 on page 321, which is the empirical rule for normally distributed random variables, and (2) Key Fact 7.6 on page 363, which gives the sampling distribution of the mean for general populations.

EXAMPLE 8.2   *Illustrates confidence intervals*

Suppose that a random sample of size $n \geq 30$ is to be taken from a population with mean μ and standard deviation σ.
a) Determine the probability that the interval from

$$\bar{x} - 2 \cdot \frac{\sigma}{\sqrt{n}} \quad \text{to} \quad \bar{x} + 2 \cdot \frac{\sigma}{\sqrt{n}}$$

will contain the population mean, μ.
b) Interpret the result from part (a) in terms of percentages.

SOLUTION   a) First we note that since the sample size is at least 30, Key Fact 7.6 implies that the random variable $\bar{x}$ is approximately normally distributed with mean $\mu_{\bar{x}} = \mu$ and standard deviation $\sigma_{\bar{x}} = \sigma/\sqrt{n}$. Consequently, by Property 2 of Key Fact 6.7, the probability is approximately 0.9544 that $\bar{x}$ will be within two standard deviations to either side of its mean:

$$P(\mu_{\bar{x}} - 2\sigma_{\bar{x}} < \bar{x} < \mu_{\bar{x}} + 2\sigma_{\bar{x}}) = 0.9544$$

Since $\mu_{\bar{x}} = \mu$ and $\sigma_{\bar{x}} = \sigma/\sqrt{n}$, we can rewrite the above equation as

$$P\left(\mu - 2 \cdot \frac{\sigma}{\sqrt{n}} < \bar{x} < \mu + 2 \cdot \frac{\sigma}{\sqrt{n}}\right) = 0.9544$$

In words, the previous equation states that the probability is 0.9544 that $\bar{x}$ will be within $2 \cdot \sigma/\sqrt{n}$ of μ. But to say that "$\bar{x}$ will be within $2 \cdot \sigma/\sqrt{n}$ of μ" is the same as saying that "μ will be within $2 \cdot \sigma/\sqrt{n}$ of $\bar{x}$." Therefore, the previous equation can be rewritten as

$$P\left(\bar{x} - 2 \cdot \frac{\sigma}{\sqrt{n}} < \mu < \bar{x} + 2 \cdot \frac{\sigma}{\sqrt{n}}\right) = 0.9544$$

(The algebra for doing these manipulations is discussed in Exercise 8.14.) This last equation provides us with the answer to our question: The probability is **0.9544** that the interval from

$$\bar{x} - 2 \cdot \frac{\sigma}{\sqrt{n}} \quad \text{to} \quad \bar{x} + 2 \cdot \frac{\sigma}{\sqrt{n}}$$

will contain μ. It is important to realize that in this and similar such statements the random variable is $\bar{x}$, not μ. The population mean, μ, is a fixed number, although it may be unknown. The sample mean, $\bar{x}$, depends on chance; namely, on which sample is selected.

b) Here we are to interpret the result obtained in part (a) in terms of percentages. One interpretation is the following: Suppose that for each possible sample of size n, we compute the sample mean, $\bar{x}$, and construct the interval from

$$\bar{x} - 2 \cdot \frac{\sigma}{\sqrt{n}} \quad \text{to} \quad \bar{x} + 2 \cdot \frac{\sigma}{\sqrt{n}}$$

Then about 95.44% of such intervals will contain the population mean, μ.   ■

Note: In the above example, we could use Property 1 or Property 3 of Key Fact 6.7 instead of Property 2. By doing so, we would get a different confidence interval with a different confidence level.

Let us now apply the results of Example 8.2 in order to obtain a confidence interval for the mean price, μ, of all new mobile homes.

EXAMPLE 8.3   *Illustrates confidence intervals*

Suppose that a random sample of 40 new mobile homes yields the price data shown in Table 8.1 on page 374. Find a 95.44% confidence interval for the mean price, μ, of all new mobile homes. Assume that the population standard deviation of the prices is $\sigma = \$7.2$ thousand dollars.[†]

[†] We might know this from previous research or from a preliminary study of prices. The more usual case where σ is unknown will be considered in the next section.

SOLUTION   Since the sample size, $n = 40$, is at least 30, we can apply the results of Example 8.2. From part (a) of that example, the probability is 0.9544 that the interval from

$$\bar{x} - 2 \cdot \frac{\sigma}{\sqrt{n}} \quad \text{to} \quad \bar{x} + 2 \cdot \frac{\sigma}{\sqrt{n}}$$

will contain μ.

We know that $\sigma = 7.2$. Also, from the data in Table 8.1, we see that $n = 40$ and $\bar{x} = 24.31$. Hence,

$$\bar{x} - 2 \cdot \frac{\sigma}{\sqrt{n}} = 24.31 - 2 \cdot \frac{7.2}{\sqrt{40}} = 22.03$$

and

$$\bar{x} + 2 \cdot \frac{\sigma}{\sqrt{n}} = 24.31 + 2 \cdot \frac{7.2}{\sqrt{40}} = 26.59$$

Thus, our confidence interval for μ is from 22.03 to 26.59. Since 95.44% of such intervals will actually contain the population mean, μ, the confidence level is 95.44%. In other words, we can be 95.44% confident that the mean price, μ, of all new mobile homes is somewhere between \$22.03 thousand and \$26.59 thousand. It is crucial to remember that this or any other 95.44% confidence interval may or may not contain the population mean, μ, but we can be 95.44% confident that it does. ∎

We need to emphasize that the confidence interval we actually get depends on the value of $\bar{x}$, which in turn depends on the sample obtained. For example, suppose that the prices of the 40 new mobile homes obtained were as given in Table 8.2 instead of as in Table 8.1.

TABLE 8.2

20.0	21.8	32.1	25.2	38.4	19.5	28.9	24.8
17.2	26.2	10.3	25.1	30.8	35.1	33.7	11.0
22.9	23.3	30.1	24.0	47.1	23.2	16.9	30.1
30.2	26.5	29.7	14.6	13.4	29.8	35.7	30.7
20.4	35.8	38.2	14.8	16.8	42.0	32.3	34.6

Then we would have $\bar{x} = 26.58$ so that

$$\bar{x} - 2 \cdot \frac{\sigma}{\sqrt{n}} = 26.58 - 2 \cdot \frac{7.2}{\sqrt{40}} = 24.30$$

and

$$\bar{x} + 2 \cdot \frac{\sigma}{\sqrt{n}} = 26.58 + 2 \cdot \frac{7.2}{\sqrt{40}} = 28.86$$

Hence, in this case, the 95.44% confidence interval for μ would be the interval from 24.30 to 28.86. We could be 95.44% confident that the mean price, μ, of all new mobile homes is somewhere between \$24.30 thousand and \$28.86 thousand.

The next example stresses the interpretation of a confidence interval. It also shows that the population mean, μ, may or may not lie in the confidence interval.

EXAMPLE 8.4   *Illustrates the interpretation of confidence intervals*

Consider once more the prices of new mobile homes. Suppose that 40 new mobile homes are to be selected at random. Then, as we have seen, the probability is 0.9544 that the interval from

$$(1) \qquad\qquad \overline{x} - 2 \cdot \frac{\sigma}{\sqrt{n}} \quad \text{to} \quad \overline{x} + 2 \cdot \frac{\sigma}{\sqrt{n}}$$

will contain the mean price, μ, of all new mobile homes. In other words, once the sample of 40 new mobile homes is obtained and their mean price, $\overline{x}$, is computed, then the interval given by Formula (1) will be a 95.44% confidence interval for μ.

To illustrate that the mean price, μ, of all new mobile homes may or may not lie in the 95.44% confidence interval obtained, we used a computer to simulate the sampling of 40 new mobile homes 20 times. For the simulation, we assumed that $\mu = \$25$ thousand. [Of course, in practice, we don't know μ. We are assuming a value for μ to illustrate a point.]

For each of the 20 samples of size $n = 40$, we did three things: First, we computed the sample mean, $\overline{x}$; second, we used the fact that $\sigma = \$7.2$ thousand and Formula (1) above to obtain the 95.44% confidence interval for μ; and third, we noted whether the population mean, $\mu = \$25$ thousand, actually lies in the confidence interval.

Figure 8.1, at the top of the next page, summarizes our results. For each sample, we have drawn a graph on the right-hand side of Figure 8.1. The dot represents the sample mean, $\overline{x}$, and the horizontal line represents the corresponding 95.44% confidence interval. As you can see, the population mean, μ, lies in the confidence interval when and only when the horizontal line crosses the dashed line.

From Figure 8.1, we observe that μ lies in the 95.44% confidence interval in 19 out of the 20 samples, that is, in 95% of the 20 samples. If instead of 20 samples, we simulated, say, 1000 samples, then we would find that the percentage of those 1000 samples for which μ lies in the 95.44% confidence interval would be even closer to 95.44%. Thus, we can be 95.44% confident that any computed 95.44% confidence interval will actually contain μ. ∎

Up to this point we have only considered confidence intervals with a confidence level of 95.44%. To obtain such confidence intervals, we applied Key Fact 7.6 (the sampling distribution of the mean for general populations) and Property 2 of Key Fact 6.7 (the empirical rule for normally distributed random variables). If instead of this second key fact, we use Key Fact 6.8 on page 323 (the general empirical rule for normally distributed random variables), then we can obtain confidence intervals with any specified confidence level. We will do that in the next section.

FIGURE 8.1 Confidence intervals for 20 random samples of size $n = 40$

Sample	$\bar{x}$	95.44% CI	μ in CI?
1	25.02	22.74 to 27.30	yes
2	22.36	20.08 to 24.64	no
3	24.80	22.52 to 27.08	yes
4	26.31	24.03 to 28.59	yes
5	24.76	22.48 to 27.04	yes
6	24.43	22.15 to 26.71	yes
7	25.61	23.33 to 27.89	yes
8	25.21	22.93 to 27.49	yes
9	25.38	23.10 to 27.66	yes
10	24.90	22.62 to 27.18	yes
11	23.89	21.61 to 26.17	yes
12	25.28	23.00 to 27.56	yes
13	23.94	21.66 to 26.22	yes
14	24.18	21.90 to 26.46	yes
15	24.80	22.52 to 27.08	yes
16	27.06	24.78 to 29.34	yes
17	24.99	22.71 to 27.27	yes
18	24.40	22.12 to 26.68	yes
19	24.96	22.68 to 27.24	yes
20	25.93	23.65 to 28.21	yes

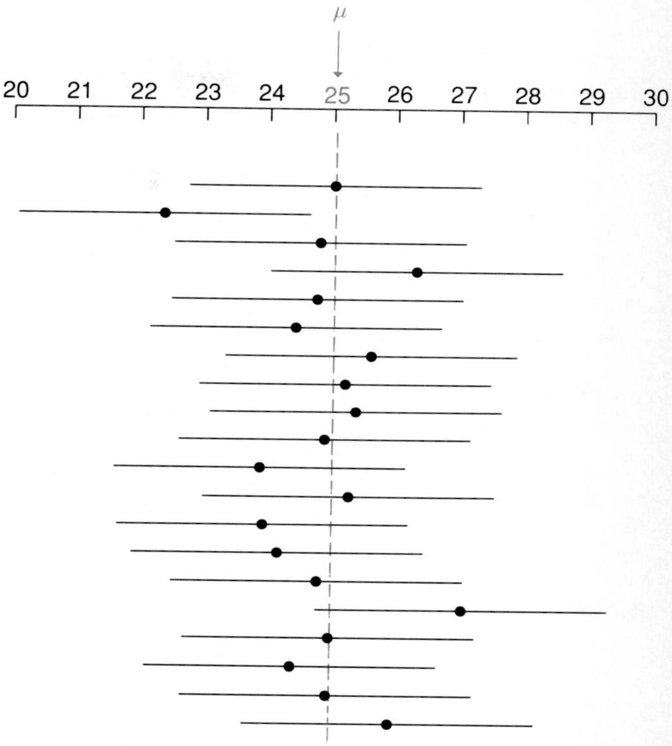

Exercises 8.1

__ **8.1** The U.S. National Center for Health Statistics compiles natality (birth) statistics and publishes the results in *Vital Statistics of the United States*. Suppose that a random sample of 35 babies yields the following birth weights, in pounds:

7.4	6.0	8.6	4.5	2.0	7.9	4.0
2.6	5.9	7.3	7.3	7.0	6.3	8.1
7.1	7.3	6.6	5.2	9.8	8.0	10.9
6.3	3.8	5.0	8.0	10.7	9.7	6.0
6.8	10.3	7.6	6.5	7.1	5.8	6.9

a) Use the data to obtain a point estimate for the population mean weight, μ, of all newborns. (*Note: The sum of the data is 240.3 lb.*)

b) Is it likely that your point estimate in part (a) is exactly equal to μ? Explain.

__ **8.2** An educational psychologist at a large university wants to estimate the mean IQ of the students in attendance. Suppose that a random sample of 30 students gives the following data on IQs:

107	99	101	93	99	103
134	132	103	109	104	103
101	128	113	106	126	103
131	106	119	102	98	116
108	103	111	119	112	105

a) Use the data to obtain a point estimate for the mean IQ, μ, of all students attending the university. (*Note: The sum of the data is 3294.*)

b) Is it likely that your point estimate in part (a) is exactly equal to μ? Explain.

___ 8.3   The R. R. Bowker Company of New York collects data on books and periodicals. Sources for the information are *Publishers Weekly, The Bowker Annual of Library and Book Trade Information,* and *Library Journal.* Suppose 40 randomly selected science books have the prices, to the nearest dollar, as given here.

70	71	56	71	69	78	87	83
74	94	52	84	65	69	62	65
73	70	63	64	79	61	69	86
70	48	75	62	90	100	53	74
64	68	90	56	60	79	73	50

Obtain a point estimate for the mean price, μ, of all science books. *(Note: $\Sigma x = \$2827$.)*

___ 8.4   The publication *Trends in Television* from the Television Bureau of Advertising, Inc., contains information on the number of TV sets owned by American households. Suppose that 50 households are selected at random and that the number of TV sets per household sampled is as follows:

1	1	1	2	6	3	3	4	2	4
3	2	1	5	2	1	3	6	2	2
3	1	1	4	3	2	2	2	2	3
0	3	1	2	1	2	3	1	1	3
3	2	1	2	1	1	3	1	5	1

Use the sample data to find a point estimate for the mean number of TV sets, μ, per American household.

___ 8.5   Define the following terms:
a) point estimate
b) confidence-interval estimate
c) confidence level

___ 8.6   Why is a confidence-interval estimate of a parameter more useful than a point estimate?

For Exercises 8.7–8.10, you may want to review what was done in Example 8.3 on pages 376–377.

___ 8.7   Refer to Exercise 8.1. Assume that the population standard deviation of weights for newborns is $\sigma = 1.9$ lb.
a) Use the data given in Exercise 8.1 to find a 95.44% confidence interval for the mean weight, μ, of all newborns.
b) Interpret your result in part (a) in words.

c) Does the true (population) mean birth weight, μ, lie in the confidence interval that you obtained in part (a)? Explain.

___ 8.8   Refer to Exercise 8.2. Assume that the standard deviation of IQs for all students attending the university is $\sigma = 12$.
a) Use the data given in Exercise 8.2 to find a 95.44% confidence interval for the mean IQ, μ, of all students attending the university.
b) Interpret your result in part (a) in words.
c) Does the mean IQ, μ, of all students attending the university lie in the confidence interval that you obtained in part (a)? Explain.

___ 8.9   Refer to Exercise 8.3. Assume that the standard deviation of prices for all science books is $15.
a) Find a 95.44% confidence interval for the mean price of all science books based on the data given in Exercise 8.3.
b) Interpret your result in part (a) in words.

___ 8.10   Refer to Exercise 8.4. Assume that $\sigma = 1.4$.
a) Find a 95.44% confidence interval for the mean number of TV sets per American household using the data given in Exercise 8.4.
b) Interpret your result in part (a) in words.

=== 8.11   Suppose a random sample of size $n \geq 30$ is to be taken from a population with mean μ and standard deviation σ.
a) Determine the probability that the interval from

$$\bar{x} - 3 \cdot \frac{\sigma}{\sqrt{n}} \quad \text{to} \quad \bar{x} + 3 \cdot \frac{\sigma}{\sqrt{n}}$$

will contain the population mean, μ. *(Hint: Refer to Example 8.2 on pages 375–376 and the note following that example.)*
b) Interpret your result from part (a) in terms of percentages.
c) Apply your result from part (a) and the data in Exercise 8.1 to obtain a 99.74% confidence interval for the mean weight, μ, of all newborns. [Recall that $\sigma = 1.9$ lb.]

=== 8.12   Suppose a random sample of size $n \geq 30$ is to be taken from a population with mean μ and standard deviation σ.
a) Determine the probability that the interval from

$$\bar{x} - 1 \cdot \frac{\sigma}{\sqrt{n}} \quad \text{to} \quad \bar{x} + 1 \cdot \frac{\sigma}{\sqrt{n}}$$

will contain the population mean, μ. *(Hint: Refer to Example 8.2 on pages 375–376 and the note following that example.)*

b) Interpret your result from part (a) in terms of percentages.

c) Apply your result from part (a) and the data in Exercise 8.2 to obtain a 68.26% confidence interval for the mean IQ, μ, of all students attending the university. [Recall that $\sigma = 12$.]

≡ **8.13**   The publication *Employment and Earnings*, released by the U.S. Bureau of Labor Statistics, contains information on the ages of persons, 16 years old and over, in the civilian labor force. Suppose that 50 such persons are randomly selected and that their ages are as displayed below.

22	58	40	42	43	32	34	45	38	19
33	16	49	29	30	43	37	19	21	62
60	41	28	35	37	51	37	65	57	26
27	31	33	24	34	28	39	43	26	38
42	40	31	34	38	35	29	33	32	33

a) Use the data to obtain a point estimate for the mean age, μ, of all people in the civilian labor force. (*Note:* $\Sigma x = 1819$.)

b) Use the data to obtain a point estimate for the standard deviation, σ, of the ages of all people in the civilian labor force. (*Note:* $\Sigma x^2 = 72{,}179$.)

c) Find a 95.44% confidence interval for the mean age, μ, of all people in the civilian labor force. (*Hint:* Use the point estimate obtained in part (b) in place of the unknown value of σ.)

≡ **8.14**   On page 376, we made the following statement: But to say that "$\bar{x}$ will be within $2 \cdot \sigma/\sqrt{n}$ of μ" is the same as saying that "μ will be within $2 \cdot \sigma/\sqrt{n}$ of $\bar{x}$." Mathematically, this means that the inequalities

$$\mu - 2 \cdot \frac{\sigma}{\sqrt{n}} < \bar{x} < \mu + 2 \cdot \frac{\sigma}{\sqrt{n}}$$

and

$$\bar{x} - 2 \cdot \frac{\sigma}{\sqrt{n}} < \mu < \bar{x} + 2 \cdot \frac{\sigma}{\sqrt{n}}$$

are equivalent. Verify that by proving the following more general result: The inequalities

$$\mu - E < \bar{x} < \mu + E$$

and

$$\bar{x} - E < \mu < \bar{x} + E$$

are equivalent.

8.2   Large-sample confidence intervals for a population mean

Recall that the percentage confidence for a confidence interval is called the *confidence level.* In the previous section, we discovered how to find a large-sample confidence interval for a population mean, μ, with a confidence level of 95.44%. Now we will learn how to obtain a large-sample confidence interval for a population mean, μ, with any prescribed confidence level.

Before we begin, it will be helpful to introduce some general notation that is used with confidence intervals. Frequently, we want to write the confidence level in the form $1 - \alpha$, where α is a number between 0 and 1. That is, if the confidence level is expressed as a decimal, then α is the number that must be subtracted from 1 to get that confidence level. For example, if the confidence level is 0.9544 (i.e., 95.44%), then we have $0.9544 = 1 - 0.0456$ and, so, $\alpha = 0.0456$. Similarly, if the confidence level is 0.90, then $\alpha = 0.10$.

Finally, let us review the z_α-notation. The symbol z_α denotes the z-value with area α to its right under the standard normal curve. Thus, $z_{0.05}$ denotes the z-value with area 0.05 to its right under the standard normal curve, $z_{0.025}$ denotes the z-value with area 0.025 to its right under the standard normal curve, and $z_{\alpha/2}$ denotes the z-value with area $\alpha/2$ to its right under the standard normal curve.

LARGE-SAMPLE CONFIDENCE INTERVALS FOR A POPULATION MEAN

To obtain a large-sample confidence interval for a population mean with any specified confidence level, we proceed as in Example 8.2 on pages 375–376 except for one thing. Instead of using Property 2 of Key Fact 6.7 (the empirical rule for normally distributed random variables), we use Key Fact 6.8 (the general empirical rule for normally distributed random variables).

Specifically, suppose that a random sample of size $n \geq 30$ is to be taken from a population with mean μ and standard deviation σ. Then by Key Fact 7.6 on page 363, the random variable $\overline{x}$ is approximately normally distributed with mean $\mu_{\overline{x}} = \mu$ and standard deviation $\sigma_{\overline{x}} = \sigma/\sqrt{n}$. Consequently, by Key Fact 6.8 on page 323, the probability is approximately $1 - \alpha$ that $\overline{x}$ will be within $z_{\alpha/2}$ standard deviations to either side of its mean:

$$P(\mu_{\overline{x}} - z_{\alpha/2} \cdot \sigma_{\overline{x}} < \overline{x} < \mu_{\overline{x}} + z_{\alpha/2} \cdot \sigma_{\overline{x}}) = 1 - \alpha$$

Since $\mu_{\overline{x}} = \mu$ and $\sigma_{\overline{x}} = \sigma/\sqrt{n}$, we can express the above equation as

$$P\left(\mu - z_{\alpha/2} \cdot \frac{\sigma}{\sqrt{n}} < \overline{x} < \mu + z_{\alpha/2} \cdot \frac{\sigma}{\sqrt{n}}\right) = 1 - \alpha$$

Using algebra, the previous equation can be rewritten as

$$P\left(\overline{x} - z_{\alpha/2} \cdot \frac{\sigma}{\sqrt{n}} < \mu < \overline{x} + z_{\alpha/2} \cdot \frac{\sigma}{\sqrt{n}}\right) = 1 - \alpha$$

This last equation shows that once the sample is taken, the interval from

$$\overline{x} - z_{\alpha/2} \cdot \frac{\sigma}{\sqrt{n}} \quad \text{to} \quad \overline{x} + z_{\alpha/2} \cdot \frac{\sigma}{\sqrt{n}}$$

will be a $(1 - \alpha)$-level confidence interval for μ. Thus, we have the following:

PROCEDURE 8.1
To find a confidence interval for a population mean, μ.

ASSUMPTION

Sample size is large $(n \geq 30)$.

STEP 1 *For a confidence level of $1 - \alpha$, use Table II to find $z_{\alpha/2}$.*
STEP 2 *The confidence interval for μ is from*

$$\overline{x} - z_{\alpha/2} \cdot \frac{s}{\sqrt{n}} \quad \text{to} \quad \overline{x} + z_{\alpha/2} \cdot \frac{s}{\sqrt{n}}$$

where $z_{\alpha/2}$ is found in Step 1, n is the sample size, and $\overline{x}$ and s are computed from the actual sample data obtained.

Note: In Step 2 of Procedure 8.1 we have used the sample standard deviation, *s*, in the formula for the confidence interval. Theoretically, we should use the population standard deviation, σ. However, σ is rarely known and so we have used *s* in its place. This is acceptable because, for large samples, the sample standard deviation is likely to be a good approximation of the population standard deviation. In the rare case where the population standard deviation is known, it should always be used in place of the sample standard deviation in Step 2 of Procedure 8.1.

EXAMPLE 8.5   *Illustrates Procedure 8.1*

The U.S. Bureau of Labor Statistics collects information on the ages of people in the civilian labor force and publishes the results in *Employment and Earnings*. Suppose that 50 people in the civilian labor force are randomly selected and that their ages are as displayed in Table 8.3.

TABLE 8.3
Ages of 50 randomly selected people in the civilian labor force

22	58	40	42	43	32	34	45	38	19
33	16	49	29	30	43	37	19	21	62
60	41	28	35	37	51	37	65	57	26
27	31	33	24	34	28	39	43	26	38
42	40	31	34	38	35	29	33	32	33

Determine a 90% confidence interval for the mean age, μ, of all people in the civilian labor force.

SOLUTION   Since the sample size is 50, which exceeds 30, we can apply Procedure 8.1 to obtain the confidence interval.

STEP 1   *For a confidence level of* $1 - \alpha$, *use Table II to find* $z_{\alpha/2}$.

We want a 90% confidence interval, so the confidence level is $0.90 = 1 - 0.10$. This means that $\alpha = 0.10$. Consulting Table II, we see that

$$z_{\alpha/2} = z_{0.10/2} = z_{0.05} = \boxed{1.645}$$

STEP 2   *The confidence interval for* μ *is from*

$$\bar{x} - z_{\alpha/2} \cdot \frac{s}{\sqrt{n}} \quad \text{to} \quad \bar{x} + z_{\alpha/2} \cdot \frac{s}{\sqrt{n}}$$

We have $n = 50$ and, from Step 1, $z_{\alpha/2} = 1.645$. To compute $\bar{x}$ and *s* for the data in Table 8.3, we apply the usual formulas:

$$\bar{x} = \frac{\Sigma x}{n} = \frac{1819}{50} = 36.38$$

and

$$s = \sqrt{\frac{n(\Sigma x^2) - (\Sigma x)^2}{n(n-1)}} = \sqrt{\frac{50(72,179) - (1819)^2}{50 \cdot 49}} = 11.07$$

Consequently, a 90% confidence interval for μ is from

$$36.38 - 1.645 \cdot \frac{11.07}{\sqrt{50}} \quad \text{to} \quad 36.38 + 1.645 \cdot \frac{11.07}{\sqrt{50}}$$

or

$$\boxed{33.8 \quad \text{to} \quad 39.0}$$

MTB

We can be 90% confident that the mean age, μ, of all people in the civilian labor force is somewhere between 33.8 and 39.0 years.   ■

RELATION OF CONFIDENCE LEVEL TO LENGTH OF CONFIDENCE INTERVAL

The confidence level of a confidence interval for a population mean, μ, signifies the confidence of the estimate. That is, it expresses the confidence we have that μ actually lies in the confidence interval. The length of the confidence interval, on the other hand, indicates the precision of the estimate. Long confidence intervals indicate poor precision, while short confidence intervals indicate good precision.

How does the confidence level affect the length of the confidence interval? Look back to Example 8.5 on page 383, where we found a 90% confidence interval for the mean age, μ, of all people in the civilian labor force. The confidence level there is 0.90, and the confidence interval we computed is from 33.8 to 39.0. We can be 90% confident that μ is somewhere between 33.8 and 39.0 years.

What happens to the length of the confidence interval if we change the confidence level from 0.90 to, say, 0.95? Then $z_{\alpha/2}$ changes from $z_{0.10/2} = z_{0.05} = 1.645$ to $z_{0.05/2} = z_{0.025} = 1.96$. The resulting confidence interval, using the same sample data (Table 8.3), is then from

$$36.38 - 1.96 \cdot \frac{11.07}{\sqrt{50}} \quad \text{to} \quad 36.38 + 1.96 \cdot \frac{11.07}{\sqrt{50}}$$

or from 33.3 to 39.4 years. We picture both the 90% and 95% confidence intervals in Figure 8.2.

FIGURE 8.2
90% and 95%
confidence
intervals for μ
using Table 8.3 data

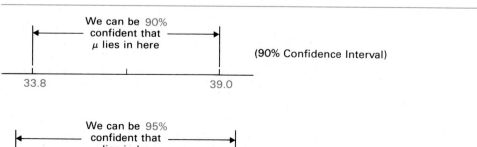

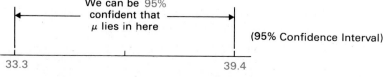

Thus, increasing the confidence level increases the length of the confidence interval. This makes sense. If we want to be more confident that μ lies in our confidence interval, then we must naturally have a more extensive interval. In summary, we have the following important fact.

KEY FACT 8.1   Confidence level and the length of a confidence interval

For a fixed sample size, the greater the confidence level, the greater the length of the confidence interval.

USING THE COMPUTER (OPTIONAL)

Procedure 8.1 on page 382 gives a step-by-step method for obtaining a confidence interval for a population mean, μ, when the sample size is large ($n \geq 30$). Minitab has a program called **ZINTERVAL** that will perform Procedure 8.1 for us.

In Example 8.5 on pages 383–384, we applied Procedure 8.1 to determine a confidence interval for the mean age of all people in the civilian labor force. Let us use ZINTERVAL to obtain that confidence interval.

EXAMPLE 8.6   *Illustrates the ZINTERVAL command*

The U.S. Bureau of Labor Statistics collects information on the ages of people in the civilian labor force and publishes the results in *Employment and Earnings*. Suppose that 50 people in the civilian labor force are randomly selected and that their ages are as displayed in Table 8.4.

TABLE 8.4
Ages of 50 randomly selected people in the civilian labor force

22	58	40	42	43	32	34	45	38	19
33	16	49	29	30	43	37	19	21	62
60	41	28	35	37	51	37	65	57	26
27	31	33	24	34	28	39	43	26	38
42	40	31	34	38	35	29	33	32	33

Use Minitab to determine a 90% confidence interval for the mean age, μ, of all people in the civilian labor force.

SOLUTION   The program ZINTERVAL requires the user to specify the population standard deviation, σ. In this case, however, σ is unknown. But since the sample size is large ($n = 50$), we can use the sample standard deviation, s, in place of σ.

We first enter the sample data in Table 8.4 into C1 using the SET command and name C1 "Ages" using the NAME command. Next, we apply the STDEV command to obtain the sample standard deviation, s, of the data. See Printout 8.1 at the top of the next page.

Now we can have Minitab determine the 90% confidence interval. We type the command ZINTERVAL, followed by the desired confidence level, the estimated value of σ, and the storage location of the sample data. In other words, we type

```
ZINTERVAL 90%, sigma=11.069, data in 'AGES'
```

This command and its results are shown in Printout 8.2 on the next page.

PRINTOUT 8.1

```
MTB > SET C1
DATA> 22 58 40 42 43 32 34 45 38 19
DATA> 33 16 49 29 30 43 37 19 21 62
DATA> 60 41 28 35 37 51 37 65 57 26
DATA> 27 31 33 24 34 28 39 43 26 38
DATA> 42 40 31 34 38 35 29 33 32 33
DATA> END
MTB > NAME C1 'AGES'
MTB > STDEV 'AGES'
    ST.DEV. =      11.069
```

PRINTOUT 8.2
Minitab output
for ZINTERVAL

```
MTB > ZINTERVAL 90%, sigma=11.069, data in 'AGES'

THE ASSUMED SIGMA =11.1

              N      MEAN    STDEV  SE MEAN   90.0 PERCENT C.I.
AGES          50     36.38   11.07    1.57   (  33.80,   38.96)
```

The first line of the output displays the value used for the population standard deviation, σ: **THE ASSUMED SIGMA =11.1**. Then we find the sample size, sample mean, sample standard deviation, and standard error of the mean $(\sigma/\sqrt{n})$. The final item is the confidence interval.

Hence, a 90% confidence interval for μ is from 33.80 to 38.96. We can be 90% confident that the mean age, μ, of all people in the civilian labor force is somewhere between 33.80 and 38.96 years. ∎

Exercises 8.2

__ **8.15** Find the confidence level and α for
a) a 90% confidence interval.
b) a 99% confidence interval.

__ **8.16** Find the confidence level and α for
a) an 85% confidence interval.
b) a 95% confidence interval.

In Exercises 8.17–8.27 use Procedure 8.1 on page 382 to determine the required confidence intervals.

__ **8.17** The Gallup Organization conducts annual national surveys on home gardening. Results are published by the National Association for Gardening in *National Gardening Survey.* Suppose that 250 households with vegetable gardens are selected at random and that the average size of their gardens is $\overline{x} = 643$ square feet.

a) Find a 90% confidence interval for the mean size, μ, of all household vegetable gardens in the United States. Assume that $\sigma = 247$ square feet.
b) Interpret your result in part (a) in words.

__ **8.18** A quality-control engineer in a bakery goods plant needs to estimate the mean weight, μ, of bags of potato chips that are packed by machine. He knows from experience that $\sigma = 0.1$ oz for this machine. He takes a random sample of 36 bags and finds the sample mean weight to be 16.01 oz.
a) Find a 99% confidence interval for μ.
b) Interpret your result in part (a) in words.

__ **8.19** The Bureau of Labor Statistics collects data on employment and hourly earnings in the aircraft industry and publishes its findings in *Employment and*

Earnings. Suppose that 30 people working in the aircraft industry are selected at random and that their hourly earnings are as follows:

$15.20	5.94	12.28	14.99	5.99
16.42	15.71	4.01	8.90	9.25
13.77	18.22	9.18	11.60	13.20
14.05	15.17	13.57	12.52	14.49
9.43	8.85	18.82	5.83	14.82
17.02	11.84	12.16	7.73	9.30

a) Find a 95% confidence interval for the mean hourly earnings, μ, of all people employed in the aircraft industry. Assume that the population standard deviation of the hourly earnings is $3.25. (*Note:* The sum of the data is $360.26.)

b) Interpret your result in part (a) in words.

__ 8.20   A sociologist wants information on the number of children per farm family in her native state of Nebraska. Forty randomly selected farm families have the following number of children:

3	5	2	1	1	0	2	3
1	1	2	1	2	0	1	5
4	1	0	1	3	1	0	1
0	1	8	0	1	2	2	2
2	1	5	3	1	4	1	0

a) Find a 90% confidence interval for the mean number of children, μ, per farm family in Nebraska. Assume $\sigma = 1.95$. (*Note:* $\Sigma x = 74$.)

b) Interpret your result in part (a) in words.

__ 8.21   The National Center for Health Statistics estimates mean weights of Americans by age, height, and sex. Those estimates can be found in the publication *Vital and Health Statistics.* Suppose 40 American women, 5 feet 4 inches tall and aged 18–24, are randomly selected and that their weights, in pounds, are as shown here.

140	136	147	138	143	122	115	125
136	152	130	134	150	153	148	132
116	159	128	136	134	126	120	146
131	167	145	132	138	137	115	145
154	139	139	147	123	154	127	116

a) Find a 90% confidence interval for the mean weight, μ, of all American women 5 feet 4 inches tall and in the age group 18–24 years. (*Note:* $\Sigma x = 5475$ and $\Sigma x^2 = 755,749$.)

b) Interpret your result in part (a) in words.

__ 8.22   A research physician wants to estimate the average age of people with diabetes. She takes a sample of 35 such people and obtains the following ages:

48	41	57	83	41	55	59
61	38	48	79	75	77	7
54	23	47	56	79	68	61
64	45	53	82	68	38	70
10	60	83	76	21	65	47

a) Determine a 95% confidence interval for the mean age, μ, of people with diabetes. (*Note:* $\Sigma x = 1939$ and $\Sigma x^2 = 120,861$.)

b) Interpret your result in part (a) in words.

__ 8.23   The *College Board* surveys colleges and universities to obtain data on the costs of attending an institution of higher education. A random sample is taken of 150 private four-year colleges and universities. The average tuition and fees for the schools selected is $\bar{x} = \$8737$ and the standard deviation is $s = \$3241$. Use these statistics to obtain a 90% confidence interval for the mean tuition and fees of all private four-year colleges and universities.

__ 8.24   A telephone company in the southwest undertook a study on various aspects of telephone usage. One item of interest was how long telephone calls last. According to the media relations manager, the company randomly selected 15,000 local telephone calls involving Phoenix residential customers. The mean duration of the calls sampled was 3.8 minutes and the standard deviation was 4.0 minutes. Use this information to determine a 95% confidence interval for the mean duration of all telephone calls made by Phoenix residential customers.

__ 8.25   Refer to Exercise 8.21.
a) Find a 99% confidence interval for μ.
b) Why is the confidence interval that you just found in part (a) longer than the one in Exercise 8.21?
c) Draw a graph similar to Figure 8.2 on page 384 that displays both confidence intervals.

__ 8.26   Refer to Exercise 8.22.
a) Determine an 80% confidence interval for μ.
b) Why is the confidence interval that you just found in part (a) shorter than the one in Exercise 8.22?
c) Draw a graph similar to Figure 8.2 on page 384 that displays both confidence intervals.

__ 8.27   The U.S. Bureau of the Census compiles data on family size and presents its findings in the publication *Current Population Reports.* Suppose that

500 American families are randomly selected in order to estimate the mean size, μ, of all American families. Further suppose that the results are as shown in the following frequency distribution:

Size of family x	Frequency f
2	198
3	118
4	101
5	59
6	12
7	3
8	8
9	1

a) Find a 95% confidence interval for the mean size, μ, of all American families. *(Hint:* Use Formula 3.2 on page 107 to compute $\bar{x}$ and Formula 3.3 on pages 108–109 to compute s.)
b) Interpret your result in part (a) in words.

Exercises 8.28–8.33 are computer exercises.

___ **8.28 (Computer exercise)** Suppose that the data in Exercise 8.20 are stored in a column named CHILDREN.
a) Which Minitab command and subcommands (if any) should be used to obtain the confidence interval required in part (a) of that exercise?
b) If you have access to Minitab, use it to obtain the confidence interval.

___ **8.29 (Computer exercise)** Suppose that the data in Exercise 8.19 are stored in a column named EARNINGS.
a) Which Minitab command and subcommands (if any) should be used to obtain the confidence interval required in part (a) of that exercise?
b) If you have access to Minitab, use it to obtain the confidence interval.

___ **8.30 (Computer exercise)** Suppose the data in Exercise 8.22 are stored in a column named AGES.
a) Which Minitab commands and subcommands (if any) should be used to obtain the confidence interval required in part (a) of that exercise?
b) If you have access to Minitab, use it to obtain the confidence interval.

___ **8.31 (Computer exercise)** Suppose the data in Exercise 8.21 are stored in a column named WEIGHTS.

a) Which Minitab commands and subcommands (if any) should be used to obtain the confidence interval required in part (a) of that exercise?
b) If you have access to Minitab, use it to obtain the confidence interval.

___ **8.32 (Computer exercise)** The manufacturer of a new model car, called the Orion, claims that a "typical" Orion gets 26 mpg (miles per gallon). An independent consumer group is skeptical of this claim and thinks that the mean gas mileage of all Orions may be less than 26 mpg. The consumer group performs mileage tests on a random sample of Orions and obtains the following (abridged) Minitab output:

THE ASSUMED SIGMA =1.50

N	MEAN	STDEV	SE MEAN	95.0 PERCENT C.I.
30	25.230	1.592	0.274	(24.692, 25.768)

Use the printout to determine
a) the (assumed) population standard deviation of the gas mileages of all Orions.
b) the standard deviation of the gas mileages of the Orions in the sample.
c) the number of Orions in the sample.
d) the mean gas mileage of the Orions in the sample.
e) a 95% confidence interval for the mean gas mileage, μ, of all Orions.
f) Does it appear that the consumer group's skepticism is justified? Explain.

___ **8.33 (Computer exercise)** A Louisiana cotton farmer is interested in a new brand of fertilizer. The farmer uses the new fertilizer on a random sample of one-acre plots and records the yield, in pounds, for each plot. Below is an abridged Minitab printout obtained by applying the ZINTERVAL command to the cotton-yield data.

THE ASSUMED SIGMA =20.8

N	MEAN	STDEV	SE MEAN	90.0 PERCENT C.I.
80	625.16	21.47	2.33	(621.33, 628.99)

From the printout, determine
a) the (assumed) population standard deviation, σ, of the cotton yields for all one-acre plots on the farm when the new fertilizer is used.
b) the standard deviation, s, of the cotton yields for the one-acre plots in the sample.
c) the number of one-acre plots in the sample.
d) the mean yield of cotton, $\bar{x}$, for the one-acre plots in the sample.

e) a 90% confidence interval for the mean yield of cotton, μ, for all one-acre plots on the farm when the new fertilizer is used.

≡ **8.34** In this exercise we will verify Key Fact 8.1 on page 385; namely, that for a fixed sample size, the greater the confidence level, the greater the length of the confidence interval. We will assume that the population standard deviation, σ, is known, so that the form of the confidence interval is

$$\bar{x} - z_{\alpha/2} \cdot \frac{\sigma}{\sqrt{n}} \quad \text{to} \quad \bar{x} + z_{\alpha/2} \cdot \frac{\sigma}{\sqrt{n}}$$

a) Show that the length of the confidence interval is

$$L = 2z_{\alpha/2} \cdot \frac{\sigma}{\sqrt{n}}$$

b) Show that increasing the confidence level, decreases the value of α.
c) Verify that decreasing the value of α, increases the value of $z_{\alpha/2}$.
d) Deduce from parts (a)–(c) that, for a fixed sample size, increasing the confidence level, increases the length of the confidence interval.

8.3 Sample size considerations

In this section we will examine in detail how the sample size affects the precision of estimating a population mean by a sample mean. We have already alluded to the fact that the larger the sample size, the greater the likelihood for small sampling error (Key Fact 7.3 on page 357). Now that we have studied confidence intervals, we can determine exactly how the sample size affects the accuracy of the estimate. Consider the following example.

EXAMPLE 8.7   *Introduces the maximum error of the estimate*

The U.S. Energy Information Administration surveys households in order to obtain data on monthly fuel expenditures for household vehicles. Results of those surveys are contained in the publication *Residential Transportation Energy Consumption Survey, Consumption Patterns of Household Vehicles.* Suppose that 30 monthly fuel expenditures for household vehicles are selected at random and that their mean is $\bar{x} = \$58.56$.

a) Find a 95% confidence interval for the mean monthly fuel expenditure, μ, per household vehicle for all household vehicles. Assume that $\sigma = \$20.65$.
b) Discuss the precision with which $\bar{x}$ estimates μ.

SOLUTION   a) For a 95% confidence interval, $\alpha = 0.05$ and so $z_{\alpha/2} = z_{0.05/2} = z_{0.025} = 1.96$. Also, we have $n = 30$, $\sigma = \$20.65$, and $\bar{x} = \$58.56$. Applying Procedure 8.1 (page 382), we find that a 95% confidence interval for μ is from

$$\bar{x} - z_{\alpha/2} \cdot \frac{\sigma}{\sqrt{n}} \quad \text{to} \quad \bar{x} + z_{\alpha/2} \cdot \frac{\sigma}{\sqrt{n}}$$

or

$$58.56 - 1.96 \cdot \frac{20.65}{\sqrt{30}} \quad \text{to} \quad 58.56 + 1.96 \cdot \frac{20.65}{\sqrt{30}}$$

or

$$58.56 - 7.39 \quad \text{to} \quad 58.56 + 7.39$$

or

$$51.17 \quad \text{to} \quad 65.95$$

We can be 95% confident that the mean monthly fuel expenditure, μ, per household vehicle is somewhere between \$51.17 and \$65.95.

b) The confidence interval obtained in part (a) is relatively long and hence provides a rather wide range for the possible values of μ. In other words, the precision of the estimate is quite poor. To improve the precision, we need to decrease the length of the confidence interval.

As we learned in Section 8.2, one way to decrease the length of the confidence interval is to lower the confidence level from 95% to some lower level. But suppose that we want to retain the same level of confidence and still narrow the confidence interval. What can be done to accomplish this? To answer that question, let us first look more closely at the confidence interval in part (a). That confidence interval is displayed graphically in Figure 8.3.

FIGURE 8.3
95% confidence interval for mean monthly fuel expenditure, μ, per household vehicle

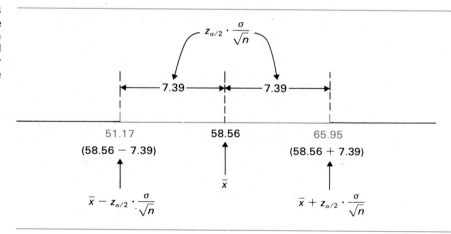

If we study Figure 8.3 or refer to the computations done in part (a), we find that the length of the confidence interval is determined by the quantity

$$E = z_{\alpha/2} \cdot \frac{\sigma}{\sqrt{n}}$$

which is just half the length of the confidence interval. In this case, $E = 7.39$. That quantity is called the **maximum error of the estimate.** We use this terminology because we can be 95% confident that μ lies in the confidence interval or, equivalently, that the maximum error made in using $\bar{x}$ to estimate μ is $E = \$7.39$. (See Figure 8.3.)

In any event, we see that to narrow the confidence interval and thereby increase the precision of the estimate, we need only decrease the value of E. Since the sample size, n, occurs in the denominator of the formula for E, we can decrease E by increasing the sample size. This makes sense, of course, because we expect to get more accurate information from larger samples. ∎

Definition 8.3 and Key Fact 8.2 summarize the preceding discussion.

DEFINITION 8.3   Maximum error of the estimate for μ

The *maximum error of the estimate*, E, is defined by

$$E = z_{\alpha/2} \cdot \frac{\sigma}{\sqrt{n}}$$

and is equal to half the length of the confidence interval. See Figure 8.4.

FIGURE 8.4
Maximum error of
the estimate, E

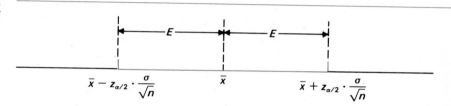

KEY FACT 8.2

The length of a confidence interval for a population mean, μ, and hence the precision with which $\bar{x}$ estimates μ, is determined by the maximum error of the estimate, E. For a given confidence level, we can increase the precision of the estimate by increasing the sample size, n.

DETERMINING THE REQUIRED SAMPLE SIZE

Oftentimes, the precision (maximum error of the estimate) and confidence level of a confidence interval for μ are specified in advance. We must then determine the sample size required to meet the given specifications. The formula for the required sample size can be obtained by solving for n in the expression $E = z_{\alpha/2} \cdot \sigma/\sqrt{n}$. The result is Formula 8.1.

FORMULA 8.1   Sample size for estimating μ

The sample size required for a $(1 - \alpha)$-level confidence interval for μ with a specified maximum error of the estimate, E, is given by

$$n = \left[\frac{z_{\alpha/2} \cdot \sigma}{E} \right]^2$$

EXAMPLE 8.8   *Illustrates Formula 8.1*

Consider again the problem of estimating the mean monthly fuel expenditure, μ, per household vehicle.

a) Determine the sample size required to insure that we can be 95% confident that the estimate, $\bar{x}$, will be within $0.50 of μ. [Recall that $\sigma = \$20.65$.]

b) Find a 95% confidence interval for μ if a sample of the size determined in part (a) has a mean of $59.02.

SOLUTION   a) To determine the required sample size, we apply Formula 8.1. In doing so, we must identify σ, E, and $z_{\alpha/2}$. First note that, by assumption, $\sigma = \$20.65$ and that the maximum error of the estimate is specified to be $E = \$0.50$. Next note that, since the confidence level is stipulated to be 0.95, we have $\alpha = 0.05$. Consulting Table II, we see that $z_{\alpha/2} = z_{0.05/2} = z_{0.025} = 1.96$. Thus, the required sample size is

$$n = \left[\frac{z_{\alpha/2} \cdot \sigma}{E}\right]^2 = \left[\frac{1.96 \cdot 20.65}{0.5}\right]^2 = 6552.58$$

Obviously, we cannot take a fractional sample size, so to be conservative, we round up to 6553. Consequently, if 6553 monthly fuel expenditures are randomly selected, then we can be 95% confident that their mean, $\bar{x}$, is within $0.50 of the true mean monthly fuel expenditure, μ. [By the way, the U.S. Energy Information Administration takes a sample of size 6841.]

b) For this part, we are to find a 95% confidence interval for μ, if a sample of the size determined in part (a), $n = 6553$, has a mean of $\bar{x} = \$59.02$. Applying Procedure 8.1, we obtain the confidence interval:

$$\bar{x} - z_{\alpha/2} \cdot \frac{\sigma}{\sqrt{n}} \quad \text{to} \quad \bar{x} + z_{\alpha/2} \cdot \frac{\sigma}{\sqrt{n}}$$

or

$$59.02 - 1.96 \cdot \frac{20.65}{\sqrt{6553}} \quad \text{to} \quad 59.02 + 1.96 \cdot \frac{20.65}{\sqrt{6553}}$$

or

$$59.02 - 0.50 \quad \text{to} \quad 59.02 + 0.50$$

or

$$58.52 \quad \text{to} \quad 59.52$$

We can be 95% confident that the mean monthly fuel expenditure, μ, per household vehicle is somewhere between $58.52 and $59.52. ∎

Note: The formula for finding the required sample size, Formula 8.1 on page 391, involves the population standard deviation, σ. If σ is unknown and we want to apply the formula, then we must first estimate σ. We can do this by taking a preliminary sample of size 30 or more. The sample standard deviation, s, of the sample obtained provides an estimate of σ and can be used in place of σ in Formula 8.1.

Exercises 8.3

— **8.35** A 90% confidence interval for the mean size, μ, of household vegetable gardens was found in Exercise 8.17 of Section 8.2 (page 386). The confidence interval is from 617.3 to 668.7 square feet.
a) Find the maximum error of the estimate, E.
b) Explain the meaning of E with regard to the estimation of μ.
c) Determine the sample size required to have the same maximum error of the estimate as in part (a) but with a 95% confidence level. [Assume that $\sigma = 247$ square feet.]

— **8.36** In Exercise 8.20 of Section 8.2, you were asked to find a 90% confidence interval for the mean number of children, μ, per farm family in Nebraska based on a sample of size 40 (given on page 387). The 90% confidence interval for μ is from 1.3 to 2.4 children per farm family.
a) Determine the maximum error of the estimate, E.
b) Explain the meaning of E, in this context, as far as the accuracy of the estimate is concerned.
c) Determine the sample size required to insure that we can be 90% confident that our estimate, $\bar{x}$, will be within 0.1 children of μ. [Recall that $\sigma = 1.95$.]
d) Find a 90% confidence interval for μ, if a sample of the size indicated in part (c) yields a mean of $\bar{x} = 1.9$ children.

— **8.37** In Exercise 8.19 of Section 8.2, you were asked to determine a 95% confidence interval for the mean hourly earnings, μ, of people employed in the aircraft industry based on a sample of size 30 (given on page 387). The 95% confidence interval is from $10.85 to $13.17.
a) Determine the maximum error of the estimate, E.
b) Explain the meaning of E, in this context, as far as the accuracy of the estimate is concerned.
c) Determine the sample size required to insure that we can be 95% confident that our estimate, $\bar{x}$, will be within $0.50 of μ. [Recall that $\sigma = 3.25$.]
d) Find a 95% confidence interval for μ, if a sample of the size indicated in part (c) has a mean of $12.87.

— **8.38** You were asked, in Exercise 8.22 of Section 8.2, to find a 95% confidence interval for the average age of people with diabetes based on a sample of size 35 (given on page 387.) The confidence interval is from 48.8 to 62.0 years.
a) Determine the maximum error of the estimate, E.

b) Find the sample size required to have a maximum error of the estimate of 0.5 years and a 95% confidence level. (*Note:* The sample standard deviation of the sample of 35 ages from Exercise 8.22 is 19.88 years.)
c) Why did you use the sample standard deviation, $s = 19.88$, in place of σ in your solution to part (b)?
d) Find a 95% confidence interval for the mean age, μ, of people with diabetes if a sample of the size indicated in part (b) gives the statistics $\bar{x} = 58.3$ years and $s = 19.0$ years.

— **8.39** In Exercise 8.21 of Section 8.2 (page 387), you were given a sample of weights obtained from the random selection of 40 American women, 5 feet 4 inches tall and aged 18–24. Based on that data, a 90% confidence interval for the mean weight, μ, of all such women is from 133.6 to 140.2 lb.
a) Determine the maximum error of the estimate, E.
b) Find the sample size required to have a maximum error of the estimate of 2.0 lb and a 99% confidence level. (*Note:* $s = 12.77$ lb for the sample data in Exercise 8.21.)
c) Why did you use the sample standard deviation, $s = 12.77$, in place of σ in your solution to part (b)?
d) Find a 99% confidence interval for the mean weight, μ, of all American women aged 18–24 who are 5 feet 4 inches tall, if a sample of the size indicated in part (b) has a mean of 134.2 lb and a standard deviation of 13.0 lb.

— **8.40** Explain how to apply the formula

$$n = \left[\frac{z_{\alpha/2} \cdot \sigma}{E}\right]^2$$

if σ is unknown.

≡ **8.41** The Bureau of the Census gives estimates for the mean value, μ, of the land and buildings per corporate farm. Those estimates are contained in the publication *Census of Agriculture*. Suppose that an estimate, $\bar{x}$, is obtained and that the maximum error of the estimate is $E = 1000. Does this indicate that the estimate is within $1000 of the true mean, μ? Explain.

≡ **8.42** Use Formula 8.1 on page 391 to establish the following fact: For a fixed confidence level, it is necessary to (approximately) quadruple the sample size, in order to double the accuracy of the estimate.

8.4   Confidence intervals for a normal population mean

In Section 8.2, we learned how to find a confidence interval for a population mean, μ, when the sample size is large ($n \geq 30$). The procedure for obtaining a large-sample confidence interval for μ is based on Key Fact 7.6, the sampling distribution of the mean for general populations. That key fact states that, for large samples, the random variable $\overline{x}$ is approximately normally distributed with mean $\mu_{\overline{x}} = \mu$ and standard deviation $\sigma_{\overline{x}} = \sigma/\sqrt{n}$. Or, equivalently, that the standardized random variable

(1)
$$z = \frac{\overline{x} - \mu}{\sigma/\sqrt{n}}$$

has approximately the standard normal distribution when the sample size is large.

However, it is quite often the case that a large sample is unavailable, extremely expensive, or undesirable. For example, tests of automobiles to analyze the impact of collisions frequently involve wrecking the cars. Clearly, it is more appropriate to employ a small sample in this and similar situations.

There are several methods for obtaining a confidence interval for a population mean that do not require a large sample. One such method applies when the population being sampled is *normally distributed*. That method is what we will study in this section.

STUDENT'S *t*-DISTRIBUTION

As we said, our next objective is to learn how to find a confidence interval for a population mean, μ, when the population being sampled is normally distributed. We will assume that the population standard deviation, σ, is unknown since that is usually the case in practice.

To begin, we need to identify the probability distribution of the random variable obtained by replacing the unknown population standard deviation, σ, in Equation (1) by the sample standard deviation, s. In other words, we need to identify the probability distribution of the random variable

$$\frac{\overline{x} - \mu}{s/\sqrt{n}}$$

In 1908, W. S. Gosset determined the probability distribution of that random variable under the assumption that the population being sampled is normally distributed. He named the probability distribution **Student's *t*-distribution.**

Gosset did much of his research while working for the Guiness Brewery of Ireland. The company prohibited its employees from publishing research, so Gosset presented his finding using the pen name *Student*. Thus the name "Student" in "Student's *t*-distribution." For brevity, we will often omit "Student's" and just use **t-distribution.**

As you know, probabilities for a random variable having a normal distribution are equal to areas under a normal curve. Probabilities for a random variable having a t-distribution are also equal to areas under a curve, suitably called a **t-curve.**

Actually, there are infinitely many t-curves, and the one that is used depends on the sample size. If the sample size is n, then we identify the t-curve in question by saying that it is the t-curve with $n-1$ **degrees of freedom.** The mathematical concepts involved in defining degrees of freedom are somewhat complex. Therefore, we define degrees of freedom simply as a number that identifies the appropriate t-curve or t-distribution. For convenience, we will usually write **df $= n - 1$** to indicate $n-1$ degrees of freedom.

Before examining the t-distribution and t-curves further, we summarize the preceding discussion in Key Fact 8.3.

KEY FACT 8.3

Suppose a random sample of size n is to be taken from a normally distributed population with mean μ. Then the random variable

$$t = \frac{\bar{x} - \mu}{s/\sqrt{n}}$$

has the t-distribution with $n-1$ degrees of freedom. Thus, probabilities for that random variable are equal to areas under the t-curve with df $= n - 1$.

Let us now look at some specific examples of t-curves and then note some general properties of such curves. As we mentioned, there is a different t-curve for each number of degrees of freedom. Nonetheless, all t-curves look quite alike and are very similar to the standard normal curve. For illustration, we have drawn in Figure 8.5 the standard normal curve and two t-curves. One t-curve has one degree of freedom and the other has six degrees of freedom.

FIGURE 8.5

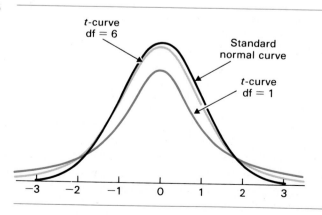

As illustrated by Figure 8.5, t-curves have the following properties:

KEY FACT 8.4   Basic properties of t-curves

PROPERTY 1   The total area under a t-curve is equal to 1.

PROPERTY 2   A t-curve extends indefinitely in both directions, approaching the horizontal axis as it does so.

PROPERTY 3   A t-curve is symmetric about 0.

PROPERTY 4   As the number of degrees of freedom gets larger, t-curves look increasingly like the standard normal curve.

USING THE t-TABLE

In Section 6.1 we learned how to find areas under the standard normal curve by using the standard-normal table, Table II. We will now see how to find areas under t-curves by using the t-table, Table III.

For our purposes, one of which is obtaining confidence intervals for a population mean, we do not need a complete t-table for each t-curve. There are only certain areas that will be important for us to know. In Table 8.5 we have reproduced part of the t-table, Table III, which can be found inside the front cover of the book.

TABLE 8.5
Values of t_α

df	$t_{0.10}$	$t_{0.05}$	$t_{0.025}$	$t_{0.01}$	$t_{0.005}$	df
.	.	.	.	.	.	.
.	.	.	.	.	.	.
.	.	.	.	.	.	.
11	1.363	1.796	2.201	2.718	3.106	11
12	1.356	1.782	2.179	2.681	3.055	12
13	1.350	1.771	2.160	2.650	3.012	13
14	1.345	1.761	2.145	2.624	2.977	14
15	1.341	1.753	2.131	2.602	2.947	15
16	1.337	1.746	2.120	2.583	2.921	16
17	1.333	1.740	2.110	2.567	2.898	17
.	.	.	.	.	.	.
.	.	.	.	.	.	.
.	.	.	.	.	.	.

The two outside columns of the table, labeled df, give the number of degrees of freedom. As you might expect, the symbol t_α denotes the t-value with area α to its right under a t-curve. Thus, the column headed $t_{0.10}$ contains t-values with area 0.10 to their right; the column headed $t_{0.05}$ contains t-values with area 0.05 to their right; and so on. We illustrate a use of Table III in Example 8.9.

EXAMPLE 8.9   *Illustrates how to find the t-value for a specified area*

For a t-curve with 13 degrees of freedom, determine $t_{0.05}$; that is, find the t-value with area 0.05 to its right. See Figure 8.6.

FIGURE 8.6

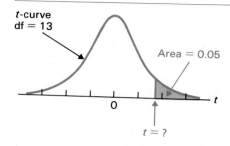

SOLUTION To find the *t*-value in question, we use Table III (or Table 8.5). Since the number of degrees of freedom is 13, we first go down the outside columns, labeled df, to "13." Then we go across that row until we are under the column headed $t_{0.05}$. The number in the body of the table there, 1.771, is the required *t*-value. That is, for a *t*-curve with df $= 13$, the *t*-value with area 0.05 to its right is $t_{0.05} = 1.771$. Figure 8.7 summarizes our results. ∎

FIGURE 8.7

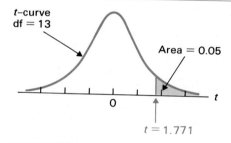

CONFIDENCE INTERVALS FOR A NORMAL POPULATION MEAN

Having discussed *t*-distributions and *t*-curves, we can now develop a procedure to find a confidence interval for a population mean, μ, when the population being sampled is normally distributed. The development is outlined below. For details, see Exercise 8.64.

So, suppose that a random sample of size n is to be taken from a normally distributed population with mean μ. Then by Key Fact 8.3 on page 395, the random variable

$$t = \frac{\bar{x} - \mu}{s/\sqrt{n}}$$

has the *t*-distribution with $n - 1$ degrees of freedom; that is, probabilities for that random variable are equal to areas under the *t*-curve with df $= n - 1$. Therefore,

$$P\left(-t_{\alpha/2} < \frac{\bar{x} - \mu}{s/\sqrt{n}} < t_{\alpha/2}\right) = 1 - \alpha$$

Using algebra, we can rewrite the previous equation as

$$P\left(\bar{x} - t_{\alpha/2} \cdot \frac{s}{\sqrt{n}} < \mu < \bar{x} + t_{\alpha/2} \cdot \frac{s}{\sqrt{n}}\right) = 1 - \alpha$$

This last equation shows that once the sample is taken, the interval from

$$\bar{x} - t_{\alpha/2} \cdot \frac{s}{\sqrt{n}} \quad \text{to} \quad \bar{x} + t_{\alpha/2} \cdot \frac{s}{\sqrt{n}}$$

will be a $(1 - \alpha)$-level confidence interval for μ. Consequently, we have the following procedure for obtaining a confidence interval for a normal population mean.

PROCEDURE 8.2

To find a confidence interval for a population mean, μ.

ASSUMPTION

Normal population.

STEP 1 *For a confidence level of $1 - \alpha$, use Table III to find $t_{\alpha/2}$ with df $= n - 1$, where n is the sample size.*

STEP 2 *The confidence interval for μ is from*

$$\bar{x} - t_{\alpha/2} \cdot \frac{s}{\sqrt{n}} \quad \text{to} \quad \bar{x} + t_{\alpha/2} \cdot \frac{s}{\sqrt{n}}$$

where $t_{\alpha/2}$ is found in Step 1, and $\bar{x}$ and s are computed from the actual sample data obtained.

We should emphasize that Procedure 8.2 applies for any sample size, large or small. The only condition for using the procedure is that the population being sampled is normally distributed. And, actually, the procedure works reasonably well even if the population being sampled is only approximately normally distributed.

Procedures that are insensitive to departures from the assumptions on which they are based are called **robust**. Thus, Procedure 8.2 is robust to moderate deviations of the normality assumption. Here now is an example that illustrates the use of Procedure 8.2.

EXAMPLE 8.10   *Illustrates Procedure 8.2*

The gestation periods of domestic dogs are normally distributed. To estimate the mean gestation period, 15 randomly selected dogs are observed during pregnancy. The gestation periods, in days, of the 15 dogs are displayed in Table 8.6.

TABLE 8.6 Gestation periods				
62.0	61.4	59.8	62.2	60.3
60.4	59.4	60.2	60.4	60.8
61.8	59.2	61.1	60.4	60.9

Determine a 95% confidence interval for the mean gestation period, μ, of the domestic dog.

SOLUTION Since the population of gestation periods is normally distributed, we can apply Procedure 8.2 to obtain the required confidence interval.

STEP 1 *For a confidence level of $1-\alpha$, use Table III to find $t_{\alpha/2}$ with df $= n-1$, where n is the sample size.*

The specified confidence level is 0.95, so that $\alpha = 0.05$. Since $n = 15$, we have df $= n - 1 = 15 - 1 = 14$. Table III shows that for df $= 14$,

$$t_{\alpha/2} = t_{0.05/2} = t_{0.025} = \mathbf{2.145}$$

STEP 2 *The confidence interval for μ is from*

$$\bar{x} - t_{\alpha/2} \cdot \frac{s}{\sqrt{n}} \quad \text{to} \quad \bar{x} + t_{\alpha/2} \cdot \frac{s}{\sqrt{n}}$$

From Step 1, $t_{\alpha/2} = 2.145$. Applying the usual formulas for $\bar{x}$ and s to the data in Table 8.6, we find that

$$\bar{x} = \frac{\Sigma x}{n} = \frac{910.3}{15} = 60.69$$

and

$$s = \sqrt{\frac{n(\Sigma x^2) - (\Sigma x)^2}{n(n-1)}} = \sqrt{0.81} = 0.90$$

Consequently, a 95% confidence interval for μ is from

$$60.69 - 2.145 \cdot \frac{0.90}{\sqrt{15}} \quad \text{to} \quad 60.69 + 2.145 \cdot \frac{0.90}{\sqrt{15}}$$

or

$$60.19 \quad \text{to} \quad 61.18$$

MTB We can be 95% confident that the mean gestation period, μ, of the domestic dog is somewhere between 60.19 and 61.18 days. ■

WHICH PROCEDURE SHOULD I USE?

We have now learned two methods for obtaining a confidence interval for a population mean, μ. If the sample size is large, we can use Procedure 8.1 on page 382, regardless of the distribution of the population being sampled. If the population being sampled is normally distributed, we can use Procedure 8.2 on page 398, regardless of the sample size. Two questions may come to mind.

First, what if both conditions are satisfied? That is, suppose we want to obtain a large-sample confidence interval for the mean of a normally distributed population. Then we can use either Procedure 8.1, which is based on the fact that, for large samples, the random variable $(\bar{x} - \mu)/(s/\sqrt{n})$ has *approximately* the standard normal distribution, regardless of the distribution of the population. Or, we can use Procedure 8.2, which is based on the fact that, for a normal population, the random variable $(\bar{x} - \mu)/(s/\sqrt{n})$ has *exactly* the t-distribution with df $= n - 1$, regardless of the size of the sample.

Strictly speaking, Procedure 8.2 is the correct procedure to use for a normal population since it is exact. Practically, however, both procedures give essentially the same results if the sample size is large. This is because, for large samples (and hence large df), there is little difference between the t-distribution and the standard normal distribution. Thus, we stopped the t-table at df $= 29$ and then added a final row with df $= \infty$. This last row actually gives values of z_α.

In summary: To obtain a confidence interval for the mean of a normally distributed population, the correct procedure to use is Procedure 8.2. To apply that procedure when the sample size exceeds 30, determine $t_{\alpha/2}$ from the last row (the ∞ row) of the t-table, Table III.

The second question that may come to mind is: What if neither the large-sample condition nor the normality condition are satisfied? That is, suppose we want to obtain a confidence interval for a population mean using a small sample and that either the population is not normally distributed or we simply do not know its distribution. Then neither Procedure 8.1 nor Procedure 8.2 is appropriate. However, under certain conditions, we can use a *nonparametric method*. Such methods are discussed in Chapter 15.

USING THE COMPUTER (OPTIONAL)

We can use Minitab to obtain a confidence interval for a population mean, μ, when the population being sampled is normally distributed. The appropriate command is called **TINTERVAL** (the first "T" stands for t-distribution).

In Example 8.10, we applied Procedure 8.2 to find a confidence interval for the mean gestation period of the domestic dog. The following example shows how to determine that confidence interval by employing the TINTERVAL command.

EXAMPLE 8.11   *Illustrates the TINTERVAL command*

The gestation periods of domestic dogs are normally distributed. To estimate the mean gestation period, 15 randomly selected dogs are observed during pregnancy. The gestation periods, in days, of the 15 dogs are displayed in Table 8.7 at the top of the next page. Use Minitab to find a 95% confidence interval for the mean gestation period, μ, of the domestic dog.

TABLE 8.7 Gestation periods				
62.0	61.4	59.8	62.2	60.3
60.4	59.4	60.2	60.4	60.8
61.8	59.2	61.1	60.4	60.9

SOLUTION   Since the population is normally distributed, we can apply TINTERVAL to obtain the 95% confidence interval. We first store the sample data in Table 8.7 into C2 using the SET command and name C2 "Gestn" using the NAME command. Then we type the command TINTERVAL, followed by the desired confidence level and the storage location of the sample data. In other words, we type

TINTERVAL 95%, data in 'GESTN'

Printout 8.3 summarizes the above discussion and also shows the resulting output.

PRINTOUT 8.3
Minitab output
for TINTERVAL

```
MTB > SET C2
DATA> 62.0 61.4 59.8 62.2 60.3
DATA> 60.4 59.4 60.2 60.4 60.8
DATA> 61.8 59.2 61.1 60.4 60.9
DATA> END
MTB > NAME C2 'GESTN'
MTB > TINTERVAL 95%, data in 'GESTN'
```

	N	MEAN	STDEV	SE MEAN	95.0 PERCENT C.I.
GESTN	15	60.687	0.898	0.232	(60.190, 61.184)

The output first displays the sample size, sample mean, sample standard deviation, and estimated standard error of the mean ($s/\sqrt{n}$). Then comes the confidence interval. From this last item, we see that a 95% confidence interval for μ is from 60.190 to 61.184. We can be 95% confident that the mean gestation period, μ, of the domestic dog is somewhere between 60.190 and 61.184 days.   ■

Exercises 8.4

—— 8.43   For a t-curve with df $= 6$, use Table III to find the following t-values:
a) $t_{0.10}$ b) $t_{0.025}$ c) $t_{0.01}$

—— 8.44   For a t-curve with df $= 17$, use Table III to find the following t-values:
a) $t_{0.05}$ b) $t_{0.025}$ c) $t_{0.005}$

—— 8.45   For a t-curve with df $= 21$, find the following t-values and illustrate your results graphically.
a) The t-value with area 0.10 to its right.
b) $t_{0.01}$.

c) The t-value with area 0.025 to its left. (Hint: A t-curve is symmetric about 0.)
d) The two t-values that divide the area under the curve into a middle 0.90 area and two outside areas of 0.05.

—— 8.46   For a t-curve with df $= 8$, find the following t-values and illustrate your results graphically.
a) The t-value with area 0.05 to its right.
b) $t_{0.10}$.
c) The t-value with area 0.01 to its left.

d) The two *t*-values that divide the area under the curve into a middle 0.95 area and two outside areas of 0.025.

__ **8.47**   According to the Salt River Project (SRP), a supplier of electricity to the greater Phoenix area, the mean annual electric bill in 1984 was $852.31. An economist wants to estimate the mean for last year. He takes a random sample of 18 SRP customers and obtains the following amounts for their last year's electric bills:

$1875	1478	2206	1740	1830	1516
1738	1486	1941	1608	1794	1828
1264	1999	1798	1794	1598	1568

a) Assume that, for a given year, annual electric bills are normally distributed. Find a 95% confidence interval for last year's mean annual electric bill, μ, for all SRP customers. *(Note:* The sample mean and sample standard deviation of the data are $\bar{x} = \$1725.61$ and $s = \$222.45$.)

b) Does it appear that the mean annual electric bill has increased from the 1984 figure of $852.31? Explain your answer.

__ **8.48**   A city planner working on bikeways needs information about local bicycle commuters. He designs a questionnaire. One of the questions asks how many minutes it takes the rider to pedal from home to his or her destination. A random sample of local bicycle commuters yields the following times:

22	19	24	31	29	29
21	15	27	23	37	31
30	26	16	26	12	
23	48	22	29	28	

a) Determine a 90% confidence interval for the mean commuting time, μ, of all local bicycle commuters in the city. Assume that the times are normally distributed. *(Note:* The sample mean and sample standard deviation of the data are $\bar{x} = 25.82$ and $s = 7.71$.)

b) Interpret your result in part (a).

__ **8.49**   As reported by the Department of Agriculture in the publication *Crop Production,* the mean yield of oats for American farms is about 58.4 bushels per acre. A farmer wants to estimate his mean yield using a newly-developed fertilizer. He uses the fertilizer on a random sample of one-acre plots and obtains the following yields, in bushels:

67	65	55	57	58
61	61	61	64	62
62	60	62	60	67

a) Find a 99% confidence interval for the mean yield per acre, μ, that this farmer will get on his land with the new fertilizer.

b) What assumption did you make in solving part (a)?

c) Does it appear that, by using the new fertilizer, the farmer can get a better mean yield than the national average? Explain your answer.

__ **8.50**   According to *Library Journal,* published by the R. R. Bowker Company of New York, the mean annual subscription rate for law periodicals was $29.66 in 1983. A random sample of 12 law periodicals yields the following annual subscription rates, to the nearest dollar, for this year:

$30	46	44	47
42	38	62	55
52	48	43	54

a) Find a 95% confidence interval for this year's mean annual subscription rate, μ, for all law periodicals.

b) What assumption did you make in solving part (a)?

c) Does your result from part (a) suggest an increase in the mean annual subscription rate over that in 1983? Justify your answer.

__ **8.51**   The *Physician's Handbook* gives statistics on heights and weights of children by age. Suppose that 20 six-year-old girls are chosen at random in order to estimate the mean height, μ, of all such girls. The resulting data are shown below (heights in inches).

44	44	47	46
38	42	46	41
50	43	40	51
47	43	47	48
48	45	41	46

a) Assuming the heights of all six-year-old girls are normally distributed, find a 95% confidence interval for the mean height, μ, of all six-year-old girls. *(Note:* $\Sigma x = 897$ and $\Sigma x^2 = 40{,}449$.)

b) Interpret your result from part (a) in words.

__ **8.52**   The length of the western rattlesnake is normally distributed. To estimate the mean length, μ, of such snakes, 10 snakes are randomly selected. Their lengths, in inches, are as follows:

40.2	43.1	45.5	44.5	39.5
40.2	41.0	41.6	43.1	44.9

a) Find a 90% confidence interval for the mean length, μ, of all western rattlesnakes. (*Note:* $\Sigma x = 423.6$ and $\Sigma x^2 = 17{,}985.62$.)

b) Interpret your result from part (a) in words.

Exercises 8.53–8.56 are computer exercises.

— 8.53 **(Computer exercise)** Suppose that the data in Exercise 8.47 are stored in a column named ELECBILL.

a) Which Minitab command and subcommands (if any) should be used to obtain the confidence interval required in part (a) of that exercise?

b) If you have access to Minitab, use it to obtain the confidence interval.

— 8.54 **(Computer exercise)** Suppose the data in Exercise 8.48 are stored in a column named TIMES.

a) Which Minitab command and subcommands (if any) should be used to obtain the confidence interval required in part (a) of that exercise?

b) If you have access to Minitab, use it to obtain the confidence interval.

— 8.55 **(Computer exercise)** A manufacturer of tobacco products intends to begin marketing a new brand of cigarette. Among other things, the manufacturer needs information about the mean tar content, μ, for this new brand. Obviously, he cannot test all such cigarettes, so he tests a sample of them. The company statistician knows that tar content is normally distributed, so she applies the TINTERVAL command to the tar-content data. Below is the (abridged) Minitab output. Tar contents are in milligrams.

N	MEAN	STDEV	SE MEAN	95.0 PERCENT C.I.
25	10.985	0.604	0.121	(10.736, 11.235)

Use the printout to find

a) the mean tar content, $\bar{x}$, of the cigarettes tested.

b) the standard deviation, s, of the tar contents of the cigarettes tested.

c) the number of cigarettes tested.

d) a 95% confidence interval for the mean tar content, μ, of all cigarettes of the new brand.

— 8.56 **(Computer exercise)** For advertising purposes, a tire company wants to estimate the mean life of a line of steel-belted radials. From past experi-

ence, it is known that tire life is normally distributed. A random sample of steel-belted radials is tested and the following (abridged) output is obtained from the application of Minitab's TINTERVAL command. Data are in thousands of miles.

N	MEAN	STDEV	SE MEAN	90.0 PERCENT C.I.
16	39.469	1.849	0.462	(38.659, 40.280)

From the printout, determine

a) the mean life of the tires sampled.

b) the estimated standard error of the mean.

c) the number of tires in the sample.

d) a 90% confidence interval for the mean life, μ, of all tires in this line.

= 8.57 Suppose that a random sample of size n is to be taken from a normally distributed population with mean μ and standard deviation σ. Determine the probability distributions of the random variables in parts (a) and (b) below.

a) $\dfrac{\bar{x} - \mu}{\sigma/\sqrt{n}}$

b) $\dfrac{\bar{x} - \mu}{s/\sqrt{n}}$

c) How do we obtain probabilities for the random variable in part (a)?

d) How do we obtain probabilities for the random variable in part (b)?

= 8.58 Intelligence quotients measured on the Stanford Revision of the *Binet-Simon Intelligence Scale* are known to be approximately normally distributed with a mean of $\mu = 100$ and a standard deviation of $\sigma = 16$. Suppose 250 IQs are to be selected at random. Find the probability distribution of each of the following random variables:

a) $\dfrac{\bar{x} - 100}{16/\sqrt{250}}$

b) $\dfrac{\bar{x} - 100}{s/\sqrt{250}}$

Referring to the random variable in part (b), we note that it has a t-distribution with df $= 249$. Thus, probabilities for that random variable are equal to areas under the t-curve with df $= 249$. But our t-table, Table III, does not have df $= 249$. So, for instance,

c) how would you find the t-value with area 0.025 to its right for a t-curve with df $= 249$?

Confidence intervals for a normal population mean when σ is known: Suppose that a random sample of size n is to be taken from a normally distributed population with mean μ and standard deviation σ. Then, by Key Fact 7.4 on page 360 (the sampling distribution of the mean for normal populations), the random variable $\bar{x}$ is normally distributed with mean $\mu_{\bar{x}} = \mu$ and standard deviation $\sigma_{\bar{x}} = \sigma/\sqrt{n}$. This is true regardless of the size of the sample. If we now use the same argument as we did just prior to Procedure 8.1 on page 382, then we find that once the sample is taken, the interval from

$$(2) \qquad \bar{x} - z_{\alpha/2} \cdot \frac{\sigma}{\sqrt{n}} \quad \text{to} \quad \bar{x} + z_{\alpha/2} \cdot \frac{\sigma}{\sqrt{n}}$$

will be a $(1 - \alpha)$-level confidence interval for μ. Use this fact to solve Exercises 8.59–8.61.

= 8.59   A certain brand of water-softener salt comes in packages marked "net weight 40 lb." The weights are known to be normally distributed with a standard deviation of $\sigma = 1.5$ lb. To ensure that the amount of salt being packaged does weigh 40 lb, on the average, 15 bags are randomly selected and their contents carefully weighed. The results are as follows:

38.1	38.6	42.5	40.7	40.0
39.9	40.4	41.0	41.6	40.4
43.2	37.9	39.5	42.3	39.4

a) Use the data to determine a 99% confidence interval for the mean net weight, μ, of all such packages of water-softener salt.
b) Is your result from part (a) in conflict with the advertised weight of 40 lb? Explain.

= 8.60   An English professor wants to estimate the average study time per week for students in introductory English courses at her school. To accomplish that, she randomly selects 25 such students and records their weekly study times. The times, in hours, are:

9	8	7	6	7
8	9	4	7	6
6	4	11	5	4
3	7	8	8	7
6	2	2	8	6

Assume that weekly study times are normally distributed with a standard deviation of 2.0 hours.
a) Find a 95% confidence interval for the mean weekly study time, μ, of introductory English students at the school.

b) Interpret your result from part (a) in words.

= 8.61   Refer to Exercise 8.49.
a) Rework part (a) of Exercise 8.49 under the assumption that σ is known and $\sigma = 3.38$ bushels.
b) Compare the confidence interval from part (a) of this exercise to the one found in Exercise 8.49.
c) What is the general effect on a confidence interval for a normal population mean, if σ is known and Formula (2) can be used to obtain the confidence interval instead of Procedure 8.2?

= 8.62   Suppose a random sample of size $n \geq 30$ is to be taken from a normally distributed population in order to obtain a confidence interval for the mean, μ, of the population. Then we can use Procedure 8.1 on page 382, which applies when the sample size is large, or we can use Procedure 8.2 on page 398, which applies when the population being sampled is normally distributed. Which is the better procedure to use? Why?

= 8.63   Let $0 < \alpha < 1$. For a t-curve, determine
a) the t-value with area α to its right.
b) the t-value with area α to its left.
c) the two t-values that divide the area under the curve into a middle $1 - \alpha$ area and two outside areas of $\alpha/2$.
d) Draw graphs to illustrate the results that you obtained in parts (a)–(c).

≡ 8.64   Suppose that a random sample of size n is to be taken from a normally distributed population with mean μ.
a) Use Key Fact 8.3 on page 395 and part (c) of Exercise 8.63 to verify that

$$P\left(-t_{\alpha/2} < \frac{\bar{x} - \mu}{s/\sqrt{n}} < t_{\alpha/2}\right) = 1 - \alpha$$

b) Use algebra to show that the result from part (a) can be expressed as

$$P\left(\bar{x} - t_{\alpha/2} \cdot \frac{s}{\sqrt{n}} < \mu < \bar{x} + t_{\alpha/2} \cdot \frac{s}{\sqrt{n}}\right) = 1 - \alpha$$

c) Use the result from part (b) to explain why once the sample is taken, the interval from

$$\bar{x} - t_{\alpha/2} \cdot \frac{s}{\sqrt{n}} \quad \text{to} \quad \bar{x} + t_{\alpha/2} \cdot \frac{s}{\sqrt{n}}$$

will be a $(1 - \alpha)$-level confidence interval for μ.

8.5 Large-sample confidence intervals for a population proportion

Many statistical studies are concerned with obtaining the proportion (percentage) of a population that has a specified attribute. For example, we might be interested in the percentage of adults who drink alcoholic beverages or in the percentage of cars in use that are imports.

In most cases, the population under consideration will be large and, hence, it will not be practical to obtain the proportion exactly by taking a census. For instance, imagine trying to interview *every* adult in order to ascertain the percentage of adults who drink alcoholic beverages. Thus, we generally resort to sampling and use the sample data to make inferences about the entire population. Let us now look at an example for the purpose of introducing some notation and terminology.

EXAMPLE 8.12 *Introduces proportion terminology*

The U.S. Energy Information Administration conducts annual surveys to obtain data on appliances possessed and generally used by U.S. households. Results are published in *Residential Energy Consumption Survey: Housing Characteristics.* For this year's survey, 6841 U.S. households are to be randomly selected and asked, among other things, whether or not they have a dishwasher. The proportion of the 6841 households sampled that have a dishwasher will then be used to estimate the proportion of all U.S. households that have a dishwasher.

We employ the letter p to denote the proportion of all U.S. households that have a dishwasher. This is the **population proportion** and is the parameter whose value is to be estimated. The proportion of the 6841 U.S. households surveyed that have a dishwasher is called the **sample proportion** and is designated by the symbol $\hat{p}$ (read "p hat"). This is the statistic that will be used to estimate the unknown population proportion, p. In summary then, we will employ the following notation:

$$p = \text{population proportion}$$
$$\hat{p} = \text{sample proportion}$$

We should emphasize that the population proportion, p, although unknown, is a fixed number. On the other hand, the sample proportion, $\hat{p}$, is a random variable. Its value depends on chance; namely, on which households are selected for the survey. For instance, if out of the 6841 households sampled, 2564 have a dishwasher, then $\hat{p} = 2564/6841 = 0.375$, or 37.5%. But, if out of the 6841 households sampled, 2448 have a dishwasher, then $\hat{p} = 2448/6841 = 0.358$, or 35.8%.

These last two calculations also indicate how the sample proportion is computed: Divide the number of households sampled that have a dishwasher, x, by the total number of households sampled, n. In symbols,

$$\hat{p} = \frac{x}{n}$$

We have used a specific example, Example 8.12, to introduce some notation and terminology that is employed when inferences are made about a population proportion. In general, we have the following definitions:

DEFINITION 8.4   Population proportion and sample proportion

Consider a population in which each member is classified as either having or not having a specified attribute, a so-called *two-category population.*

Population proportion, p: The proportion (percentage) of the entire population that has the specified attribute.

Sample proportion, $\hat{p}$: The proportion (percentage) of a sample from the population that has the specified attribute.

In Example 8.12, the specified attribute is "has a dishwasher." The population proportion, p, is the proportion of all U.S. households that have a dishwasher; the sample proportion, $\hat{p}$, is the proportion of the households sampled that have a dishwasher. As we saw in Example 8.12, a sample proportion, $\hat{p}$, is computed using the formula given below.

FORMULA 8.2   Sample proportion

A sample proportion, $\hat{p}$, is computed using the formula

$$\hat{p} = \frac{x}{n}$$

where x denotes the number of members sampled that have the specified attribute and n denotes the sample size.

Note: For convenience, we will refer to x—the number of members sampled that have the specified attribute—as the **number of successes.**

Before proceeding, it might be helpful to draw some parallels between our discussion of proportions and our previous work with means. Table 8.8 shows the correspondence between the notation for means and the notation for proportions.

TABLE 8.8
Correspondence
between notations
for means and
proportions

	Population parameter	Sample statistic
Means	μ	$\overline{x}$
Proportions	p	$\hat{p}$

As we know, a sample mean, $\overline{x}$, can be used to make inferences about a population mean, μ. Similarly, a sample proportion, $\hat{p}$, can be used to make inferences about a population proportion, p.

THE SAMPLING DISTRIBUTION OF THE PROPORTION

We have seen that in order to make inferences about a population mean, μ, it is necessary to know the sampling distribution of the mean, that is, the probability distribution of the random variable $\bar{x}$. The same is true for proportions. To make inferences about a population proportion, p, we need to know the **sampling distribution of the proportion,** that is, the probability distribution of the random variable $\hat{p}$.

   The sampling distribution of the proportion can be derived from our knowledge of the sampling distribution of the mean since a proportion can always be regarded as a mean. We present the result as Key Fact 8.5. See Exercise 8.74 or Exercise 8.75 for a verification.

KEY FACT 8.5   The sampling distribution of the proportion

Suppose that a large random sample of size n is to be taken from a two-category population with population proportion, p. Then the random variable $\hat{p}$ is approximately normally distributed and has mean $\mu_{\hat{p}} = p$ and standard deviation $\sigma_{\hat{p}} = \sqrt{p(1-p)/n}$. In other words, probabilities for $\hat{p}$ are approximately equal to areas under the normal curve with parameters p and $\sqrt{p(1-p)/n}$. The approximation is adequate provided that the number of successes, x, and the number of failures, $n - x$, are both at least 5.

LARGE-SAMPLE CONFIDENCE INTERVALS FOR A POPULATION PROPORTION

We will now present a procedure for finding a confidence interval for a population proportion, p, based on the proportion, $\hat{p}$, of a sample from the population. Basically, the procedure is just a special case of Procedure 8.1 on page 382 which provides a method for obtaining a large-sample confidence interval for a population mean, μ. See Exercise 8.76 for a detailed derivation.

PROCEDURE 8.3

To find a confidence interval for a population proportion, p.

ASSUMPTION

The number of successes, x, and the number of failures, $n - x$, are both at least 5.

STEP 1 *For a confidence level of $1 - \alpha$, use Table II to find $z_{\alpha/2}$.*
STEP 2 *The confidence interval for p is from*

$$\hat{p} - z_{\alpha/2} \cdot \sqrt{\hat{p}(1-\hat{p})/n} \quad \text{to} \quad \hat{p} + z_{\alpha/2} \cdot \sqrt{\hat{p}(1-\hat{p})/n}$$

where $z_{\alpha/2}$ is found in Step 1, n is the sample size, and $\hat{p} = x/n$ is the sample proportion.

EXAMPLE 8.13   **Illustrates Procedure 8.3**

Let us now return to the situation of Example 8.12: Suppose that 6841 U.S. households are selected at random in order to estimate the proportion, p, of all U.S. households that have a dishwasher. If 2470 of the 6841 households chosen have a dishwasher, find a 95% confidence interval for p.

SOLUTION   We will apply Procedure 8.3, but first we need to check that the conditions for its use are satisfied. The attribute in question is "has a dishwasher." The sample size is $n = 6841$. Since 2470 of the households sampled have a dishwasher,

$$x = 2470$$

and

$$n - x = 6841 - 2470 = 4371$$

Thus, both x and $n - x$ are at least 5 and, consequently, the conditions for using Procedure 8.3 are met.

STEP 1 *For a confidence level of $1 - \alpha$, use Table II to find $z_{\alpha/2}$.*

We want a 95% confidence interval. This means that $\alpha = 0.05$ and, hence,

$$z_{\alpha/2} = z_{0.05/2} = z_{0.025} = \boxed{1.96}$$

STEP 2 *The confidence interval for p is from*

$$\hat{p} - z_{\alpha/2} \cdot \sqrt{\hat{p}(1 - \hat{p})/n} \quad \text{to} \quad \hat{p} + z_{\alpha/2} \cdot \sqrt{\hat{p}(1 - \hat{p})/n}$$

We have $n = 6841$ and, from Step 1, $z_{\alpha/2} = 1.96$. Also, since 2470 of the 6841 households sampled have a dishwasher, the sample proportion is

$$\hat{p} = \frac{x}{n} = \frac{2470}{6841} = 0.361$$

Consequently, a 95% confidence interval for p is from

$$0.361 - 1.96 \cdot \sqrt{(0.361)(1 - 0.361)/6841} \quad \text{to} \quad 0.361 + 1.96 \cdot \sqrt{(0.361)(1 - 0.361)/6841}$$

or

$$\boxed{0.350 \quad \text{to} \quad 0.372}$$

We can be 95% confident that the percentage of all U.S. households that have an automatic dishwasher is somewhere between 35.0% and 37.2%. ∎

Exercises 8.5

___ **8.65** Is a population proportion, p, a parameter or a statistic? What about a sample proportion, $\hat{p}$? Justify your answers.

In each of Exercises 8.66–8.71 apply Procedure 8.3 on page 407 to determine the required confidence interval. Be sure to check that the conditions for using that procedure are met.

___ **8.66** Studies are performed to estimate the percentage of the nation's 10 million asthmatics who are allergic to sulfites. Suppose that 500 asthmatics are randomly selected and that 38 of the 500 are found to be allergic to sulfites.
a) Find a 95% confidence interval for the proportion, p, of all U.S. asthmatics who are allergic to sulfites.
b) Interpret your result from part (a) in words.

___ **8.67** A *Reader's Digest/Gallup Survey* on the drinking habits of Americans estimated the percentage of adults across the country who drink beer, wine, or hard liquor, at least occasionally. Of the 1516 adults interviewed, 985 said they drank.
a) Find a 95% confidence interval for the proportion, p, of all Americans who drink beer, wine, or hard liquor, at least occasionally.
b) Interpret your result from part (a) in words.

___ **8.68** A Gallup Poll asked public-school teachers nationwide to grade their fellow educators' performance. The poll found that 634 of the 813 teachers surveyed gave their fellow educators an *A* or *B* grade for performance.
a) Find a 90% confidence interval for the percentage of all public-school teachers who would give their fellow educators an *A* or *B* grade for performance.
b) Interpret your result from part (a) in words.

___ **8.69** A 1985 Gallup Poll estimated the support among Americans for "right to die" laws. For the survey, 1528 adults, 18 years old and older, were asked whether they were in favor of voluntary withholding of life support systems from the terminally ill. The results: 1238 said they were.
a) Determine a 99% confidence interval for the percentage of all adult Americans who are in favor of "right to die" laws.
b) Interpret your result from part (a) in words.

___ **8.70** Dr. Charles Kuntzleman directed a study to estimate the percentage of all schoolchildren who have at least one of the significant factors contributing to heart disease. As reported by the *Arizona Republic,* the three-year study involved nearly 24,000 schoolchildren and concluded that 98% of the schoolchildren examined had at least one of the significant factors contributing to heart disease.
a) Use the data to obtain a 99% confidence interval for the proportion, p, of all schoolchildren who have at least one of the significant factors contributing to heart disease.
b) Interpret your result in terms of percentages.

___ **8.71** Real Estate Research Corporation, a Los Angeles research group, publishes an annual report called *Emerging Trends in Real Estate.*
a) Suppose 200 randomly selected real-estate experts are asked whether they believe next year will be a good time to buy real estate. If 78% reply in the affirmative, determine a 90% confidence interval for the proportion of all real-estate experts who believe next year will be a good time to buy real estate.
b) Interpret your result in words.

___ **8.72** Suppose that you have been hired to estimate the percentage of adults in your state who are literate. You take a random sample of 100 adults and find that 96 of them are literate. You then determine a 95% confidence interval as follows:

$$0.96 \pm 1.96 \cdot \sqrt{(0.96)(0.04)/100}$$

or

$$0.922 \quad \text{to} \quad 0.998$$

From this you conclude that we can be 95% confident that the percentage of all adults in your state who are literate is somewhere between 92.2% and 99.8%. Is there anything wrong with your reasoning?

___ **8.73** Suppose that I have been commissioned to estimate the infant mortality rate in Norway. From a random sample of 500 live births, I find that 0.8% of them resulted in infant deaths. I then construct a 90% confidence interval for the true infant mortality rate in Norway:

$$0.008 \pm 1.645 \cdot \sqrt{(0.008)(0.992)/500}$$

or

$$0.001 \quad \text{to} \quad 0.015$$

Then I conclude: "We can be 90% confident that the proportion of infant deaths in Norway is somewhere between 0.001 and 0.015." How did I do?

Exercises 8.74 and 8.75 provide two different ways to verify Key Fact 8.5 on page 407, which gives the sampling distribution of the proportion.

≡ **8.74**   Recall from Formula 8.2 on page 406 that a sample proportion, $\hat{p}$, is computed as follows:

$$\hat{p} = \frac{x}{n}$$

where x denotes the number of successes and n denotes the sample size. The random variable x has, at least approximately, the binomial distribution with parameters n and p, where p is the population proportion. In Section 6.5 we learned that, for large n, a binomial random variable, such as x, is approximately normally distributed. From this it follows that the random variable $\hat{p}$ is also approximately normally distributed. So, to complete the verification of Key Fact 8.5, we need only show that the mean and standard deviation of the random variable $\hat{p}$ are given by

$$\mu_{\hat{p}} = p \quad \text{and} \quad \sigma_{\hat{p}} = \sqrt{p(1-p)/n}$$

Establish these last two formulas. *(Hint: First apply part (f) of Exercise 5.27 on page 250 to Formula 8.2. Then use Formulas 5.3 and 5.4 on page 275.)*

≡ **8.75**   Consider a finite, two-category population in which the proportion of members that have the specified attribute is equal to p. We can think of such a population as consisting of 1s and 0s. A member of the population is a "1" if it has the specified attribute and is a "0" otherwise.
a) If the size of the population is N, how many 1s are in the population?
b) Use part (a) and Definition 3.8 on page 124 to show that the mean of this population of 1s and 0s is p; that is, $\mu = p$.
c) Use part (b) and the shortcut formula given in Definition 3.9 on page 126 to show that the stan-

dard deviation of this population of 1s and 0s is $\sqrt{p(1-p)}$; in other words, $\sigma = \sqrt{p(1-p)}$.
d) Suppose a random sample of size n is to be taken from this population of 1s and 0s. Show that $\bar{x} = \hat{p}$.
e) Use parts (b)–(d) and the sampling distribution of the mean for general populations (Key Fact 7.6 on page 363) to verify Key Fact 8.5.

≡ **8.76**   In this exercise we will derive Procedure 8.3 on page 407, which provides a method for obtaining a confidence interval for a population proportion. So, suppose that a large random sample of size n is to be taken from a two-category population with population proportion p.
a) Apply Key Fact 8.5 on page 407 (the sampling distribution of the proportion) and Key Fact 6.8 on page 323 (the general empirical rule for normally distributed random variables) to deduce that the probability is approximately $1 - \alpha$ that the sample proportion, $\hat{p}$, will be in the interval with endpoints

$$p \pm z_{\alpha/2} \cdot \sqrt{p(1-p)/n}$$

b) Use part (a) to show that the probability is approximately $1 - \alpha$ that the interval with endpoints

$$\hat{p} \pm z_{\alpha/2} \cdot \sqrt{p(1-p)/n}$$

will contain p.
c) Deduce from part (b) that the probability is approximately $1 - \alpha$ that the interval with endpoints

$$\hat{p} \pm z_{\alpha/2} \cdot \sqrt{\hat{p}(1-\hat{p})/n}$$

will contain p.
d) Use part (c) to explain why once the sample is taken, the interval from

$$\hat{p} - z_{\alpha/2} \cdot \sqrt{\hat{p}(1-\hat{p})/n} \quad \text{to} \quad \hat{p} + z_{\alpha/2} \cdot \sqrt{\hat{p}(1-\hat{p})/n}$$

will be a $(1 - \alpha)$-level confidence interval for p.

Chapter review

KEY TERMS   confidence-interval estimate, 375
confidence level, 375
degrees of freedom (df), 395

maximum error of the estimate
for μ, 391
number of successes, 406

FORMULAS

In the formulas below,

μ = population mean
σ = population standard deviation
$\bar{x}$ = sample mean
s = sample standard deviation

n = sample size
$\alpha = 1 -$ confidence level
p = population proportion
$\hat{p}$ = sample proportion

Large-sample confidence interval for a population mean, μ, 382

$$\bar{x} - z_{\alpha/2} \cdot \frac{s}{\sqrt{n}} \quad \text{to} \quad \bar{x} + z_{\alpha/2} \cdot \frac{s}{\sqrt{n}}$$

Maximum error of the estimate for μ, 391

$$E = z_{\alpha/2} \cdot \frac{\sigma}{\sqrt{n}}$$

Sample size for estimating μ, 391

$$n = \left[\frac{z_{\alpha/2} \cdot \sigma}{E} \right]^2$$

(E = maximum error of the estimate)

Confidence interval for a normal population mean, μ, 398

$$\bar{x} - t_{\alpha/2} \cdot \frac{s}{\sqrt{n}} \quad \text{to} \quad \bar{x} + t_{\alpha/2} \cdot \frac{s}{\sqrt{n}}$$

(df $= n - 1$)

Sample proportion, 406

$$\hat{p} = \frac{x}{n}$$

(x = number of successes)

Large-sample confidence interval for a population proportion, p, 407

$$\hat{p} - z_{\alpha/2} \cdot \sqrt{\hat{p}(1 - \hat{p})/n} \quad \text{to} \quad \hat{p} + z_{\alpha/2} \cdot \sqrt{\hat{p}(1 - \hat{p})/n}$$

1. use and understand each of the preceding formulas.
2. use Table II to find $z_{\alpha/2}$ for any specified value of α.
3. find a large-sample confidence interval for a population mean, μ.
4. compute the maximum error of the estimate for μ.
5. understand the relationship between the sample size, standard deviation, confidence level, and maximum error of the estimate for a confidence interval for μ.
6. determine the sample size required for a specified confidence level and maximum error of the estimate for μ.
7. use Table III to find $t_{\alpha/2}$ for df $= n - 1$.
8. find a confidence interval for a population mean, μ, when the population being sampled is normally distributed.
9. decide which procedure to use when finding a confidence interval for a population mean, μ.
10. find a large-sample confidence interval for a population proportion, p.
11. interpret a confidence interval for a mean or proportion.
12. use the Minitab commands covered in this chapter.*
13. interpret the output obtained from the application of the Minitab commands discussed in this chapter.*

REVIEW TEST

1. Suppose that a random sample of size $n \geq 30$ is to be taken from a population with mean μ and standard deviation σ.
   a) What is the probability distribution of the random variable $\bar{x}$?
   b) Identify the probability distribution of the random variable

$$\frac{\bar{x} - \mu}{\sigma/\sqrt{n}}$$

   c) How do we find probabilities for the random variable in part (b)?

2. Use Table II to find the following z-values and draw graphs to illustrate your work.
   a) $z_{0.063}$ b) $z_{0.40}$ c) $z_{0.007}$

3. Dr. Thomas Stanley of Georgia State University has surveyed millionaires since 1973. Among other things, Dr. Stanley obtains estimates for the mean age, μ, of all American millionaires. Suppose that 36 American millionaires are randomly selected and that their ages are as follows:

31	45	79	64	48	38	39	68	52
59	68	79	42	79	53	74	66	66
71	61	52	47	39	54	67	55	71
77	64	60	75	42	69	48	57	48

Find a 95% confidence interval for the mean age, μ, of all American millionaires. (*Note:* The sample mean and sample standard deviation of the data are $\bar{x} = 58.53$ years and $s = 13.36$ years.)

4. From Problem 3, we know that "a 95% confidence interval for the mean age of all American millionaires is from 54.2 to 62.9 years." Decide which of the following provide a correct interpretation of the statement in quotes. Justify your answers.
   a) 95% of all American millionaires are between the ages of 54.2 and 62.9 years.
   b) There is a 95% chance that the mean age of all American millionaires is between 54.2 and 62.9 years.
   c) We can be 95% confident that the mean age of all American millionaires is between 54.2 and 62.9 years.
   d) The probability is 0.95 that the mean age of all American millionaires is between 54.2 and 62.9 years.

*5. (**Computer problem**) Suppose the age data in Problem 3 are stored in a column named AGES.
   a) Which Minitab commands and subcommands (if any) should be used to obtain the confidence interval required in that problem?
   b) If you have access to Minitab, use it to obtain the confidence interval.

*6. (**Computer problem**) The abridged Minitab printout below was obtained by applying the command ZINTERVAL to the age data in Problem 3.

```
THE ASSUMED SIGMA =13.4

 N    MEAN   STDEV   SE MEAN     95.0 PERCENT C.I.
36   58.53   13.36     2.23    (   54.16,   62.90)
```

From the printout, determine
   a) the (assumed) population standard deviation, σ, of the ages of all American millionaires.
   b) the sample standard deviation, s.
   c) the number of millionaires sampled.
   d) the mean age, $\bar{x}$, of the millionaires sampled.
   e) a 95% confidence interval for the mean age, μ, of all American millionaires.

7. A certain make of hand-held computer runs on four 1.5V, size AAA batteries. To estimate the average battery life, 50 computers are tested. The mean battery life for the 50 computers is found to be $\bar{x} = 60.1$ hours. Assume that $\sigma = 4.3$ hours.
   a) Obtain a 99% confidence interval for the mean battery life, μ, for this make of computer.
   b) Interpret your result from part (a) in words.

8. Refer to Problem 7.
   a) Find the maximum error of the estimate, E, for the confidence interval obtained in Problem 7.
   b) Explain the meaning of E as far as the accuracy of the estimate is concerned.
   c) Determine the sample size required to have a maximum error of the estimate of 0.5 hours and a 99% confidence level.
   d) Find a 99% confidence interval for μ, if a sample of the size indicated in part (c) yields a mean of 59.8 hours.

9. The figure below shows the standard normal curve and two t-curves. Which of the two t-curves has the larger degrees of freedom? Explain.

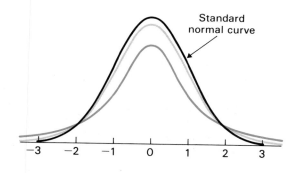

10. Suppose that a random sample of size n is to be taken from a normally distributed population with mean μ and standard deviation σ.
    a) What is the probability distribution of the random variable $\bar{x}$?
    b) Identify the probability distribution of the random variable

$$\frac{\bar{x} - \mu}{s/\sqrt{n}}$$

    c) How do we find probabilities for the random variable in part (b)?

11. For a t-curve with df $= 18$, obtain the following t-values and illustrate your results graphically.
    a) The t-value with area 0.025 to its right.
    b) $t_{0.05}$.
    c) The t-value with area 0.10 to its left.
    d) The two t-values that divide the area under the curve into a middle 0.99 area and two outside 0.005 areas.

12. The A. C. Nielsen Company collects information on television viewing by Americans. Data are published in *Nielsen Report on Television*. Suppose that a random sample of 20 American households yields the following daily viewing times in hours:

7.5	2.5	6.8	5.0	7.9
5.3	8.9	8.8	10.3	8.8
9.5	9.5	6.1	9.4	8.4
8.2	6.5	9.0	6.4	8.4

    a) Assuming daily viewing times for households are approximately normally distributed, find a 95% confidence interval for the mean daily viewing time, μ, of all American households.
    b) Interpret your result from part (a) in words.
    c) In the year 1988, the average American household watched seven hours and six minutes of TV per day. Does your result in part (a) provide evidence of an increase in average daily viewing time? Explain.

*13. (**Computer problem**) Suppose that the data displayed in Problem 12 are stored in a column named VIEWTIME.
    a) Which Minitab command and subcommands (if any) should be used to obtain the confidence interval required in that problem?
    b) If you have access to Minitab, use it to obtain the confidence interval.

*14. **(Computer problem)** The following abridged Minitab output was obtained by applying the command TINTERVAL to the data in Problem 12.

```
 N   MEAN   STDEV   SE MEAN    95.0 PERCENT C.I.
20   7.660   1.913    0.428   (  6.764,  8.556)
```

From the printout, determine
a) the mean daily viewing time of the households sampled.
b) the standard deviation of the daily viewing times for the households sampled.
c) the estimated standard error of the mean.
d) the number of households in the sample.
e) a 95% confidence interval for the mean daily viewing time, μ, of all American households.

15. On Sunday, October 7, 1984, presidential candidates Ronald Reagan and Walter Mondale debated on national television. A *Newsweek* poll, conducted by the Gallup Organization, was taken to determine which candidate the public thought did a better job in the debate. The Gallup Organization telephoned 379 registered voters who watched the debate and found that 205 thought Mondale did a better job.
a) Find a 95% confidence interval for the proportion, p, of all registered voters watching the debate who thought Mondale did a better job.
b) Interpret your result from part (a) in words.
c) Does the poll convince you that a majority of the registered voters watching the debate thought Mondale did a better job? Explain.

CHAPTER 9

HYPOTHESIS TESTS FOR ONE MEAN OR PROPORTION

In Chapter 8 we studied problems involving the estimation of certain parameters. Specifically, we examined methods for obtaining a confidence interval for one population mean or one population proportion. As we know, a confidence interval for a population mean, μ, is based on the statistic $\bar{x}$ and a confidence interval for a population proportion, p, is based on the statistic $\hat{p}$.

Now we will learn how those statistics can be used to make decisions about hypothesized values of the corresponding parameters. For example, according to the R. R. Bowker Company of New York, the mean retail price of all hardcover history books was $28.44 in 1986. We will see how the mean, $\bar{x}$, of a sample of this year's prices can be used to decide whether this year's mean price, μ, of all hardcover history books has increased over the 1986 mean of $28.44. Statistical inferences of this kind are called *hypothesis tests*. In this chapter, we will study hypothesis tests for one population mean or one population proportion.

CHAPTER OUTLINE

9.1 The nature of hypothesis testing   Presents the logic of hypothesis testing and introduces some of the notation and terminology that is used.

9.2 Terms, errors, and hypotheses   Defines some additional terms used in hypothesis testing, discusses the two types of errors that can be made in a hypothesis test, and interprets the possible conclusions for a hypothesis test.

9.3 Large-sample hypothesis tests for a population mean   Explains how to perform a hypothesis test for a population mean when the sample size is large.

9.4 *P*-values (Optional)   Introduces and applies the concept of *P*-values and shows how large-sample hypothesis tests for a population mean can be carried out using a statistical computer package.

9.5 Hypothesis tests for a normal population mean   Describes how to perform a hypothesis test for a population mean when the population being sampled is normally distributed.

9.6 Large-sample hypothesis tests for a population proportion   Discusses a procedure for performing a hypothesis test for a population proportion when the sample size is large.

9.1 The nature of hypothesis testing

We often find it necessary to employ inferential statistics in order to make decisions about the value of a parameter, such as a population mean or a population proportion. For example, we might need to decide whether the mean weight, μ, of all bags of pretzels being packaged differs from the advertised weight of 454 grams; or we might want to decide whether the mean age, μ, of juveniles being held in public custody has decreased from the 1982 mean of 15.4 years; or we might need to decide whether the proportion, p, of adult Americans who approve of the way the president is handling his job exceeds 50%.

One of the most commonly used methods for making such decisions is to perform a **hypothesis test.** A hypothesis is simply a statement that something is true. For instance, the statement "the mean weight of all bags of pretzels being packaged differs from the advertised weight of 454 grams" is a hypothesis.

Typically, there are two hypotheses in a hypothesis test. One hypothesis is called the **null hypothesis** and the other is called the **alternative hypothesis** (or **research hypothesis**). These can be defined as follows.

DEFINITION 9.1 Null and alternative hypotheses

Null hypothesis: A hypothesis to be tested. We use the symbol H_0 to stand for null hypothesis.
Alternative hypothesis: A hypothesis to be considered as an alternate to the null hypothesis. We use the symbol H_a to stand for alternative hypothesis.

For example, in the pretzel packaging illustration, the null hypothesis might be "the mean weight of all bags of pretzels being packaged equals the advertised weight of 454 grams" and the alternative hypothesis might be "the mean weight of all bags of pretzels being packaged differs from the advertised weight of 454 grams."

Originally, the term "null" in null hypothesis stood for "no difference" or "the difference is null." However, nowadays null hypothesis has come to mean simply a hypothesis to be tested. The problem in a hypothesis test is to determine whether or not the null hypothesis should be rejected in favor of the alternative hypothesis.

CHOOSING THE HYPOTHESES

The first step in setting up a hypothesis test is to decide what the null hypothesis should be and what the alternative hypothesis should be. Below we offer some guidelines for choosing those two hypotheses. Although the guidelines refer specifically to hypothesis tests for one population mean, μ, they apply to any hypothesis test concerning one parameter.

Null hypothesis: In this book, the null hypothesis for a hypothesis test concerning a population mean, μ, should always specify a single value for that parameter. This means that the null hypothesis should always be of the form $\mu = \mu_0$, where μ_0 is some specified number. In other words, there should be an equal sign (=) in the null hypothesis. We can, therefore, express the null hypothesis concisely as

$$H_0: \mu = \mu_0$$

Alternative hypothesis: The choice of the alternative hypothesis depends on and should reflect our purpose in performing the hypothesis test. There are three possibilities for the choice of the alternative hypothesis.

1. *Two-tailed test:* If we are primarily concerned with deciding whether a population mean, μ, is *different from* a specified value μ_0, then the alternative hypothesis should be $\mu \neq \mu_0$. In other words, there should be a not-equal sign ($\neq$) in the alternative hypothesis. We express such an alternative hypothesis as

$$H_a: \mu \neq \mu_0$$

A hypothesis test with an alternative hypothesis of this form is called a **two-tailed test.**

2. *Left-tailed test:* If we are primarily concerned with deciding whether a population mean, μ, is *less than* a specified value μ_0, then the alternative hypothesis should be $\mu < \mu_0$. In other words, there should be a less-than sign ($<$) in the alternative hypothesis. We express such an alternative hypothesis as

$$H_a: \mu < \mu_0$$

A hypothesis test with an alternative hypothesis of this form is called a **left-tailed test.**

3. *Right-tailed test:* If we are primarily concerned with deciding whether a population mean, μ, is *greater than* a specified value μ_0, then the alternative hypothesis should be $\mu > \mu_0$. In other words, there should be a greater-than sign ($>$) in the alternative hypothesis. We express such an alternative hypothesis as

$$H_a: \mu > \mu_0$$

A hypothesis test with an alternative hypothesis of this form is called a **right-tailed test.**

A hypothesis test is called a **one-tailed test** if it is either left-tailed or right-tailed (i.e., if it is not two-tailed). We will explain a little later the significance of the terminology "tailed." But for now, let's consider some examples that illustrate the above discussion.

EXAMPLE 9.1   *Illustrates the choice of the null and alternative hypotheses*

A snack-food company produces a 454-gram (one-pound) bag of pretzels. Although the actual net weights deviate somewhat from 454 grams and vary from one bag to another, it is nevertheless important to the company that the mean net weight of the bags be kept at 454 grams. Consequently, the quality assurance department

periodically performs a hypothesis test to decide whether the packaging machine is working properly; that is, to decide whether the mean net weight of all bags being packaged is 454 grams.
a) Determine the null hypothesis for the hypothesis test.
b) Determine the alternative hypothesis for the hypothesis test.
c) Classify the hypothesis test as two-tailed, left-tailed, or right-tailed.

SOLUTION   Let μ denote the (population) mean net weight of all bags being packaged.
a) As we said, the null hypothesis for a hypothesis test concerning a population mean, μ, should always specify a single value for that parameter. Thus, the null hypothesis for this hypothesis test is that the packaging machine is working properly; that is, the mean net weight, μ, of all bags being packaged equals 454 grams. In symbols,

$$H_0\text{: } \mu = 454 \text{ grams}$$

b) The alternative hypothesis for this hypothesis test is that the packaging machine is not working properly; that is, the mean net weight, μ, of all bags being packaged is *different from* 454 grams. In symbols,

$$H_a\text{: } \mu \neq 454 \text{ grams}$$

c) This hypothesis test is two-tailed since there is a not-equal sign ($\neq$) in the alternative hypothesis.   ∎

EXAMPLE 9.2   *Illustrates the choice of the null and alternative hypotheses*

The R. R. Bowker Company of New York collects information on the retail prices of books. Data are published in *Publishers Weekly*. In 1986, the mean retail price of all hardcover history books was $28.44. Suppose we want to perform a hypothesis test to decide whether this year's mean retail price of all hardcover history books has increased over the 1986 mean of $28.44.
a) Determine the null hypothesis for the hypothesis test.
b) Determine the alternative hypothesis for the hypothesis test.
c) Classify the hypothesis test as two-tailed, left-tailed, or right-tailed.

SOLUTION   Let μ denote this year's mean retail price of all hardcover history books.
a) Again, the null hypothesis for a hypothesis test concerning a population mean, μ, should always specify a single value for that parameter. Thus, the null hypothesis for this hypothesis test is that this year's mean retail price of all hardcover history books is the same as the 1986 mean of $28.44; that is,

$$H_0\text{: } \mu = \$28.44$$

b) We want to decide whether this year's mean retail price for hardcover history books has increased over the 1986 mean of $28.44. So, the alternative hypothesis for this hypothesis test is that this year's mean retail price of all hardcover history books is *greater than* $28.44; that is,

$$H_a: \mu > \$28.44$$

c) This hypothesis test is right-tailed since there is a greater-than sign ($>$) in the alternative hypothesis. ∎

EXAMPLE 9.3   *Illustrates the choice of the null and alternative hypotheses*

Calcium is the most abundant and one of the most important minerals in the body. It works with phosphorus to build and maintain bones and teeth. According to the Food and Nutrition Board of the National Academy of Sciences, the recommended daily allowance (RDA) of calcium for adults is 800 mg (milligrams). A nutritionist thinks that the average person with an income below the poverty level gets less than the RDA of 800 mg. She plans to perform a hypothesis test in order to decide whether her suspicion is correct.
a) Determine the null hypothesis for the hypothesis test.
b) Determine the alternative hypothesis for the hypothesis test.
c) Classify the hypothesis test as two-tailed, left-tailed, or right-tailed.

SOLUTION   Let μ denote the mean daily intake of calcium of all people with incomes below the poverty level.
a) The null hypothesis must specify a single value for the parameter μ. Hence, the null hypothesis for this hypothesis test is that the mean calcium intake of all people with incomes below the poverty level is 800 mg per day:

$$H_0: \mu = 800 \text{ mg}$$

b) The nutritionist's suspicion is that the mean calcium intake of all people with incomes below the poverty level is *less than* the RDA of 800 mg per day. Thus, the alternative hypothesis for this hypothesis test is:

$$H_a: \mu < 800 \text{ mg}$$

c) This hypothesis test is left-tailed since there is a less-than sign ($<$) in the alternative hypothesis. ∎

THE LOGIC OF HYPOTHESIS TESTING

We have now seen how to choose the null and alternative hypotheses for a hypothesis test. The next question is: How do we decide which of the two hypotheses is true; that is, how do we decide whether or not to reject the null hypothesis in favor of the alternative hypothesis?

Very roughly, the procedure for deciding is the following: Take a random sample from the population. If the sample data are consistent with the null hypothesis, then do not reject the null hypothesis. On the other hand, if the sample data are inconsistent with the null hypothesis, then reject the null hypothesis and conclude that the alternative hypothesis is true.

Of course, in practice, we must give a precise criterion for deciding whether or not to reject the null hypothesis. In the next example, we will illustrate how such a criterion can be devised for a two-tailed hypothesis test concerning a population mean, μ. The example is being presented to introduce the logic of hypothesis testing and some of the terminology of hypothesis testing. General procedures for performing hypothesis tests will be given later.

EXAMPLE 9.4   *Introduces hypothesis testing*

Consider again the situation of Example 9.1. A company that produces snack foods uses a packaging machine to package 454-gram bags of pretzels. To check whether the machine is working properly, the quality assurance department takes a random sample of 50 bags of pretzels. The net weights, in grams, of the 50 bags of pretzels obtained are displayed in Table 9.1.

TABLE 9.1
Weights of the
50 bags of pretzels

464	450	450	456	452	433	446	446	450	447
442	438	452	447	460	450	453	456	446	433
448	450	439	452	459	454	456	454	452	449
463	449	447	466	446	447	450	449	457	464
468	447	433	464	469	457	454	451	453	443

Do the data provide sufficient evidence to conclude that the packaging machine is not working properly? Use the following steps to answer the question:
a) State the null and alternative hypotheses for the hypothesis test.
b) Discuss the basic idea for carrying out the hypothesis test.
c) Obtain a precise criterion for deciding whether or not to reject the null hypothesis in favor of the alternative hypothesis.
d) Apply the criterion in part (c) to the sample data and state the conclusion.

SOLUTION   Let μ denote the mean net weight of all bags being packaged.
a) The null and alternative hypotheses for the hypothesis test were found in Example 9.1. They are:

H_0: $\mu = 454$ grams (the machine is working properly)
H_a: $\mu \neq 454$ grams (the machine is not working properly)

b) Basically, the idea for carrying out the hypothesis test is this: If the null hypothesis is true (i.e., if $\mu = 454$ grams), then the mean weight, $\bar{x}$, of the sample of 50 bags of pretzels should be approximately equal to 454 grams. Of course, we cannot expect a sample mean to be exactly equal to a population mean—some sampling error is to be anticipated. However, if the sample mean weight, $\bar{x}$, differs by too much from 454 grams, then we would be inclined to reject the null hypothesis and conclude that the alternative hypothesis is true.

From Table 9.1, we can compute the mean weight of the sample of 50 bags of pretzels:

$$\overline{x} = \frac{\Sigma x}{n} = \frac{22{,}561}{50} = \boxed{451.2 \text{ grams}}$$

The question now is whether the difference of 2.8 grams between the sample mean of 451.2 grams and the hypothesized population mean of 454 grams can reasonably be attributed to sampling error or whether the difference is large enough to indicate that the population mean is not 454 grams. To answer this question, we need to know how likely it would be to get such a difference if, in fact, the null hypothesis, $\mu = 454$ grams, is true. That likelihood can be determined by using our knowledge of the sampling distribution of the mean, as explained in part (c).

c) For this part, we are to obtain a precise criterion for deciding whether or not to reject the null hypothesis in favor of the alternative hypothesis. We will obtain that criterion by applying Key Fact 7.6 on page 363, the sampling distribution of the mean for general populations. Since the sample size here is large ($n = 50$), Key Fact 7.6 implies that the random variable $\overline{x}$ is approximately normally distributed and has mean $\mu_{\overline{x}} = \mu$ and standard deviation $\sigma_{\overline{x}} = \sigma/\sqrt{n}$.

Knowing that $\overline{x}$ is approximately normally distributed, we can, for instance, use Property 2 of Key Fact 6.7 on page 321 (the empirical rule for normally distributed random variables) to deduce the following: The probability is about 0.9544 that $\overline{x}$ will be within two standard deviations of its mean, $\mu_{\overline{x}}$, which is the same as the population mean, μ. See Figure 9.1.

FIGURE 9.1

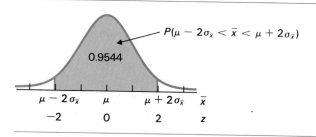

Thus, the probability is only 0.0456 ($= 1 - 0.9544$) that $\overline{x}$ will be more than two standard deviations (i.e., $2\sigma_{\overline{x}}$) away from μ. It is quite unlikely that the sample mean, $\overline{x}$, will be more than two standard deviations away from the population mean, μ. We can employ this last fact as a basis for deciding whether or not to reject the null hypothesis, $\mu = 454$ grams.

Specifically, if the mean weight, $\overline{x}$, of the 50 bags of pretzels sampled is within two standard deviations of 454 grams, then we can reasonably attribute the difference between $\overline{x}$ and 454 grams to sampling error and, thereby, accept the null hypothesis. However, if $\overline{x}$ is more than two standard deviations away from 454 grams, then either (1) the null hypothesis, $\mu = 454$ grams, is true and an extremely unlikely event occurred or (2) the null hypothesis is false and the alternative hypothesis, $\mu \neq 454$ grams, is true. Surely we would select (2) as the more reasonable conclusion.

In summary then, we have obtained the following precise criterion for deciding whether or not to reject the null hypothesis:

> If the mean weight, $\bar{x}$, of the 50 bags of pretzels sampled is more than two standard deviations away from 454 grams, then reject the null hypothesis, $\mu = 454$ grams, and conclude that the alternative hypothesis, $\mu \neq 454$ grams, is true. Otherwise, do not reject the null hypothesis.

We can picture this criterion graphically as shown in Figure 9.2.

FIGURE 9.2

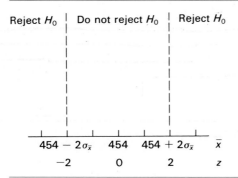

To see the implications of our decision criterion if in fact the null hypothesis, $\mu = 454$ grams, is true, we superimpose on Figure 9.2 the normal curve for $\bar{x}$ under that condition. That is, we superimpose the normal curve with parameters $\mu_{\bar{x}} = \mu = 454$ and $\sigma_{\bar{x}} = \sigma/\sqrt{n}$. This is done in Figure 9.3.

FIGURE 9.3

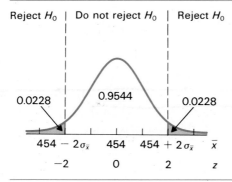

Figure 9.3 shows that, using our decision criterion, the probability is only 0.0456 ($= 1 - 0.9544 = 0.0228 + 0.0228$) of rejecting the null hypothesis, if it is in fact true. That probability is called the *significance level* of the hypothesis test.

d) Finally, we are to apply the criterion in part (c) t[...]
the conclusion. We will assume as known that $\sigma =$
the criterion in part (c), we need to determine how
the sample mean, $\bar{x}$, is away from 454 grams. As we k[...]
by computing the z-score for $\bar{x}$; that is, by finding the [...]
random variable

$$ z = \frac{\bar{x} - 454}{\sigma_{\bar{x}}} = \frac{\bar{x} - 454}{\sigma/\sqrt{n}} $$

We have $\sigma = 7.9$ grams, $n = 50$, and, from part (b), we know that the mean
weight of the 50 bags of pretzels sampled is $\bar{x} = 451.2$ grams. Thus,

$$ z = \frac{\bar{x} - 454}{\sigma/\sqrt{n}} = \frac{451.2 - 454}{7.9/\sqrt{50}} = -2.51 $$

Consequently, the sample mean, $\bar{x}$, is 2.51 standard deviations below the null-
hypothesis mean of 454 grams. See Figure 9.4.

FIGURE 9.4

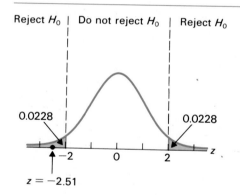

Reject H_0 | Do not reject H_0 | Reject H_0

0.0228

0.0228

z

−2   0   2

$z = -2.51$

Since the mean weight, $\bar{x}$, of the 50 bags of pretzels sampled is more than
two standard deviations away from 454 grams, we reject the null hypothesis,
$\mu = 454$ grams, and conclude that the alternative hypothesis, $\mu \neq 454$ grams, is
true. In other words, the data provide sufficient evidence to conclude that the
packaging machine is not working properly. ∎

Example 9.4 contains all the elements of a hypothesis test. But don't worry
too much about the details at this point. We will cover them systematically in
future sections. What you should understand at this point is how to choose the null
and alternative hypotheses for a hypothesis test and the logic behind performing a
hypothesis test.

...ses 9.1

9.1   Explain the meaning of the term *hypothesis* as used in inferential statistics.

___ **9.2**   What role does the decision criterion play in a hypothesis test?

In each of Exercises 9.3–9.10, a hypothesis test will be proposed. For each hypothesis test,
a) determine the null hypothesis.
b) determine the alternative hypothesis.
c) classify the hypothesis test as a two-tailed test, a left-tailed test, or a right-tailed test.

___ **9.3**   According to the Census Bureau publication *Construction Reports,* the mean expenditure per residential property owner for maintenance and repairs was $280 in 1983. Suppose that we want to perform a hypothesis test to decide whether last year's mean amount spent for maintenance and repairs per residential property owner has increased over the 1983 mean of $280.

___ **9.4**   *The World Almanac, 1985,* reports that the mean travel time to work in 1980 for all South Dakota residents was 13 minutes. A transportation official wants to use a sample of this year's travel times for South Dakota residents to decide whether the mean travel time to work for all South Dakota residents has changed from the 1980 mean of 13 minutes.

___ **9.5**   The Food and Nutrition Board of the National Academy of Sciences states that the recommended daily allowance (RDA) of iron for adult females under the age of 51 is 18 mg. A hypothesis test is to be performed to decide whether adult females under the age of 51 are, on the average, getting less than the RDA of 18 mg of iron.

___ **9.6**   As reported by the U.S. Office of Juvenile Justice and Delinquency Prevention in *Children in Custody,* the mean age of all juveniles held in public custody in 1982 was 15.4 years. Suppose that the ages of a random sample of juveniles currently being held in public custody are to be used to decide whether this year's mean age of all juveniles being held in public custody has decreased since 1982.

___ **9.7**   A dog-food manufacturer sells "50-lb" bags of dog food. Seventy-five bags are randomly selected and carefully weighed. Using the weights obtained, a hypothesis test is to be performed in order to decide whether the mean weight of all bags of this dog food differs from the advertised weight of 50 lb.

___ **9.8**   The Health Insurance Association of America reports in *Survey of Hospital Semi-Private Room Charges* that the mean daily charge for a semi-private room in American hospitals was $253 in 1988. In that same year, the semi-private room rates were obtained for a random sample of 30 Massachusetts hospitals. A hypothesis test was then performed to decide whether the mean semi-private room rate in Massachusetts hospitals exceeded the national mean.

___ **9.9**   The manufacturer of a new model car, called the Orion, claims that a typical car gets 26 mpg (miles per gallon). An independent consumer group is somewhat skeptical of this claim and thinks that the mean gas mileage of all Orions may very well be less than 26 mpg. To try to justify its contention, the consumer group plans to conduct mileage tests on 30 randomly selected Orions and use the results to perform a hypothesis test.

___ **9.10**   A Louisiana cotton farmer has used a certain brand of fertilizer for the past five years. His experience with the fertilizer indicates that the mean yield is about 623 lb/acre. Recently, a new brand of fertilizer came out on the market that will supposedly increase cotton yield. The farmer intends to try the new fertilizer on 80 one-acre plots in order to decide for himself whether it increases the mean yield of cotton on his land.

For Exercises 9.11 and 9.12, use the same method as that in Example 9.4 on pages 420–423 to perform the indicated hypothesis tests.

=== **9.11**   According to *Food Consumption, Prices, and Expenditures,* published by the U.S. Department of Agriculture, the mean consumption of beef per person in 1983 was 106.5 lb. Suppose that last year's beef consumptions, in pounds, for 40 randomly selected people are as shown below.

119	91	81	106	119	129	142	119
138	115	102	79	48	148	107	122
96	114	102	126	110	130	118	109
113	102	99	83	101	100	132	96
116	134	120	109	135	93	118	85

Do the data provide sufficient evidence to conclude that last year's mean beef consumption per person has changed from the 1983 mean of 106.5 lb? Use the following steps to answer the question:

a) State the null and alternative hypotheses for the hypothesis test.

b) Discuss the basic idea for carrying out the hypothesis test.

c) Obtain a precise criterion for deciding whether or not to reject the null hypothesis in favor of the alternative hypothesis.

d) Apply the criterion in part (c) to the sample data and state the conclusion. (Assume as known that the population standard deviation, σ, of last year's beef consumptions is 16.7 lb.)

— 9.12   The U.S. Energy Information Administration compiles data on energy consumption and publishes its findings in *Residential Energy Consumption Survey: Consumption and Expenditures*. In 1982, the mean energy consumed per American household was 114 million BTU. For that same year, 50 randomly selected households in the South had the following energy consumptions:

130	55	45	64	155	66	60	80	102	62
58	101	75	111	151	139	81	55	66	90
97	77	51	67	125	50	136	55	83	91
54	86	100	78	93	113	111	104	96	113
96	87	129	109	69	94	99	97	83	97

Do the data provide sufficient evidence to conclude that in 1982 the mean energy consumed by southern households differed from that of all American households? Use the following steps to answer the question:

a) State the null and alternative hypotheses for the hypothesis test.

b) Discuss the basic idea for carrying out the hypothesis test.

c) Obtain a precise criterion for deciding whether or

not to reject the null hypothesis in favor of the alternative hypothesis.

d) Apply the criterion in part (c) to the sample data and state the conclusion. (Assume that, for 1982, the standard deviation of energy consumptions of all southern households was $\sigma = 25$ million BTU.)

— 9.13   Refer to the pretzel-packaging illustration of Example 9.4. Suppose that in the second paragraph of the solution to part (c) (page 421), we use Property 1 of Key Fact 6.7 on page 321 instead of Property 2.

a) Determine the resulting decision criterion and portray that criterion graphically using a graph similar to the one in Figure 9.2.

b) Construct a graph similar to the one in Figure 9.3 that shows the implications of the decision criterion in part (a) if, in fact, the null hypothesis is true.

c) Obtain the significance level of the hypothesis test.

d) Apply the criterion in part (a) to the sample data in Table 9.1 on page 420 and state the conclusion. (*Note:* Recall that the sample mean of the data is $\bar{x} = 451.2$ grams and that we are assuming as known that $\sigma = 7.9$ grams.)

— 9.14   Refer to the pretzel-packaging illustration of Example 9.4. Suppose that in the second paragraph of the solution to part (c) (page 421), we use Property 3 of Key Fact 6.7 on page 321 instead of Property 2.

a) Determine the resulting decision criterion and portray that criterion graphically using a graph similar to the one in Figure 9.2.

b) Construct a graph similar to the one in Figure 9.3 that shows the implications of the decision criterion in part (a) if, in fact, the null hypothesis is true.

c) Obtain the significance level of the hypothesis test.

d) Apply the criterion in part (a) to the sample data in Table 9.1 on page 420 and state the conclusion. (*Note:* Recall that the sample mean of the data is $\bar{x} = 451.2$ grams and that we are assuming as known that $\sigma = 7.9$ grams.)

9.2   Terms, errors, and hypotheses

To fully understand the nature of hypothesis testing, we need to learn some additional terms and concepts. In this section we will define several more terms that are used in hypothesis testing, discuss the two types of errors that can be made in a hypothesis test, and interpret the possible conclusions for a hypothesis test.

SOME ADDITIONAL TERMINOLOGY

In order to introduce some additional terminology that is used in hypothesis testing, we will refer to the pretzel-packaging hypothesis test of Example 9.4 on page 420. Recall that the null and alternative hypotheses for that hypothesis test are

$$H_0: \mu = 454 \text{ grams (the machine is working properly)}$$
$$H_a: \mu \neq 454 \text{ grams (the machine is not working properly)}$$

where μ is the mean net weight of all bags of pretzels being packaged.

As a basis for deciding whether to reject the null hypothesis, we employed in part (d) of Example 9.4 the random variable

$$z = \frac{\bar{x} - 454}{\sigma/\sqrt{n}}$$

That random variable is called the **test statistic** for the hypothesis test.

Figure 9.4 on page 423 graphically portrays the criterion used to decide whether or not the null hypothesis should be rejected. For convenience, we repeat that figure here as Figure 9.5.

FIGURE 9.5

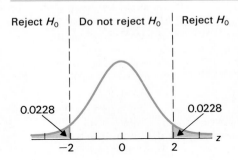

The set of values for the test statistic that lead us to reject the null hypothesis is called the **rejection region.** In this case, the rejection region consists of all z-values that lie either to the left of -2 or to the right of 2, that part of the horizontal axis under the shaded area in Figure 9.5.

On the other hand, the set of values for the test statistic that lead us not to reject the null hypothesis is called the **nonrejection region** or **acceptance region.** In this case, the nonrejection region consists of all z-values that lie between -2 and 2, that part of the horizontal axis under the unshaded area in Figure 9.5.

Finally, the values of the test statistic that separate the rejection and non-rejection regions are called the **critical values.** In this case, the critical values are $z = \pm 2$, as we see from Figure 9.5. We summarize the preceding discussion in Figure 9.6.

FIGURE 9.6
Rejection region,
nonrejection region,
and critical values
for pretzel-packaging
illustration

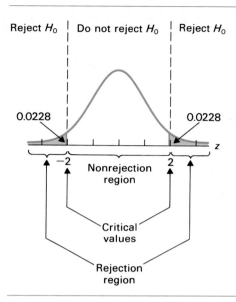

The terminology introduced above with reference to Example 9.4 is presented formally in Definition 9.2. We should emphasize that the terminology applies to any hypothesis test, not just to hypothesis tests for a population mean.

DEFINITION 9.2 Test statistic, rejection region, nonrejection region, critical values

Test statistic: The statistic used as a basis for deciding whether the null hypothesis should be rejected.
Rejection region: The set of values for the test statistic that lead to rejection of the null hypothesis.
Nonrejection region: The set of values for the test statistic that lead to nonrejection of the null hypothesis.
Critical values: The values of the test statistic that separate the rejection and nonrejection regions.

For a two-tailed test, as in the pretzel-packaging illustration, the null hypothesis will be rejected if the test statistic is either too small or too large. Consequently, the rejection region for such a test will consist of two parts, one on the left and one on the right, as illustrated in Figure 9.6.

For a left-tailed test, as in Example 9.3 on page 419, the null hypothesis will only be rejected if the test statistic is too small. Thus, the rejection region for such a test will consist of only one part and that part will be on the left.

For a right-tailed test, as in Example 9.2 on page 418, the null hypothesis will only be rejected if the test statistic is too large. Hence, the rejection region for such a test will consist of only one part and that part will be on the right.

Table 9.2 summarizes the discussion in the previous three paragraphs. A graphical summary is provided in Figure 9.7. By examining Figure 9.7, we can understand why the terminology "tailed" is used. Indeed, the rejection region is in both tails for a two-tailed test, is in the left tail for a left-tailed test, and is in the right tail for a right-tailed test.

TABLE 9.2

	Two-tailed test	Left-tailed test	Right-tailed test
Sign in H_a	$\neq$	$<$	$>$
Rejection region	Both sides	Left side	Right side

FIGURE 9.7

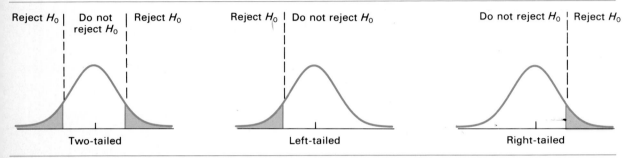

TYPE I AND TYPE II ERRORS

Whenever statistical inference methods are employed, it is always possible that the decision reached will be incorrect. This is because partial information, obtained from the sample, is used to draw conclusions about the entire population.

In hypothesis testing, there are four outcomes possible, two of which lead to incorrect decisions. The four possible outcomes are depicted in Table 9.3.

TABLE 9.3
The four possible outcomes for a hypothesis test

		H_0 is:	
		True	False
Decision:	Do not reject H_0	Correct decision	Type II error
	Reject H_0	Type I error	Correct decision

As we see from Table 9.3, an incorrect decision occurs if either a true null hypothesis is rejected or a false null hypothesis is not rejected. The first incorrect decision is called a **Type I error** and the second a **Type II error.**

DEFINITION 9.3   Type I and Type II errors

Type I error: Rejecting the null hypothesis when it is in fact true.
Type II error: Not rejecting the null hypothesis when it is in fact false.

We should emphasize that in a hypothesis test, it is presumed that the null and alternative hypotheses are exhaustive. In other words, if the null hypothesis is false, then the alternative hypothesis is true; and, vice-versa, if the alternative hypothesis is false, then the null hypothesis is true.

EXAMPLE 9.5   *Illustrates Type I and Type II errors*

Consider once more the pretzel-packaging illustration. The null and alternative hypotheses are

$$H_0: \mu = 454 \text{ grams (the machine is working properly)}$$
$$H_a: \mu \neq 454 \text{ grams (the machine is not working properly)}$$

where μ is the mean net weight of all bags of pretzels being packaged. Explain what each of the following would mean:
a) A Type I error.
b) A Type II error.
c) A correct decision.

Now, recall from Example 9.4 on pages 420–423 that the results of sampling 50 bags of pretzels lead to rejection of the null hypothesis, $\mu = 454$ grams; that is, to the conclusion that $\mu \neq 454$ grams. Classify that conclusion by error type or as a correct decision, if in fact the mean net weight, μ, being packaged
d) is 454 grams.
e) is not 454 grams.

SOLUTION   a) A Type I error occurs when a true null hypothesis is rejected. In this case, a Type I error would occur if in fact $\mu = 454$ grams but the results of the sampling lead to the conclusion that $\mu \neq 454$ grams.
b) A Type II error occurs when a false null hypothesis is not rejected. In this case, a Type II error would occur if in fact $\mu \neq 454$ grams but the results of the sampling fail to lead to that conclusion.
c) A correct decision can occur in either of two ways. First, a correct decision occurs when a true null hypothesis is not rejected. In the present situation, this would happen if in fact $\mu = 454$ grams and the results of the sampling do not lead to the rejection of that fact. A correct decision also occurs when a false null hypothesis is rejected. In the present situation, this would happen if in fact $\mu \neq 454$ grams and the results of the sampling lead to that conclusion.
d) If in fact $\mu = 454$ grams, then the null hypothesis is true. So, by rejecting the null hypothesis, $\mu = 454$ grams, a Type I error has been made—a true null hypothesis has been rejected.
e) If in fact $\mu \neq 454$ grams, then the null hypothesis is false. Consequently, by rejecting the null hypothesis, $\mu = 454$ grams, a correct decision has been made—a false null hypothesis has been rejected.   ∎

PROBABILITIES OF TYPE I AND TYPE II ERRORS

The probability of making a Type I error is the probability of rejecting a true null hypothesis. In other words, it is the probability that the test statistic will be in the rejection region if, in fact, the null hypothesis is true.

To illustrate, let us find the probability of making a Type I error in the pretzel-packaging hypothesis test of Example 9.4. Figure 9.3 on page 422 (or Figure 9.5 on page 426) provides a graphical display of the criterion used to decide whether or not to reject the null hypothesis, $\mu = 454$ grams. It also shows the implications of that criterion when in fact the null hypothesis is true.

As we see from Figure 9.3, the probability is 0.0456 ($= 0.0228 + 0.0228$) that the test statistic, z, will be in the rejection region if the null hypothesis is in fact true. Thus, the probability of a Type I error for this hypothesis test is 0.0456. There is only a 4.56% chance of concluding that $\mu \neq 454$ grams when in fact $\mu = 454$ grams.

The probability of making a Type I error is called the **significance level** of the hypothesis test and is denoted by the Greek letter $\boldsymbol{\alpha}$. Hence, the significance level of the pretzel-packaging hypothesis test is $\alpha = 0.0456$. We summarize the discussion on the probability of a Type I error in Definition 9.4.

DEFINITION 9.4   Significance level

The *significance level*, α, of a hypothesis test is defined to be the probability of making a Type I error; that is, the probability of rejecting a true null hypothesis.

The probability of making a Type II error is the probability of not rejecting a false null hypothesis. In other words, it is the probability that the test statistic will be in the nonrejection region if, in fact, the null hypothesis is false. We use the Greek letter $\boldsymbol{\beta}$ (beta) to denote the probability of a Type II error. The probability, β, of a Type II error depends on the true value of μ. A thorough examination of Type II error probabilities will be considered in Chapter 16.

Ideally, we would like both Type I and Type II errors to have small probabilities. For then the chances of making an incorrect decision would be small, regardless of whether the null hypothesis is true. As we will see, it is always possible to design a hypothesis test with any desired significance level (i.e., Type I error probability). Thus, if it is important not to reject a true null hypothesis, then we should specify a small value for the significance level, α. However, in making our choice for α, we should keep the following fact in mind:

KEY FACT 9.1

For a fixed sample size, the smaller the Type I error probability, α, of rejecting a true null hypothesis, the larger the Type II error probability, β, of not rejecting a false null hypothesis; and vice versa.

Consequently, it is always necessary to assess the risks involved in committing both types of errors and to use that assessment as a method for balancing the Type I and Type II error probabilities.

POSSIBLE CONCLUSIONS FOR A HYPOTHESIS TEST

It is important to be aware of the following points regarding the possible conclusions for a hypothesis test. First recall that the significance level, α, is the probability of making a Type I error, that is, of rejecting a true null hypothesis. Thus, if the hypothesis test is performed at a small significance level (e.g., $\alpha = 0.05$), then it is unlikely that the null hypothesis will be rejected when it is in fact true.

Since, in this text, we will generally specify a small significance level, we can make the following statement concerning a hypothesis test: If we do reject the null hypothesis in a hypothesis test, then we can be reasonably confident that the alternative hypothesis is true.

On the other hand, we will usually not know the probability, β, of making a Type II error, that is, of not rejecting a false null hypothesis. Therefore, if we do not reject the null hypothesis in a hypothesis test, then we simply reserve judgement about which hypothesis is true. In other words, if we do not reject the null hypothesis, we conclude only that the data did not provide sufficient evidence to support the alternative hypothesis. We do not conclude that the data provided sufficient evidence to support the null hypothesis. In summary:

KEY FACT 9.2 Possible conclusions for a hypothesis test

The possible conclusions for a hypothesis test are:

1. If the null hypothesis is rejected, conclude that the alternative hypothesis is probably true.

2. If the null hypothesis is not rejected, conclude that the data do not provide sufficient evidence to support the alternative hypothesis.

Exercises 9.2

In each of Exercises 9.15–9.20, we have presented a graph that portrays the decision criterion for a hypothesis test concerning a population mean, μ. The null hypothesis for each test is H_0: $\mu = \mu_0$ and the test statistic is

$$z = \frac{\bar{x} - \mu_0}{\sigma/\sqrt{n}}$$

Also, the curve in each graph shows the implications of the decision criterion if, in fact, the null hypothesis is true. For each exercise, determine the
a) rejection region.
b) nonrejection region.
c) critical value(s).
d) significance level.
e) Construct a graph similar to Figure 9.6 on page 427 that depicts your results from parts (a)–(d).

f) Identify the hypothesis test as a two-tailed test, a left-tailed test, or a right-tailed test.

___ 9.15 The graphical display of the decision criterion is:

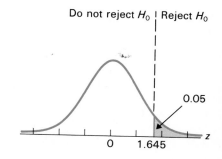

___ **9.16** The graphical display of the decision criterion is:

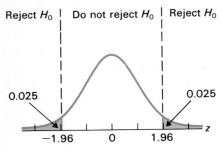

___ **9.17** The graphical display of the decision criterion is:

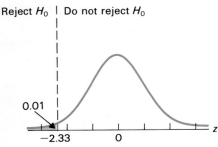

___ **9.18** The graphical display of the decision criterion is:

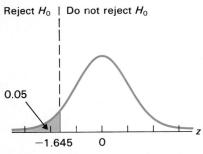

___ **9.19** The graphical display of the decision criterion is:

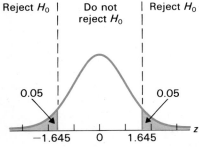

___ **9.20** The graphical display of the decision criterion is:

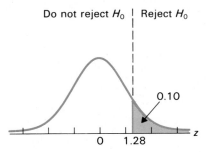

___ **9.21** According to the Census Bureau publication *Construction Reports,* the mean expenditure per residential property owner for maintenance and repairs was $280 in 1983. Suppose we want to perform a hypothesis test to decide whether last year's mean amount spent for maintenance and repairs per residential property owner has increased over the 1983 mean of $280. Then the null and alternative hypotheses are

$$H_0: \mu = \$280$$
$$H_a: \mu > \$280$$

where μ is last year's mean amount spent for maintenance and repairs per residential property owner. Explain what each of the following would mean:
a) A Type I error.
b) A Type II error.
c) A correct decision.

Now, suppose the results of carrying out the hypothesis test lead to nonrejection of the null hypothesis. Classify that conclusion by error type or as a correct decision if, in fact, last year's mean amount spent for maintenance and repairs per residential property owner
d) is equal to the 1983 mean of $280.
e) is greater than the 1983 mean of $280.

___ **9.22** *The World Almanac, 1985,* reports that the mean travel time to work in 1980 for all South Dakota residents was 13 minutes. A transportation official wants to use a sample of this year's travel times for South Dakota residents to decide whether the mean travel time to work for all South Dakota residents has changed from the 1980 mean of 13 minutes. The null and alternative hypotheses for the hypothesis test are

$$H_0: \mu = 13 \text{ minutes}$$
$$H_a: \mu \neq 13 \text{ minutes}$$

where μ is this year's mean travel time to work for all South Dakota residents. Explain what each of the following would mean:

a) A Type I error.
b) A Type II error.
c) A correct decision.

Now, suppose that the results of the sampling lead to nonrejection of the null hypothesis. Classify that conclusion by error type or as a correct decision if, in fact, this year's mean travel time to work, μ, for all South Dakota residents

d) has not changed from the 1980 mean of 13 minutes.
e) has changed from the 1980 mean of 13 minutes.

— **9.23**   The Food and Nutrition Board of the National Academy of Sciences states that the recommended daily allowance (RDA) of iron for adult females under the age of 51 is 18 mg. A hypothesis test is to be performed to decide whether adult females under the age of 51 are, on the average, getting less than the RDA of 18 mg of iron. The null and alternative hypotheses for the hypothesis test are

$$H_0: \mu = 18 \text{ mg}$$
$$H_a: \mu < 18 \text{ mg}$$

where μ is the mean daily iron intake of all adult females under the age of 51. Explain what each of the following would mean:

a) A Type I error.
b) A Type II error.
c) A correct decision.

Now, suppose that the results of carrying out the hypothesis test lead to rejection of the null hypothesis, $\mu = 18$ mg; that is, to the conclusion that $\mu < 18$ mg. Classify that conclusion by error type or as a correct decision if, in fact, the mean iron intake, μ, of all adult females under the age of 51

d) is not less than the RDA of 18 mg per day.
e) is less than the RDA of 18 mg per day.

— **9.24**   As reported by the U.S. Office of Juvenile Justice and Delinquency Prevention in *Children in Custody*, the mean age of all juveniles held in public custody in 1982 was 15.4 years. Suppose that the ages of a random sample of juveniles currently being held in public custody are to be used to decide whether this year's mean age of all juveniles being held in public custody has decreased since 1982. The null and alternative hypotheses for the hypothesis test are

$$H_0: \mu = 15.4 \text{ years}$$
$$H_a: \mu < 15.4 \text{ years}$$

where μ is this year's mean age of all juveniles being held in public custody. Explain what each of the following would mean:

a) A Type I error.
b) A Type II error.
c) A correct decision.

Now, suppose that the results of carrying out the hypothesis test lead to rejection of the null hypothesis, $\mu = 15.4$ years; that is, to the conclusion that $\mu < 15.4$ years. Classify that conclusion by error type or as a correct decision if, in fact, this year's mean age, μ, of all juveniles being held in public custody

d) is 15.4 years.
e) is less than 15.4 years.

— **9.25**   A dog-food manufacturer sells "50-lb" bags of dog food. Seventy-five bags of this brand of dog food are randomly selected and carefully weighed. Using the weights obtained, a hypothesis test is to be performed in order to decide whether the mean weight of all bags of this dog food differs from the advertised weight of 50 lb. The null and alternative hypotheses for the hypothesis test are

$$H_0: \mu = 50 \text{ lb}$$
$$H_a: \mu \neq 50 \text{ lb}$$

where μ is the actual mean weight of all "50-lb" bags of this dog food. Explain what each of the following would mean:

a) A Type I error.
b) A Type II error.
c) A correct decision.

Now, suppose that the weights of the 75 bags sampled lead to nonrejection of the null hypothesis. Classify that conclusion by error type or as a correct decision if, in fact, the mean weight, μ, of all "50-lb" bags of this dog food

d) equals the advertised weight of 50 lb.
e) does not equal the advertised weight of 50 lb.

— **9.26**   The Health Insurance Association of America reports in *Survey of Hospital Semi-Private Room Charges* that the mean daily charge for a semi-private room in American hospitals was $253 in 1988. In that same year, the daily semi-private room charges were obtained for a sample of 30 Massachusetts hospitals. A hypothesis test was then performed to decide whether the mean daily semi-private room charge in Massachusetts hospitals exceeded the national mean. The null and alternative hypotheses for the hypothesis test performed are

$$H_0: \mu = \$253$$
$$H_a: \mu > \$253$$

where μ is the 1988 mean daily semi-private room charge in Massachusetts hospitals. Explain what each of the following would mean:
a) A Type I error.
b) A Type II error.
c) A correct decision.

Now, suppose that the results of the sampling led to rejection of the null hypothesis, $\mu = \$253$; that is, to the conclusion that $\mu > \$253$. Classify that conclusion by error type or as a correct decision if, in fact, the 1988 mean daily semi-private room charge in Massachusetts hospitals
d) did not exceed the national mean.
e) did exceed the national mean.

___ 9.27   The manufacturer of a new model car, called the Orion, claims that a typical car gets 26 mpg (miles per gallon). An independent consumer group is somewhat skeptical of this claim and thinks that the mean gas mileage of all Orions may very well be less than 26 mpg. To try to justify its contention, the consumer group plans to conduct mileage tests on 30 randomly selected Orions and use the results to perform a hypothesis test. The null and alternative hypotheses for the hypothesis test are

$$H_0: \mu = 26 \text{ mpg}$$
$$H_a: \mu < 26 \text{ mpg}$$

where μ is the mean gas mileage of all Orions. Explain what each of the following would mean:
a) A Type I error.
b) A Type II error.
c) A correct decision.

Now, suppose that the mileages of the 30 Orions tested lead to rejection of the null hypothesis, $\mu = 26$ mpg; that is, to the conclusion that $\mu < 26$ mpg. Classify that conclusion by error type or as a correct decision if, in fact,
d) the consumer group's conjecture is correct.
e) the manufacturer's claim is true.

___ 9.28   A Louisiana cotton farmer has used a certain brand of fertilizer for the past five years. His experience with the fertilizer indicates that the mean yield is about 623 lb/acre. Recently, a new brand of fertilizer came out on the market that will supposedly increase cotton yield. The farmer intends to try the new fertilizer on 80 one-acre plots in order to decide for himself whether it increases the mean yield of cotton on his land. The null and alternative hypotheses for the hypothesis test are

$$H_0: \mu = 623 \text{ lb/acre}$$
$$H_a: \mu > 623 \text{ lb/acre}$$

where μ is the actual mean yield of cotton that the farmer will get on his land using the new fertilizer. Explain what each of the following would mean:
a) A Type I error.
b) A Type II error.
c) A correct decision.

Now, suppose that the yields on the 80 one-acre test plots lead to nonrejection of the null hypothesis. Classify that conclusion by error type or as a correct decision if, in fact, the new fertilizer
d) will increase the mean yield.
e) will not increase the mean yield.

___ 9.29   True or False: If it is important not to reject a true null hypothesis, then the hypothesis test should be performed at a small significance level.

___ 9.30   True or False: For a fixed sample size, decreasing the significance level of a hypothesis test results in an increase in the probability of making a Type II error.

=== 9.31   Suppose we choose the significance level of a hypothesis test to be $\alpha = 0$.
a) What is the probability of a Type I error, that is, of rejecting a true null hypothesis?
b) What is the probability of a Type II error, that is, of not rejecting a false null hypothesis?

=== 9.32   Find an exercise in this section for which it is important to have
a) a small α probability.
b) a small β probability.
c) both α and β probabilities small.

=== 9.33   Suppose that you are performing a statistical test to decide whether a nuclear reactor should be approved for use. Further suppose that failing to reject the null hypothesis corresponds to approval. What property would you want the Type II error probability, β, to have?

=== 9.34   In the United States court system, a defendant is assumed innocent until proven guilty. Suppose we regard a court trial as a hypothesis test with null and alternative hypotheses given by

H_0: Defendant is innocent.

H_a: Defendant is guilty.

a) Explain the meaning of a Type I error.

b) Explain the meaning of a Type II error.
c) If you were the defendant, what kind of value would you want for α? Why?
d) If you were the prosecuting attorney, what kind of value would you want for β? Why?
e) What are the consequences to our court system if we make $\alpha = 0$? $\beta = 0$?

9.3 Large-sample hypothesis tests for a population mean

In this section, we will learn a procedure for performing a large-sample hypothesis test for a population mean, μ, at any prescribed significance level. The only preliminary topic that we still need to discuss is how to obtain the critical value(s) for such a hypothesis test when the significance level, α, is specified in advance. We now commence with that discussion.

OBTAINING THE CRITICAL VALUE(S) FOR A SPECIFIED SIGNIFICANCE LEVEL

Recall that the significance level, α, of a hypothesis test is the probability of making a Type I error; that is, the probability of rejecting a true null hypothesis. Equivalently, α is the probability that the test statistic will be in the rejection region if, in fact, the null hypothesis is true. Thus, we have the following key fact which holds for any hypothesis test:

KEY FACT 9.3 Choosing the critical value(s) for a specified significance level

Suppose a hypothesis test is to be performed at a specified significance level, α. Then the critical value(s) must be chosen so that, if the null hypothesis is true, the probability is equal to α that the test statistic will be in the rejection region.

Let us now apply Key Fact 9.3 to large-sample hypothesis tests for a population mean. As we have seen, the null hypothesis for a hypothesis test concerning one population mean, μ, is of the form $H_0: \mu = \mu_0$, where μ_0 is some number. If the sample size is large, then we use the random variable

$$(1) \qquad z = \frac{\bar{x} - \mu_0}{\sigma/\sqrt{n}}$$

as the test statistic for the hypothesis test. That test statistic tells us how many standard deviations the observed sample mean, $\bar{x}$, is away from μ_0.

From Key Fact 7.6 on page 363 (the sampling distribution of the mean for general populations), we know that for large samples, the random variable $\bar{x}$ is approximately normally distributed and has mean $\mu_{\bar{x}} = \mu$ and standard deviation $\sigma_{\bar{x}} = \sigma/\sqrt{n}$. Therefore, if the null hypothesis, $\mu = \mu_0$, is true, then the test statistic in Equation (1) has approximately the standard normal distribution. In other words, if the null hypothesis is true, then probabilities for that test statistic are equal to areas under the standard normal curve.

Thus, in view of Key Fact 9.3, we see that for a specified significance level, α, we need to choose the critical value(s) so that the area under the standard normal curve that lies above the rejection region is equal to α. Here is an example.

EXAMPLE 9.6   *Illustrates obtaining the critical values for a specified significance level*

Suppose we want to perform a large-sample hypothesis test for a population mean, μ, with null hypothesis H_0: $\mu = \mu_0$. If the hypothesis test is to be performed at the 5% significance level (i.e., $\alpha = 0.05$), find the critical value(s) for
a) a two-tailed test; that is, for a test with alternative hypothesis H_a: $\mu \neq \mu_0$.
b) a left-tailed test; that is, for a test with alternative hypothesis H_a: $\mu < \mu_0$.
c) a right-tailed test; that is, for a test with alternative hypothesis H_a: $\mu > \mu_0$.

SOLUTION   Since $\alpha = 0.05$, we need to choose the critical value(s) so that the area under the standard normal curve that lies above the rejection region is equal to 0.05.
a) For a two-tailed test, the rejection region will be on both the left and right. So, for a test with $\alpha = 0.05$, the critical values will be the two z-values that divide the area under the standard normal curve into a middle 0.95 area and two outside areas of 0.025. In other words, the critical values will be $\pm z_{0.025}$. Consulting Table II, we find that $\pm z_{0.025} = \pm 1.96$. See Figure 9.8(a).
b) For a left-tailed test, the rejection region will be on the left. So, for a test with $\alpha = 0.05$, the critical value will be the z-value with area 0.05 to its left under the standard normal curve. Since the standard normal curve is symmetric about 0, that z-value is the negative of the z-value with area 0.05 to its right. In other words, the critical value will be $-z_{0.05}$. Consulting Table II, we find that $-z_{0.05} = -1.645$. See Figure 9.8(b).
c) For a right-tailed test, the rejection region will be on the right. So, for a test with $\alpha = 0.05$, the critical value will be the z-value with area 0.05 to its right under the standard normal curve. In other words, the critical value will be $z_{0.05}$. Consulting Table II, we find that $z_{0.05} = 1.645$. See Figure 9.8(c).   ∎

FIGURE 9.8   Critical value(s) for a hypothesis test at the 5% significance level, if the hypothesis test is (a) two-tailed, (b) left-tailed, and (c) right-tailed

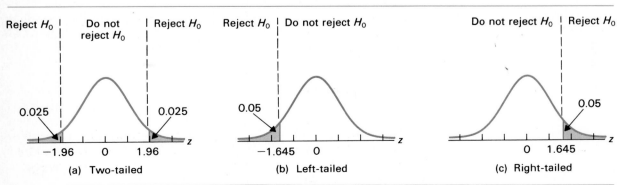

(a) Two-tailed   (b) Left-tailed   (c) Right-tailed

By arguing as we did in Example 9.6, we can obtain any specified significance level, α. For a two-tailed test, t $\pm z_{\alpha/2}$; for a left-tailed test, the critical value will be $-$ test, the critical value will be z_α. A graphical portrayal o in Figure 9.9.

FIGURE 9.9 Critical value(s) for a hypothesis test at the significance level α, if the hypothesis test is (a) two-tailed, (b) left-tailed, and (c) right-tailed

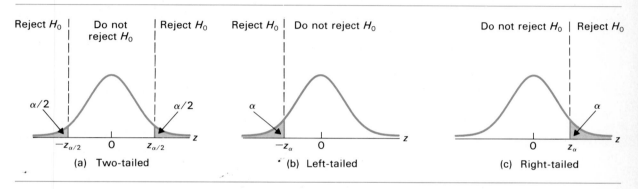

The most commonly used significance levels for hypothesis tests are 0.10, 0.05, and 0.01. If we consider both one-tailed and two-tailed hypothesis tests, then these three significance levels give rise to five "tail areas." Using Table II, we obtained the value of z_α corresponding to each of those five tail areas. Table 9.4 provides a summary of our work.

TABLE 9.4
Some important values of z_α

$z_{0.10}$	$z_{0.05}$	$z_{0.025}$	$z_{0.01}$	$z_{0.005}$
1.28	1.645	1.96	2.33	2.575

Alternatively, the five values of z_α presented in Table 9.4 can be found in the last row (the row labeled ∞) of the t-table, Table III. In this latter table, the values of z_α are displayed to three decimal places. (Can you explain the slight discrepancy between the values given for $z_{0.005}$ in the two tables?)

LARGE-SAMPLE HYPOTHESIS TESTS FOR A POPULATION MEAN

We can now write down a quick and simple procedure for performing a hypothesis test concerning a population mean when the sample size is large. The procedure is obtained by carefully studying what we have done in the previous two sections and what we have just done in this section.

PROCEDURE 9.1

To perform a hypothesis test for a population mean with null hypothesis $H_0: \mu = \mu_0$.

ASSUMPTION

Sample size is large ($n \geq 30$).

STEP 1 *State the null and alternative hypotheses.*
STEP 2 *Decide on the significance level, α.*
STEP 3 *The critical value(s)*
      a) *for a two-tailed test are $\pm z_{\alpha/2}$.*
      b) *for a left-tailed test is $-z_\alpha$.*
      c) *for a right-tailed test is z_α.*
      Use Table II to find the critical value(s).

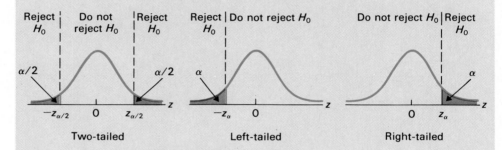

    Two-tailed   Left-tailed   Right-tailed

STEP 4 *Compute the value of the test statistic*

$$z = \frac{\overline{x} - \mu_0}{s/\sqrt{n}}$$

STEP 5 *If the value of the test statistic falls in the rejection region, then reject H_0; otherwise, do not reject H_0.*
STEP 6 *State the conclusion in words.*

Note: In Step 4 of Procedure 9.1 we have used the sample standard deviation, *s*, in the denominator of the test statistic. Theoretically, we should use the population standard deviation, σ. However, σ is rarely known and so we have used *s* in its place. This is acceptable because, for large samples, the sample standard deviation is likely to be a good approximation of the population standard deviation. In the rare case where σ is known, it should always be used in place of *s* in Step 4 of Procedure 9.1.

EXAMPLE 9.7   **Illustrates Procedure 9.1**

The R. R. Bowker Company of New York collects information on the retail prices of books. Data are published in *Publishers Weekly*. In 1986, the mean retail price of all hardcover history books was \$28.44. This year's retail prices for 40 randomly selected history books are presented, to the nearest dollar, in Table 9.5.

TABLE 9.5
This year's prices ($) for 40 history books

35	37	33	26	50	32	30	39
32	33	48	27	20	24	33	31
39	25	28	31	36	32	26	41
33	25	35	23	41	36	45	27
18	28	32	36	22	34	26	21

Do the data provide sufficient evidence to conclude that this year's mean retail price of all hardcover history books has increased over the 1986 mean of $28.44? Perform the appropriate hypothesis test at the 1% significance level.

SOLUTION   Since the sample size, $n = 40$, is large, we apply Procedure 9.1 to perform the hypothesis test.

STEP 1 *State the null and alternative hypotheses.*

Let μ denote this year's mean retail price of all hardcover history books. The null and alternative hypotheses were stated in Example 9.2 on page 418. They are:

H_0: $\mu = \$28.44$ (mean price has not increased)
H_a: $\mu > \$28.44$ (mean price has increased)

Note that the hypothesis test is right-tailed since there is a greater-than sign ($>$) in the alternative hypothesis.

STEP 2 *Decide on the significance level, α.*

We are to perform the test at the 1% significance level. Thus, $\alpha = 0.01$.

STEP 3 *The critical value for a right-tailed test is z_α.*

Since $\alpha = 0.01$, the critical value is $z_{0.01}$. From Table II (or Table 9.4), we see that $z_{0.01} = 2.33$. So, the critical value is $z_{0.01} = 2.33$. See Figure 9.10.

FIGURE 9.10

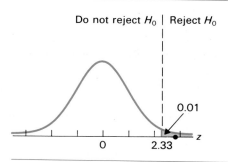

Do not reject H_0 | Reject H_0

0.01

0   2.33   z

STEP 4 *Compute the value of the test statistic*

$$z = \frac{\bar{x} - \mu_0}{s/\sqrt{n}}$$

We have $\mu_0 = \$28.44$ and $n = 40$. Applying the usual formulas for computing $\bar{x}$ and s, we find from Table 9.5 that $\bar{x} = \$31.75$ and $s = \$7.35$. Thus, the value of the test statistic is

$$z = \frac{\bar{x} - \mu_0}{s/\sqrt{n}} = \frac{31.75 - 28.44}{7.35/\sqrt{40}} = 2.85$$

This value of z is marked with a dot in Figure 9.10.

STEP 5　*If the value of the test statistic falls in the rejection region, reject H_0; otherwise, do not reject H_0.*

The value of the test statistic, found in Step 4, is $z = 2.85$. As we see from Figure 9.10, this falls in the rejection region and so we reject H_0.

STEP 6　*State the conclusion in words.*

The data provide sufficient evidence to conclude that this year's mean retail price of all hardcover history books has increased over the 1986 mean of $28.44. ■

EXAMPLE 9.8　**Illustrates Procedure 9.1**

Calcium is the most abundant and one of the most important minerals in the body. It works with phosphorus to build and maintain bones and teeth. According to the Food and Nutrition Board of the National Academy of Sciences, the recommended daily allowance (RDA) of calcium for adults is 800 mg (milligrams).

　　A nutritionist thinks that the average person with an income below the poverty level gets less than the RDA of 800 mg. To test her claim, she obtains the daily intakes of calcium for a random sample of 35 people with incomes below the poverty level. The results are displayed in Table 9.6. Data are in milligrams.

TABLE 9.6
Daily calcium intakes

879	1096	701	986	828	1077	703
555	422	997	473	702	508	530
513	720	944	673	574	707	864
1199	743	1325	655	1043	599	1008
705	180	287	542	893	1052	473

At the 5% significance level, do the data provide sufficient evidence to conclude that the mean calcium intake of all people with incomes below the poverty level is less than the RDA of 800 mg per day?

SOLUTION　Since the sample size, $n = 35$, is large, we apply Procedure 9.1 to perform the hypothesis test.

STEP 1　*State the null and alternative hypotheses.*

　　Let μ denote the mean daily intake of calcium of all people with incomes below the poverty level. The null and alternative hypotheses were obtained in Example 9.3 on page 419. They are:

H_0: $\mu = 800$ mg (mean calcium intake is not less than the RDA)
H_a: $\mu < 800$ mg (mean calcium intake is less than the RDA)

Note that the hypothesis test is left-tailed since there is a less-than sign ($<$) in the alternative hypothesis.

STEP 2 *Decide on the significance level, α.*

We are to perform the test at the 5% significance level. Thus, $\alpha = 0.05$.

STEP 3 *The critical value for a left-tailed test is $-z_\alpha$.*

Since $\alpha = 0.05$, the critical value is $-z_{0.05}$. From Table II (or Table 9.4 or Table III), we see that $z_{0.05} = 1.645$. Hence, the critical value is $-z_{0.05} = -1.645$. See Figure 9.11.

FIGURE 9.11

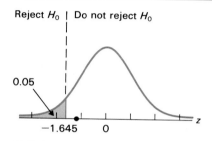

Reject H_0 | Do not reject H_0

0.05

−1.645   0

z

STEP 4 *Compute the value of the test statistic*

$$z = \frac{\bar{x} - \mu_0}{s/\sqrt{n}}$$

We have $\mu_0 = 800$ mg and $n = 35$. From the data in Table 9.6, we find that

$$\bar{x} = \frac{\Sigma x}{n} = \frac{26{,}156}{35} = 747.3 \text{ mg}$$

and

$$s = \sqrt{\frac{n(\Sigma x^2) - (\Sigma x)^2}{n(n-1)}} = \sqrt{\frac{35(21{,}884{,}700) - (26{,}156)^2}{35 \cdot 34}} = 262.2 \text{ mg}$$

Thus, the value of the test statistic is

$$z = \frac{\bar{x} - \mu_0}{s/\sqrt{n}} = \frac{747.3 - 800}{262.2/\sqrt{35}} = -1.19$$

This value of z is marked with a dot in Figure 9.11.

STEP 5 *If the value of the test statistic falls in the rejection region, reject H_0; otherwise, do not reject H_0.*

The value of the test statistic, found in Step 4, is $z = -1.19$. As we see from Figure 9.11, this does not fall in the rejection region and so we do not reject H_0.

STEP 6 *State the conclusion in words.*

The sample of 35 calcium intakes does not provide sufficient evidence to conclude that the mean calcium intake, μ, of all people with incomes below the poverty level is less than the RDA of 800 mg per day.   ■

EXAMPLE 9.9   *Illustrates Procedure 9.1*

The general partner of a limited partnership firm has told a potential investor that the mean monthly rent for three-bedroom apartments in the city is $587. To check out this claim, the investor randomly selects 30 three-bedroom apartments in the city and determines their monthly rents. Table 9.7 displays the monthly rents, in dollars, for the 30 three-bedroom apartments selected.

TABLE 9.7
Monthly rents ($) for
30 three-bedroom
apartments

348	567	475	688	710	529
566	628	496	668	617	700
619	567	563	662	504	669
708	620	682	594	475	554
669	547	623	696	603	638

Do the data suggest that the general partner's claim is in error? Perform the appropriate hypothesis test at the 0.05 level of significance.

SOLUTION   We apply Procedure 9.1.

STEP 1 *State the null and alternative hypotheses.*

Let μ denote the mean monthly rent of all three-bedroom apartments in the city. Then the null and alternative hypotheses are

$$H_0: \mu = \$587 \text{ (general partner's claim is correct)}$$
$$H_a: \mu \neq \$587 \text{ (general partner's claim is not correct)}$$

Note that the hypothesis test is two-tailed since there is a not-equal sign ($\neq$) in the alternative hypothesis.

STEP 2 *Decide on the significance level, α.*

We are to perform the hypothesis test at the 0.05 level of significance. Therefore, $\alpha = 0.05$.

STEP 3 *The critical values for a two-tailed test are $\pm z_{\alpha/2}$.*

Since $\alpha = 0.05$, we see from Table II that the critical values are $\pm z_{0.05/2} = \pm z_{0.025} = \pm 1.96$. See Figure 9.12.

FIGURE 9.12

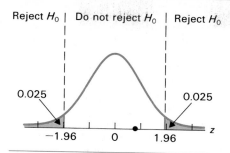

STEP 4 *Compute the value of the test statistic*

$$z = \frac{\overline{x} - \mu_0}{s/\sqrt{n}}$$

We have $\mu_0 = \$587$, $n = 30$, and from the data in Table 9.7 we find that $\overline{x} = \$599.50$ and $s = \$84.84$. Thus, the value of the test statistic is

$$z = \frac{\overline{x} - \mu_0}{s/\sqrt{n}} = \frac{599.50 - 587}{84.84/\sqrt{30}} = 0.81$$

This value of z is marked with a dot in Figure 9.12.

STEP 5 *If the value of the test statistic falls in the rejection region, reject H_0; otherwise, do not reject H_0.*

From Step 4, the value of the test statistic is $z = 0.81$. This does not fall in the rejection region, as we see from Figure 9.12. Hence we do not reject H_0.

STEP 6 *State the conclusion in words.*

The data do not provide sufficient evidence to conclude that the mean monthly rent, μ, of all three-bedroom apartments in the city differs from the general partner's claim of $587.

■

THE CLASSICAL APPROACH TO HYPOTHESIS TESTING

Procedure 9.1 on page 438 gives a step-by-step method that applies specifically to a large-sample hypothesis test for a population mean. The steps used in that procedure are typical of those employed in the so-called **classical approach to hypothesis testing.**

We will usually provide procedures that are specific to a particular kind of inference (e.g., large-sample hypothesis test for a population mean). Nonetheless, it is a good idea to be aware of the elements common to all hypothesis-testing procedures that are based on the classical approach. Here are those elements:

	PROCEDURE 9.2
	To perform a hypothesis test (classical approach):

STEP 1 *State the null and alternative hypotheses.*
STEP 2 *Decide on the significance level, α.*
STEP 3 *Determine the critical value(s).*
STEP 4 *Compute the value of the test statistic.*
STEP 5 *If the value of the test statistic falls in the rejection region, then reject H_0; otherwise, do not reject H_0.*
STEP 6 *State the conclusion in words.*

Exercises 9.3

In Exercises 9.35–9.40, suppose that a large-sample hypothesis test is to be performed for a population mean, μ, with null hypothesis H_0: $\mu = \mu_0$. Further suppose that the test statistic used will be

$$z = \frac{\bar{x} - \mu_0}{\sigma/\sqrt{n}}$$

For each exercise, obtain the required critical value(s) and draw a graph that illustrates your results.

__ **9.35**   A right-tailed test with $\alpha = 0.01$.

__ **9.36**   A left-tailed test with $\alpha = 0.10$.

__ **9.37**   A two-tailed test with $\alpha = 0.10$.

__ **9.38**   A right-tailed test with $\alpha = 0.05$.

__ **9.39**   A left-tailed test with $\alpha = 0.05$.

__ **9.40**   A two-tailed test with $\alpha = 0.01$.

In each of Exercises 9.41–9.48, apply Procedure 9.1 on page 438 to perform the required hypothesis test.

__ **9.41**   According to the Census Bureau publication *Construction Reports*, the mean expenditure per residential property owner for maintenance and repairs was $280 in 1983. For last year, a random sample of 40 residential property owners revealed the following expenditures:

$158	110	185	136	167	84	437	420
230	57	942	287	259	123	712	170
531	347	505	148	111	442	145	38
49	188	107	134	223	36	360	253
321	821	89	202	759	151	1176	31

Do the data provide sufficient evidence to conclude that last year's mean amount spent for maintenance and repairs has increased over the 1983 mean of $280? Assume $\sigma = \$265$ and perform the hypothesis test at the 5% significance level. (*Note:* The sum of the data is $11,644.)

__ **9.42**   *The World Almanac, 1985,* reports that the mean travel time to work in 1980 for all South Dakota residents was 13 minutes. A transportation official obtained this year's travel times, in minutes, for a random sample of 35 South Dakota residents. The data are presented below.

29	40	0	12	10	6	41
25	21	5	4	19	2	7
10	8	3	6	52	4	12
0	33	6	2	17	21	8
38	2	13	8	14	11	2

At the 5% significance level, do the data provide sufficient evidence to conclude that the mean travel time to work for all South Dakota residents has changed from the 1980 mean of 13 minutes? Assume $\sigma = 11.6$ minutes. (*Note:* The sum of the data is 491.)

__ **9.43**   The Food and Nutrition Board of the National Academy of Sciences states that the recommended daily allowance (RDA) of iron for adult females under the age of 51 is 18 mg. The iron intakes, in milligrams, during a 24-hour period were obtained for 45 randomly selected adult females under the age of 51. Here are the data:

15.0	18.1	14.4	14.6	10.9	18.1	18.2	18.3	15.0
16.0	12.6	16.6	20.7	19.8	11.6	12.8	15.6	11.0
15.3	9.4	19.5	18.3	14.5	16.6	11.5	16.4	12.5
14.6	11.9	12.5	18.6	13.1	12.1	10.7	17.3	12.4
17.0	6.3	16.8	12.5	16.3	14.7	12.7	16.3	11.5

At the 1% significance level, do the data suggest that adult females under the age of 51 are, on the average, getting less than the RDA of 18 mg of iron? *(Note: $\Sigma x = 660.6$ and $\Sigma x^2 = 10,115.88$.)*

___ **9.44** As reported by the U.S. Office of Juvenile Justice and Delinquency Prevention in *Children in Custody,* the mean age of all juveniles held in public custody in 1982 was 15.4 years. The mean age of 250 randomly selected juveniles currently being held in public custody is 15.26 years; and the standard deviation of the ages is 1.01 years. Does it appear that the mean age, μ, of all juveniles being held in public custody this year has decreased since 1982? Perform the appropriate hypothesis test at the 0.10 level of significance.

___ **9.45** A dog-food manufacturer sells "50-lb" bags of dog food. Suppose that you randomly select 75 of these bags and find that $\bar{x} = 50.11$ lb and $s = 0.84$ lb.
a) Would you be inclined to believe that the actual mean weight, μ, of all "50-lb" bags of this dog food differs from the advertised weight of 50 lb? Perform your hypothesis test at the 5% significance level.
b) Repeat part (a) if the mean weight of the 75 bags is 50.21 lb instead of 50.11 lb.

___ **9.46** The Health Insurance Association of America reports in *Survey of Hospital Semi-Private Room Charges* that the mean daily charge for a semi-private room in American hospitals was $253 in 1988. In that same year, a random sample of 30 Massachusetts hospitals yielded a mean daily semi-private room charge of $260.68 with a standard deviation of $12.77. Do the data provide sufficient evidence to conclude that, for 1988, the mean daily semi-private room charge in Massachusetts hospitals exceeded the national mean of $253? Perform the required hypothesis test using $\alpha = 0.05$.

___ **9.47** The manufacturer of a new model car, called the Orion, claims that a typical car gets 26 mpg (miles per gallon). An independent consumer group is somewhat skeptical of this claim and thinks that the mean gas mileage of all Orions may very well be less than 26 mpg. To try to justify its contention, the consumer group conducts mileage tests on 30 randomly selected Orions and obtains the following data:

25.3	25.1	29.6	24.6	26.0	26.0
26.3	23.6	26.0	25.4	26.1	23.8
25.1	24.1	25.8	26.4	23.4	24.8
22.6	26.6	25.1	26.6	28.0	23.3
23.8	25.4	26.2	25.1	25.3	21.5

At the 5% significance level, do the data support the consumer group's conjecture?

___ **9.48** A Louisiana cotton farmer has used a certain brand of fertilizer for the past five years. His experience with the fertilizer indicates that the mean yield is about 623 lb/acre. Recently, a new brand of fertilizer came out on the market that will supposedly increase cotton yield. The farmer uses the new fertilizer on 80 of his one-acre plots. Here are the resulting cotton yields, in pounds:

639	653	631	590	628	638	618	602
622	667	614	644	611	652	598	627
637	613	591	627	637	604	620	618
621	642	654	643	634	630	621	560
637	634	645	630	605	617	598	641
636	599	616	578	620	639	608	615
587	601	629	627	626	613	568	638
599	627	630	615	620	658	629	629
626	670	642	647	637	593	617	654
627	645	646	643	648	654	651	613

a) At the 10% significance level, do the data provide sufficient evidence to conclude that the new fertilizer increases the mean yield of cotton on the farmer's land? *(Note: $\bar{x} = 625.2$ lb/acre and $s = 21.5$ lb/acre.)*
b) If the new fertilizer costs more than the one the farmer presently uses, would you buy the new fertilizer if you were the farmer? Why?

___ **9.49** Explain why it is permissible to use the sample standard deviation, s, in place of an unknown population standard deviation, σ, when performing a large-sample hypothesis test for a population mean, μ.

=== **9.50** The average passenger vehicle was driven 8.9 thousand miles in 1982, as reported by the U.S. Federal Highway Administration in the publication *Highway Statistics.* A random sample of 500 passenger vehicles gave a mean of 8.7 thousand miles driven for last year, with a standard deviation of 6.0 thousand miles. Let μ denote last year's mean distance driven for all passenger vehicles.

a) Perform the hypothesis test

$$H_0: \mu = 8.9 \text{ thousand miles}$$
$$H_a: \mu \neq 8.9 \text{ thousand miles}$$

at the 5% significance level.

b) Use Procedure 8.1 on page 382 to find a 95% confidence interval for μ.

c) Does the value of 8.9 thousand miles hypothesized for the mean, μ, in the null hypothesis of part (a) lie in your confidence interval from part (b)?

d) Repeat parts (a)–(c) if the 500 passenger vehicles sampled were driven an average of 9.5 thousand miles last year.

e) Based on your observations in parts (a)–(d), complete the following statements concerning the relationship between a two-tailed hypothesis test

$$H_0: \mu = \mu_0$$
$$H_a: \mu \neq \mu_0$$

at the significance level α and a $(1 - \alpha)$-level confidence interval for μ:

(i) If μ_0 lies in the $(1-\alpha)$-level confidence interval for μ, then the null hypothesis *(will, will not)* be rejected.

(ii) If μ_0 lies outside the $(1 - \alpha)$-level confidence interval for μ, then the null hypothesis *(will, will not)* be rejected.

≡ 9.51   **The relationship between hypothesis tests and confidence intervals:** In this exercise, we will examine the relationship between a large-sample, two-tailed, hypothesis test for a population mean, μ, and a large-sample confidence-interval estimate for a population mean, μ.

a) Show that the inequalities

$$\bar{x} - z_{\alpha/2} \cdot \frac{s}{\sqrt{n}} < \mu_0 < \bar{x} + z_{\alpha/2} \cdot \frac{s}{\sqrt{n}}$$

are equivalent to

$$-z_{\alpha/2} < \frac{\bar{x} - \mu_0}{s/\sqrt{n}} < z_{\alpha/2}$$

b) Deduce the following fact from part (a): For a two-tailed hypothesis test

$$H_0: \mu = \mu_0$$
$$H_a: \mu \neq \mu_0$$

at the significance level α, the null hypothesis will not be rejected if μ_0 lies in the $(1 - \alpha)$-level confidence interval for μ and, conversely, the null hypothesis will be rejected if μ_0 does not lie in the $(1 - \alpha)$-level confidence interval for μ.

9.4 *P*-values (Optional)

Procedure 9.2 on page 444 lists the steps for performing any hypothesis test using the classical approach to hypothesis testing. In the classical approach to hypothesis testing, the significance level is specified in advance and then the conclusion is stated in terms of rejecting or not rejecting the null hypothesis.

There are some disadvantages with this approach to hypothesis testing. First, it does not permit people reading only the conclusions of the study to make their own evaluation (i.e., to choose their own significance level). Second, it does not provide the reader with the actual significance of the test.

To alleviate these problems, many researchers and most computer packages include the **P-value** of the hypothesis test in their reports. Roughly speaking, the *P*-value of a hypothesis test indicates the likelihood of observing the value obtained for the test statistic, if the null hypothesis is true.

A large *P*-value indicates that it would not be unlikely to observe the value obtained for the test statistic, if the null hypothesis were true. In other words, a large *P*-value does not provide evidence that the null hypothesis is false. On the other hand, a small *P*-value indicates that it would be unlikely to observe the value obtained for the test statistic, if the null hypothesis were true. In other words, a small *P*-value provides evidence that the null hypothesis is false.

We now present the precise definition of the *P*-value of a hypothesis test. Following that definition, we will consider several illustrations.

DEFINITION 9.5 *P*-value

The *P-value* of a hypothesis test is the probability of observing a value of the test statistic that is at least as inconsistent with the null hypothesis as the value of the test statistic actually observed.

Note: The *P*-value is also referred to as the **observed significance level** or the **probability value.**

If we are considering a large-sample hypothesis test for a population mean, μ, then the *P*-value can be obtained as follows: For a right-tailed test, the *P*-value is the probability, if the null hypothesis is true, of observing a value of the test statistic, z, that is at least as large as the value actually observed. For a left-tailed test, the *P*-value is the probability, if the null hypothesis is true, of observing a value of the test statistic, z, that is at least as small as the value actually observed. For a two-tailed test, the *P*-value is the probability, if the null hypothesis is true, of observing a value of the test statistic, z, that is at least as large in magnitude as the value actually observed. See Figure 9.13.

FIGURE 9.13 *P*-value for a large-sample hypothesis test concerning a population mean, μ

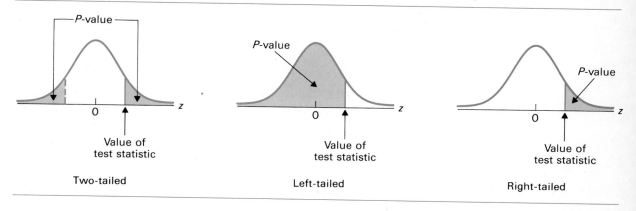

Two-tailed   Left-tailed   Right-tailed

The procedure for obtaining the *P*-value of any hypothesis test is similar to the one given in the previous paragraph. See Exercise 9.75 for the details.

The easiest way to understand *P*-values is to look at some examples. We will determine the *P*-values for two of the hypothesis tests that we performed in the previous section.

EXAMPLE 9.10   **Illustrates P-values**

Consider again the history-book illustration of Example 9.7 on pages 438–440. The null and alternative hypotheses for the hypothesis test in that example are

$$H_0: \mu = \$28.44 \text{ (mean price has not increased)}$$

$$H_a: \mu > \$28.44 \text{ (mean price has increased)}$$

where μ is this year's mean retail price of all hardcover history books. Note that the hypothesis test is right-tailed since there is a greater-than sign ($>$) in the alternative hypothesis.

Table 9.5 on page 439 gives this year's prices for 40 randomly selected history books. Using that data, we found the value of the test statistic to be $z = 2.85$. Determine and interpret the P-value of the hypothesis test.

SOLUTION   The P-value of a hypothesis test is the probability of observing a value of the test statistic that is at least as inconsistent with the null hypothesis as the value of the test statistic actually observed. Since, in this case, the test is right-tailed, the P-value is the probability of observing a value of z of 2.85 or greater, $P(z \geq 2.85)$, if the null hypothesis is true. See Figure 9.14.

FIGURE 9.14
P-value for history-book illustration

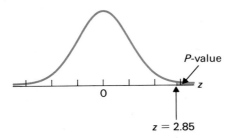

Now, if the null hypothesis is true, then the test statistic, z, has (approximately) the standard normal distribution. Therefore, if the null hypothesis is true, $P(z \geq 2.85)$ is equal to the area under the standard normal curve to the right of 2.85, the shaded area in Figure 9.14. From Table II, we find that area to be $0.5000 - 0.4978 = 0.0022$.

Thus, the P-value of this hypothesis test is 0.0022. This means that, if the null hypothesis is true, then the probability is only 0.0022 that the test statistic will be 2.85 or greater. Consequently, observing a value of 2.85 for the test statistic provides strong evidence that the null hypothesis is false.   ■

EXAMPLE 9.11   **Illustrates P-values**

Refer to the monthly-rental illustration of Example 9.9 on pages 442–443. The null and alternative hypotheses for the hypothesis test in that example are

$$H_0: \mu = \$587 \text{ (general partner's claim is correct)}$$

$$H_a: \mu \neq \$587 \text{ (general partner's claim is not correct)}$$

where μ is the mean monthly rent of all three-bedroom apartments in the city. Note that the hypothesis test is two-tailed since there is a not-equal sign ($\neq$) in the alternative hypothesis.

Table 9.7 on page 442 gives the monthly rents of a random sample of 30 three-bedroom apartments in the city. The value of the test statistic for that data was found to be $z = 0.81$. Obtain and interpret the *P*-value of the hypothesis test.

SOLUTION   Again, the *P*-value of a hypothesis test is the probability of observing a value of the test statistic that is at least as inconsistent with the null hypothesis as the value of the test statistic actually observed. Since, in this case, the test is two-tailed, the *P*-value is the probability of observing a value of z of 0.81 or greater in magnitude, $P(z \leq -0.81 \text{ or } z \geq 0.81)$, if the null hypothesis is true. See Figure 9.15.

FIGURE 9.15
P-value for monthly-rents illustration

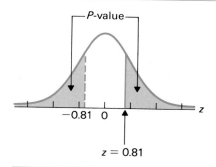

If the null hypothesis is true, z has (approximately) the standard normal distribution. So, under that condition, $P(z \leq -0.81 \text{ or } z \geq 0.81)$ is equal to the area under the standard normal curve that lies either to the left of -0.81 or to the right of 0.81, the shaded area in Figure 9.15. From Table II, we find that area to be $2 \cdot (0.5000 - 0.2910) = 2 \cdot 0.2090 = 0.4180$.

Hence, the *P*-value of this hypothesis test is 0.4180. In other words, if the null hypothesis is true, then the probability is 0.4180 that the test statistic will be 0.81 or greater in magnitude. Therefore, observing a value of 0.81 for the test statistic certainly does not provide evidence that the null hypothesis is false. ∎

THE *P*-VALUE APPROACH TO HYPOTHESIS TESTING

The *P*-value can be interpreted as the observed significance level of a hypothesis test. To illustrate, suppose that the *P*-value of a right-tailed, large-sample, hypothesis test for a population mean turns out to be 0.03, as depicted by the shaded area in Figure 9.16 at the top of the next page.

Then, as we see from Figure 9.16, the null hypothesis would be rejected for a test at the 0.05 significance level but would not be rejected for a test at the 0.01 significance level. In fact, the figure makes it clear that the *P*-value is precisely the smallest significance level at which the null hypothesis would be rejected.

FIGURE 9.16

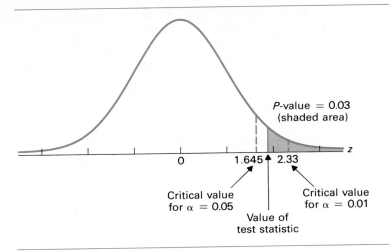

KEY FACT 9.4   *P*-value as the observed significance level
The *P*-value of a hypothesis test is equal to the smallest significance level at which the null hypothesis can be rejected.

In view of Key Fact 9.4, we have the following criterion for deciding whether or not the null hypothesis should be rejected in favor of the alternative hypothesis:

KEY FACT 9.5   Decision criterion for a hypothesis test using the *P*-value
If the *P*-value is less than or equal to the specified significance level, α, then reject the null hypothesis. Otherwise, do not reject the null hypothesis.

Key Fact 9.5 provides us with a means for employing the *P*-value as a basis for performing a hypothesis test. A general method for doing that is presented in Procedure 9.3.

PROCEDURE 9.3
To perform a hypothesis test (*P*-value approach):

STEP 1 *State the null and alternative hypotheses.*
STEP 2 *Decide on the significance level, α.*
STEP 3 *Compute the value of the test statistic.*
STEP 4 *Determine the P-value.*
STEP 5 *If the P-value is less than or equal to α, reject H_0; otherwise, do not reject H_0.*
STEP 6 *State the conclusion in words.*

In Example 9.8 on pages 440–442, we performed a hypothesis test concerning the mean intake of calcium, μ, of all people with incomes below the poverty level. To carry out that hypothesis test, we used the classical approach to hypothesis testing. Now we will perform that same hypothesis test using the *P*-value approach to hypothesis testing.

EXAMPLE 9.12 *Illustrates Procedure 9.3*

Calcium is the most abundant and one of the most important minerals in the body. It works with phosphorus to build and maintain bones and teeth. According to the Food and Nutrition Board of the National Academy of Sciences, the recommended daily allowance (RDA) of calcium for adults is 800 mg (milligrams).

A nutritionist thinks that the average person with an income below the poverty level gets less than the RDA of 800 mg. To test her claim, she obtains the daily intakes of calcium for a random sample of 35 people with incomes below the poverty level. The results are displayed in Table 9.8. Data are in milligrams.

TABLE 9.8
Daily calcium intakes

879	1096	701	986	828	1077	703
555	422	997	473	702	508	530
513	720	944	673	574	707	864
1199	743	1325	655	1043	599	1008
705	180	287	542	893	1052	473

At the 5% significance level, do the data provide sufficient evidence to conclude that the mean calcium intake of all people with incomes below the poverty level is less than the RDA of 800 mg per day?

SOLUTION We will apply Procedure 9.3 to perform the hypothesis test.

STEP 1 *State the null and alternative hypotheses.*

Let μ denote the mean daily intake of calcium of all people with incomes below the poverty level. The null and alternative hypotheses are:

H_0: $\mu = 800$ mg (mean calcium intake is not less than the RDA)
H_a: $\mu < 800$ mg (mean calcium intake is less than the RDA)

Note that the hypothesis test is left-tailed since there is a less-than sign ($<$) in the alternative hypothesis.

STEP 2 *Decide on the significance level, α.*

We are to perform the test at the 5% significance level. Thus, $\alpha = 0.05$.

STEP 3 *Compute the value of the test statistic.*

Since the sample size, $n = 35$, is large, we use the test statistic

$$z = \frac{\bar{x} - \mu_0}{s/\sqrt{n}}$$

We have $\mu_0 = 800$ mg and $n = 35$. From the data in Table 9.8 we find that $\bar{x} = 747.3$ mg and $s = 262.2$ mg. Thus, the value of the test statistic is

$$z = \frac{\bar{x} - \mu_0}{s/\sqrt{n}} = \frac{747.3 - 800}{262.2/\sqrt{35}} = -1.19$$

See Figure 9.17.

FIGURE 9.17

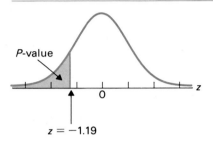

STEP 4 *Determine the P-value.*

Since the test is left-tailed, the *P*-value is the probability of observing a value of z of -1.19 or less, $P(z \leq -1.19)$, if the null hypothesis is true. That probability equals the shaded area in Figure 9.17 which, by Table II, is $0.5000 - 0.3830 = 0.1170$. Hence, the *P*-value of the hypothesis test equals 0.1170.

STEP 5 *If the P-value is less than or equal to α, reject H_0; otherwise, do not reject H_0.*

From Step 4, the *P*-value is 0.1170. Since this exceeds the specified significance level of $\alpha = 0.05$, we do not reject H_0.

STEP 6 *State the conclusion in words.*

The sample of 35 calcium intakes does not provide sufficient evidence to conclude that the mean calcium intake, μ, of all people with incomes below the poverty level is less than the RDA of 800 mg per day. ∎

MTB

USING THE COMPUTER (OPTIONAL)

We can use Minitab to perform a large-sample hypothesis test for a population mean, μ. The appropriate command is called **ZTEST**. But before we can apply the ZTEST command, we need to discuss the subcommands that are used to specify whether the hypothesis test is two-tailed, left-tailed, or right-tailed.

Minitab assumes that the hypothesis test is two-tailed unless told otherwise. For a left-tailed test, the subcommand **ALTERNATIVE** = -1 is used; and for a right-tailed test the subcommand **ALTERNATIVE = 1** is used. Table 9.9 records these subcommands for future reference.

TABLE 9.9
Subcommands
for specifying the
alternative hypothesis

Type of test	Subcommand
Two-tailed	None
Left-tailed	ALTERNATIVE $= -1$
Right-tailed	ALTERNATIVE $= 1$

Now we are ready to see how ZTEST can be used to perform a large-sample hypothesis test for a population mean, μ. To explain the details, we return once more to the hypothesis test concerning the mean calcium intake of all people with incomes below the poverty level.

EXAMPLE 9.13   **Illustrates the ZTEST command**

Refer to Example 9.12 on pages 451–452. Use Minitab to perform the hypothesis test considered in that example.

SOLUTION   Let μ denote the mean daily intake of calcium of all people with incomes below the poverty level. The problem is to perform the hypothesis test

$$H_0: \mu = 800 \text{ mg (mean calcium intake is not less than the RDA)}$$
$$H_a: \mu < 800 \text{ mg (mean calcium intake is less than the RDA)}$$

at the 5% significance level ($\alpha = 0.05$). Note that the hypothesis test is left-tailed since there is a less-than sign ($<$) in the alternative hypothesis.

The program ZTEST requires the user to specify the population standard deviation, σ, just like the program ZINTERVAL does. In this case, however, we do not know the value of σ. But since the sample size is large ($n = 35$), we can use the sample standard deviation, s, in place of σ.

To begin, we enter the sample data from Table 9.8 on page 451 into C1 using the SET command and name C1 "Calcium" using the NAME command. Next, we apply the STDEV command to obtain the sample standard deviation, s, of the data. See Printout 9.1.

PRINTOUT 9.1

```
MTB > SET C1
DATA> 879 1096 701 986 828 1077 703
DATA> 555 422 997 473 702 508 530
DATA> 513 720 944 673 574 707 864
DATA> 1199 743 1325 655 1043 599 1008
DATA> 705 180 287 542 893 1052 473
DATA> END
MTB > NAME C1 'CALCIUM'
MTB > STDEV 'CALCIUM'
   ST.DEV. =      262.23
```

Now we can employ Minitab to perform the required hypothesis test. We first type the command ZTEST, followed by the null hypothesis, the estimated value of σ, and the storage location of the sample data. In other words, we type the following command:

<u>ZTEST of mu=800, sigma=262.23, data in 'CALCIUM';</u>

Since the test is left-tailed, we must also use the subcommand <u>ALTERNATIVE=-1.</u>, as indicated in Table 9.9. These commands and the resulting output are displayed in Printout 9.2.

```
MTB > ZTEST of mu=800, sigma=262.23, data in 'CALCIUM';
SUBC> ALTERNATIVE=-1.

TEST OF MU = 800.000 VS MU L.T.  800.000
THE ASSUMED SIGMA = 262

                N      MEAN     STDEV    SE MEAN         Z     P VALUE
CALCIUM        35   747.314   262.227    44.325     -1.19        0.12
```

The output first displays a statement of the null and alternative hypotheses for the hypothesis test: TEST OF MU = 800.000 VS MU L.T. 800.000 (L.T. stands for "less than"). Next we find the value used for the population standard deviation, σ: THE ASSUMED SIGMA = 262. Then the output shows the sample size, sample mean, sample standard deviation, and standard error of the mean. The next to last entry, Z, gives the value of the test statistic

$$z = \frac{\bar{x} - \mu_0}{s/\sqrt{n}} = \frac{747.314 - 800}{262.23/\sqrt{35}}$$

So, $z = -1.19$.

The final entry shown in the output in Printout 9.2 is P VALUE. This is really the only quantity that we need in order to decide whether the null hypothesis should be rejected. From Key Fact 9.4, we know that the P-value is equal to the smallest significance level at which the null hypothesis can be rejected. Therefore, if the P-value is less than or equal to the specified significance level, then we reject H_0; otherwise, we do not reject H_0.

From Printout 9.2, we see that the P-value for this hypothesis test equals 0.12. Since this P-value exceeds the specified significance level of $\alpha = 0.05$, we do not reject H_0. The data do not provide sufficient evidence to conclude that the mean calcium intake of all people with incomes below the poverty level is less than the RDA of 800 mg per day.   ∎

Exercises 9.4

___ **9.52** State two reasons why it is a good idea to include the *P*-value in a report concerning a hypothesis test.

In Exercises 9.53–9.58, we have given the value obtained for the test statistic

$$z = \frac{\bar{x} - \mu_0}{\sigma/\sqrt{n}}$$

in a large-sample hypothesis test concerning a population mean, μ. We have also specified whether the test is two-tailed, left-tailed, or right-tailed. Determine the P-value corresponding to each z-value.

___ **9.53** Right-tailed test:
a) $z = 2.03$.   b) $z = -0.31$.

___ **9.54** Left-tailed test:
a) $z = -1.84$.   b) $z = 1.25$.

___ **9.55** Left-tailed test:
a) $z = -0.74$.   b) $z = 1.16$.

___ **9.56** Two-tailed test:
a) $z = 3.08$.   b) $z = -2.42$.

___ **9.57** Two-tailed test:
a) $z = -1.66$.   b) $z = 0.52$.

___ **9.58** Right-tailed test:
a) $z = 1.24$.   b) $z = -0.69$.

In each of Exercises 9.41–9.48 of Section 9.3, you were asked to perform a large-sample hypothesis test for a population mean, μ, using the classical approach to hypothesis testing. Now, in Exercises 9.59–9.66, you are asked to perform those same hypothesis tests using the P-value approach to hypothesis testing (Procedure 9.3 on page 450).

___ **9.59** According to the Census Bureau publication *Construction Reports*, the mean expenditure per residential property owner for maintenance and repairs was $280 in 1983. For last year, a random sample of 40 residential property owners revealed the following expenditures:

$158	110	185	136	167	84	437	420
230	57	942	287	259	123	712	170
531	347	505	148	111	442	145	38
49	188	107	134	223	36	360	253
321	821	89	202	759	151	1176	31

Do the data provide sufficient evidence to conclude that last year's mean amount spent for maintenance and repairs has increased over the 1983 mean of $280? Assume $\sigma = \$265$ and perform the hypothesis test at the 5% significance level. (*Note:* The sum of the data is $11,644.)

___ **9.60** *The World Almanac, 1985*, reports that the mean travel time to work in 1980 for all South Dakota residents was 13 minutes. A transportation official obtained this year's travel times, in minutes, for a random sample of 35 South Dakota residents. The data are presented below.

29	40	0	12	10	6	41
25	21	5	4	19	2	7
10	8	3	6	52	4	12
0	33	6	2	17	21	8
38	2	13	8	14	11	2

At the 5% significance level, do the data provide sufficient evidence to conclude that the mean travel time to work for all South Dakota residents has changed from the 1980 mean of 13 minutes? Assume $\sigma = 11.6$ minutes. (*Note:* The sum of the data is 491.)

___ **9.61** The Food and Nutrition Board of the National Academy of Sciences states that the recommended daily allowance (RDA) of iron for adult females under the age of 51 is 18 mg. The iron intakes, in milligrams, during a 24-hour period were obtained for 45 randomly selected adult females under the age of 51. Here are the data:

15.0	18.1	14.4	14.6	10.9	18.1	18.2	18.3	15.0
16.0	12.6	16.6	20.7	19.8	11.6	12.8	15.6	11.0
15.3	9.4	19.5	18.3	14.5	16.6	11.5	16.4	12.5
14.6	11.9	12.5	18.6	13.1	12.1	10.7	17.3	12.4
17.0	6.3	16.8	12.5	16.3	14.7	12.7	16.3	11.5

At the 1% significance level, do the data suggest that adult females under the age of 51 are, on the average, getting less than the RDA of 18 mg of iron? (*Note:* $\Sigma x = 660.6$ and $\Sigma x^2 = 10,115.88$.)

___ **9.62** As reported by the U.S. Office of Juvenile Justice and Delinquency Prevention in *Children in Custody*, the mean age of all juveniles held in public custody in 1982 was 15.4 years. The mean age of 250 randomly selected juveniles currently being held in public custody is 15.26 years; and the standard deviation of the ages is 1.01 years. Does it appear that the mean age, μ, of all juveniles being held in public custody this year has decreased since 1982? Perform the appropriate hypothesis test at the 0.10 level of significance.

___ **9.63** A dog-food manufacturer sells "50-lb" bags of dog food. Suppose that you randomly select 75 of these bags and find that $\bar{x} = 50.11$ lb and $s = 0.84$ lb.

a) Would you be inclined to believe that the actual mean weight, μ, of all "50-lb" bags of this dog food differs from the advertised weight of 50 lb? Perform your hypothesis test at the 5% significance level.

b) Repeat part (a) if the mean weight of the 75 bags is 50.21 lb instead of 50.11 lb.

___ **9.64** The Health Insurance Association of America reports in *Survey of Hospital Semi-Private Room Charges* that the mean daily charge for a semi-private room in American hospitals was $253 in 1988. In that same year, a random sample of 30 Massachusetts hospitals yielded a mean daily semi-private room charge of $260.68 with a standard deviation of $12.77. Do the data provide sufficient evidence to conclude that, for 1988, the mean daily semi-private room charge in Massachusetts hospitals exceeded the national mean of $253? Perform the required hypothesis test using $\alpha = 0.05$.

___ **9.65** The manufacturer of a new model car, called the Orion, claims that a typical car gets 26 mpg (miles per gallon). An independent consumer group is somewhat skeptical of this claim and thinks that the mean gas mileage of all Orions may very well be less than 26 mpg. To try to justify its contention, the consumer group conducts mileage tests on 30 randomly selected Orions and obtains the following data:

25.3	25.1	29.6	24.6	26.0	26.0
26.3	23.6	26.0	25.4	26.1	23.8
25.1	24.1	25.8	26.4	23.4	24.8
22.6	26.6	25.1	26.6	28.0	23.3
23.8	25.4	26.2	25.1	25.3	21.5

At the 5% significance level, do the data support the consumer group's conjecture?

___ **9.66** A Louisiana cotton farmer has used a certain brand of fertilizer for the past five years. His experience with the fertilizer indicates that the mean yield is about 623 lb/acre. Recently, a new brand of fertilizer came out on the market that will supposedly increase cotton yield. The farmer uses the new fertilizer on 80 of his one-acre plots. Here are the resulting cotton yields, in pounds:

639	653	631	590	628	638	618	602
622	667	614	644	611	652	598	627
637	613	591	627	637	604	620	618
621	642	654	643	634	630	621	560
637	634	645	630	605	617	598	641
636	599	616	578	620	639	608	615
587	601	629	627	626	613	568	638
599	627	630	615	620	658	629	629
626	670	642	647	637	593	617	654
627	645	646	643	648	654	651	613

a) At the 10% significance level, do the data provide sufficient evidence to conclude that the new fertilizer increases the mean yield of cotton on the farmer's land? (*Note:* $\bar{x} = 625.2$ lb/acre and $s = 21.5$ lb/acre.)

b) If the new fertilizer costs more than the one the farmer presently uses, would you buy the new fertilizer if you were the farmer? Why?

Exercises 9.67–9.72 are computer exercises.

___ **9.67 (Computer exercise)** Suppose that the data in Exercise 9.59 are stored in a column named EXPENDIT.

a) Which Minitab command and subcommands (if any) should be used to carry out the hypothesis test considered in that exercise?

b) If you have access to Minitab, use it to perform the hypothesis test.

___ **9.68 (Computer exercise)** Suppose that the data in Exercise 9.60 are stored in a column named TRAVTIME.

a) Which Minitab command and subcommands (if any) should be used to carry out the hypothesis test considered in that exercise?

b) If you have access to Minitab, use it to perform the hypothesis test.

___ **9.69 (Computer exercise)** Assume the data in Exercise 9.61 are stored in a column named IRON.

a) Which Minitab commands and subcommands (if any) should be used to perform the hypothesis test considered in that exercise?

b) If you have access to Minitab, use it to carry out the hypothesis test.

___ **9.70 (Computer exercise)** Assume the data in Exercise 9.66 are stored in a column named YIELD.

a) Which Minitab commands and subcommands (if any) should be used to perform the hypothesis test considered in part (a) of that exercise?
b) If you have access to Minitab, use it to carry out the hypothesis test.

___ **9.71 (Computer exercise)** The U.S. National Center for Health Statistics compiles information on the length of stay in hospitals by patients. Data are reported in *Vital and Health Statistics*. According to that publication, the mean hospital stay was 7.1 days in 1982. A researcher thinks that this year's mean will be less. She first determines the stays, in days, of a random sample of patients who were discharged from the hospital this year. Then she applies Minitab's ZTEST command and obtains the following (abridged) computer output:

```
TEST OF MU = 7.100 VS MU L.T. 7.100
THE ASSUMED SIGMA = 7.70
```

N	MEAN	STDEV	SE MEAN	Z	P VALUE
40	6.850	7.011	1.217	-0.21	0.42

Use the printout to determine the
a) null and alternative hypotheses for the researcher's hypothesis test.
b) subcommand used by the researcher.
c) (assumed) population standard deviation of this year's hospital stays.
d) standard deviation of the hospital stay's in the sample obtained by the researcher.
e) number of patients sampled by the researcher.
f) mean hospital stay of the patients sampled.
g) value obtained for the test statistic, z.
h) *P*-value of the hypothesis test.
i) smallest significance level at which the null hypothesis can be rejected.
j) conclusion if the test is performed at the 5% significance level.

___ **9.72 (Computer exercise)** A few years ago, the owner of a menswear store decided to advertise in local papers in an attempt to improve sales. From past records he knows that, without advertising, average weekly sales had been $1700. Of course, he does not wish to continue spending money on advertising if sales have not increased and consequently decides to perform a hypothesis test. The owner determines the weekly sales for a random sample of weeks in which advertising was used and obtains the following abridged Minitab output:

```
TEST OF MU = 1700.000 VS MU G.T. 1700.000
THE ASSUMED SIGMA = 250
```

N	MEAN	STDEV	SE MEAN	Z	P VALUE
32	1801.469	395.699	44.194	2.30	0.011

Use the printout to determine the
a) null and alternative hypotheses for the test.
b) subcommand used by the store owner.
c) (assumed) population standard deviation of weekly sales with advertising.
d) standard deviation of the weekly sales in the sample obtained by the store owner.
e) number of weeks sampled by the store owner.
f) mean sales of the weeks sampled.
g) value obtained for the test statistic, z.
h) *P*-value of the hypothesis test.
i) conclusion if the test is performed with $\alpha = 0.05$.

≡ **9.73** Suppose that for a large-sample hypothesis test concerning a population mean, μ, the observed value of the test statistic z is z_0. If the test is right-tailed, then the *P*-value of the hypothesis test can be expressed as $P(z \geq z_0)$. Determine the corresponding expression for the *P*-value if the test is
a) left-tailed.   b) two-tailed.

≡ **9.74** The symbol $\Phi(z)$ is often used to denote the area under the standard normal curve that lies to the left of a specified value of z. Suppose that for a large-sample hypothesis test concerning a population mean, μ, the observed value of the test statistic is z_0. Express the *P*-value of the hypothesis test in terms of Φ if the test is
a) left-tailed.   b) right-tailed.   c) two-tailed.

≡ **9.75 Obtaining the *P*-value:** This exercise discusses the general procedure for obtaining the *P*-value of a hypothesis test. Let x denote the test statistic for a hypothesis test and let x_0 denote the value of the test statistic actually observed. Then the *P*-value of the hypothesis test equals

1. $P(x \geq x_0)$, for a right-tailed test,
2. $P(x \leq x_0)$, for a left-tailed test,
3. $2 \cdot \min \{P(x \leq x_0), P(x \geq x_0)\}$, for a two-tailed test,

where the probabilities are computed under the condition that the null hypothesis is true.

Suppose now that we are considering a large-sample hypothesis test for a population mean using the test statistic z. Verify that the probability expressions in 1–3 are equivalent to those obtained in Exercise 9.73.

9.5 Hypothesis tests for a normal population mean

In Section 9.3, we learned how to perform a hypothesis test concerning a population mean, μ, when the sample size is large ($n \geq 30$). However, as mentioned previously, there are many instances in which a large sample is unavailable, extremely expensive, or undesirable.

There are several methods for performing a hypothesis test concerning a population mean that do not require a large sample. One such method applies when the population being sampled is *normally distributed*. That method is what we will study in this section. We will assume that the population standard deviation, σ, is unknown since, in practice, that is usually the case.

To develop a hypothesis-testing procedure for a normal population mean, we begin by recalling Key Fact 8.3 on page 395: Suppose that a random sample of size n is to be taken from a normally distributed population with mean μ. Then the random variable

$$t = \frac{\overline{x} - \mu}{s/\sqrt{n}}$$

has the t-distribution with $n - 1$ degrees of freedom. In other words, probabilities for that random variable are equal to areas under the t-curve with df $= n - 1$.

Consequently, when the population being sampled is normally distributed, we can perform a hypothesis test with null hypothesis $H_0: \mu = \mu_0$ by employing the random variable

$$t = \frac{\overline{x} - \mu_0}{s/\sqrt{n}}$$

as our test statistic and using the t-table, Table III, to obtain the critical value(s). Procedure 9.4 supplies the steps.

PROCEDURE 9.4

To perform a hypothesis test for a population mean with null hypothesis $H_0: \mu = \mu_0$.

ASSUMPTION

Normal population.

STEP 1 *State the null and alternative hypotheses.*
STEP 2 *Decide on the significance level, α.*
STEP 3 *The critical value(s)*
    a) *for a two-tailed test are $\pm t_{\alpha/2}$,*
    b) *for a left-tailed test is $-t_{\alpha}$,*
    c) *for a right-tailed test is t_{α},*
    with df $= n - 1$. Use Table III to find the critical value(s).

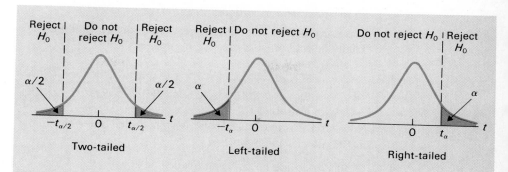

STEP 4 *Compute the value of the test statistic*

$$t = \frac{\bar{x} - \mu_0}{s/\sqrt{n}}$$

STEP 5 *If the value of the test statistic falls in the rejection region, then reject H_0; otherwise, do not reject H_0.*
STEP 6 *State the conclusion in words.*

EXAMPLE 9.14 *Illustrates Procedure 9.4*

The U.S. Energy Information Administration surveys households to obtain data on residential energy consumption and expenditures. Results of the surveys can be found in *Residential Energy Consumption Survey: Consumption and Expenditures.* According to that publication, the mean residential energy expenditure of all American families was $1123 in 1985. That same year, 15 randomly selected upper-income families reported the energy expenditures shown in Table 9.10. The expenditures are given to the nearest dollar.

TABLE 9.10
Energy expenditures for 15 upper-income families

$1254	1350	1227	1154	1790
1615	1521	908	1231	1369
1711	1293	1205	1351	1185

At the 5% significance level, do the data indicate that, in 1985, upper-income families spent more, on the average, for energy than the national average of $1123? [Assume that the 1985 energy expenditures of all upper-income families are normally distributed.]

SOLUTION Since the population under consideration is normally distributed, we can apply Procedure 9.4 to perform the required hypothesis test.

STEP 1 *State the null and alternative hypotheses.*

Let μ denote the mean energy expenditure of all upper-income families in 1985. Then the null and alternative hypotheses are

$$H_0: \mu = \$1123 \text{ (mean was not greater than the national mean)}$$
$$H_a: \mu > \$1123 \text{ (mean was greater than the national mean)}$$

Note that the hypothesis test is right-tailed since there is a greater-than sign ($>$) in the alternative hypothesis.

STEP 2 *Decide on the significance level, α.*

We are to perform the hypothesis test at the 5% significance level. Thus, $\alpha = 0.05$.

STEP 3 *The critical value for a right-tailed test is t_α, with df $= n - 1$.*

Here $n = 15$ and $\alpha = 0.05$. Table III shows that for df $= 15 - 1 = 14$, $t_{0.05} = 1.761$. See Figure 9.18.

FIGURE 9.18

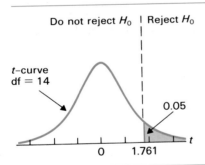

STEP 4 *Compute the value of the test statistic*

$$t = \frac{\bar{x} - \mu_0}{s/\sqrt{n}}$$

We have $\mu_0 = \$1123$ and $n = 15$. Furthermore, the mean and standard deviation of the sample data in Table 9.10 are

$$\bar{x} = \frac{\Sigma x}{n} = \frac{20{,}164}{15} = \$1344.27$$

and

$$s = \sqrt{\frac{n(\Sigma x^2) - (\Sigma x)^2}{n(n-1)}} = \sqrt{\frac{15(27{,}852{,}834) - (20{,}164)^2}{15 \cdot 14}} = \$231.00$$

Consequently, the value of the test statistic is

$$t = \frac{\bar{x} - \mu_0}{s/\sqrt{n}} = \frac{1344.27 - 1123}{231.00/\sqrt{15}} = \boxed{3.710}$$

STEP 5 *If the value of the test statistic falls in the rejection region, reject H_0; otherwise, do not reject H_0.*

The value of the test statistic, found in Step 4, is $t = 3.710$. As we see from Figure 9.18, this falls in the rejection region. Hence, we **reject H_0.**

STEP 6 *State the conclusion in words.*

The sample data provide sufficient evidence to conclude that, in the year 1985, upper-income families spent more, on the average, for energy than the national average of $1123.

MTB

USING THE COMPUTER (OPTIONAL)

Procedure 9.4 on page 458 gives a step-by-step method for performing a hypothesis test concerning a population mean, μ, when the population being sampled is normally distributed. Alternatively, we can use Minitab to carry out such a hypothesis test. The appropriate command is called **TTEST**. Example 9.15 explains in detail how to apply TTEST.

EXAMPLE 9.15   *Illustrates the TTEST command*

Refer to Example 9.14 on page 459. Use Minitab to perform the hypothesis test considered in that example.

SOLUTION   Let μ denote the mean energy expenditure of all upper-income families in 1985. The problem is to perform the hypothesis test

H_0: $\mu = \$1123$ (mean was not greater than the national mean)

H_a: $\mu > \$1123$ (mean was greater than the national mean)

at the 5% significance level ($\alpha = 0.05$). Note that the hypothesis test is right-tailed since there is a greater-than sign ($>$) in the alternative hypothesis.

Since, by assumption, the population under consideration is normally distributed, we can apply Minitab's TTEST to carry out the hypothesis test. We first enter the sample data from Table 9.10 on page 459 into C2 using the SET command and name C2 "Energy$" using the NAME command. Then we type the command TTEST, followed by the null hypothesis and the storage location of the sample data. That is, we type

```
TTEST of mu=1123, data in 'ENERGY$';
```

Because the test is right-tailed, we also type the subcommand <u>ALTERNATIVE=1</u>. (see Table 9.9 on page 453). Printout 9.3 displays the above commands and the output that results.

PRINTOUT 9.3
Minitab output
for TTEST

```
MTB > SET C2
DATA> 1254 1350 1227 1154 1790
DATA> 1615 1521  908 1231 1369
DATA> 1711 1293 1205 1351 1185
DATA> END
MTB > NAME C2 'ENERGY$'
MTB > TTEST of mu=1123, data in 'ENERGY$';
SUBC> ALTERNATIVE=1.

TEST OF MU = 1123.000 VS MU G.T. 1123.000

              N     MEAN    STDEV   SE MEAN      T    P VALUE
ENERGY$      15  1344.267  230.998   59.643    3.71    0.0012
```

On the first line of the output, we find a statement of the null and alternative hypotheses: TEST OF MU = 1123.000 VS MU G.T. 1123.000 (G.T. stands for "greater than"). Then we find the sample size, sample mean, sample standard deviation, and estimated standard error of the mean. The next-to-last entry, T, gives the value of the test statistic

$$t = \frac{\bar{x} - \mu_0}{s/\sqrt{n}} = \frac{1344.267 - 1123}{230.998/\sqrt{15}}$$

which, as we see from Printout 9.3, equals 3.71.

The final entry in Printout 9.3 shows the P-value of the hypothesis test. Since the P-value of 0.0012 is less than the specified significance level of $\alpha = 0.05$, we reject H_0. In other words, the data provide sufficient evidence to conclude that upper-income families spent more, on the average, for energy in 1985 than the national average of $1123. ∎

Exercises 9.5

__ 9.76 Is there any restriction on sample size for using Procedure 9.4? Provide a detailed explanation for your answer.

In each of Exercises 9.77–9.82, apply Procedure 9.4 on page 458 to perform the required hypothesis test.

__ 9.77 A paint manufacturer claims that the average drying time for its new latex paint is two hours. To test this claim, the drying times are obtained for 20 randomly selected cans of paint. The results are displayed in the table at the top of the following column. The data are given in minutes.

123	109	115	121	130
127	106	120	116	136
131	128	139	110	133
122	133	119	135	109

Assuming the drying times are normally distributed, do the sample data provide sufficient evidence to conclude that the mean drying time is actually greater than the manufacturer's claim of 120 minutes? Use $\alpha = 0.05$. (*Note:* The sample mean and sample standard deviation of the data are $\bar{x} = 123.1$ minutes and $s = 10.0$ minutes, respectively.)

— **9.78** The U.S. Energy Information Administration compiles data on household motor fuel expenditures and publishes the results in *Residential Transportation Energy Consumption Survey, Consumption Patterns of Household Vehicles*. According to that publication, the mean annual motor fuel expenditure per American household was $1317 in 1983. That same year, a sample of 16 households within metropolitan areas provided the following data on annual motor fuel expenditures:

$1390	1459	2043	1551
415	1359	1778	1537
1167	1716	1463	904
560	1592	1710	638

Assume that the annual motor fuel expenditures for households within metropolitan areas are normally distributed. At the 5% significance level, do the data provide evidence that the 1983 mean annual fuel expenditure for households within metropolitan areas differed from the national mean of $1317? *(Note:* The sample mean and sample standard deviation of the data are $\bar{x} = \$1330.13$ and $s = \$470.36$.)

— **9.79** A battery retailer has received a large shipment of automobile batteries from a supplier. The supplier claims that the batteries have a mean life of 36 months. A test on 10 batteries randomly sampled from the shipment yields the following lifetimes in months:

27.6	28.7	34.7	29.0	22.9
29.6	29.4	30.2	36.5	34.7

a) Do the data indicate that the mean life of the supplier's batteries is less than the claimed 36 months? Perform the hypothesis test at the 1% significance level. *(Note:* $\Sigma x = 303.3$ and $\Sigma x^2 = 9343.85$.)
b) What assumption must you make about battery life in order to carry out the hypothesis test in the way that you did in part (a)?

— **9.80** As reported by the College Entrance Examination Board in *National College-Bound Senior,* the mean verbal score on the Scholastic Aptitude Test in 1987 was 430 points out of a possible 800. The scores on the SAT are approximately normally distributed. A random sample of 25 verbal scores for last year yielded the following data:

346	496	352	378	315
491	360	385	500	558
381	303	434	562	496
420	485	446	479	422
494	289	436	516	615

At the 10% significance level, does it appear that last year's mean for verbal SAT scores has increased over the 1987 mean of 430 points? *(Note:* $\Sigma x = 10{,}959$ and $\Sigma x^2 = 4{,}979{,}401$.)

— **9.81** According to *Food Cost Review,* published by the U.S. Department of Agriculture, the average retail price for oranges in 1983 was 38.5 cents per pound. Recently, a random sample of 15 markets reported the following prices for oranges in cents per pound:

43.0	40.0	42.6	40.2	37.5
44.1	45.2	41.8	35.6	34.6
37.9	44.2	44.5	38.2	42.4

Assuming the retail prices for oranges are normally distributed, can we conclude that the mean retail price for oranges now is different from the 1983 mean of 38.5 cents per pound? Use $\alpha = 0.05$.

— **9.82** Atlas Fishing Line, Inc., manufactures a 10-lb test line. Twelve randomly selected spools are subjected to tensile-strength tests. The results are:

9.8	10.2	9.8	9.4
9.7	9.7	10.1	10.1
9.8	9.6	9.1	9.7

Use the fact that tensile strengths are normally distributed to decide whether Atlas Fishing Line's 10-lb test line is not up to specifications. Perform the test at the 5% significance level.

Exercises 9.83–9.86 are computer exercises.

— **9.83** (**Computer exercise**) Suppose that the data in Exercise 9.77 are stored in a column named DRYTIMES.
a) Which Minitab command and subcommands (if any) should be used to perform the hypothesis test considered in that exercise?
b) If you have access to Minitab, use it to carry out the hypothesis test.

— **9.84** (**Computer exercise**) Suppose the data in Exercise 9.78 are stored in a column named FUEL$.

a) Which Minitab command and subcommands (if any) should be used to perform the hypothesis test considered in that exercise?

b) If you have access to Minitab, use it to carry out the hypothesis test.

___ **9.85 (Computer exercise)** An automobile manufacturer is experimenting with a new bumper that could reduce repair costs resulting from front-end collisions at low speeds. Experience with the presently used bumper indicates that, at 10 mph, the mean cost of repair resulting from a front-end collision is $550. The manufacturer equips a sample of cars with the new bumper and has these cars undergo front-end collisions at 10 mph. Following the collisions, the cars are repaired and the repair costs recorded. Below is a Minitab printout obtained by applying the TTEST command to the repair-cost data.

```
TEST OF MU = 550.000 VS MU L.T. 550.000


 N     MEAN    STDEV   SE MEAN     T   P VALUE
15   496.667  68.624   17.719   -3.01  0.0047
```

Using the printout, determine the

a) null and alternative hypotheses for the manufacturer's hypothesis test.

b) subcommand used.

c) sample standard deviation of the repair costs.

d) number of cars in the sample.

e) sample mean repair cost.

f) value obtained for the test statistic, t.

g) P-value of the hypothesis test.

h) smallest significance level at which the null hypothesis can be rejected.

i) conclusion if the test is performed at the 5% significance level.

___ **9.86 (Computer exercise)** A company produces cans of stewed tomatoes with an advertised weight of 14 ounces. Recently, the company hired a quality-control engineer. The engineer wants to check whether, on the average, the cans do contain 14 ounces of stewed tomatoes. For an initial test, she takes a random sample of cans of stewed tomatoes, finds their net weights, and obtains the following Minitab printout by applying the TTEST command to the net weights:

```
TEST OF MU = 14.0000 VS MU N.E. 14.0000


 N    MEAN    STDEV   SE MEAN     T   P VALUE
20  13.9505  0.3242   0.0725   -0.68   0.50
```

From the printout, determine the

a) null and alternative hypotheses for the quality-control engineer's hypothesis test.

b) subcommand used by the quality-control engineer.

c) standard deviation of the net weights of the cans sampled by the quality-control engineer.

d) number of cans sampled.

e) mean net weight of the cans sampled.

f) value obtained for the test statistic, t.

g) P-value of the hypothesis test.

h) smallest significance level at which the null hypothesis can be rejected.

i) conclusion if the test is performed at the 5% significance level.

=== **9.87** A manufacturer of light bulbs produces a 60-watt bulb with a mean life of 1000 hours. The research and development (R&D) department has developed a new bulb that it claims will, on the average, outlast the present bulb. To try to justify its claim, R&D tests 10 new bulbs. The results show a mean life of $\bar{x} = 1050.2$ hours for the 10 bulbs, with a standard deviation of $s = 65.8$ hours.

a) At the 1% significance level, do the data obtained by R&D support its claim? [Assume that bulb life is normally distributed.]

b) Suppose that in part (a) you had mistakenly concluded that the test statistic

$$\frac{\bar{x} - 1000}{s/\sqrt{10}}$$

has the standard normal distribution.

(i) What critical value (z-value) would you then have used?

(ii) What critical value (t-value) did you actually use?

(iii) In general, does the mistaken use of a z critical value, when a t critical value should be used, make it more or less likely that the null hypothesis will be rejected? Provide a detailed explanation for your answer.

Hypothesis tests for a normal population mean when σ is known: Suppose that a random sample of size n is to be taken from a normally distributed population with mean μ and standard deviation σ. Then, by Key Fact 7.4 on page 360 (the sampling distribution of the mean for normal populations), the random variable $\bar{x}$ is normally distributed with mean $\mu_{\bar{x}} = \mu$

and standard deviation $\sigma_{\bar{x}} = \sigma/\sqrt{n}$. Or, equivalently, the standardized random variable

$$z = \frac{\bar{x} - \mu}{\sigma/\sqrt{n}}$$

has the standard normal distribution. Consequently, if the population being sampled is normally distributed and σ is known, then we can perform a hypothesis test with null hypothesis H_0: $\mu = \mu_0$ by employing the random variable

$$(2) \qquad\qquad z = \frac{\bar{x} - \mu_0}{\sigma/\sqrt{n}}$$

as the test statistic and using the standard-normal table, Table II, to obtain the critical value(s). We will discuss this fact in Exercises 9.88–9.92.

═ 9.88 Suppose that we want to perform a hypothesis test for a normal population mean with null hypothesis H_0: $\mu = \mu_0$ and that the population standard deviation, σ, is known. Formulate a step-by-step procedure for the hypothesis test that uses the test statistic given in Equation (2).

═ 9.89 A consumer group thinks that Wheat Flakes brand cereal contains, on the average, less than the advertised weight of 15 ounces per box. A random sample of 40 boxes of Wheat Flakes gives the following weights, in ounces:

15.8	15.1	15.2	15.4	14.8	15.6	15.7	14.5
14.8	15.4	15.3	15.5	15.2	14.6	15.4	15.4
15.5	14.7	14.7	15.1	14.7	15.3	15.3	15.5
14.0	14.2	14.6	15.0	15.1	14.9	14.9	15.8
15.0	14.4	15.4	14.3	15.4	15.9	15.2	15.6

Assume that the weights are normally distributed and that $\sigma = 0.5$ oz. Use your procedure from Exercise 9.88 to test, at the 5% significance level, whether the data provide sufficient evidence to conclude that the consumer group's conjecture is correct. (Note: The sum of the data is 604.2 oz.)

═ 9.90 When performing a hypothesis test for the population mean, μ, of a normally distributed population, why can't we always use the random variable in Equation (2) as the test statistic?

═ 9.91 Brown Swiss Dairy sells "half-gallon" cartons of milk. The contents of the cartons are known to be normally distributed with a standard deviation of $\sigma = 1.1$ fluid ounces. Suppose 15 randomly selected cartons have a mean content of 64.48 fluid ounces.

a) Do the data provide sufficient evidence to infer that the cartons actually contain more milk, on the average, than 64 fluid ounces? Perform the appropriate hypothesis test at the 0.05 level of significance using your procedure from Exercise 9.88.

b) Suppose that in part (a) you had incorrectly concluded that the test statistic

$$\frac{\bar{x} - 64}{1.1/\sqrt{15}}$$

has the t-distribution with df $= 14$.
 (i) What critical value (t-value) would you then have used?
 (ii) What critical value (z-value) did you actually use?
 (iii) In general, does the mistaken use of a t critical value, when a z critical value should be used, make it more or less likely that the null hypothesis will be rejected?

═ 9.92 Refer to Exercise 9.91.

a) Suppose you mistakenly use the t-table (Table III) instead of the standard-normal table (Table II) to obtain the critical value for the hypothesis test in part (a) of Exercise 9.91. What will be the significance level of the resulting test? Compare this with the desired significance level of 0.05.

b) More generally, suppose that you are performing a hypothesis test, at the significance level α, for the population mean, μ, of a normally distributed population. Further suppose that the population standard deviation, σ, is known so that the appropriate hypothesis testing procedure is the z-test of Exercise 9.88. If you mistakenly use the t-table instead of the standard-normal table to obtain the critical value(s), will the actual significance level of the resulting test be higher or lower than α? Justify your answer.

═ 9.93 Suppose that you are performing a hypothesis test, at the significance level α, for the population mean, μ, of a normally distributed population. Further suppose that the population standard deviation, σ, is unknown so that the appropriate hypothesis testing procedure is the t-test (Procedure 9.4 on page 458). If you mistakenly use the standard-normal table (Table II) instead of the t-table (Table III) to obtain the critical value(s), will the actual significance level of the resulting test be higher or lower than α? Provide a detailed explanation to justify your answer.

9.6 Large-sample hypothesis tests for a population proportion

Large-sample hypothesis tests for a population proportion follow essentially the same pattern as those for a population mean. Before proceeding, however, let us review the terminology and notation used in studying proportions.

Recall that a *two-category population* is one in which each member of the population is classified as either having or not having a specified attribute. For a human population, that attribute might be "female," it might be "owning a home," or it might be "over 40."

The proportion (percentage) of the entire population that has the specified attribute is called the *population proportion* and is denoted by the letter p. The proportion of a sample from the population that has the specified attribute is called a *sample proportion* and is denoted by the symbol $\hat{p}$. Thus,

$$p = \text{population proportion}$$
$$\hat{p} = \text{sample proportion}$$

For a sample of size n, the sample proportion is computed from the formula

$$\hat{p} = \frac{x}{n}$$

where x is the number of members sampled that have the specified attribute (the number of "successes").

Now that we have reviewed the terminology and notation used in statistical inferences for a population proportion, we return to the development of a large-sample hypothesis-testing procedure. The essential fact used in designing such a procedure is Key Fact 8.5 on page 407, the sampling distribution of the proportion: Suppose that a large random sample of size n is to be taken from a two-category population with population proportion p. Then the random variable $\hat{p}$ is approximately normally distributed and has mean $\mu_{\hat{p}} = p$ and standard deviation $\sigma_{\hat{p}} = \sqrt{p(1-p)/n}$. Or, equivalently, the standardized random variable

$$z = \frac{\hat{p} - p}{\sqrt{p(1-p)/n}}$$

has approximately the standard normal distribution.

Consequently, to perform a large-sample hypothesis test with null hypothesis H_0: $p = p_0$, we can use the random variable

$$z = \frac{\hat{p} - p_0}{\sqrt{p_0(1-p_0)/n}}$$

as the test statistic and obtain the critical value(s) from the standard-normal table, Table II.

PROCEDURE 9.5

To perform a hypothesis test for a population proportion with null hypothesis $H_0: p = p_0$.

ASSUMPTION

Both np_0 and $n(1 - p_0)$ are at least 5.

STEP 1 *State the null and alternative hypotheses.*
STEP 2 *Decide on the significance level, α.*
STEP 3 *The critical value(s)*
    a) *for a two-tailed test are $\pm z_{\alpha/2}$.*
    b) *for a left-tailed test is $-z_\alpha$.*
    c) *for a right-tailed test is z_α.*
    Use Table II to find the critical value(s).

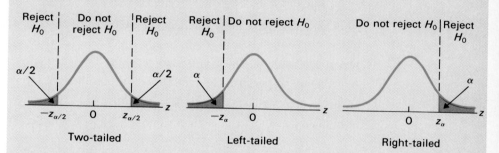

Two-tailed   Left-tailed   Right-tailed

STEP 4 *Compute the value of the test statistic*

$$z = \frac{\hat{p} - p_0}{\sqrt{p_0(1 - p_0)/n}}$$

STEP 5 *If the value of the test statistic falls in the rejection region, then reject H_0; otherwise, do not reject H_0.*
STEP 6 *State the conclusion in words.*

EXAMPLE 9.16   **Illustrates Procedure 9.5**

In the final days of the 1988 presidential campaign, the Gallup Organization conducted a telephone poll of 3079 registered voters nationwide. Of those surveyed, 1632 said that they planned to vote for George Bush. At the 5% significance level, did the data provide sufficient evidence to conclude that George Bush would receive a majority of the popular vote?

SOLUTION   We will apply Procedure 9.5 to perform the required hypothesis test. But first it is necessary to check that the conditions for its use are satisfied. We have $n = 3079$

and $p_0 = 0.50$ (50%). Therefore,

$$np_0 = 3079 \cdot 0.50 = 1539.5$$

and

$$n(1 - p_0) = 3079 \cdot (1 - 0.50) = 1539.5$$

Since both np_0 and $n(1 - p_0)$ are at least 5, we can use Procedure 9.5 to perform the hypothesis test.

STEP 1 *State the null and alternative hypotheses.*

> $H_0: p = 0.50$ (George Bush would not receive a majority of the votes)
> $H_a: p > 0.50$ (George Bush would receive a majority of the votes)

Note that the hypothesis test is right-tailed since there is a greater-than sign (>) in the alternative hypothesis.

STEP 2 *Decide on the significance level, α.*

We are to perform the test at the 5% significance level. So, $\alpha = 0.05$.

STEP 3 *The critical value for a right-tailed test is z_α.*

Since $\alpha = 0.05$, the critical value is $z_{0.05} = 1.645$. See Figure 9.19.

FIGURE 9.19

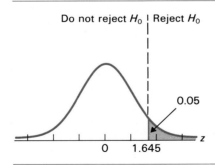

STEP 4 *Compute the value of the test statistic*

$$z = \frac{\hat{p} - p_0}{\sqrt{p_0(1 - p_0)/n}}$$

We have $n = 3079$ and $p_0 = 0.50$. The number of registered voters surveyed who planned to vote for George Bush is 1632. Therefore, the proportion of those surveyed who planned to vote for George Bush is

$$\hat{p} = \frac{x}{n} = \frac{1632}{3079} = 0.530 \ (53.0\%)$$

Consequently, the value of the test statistic is

$$z = \frac{\hat{p} - p_0}{\sqrt{p_0(1 - p_0)/n}} = \frac{0.530 - 0.50}{\sqrt{(0.50)(1 - 0.50)/3079}} = \boxed{3.33}$$

STEP 5 *If the value of the test statistic falls in the rejection region, reject H_0; otherwise, do not reject H_0.*

From Step 4, the value of the test statistic is $z = 3.33$. A glance at Figure 9.19 shows that this falls in the rejection region. Thus, we reject H_0.

STEP 6 *State the conclusion in words.*

The survey done by the Gallup Organization provided sufficient evidence to conclude that George Bush would receive a majority of the popular vote. ∎

Exercises 9.6

In Exercises 9.94–9.102, use Procedure 9.5 on page 467 to perform each of the hypothesis tests. Be sure to check whether the conditions for applying that procedure are met.

___ **9.94**   The Motor Vehicle Manufacturers Association of the United States reports in *Motor Vehicle Facts and Figures* that, in 1962, foreign cars made up 4.8% of all U.S. car sales. In 1983, the percentage was 27.8%. U.S. automakers have implemented several policies to cut costs in order to challenge the imports. From a random sample of 500 of this year's car sales, it is found that 128 are imports. Do the data suggest, at the 5% significance level, that the percentage of foreign car sales has declined from the 1983 figure of 27.8%?

___ **9.95**   In 1988, when a political incumbent ran for city council, 63.8% of the voters thought that repair of city streets was an important priority. Since then, repairs have been effected in several areas of the city. The councillor wonders if attitudes on the issue have changed. Out of a sample of 350 voters, 204 still think that repair of city streets is an important issue. Perform the appropriate hypothesis test at the 5% significance level.

___ **9.96**   According to the Arizona Real Estate Commission, in 1982, only 10% of the people holding real estate licenses were active in the industry. An independent agency has been asked by the Commission to determine whether this year's percentage is higher.

A random sample of 150 licensed people reveals that 24 are currently active. Perform the appropriate hypothesis test at the 1% significance level.

___ **9.97**   As reported by the American Medical Association in *Physician Characteristics and Distribution in the U.S.*, 9.0% of all physicians were women in 1975; in 1980, 11.6% were women; and in 1981, 12.2% were women. For this year, out of a sample of 125 randomly selected physicians, 19 are women. Do the data indicate, at the 10% significance level, that the proportion of female physicians is higher now than in 1981?

___ **9.98**   A 1982 study, conducted by the U.S. National Center for Health Statistics and published in *Vital and Health Statistics,* revealed that about 39.5% of young adults (18–25 years old) smoked cigarettes. In 1985, 37.2% of 1283 randomly selected young adults smoked cigarettes. Do the data provide sufficient evidence to conclude that the percentage of young adult smokers declined between 1982 and 1985? Use a significance level of 0.05.

___ **9.99**   A direct-mail firm wants to conduct a test-market offering for one of its new products. A random sample of 800 people is chosen to receive advertising material describing the new product. It is decided that additional advertising and promotion will occur only if the sample results provide strong evidence that the actual (population) response rate, *p,* will exceed 6.5%. What decision will be made if 70 out of the 800 people make a purchase? Use $\alpha = 0.01$.

__ **9.100**   Of the 38 numbers on a roulette wheel, 18 are red, 18 are black, and two are green. If the wheel is balanced, the probability of the ball landing on red is $18/38 = 0.474$. A gambler has been studying a roulette wheel. If the wheel is out of balance, then the gambler can improve his odds of winning. The gambler observes 200 spins of the wheel and finds that, out of those 200 trials, the ball lands on red 93 times. At the 10% significance level, do the data provide sufficient evidence to conclude that the ball is not landing on red the correct percentage of the time?

__ **9.101**   In 1983, 12.3% of all families earned incomes below the poverty level, as reported by the U.S. Bureau of the Census in the publication *Current Population Reports*. During that same year, a sociologist randomly selected 250 families in the Northeast. Twenty-eight of these 250 families earned incomes below the poverty level. Does the study provide sufficient evidence to conclude that, in 1983, the proportion of northeastern families below the poverty level was less than the national proportion? Use $\alpha = 0.05$.

__ **9.102**   On Sunday, October 7, 1984, presidential candidates Ronald Reagan and Walter Mondale debated on national television. A *Newsweek* poll, which was conducted by the Gallup Organization, was taken to determine which candidate the public thought did a better job in the debate. The Gallup Organization telephoned 379 registered voters who watched the debate and found that 205 of the 379 thought that Mondale did a better job. Can we infer from the poll, at the 5% significance level, that a majority of all registered voters who watched the debate thought Mondale did a better job?

≡ **9.103**   The test statistic used in Procedure 9.5 is

(3)
$$z = \frac{\hat{p} - p_0}{\sqrt{p_0(1 - p_0)/n}}$$

In this exercise we will show that the test statistic in Equation (3) is the familiar

(4)
$$z = \frac{\bar{x} - \mu_0}{\sigma/\sqrt{n}}$$

when the latter is specialized to the situation of proportions. In other words, we will prove that Procedure 9.5 is really just a special case of Procedure 9.1.

Consider now a finite, two-category population in which the proportion of members with the specified attribute is equal to p. We can think of such a population as consisting of 1s and 0s. A member of the population is a "1" if it has the specified attribute and is a "0" otherwise.

a) If the size of the population is N, how many 1s are in the population?

b) Use part (a) and Definition 3.8 on page 124 to show that the mean of this population of 1s and 0s is p; that is, $\mu = p$.

c) Use part (b) and the shortcut formula in Definition 3.9 on page 126 to show that the standard deviation of this population of 1s and 0s is $\sqrt{p(1-p)}$; in other words, $\sigma = \sqrt{p(1-p)}$.

d) Suppose a random sample of size n is to be taken from this population of 1s and 0s. Show that $\bar{x} = \hat{p}$.

e) Deduce from parts (b)–(d) that the test statistic in Equation (4) becomes the one in Equation (3), when specialized to the situation of proportions.

Chapter review

FORMULAS   In the formulas below,

μ = population mean n = sample size
$\bar{x}$ = sample mean p = population proportion
s = sample standard deviation $\hat{p}$ = sample proportion

Test statistic for H_0: $\mu = \mu_0$ (large sample), 438

$$z = \frac{\bar{x} - \mu_0}{s/\sqrt{n}}$$

Test statistic for H_0: $\mu = \mu_0$ (normal population), 459

$$t = \frac{\bar{x} - \mu_0}{s/\sqrt{n}}$$

with df = $n - 1$.

Test statistic for H_0: $p = p_0$ (large sample), 467

$$z = \frac{\hat{p} - p_0}{\sqrt{p_0(1 - p_0)/n}}$$

YOU SHOULD   1. use and understand the preceding formulas.
BE ABLE TO   2. define the terms associated with hypothesis testing.
              3. choose the null and alternative hypotheses for a hypothesis test.
              4. explain the logic behind hypothesis testing.
              5. identify the test statistic, rejection region, nonrejection region, and critical value(s) for a hypothesis test.
              6. define and apply the concepts of Type I and Type II errors.
              7. state the possible conclusions for a hypothesis test.
              8. obtain the critical value(s) for a specified significance level.
              9. perform a large-sample hypothesis test for a population mean, μ.
             10. state and apply the steps for performing a hypothesis test using the classical approach to hypothesis testing.
             11. obtain the P-value of a hypothesis test.*
             12. state and apply the steps for performing a hypothesis test using the P-value approach to hypothesis testing.*
             13. perform a hypothesis test for a population mean, μ, when the population being sampled is normally distributed.
             14. perform a large-sample hypothesis test for a population proportion, p.
             15. use the Minitab commands covered in this chapter.*
             16. interpret the output obtained from the application of the Minitab commands discussed in this chapter.*

REVIEW TEST

1. The U.S. Department of Agriculture reports in *Food Consumption, Prices, and Expenditures* that the average American consumed 20.6 lb of cheese in 1983. There has been a steady increase in cheese consumption since 1960, when the average American ate only 8.3 lb of cheese. A researcher thinks that the trend of increasing cheese consumption is still continuing. He wants to perform a hypothesis test to decide whether last year's mean cheese consumption is greater than the 1983 mean of 20.6 lb.
   a) Determine the null hypothesis for the researcher's hypothesis test.
   b) Determine the alternative hypothesis for the researcher's hypothesis test.
   c) Classify the hypothesis test as a two-tailed test, a left-tailed test, or a right-tailed test.

2. Below is a graph that portrays the decision criterion for a hypothesis test concerning a population mean, μ. The null hypothesis for the hypothesis test is $H_0: \mu = \mu_0$ and the test statistic is

$$z = \frac{\bar{x} - \mu_0}{\sigma/\sqrt{n}}$$

   Also, the curve in the graph shows the implications of the decision criterion if, in fact, the null hypothesis is true.

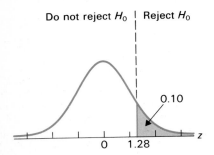

   Do not reject H_0 | Reject H_0

   0.10

   0   1.28

   Determine the
   a) rejection region.
   b) nonrejection region.
   c) critical value(s).
   d) significance level.
   e) Construct a graph that depicts your results from parts (a)–(d).
   f) Identify the hypothesis test as a two-tailed test, a left-tailed test, or a right-tailed test.

3. The null and alternative hypotheses for the hypothesis test in Problem 1 are

   $H_0: \mu = 20.6$ lb (mean has not increased)
   $H_a: \mu > 20.6$ lb (mean has increased)

   where μ is last year's mean cheese consumption for all Americans. Explain what each of the following would mean:
   a) A Type I error.
   b) A Type II error.
   c) A correct decision.

   Now, suppose the results of carrying out the hypothesis test lead to rejection of the null hypothesis. Classify that decision by error type or as a correct decision if, in fact, last year's mean consumption of cheese for all Americans
   d) has not increased over the 1983 mean of 20.6 lb.
   e) has increased over the 1983 mean of 20.6 lb.

4. Suppose that the researcher in Problem 1 randomly selects 35 Americans and obtains the following data, in pounds, on their cheese consumption for last year:

33	16	20	25	29	23	27
21	20	19	23	9	15	34
13	29	23	31	19	32	11
26	15	23	20	31	20	13
24	16	14	18	23	24	24

   a) At the 10% significance level, do the data provide evidence to conclude that last year's mean cheese consumption for all Americans has increased over the 1983 mean of 20.6 lb? Assume as known that $\sigma = 6.9$ lb. (*Note:* The sum of the data is 763 lb.)
   b) Given the conclusion in part (a), if an error has been made, what type must it be? Explain.

*5. (**Computer problem**) Suppose that the data in Problem 4 are stored in a column named CHEESE.
   a) Which Minitab command and subcommands (if any) should be used to carry out the hypothesis test considered in part (a) of that problem?
   b) If you have access to Minitab, use it to perform the hypothesis test.

*6. (**Computer problem**) The abridged Minitab printout at the top of the first column on the next page was obtained by applying the command ZTEST to the data in Problem 4.

```
TEST OF MU = 20.600 VS MU G.T. 20.600
THE ASSUMED SIGMA = 6.90


N     MEAN   STDEV   SE MEAN      Z   P VALUE
35   21.800  6.480    1.166    1.03    0.15
```

Using the printout, determine

a) the null and alternative hypotheses.

b) the subcommands used (if any).

c) the (assumed) population standard deviation, σ, for last year's cheese consumptions.

d) the sample standard deviation, s.

e) the number of people in the sample.

f) last year's mean cheese consumption, $\bar{x}$, for the people in the sample.

g) the value of the test statistic, z.

h) the P-value of the hypothesis test.

i) the smallest significance level at which the null hypothesis can be rejected.

j) the conclusion if the hypothesis test is performed at the 10% significance level.

7. Between May 17 and 20, 1985, the Gallup Organization conducted a survey to estimate then President Ronald Reagan's overall standing with the public. Out of a random sample of 1528 adults, 840 approved of the way Ronald Reagan was handling his job as president. Do these results indicate, at the 1% significance level, that a majority of all Americans were satisfied with President Reagan's performance?

8. Each year, manufacturers perform mileage tests on new car models and submit the results to the Environmental Protection Agency. The EPA then tests the vehicles to determine whether the manufacturers are correct. In 1989, one company reported that a particular model equipped with a four-speed manual transmission averaged 29 miles per gallon (mpg) on the highway. Let us suppose that the EPA tested 15 of the cars and obtained the gas mileages given below.

27.3	31.2	29.4	31.6	28.6
30.9	29.7	28.5	27.8	27.3
25.9	28.8	28.9	27.8	27.6

a) At the 0.05 level of significance, what decision would you make regarding the company's report on the gas mileage of the car? [Assume that the gas mileages are normally distributed.]

b) Is it necessary to assume that the gas mileages are normally distributed in order to carry out the hypothesis test in part (a) in the way that you did? Explain.

*9. (Computer problem) Assume that the mileage data displayed in Problem 8 are stored in a column named MILEAGE.

a) Which Minitab command and subcommands (if any) should be used to carry out the hypothesis test considered in part (a) of that problem?

b) If you have access to Minitab, use it to perform the hypothesis test.

*10. (Computer problem) The following abridged Minitab printout was obtained by applying the TTEST command to the data in Problem 8:

```
TEST OF MU = 29.000 VS MU N.E. 29.000


N     MEAN   STDEV   SE MEAN      T   P VALUE
15   28.753  1.595    0.412   -0.60    0.56
```

Employ the printout to determine

a) the null and alternative hypotheses for the hypothesis test.

b) the subcommands used (if any).

c) the standard deviation of the mileages of the cars tested.

d) the number of cars tested.

e) the mean mileage of the cars tested.

f) the value of the test statistic, t.

g) the P-value of the hypothesis test.

h) the smallest significance level at which the null hypothesis can be rejected.

i) the conclusion if the hypothesis test is performed at the 5% significance level.

11. According to *Crime in the United States*, published by the U.S. Federal Bureau of Investigation (FBI), the mean value lost due to purse snatching was $178 in 1983. For this year, 40 randomly selected purse-snatching offenses have a mean value lost of $159 with a standard deviation of $84. Do the data provide sufficient evidence to conclude that the mean value lost due to purse snatching has decreased from the 1983 mean of $178? Use $\alpha = 0.05$.

*12. Refer to Problem 11.

a) Determine the P-value of the hypothesis test.

b) Perform the hypothesis test using the P-value approach to hypothesis testing.

CHAPTER 10

INFERENCES FOR TWO MEANS OR PROPORTIONS

In Chapters 8 and 9 we learned how to obtain confidence intervals and perform hypothesis tests for one population mean, μ, or one population proportion, p. Frequently, however, inferential statistics is used to compare the means or proportions of two populations.

For instance, we might perform a hypothesis test to decide whether the mean starting salary of business majors is greater than the mean starting salary of liberal arts majors. Or we might need to find a confidence interval for the difference between the percentages of unemployed white-collar workers and unemployed blue-collar workers. In this chapter, we will study methods of making statistical inferences for two population means or two population proportions.

CHAPTER OUTLINE

10.1 Large-sample inferences for two population means using independent samples

In this section we will examine methods of making statistical inferences for the means of two populations when dealing with large samples. The methods we will consider here require not only that the samples selected from the two populations are random but also that they are *independent*. Two samples are **independent** if the sample selected from one of the populations has no effect on the sample selected from the other population.

We begin by discussing large-sample hypothesis tests for comparing the means of two populations using independent samples. The following example introduces the basic ideas involved in performing such hypothesis tests.

EXAMPLE 10.1   *Introduces hypothesis tests for comparing two population means*

Suppose we want to perform a hypothesis test in order to decide whether there is a difference between the mean salary of faculty teaching in public institutions and the mean salary of faculty teaching in private institutions. We can formulate the problem statistically in the following way: First, let the two populations in question be designated as Population 1 and Population 2:

Population 1: All salaries of faculty teaching in public institutions.
Population 2: All salaries of faculty teaching in private institutions.

Next, denote the mean salary of all faculty teaching in public institutions by μ_1 and the mean salary of all faculty teaching in private institutions by μ_2. Then the hypothesis test that we want to perform can be stated as

$$H_0: \mu_1 = \mu_2 \text{ (mean salaries are the same)}$$
$$H_a: \mu_1 \neq \mu_2 \text{ (mean salaries are different)}$$

Roughly speaking, the hypothesis test can be carried out as follows:

1. Take a random sample from each of the two populations.
2. Compute the mean, $\overline{x}_1$, of the sample from Population 1 and the mean, $\overline{x}_2$, of the sample from Population 2.
3. Reject the null hypothesis if $\overline{x}_1$ and $\overline{x}_2$ differ by too much; otherwise, do not reject the null hypothesis.

This process is pictured in Figure 10.1 at the top of the next page.

Suppose we obtain a sample of 30 salaries from Population 1 (public institution salaries) and a sample of 35 salaries from Population 2 (private institution salaries); and that the results are as depicted in Table 10.1.

FIGURE 10.1

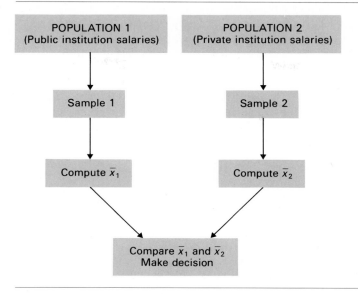

TABLE 10.1   Annual salaries ($) for 30 randomly selected faculty in public institutions and 35 randomly selected faculty in private institutions

Sample 1 (public institutions)						Sample 2 (private institutions)							
27,565	30,497	47,309	45,419	32,958	47,321	28,666	50,108	31,258	79,204	41,502	39,401	27,968
25,789	69,433	34,238	24,240	17,227	45,816	52,015	39,741	51,872	52,341	30,843	37,300	69,725
23,706	31,131	43,757	25,958	44,748	63,877	24,706	31,303	36,691	28,611	36,813	27,486	19,490
35,650	55,693	24,390	31,268	56,503	16,920	43,217	43,494	27,607	35,853	17,211	15,775	21,845
30,833	34,755	30,565	49,337	40,400	32,753	35,687	68,767	40,566	40,633	43,740	48,927	64,198

The sample means for the two data sets in Table 10.1 are

$$\overline{x}_1 = \frac{\Sigma x}{n_1} = \frac{\$1{,}120{,}056}{30} = \$37{,}335.20$$

$$\overline{x}_2 = \frac{\Sigma x}{n_2} = \frac{\$1{,}384{,}564}{35} = \$39{,}558.97$$

The question now is whether the difference of $2223.77 between these two sample means can reasonably be attributed to sampling error or whether the difference is large enough to indicate that the two populations have different means. To answer this question, we need to know how likely it would be to get such a difference in sample means if, in fact, the two population means are equal. And that likelihood can be determined once we have obtained the probability distribution of the difference, $\overline{x}_1 - \overline{x}_2$, between two sample means—the **sampling distribution of the difference between two means.** Let us therefore examine that sampling distribution. We will complete the solution to the salary problem shortly.   ■

ε SAMPLING DISTRIBUTION OF THE DIFFERENCE BETWEEN NO MEANS (LARGE AND INDEPENDENT SAMPLES)

First we need to become familiar with the notation for parameters and statistics when analyzing two populations. Let us call the two populations under consideration Population 1 and Population 2. Then, as indicated in Example 10.1, we use a subscript "1" when referring to parameters or statistics for Population 1 and we use a subscript "2" when referring to parameters or statistics for Population 2. This notation is displayed in Table 10.2.

TABLE 10.2
Notation for parameters and statistics when dealing with two populations

	Population 1	Population 2
Population mean	μ_1	μ_2
Population std. dev.	σ_1	σ_2
Sample mean	$\bar{x}_1$	$\bar{x}_2$
Sample std. dev.	s_1	s_2
Sample size	n_1	n_2

Next we will obtain formulas that relate the mean and standard deviation of the random variable $\bar{x}_1 - \bar{x}_2$ to the means, μ_1 and μ_2, and standard deviations, σ_1 and σ_2, of the two populations. It can be shown that the mean of the difference of the random variables $\bar{x}_1$ and $\bar{x}_2$ is simply the difference of the means:

$$\mu_{\bar{x}_1 - \bar{x}_2} = \mu_{\bar{x}_1} - \mu_{\bar{x}_2}$$

Additionally, it can be shown that for independent samples, the standard deviation of the random variable $\bar{x}_1 - \bar{x}_2$ is related to the standard deviations of the random variables $\bar{x}_1$ and $\bar{x}_2$ by the equation

$$\sigma_{\bar{x}_1 - \bar{x}_2} = \sqrt{\sigma_{\bar{x}_1}^2 + \sigma_{\bar{x}_2}^2}$$

Combining the previous two formulas with the formulas $\mu_{\bar{x}} = \mu$ and $\sigma_{\bar{x}} = \sigma/\sqrt{n}$ from Section 7.3, we obtain the formulas

$$\mu_{\bar{x}_1 - \bar{x}_2} = \mu_1 - \mu_2$$

and

$$\sigma_{\bar{x}_1 - \bar{x}_2} = \sqrt{(\sigma_1^2/n_1) + (\sigma_2^2/n_2)}$$

These last two formulas express the mean and standard deviation of the random variable $\bar{x}_1 - \bar{x}_2$ in terms of the population parameters and the sample sizes. See Exercise 10.15 for details.

Finally, by the Central Limit Theorem (Key Fact 7.5 on page 363), we know that the random variable $\bar{x}$ is approximately normally distributed for large samples.

Thus, if both n_1 and n_2 are large, the random variables $\bar{x}_1$ and $\bar{x}_2$ are both approximately normally distributed. From this it can be proved mathematically that, for independent samples, the random variable $\bar{x}_1 - \bar{x}_2$ is also approximately normally distributed. Consequently, we have the following fundamental fact:

KEY FACT 10.1   The sampling distribution of the difference between two means (large and independent samples)

Suppose that a random sample of size $n_1 \geq 30$ is to be taken from a population with mean μ_1 and standard deviation σ_1; and that a random sample of size $n_2 \geq 30$ is to be taken from a population with mean μ_2 and standard deviation σ_2. Further suppose that the two samples are to be selected independently of one another. Then the random variable $\bar{x}_1 - \bar{x}_2$ is approximately normally distributed and has mean $\mu_{\bar{x}_1 - \bar{x}_2} = \mu_1 - \mu_2$ and standard deviation $\sigma_{\bar{x}_1 - \bar{x}_2} = \sqrt{(\sigma_1^2/n_1) + (\sigma_2^2/n_2)}$. Thus, the standardized random variable

$$z = \frac{(\bar{x}_1 - \bar{x}_2) - (\mu_1 - \mu_2)}{\sqrt{(\sigma_1^2/n_1) + (\sigma_2^2/n_2)}}$$

has approximately the standard normal distribution.

LARGE-SAMPLE HYPOTHESIS TESTS FOR TWO POPULATION MEANS USING INDEPENDENT SAMPLES

Now that we have obtained the sampling distribution of the difference between two means when dealing with large and independent samples, we can develop a procedure for performing a hypothesis test. The null hypothesis will be

$$H_0: \mu_1 = \mu_2 \text{ (population means are equal)}$$

Note that if the null hypothesis is true, then $\mu_1 - \mu_2 = 0$, and so the standardized random variable in Key Fact 10.1 becomes simply

$$(1) \qquad\qquad z = \frac{\bar{x}_1 - \bar{x}_2}{\sqrt{(\sigma_1^2/n_1) + (\sigma_2^2/n_2)}}$$

This last random variable can serve as the test statistic for a hypothesis test concerning two population means in the same way as the random variable

$$z = \frac{\bar{x} - \mu_0}{\sigma/\sqrt{n}}$$

does for a hypothesis test concerning one population mean. The numerator of the test statistic for two means, given in Equation (1), measures the difference between the sample means. The denominator standardizes the test statistic so that the standard-normal table, Table II, can be used to obtain the critical value(s).

Let us now discuss the alternative hypothesis for a hypothesis test concerning the means of two populations. As we know, there are three possibilities for the alternative hypothesis:

1. *Two-tailed test:* If we want to decide whether the means of Population 1 and Population 2 are different, then the alternative hypothesis will be

$$H_a: \mu_1 \neq \mu_2$$

If the null hypothesis, $H_0: \mu_1 = \mu_2$, is true, then we would expect the sample means $\overline{x}_1$ and $\overline{x}_2$ to be roughly equal (why?). Thus, the null hypothesis will be rejected in favor of the alternative hypothesis if $\overline{x}_1$ and $\overline{x}_2$ differ by too much, that is, if the test statistic in Equation (1) is either too small or too large. In other words, the rejection region will consist of two "tails" of the standard normal distribution, just as in two-tailed tests for one population mean.

2. *Left-tailed test:* If we want to decide whether the mean of Population 1 is smaller than the mean of Population 2, then the alternative hypothesis will be

$$H_a: \mu_1 < \mu_2$$

In this case, the null hypothesis, $H_0: \mu_1 = \mu_2$, will be rejected in favor of the alternative hypothesis only if $\overline{x}_1$ is too much smaller than $\overline{x}_2$, that is, if the test statistic in Equation (1) is too small. In other words, the rejection region will consist of a left "tail" of the standard normal distribution, just as in left-tailed tests for one population mean.

3. *Right-tailed test:* If we want to decide whether the mean of Population 1 is larger than the mean of Population 2, then the alternative hypothesis will be

$$H_a: \mu_1 > \mu_2$$

In this case, the null hypothesis, $H_0: \mu_1 = \mu_2$, will be rejected in favor of the alternative hypothesis only if $\overline{x}_1$ is too much larger than $\overline{x}_2$, that is, if the test statistic in Equation (1) is too large. In other words, the rejection region will consist of a right "tail" of the standard normal distribution, just as in right-tailed tests for one population mean.

With the above discussion in mind, we can now write down a method for performing a large-sample hypothesis test to compare two population means using independent samples. Procedure 10.1 supplies the steps.

PROCEDURE 10.1
To perform a hypothesis test for two population means with null hypothesis $H_0: \mu_1 = \mu_2$.

ASSUMPTIONS

1. Independent samples.
2. Large samples ($n_1 \geq 30$, $n_2 \geq 30$).

STEP 1 *State the null and alternative hypotheses.*
STEP 2 *Decide on the significance level, α.*

STEP 3 *The critical value(s)*
      a) *for a two-tailed test are* $\pm z_{\alpha/2}$.
      b) *for a left-tailed test is* $-z_{\alpha}$.
      c) *for a right-tailed test is* z_{α}.
      Use Table II to find the critical value(s).

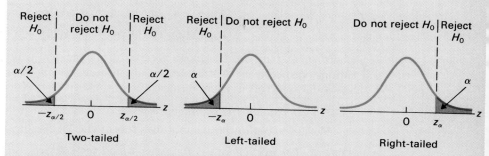

STEP 4 *Compute the value of the test statistic*

$$z = \frac{\bar{x}_1 - \bar{x}_2}{\sqrt{(s_1^2/n_1) + (s_2^2/n_2)}}$$

STEP 5 *If the value of the test statistic falls in the rejection region, then reject H_0; otherwise, do not reject H_0.*
STEP 6 *State the conclusion in words.*

Note: In Step 4 of Procedure 10.1 we have used the sample standard deviations, s_1 and s_2, in the denominator of the test statistic. Theoretically, we should use the population standard deviations, σ_1 and σ_2. However, σ_1 and σ_2 are rarely known and so we have used s_1 and s_2 in their place. This is acceptable because for large samples the sample standard deviations are likely to be a good approximation of the population standard deviations. In the rare case where the population standard deviations are known, they should always be used in place of the sample standard deviations in Step 4 of Procedure 10.1.

EXAMPLE 10.2   *Illustrates Procedure 10.1*

We now return to the salary problem posed in Example 10.1. Recall that we want to perform a hypothesis test to decide whether there is a difference between the mean salaries of faculty teaching in public and private institutions.
    Random samples of 30 faculty teaching in public institutions and 35 faculty teaching in private institutions yield the data in Table 10.1 on page 477. Do the data suggest, at the 5% significance level, that there is a difference in mean salaries for faculty teaching in public and private institutions?

SOLUTION   First observe that both sample sizes are large: $n_1 = 30$ and $n_2 = 35$. Next observe that the description of the problem implies that the samples are independent. Thus, we can employ Procedure 10.1 to perform the hypothesis test.

STEP 1 *State the null and alternative hypotheses.*

The null and alternative hypotheses are

$$H_0\text{: } \mu_1 = \mu_2 \text{ (mean salaries are the same)}$$
$$H_\text{a}\text{: } \mu_1 \neq \mu_2 \text{ (mean salaries are different)}$$

where μ_1 and μ_2 are the mean salaries of all faculty in public and private institutions, respectively. Note that the hypothesis test is two-tailed since there is a not-equal sign ($\neq$) in the alternative hypothesis.

STEP 2 *Decide on the significance level, α.*

We are to perform the hypothesis test at the 5% significance level. So, $\alpha = 0.05$.

STEP 3 *The critical values for a two-tailed test are $\pm z_{\alpha/2}$.*

Since $\alpha = 0.05$, we find from Table II (or the last row of Table III) that the critical values are $\pm z_{0.05/2} = \pm z_{0.025} = \pm 1.96$. See Figure 10.2.

FIGURE 10.2

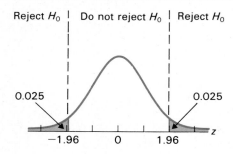

STEP 4 *Compute the value of the test statistic*

$$z = \frac{\bar{x}_1 - \bar{x}_2}{\sqrt{(s_1^2/n_1) + (s_2^2/n_2)}}$$

We have $n_1 = 30$ and $n_2 = 35$. To obtain the sample mean and sample standard deviation for each of the two data sets in Table 10.1, we proceed in the usual way:

$$\bar{x}_1 = \frac{\Sigma x}{n_1} = \$37{,}335.20, \quad s_1 = \sqrt{\frac{n_1(\Sigma x^2) - (\Sigma x)^2}{n_1(n_1 - 1)}} = \$13{,}129.09$$

and

$$\bar{x}_2 = \frac{\Sigma x}{n_2} = \$39{,}558.97, \quad s_2 = \sqrt{\frac{n_2(\Sigma x^2) - (\Sigma x)^2}{n_2(n_2 - 1)}} = \$14{,}940.88$$

Thus, the value of the test statistic is

$$z = \frac{\overline{x}_1 - \overline{x}_2}{\sqrt{(s_1^2/n_1) + (s_2^2/n_2)}} = \frac{37{,}335.20 - 39{,}558.97}{\sqrt{(13{,}129.09^2/30) + (14{,}940.88^2/35)}} = \boxed{-0.64}$$

STEP 5 *If the value of the test statistic falls in the rejection region, reject H_0; otherwise, do not reject H_0.*

From Step 4, we see that the value of the test statistic is $z = -0.64$, which does not fall in the rejection region. Hence, we do not reject H_0.

STEP 6 *State the conclusion in words.*

MTB

Based on the sample data, we have insufficient evidence to conclude that there is a difference in mean salaries for faculty in public and private institutions. ∎

LARGE-SAMPLE CONFIDENCE INTERVALS FOR THE DIFFERENCE BETWEEN TWO POPULATION MEANS USING INDEPENDENT SAMPLES

We can also use Key Fact 10.1 on page 479 to derive the following confidence-interval procedure for the difference between two population means when dealing with large, independent samples. See Exercise 10.16 for details of the derivation.

PROCEDURE 10.2

To find a confidence interval for the difference between two population means.

ASSUMPTIONS

1. Independent samples.
2. Large samples ($n_1 \geq 30$, $n_2 \geq 30$).

STEP 1 *For a confidence level of $1 - \alpha$, use Table II to find $z_{\alpha/2}$.*
STEP 2 *The endpoints of the confidence interval for $\mu_1 - \mu_2$ are*

$$(\overline{x}_1 - \overline{x}_2) \pm z_{\alpha/2} \cdot \sqrt{(s_1^2/n_1) + (s_2^2/n_2)}$$

EXAMPLE 10.3   *Illustrates Procedure 10.2*

Consider once more the situation of Example 10.1. Find a 95% confidence interval for the difference, $\mu_1 - \mu_2$, between the mean salaries of faculty teaching in public and private institutions.

SOLUTION   We apply Procedure 10.2.

STEP 1 *For a confidence level of $1 - \alpha$, use Table II to find $z_{\alpha/2}$.*

For a 95% confidence interval, the confidence level is $0.95 = 1 - 0.05$. So, $\alpha = 0.05$. From Table II, we find that $z_{\alpha/2} = z_{0.05/2} = z_{0.025} = 1.96$.

STEP 2 *The endpoints of the confidence interval for $\mu_1 - \mu_2$ are*

$$(\bar{x}_1 - \bar{x}_2) \pm z_{\alpha/2} \cdot \sqrt{(s_1^2/n_1) + (s_2^2/n_2)}$$

From Step 1, $z_{\alpha/2} = 1.96$. Also, $n_1 = 30$, $n_2 = 35$ and, from Example 10.2, we know that $\bar{x}_1 = \$37,335.20$, $s_1 = \$13,129.09$ and $\bar{x}_2 = \$39,558.97$, $s_2 = \$14,940.88$. Hence, the endpoints of the confidence interval for $\mu_1 - \mu_2$ are

$$(37,335.20 - 39,558.97) \pm 1.96 \cdot \sqrt{(13,129.09^2/30) + (14,940.88^2/35)}$$

or

$$-2223.77 \pm 6824.56$$

Thus, a 95% confidence interval for $\mu_1 - \mu_2$ is from

$$-9048.33 \quad \text{to} \quad 4600.79$$

We can be 95% confident that the difference, $\mu_1 - \mu_2$, between the mean salaries of faculty teaching in public institutions and faculty teaching in private institutions is somewhere between $-\$9,048.33$ and $\$4,600.79$. ∎

MTB

Exercises 10.1

___ **10.1** Consider the quantities μ_1, σ_1, $\bar{x}_1$, s_1, μ_2, σ_2, $\bar{x}_2$, s_2.
a) Which quantities represent parameters and which represent statistics?
b) Which quantities are fixed numbers and which are random variables?

In Exercises 10.2–10.7, use Procedure 10.1 on page 480 to perform each of the hypothesis tests.

___ **10.2** Surveys are conducted by the Northwestern University Placement Center, Evanston, IL, on starting salaries for college graduates. Results of the surveys can be found in the publication *The Northwestern Lindquist-Endicott Report*. Independent random samples of 32 accounting graduates and 35 liberal arts graduates yield the starting annual salaries shown in the table at the top of the next column. The data are in thousands of dollars, rounded to the nearest hundred dollars.

Accounting				Liberal arts				
24.9	20.4	23.3	20.9	20.0	16.8	21.3	20.6	21.0
23.4	25.3	26.8	22.6	18.7	23.8	18.3	17.7	18.0
24.0	22.0	21.7	23.0	19.1	21.1	19.6	19.0	18.7
26.2	25.9	24.8	25.3	20.8	20.4	17.1	19.5	
25.5	20.4	22.4	24.5	19.9	19.2	19.1	17.2	
24.9	25.4	25.8	24.2	18.3	22.7	20.0	19.2	
22.2	22.1	21.0	24.7	20.9	19.1	20.8	20.5	
21.5	22.5	25.4	26.6	21.4	16.3	16.3	20.5	

At the 5% significance level, does it appear that accounting graduates have a higher mean starting salary than liberal arts graduates? Assume that the population standard deviation of all starting salaries for accounting graduates is $1.73 thousand and that for liberal arts graduates is $1.82 thousand. *(Note: The sum of the accounting data is $759.6 thousand and the sum of the liberal arts data is $682.9 thousand.)*

__ **10.3** The U.S. National Center for Health Statistics compiles data on the length of stay by patients in short-term hospitals and publishes its findings in *Vital and Health Statistics*. Independent samples of 40 male patients and 35 female patients gave the following data on length of stay. The data are in days.

Male								Female						
4	4	12	18	9	6	12	10	14	7	15	1	12	1	3
3	6	15	7	3	55	1	2	7	21	4	1	5	4	4
10	13	5	7	1	23	9	2	3	5	18	12	5	1	7
1	17	2	24	11	14	6	2	7	2	15	4	9	10	7
1	8	1	3	19	3	1	13	3	6	5	9	6	2	14

At the 10% significance level, do the data suggest that, on the average, males stay in the hospital longer than females? Assume $\sigma_1 = 7.5$ days and $\sigma_2 = 6.8$ days. (*Note:* The sum of the male data is 363 days and the sum of the female data is 249 days.)

__ **10.4** An agronomist wants to determine whether a larger corn crop can be obtained if sterilized males of an insect pest are introduced to control the pest instead of using insecticides. The insecticide is used on 40 randomly selected one-acre plots of corn and the sterilized male insects on another 40 randomly selected one-acre plots of corn. The yields, in bushels, are as shown in the following table:

Insecticide					Sterilized males				
109	101	97	89	100	105	109	110	118	109
98	98	94	99	104	113	111	111	99	112
103	88	108	102	106	106	117	99	107	119
97	105	102	104	101	110	111	103	110	108
101	100	105	110	96	104	102	111	114	114
102	95	100	95	109	122	117	101	109	109
91	98	113	91	95	102	109	103	109	106
106	98	101	99	96	107	107	111	128	109

Do the data provide sufficient evidence to conclude that the use of sterilized male insects is more effective than insecticides in controlling the insect pest? Use $\alpha = 0.01$. (*Note:* For the insecticide data, $\Sigma x = 4006$ and $\Sigma x^2 = 402,480$; and for the sterilized-male data, $\Sigma x = 4381$ and $\Sigma x^2 = 481,261$.)

__ **10.5** The U.S. Energy Information Administration publishes data on residential energy consumption and expenditures in *Residential Energy Consumption Survey: Consumption and Expenditures*. Suppose that you want to decide whether last year's mean annual fuel expenditure for households using natural gas is different from that for households using only electricity. What conclusion would you draw given the data below? Use $\alpha = 0.05$.

Natural gas			Electricity			
$2002	1456	1394	$1376	1452	1235	1480
1541	1321	1338	1185	1327	1059	1400
1495	1526	1358	1227	1102	1168	1070
1801	1478	1376	1180	1221	1351	1014
1579	1375	1664	1461	1102	976	1394
1305	1458	1369	1379	987	1002	1532
1495	1507	1636	1450	1177	1150	
1698	1249	1377	1352	1266	1109	
1648	1557	1491	949	1351	1259	
1505	1355	1574	1179	1393	1456	

(*Note:* For the natural gas data, $\Sigma x = 44,928$ and $\Sigma x^2 = 68,029,844$; and for the electricity data, $\Sigma x = 44,771$ and $\Sigma x^2 = 56,633,389$.)

__ **10.6** Researchers in obesity wanted to compare the effectiveness of dieting with exercise against dieting without exercise. Seventy-three patients were randomly divided into two groups. Group 1, numbering 37 patients, was put on a program of dieting with exercise. Group 2, numbering 36 patients, dieted only. The results for weight loss, in pounds, after two months are summarized in the following table:

Diet-with-exercise group	Diet-only group
$\bar{x}_1 = 16.8$ lb	$\bar{x}_2 = 17.1$ lb
$s_1 = 3.5$ lb	$s_2 = 5.2$ lb

Determine, at the 0.05 significance level, whether there is a difference between the two treatments.

__ **10.7** A regional sales manager chooses two similar offices to study the effectiveness of a new training program aimed at increasing sales. One office institutes the training program and the other does not. The office that does not, Office 1, has 47 sales people. For this office, the mean sales per person over the next month is $3,197 with a standard deviation of $102. Office 2, the office that does institute the training program, has 51 sales people. For this office, the mean sales per person over the next month is $3,229 with a standard deviation of $107.

a) At the 5% significance level, does the training program appear to increase sales?
b) Repeat part (a) at the 10% significance level.

In each of Exercises 10.8–10.13, use Procedure 10.2 on page 483 to determine the required confidence interval.

___ **10.8**   Refer to Exercise 10.2.
a) Determine a 90% confidence interval for the difference, $\mu_1 - \mu_2$, between the mean starting salaries of accounting and liberal arts graduates.
b) Interpret your results in words.

___ **10.9**   Refer to Exercise 10.3.
a) Determine an 80% confidence interval for the difference, $\mu_1 - \mu_2$, between the mean lengths of stay in short-term hospitals by males and females.
b) Interpret your results in words.

___ **10.10**   Refer to Exercise 10.4.
a) Determine a 98% confidence interval for the difference, $\mu_1 - \mu_2$, between the mean yields of corn when the insecticide is used to control the insect pest and when sterilized males are used.
b) Interpret your results in words.

___ **10.11**   Refer to Exercise 10.5.
a) Find a 95% confidence interval for the difference, $\mu_1 - \mu_2$, between last year's mean fuel expenditures for households using natural gas and those using only electricity.
b) Interpret your results in words.

___ **10.12**   Refer to Exercise 10.6.
a) Find a 95% confidence interval for the difference, $\mu_1 - \mu_2$, between the mean weight losses after two months using the diet-with-exercise method and the diet-only method.
b) Interpret your results in words.

___ **10.13**   Refer to Exercise 10.7.
a) Determine a 90% confidence interval for the difference, $\mu_1 - \mu_2$, between the mean sales per person per month for those that do not take the training program and those that do.
b) Repeat part (a) using an 80% confidence level.

▦ **10.14**   Use your results from Exercises 10.5, 10.6, 10.11, and 10.12 to complete the following statement: A hypothesis test of H_0: $\mu_1 = \mu_2$ versus H_a: $\mu_1 \neq \mu_2$, at the significance level α, will lead to rejection of the null hypothesis if and only if the number _____ does not lie in the $(1 - \alpha)$-level confidence interval for $\mu_1 - \mu_2$.

▦ **10.15**   In this exercise, we will derive the formulas given on page 478 for the mean and standard deviation of the random variable $\bar{x}_1 - \bar{x}_2$. So, suppose that independent random samples of sizes n_1 and n_2 are to be taken, respectively, from populations with means μ_1 and μ_2 and standard deviations σ_1 and σ_2.
a) Use the results of Exercise 5.26(e) on page 249 and Exercise 5.27(f) on page 250 to show that

$$\mu_{\bar{x}_1 - \bar{x}_2} = \mu_{\bar{x}_1} - \mu_{\bar{x}_2}$$

and

$$\sigma_{\bar{x}_1 - \bar{x}_2} = \sqrt{\sigma_{\bar{x}_1}^2 + \sigma_{\bar{x}_2}^2}$$

b) Apply the formulas $\mu_{\bar{x}} = \mu$ and $\sigma_{\bar{x}} = \sigma / \sqrt{n}$ to the results in part (a) to derive the formulas

$$\mu_{\bar{x}_1 - \bar{x}_2} = \mu_1 - \mu_2$$

and

$$\sigma_{\bar{x}_1 - \bar{x}_2} = \sqrt{(\sigma_1^2/n_1) + (\sigma_2^2/n_2)}$$

▦ **10.16**   This exercise provides a verification of Procedures 10.1 and 10.2. Suppose that independent random samples of sizes $n_1 \geq 30$ and $n_2 \geq 30$ are to be taken, respectively, from populations with means, μ_1 and μ_2, and standard deviations, σ_1 and σ_2. Assume as known that the random variable $\bar{x}_1 - \bar{x}_2$ is approximately normally distributed.
a) Use the results of part (b) of Exercise 10.15 to show that the random variable

$$z = \frac{(\bar{x}_1 - \bar{x}_2) - (\mu_1 - \mu_2)}{\sqrt{(\sigma_1^2/n_1) + (\sigma_2^2/n_2)}}$$

has, at least approximately, the standard normal distribution.
b) Deduce from part (a) that the random variable

$$z = \frac{(\bar{x}_1 - \bar{x}_2) - (\mu_1 - \mu_2)}{\sqrt{(s_1^2/n_1) + (s_2^2/n_2)}}$$

also has approximately the standard normal distribution. [This justifies Procedure 10.1.]
c) Use part (b) to show that

$$P\left(-z_{\alpha/2} < \frac{(\bar{x}_1 - \bar{x}_2) - (\mu_1 - \mu_2)}{\sqrt{(s_1^2/n_1) + (s_2^2/n_2)}} < z_{\alpha/2}\right) \approx 1 - \alpha$$

d) Use part (c) to verify Procedure 10.2.

10.2   Inferences for the means of two normal populations with equal standard deviations using independent samples

In the previous section we learned how to perform inferences to compare the means of two populations when the sample sizes are large and the samples are chosen independently. There are several methods for performing such inferences that do not require large samples.

Included among those methods, are two that apply when the populations being sampled are *normally distributed* and have unknown standard deviations. One requires that the populations under consideration have equal standard deviations and the other does not. We will study the former method in this section and the latter method in the next section.

THE SAMPLING DISTRIBUTION OF THE DIFFERENCE BETWEEN TWO MEANS (NORMAL POPULATIONS AND INDEPENDENT SAMPLES)

To begin, we need to determine the probability distribution of the random variable $\bar{x}_1 - \bar{x}_2$ when independent samples are taken from two normally distributed populations. From Key Fact 7.4 on page 360 (the sampling distribution of the mean for normal populations), we know that when sampling from a normally distributed population, the random variable $\bar{x}$ is normally distributed, regardless of the sample size, n.

Using that fact it can be proved that, when independent sampling is done from two normally distributed populations, the random variable $\bar{x}_1 - \bar{x}_2$ is also normally distributed, regardless of the sample sizes, n_1 and n_2. Combining this last result with the formulas given on page 478 for the mean and standard deviation of $\bar{x}_1 - \bar{x}_2$, we obtain the following:

KEY FACT 10.2   The sampling distribution of the difference between two means (normal populations and independent samples)

Suppose that a random sample of size n_1 is to be taken from a normally distributed population with mean μ_1 and standard deviation σ_1; and that a random sample of size n_2 is to be taken from a normally distributed population with mean μ_2 and standard deviation σ_2. Further suppose that the two samples are to be selected independently. Then the random variable $\bar{x}_1 - \bar{x}_2$ is also normally distributed and has mean $\mu_{\bar{x}_1 - \bar{x}_2} = \mu_1 - \mu_2$ and standard deviation $\sigma_{\bar{x}_1 - \bar{x}_2} = \sqrt{(\sigma_1^2/n_1) + (\sigma_2^2/n_2)}$. Thus, the standardized random variable

$$z = \frac{(\bar{x}_1 - \bar{x}_2) - (\mu_1 - \mu_2)}{\sqrt{(\sigma_1^2/n_1) + (\sigma_2^2/n_2)}}$$

has the standard normal distribution.

HYPOTHESIS TESTS FOR THE MEANS OF TWO NORMAL POPULATIONS WITH EQUAL STANDARD DEVIATIONS USING INDEPENDENT SAMPLES

We will now develop a procedure for performing a hypothesis test to compare the means of two normally distributed populations with equal, but unknown, standard deviations using independent samples. Our immediate goal is to find a test statistic for such a hypothesis test.

Let us use σ to denote the common standard deviation of the two populations. According to Key Fact 10.2, when independent samples are taken from two normally distributed populations, the random variable

$$z = \frac{(\bar{x}_1 - \bar{x}_2) - (\mu_1 - \mu_2)}{\sqrt{(\sigma_1^2/n_1) + (\sigma_2^2/n_2)}}$$

has the standard normal distribution. Replacing σ_1 and σ_2 in the above expression by their common value σ and using some algebra, we obtain the random variable

$$(2) \qquad z = \frac{(\bar{x}_1 - \bar{x}_2) - (\mu_1 - \mu_2)}{\sigma\sqrt{(1/n_1) + (1/n_2)}}$$

Of course, this random variable cannot be used as a basis for obtaining the required test statistic since the common population standard deviation, σ, is unknown.

Consequently, we need to use sample information to estimate the unknown population standard deviation, σ, or, equivalently, the population variance, σ^2. The best way to do this is to regard the sample variances, s_1^2 and s_2^2, as two estimates of σ^2 and then **pool** those estimates by weighting them according to sample size (actually by degrees of freedom). Thus, our estimate for the common population variance, σ^2, is

$$s_p^2 = \frac{(n_1 - 1)s_1^2 + (n_2 - 1)s_2^2}{n_1 + n_2 - 2}$$

and, hence, for the common population standard deviation, σ, is

$$s_p = \sqrt{\frac{(n_1 - 1)s_1^2 + (n_2 - 1)s_2^2}{n_1 + n_2 - 2}}$$

The subscript "p" stands for "pooled," and the quantity s_p is called the **pooled sample standard deviation.** In summary, we use the pooled sample standard deviation, s_p, as our estimate for the common, but unknown, standard deviation, σ, of the two populations.

Replacing the unknown population standard deviation, σ, in Equation (2) by its estimate, s_p, we get the random variable

$$\frac{(\bar{x}_1 - \bar{x}_2) - (\mu_1 - \mu_2)}{s_p\sqrt{(1/n_1) + (1/n_2)}}$$

which *can* be used as a basis for obtaining the required test statistic. However, unlike the random variable in Equation (2), this last random variable does not have the standard normal distribution. But, its distribution is one with which we are familiar—a t-distribution.

KEY FACT 10.3

Suppose that independent random samples of sizes n_1 and n_2 are to be taken from two normally distributed populations with means μ_1 and μ_2, respectively. Further suppose that the standard deviations of the two populations are equal. Then the random variable

$$ t = \frac{(\bar{x}_1 - \bar{x}_2) - (\mu_1 - \mu_2)}{s_{\mathrm{p}}\sqrt{(1/n_1) + (1/n_2)}} $$

has the t-distribution with df $= n_1 + n_2 - 2$.

In view of Key Fact 10.3, we see that for a hypothesis test with null hypothesis H_0: $\mu_1 = \mu_2$, we can use the random variable

$$ t = \frac{\bar{x}_1 - \bar{x}_2}{s_{\mathrm{p}}\sqrt{(1/n_1) + (1/n_2)}} $$

as the test statistic and obtain the critical value(s) from the t-table, Table III.

PROCEDURE 10.3

To perform a hypothesis test for two population means with null hypothesis H_0: $\mu_1 = \mu_2$.

ASSUMPTIONS

1. Independent samples.
2. Normal populations.
3. Equal, but unknown, population standard deviations.[†]

STEP 1 *State the null and alternative hypotheses.*
STEP 2 *Decide on the significance level, α.*
STEP 3 *The critical value(s)*
    a) *for a two-tailed test are $\pm t_{\alpha/2}$,*
    b) *for a left-tailed test is $-t_{\alpha}$,*
    c) *for a right-tailed test is t_{α},*
    with df $= n_1 + n_2 - 2$. Use Table III to find the critical value(s).

[†] Although the standard deviations of the populations are unknown, they may be known to be equal on the basis of previous studies or theoretical considerations.

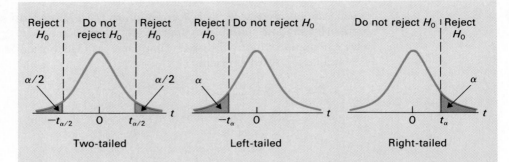

STEP 4 *Compute the value of the test statistic*

$$t = \frac{\bar{x}_1 - \bar{x}_2}{s_\mathrm{p}\sqrt{(1/n_1) + (1/n_2)}}$$

where

$$s_\mathrm{p} = \sqrt{\frac{(n_1 - 1)s_1^2 + (n_2 - 1)s_2^2}{n_1 + n_2 - 2}}$$

STEP 5 *If the value of the test statistic falls in the rejection region, then reject H_0; otherwise, do not reject H_0.*

STEP 6 *State the conclusion in words.*

Note: In Step 4 of Procedure 10.3, we need to calculate the pooled sample standard deviation, s_p. The pooled sample standard deviation always lies between the two sample standard deviations, s_1 and s_2. If you calculate s_p and it does not lie between s_1 and s_2, then you made an error.

EXAMPLE 10.4   *Illustrates Procedure 10.3*

The U.S. National Center for Health Statistics collects data on the daily intake of selected nutrients by race and income level. Results are published in *Vital and Health Statistics*. Suppose that we want to compare the mean protein intake of all people with incomes below the poverty level to that of all people with incomes above the poverty level. The data in Table 10.3 give the protein intakes, in grams, over a 24-hour period for independent random samples of 15 people with incomes below the poverty level and 10 people with incomes above the poverty level.

TABLE 10.3
Daily protein intakes

Below poverty level			Above poverty level	
51.4	49.7	72.0	86.0	69.0
76.7	65.8	55.0	59.7	80.2
73.7	62.1	79.7	68.6	78.1
66.2	75.8	65.4	98.6	69.8
65.5	62.0	73.3	87.7	77.2

At the 5% significance level, do the data provide sufficient evidence to conclude that the mean protein intake of all people with incomes below the poverty level is less than the mean protein intake of all people with incomes above the poverty level? [Assume that the daily protein intakes of all people with incomes below the poverty level are normally distributed, that the daily protein intakes of all people with incomes above the poverty level are normally distributed, and that the standard deviations of both of these populations are equal.]

SOLUTION   We apply Procedure 10.3.

STEP 1 *State the null and alternative hypotheses.*

Let μ_1 denote the mean daily intake of protein of all people with incomes below the poverty level and μ_2 denote the mean daily intake of protein of all people with incomes above the poverty level. Then the null and alternative hypotheses for the hypothesis test are

H_0: $\mu_1 = \mu_2$ (below-poverty mean is not less than the above-poverty mean)
H_a: $\mu_1 < \mu_2$ (below-poverty mean is less than the above-poverty mean)

Note that the hypothesis test is left-tailed since there is a less-than sign ($<$) in the alternative hypothesis.

STEP 2 *Decide on the significance level, α.*

The hypothesis test is to be performed at the 5% significance level and, so, $\alpha = 0.05$.

STEP 3 *The critical value for a left-tailed test is $-t_\alpha$ with df $= n_1 + n_2 - 2$.*

From Step 2, $\alpha = 0.05$. Also, we see from Table 10.3 that $n_1 = 15$ and $n_2 = 10$. Hence, df $= 15 + 10 - 2 = 23$. Consulting Table III, we find that the critical value is $-t_{0.05} = -1.714$. See Figure 10.3.

FIGURE 10.3

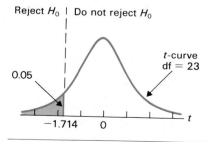

Reject H_0 | Do not reject H_0

0.05

t-curve
df $= 23$

-1.714 0 t

STEP 4 *Compute the value of the test statistic*

$$t = \frac{\bar{x}_1 - \bar{x}_2}{s_{\mathrm{p}}\sqrt{(1/n_1) + (1/n_2)}}$$

where

$$s_{\mathrm{p}} = \sqrt{\frac{(n_1 - 1)s_1^2 + (n_2 - 1)s_2^2}{n_1 + n_2 - 2}}$$

We first determine the pooled sample standard deviation, s_{p}. The sample standard deviations, s_1 and s_2, of the two data sets in Table 10.3 are calculated in the usual way. We find that $s_1 = 9.17$ g and $s_2 = 11.34$ g. Consequently, since $n_1 = 15$ and $n_2 = 10$,

$$s_{\mathrm{p}} = \sqrt{\frac{(15 - 1) \cdot 9.17^2 + (10 - 1) \cdot 11.34^2}{15 + 10 - 2}} = 10.07 \text{ g}$$

The sample means of the two data sets in Table 10.3 are $\bar{x}_1 = 66.29$ g and $\bar{x}_2 = 77.49$ g. Therefore, the value of the test statistic is

$$t = \frac{\bar{x}_1 - \bar{x}_2}{s_{\mathrm{p}}\sqrt{(1/n_1) + (1/n_2)}} = \frac{66.29 - 77.49}{10.07\sqrt{(1/15) + (1/10)}} = \boxed{-2.724}$$

STEP 5 *If the value of the test statistic falls in the rejection region, reject H_0; otherwise, do not reject H_0.*

From Step 4, the value of the test statistic is $t = -2.724$, which falls in the rejection region (see Figure 10.3). Thus, we reject H_0.

STEP 6 *State the conclusion in words.*

The data provide sufficient evidence to conclude that the mean protein intake of all people with incomes below the poverty level is less than the mean protein intake of all people with incomes above the poverty level. In other words, it appears that the average person with an income below the poverty level gets less protein than the average person with an income above the poverty level.

MTB

CONFIDENCE INTERVALS FOR THE DIFFERENCE BETWEEN THE MEANS OF TWO NORMAL POPULATIONS WITH EQUAL STANDARD DEVIATIONS USING INDEPENDENT SAMPLES

We can also use Key Fact 10.3 on page 489 to derive the following confidence-interval procedure for the difference between two means. Note carefully the conditions required for using this procedure.

> ## PROCEDURE 10.4
> **To find a confidence interval for the difference between two population means.**
>
> *ASSUMPTIONS*
> 1. Independent samples.
> 2. Normal populations.
> 3. Equal, but unknown, population standard deviations.
>
> **STEP 1** *For a confidence level of $1 - \alpha$, use Table III to find $t_{\alpha/2}$ with df = $n_1 + n_2 - 2$.*
>
> **STEP 2** *The endpoints of the confidence interval for $\mu_1 - \mu_2$ are*
>
> $$(\bar{x}_1 - \bar{x}_2) \pm t_{\alpha/2} \cdot s_{\mathrm{p}} \sqrt{(1/n_1) + (1/n_2)}$$

EXAMPLE 10.5 *Illustrates Procedure 10.4*

Refer to Example 10.4. Use the sample data displayed in Table 10.3 on page 490 to obtain a 95% confidence interval for the difference, $\mu_1 - \mu_2$, between the mean daily protein intake of all people with incomes below the poverty level and the mean daily protein intake of all people with incomes above the poverty level.

SOLUTION We apply Procedure 10.4.

STEP 1 *For a confidence level of $1 - \alpha$, use Table III to find $t_{\alpha/2}$ with df = $n_1 + n_2 - 2$.*

For a 95% confidence interval, the confidence level is $0.95 = 1 - 0.05$. So, $\alpha = 0.05$. From Table 10.3, we see that $n_1 = 15$ and $n_2 = 10$, which makes df $= n_1 + n_2 - 2 = 15 + 10 - 2 = 23$. Consulting Table III, we find that for df = 23, $t_{\alpha/2} = t_{0.05/2} = t_{0.025} = $ **2.069.**

STEP 2 *The endpoints of the confidence interval for $\mu_1 - \mu_2$ are*

$$(\bar{x}_1 - \bar{x}_2) \pm t_{\alpha/2} \cdot s_{\mathrm{p}} \sqrt{(1/n_1) + (1/n_2)}$$

From Step 1, $t_{\alpha/2} = 2.069$. Also, $n_1 = 15$, $n_2 = 10$ and, from Example 10.4 (page 492), we know that $\bar{x}_1 = 66.29$ g, $\bar{x}_2 = 77.49$ g, and $s_{\mathrm{p}} = 10.07$ g. Consequently, the endpoints of the confidence interval for $\mu_1 - \mu_2$ are

$$(66.29 - 77.49) \pm 2.069 \cdot 10.07 \sqrt{(1/15) + (1/10)}$$

or

$$-11.20 \pm 8.51$$

Thus, a 95% confidence interval for $\mu_1 - \mu_2$ is from

$$-19.71 \quad \text{to} \quad -2.69$$

We can be 95% confident that the difference, $\mu_1 - \mu_2$, between the mean daily protein intakes of all people with incomes below the poverty level and all people with incomes above the poverty level is somewhere between -19.71 and -2.69 grams. In particular, we can be 95% confident that the average person with an income above the poverty level gets at least 2.69 grams more protein per day than the average person with an income below the poverty level. ■

MTB

USING THE COMPUTER (OPTIONAL)

Procedure 10.3 on page 489 gives a step-by-step method for performing a hypothesis test to compare the means of two normally distributed populations with equal standard deviations using independent samples. Procedure 10.4 on page 493 provides a step-by-step method for obtaining a confidence interval for the difference between the means of two normally distributed populations with equal standard deviations using independent samples.

Alternatively, we can use Minitab to carry out both of those procedures simultaneously. The appropriate command is **TWOSAMPLE T** with the subcommand **POOLED**. The next example illustrates the use of TWOSAMPLE T; POOLED.

EXAMPLE 10.6   *Illustrates the TWOSAMPLE T; POOLED command*

Use Minitab to simultaneously perform the hypothesis test considered in Example 10.4 and obtain the confidence interval required in Example 10.5.

SOLUTION   Table 10.3 gives the protein intakes, in grams, over a 24-hour period for independent random samples of 15 people with incomes below the poverty level and 10 people with incomes above the poverty level. We repeat that table here as Table 10.4.

TABLE 10.4
Daily protein intakes

Below poverty level			Above poverty level	
51.4	49.7	72.0	86.0	69.0
76.7	65.8	55.0	59.7	80.2
73.7	62.1	79.7	68.6	78.1
66.2	75.8	65.4	98.6	69.8
65.5	62.0	73.3	87.7	77.2

Let μ_1 denote the mean daily intake of protein of all people with incomes below the poverty level and μ_2 denote the mean daily intake of protein of all people with incomes above the poverty level. The problem in Example 10.4 is to perform the hypothesis test

$H_0: \mu_1 = \mu_2$ (below-poverty mean is not less than the above-poverty mean)

$H_a: \mu_1 < \mu_2$ (below-poverty mean is less than the above-poverty mean)

at the 5% significance level; and the problem in Example 10.5 is to obtain a 95% confidence interval for $\mu_1 - \mu_2$.

By assumption, the samples are independent and the populations are normally distributed with equal standard deviations. Hence, we can apply TWOSAMPLE T; POOLED to simultaneously carry out the hypothesis test and obtain the required confidence interval.

To begin, we enter the two sets of sample data from Table 10.4 into, say, C1 and C2 using the SET command and name C1 "Belowpov" and C2 "Abovepov" using the NAME command. Next we type the command TWOSAMPLE T, the desired confidence level for the confidence interval, and the storage locations of the sample data. That is, we type

TWOSAMPLE T, 95% confidence, for 'BELOWPOV' vs 'ABOVEPOV';

Then we type the subcommand <u>POOLED;</u>. Finally, since the hypothesis test is left-tailed (H_a: $\mu_1 < \mu_2$), we must also type the subcommand <u>ALTERNATIVE=-1</u>.. Printout 10.1 summarizes the above discussion and also displays the resulting output.

PRINTOUT 10.1
Minitab output for
TWOSAMPLE T;
POOLED

```
MTB > SET C1
DATA> 51.4 49.7 72.0 76.7 65.8
DATA> 55.0 73.7 62.1 79.7 66.2
DATA> 75.8 65.4 65.5 62.0 73.3
DATA> END
MTB > SET C2
DATA> 86.0 69.0 59.7 80.2 68.6
DATA> 78.1 98.6 69.8 87.7 77.2
DATA> END
MTB > NAME C1 'BELOWPOV' C2 'ABOVEPOV'
MTB > TWOSAMPLE T, 95% confidence, for 'BELOWPOV' vs 'ABOVEPOV';
SUBC> POOLED;
SUBC> ALTERNATIVE=-1.


TWOSAMPLE T FOR BELOWPOV VS ABOVEPOV
              N       MEAN     STDEV    SE MEAN
BELOWPOV     15      66.29      9.17      2.4
ABOVEPOV     10      77.5      11.3       3.6


95 PCT CI FOR MU BELOWPOV - MU ABOVEPOV: (-19.7, -2.7)


TTEST MU BELOWPOV = MU ABOVEPOV (VS LT): T= -2.72  P=0.0060  DF=  23


POOLED STDEV =        10.1
```

The first line of the output describes the test being performed: TWOSAMPLE T FOR BELOWPOV VS ABOVEPOV. The next three lines give the sample size, sample mean, sample standard deviation, and estimated standard error of the mean for each of the two samples.

On the fifth line, we find the required 95% confidence interval for the difference between the two population means. Thus, we can be 95% confident that the difference, $\mu_1 - \mu_2$, between the mean daily protein intake of all people with incomes below the poverty level and the mean daily protein intake of all people with incomes above the poverty level is somewhere between -19.7 and -2.7 grams.

The next-to-last line of the output gives a statement of the null and alternative hypotheses, followed by the value of the test statistic (T= -2.72), the *P*-value (P=0.0060), and the degrees of freedom (DF= 23). On the last line, we find the value of the pooled sample standard deviation, s_p.

We see, in particular, that the *P*-value is 0.0060. Since this is less than the specified significance level of $\alpha = 0.05$, we reject H_0. In other words, the data provide sufficient evidence to conclude that the mean protein intake of all people with incomes below the poverty level is less than the mean protein intake of all people with incomes above the poverty level.   ■

Exercises 10.2

In each of Exercises 10.17–10.22, assume the two populations being sampled are normally distributed and have equal standard deviations. Use Procedure 10.3 on page 489 to perform each of the hypothesis tests.

___ **10.17** A highway official wants to compare two brands of paint used for striping roads. Ten stripes of each paint are run across the highway. The number of months that each stripe lasts is presented in the following table:

Brand *A*		Brand *B*	
35.6	36.1	37.2	36.4
37.0	35.8	39.7	37.5
34.9	34.9	37.2	40.5
36.0	38.8	38.8	38.2
36.6	36.5	37.7	36.6

Based on the sample data, does there appear to be a difference in mean lasting times for the two brands of paint? Use $\alpha = 0.05$. (*Note:* For Brand *A*, the sum of the data is 362.2 and the sum of the squares of the data is 13,130.48. For Brand *B*, the sum of the data is 379.8 and the sum of the squares of the data is 14,440.76.)

___ **10.18** In a packing plant, a machine packs cartons with jars. A salesperson claims that the machine she is selling will pack faster. To test that claim, the time it takes each machine to pack 10 cartons is recorded. The results, in seconds, are shown in the table at the top of the next column.

New machine		Present machine	
42.0	41.0	42.7	43.6
41.3	41.8	43.8	43.3
42.4	42.8	42.5	43.5
43.2	42.3	43.1	41.7
41.8	42.7	44.0	44.1

Do the data provide sufficient evidence to conclude that, on the average, the new machine packs faster? Use $\alpha = 0.05$. (*Note:* For the new machine, the sum of the data is 421.3 and the sum of the squares of the data is 17,753.59; and for the present machine, the sum of the data is 432.3 and the sum of the squares of the data is 18,693.39.)

___ **10.19** The U.S. Bureau of Labor Statistics conducts monthly surveys to estimate hourly earnings of nonsupervisory employees in various industry groups. Results of the surveys can be found in the publication *Employment and Earnings.* Suppose that independent random samples of 14 mine workers and 17 construction workers yield the following statistics:

Mining	Construction
$\bar{x}_1 = \$12.93$	$\bar{x}_2 = \$13.42$
$s_1 = \$2.25$	$s_2 = \$2.36$

At the 5% significance level, do the data provide sufficient evidence to conclude that mine workers earn less on the average than construction workers?

10.20   The U.S. Energy Information Administration compiles data on the miles driven annually by American households and publishes its findings in *Residential Transportation Energy Consumption Survey, Consumption Patterns of Household Vehicles.* Suppose that independent random samples of 15 midwestern households and 14 southern households provide the statistics on the number of miles driven last year as shown in the following table:

Midwest	South
$\bar{x}_1 = 16{,}229$ mi	$\bar{x}_2 = 17{,}689$ mi
$s_1 = 4{,}057$ mi	$s_2 = 4{,}420$ mi

At the 5% significance level, does there appear to be a difference in the average number of miles driven by midwestern and southern households?

10.21   The publication *Vital and Health Statistics,* from the National Center for Health Statistics, provides information on heights and weights of Americans, by age and sex. Suppose that independent samples of 10 males aged 25–34 years and 15 males aged 45–54 years yield the following heights, in inches:

25–34		45–54		
73.3	70.4	73.2	69.5	64.7
64.8	66.8	68.5	74.5	73.0
72.1	70.7	62.4	70.6	66.7
68.9	74.4	65.5	69.3	68.1
68.7	71.8	71.3	67.1	64.3

At the 5% significance level, do the data provide sufficient evidence to conclude that males in the age group 25–34 years are, on the average, taller than those who were in that age group 20 years ago?

10.22   A supervisor records the time required by each of two workers to perform an assembly-line task. Each worker is observed on six randomly selected occasions. The times, to the nearest minute, are as shown in the following table:

Sigmund		Marcia	
8	9	8	9
10	11	9	8
9	10	9	10

Do the data provide sufficient evidence to conclude, at the 10% significance level, that there is a difference in the mean times required to complete the assembly-line task for the two workers?

In each of Exercises 10.23–10.28, apply Procedure 10.4 on page 493 to obtain the required confidence interval.

10.23   Refer to Exercise 10.17.
a) Determine a 95% confidence interval for the difference, $\mu_1 - \mu_2$, between the mean lasting times of Brand A and Brand B paints.
b) Interpret your results in words.

10.24   Refer to Exercise 10.18.
a) Find a 90% confidence interval for the difference, $\mu_1 - \mu_2$, between the mean time it takes the new machine to pack 10 cartons and the mean time it takes the present machine to pack 10 cartons.
b) Interpret your results in words.

10.25   Refer to Exercise 10.19.
a) Determine a 90% confidence interval for the difference, $\mu_1 - \mu_2$, between the mean hourly earnings of nonsupervisory mine and construction workers.
b) Interpret your results in words.

10.26   Refer to Exercise 10.20.
a) Find a 95% confidence interval for the difference, $\mu_1 - \mu_2$, between last year's mean number of miles driven by midwestern and southern households.
b) Interpret your results in words.

10.27   Refer to Exercise 10.21.
a) Obtain a 90% confidence interval for the difference between the mean height of males in the age group 25–34 years and the mean height of males in the age group 45–54 years.
b) Interpret your results in words.

10.28   Refer to Exercise 10.22.
a) Obtain a 90% confidence interval for the difference between the mean time it takes Sigmund to complete the assembly-line task and the mean time it takes Marcia to complete the assembly-line task.
b) Interpret your results in words.

Exercises 10.29–10.32 are computer exercises.

10.29 (Computer exercise) Suppose that the data in Exercise 10.17 are stored in columns named BRAND A and BRAND B.
a) Which Minitab command and subcommands (if any) should be used to simultaneously perform the hypothesis test considered in Exercise 10.17 and

obtain the confidence interval required in Exercise 10.23?

b) If you have access to Minitab, use it to perform the hypothesis test and obtain the confidence interval.

___ 10.30 (Computer exercise) Suppose the data in Exercise 10.18 are stored in columns named NEW and PRESENT.

a) Which Minitab command and subcommands (if any) should be used to simultaneously perform the hypothesis test considered in Exercise 10.18 and obtain the confidence interval required in Exercise 10.24?

b) If you have access to Minitab, use it to perform the hypothesis test and obtain the confidence interval.

___ 10.31 (Computer exercise) The research and development (R&D) department of a light-bulb manufacturing company claims to have developed a new bulb that, on the average, will outlast the bulb currently produced. To try to justify the claim, R&D takes independent random samples of the current bulb and the new bulb. After the lifetimes of the bulbs sampled are determined, Minitab's TWOSAMPLE T; POOLED command is applied to get the output shown in Printout 10.2 at the top of the next page. Use the output to obtain

a) the null and alternative hypotheses for R&D's hypothesis test.

b) the command and subcommand(s) used.

c) the sample standard deviations of the lifetimes for both the current bulb and the new bulb.

d) the number of current bulbs sampled and the number of new bulbs sampled.

e) the sample mean lifetimes for both the current and new bulbs.

f) the value of the test statistic, t.

g) the P-value of the hypothesis test.

h) the smallest significance level at which the null hypothesis can be rejected.

i) the conclusion if the hypothesis test is performed at the 1% significance level.

j) a 99% confidence interval for the difference, $\mu_1 - \mu_2$, between the mean lifetimes of the current bulb and the new bulb.

k) the assumptions made by R&D in using the Minitab program TWOSAMPLE T; POOLED.

l) the value of s_p.

___ 10.32 (Computer exercise) A researcher in nutrition wants to compare the protein intakes of whites and blacks. He takes independent random samples of

10 whites and 12 blacks and determines their 24-hour protein intakes, in grams. Following that, he applies Minitab's TWOSAMPLE T; POOLED command to the resulting data and obtains the output shown in Printout 10.3 on the next page. From the output, determine

a) the null and alternative hypotheses.

b) the command and subcommand(s) used.

c) the sample standard deviation of each of the two sets of protein-intake data.

d) the mean daily protein intake of the whites sampled and the mean daily protein intake of the blacks sampled.

e) the value of the test statistic, t.

f) the P-value of the hypothesis test.

g) the smallest significance level at which the null hypothesis can be rejected.

h) the conclusion if the hypothesis test is performed at the 5% significance level.

i) a 95% confidence interval for the difference, $\mu_1 - \mu_2$, between the mean daily protein intakes of whites and blacks.

j) the assumptions made by the researcher in order to use TWOSAMPLE T; POOLED.

k) the value of s_p.

≡ 10.33 Suppose that we want to perform a hypothesis test to compare the means of two normally distributed populations using independent samples. Further suppose that the standard deviations of the two populations are known.

a) What test statistic should be used to perform the hypothesis test? (Hint: Refer to Key Fact 10.2 on page 487.)

b) Which table should be consulted in order to obtain the critical value(s)?

c) Formulate a step-by-step procedure for the hypothesis test that uses the test statistic from part (a).

≡ 10.34 Let

$$z = \frac{(\bar{x}_1 - \bar{x}_2) - (\mu_1 - \mu_2)}{\sqrt{(\sigma_1^2/n_1) + (\sigma_2^2/n_2)}}$$

Show that if $\sigma_1 = \sigma_2 = \sigma$, then we can rewrite the expression for z as

$$z = \frac{(\bar{x}_1 - \bar{x}_2) - (\mu_1 - \mu_2)}{\sigma\sqrt{(1/n_1) + (1/n_2)}}$$

[This verifies Equation (2) on page 488.]

PRINTOUT 10.2   Minitab output for Exercise 10.31

```
TWOSAMPLE T FOR CURRENT VS NEW
              N      MEAN    STDEV   SE MEAN
CURRENT   20      1023.2     55.3       12
NEW       10      1093.0     44.6       14

99 PCT CI FOR MU CURRENT - MU NEW: (-126, -14)

TTEST MU CURRENT = MU NEW (VS LT): T= -3.46  P=0.0009  DF=  28

POOLED STDEV =        52.1
```

PRINTOUT 10.3   Minitab output for Exercise 10.32

```
TWOSAMPLE T FOR WHITE VS BLACK
            N      MEAN    STDEV   SE MEAN
WHITE   10      77.4     16.4       5.2
BLACK   12      68.8     15.2       4.4

95 PCT CI FOR MU WHITE - MU BLACK: (-5.4, 22.8)

TTEST MU WHITE = MU BLACK (VS NE): T= 1.29  P=0.21  DF=  20

POOLED STDEV =        15.7
```

≡ **10.35**   The formula given on page 488 for the pooled variance is

$$s_{\mathrm{p}}^2 = \frac{(n_1 - 1)s_1^2 + (n_2 - 1)s_2^2}{n_1 + n_2 - 2}$$

Show that, if the sample sizes, n_1 and n_2, are equal, then s_{p}^2 is just the mean of s_1^2 and s_2^2.

≡ **10.36**   Suppose that independent random samples of sizes n_1 and n_2 are to be taken from two nor-

mally distributed populations with means μ_1 and μ_2, respectively. Further suppose that the standard deviations of the two populations are equal.

a) Use Key Fact 10.3 on page 489 to show that

$$P\left(-t_{\alpha/2} < \frac{(\bar{x}_1 - \bar{x}_2) - (\mu_1 - \mu_2)}{s_{\mathrm{P}}\sqrt{(1/n_1) + (1/n_2)}} < t_{\alpha/2}\right) = 1 - \alpha$$

b) Use part (a) to verify Procedure 10.4 on page 493.

10.3   Inferences for the means of two normal populations using independent samples (Optional)

We examined, in Section 10.2, methods of performing inferences to compare the means of two normally distributed populations using independent samples. The methods discussed there require the standard deviations of the two populations under consideration to be equal.

In this section, we will develop inferential procedures to compare the means of two normally distributed populations using independent samples that do not require the population standard deviations to be equal, although they may be. As before, we will assume that the population standard deviations are unknown because that is usually the case in practice.

HYPOTHESIS TESTS FOR THE MEANS OF TWO NORMAL POPULATIONS USING INDEPENDENT SAMPLES

Let us begin by finding a test statistic. From Key Fact 10.2 on page 487, we know that when independent samples are taken from two normally distributed populations, the random variable

$$z = \frac{(\overline{x}_1 - \overline{x}_2) - (\mu_1 - \mu_2)}{\sqrt{(\sigma_1^2/n_1) + (\sigma_2^2/n_2)}}$$

has the standard normal distribution. Since we are assuming that the population standard deviations, σ_1 and σ_2, are unknown, we cannot use the previous random variable as a basis for obtaining the required test statistic. We therefore replace σ_1 and σ_2 by their sample estimates, s_1 and s_2, and obtain the random variable

$$\frac{(\overline{x}_1 - \overline{x}_2) - (\mu_1 - \mu_2)}{\sqrt{(s_1^2/n_1) + (s_2^2/n_2)}}$$

which *can* be used as a basis for obtaining the required test statistic. This last random variable does not have the standard normal distribution but, as you might expect, has a *t*-distribution.

KEY FACT 10.4

Suppose that independent random samples of sizes n_1 and n_2 are to be taken from two normally distributed populations with means μ_1 and μ_2, respectively. Then the random variable

$$t = \frac{(\overline{x}_1 - \overline{x}_2) - (\mu_1 - \mu_2)}{\sqrt{(s_1^2/n_1) + (s_2^2/n_2)}}$$

has approximately the *t*-distribution with degrees of freedom given by

$$df = \frac{\left[(s_1^2/n_1) + (s_2^2/n_2)\right]^2}{\dfrac{(s_1^2/n_1)^2}{n_1 - 1} + \dfrac{(s_2^2/n_2)^2}{n_2 - 1}}$$

rounded down to the nearest integer.

From Key Fact 10.4, we see that for a hypothesis test with null hypothesis $H_0: \mu_1 = \mu_2$, we can use the random variable

$$t = \frac{\overline{x}_1 - \overline{x}_2}{\sqrt{(s_1^2/n_1) + (s_2^2/n_2)}}$$

as the test statistic and obtain the critical value(s) from the t-table, Table III. Thus, we have the following procedure for performing a hypothesis test to compare the means of two normally distributed populations using independent samples.

PROCEDURE 10.5

To perform a hypothesis test for two population means with null hypothesis
$H_0: \mu_1 = \mu_2$.

ASSUMPTIONS

1. Independent samples.
2. Normal populations.

STEP 1 *State the null and alternative hypotheses.*
STEP 2 *Decide on the significance level, α.*
STEP 3 *The critical value(s)*
    a) *for a two-tailed test are $\pm t_{\alpha/2}$,*
    b) *for a left-tailed test is $-t_{\alpha}$,*
    c) *for a right-tailed test is t_{α},*
    with degrees of freedom given by

$$df = \frac{\left[(s_1^2/n_1) + (s_2^2/n_2)\right]^2}{\dfrac{(s_1^2/n_1)^2}{n_1 - 1} + \dfrac{(s_2^2/n_2)^2}{n_2 - 1}}$$

    rounded down to the nearest integer. Use Table III to find the critical value(s).

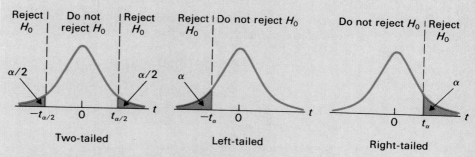

Two-tailed   Left-tailed   Right-tailed

STEP 4 *Compute the value of the test statistic*

$$t = \frac{\overline{x}_1 - \overline{x}_2}{\sqrt{(s_1^2/n_1) + (s_2^2/n_2)}}$$

STEP 5 *If the value of the test statistic falls in the rejection region, then reject H_0; otherwise, do not reject H_0.*
STEP 6 *State the conclusion in words.*

EXAMPLE 10.7   *Illustrates Procedure 10.5*

A general contractor wants to compare the lifetimes of two major brands of electric water heaters, Eagle and National. Using independent samples, the contractor obtains the data displayed in Table 10.5.

TABLE 10.5
Lifetimes, in
years, of water
heaters sampled

Eagle				National				
6.9	7.3	7.8	7.4	8.7	7.0	8.7	6.7	7.8
7.2	6.6	6.2	8.2	8.6	6.1	7.5	7.7	7.5
7.6	5.7	5.5	6.9	11.2	6.1	6.3	7.0	10.7

Determine, at the 5% significance level, whether the two brands of water heaters have different mean lifetimes. [Assume that the lifetimes of each brand are normally distributed.]

SOLUTION   We will apply Procedure 10.5. Before doing so, however, we give the values of the statistics that are needed to carry out the hypothesis test. These are obtained from the sample data in Table 10.5 in the usual way. See Table 10.6.

TABLE 10.6

Eagle	National
$\bar{x}_1 = 6.94$ yr	$\bar{x}_2 = 7.84$ yr
$s_1 = 0.82$ yr	$s_2 = 1.53$ yr
$n_1 = 12$	$n_2 = 15$

STEP 1 *State the null and alternative hypotheses.*

Let μ_1 denote the mean lifetime of all Eagle water heaters and μ_2 denote the mean lifetime of all National water heaters. Then the null and alternative hypotheses are

$$H_0: \mu_1 = \mu_2 \text{ (mean lifetimes are the same)}$$
$$H_a: \mu_1 \neq \mu_2 \text{ (mean lifetimes are different)}$$

Note that the hypothesis test is two-tailed since there is a not-equal sign ($\neq$) in the alternative hypothesis.

STEP 2 *Decide on the significance level, α.*

The test is to be performed at the 5% significance level; thus, $\alpha = 0.05$.

STEP 3 *The critical values for a two-tailed test are $\pm t_{\alpha/2}$ with degrees of freedom given by*

$$df = \frac{\left[(s_1^2/n_1) + (s_2^2/n_2)\right]^2}{\dfrac{\left(s_1^2/n_1\right)^2}{n_1 - 1} + \dfrac{\left(s_2^2/n_2\right)^2}{n_2 - 1}}$$

rounded down to the nearest integer.

From Step 2, $\alpha = 0.05$. Also, referring to Table 10.6, we find that

$$df = \frac{\left[(0.82^2/12) + (1.53^2/15)\right]^2}{\dfrac{(0.82^2/12)^2}{12 - 1} + \dfrac{(1.53^2/15)^2}{15 - 1}} = 22 \text{ (rounded down)}$$

So, the critical values are $\pm t_{\alpha/2} = \pm t_{0.05/2} = \pm t_{0.025} = \pm 2.074.$ See Figure 10.4.

FIGURE 10.4

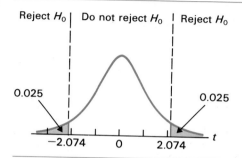

STEP 4 *Compute the value of the test statistic*

$$t = \frac{\bar{x}_1 - \bar{x}_2}{\sqrt{(s_1^2/n_1) + (s_2^2/n_2)}}$$

The required statistics for obtaining the value of the test statistic have already been calculated and are displayed in Table 10.6. We have

$$t = \frac{\bar{x}_1 - \bar{x}_2}{\sqrt{(s_1^2/n_1) + (s_2^2/n_2)}} = \frac{6.94 - 7.84}{\sqrt{(0.82^2/12) + (1.53^2/15)}} = -1.954$$

STEP 5 *If the value of the test statistic falls in the rejection region, reject H_0; otherwise, do not reject H_0.*

From Step 4, the value of the test statistic is $t = -1.954$ which, as we see from Figure 10.4, does not fall in the rejection region. Thus, we do not reject H_0.

STEP 6 *State the conclusion in words.*

MTB   The data do not provide sufficient evidence to conclude that the two brands of water heaters have different mean lifetimes.   ∎

CONFIDENCE INTERVALS FOR THE DIFFERENCE BETWEEN THE MEANS OF TWO NORMAL POPULATIONS USING INDEPENDENT SAMPLES

Suppose we want to obtain a confidence interval for the difference between the means of two normally distributed populations using independent samples. As we have seen, if the standard deviations of the two populations are equal, then we can apply Procedure 10.4 on page 493.

However, if the standard deviations of the two populations are not equal or if we simply do not know whether they are equal, then Procedure 10.4 should not be used. Instead, the following procedure should be employed. This procedure can be derived from Key Fact 10.4 on page 500.

PROCEDURE 10.6

To find a confidence interval for the difference between two population means.

ASSUMPTIONS

1. Independent samples.
2. Normal populations.

STEP 1 *For a confidence level of $1 - \alpha$, use Table III to find $t_{\alpha/2}$ with*

$$df = \frac{\left[\left(s_1^2/n_1\right) + \left(s_2^2/n_2\right)\right]^2}{\dfrac{\left(s_1^2/n_1\right)^2}{n_1 - 1} + \dfrac{\left(s_2^2/n_2\right)^2}{n_2 - 1}}$$

rounded down to the nearest integer.

STEP 2 *The endpoints of the confidence interval for $\mu_1 - \mu_2$ are*

$$(\bar{x}_1 - \bar{x}_2) \pm t_{\alpha/2} \cdot \sqrt{\left(s_1^2/n_1\right) + \left(s_2^2/n_2\right)}$$

EXAMPLE 10.8   *Illustrates Procedure 10.6*

Refer to Example 10.7. Use the sample data in Table 10.5 on page 502 to obtain a 95% confidence interval for the difference, $\mu_1 - \mu_2$, between the mean lifetimes of Eagle and National water heaters.

SOLUTION   We apply Procedure 10.6.

STEP 1 *For a confidence level of $1 - \alpha$, use Table III to find $t_{\alpha/2}$ with*

$$df = \frac{\left[\left(s_1^2/n_1\right) + \left(s_2^2/n_2\right)\right]^2}{\dfrac{\left(s_1^2/n_1\right)^2}{n_1 - 1} + \dfrac{\left(s_2^2/n_2\right)^2}{n_2 - 1}}$$

rounded down to the nearest integer.

For a 95% confidence interval, $\alpha = 0.05$. As we saw in Example 10.7, df = 22. Consulting Table III, we find that for df = 22, $t_{\alpha/2} = t_{0.05/2} = t_{0.025} = $ **2.074.**

STEP 2 *The endpoints of the confidence interval for $\mu_1 - \mu_2$ are*

$$(\bar{x}_1 - \bar{x}_2) \pm t_{\alpha/2} \cdot \sqrt{(s_1^2/n_1) + (s_2^2/n_2)}$$

From Step 1, $t_{\alpha/2} = 2.074$. Referring to Table 10.6 on page 502, we conclude that the endpoints of the confidence interval for $\mu_1 - \mu_2$ are

$$(6.94 - 7.84) \pm 2.074 \cdot \sqrt{(0.82^2/12) + (1.53^2/15)}$$

or

$$-0.90 \pm 0.96$$

Thus, a 95% confidence interval for $\mu_1 - \mu_2$ is from

$$\boxed{-1.86 \quad \text{to} \quad 0.06}$$

We can be 95% confident that the difference, $\mu_1 - \mu_2$, between the mean lifetime of Eagle water heaters and the mean lifetime of National water heaters is somewhere between -1.86 and 0.06 years.

MTB

POOLED VERSUS NON-POOLED

Suppose that we want to perform a hypothesis test to compare the means, μ_1 and μ_2, of two normally distributed populations with unknown standard deviations. If we take independent samples, then we need to use either the pooled procedure (Procedure 10.3 of Section 10.2) or the non-pooled procedure (Procedure 10.5 of this section).

The pooled procedure is to be used when the population standard deviations, σ_1 and σ_2, are equal. What if the pooled procedure is employed when, in fact, the population standard deviations are not equal? The answer to this question depends on several factors. If the population standard deviations are unequal, but not too unequal, and the sample sizes, n_1 and n_2, are about the same, then using the pooled procedure will not cause any serious difficulties. However, if the population standard deviations are actually quite different, then applying the pooled procedure can result in a significantly larger Type I error probability than the one specified.

On the other hand, the non-pooled procedure does not require the population standard deviations to be equal; it applies whether or not they are equal. Then why use the pooled procedure at all? The reason is as follows: If the population standard deviations are equal, then, on the average, the pooled procedure is slightly more powerful—there is a somewhat smaller probability of making a Type II error.

Thus, we see that for a hypothesis test to compare the means of two normally distributed populations using independent samples, the pooled procedure should be used only when we are quite sure that the two populations have equal standard deviations. Similar remarks apply to confidence intervals.

With the above discussion in mind, we can now suggest a method for deciding between a pooled and non-pooled procedure. This is done in the following key fact:

KEY FACT 10.5   Choosing between a pooled and a non-pooled procedure

Suppose you want to compare the means of two normally distributed populations using independent samples. If you are sure that the populations have equal
standard deviations, use a pooled procedure (i.e, a procedure from Section 10.2).
Otherwise, use a non-pooled procedure (i.e., a procedure from this section).

USING THE COMPUTER (OPTIONAL)

In this section, we have learned how to perform inferences to compare the means of
two normally distributed populations using independent samples without making
any assumptions about the relationship between the standard deviations of the
two populations. Procedure 10.5 on page 501 provides a step-by-step method for
performing a hypothesis test and Procedure 10.6 on page 504 gives a step-by-step
method for obtaining a confidence interval.

Minitab has a program that will carry out both Procedures 10.5 and 10.6 simultaneously. The appropriate command is **TWOSAMPLE T** *without* the POOLED
subcommand. (Recall from Section 10.2 that the POOLED subcommand is used
when the standard deviations of the two populations are known to be equal.) We
will use the command TWOSAMPLE T to simultaneously perform the hypothesis test considered in Example 10.7 and obtain the confidence interval required in
Example 10.8.

EXAMPLE 10.9   *Illustrates the TWOSAMPLE T command*

A general contractor wants to compare the lifetimes of two brands of water heaters,
Eagle and National. Using independent samples, the contractor obtains the data
shown in Table 10.7.

TABLE 10.7
Lifetimes, in
years, of water
heaters sampled

	Eagle				National			
6.9	7.3	7.8	7.4	8.7	7.0	8.7	6.7	7.8
7.2	6.6	6.2	8.2	8.6	6.1	7.5	7.7	7.5
7.6	5.7	5.5	6.9	11.2	6.1	6.3	7.0	10.7

Apply Minitab to determine, at the 5% significance level, whether the two brands of
water heaters have different mean lifetimes and to obtain a 95% confidence interval
for the difference between the mean lifetimes. [Assume that the lifetimes of each
brand are normally distributed.]

SOLUTION   Let μ_1 denote the mean lifetime of all Eagle water heaters and μ_2 denote the mean
lifetime of all National water heaters. Then we want to use Minitab to simultaneously perform the hypothesis test

$$H_0: \mu_1 = \mu_2 \text{ (mean lifetimes are the same)}$$
$$H_a: \mu_1 \neq \mu_2 \text{ (mean lifetimes are different)}$$

with $\alpha = 0.05$ and obtain a 95% confidence interval for $\mu_1 - \mu_2$.

By assumption, the populations are normally distributed and the samples are independent. Also, there is no indication that the population standard deviations are equal. So, we will employ TWOSAMPLE T, without the POOLED subcommand, to carry out the hypothesis test and obtain the required confidence interval.

We begin by entering the sample data in Table 10.7 into, say, C3 and C4 using the SET command. Next we name C3 "Eagle" and C4 "National" using the NAME command. Then we type the command TWOSAMPLE T, the desired confidence level for the confidence interval, and the storage locations of the sample data:

<u>TWOSAMPLE T, 95% confidence, for 'EAGLE' vs 'NATIONAL'</u>

Since the hypothesis test is two-tailed (H_a: $\mu_1 \neq \mu_2$), we do not need to use the ALTERNATIVE subcommand. Printout 10.4 displays the above commands and the output that results.

PRINTOUT 10.4
Minitab output for
TWOSAMPLE T

```
MTB > SET C3
DATA> 6.9 7.3 7.8 7.4 7.2 6.6
DATA> 6.2 8.2 7.6 5.7 5.5 6.9
DATA> END
MTB > SET C4
DATA> 8.7 7.0 8.7 6.7 7.8
DATA> 8.6 6.1 7.5 7.7 7.5
DATA> 11.2 6.1 6.3 7.0 10.7
DATA> END
MTB > NAME C3 'EAGLE' C4 'NATIONAL'
MTB > TWOSAMPLE T, 95% confidence, for 'EAGLE' vs 'NATIONAL'

TWOSAMPLE T FOR EAGLE VS NATIONAL
              N      MEAN     STDEV    SE MEAN
EAGLE        12      6.942    0.823    0.24
NATIONAL     15      7.84     1.53     0.40

95 PCT CI FOR MU EAGLE - MU NATIONAL: (-1.86, 0.06)

TTEST MU EAGLE = MU NATIONAL (VS NE): T= -1.95  P=0.065  DF=  22
```

The output provides the same information as TWOSAMPLE T with the POOLED subcommand except, of course, no pooled sample standard deviation is printed.

For the hypothesis test, we refer to the last line of the output in Printout 10.4. We find that the P-value is 0.065. Since this exceeds the designated significance level of $\alpha = 0.05$, we do not reject H_0. In other words, the data do not provide sufficient evidence to conclude that the two brands of water heaters have different mean lifetimes.

The required confidence interval can be found on the next-to-last line of the output in Printout 10.4. From that line, we see that a 95% confidence interval for the difference, $\mu_1 - \mu_2$, between the mean lifetimes of Eagle and National water heaters is somewhere between -1.86 and 0.06 years. ∎

Exercises 10.3

In each of Exercises 10.37–10.42 assume that the two populations being sampled are normally distributed but do not assume that their standard deviations are equal, although they may be.

___ **10.37**   Independent random samples of 17 sophomores and 13 juniors attending a large state university gave the statistics below for cumulative grade point average (GPA).

Sophomores	Juniors
$\bar{x}_1 = 2.54$	$\bar{x}_2 = 2.68$
$s_1 = 0.42$	$s_2 = 0.38$

Can you conclude from these data that there is a difference between the mean GPAs of sophomores and juniors at the university? Use $\alpha = 0.05$.

___ **10.38**   According to *High School Profile Report*, in past years, college-bound males have outperformed college-bound females on the mathematics portion of tests given by the American College Testing (ACT) Program. Independent random samples of this year's scores yield the following statistics:

Males	Females
$\bar{x}_1 = 18.3$	$\bar{x}_2 = 16.2$
$s_1 = 3.8$	$s_2 = 4.0$
$n_1 = 15$	$n_2 = 15$

Does it appear that college-bound males are, on the average, still outperforming college-bound females on the mathematics portion of ACT tests?
a) Use $\alpha = 0.05$.
b) Use $\alpha = 0.10$.

___ **10.39**   The owner of a chain of car washes needs to decide between two brands of hot wax. One of the brands, Sureglow, costs less than the other brand, Mirror-sheen. So, unless there is strong evidence that Mirror-sheen outlasts Sureglow, the owner will purchase Sureglow. With the cooperation of several local automobile dealers, 30 cars are randomly selected to take part in a test. Fifteen of the 30 cars are waxed with Sureglow and 15 with Mirror-sheen. The cars are then exposed to the same environmental conditions.

In the following table, you will find the data obtained on effectiveness times, in days.

Sureglow			Mirror-sheen		
87	90	88	92	92	91
93	90	92	91	92	91
91	89	93	93	93	92
88	87	89	92	93	94
91	91	90	93	94	91

(Note: For the Sureglow data $\Sigma x = 1349$ and $\Sigma x^2 = 121{,}373$; and for the Mirror-sheen data $\Sigma x = 1384$ and $\Sigma x^2 = 127{,}712$.)
a) At the 1% level of significance, does Mirror-sheen seem to have a longer effectiveness time, on the average, than Sureglow?
b) Do the data provide strong evidence that Mirror-sheen outlasts Sureglow? Explain.

___ **10.40**   The marketing manager of a firm that produces laundry products decides to test market a new laundry product in each of the firm's two sales regions. He wants to determine whether there will be a difference in mean sales per market per month between the two regions. Supermarkets from each region are independently and randomly selected to take part in the test, 10 from Region 1 and 15 from Region 2. The following data give the number of cases sold in each store during the testing month:

Region 1		Region 2		
74	87	84	86	95
96	94	87	89	94
78	77	92	93	88
83	83	85	81	93
86	80	92	92	85

At the 10% significance level, does the test marketing reveal a difference in potential mean sales per market in the two regions? *(Note:* For the Region 1 data, $\Sigma x = 838$ and $\Sigma x^2 = 70{,}684$; and for the Region 2 data, $\Sigma x = 1336$ and $\Sigma x^2 = 119{,}248$.)

___ **10.41**   The U.S. Department of Agriculture compiles information on acreage, production, and value of potatoes, by state, and publishes its findings in *Agricultural Statistics*. The yield of potatoes is measured in hundreds of pounds (cwt) per acre. A random sample of 40 one-acre plots of potatoes is taken from Idaho and a random sample of 32 one-acre plots of potatoes is taken from Nevada. The yields of the plots sampled are displayed in the following table:

Idaho					Nevada			
229	267	326	309	231	283	254	328	292
283	344	310	258	316	315	336	378	314
241	281	218	284	311	312	328	272	307
254	217	267	299	266	348	233	354	400
264	264	290	312	298	341	313	309	308
305	244	303	299	285	340	300	316	268
308	260	204	291	242	259	276	271	362
329	315	246	322	293	340	339	300	333

At the 5% significance level, do the data provide sufficient evidence to conclude that Idaho has a smaller mean potato yield than Nevada? (*Note:* The sample mean and sample standard deviation of the Idaho data are 279.6 cwt and 34.6 cwt, respectively; and the sample mean and sample standard deviation of the Nevada data are 313.4 cwt and 37.2 cwt, respectively.)

___ **10.42** Figures on the costs to community hospitals per patient per day are reported by the American Hospital Association in the publication *Hospital Statistics*. Independent random samples of 12 such costs in California and 10 such costs in New York gave the following data, in dollars:

California		New York	
1186	1257	702	987
1076	1560	435	1016
1394	1399	639	906
965	1040	876	868
690	1227	669	522
933	1441		

Do the data provide sufficient evidence to conclude that the mean cost to community hospitals per patient per day is greater in California than in New York? Perform the appropriate hypothesis test at the 0.01 level of significance.

In each of Exercises 10.43–10.48, apply Procedure 10.6 on page 504 to obtain the required confidence interval.

___ **10.43** Refer to Exercise 10.37.

a) Determine a 95% confidence interval for the difference, $\mu_1 - \mu_2$, between the mean GPAs of sophomores and juniors at the university.
b) Interpret your results in words.

___ **10.44** Refer to Exercise 10.38.

a) Obtain a 90% confidence interval for the difference, $\mu_1 - \mu_2$, between this year's mean mathematics ACT score for males and this year's mean mathematics ACT score for females.
b) Repeat part (a) using a confidence level of 80%.

___ **10.45** Refer to Exercise 10.39.

a) Find a 98% confidence interval for the difference, $\mu_1 - \mu_2$, between the mean effectiveness times of Sureglow and Mirror-sheen.
b) Interpret your results in words.

___ **10.46** Refer to Exercise 10.40.

a) Find a 90% confidence interval for the difference, $\mu_1 - \mu_2$, between potential mean monthly sales per market in Region 1 and Region 2.
b) Interpret your results in words. [Be extra careful here!]

___ **10.47** Refer to Exercise 10.41.

a) Obtain a 90% confidence interval for the difference, $\mu_1 - \mu_2$, between the mean yields per acre of potatoes for Idaho and Nevada.
b) Interpret your results in words.

___ **10.48** Refer to Exercise 10.42.

a) Obtain a 98% confidence interval for the difference, $\mu_1 - \mu_2$, between the mean cost to community hospitals per patient per day in California and the mean cost to community hospitals per patient per day in New York.
b) Interpret your results in words.

Exercises 10.49–10.52 are computer exercises.

___ **10.49 (Computer exercise)** Suppose that the data in Exercise 10.39 are stored in columns named SUREGLOW and MIRROR.

a) Which Minitab command and subcommands (if any) should be used to simultaneously perform the hypothesis test considered in Exercise 10.39 and obtain the confidence interval required in Exercise 10.45?
b) If you have access to Minitab, use it to perform the hypothesis test and obtain the confidence interval.

___ **10.50 (Computer exercise)** Suppose that the data in Exercise 10.40 are stored in columns named REGION 1 and REGION 2.

a) Which Minitab command and subcommands (if any) should be used to simultaneously perform the hypothesis test considered in Exercise 10.40 and obtain the confidence interval required in Exercise 10.46?

b) If you have access to Minitab, use it to perform the hypothesis test and obtain the confidence interval.

___ **10.51 (Computer exercise)** The U.S. National Center for Health Statistics compiles information on divorces in the document *Vital Statistics of the United States*. Suppose that independent random samples are taken of divorced males and divorced females in order to decide whether the mean age at the time of first divorce for males is greater than that for females. Further suppose that the output obtained by applying the TWOSAMPLE T command to the data is as shown in Printout 10.5 at the top of the next page. From the output, determine

a) the null and alternative hypotheses for the hypothesis test.

b) the command used and the subcommands used (if any) to obtain the Minitab output.

c) the standard deviation of the ages for each of the two samples.

d) the number of males sampled and the number of females sampled.

e) the mean age at first divorce for both the males sampled and the females sampled.

f) the value of the test statistic, t.

g) the P-value of the hypothesis test.

h) the smallest significance level at which the null hypothesis can be rejected.

i) the conclusion if the hypothesis test is performed at the 5% significance level.

j) a 90% confidence interval for the difference, $\mu_1 - \mu_2$, between the mean age at first divorce for males and the mean age at first divorce for females.

k) the conditions that must be satisfied in order to apply TWOSAMPLE T to the age data.

___ **10.52 (Computer exercise)** A transportation official wants to compare the mean number of miles that cars were driven last year to the mean number of miles that trucks were driven last year. He takes independent random samples of cars and trucks and records the number of miles that each vehicle was driven last year. Then he applies the TWOSAMPLE T command to the data and obtains the output displayed in Printout 10.6 on the next page. Use the output to determine

a) the null and alternative hypotheses for the official's hypothesis test.

b) the command used and the subcommands used (if any) to obtain the Minitab output.

c) the standard deviations of the number of miles driven last year for the cars sampled and for the trucks sampled.

d) the number of cars sampled and the number of trucks sampled.

e) the mean number of miles driven last year for both the cars sampled and the trucks sampled.

f) the value of the test statistic, t.

g) the P-value of the hypothesis test.

h) the smallest significance level at which the null hypothesis can be rejected.

i) the conclusion if the hypothesis test is performed at the 1% significance level.

j) a 99% confidence interval for the difference, $\mu_1 - \mu_2$, between the mean number of miles that cars were driven last year and the mean number of miles that trucks were driven last year.

k) the assumptions made by the official in using the Minitab program TWOSAMPLE T.

≡ **10.53** We have presented two procedures for performing a hypothesis test to compare the means of two normally distributed populations using independent samples. One procedure, Procedure 10.3 on page 489, applies when the standard deviations of the two populations are unknown but assumed equal. The test statistic used in that procedure is

$$t = \frac{\bar{x}_1 - \bar{x}_2}{s_\mathrm{p}\sqrt{(1/n_1) + (1/n_2)}}$$

The other procedure, Procedure 10.5 on page 501, applies when the standard deviations of the two populations are unknown but not assumed equal. The test statistic used in that procedure is

$$t = \frac{\bar{x}_1 - \bar{x}_2}{\sqrt{(s_1^2/n_1) + (s_2^2/n_2)}}$$

a) Show that, if the sample sizes, n_1 and n_2, are equal, then the values of the two test statistics above will be exactly the same. (*Hint:* Refer to Exercise 10.35 on page 499.)

b) Does part (a) imply that the two t-tests are equivalent when the sample sizes are equal? Provide a detailed explanation for your answer.

PRINTOUT 10.5   Minitab output for Exercise 10.51

```
TWOSAMPLE T FOR MALES VS FEMALES
             N      MEAN     STDEV    SE MEAN
MALES       11     35.69     4.62      1.4
FEMALES     12     32.81     7.25      2.1

90 PCT CI FOR MU MALES - MU FEMALES: (-1.5, 7.2)

TTEST MU MALES = MU FEMALES (VS GT): T= 1.14   P=0.13   DF= 18
```

PRINTOUT 10.6   Minitab output for Exercise 10.52

```
TWOSAMPLE T FOR CARS VS TRUCKS
             N      MEAN     STDEV    SE MEAN
CARS        15      9.31     2.78      0.72
TRUCKS      10     10.93     4.51      1.4

99 PCT CI FOR MU CARS - MU TRUCKS: (-6.44, 3.2)

TTEST MU CARS = MU TRUCKS (VS NE): T= -1.02   P=0.33   DF=  13
```

10.4   Inferences for two population means using paired samples

Up to this point, the methods that we have studied for comparing the means of two populations rely on independent samples. In this section, we will examine methods for comparing the means of two populations that use paired samples.

Suppose, for instance, that we want to decide whether a newly developed gasoline additive increases gas mileage. Let μ_1 denote the mean gas mileage of all cars when the additive is used and μ_2 denote the mean gas mileage of all cars when the additive is not used. Then we want to perform the hypothesis test

H_0: $\mu_1 = \mu_2$ (mean gas mileage with additive is not greater)

H_a: $\mu_1 > \mu_2$ (mean gas mileage with additive is greater)

One method that can be used to carry out the hypothesis test is this: Randomly and independently select two groups of, say, 10 cars each; have one group driven with the additive and the other group driven without the additive; and then apply a hypothesis-testing procedure, as described in Section 10.2 or 10.3, to the two samples of 10 mileages obtained. This method employs *independent samples*.

However, it is probably more appropriate to use the following method to carry out the hypothesis test: Randomly select a single group of 10 cars; have each of the 10 cars driven both with and without the additive; and then apply a hypothesis-testing procedure, as will be described in this section, to the 10 *pairs* of mileages obtained. This method employs **paired samples.** Each piece of sample data consists of a pair of numbers, the gas mileage of a particular car both with and without the additive.

By pairing the samples, we can remove extraneous sources of variation, such as the variation due to cars and drivers. As a consequence, the sampling error made in estimating the difference between the population means will generally be smaller. This fact, in turn, makes it more likely that we will detect differences between the population means when such differences exist.

We will now use the gas-mileage illustration to explain the logic behind hypothesis tests that employ paired samples. In doing so, we will also introduce some of the notation and terminology that is used in such hypothesis tests.

EXAMPLE 10.10   *Introduces hypothesis tests for two means using paired samples*

A major oil company has developed a new gasoline additive that is supposed to increase mileage. To test that hypothesis, 10 cars are randomly selected. Each car sampled is driven both with and without the additive. The resulting gas mileages, in miles per gallon, are displayed in the second and third columns of Table 10.8.

TABLE 10.8
Gas mileages, with and without additive, for 10 randomly selected cars

Car	With additive x_1	Without additive x_2	Paired difference $d = x_1 - x_2$
1	25.7	24.9	0.8
2	20.0	18.8	1.2
3	28.4	27.7	0.7
4	13.7	13.0	0.7
5	18.8	17.8	1.0
6	12.5	11.3	1.2
7	28.4	27.8	0.6
8	8.1	8.2	−0.1
9	23.1	23.1	0.0
10	10.4	9.9	0.5
			6.6

In the last column of Table 10.8, we have recorded the difference, *d,* between the gas mileages, with and without the additive, for each of the 10 cars sampled. Each difference is referred to as a **paired difference** since it is the difference of a pair of numbers. For instance, the first car got 25.7 mpg with the additive and 24.9 mpg without the additive, giving a paired difference of $d = 25.7 - 24.9 = 0.8$ mpg, an increase in gas mileage of 0.8 mpg with the additive.

We want to use the paired differences, *d,* in the last column of Table 10.8 to perform the hypotheses test

$$H_0: \mu_1 = \mu_2 \text{ (mean gas mileage with additive is not greater)}$$
$$H_a: \mu_1 > \mu_2 \text{ (mean gas mileage with additive is greater)}$$

where μ_1 denotes the mean gas mileage of all cars when the additive is used and μ_2 denotes the mean gas mileage of all cars when the additive is not used.

The logic behind performing the hypothesis test using the paired differences is as follows: If the null hypothesis is true, then the paired differences of the gas mileages for the cars sampled should average out to about zero. In other words, we would expect the sample mean, $\bar{d}$, of the paired differences in the final column of Table 10.8 to be close to zero. To put it another way, if $\bar{d}$ is too much greater than zero, then we would take this as evidence that the null hypothesis is false and conclude that the mean gas mileage, μ_1, with the additive is greater than the mean gas mileage, μ_2, without the additive.

From the last column of Table 10.8, we find that the sample mean of the paired differences is

$$\bar{d} = \frac{\Sigma d}{n} = \frac{6.6}{10} = 0.66 \text{ mpg},$$

a mean increase in gas mileage for the cars sampled of 0.66 mpg when the additive is used. The question now is whether that mean increase in gas mileage can reasonably be attributed to sampling error or whether it is large enough to indicate that, on the average, the additive improves gas mileage. To answer the question, we need to know the probability distribution of the random variable $\bar{d}$. We will discuss that probability distribution and then return to solve the gas-mileage problem. ■

THE SAMPLING DISTRIBUTION OF THE DIFFERENCE BETWEEN TWO MEANS (NORMAL DIFFERENCES AND PAIRED SAMPLES)

Consider two populations, say Population 1 and Population 2, whose members can be paired. To each pair, there corresponds a single number, obtained by subtracting the Population 2 value in the pair from the Population 1 value in the pair. Thus, from the two populations, we can form a single population consisting of the differences of all the pairs.

Denote the mean of that population of paired differences by μ_d. Then it can be shown that

(3) $$\mu_d = \mu_1 - \mu_2$$

where μ_1 is the mean of Population 1 and μ_2 is the mean of Population 2. In other words, the mean of the population of paired differences is equal to the difference between the two population means. (See Exercise 10.69 for a detailed derivation.)

Now, suppose that a random sample of n pairs is to be taken from the two populations. Let $\bar{d}$ denote the sample mean paired difference of the pairs obtained, where each difference is computed by subtracting the second number in the pair from the first number in the pair ($d = x_1 - x_2$). Also, let s_d denote the sample standard deviation of the paired differences.

We can think of the paired differences of the pairs sampled as a random sample from the population of all possible paired differences. If that population is *normally distributed*, then we can apply Key Fact 8.3 on page 395 and Equation (3) to obtain the following result:

KEY FACT 10.6

Suppose that a random sample of n pairs is to be taken from populations with means μ_1 and μ_2. Further suppose that the population of all paired differences is normally distributed. Then the random variable

$$t = \frac{\overline{d} - (\mu_1 - \mu_2)}{s_d/\sqrt{n}}$$

has the t-distribution with df $= n - 1$.

Note: We will use the phrase **normal differences** as an abbreviation of "the population of paired differences is normally distributed."

HYPOTHESIS TESTS FOR TWO POPULATION MEANS USING PAIRED SAMPLES

We can now present a hypothesis-testing procedure for comparing the means of two populations using paired samples when the population of all paired differences is normally distributed. In view of Key Fact 10.6, we see that for a hypothesis test with null hypothesis H_0: $\mu_1 = \mu_2$, we can use the random variable

$$t = \frac{\overline{d}}{s_d/\sqrt{n}}$$

as the test statistic and obtain the critical value(s) from the t-table, Table III. Thus, we have the following procedure:

PROCEDURE 10.7

To perform a hypothesis test for two population means with null hypothesis H_0: $\mu_1 = \mu_2$.

ASSUMPTIONS

1. Paired samples.
2. Normal differences.

STEP 1 *State the null and alternative hypotheses.*
STEP 2 *Decide on the significance level, α.*
STEP 3 *The critical value(s)*
      a) for a two-tailed test are $\pm t_{\alpha/2}$,
      b) for a left-tailed test is $-t_\alpha$,
      c) for a right-tailed test is t_α,
      with df $= n - 1$. Use Table III to find the critical value(s).

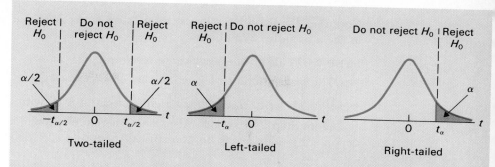

STEP 4 *Calculate the paired differences, $d = x_1 - x_2$, of the sample pairs.*
STEP 5 *Compute the value of the test statistic*

$$t = \frac{\overline{d}}{s_d/\sqrt{n}}$$

STEP 6 *If the value of the test statistic falls in the rejection region, then reject H_0; otherwise, do not reject H_0.*
STEP 7 *State the conclusion in words.*

EXAMPLE 10.11   *Illustrates Procedure 10.7*

We now return to the gas-mileage problem posed in Example 10.10. Recall that a major oil company has developed a new gasoline additive that is supposed to increase gas mileage. To test that hypothesis, the gas mileages are obtained, both with and without the additive, for each of 10 randomly selected cars. The results are displayed in the second and third columns of Table 10.8 on page 512.

Do the data provide sufficient evidence to conclude that, on the average, the gasoline additive improves gas mileage? Perform the appropriate hypothesis test at the 5% significance level. [Assume that changes in gas mileage due to the gasoline additive are normally distributed.]

SOLUTION   Note that we are dealing here with paired samples. Each pair consists of the gas mileage of a car both with and without the additive. Since, by assumption, the population of all possible paired differences is normally distributed, we can apply Procedure 10.7 to perform the hypothesis test.

STEP 1 *State the null and alternative hypotheses.*

Let μ_1 denote the mean gas mileage of all cars when the additive is used and μ_2 denote the mean gas mileage of all cars when the additive is not used. Then the null and alternative hypotheses are

H_0: $\mu_1 = \mu_2$ (mean gas mileage with additive is not greater)
H_a: $\mu_1 > \mu_2$ (mean gas mileage with additive is greater)

Note that the hypothesis test is right-tailed since there is a greater-than sign ($>$) in the alternative hypothesis.

STEP 2 *Decide on the significance level, α.*

The test is to be performed at the 5% significance level. Thus, $\alpha = 0.05.$

STEP 3 *The critical value for a right-tailed test is t_α with df $= n - 1$.*

From Step 2, $\alpha = 0.05$. Also, since there are 10 pairs in the sample, we have df $= n - 1 = 10 - 1 = 9$. So, the critical value is $t_{0.05} = 1.833.$ See Figure 10.5.

FIGURE 10.5

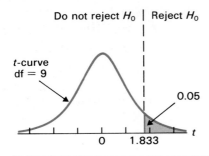

STEP 4 *Calculate the paired differences, $d = x_1 - x_2$, of the sample pairs.*

We have already done this in the last column of Table 10.8 on page 512.

STEP 5 *Compute the value of the test statistic*

$$t = \frac{\overline{d}}{s_d/\sqrt{n}}$$

We first need to determine the sample mean and sample standard deviation of the d-values in Table 10.8. This is accomplished in the usual manner:

$$\overline{d} = \frac{\Sigma d}{n} = \frac{6.6}{10} = 0.66$$

and

$$s_d = \sqrt{\frac{n(\Sigma d^2) - (\Sigma d)^2}{n(n-1)}} = \sqrt{\frac{10(6.12) - (6.6)^2}{10 \cdot 9}} = 0.44$$

Consequently, the value of the test statistic is

$$t = \frac{\overline{d}}{s_d/\sqrt{n}} = \frac{0.66}{0.44/\sqrt{10}} = 4.74$$

STEP 6 *If the value of the test statistic falls in the rejection region, reject H_0; otherwise, do not reject H_0.*

From Step 5, the value of the test statistic is $t = 4.74$, which falls in the rejection region. Hence, we **reject H_0.**

STEP 7 *State the conclusion in words.*

The data provide sufficient evidence to conclude that the mean gas mileage of all cars when the additive is used is greater than the mean gas mileage of all cars when the additive is not used. In other words, it appears that the additive is effective in increasing gas mileage. ∎

CONFIDENCE INTERVALS FOR THE DIFFERENCE BETWEEN TWO POPULATION MEANS USING PAIRED SAMPLES

We can also use Key Fact 10.6 on page 514 to derive the following confidence-interval procedure for the difference between two population means. The procedure applies when the samples are paired and the population of all paired differences is normally distributed.

PROCEDURE 10.8

To find a confidence interval for the difference between two population means.

ASSUMPTIONS

1. Paired samples.
2. Normal differences.

STEP 1 *For a confidence level of $1 - \alpha$, use Table III to find $t_{\alpha/2}$ with $df = n - 1$.*

STEP 2 *The endpoints of the confidence interval for $\mu_1 - \mu_2$ are*

$$\overline{d} \pm t_{\alpha/2} \cdot \frac{s_d}{\sqrt{n}}$$

EXAMPLE 10.12   *Illustrates Procedure 10.8*

Consider once more the gas-mileage illustration of Example 10.10. Use the sample data in Table 10.8 on page 512 to obtain a 90% confidence interval for the difference, $\mu_1 - \mu_2$, between the mean gas mileage of all cars when the additive is used and the mean gas mileage of all cars when the additive is not used.

SOLUTION   We apply Procedure 10.8.

STEP 1 *For a confidence level of $1 - \alpha$, use Table III to find $t_{\alpha/2}$ with $df = n - 1$.*

For a 90% confidence interval, we have $\alpha = 0.10$. From Table III, we find that for $df = n - 1 = 10 - 1 = 9$, $t_{\alpha/2} = t_{0.10/2} = t_{0.05} = $ **1.833.**

STEP 2 *The endpoints of the confidence interval for $\mu_1 - \mu_2$ are*

$$\bar{d} \pm t_{\alpha/2} \cdot \frac{s_d}{\sqrt{n}}$$

From Step 1, $t_{\alpha/2} = 1.833$. Also, $n = 10$ and, from Example 10.11, we know that $\bar{d} = 0.66$ and $s_d = 0.44$. Consequently, the endpoints of the confidence interval for $\mu_1 - \mu_2$ are

$$0.66 \pm 1.833 \cdot \frac{0.44}{\sqrt{10}}$$

or

$$0.66 \pm 0.26$$

Thus, a 90% confidence interval for $\mu_1 - \mu_2$ is from

$$0.40 \quad \text{to} \quad 0.92$$

We can be 90% confident that the difference, $\mu_1 - \mu_2$, between the mean gas mileage of all cars when the additive is used and the mean gas mileage of all cars when the additive is not used is somewhere between 0.40 mpg and 0.92 mpg. In particular, we can be 90% confident that, on the average, the additive increases gas mileage by at least 0.40 mpg.

MTB

LARGE-SAMPLE INFERENCES FOR TWO POPULATION MEANS USING PAIRED SAMPLES

The inferential procedures that we have discussed in this section provide methods for comparing the means of two populations using paired samples. An assumption for the use of those procedures is that the population of all paired differences is normally distributed (normal differences).

We can also derive methods for comparing the means of two populations using paired samples that do not require normal differences. For example, if the sample size is large, then we can develop inferential procedures that rely on paired samples but do not make any assumptions about the distribution of the population of paired differences. We will investigate these large-sample procedures in the exercises.

Exercises 10.4

In each of Exercises 10.54–10.59, assume that the population of paired differences is normally distributed.

___ **10.54** A pediatrician measured the blood cholesterol levels of her young patients. She was surprised to find that many of them had levels over 200 milligrams per 100 milliliters, indicating increased risk of artery disease. Ten such patients were randomly selected to take part in a nutritional program designed to lower blood cholesterol. Two months following the commencement of the program, the pediatrician measured the blood cholesterol levels of the 10 patients again. Here are the data that were obtained:

Patient	Before program	After program
1	210	212
2	217	210
3	208	210
4	215	213
5	202	200
6	209	208
7	207	203
8	210	199
9	221	218
10	218	214

Do the data suggest that the nutritional program is, on the average, effective in reducing cholesterol levels? Perform the appropriate hypothesis test at the 1% level of significance.

— **10.55** An exercise physiologist wants to decide whether a certain type of running program will reduce heart rates. He measures the heart rates of 15 randomly selected people who are then placed on the running program. One year later the exercise physiologist again measures the heart rates of the 15 people. The heart rates both before and after the running program are displayed in the table below.

Person	Before program	After program
1	68	67
2	76	77
3	74	74
4	71	74
5	71	69
6	72	70
7	75	71
8	83	77
9	75	71
10	74	74
11	76	73
12	77	68
13	78	71
14	75	72
15	75	77

Do the data provide sufficient evidence to conclude that the running program will, on the average, reduce heart rates? Use $\alpha = 0.01$.

— **10.56** The A. C. Nielsen Company collects data on the TV viewing habits of Americans and publishes the information in *Nielsen Report on Television.* Suppose that 20 married couples are randomly selected and that their weekly viewing times, in hours, are as given in the following table:

Husband	Wife	Husband	Wife	Husband	Wife
21	24	38	45	36	35
56	55	27	29	20	34
34	55	30	41	43	32
30	34	31	37	4	13
41	32	30	35	16	9
35	38	32	48	21	23
26	38	15	17		

At the 5% level of significance, does it appear that married men watch less TV, on the average, than married women? (*Note:* $\bar{d} = -4.4$ hr and $s_d = 8.15$ hr.)

— **10.57** In Exercise 10.19, we performed a hypothesis test using independent samples to decide whether nonsupervisory mine workers earn a smaller average hourly wage than nonsupervisory construction workers. Now we will perform that same hypothesis test using paired samples. Suppose that nonsupervisory mine workers and nonsupervisory construction workers are paired by matching workers with similar experience and job classification. Further suppose that a random sample of 15 pairs yields the following paired differences for hourly wages (mining *minus* construction).

0.80	1.03	0.57
−2.38	0.89	−2.16
−1.36	−0.05	−1.89
−0.63	1.20	−1.40
−0.67	−1.23	0.47

Use these data to decide whether nonsupervisory mine workers earn a smaller average hourly wage than nonsupervisory construction workers. Perform the appropriate hypothesis test at the 5% significance level. (*Note:* $\Sigma d = -6.81$ and $\Sigma d^2 = 24.5517$.)

— **10.58** An algebra teacher wants to compare two methods of teaching college algebra. One is the lecture method and the other is the personalized system of instruction (PSI) method. Students are paired by matching those with similar mathematics background and performance. A random sample of 11 pairs is selected. From each pair, one student is randomly chosen to take the lecture course; the other student takes the PSI course. Both courses are taught by the algebra teacher. The final grades for the 11 pairs of students are found to be the following:

Lecture	PSI	Lecture	PSI
66	67	80	79
93	93	73	70
36	35	74	67
84	85	83	79
60	64	52	50
66	57		

Do the data provide sufficient evidence to conclude that there is a difference in mean student performance between the two instructional methods? Use a significance level of $\alpha = 0.05$.

___ **10.59**   The publication *Current Population Reports*, released by the U.S. Bureau of the Census, presents data on the ages of married people. Suppose that 10 married couples are randomly selected and have the ages given here.

Husband	Wife	Husband	Wife
54	53	33	35
21	22	68	67
32	33	32	28
78	74	54	41
70	64	52	44

Do the data suggest that the mean age of married men is greater than the mean age of married women? Perform the appropriate hypothesis test at the 5% significance level.

In each of Exercises 10.60–10.65, use Procedure 10.8 on page 517 to obtain the required confidence interval.

___ **10.60**   Refer to Exercise 10.54.
a) Determine a 98% confidence interval for the difference, $\mu_1 - \mu_2$, between the mean blood cholesterol levels of high-level patients before and after the nutritional program.
b) Interpret your results in words.

___ **10.61**   Refer to Exercise 10.55.
a) Determine a 98% confidence interval for the difference, $\mu_1 - \mu_2$, between the mean heart rates of all people before and after the running program.
b) Interpret your results in words.

___ **10.62**   Refer to Exercise 10.56.
a) Find a 90% confidence interval for the difference, $\mu_1 - \mu_2$, between the mean weekly TV viewing times of married men and married women.
b) Interpret your results in words.

___ **10.63**   Refer to Exercise 10.57.
a) Obtain a 90% confidence interval for the difference, $\mu_1 - \mu_2$, between the mean hourly earnings of nonsupervisory mine and construction workers.
b) Interpret your results in words.

___ **10.64**   Refer to Exercise 10.58.
a) Obtain a 95% confidence interval for the difference, $\mu_1 - \mu_2$, between the mean final grade of all students who take college algebra from the teacher using the lecture method and the mean final grade of all students who take college algebra from the teacher using the PSI method.
b) Interpret your results in words.

___ **10.65**   Refer to Exercise 10.59.
a) Find a 90% confidence interval for the difference, $\mu_1 - \mu_2$, between the mean age of married men and the mean age of married women.
b) Interpret your results in words.

=== **10.66**   This exercise shows what can happen when a hypothesis-testing procedure designed for use with independent samples is applied to perform a hypothesis test in which the samples are paired. In Example 10.11 on page 515, we applied Procedure 10.7 to perform a hypothesis test using paired samples to decide whether a gasoline additive is effective in increasing gas mileage. Specifically, if we let μ_1 and μ_2 denote, respectively, the mean gas mileage of all cars when the additive is and is not used, then the hypothesis test is

$$H_0: \mu_1 = \mu_2 \text{ (additive is not effective)}$$
$$H_a: \mu_1 > \mu_2 \text{ (additive is effective)}$$

a) Apply Procedure 10.5 on page 501 to the sample data in the second and third columns of Table 10.8 on page 512 to perform the hypothesis test. Use $\alpha = 0.05$.
b) Why is it inappropriate to perform the hypothesis test in the way that you did in part (a)?
c) Compare your result in part (a) to the one obtained in Example 10.11.

Large-sample inferences for the means of two populations using paired samples: We will consider, in Exercises 10.67 and 10.68, large-sample inferences to compare the means of two populations using paired samples. The procedures used for such inferences are similar to those that apply when the population of paired differences is normally distributed. However, for large samples, (1) no assumptions need be made about the distribution of the population of paired differences and (2) the standard-normal table, Table II, is used instead of the t-table, Table III. Thus, the test statistic for a hypothesis test with null hypothesis $H_0: \mu_1 = \mu_2$ is

$$z = \frac{\bar{d}}{s_d/\sqrt{n}}$$

And the endpoints of a $(1-\alpha)$-level confidence interval for $\mu_1 - \mu_2$ are

$$\bar{d} \pm z_{\alpha/2} \cdot \frac{s_d}{\sqrt{n}}$$

= **10.67** A tire company has developed two new processes, Process A and Process B, for making longer-wearing steel-belted radials. To compare the two processes, 50 tires made using Process A and 50 tires made using Process B are randomly selected. Each of the 50 Process-A tires is randomly assigned to either the front left or front right of one of 50 cars. If, for a given car, a Process-A tire gets assigned to be the left front tire, then a Process-B tire gets assigned to be the right front tire, and vice-versa. The rear of each of the 50 cars is equipped with two tires currently manufactured by the tire company. The differences in tire life (Process-A tire life *minus* Process-B tire life) for the 50 pairs of tires are given below in thousands of miles, to the nearest hundred miles.

−0.8	−2.4	−0.2	1.6	−3.6
−0.6	−1.7	−1.4	−2.5	−1.8
−0.9	0.9	2.8	−2.7	−0.6
0.0	−2.0	1.8	−1.5	1.9
−4.2	−2.1	3.0	0.0	0.4
0.9	0.9	1.6	−2.7	0.9
1.0	−2.3	−0.4	0.5	2.4
−0.3	0.9	−0.6	0.8	3.6
−0.4	0.5	1.3	−4.0	2.4
−4.3	2.4	−0.5	1.5	3.1

a) At the 10% significance level, does there appear to be a difference in the mean lifetimes of Process-A and Process-B tires? *(Note: The sum of the data is −7.4 and the sum of the squares of the data is 197.90.)*

b) Find a 90% confidence interval for the difference, $\mu_1 - \mu_2$, between the mean lifetime of Process-A tires and the mean lifetime of Process-B tires.

= **10.68** In Example 10.2 (pages 481–483), we performed a hypothesis test using independent samples to decide whether there is a difference in mean salaries for faculty teaching in public and private institutions. Suppose we want to perform that same hypothesis test using paired samples. Pairs are formed by matching faculty in public and private institutions by rank and specialty. A random sample of 30 pairs yields the following annual salaries, in dollars:

Public	Private	Public	Private	Public	Private
42,080	39,752	70,380	69,296	25,139	26,434
36,104	46,431	39,496	38,656	27,152	27,867
33,875	37,496	24,426	29,824	26,750	28,559
30,564	34,184	45,638	47,208	25,749	29,032
57,326	68,784	25,724	37,345	32,893	31,908
39,568	38,875	33,877	36,636	24,889	25,349
43,190	46,421	36,738	32,158	28,906	36,198
31,689	32,008	37,294	37,744	36,546	38,247
29,606	29,770	20,406	20,720	52,669	51,045
48,025	54,914	55,607	56,350	61,468	60,695

a) Do the data provide sufficient evidence to conclude, at the 5% significance level, that there is a difference in mean salaries for faculty teaching in public and private institutions?

b) Compare your result in part (a) to the one obtained in Example 10.2.

c) Use the above data to find a 95% confidence interval for the difference between the mean salaries of faculty teaching in public and private institutions.

d) Compare your result in part (c) to the one obtained in Example 10.3 on pages 483–484.

≡ **10.69** On page 513, we gave the formula

$$(4) \qquad \mu_d = \mu_1 - \mu_2$$

which states that the mean of the population of paired differences is equal to the difference between the two population means.

a) Suppose that the two populations under consideration are finite and that each has N members. Prove Formula (4).

b) In general, let (x_1, x_2) denote a randomly selected pair from the population of all pairs and let $d = x_1 - x_2$ denote the paired difference. Note that x_1, x_2, and d are random variables. Use the results of Exercises 5.26(e) and 5.27(f) on pages 249 and 250, respectively, to prove Formula (4).

10.5   Which procedure should I use?

At this point, it might be helpful to summarize the various inferential procedures that we have presented for comparing the means, μ_1 and μ_2, of two populations. This is done in Table 10.9.

TABLE 10.9   Summary of inferential procedures for comparing two population means, μ_1 and μ_2 (the null hypothesis for all hypothesis tests is $H_0: \mu_1 = \mu_2$)

Type	Assumptions	Test statistic for a hypothesis test	Endpoints of a confidence interval	Procedures to use
Two-sample z	1. Independent samples 2. Large samples	$z = \dfrac{\overline{x}_1 - \overline{x}_2}{\sqrt{(s_1^2/n_1) + (s_2^2/n_2)}}$	$(\overline{x}_1 - \overline{x}_2) \pm z_{\alpha/2}$ $\cdot\sqrt{(s_1^2/n_1) + (s_2^2/n_2)}$	10.1 (p. 480) 10.2 (p. 483)
Pooled t	1. Independent samples 2. Normal populations 3. σs assumed equal	$t = \dfrac{\overline{x}_1 - \overline{x}_2}{s_P\sqrt{(1/n_1) + (1/n_2)}}$ $(\text{df} = n_1 + n_2 - 2)$	$(\overline{x}_1 - \overline{x}_2) \pm t_{\alpha/2}$ $\cdot s_P\sqrt{(1/n_1) + (1/n_2)}$ $(\text{df} = n_1 + n_2 - 2)$	10.3 (p. 489) 10.4 (p. 493)
Non-pooled t	1. Independent samples 2. Normal populations	$t = \dfrac{\overline{x}_1 - \overline{x}_2}{\sqrt{(s_1^2/n_1) + (s_2^2/n_2)}}$ †	$(\overline{x}_1 - \overline{x}_2) \pm t_{\alpha/2}$ $\cdot\sqrt{(s_1^2/n_1) + (s_2^2/n_2)}$ †	10.5 (p. 501) 10.6 (p. 504)
Paired t	1. Paired samples 2. Normal differences	$t = \dfrac{\overline{d}}{s_d/\sqrt{n}}$ $(\text{df} = n - 1)$	$\overline{d} \pm t_{\alpha/2} \cdot \dfrac{s_d}{\sqrt{n}}$ $(\text{df} = n - 1)$	10.7 (p. 514) 10.8 (p. 517)
Paired z	1. Paired samples 2. Large sample	$z = \dfrac{\overline{d}}{s_d/\sqrt{n}}$	$\overline{d} \pm z_{\alpha/2} \cdot \dfrac{s_d}{\sqrt{n}}$	Discussed in the exercises (p. 520)

† $\text{df} = [(s_1^2/n_1) + (s_2^2/n_2)]^2/[(s_1^2/n_1)^2/(n_1 - 1) + (s_2^2/n_2)^2/(n_2 - 1)]$

Let us examine Table 10.9. Look, for instance, at the row of the table that begins with "Pooled t" (this is the name that we have given to the type of procedures discussed in Section 10.2). In the second column of that row are the conditions required for using a pooled-t procedure.

The third column of the row that begins with "Pooled t" displays the test statistic for a pooled-t hypothesis test and the fourth column shows the endpoints of a pooled-t confidence interval. Finally, the fifth column gives the procedures and their page numbers. The first procedure, in this case Procedure 10.3, is for a hypothesis test and the second procedure, in this case Procedure 10.4, is for a confidence interval.

From Table 10.9, we can obtain the flowchart depicted in Figure 10.6 on the next page. The flowchart provides an organized strategy for choosing the correct inferential procedure to compare the means of two populations. We will illustrate the use of the flowchart in Example 10.13.

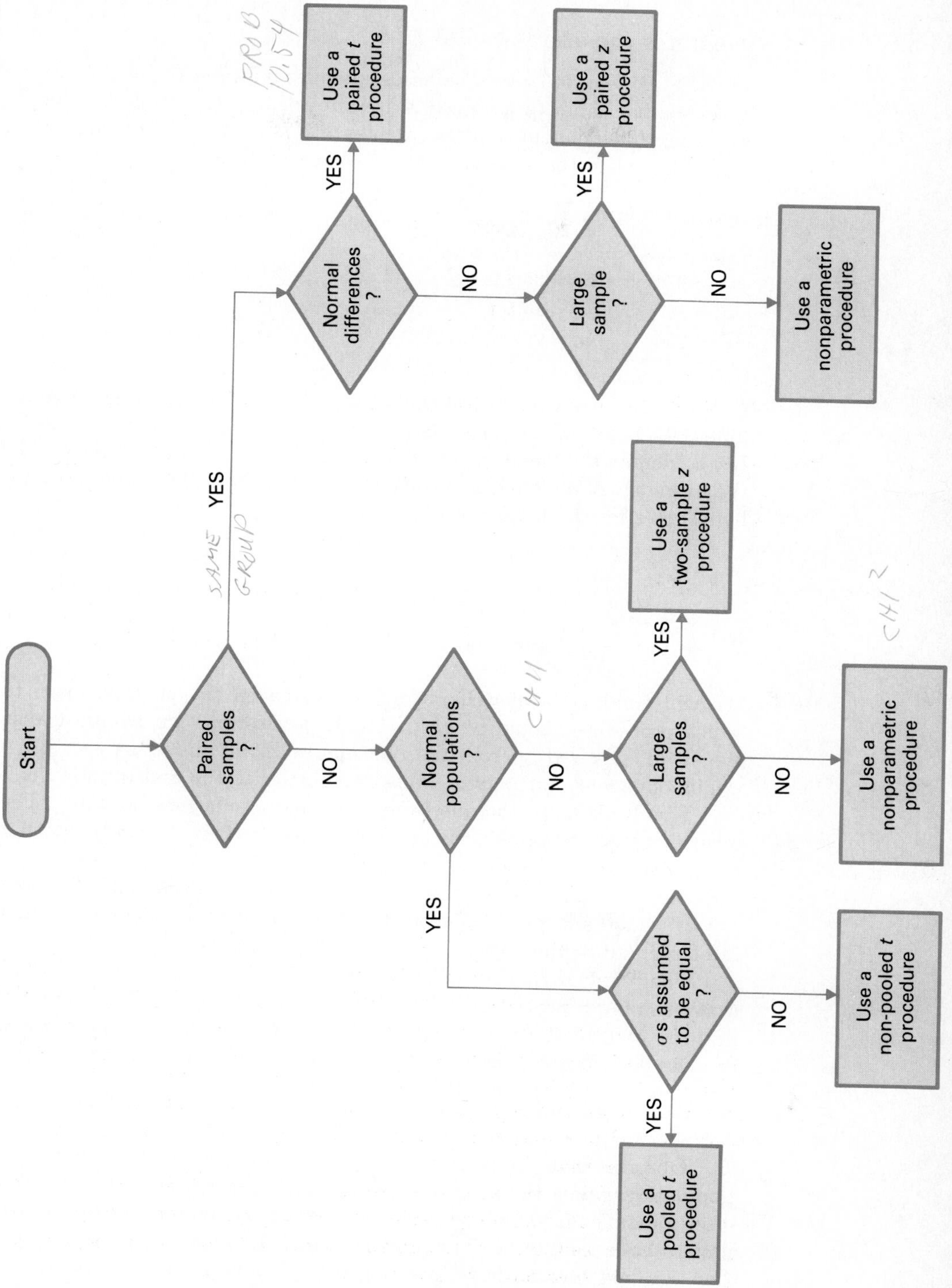

FIGURE 10.6   Flowchart for choosing the correct procedure to compare two population means

EXAMPLE 10.13   *Illustrates the use of Figure 10.6*

The U.S. Bureau of Labor Statistics compiles information on age and sex of people in the civilian labor force. Results are published in the document *Employment and Earnings*. Independent random samples of 250 males and 200 females in the civilian labor force yield the statistics on age shown in Table 10.10.

TABLE 10.10

Male	Female
$\bar{x}_1 = 37.5$ yr	$\bar{x}_2 = 36.4$ yr
$s_1 = 13.7$ yr	$s_2 = 13.5$ yr
$n_1 = 250$	$n_2 = 200$

Suppose that we want to use the data in Table 10.10 to decide whether there is a difference between the mean ages of males and females in the civilian labor force. Let μ_1 denote the mean age of all males in the civilian labor force and let μ_2 denote the mean age of all females in the civilian labor force. Then the null and alternative hypotheses for the hypothesis test are

$$H_0: \mu_1 = \mu_2 \text{ (mean ages are the same)}$$
$$H_a: \mu_1 \neq \mu_2 \text{ (mean ages are different)}$$

Which procedure should be used to perform the hypothesis test?

SOLUTION   We will employ the flowchart in Figure 10.6 on the previous page to determine which procedure should be used. The first question that we must answer is: "Are the samples paired?" From the information given above, we see that the samples are independent, not paired. Thus, the answer to the first question is "No."

     This leads us to the question: "Are the populations normal?" There is no indication that the populations under consideration are normally distributed. Hence, the answer to this second question is also "No."

     Next, we must answer the question: "Are the samples large?" The sample sizes are $n_1 = 250$ and $n_2 = 200$, both of which exceed 30. Consequently, the samples are large and so the answer to this third question is "Yes."

     The "Yes" answer to the previous question leads us to the box that states "Use a two-sample z procedure." The two-sample z procedures are summarized in the first row of Table 10.9 (page 522). Since the problem here is to perform a hypothesis test, we see from the last column of that row that the correct procedure to use is Procedure 10.1 on page 480. ∎

Observe that Table 10.9 contains no procedures that apply to non-normal populations when the sample sizes are small. As we mentioned in Chapter 8, it is often possible to use *nonparametric methods* (Chapter 15) to perform inferences under those conditions. This fact is reflected in two of the boxes in Figure 10.6. Can you find the two boxes starting from the top of the flowchart?

10.6   Large-sample inferences for two population proportions using independent samples

Recall that a *two-category population* is one in which each member is classified as either having or not having a specified attribute. The proportion (percentage) of the entire population that has the specified attribute is called the *population proportion* and is denoted by the letter p. The proportion of a sample from the population that has the specified attribute is called a *sample proportion* and is denoted by the symbol $\hat{p}$.

In this section, we will examine methods for performing inferences to compare the proportions, p_1 and p_2, of two two-category populations using independent samples. We introduce such inferences in Example 10.14.

EXAMPLE 10.14   *Introduces hypothesis tests for two population proportions*

A 1989 Associated Press article appearing in the *Arizona Republic* reported on a national smoking study conducted by the Centers for Disease Control. A face-to-face survey was taken of about 44,000 adult Americans. The following data are based on the results obtained in that survey.

Independent random samples of 21,150 American males and 22,852 American females were selected in order to compare the percentage of males who smoke cigarettes to the percentage of females who smoke cigarettes. Of the males sampled, 6598 were found to be cigarette smokers; and of the females sampled, 6147 were found to be cigarette smokers. Do the data provide sufficient evidence to conclude that the percentage of all American males who smoke cigarettes exceeds the percentage of all American females who smoke cigarettes?

SOLUTION   Note that the specified attribute here is "smokes cigarettes." Let p_1 denote the proportion of all American males who smoke cigarettes and let p_2 denote the proportion of all American females who smoke cigarettes. Then we want to perform the hypothesis test

$$H_0: p_1 = p_2 \text{ (percentage of male smokers is not higher)}$$
$$H_a: p_1 > p_2 \text{ (percentage of male smokers is higher)}$$

Roughly speaking, the hypothesis test can be carried out as follows:

1. Compute the proportion, $\hat{p}_1$, of the males sampled who smoke cigarettes and the proportion, $\hat{p}_2$, of the females sampled who smoke cigarettes.
2. If $\hat{p}_1$ is too much larger than $\hat{p}_2$, reject H_0; otherwise, do not reject H_0.

The first step is easy. Since 6598 of the 21,150 males sampled were found to be smokers and 6147 of the 22,852 females sampled were found to be smokers, we have

$$\hat{p}_1 = \frac{x_1}{n_1} = \frac{6598}{21,150} = 0.312 \ (31.2\%)$$

and

$$\hat{p}_2 = \frac{x_2}{n_2} = \frac{6147}{22{,}852} = 0.269 \ (26.9\%)$$

For the second step, we need to decide whether the sample proportion, $\hat{p}_1 = 0.312$, exceeds the sample proportion, $\hat{p}_2 = 0.269$, by a sufficient amount to warrant rejection of the null hypothesis in favor of the alternative hypothesis. In other words, we need to decide whether the difference between the two sample proportions can reasonably be attributed to sampling error or whether it indicates that the proportion of all American males who smoke cigarettes exceeds the proportion of all American females who smoke cigarettes.

To make that decision, we need to know the probability distribution of the difference, $\hat{p}_1 - \hat{p}_2$, between two sample proportions—the **sampling distribution of the difference between two proportions.** We will discuss that sampling distribution and then return to complete the hypothesis test.   ■

THE SAMPLING DISTRIBUTION OF THE DIFFERENCE BETWEEN TWO PROPORTIONS (LARGE AND INDEPENDENT SAMPLES)

To begin our discussion of the sampling distribution of the difference between two proportions, we provide a summary of the required notation in Table 10.11.

TABLE 10.11
Notation for parameters and statistics when considering two two-category populations

	Population 1	Population 2
Population proportion	p_1	p_2
Sample size	n_1	n_2
Number of successes	x_1	x_2
Sample proportion	$\hat{p}_1$	$\hat{p}_2$

Recall that the *number of successes* refers to the number of members sampled that have the specified attribute. Consequently, the sample proportions are computed from the formulas

$$\hat{p}_1 = \frac{x_1}{n_1} \quad \text{and} \quad \hat{p}_2 = \frac{x_2}{n_2}$$

Next we will present formulas that relate the mean and standard deviation of the random variable $\hat{p}_1 - \hat{p}_2$ to the population proportions, p_1 and p_2, and the sample sizes, n_1 and n_2. We have

$$\mu_{\hat{p}_1-\hat{p}_2} = p_1 - p_2$$

and

$$\sigma_{\hat{p}_1-\hat{p}_2} = \sqrt{p_1(1-p_1)/n_1 + p_2(1-p_2)/n_2}$$

These two formulas are derived in almost exactly the same way as the formulas for the mean and standard deviation of the random variable $\bar{x}_1 - \bar{x}_2$. See Exercise 10.84 for details.

Finally, as we know from Key Fact 8.5 on page 407 (the sampling distribution of the proportion), the random variables $\hat{p}_1$ and $\hat{p}_2$ are both approximately normally distributed for large sample sizes. From this it can be shown that, for independent samples, the random variable $\hat{p}_1 - \hat{p}_2$ is also approximately normally distributed. Hence, we can state the following fact:

KEY FACT 10.7 **The sampling distribution of the difference between two proportions (large and independent samples)**

Suppose that a random sample of size n_1 is to be taken from a two-category population with population proportion p_1; and that a random sample of size n_2 is to be taken from a two-category population with population proportion p_2. Further suppose that the two samples are to be selected independently of one another. Then, for large samples, the random variable $\hat{p}_1 - \hat{p}_2$ is approximately normally distributed and has mean $\mu_{\hat{p}_1 - \hat{p}_2} = p_1 - p_2$ and standard deviation $\sigma_{\hat{p}_1 - \hat{p}_2} = \sqrt{p_1(1 - p_1)/n_1 + p_2(1 - p_2)/n_2}$. Thus, the standardized random variable

$$z = \frac{(\hat{p}_1 - \hat{p}_2) - (p_1 - p_2)}{\sqrt{p_1(1 - p_1)/n_1 + p_2(1 - p_2)/n_2}}$$

has approximately the standard normal distribution.

Key Fact 10.7 provides the necessary theory for obtaining inferential procedures to compare the proportions of two two-category populations. We will use that key fact to derive both a hypothesis-testing procedure and a confidence-interval procedure. Let's start with hypothesis tests.

LARGE-SAMPLE HYPOTHESIS TESTS FOR TWO POPULATION PROPORTIONS USING INDEPENDENT SAMPLES

We will now develop a hypothesis-testing procedure for comparing the proportions of two two-category populations. Our immediate goal is to use Key Fact 10.7 to find a random variable that can be employed as the test statistic.

The null hypothesis for a hypothesis test to compare the proportions of two two-category populations will be

$$H_0: p_1 = p_2 \text{ (population proportions are equal)}$$

Note that if the null hypothesis is true, then $p_1 - p_2 = 0$, and so the standardized random variable in Key Fact 10.7 becomes

$$z = \frac{\hat{p}_1 - \hat{p}_2}{\sqrt{p(1 - p)/n_1 + p(1 - p)/n_2}}$$

where p denotes the common value of p_1 and p_2. Factoring $p(1 - p)$ out of the denominator of the previous expression gives the random variable

$$(5) \qquad z = \frac{\hat{p}_1 - \hat{p}_2}{\sqrt{p(1 - p)}\sqrt{(1/n_1) + (1/n_2)}}$$

The random variable in Equation (5) cannot be used as the test statistic since p is unknown. Consequently, we must estimate p using sample information. The best estimate of p is obtained by pooling the data to get the proportion of successes in both samples combined. That is, we estimate p by

$$\hat{p}_{\mathrm{p}} = \frac{x_1 + x_2}{n_1 + n_2}$$

We call $\hat{p}_{\mathrm{p}}$ the **pooled sample proportion.**

Replacing the p in Equation (5) by its estimate, $\hat{p}_{\mathrm{p}}$, yields the random variable

$$\frac{\hat{p}_1 - \hat{p}_2}{\sqrt{\hat{p}_{\mathrm{p}}(1 - \hat{p}_{\mathrm{p}})}\sqrt{(1/n_1) + (1/n_2)}}$$

This last random variable *can* be used as the test statistic and, like the random variable in Equation (5), has approximately the standard normal distribution for large samples if the null hypothesis is true. Therefore, we have the following procedure:

PROCEDURE 10.9
To perform a hypothesis test for two population proportions with null hypothesis $H_0\colon p_1 = p_2$.

ASSUMPTIONS

1. Independent samples.
2. Large samples.

STEP 1 *State the null and alternative hypotheses.*
STEP 2 *Decide on the significance level, α.*
STEP 3 *The critical value(s)*
       a) *for a two-tailed test are $\pm z_{\alpha/2}$.*
       b) *for a left-tailed test is $-z_{\alpha}$.*
       c) *for a right-tailed test is z_{α}.*
       Use Table II to find the critical value(s).

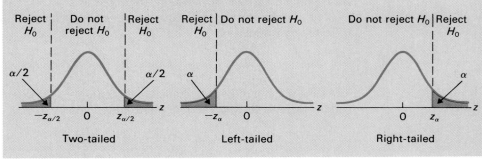

STEP 4 *Compute the value of the test statistic*

$$z = \frac{\hat{p}_1 - \hat{p}_2}{\sqrt{\hat{p}_p(1-\hat{p}_p)}\sqrt{(1/n_1) + (1/n_2)}}$$

where

$$\hat{p}_p = \frac{x_1 + x_2}{n_1 + n_2}$$

STEP 5 *If the value of the test statistic falls in the rejection region, then reject H_0; otherwise, do not reject H_0.*

STEP 6 *State the conclusion in words.*

EXAMPLE 10.15   *Illustrates Procedure 10.9*

We now return to the problem posed in Example 10.14. Independent random samples of 21,150 American males and 22,852 American females were selected in order to compare the percentage of males who smoke cigarettes to the percentage of females who smoke cigarettes. Of the males sampled, 6598 were found to be cigarette smokers; and of the females sampled, 6147 were found to be cigarette smokers.

Do the data provide sufficient evidence to conclude that the percentage of all American males who smoke cigarettes exceeds the percentage of all American females who smoke cigarettes? Perform the appropriate hypothesis test at the 5% significance level.

SOLUTION   We apply Procedure 10.9.

STEP 1 *State the null and alternative hypotheses.*

Let p_1 denote the proportion of all American males who smoke cigarettes and let p_2 denote the proportion of all American females who smoke cigarettes. Then the null and alternative hypotheses are

H_0: $p_1 = p_2$ (percentage of male smokers is not higher)
H_a: $p_1 > p_2$ (percentage of male smokers is higher)

Note that the hypothesis test is right-tailed since there is a greater-than sign ($>$) in the alternative hypothesis.

STEP 2 *Decide on the significance level, α.*

The test is to be performed at the 5% significance level. Thus, $\alpha = 0.05$.

STEP 3 *The critical value for a right-tailed test is z_α.*

Since $\alpha = 0.05$, the critical value is $z_{0.05} = 1.645$. See Figure 10.7.

FIGURE 10.7

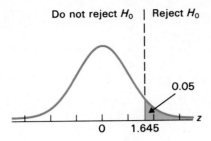

STEP 4 *Compute the value of the test statistic*

$$z = \frac{\hat{p}_1 - \hat{p}_2}{\sqrt{\hat{p}_\text{p}(1 - \hat{p}_\text{p})}\sqrt{(1/n_1) + (1/n_2)}}$$

where

$$\hat{p}_\text{p} = \frac{x_1 + x_2}{n_1 + n_2}$$

We first obtain $\hat{p}_1$, $\hat{p}_2$, and $\hat{p}_\text{p}$. Since 6598 of the 21,150 males sampled and 6147 of the 22,852 females sampled were found to be cigarette smokers, we have $x_1 = 6598$, $n_1 = 21,150$ and $x_2 = 6147$, $n_2 = 22,852$. Therefore,

$$\hat{p}_1 = \frac{x_1}{n_1} = \frac{6598}{21,150} = 0.312$$

$$\hat{p}_2 = \frac{x_2}{n_2} = \frac{6147}{22,852} = 0.269$$

and

$$\hat{p}_\text{p} = \frac{x_1 + x_2}{n_1 + n_2} = \frac{6598 + 6147}{21,150 + 22,852} = \frac{12,745}{44,002} = 0.290$$

Consequently, the value of the test statistic is

$$z = \frac{\hat{p}_1 - \hat{p}_2}{\sqrt{\hat{p}_\text{p}(1 - \hat{p}_\text{p})}\sqrt{(1/n_1) + (1/n_2)}}$$

$$= \frac{0.312 - 0.269}{\sqrt{(0.290)(1 - 0.290)}\sqrt{(1/21,150) + (1/22,852)}} = \boxed{9.93}$$

STEP 5 *If the value of the test statistic falls in the rejection region, reject H_0; otherwise, do not reject H_0.*

From Step 4, the value of the test statistic is $z = 9.93$ which, as we see from Figure 10.7, falls in the rejection region. Thus, we reject H_0.

STEP 6 *State the conclusion in words.*

The data provide sufficient evidence to conclude that the percentage of all American males who smoke cigarettes is greater than the percentage of all American females who smoke cigarettes. ∎

LARGE-SAMPLE CONFIDENCE INTERVALS FOR THE DIFFERENCE BETWEEN TWO POPULATION PROPORTIONS USING INDEPENDENT SAMPLES

Key Fact 10.7 on page 527 can also be used to derive the following confidence-interval procedure for the difference between two population proportions. For details of the derivation of the procedure, see Exercise 10.85.

PROCEDURE 10.10

To find a confidence interval for the difference between two population proportions.

ASSUMPTIONS

1. Independent samples.
2. Large samples.

STEP 1 *For a confidence level of $1 - \alpha$, use Table II to find $z_{\alpha/2}$.*

STEP 2 *The endpoints of the confidence interval for $p_1 - p_2$ are*

$$(\hat{p}_1 - \hat{p}_2) \pm z_{\alpha/2} \cdot \sqrt{\hat{p}_1(1 - \hat{p}_1)/n_1 + \hat{p}_2(1 - \hat{p}_2)/n_2}$$

EXAMPLE 10.16   *Illustrates Procedure 10.10*

Consider again the smoking illustration of Example 10.14. Use the sample data to obtain a 90% confidence interval for the difference, $p_1 - p_2$, between the proportion of all American males who smoke cigarettes and the proportion of all American females who smoke cigarettes.

SOLUTION   We apply Procedure 10.10.

STEP 1 *For a confidence level of $1 - \alpha$, use Table II to find $z_{\alpha/2}$.*

For a 90% confidence interval, $\alpha = 0.10$. Consulting Table II, we find that $z_{\alpha/2} = z_{0.10/2} = z_{0.05} = $ 1.645.

STEP 2 *The endpoints of the confidence interval for $p_1 - p_2$ are*

$$(\hat{p}_1 - \hat{p}_2) \pm z_{\alpha/2} \cdot \sqrt{\hat{p}_1(1 - \hat{p}_1)/n_1 + \hat{p}_2(1 - \hat{p}_2)/n_2}$$

From Step 1, $z_{\alpha/2} = 1.645$. Referring to Example 10.15 on page 530, we see that $\hat{p}_1 = 0.312$, $n_1 = 21,150$ and $\hat{p}_2 = 0.269$, $n_2 = 22,852$. Therefore, the endpoints of the confidence interval for $p_1 - p_2$ are

$$(0.312 - 0.269) \pm 1.645 \cdot \sqrt{0.312(1 - 0.312)/21{,}150 + 0.269(1 - 0.269)/22{,}852}$$

or

$$0.043 \pm 0.007$$

Thus, a 90% confidence interval for $p_1 - p_2$ is from

$$\boxed{0.036 \quad \text{to} \quad 0.050}$$

We can be 90% confident that the difference, $p_1 - p_2$, between the proportion of all American males who smoke cigarettes and the proportion of all American females who smoke cigarettes is somewhere between 0.036 and 0.050. In other words, we can be 90% confident that the percentage of all American males who smoke cigarettes exceeds the percentage of all American females who smoke cigarettes by at least 3.6% but by no more than 5.0%. ■

Exercises 10.6

___ **10.70** Consider a hypothesis test for two population proportions with null hypothesis H_0: $p_1 = p_2$. What parameter is being estimated by the pooled sample proportion, $\hat{p}_p$?

___ **10.71** Consider the quantities p_1, p_2, x_1, x_2, $\hat{p}_1$, $\hat{p}_2$, and $\hat{p}_p$.
a) Which quantities represent parameters and which represent statistics?
b) Which quantities are fixed numbers and which are random variables?

___ **10.72** The Gallup Organization periodically conducts surveys in order to estimate the percentage of Americans who approve of the way the president is doing his job. In April of 1985, 795 adults out of a random sample of 1528 adults said that they approved of the way then President Ronald Reagan was handling his job. In May of 1985, 840 adults out of another random sample of 1528 adults said that they approved. At the 5% significance level, do the data suggest that the percentage of Americans who approved of the way Ronald Reagan was handling his job as president increased from April to May of 1985?

___ **10.73** The U.S. Energy Information Administration performs surveys to estimate the percentage of American households that own various appliances. Re-

sults of those surveys can be found in *Residential Energy Consumption Survey: Housing Characteristics*. Euromonitor Publications Limited, London, England, conducts similar surveys for other countries and publishes its findings in *European Marketing Data and Statistics*. Suppose that out of a random sample of 500 American households, 370 own washing machines; and that out of a random sample of 450 French households, 365 own washing machines. Do the data provide sufficient evidence to conclude that there is a difference between the percentages of American and French households that own washing machines? Use $\alpha = 0.01$.

___ **10.74** The Organization for Economic Cooperation and Development, Paris, France, summarizes data on labor-force participation rates in the publication *Labour Force Statistics*. Suppose 300 American women and 250 Canadian women are randomly selected and that 184 of the American women and 148 of the Canadian women are in their respective labor forces. At the 5% significance level, do the data suggest that there is a difference between the labor-force participation rates of American and Canadian women?

___ **10.75** Annual surveys are performed by the U.S. Bureau of the Census to obtain estimates of the percentage of the voting-age population that has registered to vote. The information from those surveys are published in *Current Population Reports*. Suppose that 400 employed persons and 450 unemployed persons are independently and randomly selected. Further suppose that 262 of the employed persons and 224 of the unemployed persons have registered to vote. Can we conclude that the percentage of employed workers who have registered to vote exceeds the percentage of unemployed workers who have registered to vote? Use $\alpha = 0.05$.

— **10.76** Suppose we tell you that the percentage of adult American males who are married exceeds the percentage of adult American females who are married. Further suppose that to check this claim you randomly select 550 adult American males and 575 adult American females. You find that 367 of the males selected are married and 353 of the females selected are married. Do your data provide sufficient evidence, at the 5% significance level, to support our claim? Explain your answer.

— **10.77** Information on American physicians and dentists are compiled, respectively, by the American Medical Association and the American Dental Association. A random sample of 300 American dentists contains 77 who practice in the Northeast; and a random sample of 300 American physicians contains 82 who practice in the Northeast. Do these data suggest that the percentage of American dentists who practice in the Northeast is smaller than the percentage of American physicians who practice in the Northeast? Perform the required hypothesis test using $\alpha = 0.10$.

— **10.78** Refer to Exercise 10.72.
a) Determine a 90% confidence interval for the difference, $p_1 - p_2$, between the proportion of Americans who approved of the way Ronald Reagan was handling his job as president in April of 1985 and the proportion of Americans who approved of the way Ronald Reagan was handling his job as president in May of 1985.
b) Interpret your results in words.

— **10.79** Refer to Exercise 10.73.
a) Determine a 99% confidence interval for the difference, $p_1 - p_2$, between the proportions of American and French households that own washing machines.
b) Interpret your results in words.

— **10.80** Refer to Exercise 10.74.
a) Find a 95% confidence interval for the difference, $p_1 - p_2$, between the labor-force participation rates of American and Canadian women.
b) Interpret your results in words.

— **10.81** Refer to Exercise 10.75.
a) Obtain a 90% confidence interval for the difference, $p_1 - p_2$, between the proportions of employed and unemployed workers who have registered to vote.
b) Interpret your results in words.

— **10.82** Refer to Exercise 10.76.
a) Obtain a 90% confidence interval for the difference, $p_1 - p_2$, between the proportion of adult American males who are married and the proportion of adult American females who are married.
b) Interpret your results in words.

— **10.83** Refer to Exercise 10.77.
a) Determine an 80% confidence interval for the difference, $p_1 - p_2$, between the proportion of American dentists who practice in the Northeast and the proportion of American physicians who practice in the Northeast.
b) Interpret your results in words.

≡ **10.84** In this exercise, we will establish the formulas presented on page 526 for the mean and standard deviation of the random variable $\hat{p}_1 - \hat{p}_2$.
a) Use the results of Exercises 5.26(e) and 5.27(f) on pages 249 and 250, respectively, to show that

$$\mu_{\hat{p}_1 - \hat{p}_2} = \mu_{\hat{p}_1} - \mu_{\hat{p}_2}$$

and

$$\sigma_{\hat{p}_1 - \hat{p}_2} = \sqrt{\sigma_{\hat{p}_1}^2 + \sigma_{\hat{p}_2}^2}$$

b) Apply the formulas $\mu_{\hat{p}} = p$ and $\sigma_{\hat{p}} = \sqrt{p(1-p)/n}$ from Section 8.5 to the results in part (a) to derive the formulas

$$\mu_{\hat{p}_1 - \hat{p}_2} = p_1 - p_2$$

and

$$\sigma_{\hat{p}_1 - \hat{p}_2} = \sqrt{p_1(1-p_1)/n_1 + p_2(1-p_2)/n_2}$$

≡ **10.85** This exercise justifies Procedure 10.10.
a) Use Key Fact 10.7 on page 527 to explain why the random variable

$$\frac{(\hat{p}_1 - \hat{p}_2) - (p_1 - p_2)}{\sqrt{\hat{p}_1(1-\hat{p}_1)/n_1 + \hat{p}_2(1-\hat{p}_2)/n_2}}$$

has approximately the standard normal distribution for large samples.
b) Use part (a) to derive the confidence-interval formula in Step 2 of Procedure 10.10 on page 531.

Chapter review

KEY TERMS

independent samples, 476
normal differences, 514
paired difference, 512
paired samples, 512
pool, 488
POOLED,* 494
pooled sample proportion $(\hat{p}_\mathrm{p})$, 528

pooled sample standard deviation
  (s_p), 488
sampling distribution of the difference
  between two means, 479, 487
sampling distribution of the difference
  between two proportions, 527
TWOSAMPLE T,* 494, 506

FORMULAS

In the formulas below,

μ_1, μ_2 = population means
$\overline{x}_1, \overline{x}_2$ = sample means
s_1, s_2 = sample standard deviations
n_1, n_2 = sample sizes

p_1, p_2 = population proportions
$\hat{p}_1, \hat{p}_2$ = sample proportions
x_1, x_2 = numbers of successes

and

$\overline{d}$ = sample mean of paired differences
n = number of sample pairs

s_d = sample standard deviation of
  paired differences

Test statistic for H_0: $\mu_1 = \mu_2$ (large and independent samples), 481

$$z = \frac{\overline{x}_1 - \overline{x}_2}{\sqrt{(s_1^2/n_1) + (s_2^2/n_2)}}$$

Confidence interval for $\mu_1 - \mu_2$ (large and independent samples), 483

$$(\overline{x}_1 - \overline{x}_2) \pm z_{\alpha/2} \cdot \sqrt{(s_1^2/n_1) + (s_2^2/n_2)}$$

Pooled sample standard deviation, 488

$$s_\mathrm{p} = \sqrt{\frac{(n_1 - 1)s_1^2 + (n_2 - 1)s_2^2}{n_1 + n_2 - 2}}$$

Test statistic for H_0: $\mu_1 = \mu_2$ (normal populations, independent samples, and σs assumed equal), 490

$$t = \frac{\overline{x}_1 - \overline{x}_2}{s_\mathrm{p}\sqrt{(1/n_1) + (1/n_2)}}$$

with df $= n_1 + n_2 - 2$.

Confidence interval for $\mu_1 - \mu_2$ (normal populations, independent samples, and σs assumed equal), 493

$$(\overline{x}_1 - \overline{x}_2) \pm t_{\alpha/2} \cdot s_\mathrm{p}\sqrt{(1/n_1) + (1/n_2)}$$

with df $= n_1 + n_2 - 2$.

Test statistic for H_0: $\mu_1 = \mu_2$ (normal populations, independent samples, and σs not assumed equal),* 501

$$t = \frac{\bar{x}_1 - \bar{x}_2}{\sqrt{(s_1^2/n_1) + (s_2^2/n_2)}}$$

with df $= [(s_1^2/n_1) + (s_2^2/n_2)]^2/[(s_1^2/n_1)^2/(n_1 - 1) + (s_2^2/n_2)^2/(n_2 - 1)]$ rounded down to the nearest integer.

Confidence interval for $\mu_1 - \mu_2$ (normal populations, independent samples, and σs not assumed equal),* 504

$$(\bar{x}_1 - \bar{x}_2) \pm t_{\alpha/2} \cdot \sqrt{(s_1^2/n_1) + (s_2^2/n_2)}$$

with df $= [(s_1^2/n_1) + (s_2^2/n_2)]^2/[(s_1^2/n_1)^2/(n_1 - 1) + (s_2^2/n_2)^2/(n_2 - 1)]$ rounded down to the nearest integer.

Test statistic for H_0: $\mu_1 = \mu_2$ (normal differences and paired samples), 515

$$t = \frac{\bar{d}}{s_d/\sqrt{n}}$$

with df $= n - 1$.

Confidence interval for $\mu_1 - \mu_2$ (normal differences and paired samples), 517

$$\bar{d} \pm t_{\alpha/2} \cdot \frac{s_d}{\sqrt{n}}$$

with df $= n - 1$.

Pooled sample proportion, 528

$$\hat{p}_{\mathrm{P}} = \frac{x_1 + x_2}{n_1 + n_2}$$

Test statistic for H_0: $p_1 = p_2$ (large and independent samples), 529

$$z = \frac{\hat{p}_1 - \hat{p}_2}{\sqrt{\hat{p}_{\mathrm{P}}(1 - \hat{p}_{\mathrm{P}})}\sqrt{(1/n_1) + (1/n_2)}}$$

Confidence interval for $p_1 - p_2$ (large and independent samples), 531

$$(\hat{p}_1 - \hat{p}_2) \pm z_{\alpha/2} \cdot \sqrt{\hat{p}_1(1 - \hat{p}_1)/n_1 + \hat{p}_2(1 - \hat{p}_2)/n_2}$$

YOU SHOULD
BE ABLE TO

1. use and understand the preceding formulas.
2. perform large-sample inferences to compare the means of two populations using independent samples.

3. perform inferences to compare the means of two normally distributed populations using independent samples when the population standard deviations are unknown but assumed equal.
4. perform inferences to compare the means of two normally distributed populations using independent samples when the population standard deviations are unknown and not assumed equal.*
5. perform inferences to compare the means of two populations using paired samples when the population of paired differences is normally distributed.
6. perform large-sample inferences to compare the proportions of two two-category populations using independent samples.
7. use the Minitab commands covered in this chapter.*
8. interpret the output obtained from the application of the Minitab commands discussed in this chapter.*

REVIEW TEST

1. *Better Homes and Gardens* conducts semi-annual surveys of housing in 100 cities across the United States. One item of considerable interest is the average price for resale homes. Suppose that 50 randomly selected resale home prices in Phoenix, Arizona have a mean of $81,525, with a standard deviation of $29,670; and that 75 randomly selected resale home prices in Flint, Michigan have a mean of $47,603, with a standard deviation of $12,466. Do the data suggest that the mean resale price for homes in Phoenix, Arizona exceeds the mean resale price for homes in Flint, Michigan? Perform the hypothesis test at the 1% significance level.

2. Refer to Problem 1. Obtain a 98% confidence interval for the difference, $\mu_1 - \mu_2$, between the mean resale price of homes in Phoenix, Arizona and the mean resale price of homes in Flint, Michigan.

3. Two speed reading programs are to be compared to see whether there is a difference in results. For the comparison, 10 pairs of people are randomly selected, where each pair consists of people whose present reading speeds are nearly identical. From each pair, one person is randomly selected to take Program 1 and the other person takes Program 2. After the completion of the speed reading programs, the speeds for the 10 pairs are recorded. The speeds, in words per minute, are displayed in the table at the top of the next column.
   a) At the 10% significance level, can we conclude that there is a difference in mean results for the two speed reading programs?
   b) What assumption must you make in order to perform the hypothesis test in part (a)?

Pair	Program 1	Program 2
1	1114	1032
2	996	1148
3	979	1074
4	1125	1076
5	910	959
6	1056	1094
7	1091	1091
8	1053	1096
9	996	1032
10	894	1012

4. Refer to Problem 3. Obtain a 90% confidence interval for the difference, $\mu_1 - \mu_2$, between the mean reading speed of people using Program 1 and the mean reading speed of people using Program 2.

5. The National Science Foundation surveys doctoral scientists for selected characteristics and publishes the information obtained in *U.S. Scientists and Engineers*. Suppose that 250 male doctoral scientists and 175 female doctoral scientists are selected at random. Further suppose that 15 of the males sampled and 19 of the females sampled are unemployed. Does it appear that the percentage of unemployed male doctoral scientists is lower than the percentage of unemployed female doctoral scientists? Perform the appropriate hypothesis test using $\alpha = 0.05$.

6. Refer to Problem 5. Obtain a 90% confidence interval for the difference between the proportion of unemployed male doctoral scientists and the proportion of unemployed female doctoral scientists.

7. A psychology professor, teaching at a large university in the northeast, wants to know whether there is a difference between the mean IQs of male and female students in attendance. She randomly and independently selects 20 female students and 20 male students and has them take IQ tests. The resulting IQ data are as follows:

Female				Male			
130	109	126	116	106	131	109	116
117	100	118	122	114	133	101	120
124	118	131	115	134	134	144	119
120	127	104	129	120	122	122	114
125	130	130	117	107	111	124	110

Do the data indicate, at the 5% significance level, that there is a difference between the mean IQs of male and female students at the university? Assume that IQs of female and male students at the university are both approximately normally distributed with equal standard deviations. (*Note:* For the female IQ data, $\Sigma x = 2408$ and $\Sigma x^2 = 291{,}396$; and for the male IQ data, $\Sigma x = 2391$ and $\Sigma x^2 = 288{,}239$.)

8. Refer to Problem 7. Determine a 95% confidence interval for the difference between the mean IQs of female and male students at the university.

*9. (**Computer problem**) Suppose that the data in Problem 7 are stored in columns named FEMALE and MALE.
   a) Which Minitab command and subcommands (if any) should be used to simultaneously perform the hypothesis test considered in Problem 7 and obtain the confidence interval required in Problem 8?
   b) If you have access to Minitab, use it to perform the hypothesis test and obtain the confidence interval.

*10. (**Computer problem**) Printout 10.7 on the following page supplies the output obtained by applying Minitab's TWOSAMPLE T; POOLED command to the data in Problem 7. From the output, determine
   a) the null and alternative hypotheses for the hypothesis test.
   b) the command and subcommand(s) used.
   c) the sample standard deviation of each of the two sets of IQ data.

   d) the mean IQ of the females sampled and the mean IQ of the males sampled.
   e) the value of the test statistic, t.
   f) the P-value of the hypothesis test.
   g) the smallest significance level at which the null hypothesis can be rejected.
   h) the conclusion if the hypothesis test is performed at the 5% significance level.
   i) a 95% confidence interval for the difference, $\mu_1 - \mu_2$, between the mean IQs of female and male students at the university.
   j) the assumptions that must be made in order to use TWOSAMPLE T; POOLED.
   k) the value of s_p.

*11. Euromonitor Publications Limited conducts surveys in various countries to obtain data on food consumption of major food commodities. Results of those surveys can be found in the publication *European Marketing Data and Statistics*. Suppose that independent random samples of 10 Germans and 15 Russians consumed the following quantities of fish, in kilograms, during last year:

Germans		Russians		
17	17	16	21	12
1	9	11	5	23
15	6	19	19	22
10	13	16	23	12
14	11	18	7	17

Use the data to decide whether Germans consume less fish than Russians, on the average. Take $\alpha = 0.05$. Assume that fish consumptions in each country are approximately normally distributed. (*Note:* The mean and standard deviation of the German fish-consumption data are 11.30 kg and 5.06 kg, respectively; and the mean and standard deviation of the Russian fish-consumption data are 16.07 kg and 5.61 kg, respectively.)

*12. Refer to Problem 11. Find a 90% confidence interval for the difference, $\mu_1 - \mu_2$, between last year's mean fish consumption by Germans and last year's mean fish consumption by Russians.

*13. (**Computer problem**) Suppose that the data displayed in Problem 11 are stored in columns named GERMANS and RUSSIANS.
   a) Which Minitab command and subcommands (if any) should be used to simultaneously per-

form the hypothesis test considered in Problem 11 and obtain the confidence interval required in Problem 12?

b) If you have access to Minitab, use it to perform the hypothesis test and obtain the confidence interval.

*14. **(Computer problem)** Printout 10.8 at the bottom of the page shows the Minitab output obtained by applying the TWOSAMPLE T command to the data in Problem 11. Employ the output to determine

a) the null and alternative hypotheses for the hypothesis test.

b) the command used and the subcommands used (if any) to obtain the Minitab output.

c) the standard deviation of the fish consumptions for each of the two samples.

d) the number of Germans sampled and the number of Russians sampled.

e) the mean fish consumptions for both the Germans and Russians sampled.

f) the value of the test statistic, t.

g) the P-value of the hypothesis test.

h) the smallest significance level at which the null hypothesis can be rejected.

i) the conclusion if the hypothesis test is performed at the 5% significance level.

j) a 90% confidence interval for the difference, $\mu_1 - \mu_2$, between last year's mean fish consumption by Germans and last year's mean fish consumption by Russians.

k) the conditions that must be satisfied in order to apply the TWOSAMPLE T command to the fish-consumption data.

PRINTOUT 10.7   Minitab output for Problem 10

```
TWOSAMPLE T FOR FEMALE VS MALE
           N      MEAN     STDEV    SE MEAN
FEMALE    20    120.40      8.80      2.0
MALE      20    119.6      11.2       2.5

95 PCT CI FOR MU FEMALE - MU MALE: (-5.6, 7.3)

TTEST MU FEMALE = MU MALE (VS NE): T= 0.27  P=0.79  DF=  38

POOLED STDEV =        10.1
```

PRINTOUT 10.8   Minitab output for Problem 14

```
TWOSAMPLE T FOR GERMANS VS RUSSIANS
            N      MEAN     STDEV    SE MEAN
GERMANS    10    11.30      5.06      1.6
RUSSIANS   15    16.07      5.61      1.4

90 PCT CI FOR MU GERMANS - MU RUSSIANS: (-8.5, -1.0)

TTEST MU GERMANS = MU RUSSIANS (VS LT): T= -2.21  P=0.020  DF=  20
```

CHAPTER 11

CHI-SQUARE PROCEDURES

The statistical-inference techniques that we have presented thus far have dealt exclusively with hypothesis tests and confidence intervals for means and proportions. Now we will consider some other widely used inferential procedures. These include hypothesis tests concerning the percentage distribution of a population, hypothesis tests for deciding whether two characteristics of a population are statistically dependent, and inferences for the standard deviation of a population.

All of the inferential procedures that we will examine in this chapter rely on a continuous probability distribution called the *chi-square distribution*. Consequently, those procedures are often referred to as **chi-square procedures.**

CHAPTER OUTLINE

11.1 The chi-square distribution

The statistical-inference procedures discussed in this chapter all rely on a class of continuous probability distributions called **chi-square distributions.** Thus, we will begin this chapter by studying that class of probability distributions.

As with normal distributions and t-distributions, probabilities for a random variable having a chi-square distribution are equal to areas under a curve, suitably called a χ^2 **(chi-square) curve.** Actually, there are infinitely many χ^2-curves and we identify the χ^2-curve in question by giving its number of degrees of freedom, just like for t-curves. Figure 11.1 shows three different χ^2-curves.

FIGURE 11.1
χ^2-curves for
df = 5, 10, and 19

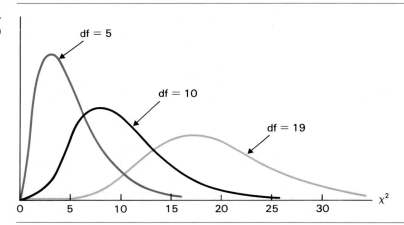

In this figure, we can observe some basic properties of χ^2-curves. These are presented in Key Fact 11.1.

KEY FACT 11.1   Basic properties of χ^2-curves

PROPERTY 1   The total area under a χ^2-curve is equal to 1.

PROPERTY 2   A χ^2-curve starts at 0 on the horizontal axis and extends indefinitely to the right, approaching the horizontal axis as it does so.

PROPERTY 3   A χ^2-curve is not symmetrical. It climbs to its high point rapidly and comes back to the axis more slowly—it is *skewed* to the right.

PROPERTY 4   As the number of degrees of freedom gets larger, χ^2-curves look increasingly like normal curves.

USING THE χ^2-TABLE

To perform a hypothesis test or obtain a confidence interval that is based on a chi-square distribution, we need to be able to determine the χ^2-value corresponding to a specified area under a χ^2-curve. Table IV, which can be found inside the back cover of the book, provides a table of χ^2-values corresponding to several areas for various degrees of freedom.

The χ^2-table, Table IV, is quite similar to the t-table, Table III. The two outside columns of Table IV, labeled df, give the number of degrees of freedom. As you might expect, the symbol χ^2_α denotes the χ^2-value with area α to its right under a χ^2-curve. Thus, the column headed $\chi^2_{0.995}$ contains χ^2-values with area 0.995 to their right; the column headed $\chi^2_{0.99}$ contains χ^2-values with area 0.99 to their right; and so on. We will illustrate the use of the χ^2-table, Table IV, in the next three examples.

EXAMPLE 11.1 *Illustrates how to find the χ^2-value for a specified area*

For a χ^2-curve with 12 degrees of freedom, find $\chi^2_{0.025}$; that is, find the χ^2-value with area 0.025 to its right. See Figure 11.2.

FIGURE 11.2

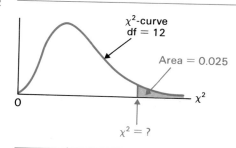

SOLUTION To find the χ^2-value in question, we use Table IV. Since the number of degrees of freedom is 12, we first go down the outside columns, labeled df, to "12." Then we go across that row until we are under the column headed $\chi^2_{0.025}$. The number in the body of the table there, 23.337, is the required χ^2-value. That is, for a χ^2-curve with df = 12, the χ^2-value with area 0.025 to its right is $\chi^2_{0.025} = \mathbf{23.337}$. Figure 11.3 summarizes our results. ∎

FIGURE 11.3

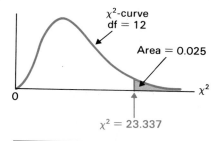

EXAMPLE 11.2   *Illustrates how to find the χ^2-value for a specified area*

Determine the χ^2-value with area 0.05 to its left for a χ^2-curve with df = 7. See Figure 11.4.

FIGURE 11.4

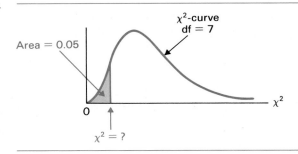

SOLUTION   We first note that since the total area under a χ^2-curve is equal to 1 (Property 1 of Key Fact 11.1), the unshaded area in Figure 11.4 must equal $1 - 0.05 = 0.95$. Thus, the area to the right of the required χ^2-value is 0.95. This means that the required χ^2-value is $\chi^2_{0.95}$. From Table IV, we find that for df = 7, $\chi^2_{0.95} = 2.167$. Therefore, for a χ^2-curve with df = 7, the χ^2-value with area 0.05 to its left is **2.167**. See Figure 11.5.   ∎

FIGURE 11.5

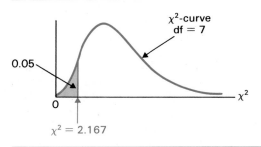

EXAMPLE 11.3   *Illustrates how to find the χ^2-values for a specified area*

For a χ^2-curve with df = 20, determine the two χ^2-values that divide the area under the curve into a middle 0.95 area and two outside 0.025 areas. See Figure 11.6 at the top of the next page.

SOLUTION   First we obtain the χ^2-value on the right in Figure 11.6. From the figure, we see that the shaded area on the right-hand side is 0.025. This means that the χ^2-value on the right is $\chi^2_{0.025}$. Consulting Table IV, we find that for df = 20, $\chi^2_{0.025} = 34.170$.

Next we obtain the χ^2-value on the left in Figure 11.6. Because the area to the left of that χ^2-value is 0.025, the area to its right is $1 - 0.025 = 0.975$. Hence, the χ^2-value on the left is $\chi^2_{0.975}$, which by Table IV equals 9.591 for df = 20.

FIGURE 11.6

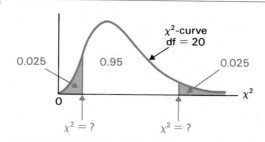

Consequently, for a χ^2-curve with df $= 20$, the two χ^2-values that divide the area under the curve into a middle 0.95 area and two outside 0.025 areas are 9.591 and 34.170. See Figure 11.7. ■

MTB

FIGURE 11.7

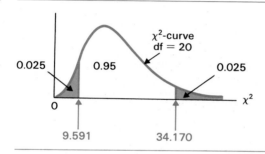

Exercises 11.1

In each of Exercises 11.1–11.8, use the χ^2-table, Table IV, to determine the required χ^2-values. Illustrate your work graphically.

___ **11.1** For a χ^2-curve with 19 degrees of freedom, find the χ^2-value with
a) area 0.025 to its right.
b) area 0.95 to its right.

___ **11.2** For a χ^2-curve with 22 degrees of freedom, find the χ^2-value with
a) area 0.01 to its right.
b) area 0.995 to its right.

___ **11.3** For a χ^2-curve with df $= 10$, determine
a) $\chi^2_{0.05}$ b) $\chi^2_{0.975}$

___ **11.4** For a χ^2-curve with df $= 4$, determine
a) $\chi^2_{0.005}$ b) $\chi^2_{0.99}$

___ **11.5** Consider a χ^2-curve with df $= 8$. Obtain the χ^2-value with
a) area 0.01 to its left.
b) area 0.95 to its left.

___ **11.6** Consider a χ^2-curve with df $= 16$. Obtain the χ^2-value with
a) area 0.025 to its left.
b) area 0.975 to its left.

___ **11.7** Determine the two χ^2-values that divide the area under the curve into a middle 0.95 area and two outside 0.025 areas for a χ^2-curve with
a) df $= 5$ b) df $= 20$

___ **11.8** Determine the two χ^2-values that divide the area under the curve into a middle 0.90 area and two outside 0.05 areas for a χ^2-curve with
a) df $= 11$ b) df $= 28$

11.2   Chi-square goodness-of-fit test

The first chi-square procedure that we will discuss is known as the **chi-square goodness-of-fit test.** Among other things, that procedure can be used to perform hypothesis tests about the percentage distribution of a population or the probability distribution of a random variable. Let us begin by looking at an example that illustrates the reasoning behind the chi-square goodness-of-fit test.

EXAMPLE 11.4   *Introduces the chi-square goodness-of-fit test*

The American Medical Association compiles information on American physicians and publishes its findings in *Physician Characteristics and Distribution in the U.S.* Physicians can be broadly classified into four categories: general practice, medical, surgical, and other. Table 11.1 gives a percentage distribution and probability (relative-frequency) distribution for American physicians in 1986.

TABLE 11.1
Specialty distribution
of American
physicians, 1986

Specialty	Percent	Probability, *p*
General practice	18.0	0.180
Medical	33.9	0.339
Surgical	27.0	0.270
Other	21.1	0.211
	100.0	1.000

The table shows, for instance, that in 1986, 18.0% of all American physicians were in general practice; or, equivalently, that the probability is 0.180 that a randomly selected 1986 American physician was in general practice.

A researcher wants to know whether this year's specialty distribution of American physicians has changed from the 1986 distribution. To begin the study, he randomly selects 500 American physicians who are currently practicing medicine. Table 11.2 provides a frequency distribution and percentage distribution for the specialties of the physicians obtained.

TABLE 11.2
Sample results
for specialties of
500 randomly
selected American
physicians currently
practicing medicine

Specialty	Frequency	Percent
General practice	80	16.0
Medical	162	32.4
Surgical	156	31.2
Other	102	20.4
	500	100.0

The researcher plans to use the sample data displayed in Table 11.2 to perform the hypothesis test

H_0: The current specialty distribution of American physicians is the same as the 1986 distribution.

H_a: The current specialty distribution of American physicians is different from the 1986 distribution.

The basic idea behind the hypothesis test is this: Compare the observed frequencies in the second column of Table 11.2 to the frequencies that would be expected if the current specialty distribution is the same as the 1986 specialty distribution. If the observed and expected frequencies match up fairly well, do not reject the null hypothesis; otherwise, reject the null hypothesis.

To transform the idea in the previous paragraph into a precise procedure for performing the hypothesis test, we need to answer the following two questions: First, what frequencies should we expect to get from a random sample of 500 American physicians who are currently practicing medicine, if the current specialty distribution is, in fact, the same as the 1986 specialty distribution? Second, how do we decide whether the frequencies actually obtained match up reasonably well with those that we would expect?

The first question is easy to answer. If the current specialty distribution is the same as the 1986 specialty distribution, then, for instance, 33.9% of all current American physicians will be in the medical category (see Table 11.1). Therefore, in a sample of 500 physicians, we would expect about 33.9% of the 500, or 169.5, to be in the medical category.

In general, each expected frequency, denoted by the letter E, is computed using the formula

$$E = np$$

where n is the sample size (in this case 500) and p is the appropriate probability from the third column of Table 11.1. For instance, the expected frequency for the medical category is

$$E = np = 500 \cdot 0.339 = 169.5,$$

as we have already seen. The expected frequencies for all four specialties are obtained in Table 11.3.

TABLE 11.3
Expected frequencies

Specialty	Probability p	Expected frequency $np = E$
General practice	0.180	$500 \cdot 0.180 = 90.0$
Medical	0.339	$500 \cdot 0.339 = 169.5$
Surgical	0.270	$500 \cdot 0.270 = 135.0$
Other	0.211	$500 \cdot 0.211 = 105.5$

The third column of Table 11.3 provides us with the answer to our first question. It gives the frequencies that we would expect to get if the current specialty distribution of American physicians is the same as the 1986 specialty distribution.

Our second question, whether the observed frequencies match up reasonably well with the expected frequencies, is harder to answer. We would like to calculate a number that measures how good the fit is.

In the second column of Table 11.4, we have repeated the **observed frequencies** from the second column of Table 11.2; that is, the frequencies actually obtained from a random sample of 500 American physicians who are currently practicing medicine. In the third column of Table 11.4, we have repeated the **expected frequencies** from the third column of Table 11.3; that is, the frequencies that we would expect to get from a random sample of 500 American physicians who are currently practicing medicine, if the current specialty distribution is the same as the 1986 specialty distribution.

TABLE 11.4

Specialty	Observed frequency O	Expected frequency E	Difference $O - E$	Square of difference $(O - E)^2$	$(O - E)^2/E$
General practice	80	90.0	−10.0	100.00	1.111
Medical	162	169.5	−7.5	56.25	0.332
Surgical	156	135.0	21.0	441.00	3.267
Other	102	105.5	−3.5	12.25	0.116
	500	500.0	0		4.826

To try to measure how well the observed and expected frequencies match up, it is logical to look at their differences, $O - E$. These differences are shown in the fourth column of Table 11.4. As you can see, summing those differences in order to obtain a "total difference" is not very useful since the sum equals zero.

Instead, each difference, $O - E$, is squared (fifth column of Table 11.4) and then divided by the corresponding expected frequency. This gives the values, $(O - E)^2/E$, shown in the sixth column of Table 11.4. The sum of those values,

$$\Sigma(O - E)^2/E = 4.826,$$

is the statistic that is used to measure how well or poorly the observed and expected frequencies match up.

If the null hypothesis is true, then the observed and expected frequencies should be about the same, thus resulting in a small value of the test statistic $\Sigma(O - E)^2/E$. In other words, if $\Sigma(O - E)^2/E$ is too large, then this gives us evidence that the null hypothesis is false.

As we have seen in Table 11.4, $\Sigma(O - E)^2/E = 4.826$. Can this value be reasonably attributed to sampling error or is it large enough to indicate that the null hypothesis is false? To answer this question, we need to know the probability distribution of the test statistic, $\Sigma(O - E)^2/E$, if the null hypothesis is true. That probability distribution is identified in Key Fact 11.2.   ■

KEY FACT 11.2

Consider a chi-square goodness-of-fit test where the null hypothesis specifies the percentage distribution of a population or the probability distribution of a random variable. Suppose that the sample size is large. Then, if the null hypothesis is true, the random variable

$$\chi^2 = \Sigma (O - E)^2 / E$$

has approximately a chi-square distribution. The number of degrees of freedom is one less than the number of categories in the distribution.

In Example 11.4, there are four categories (the four specialties) and so the number of degrees of freedom is $4 - 1 = 3$.

THE CHI-SQUARE GOODNESS-OF-FIT TEST

We can now state a procedure for performing a chi-square goodness-of-fit test. The null hypothesis is that a population (or random variable) has a specified distribution and the alternative hypothesis is that the population (or random variable) has a distribution different from the one specified in the null hypothesis. Since the null hypothesis will be rejected only when the value of the test statistic is too large, the rejection region is always on the right; that is, the hypothesis test is always right-tailed.

PROCEDURE 11.1

To perform a chi-square goodness-of-fit test.

ASSUMPTIONS

1. All expected frequencies are at least 1.
2. At most 20% of the expected frequencies are less than 5.[†]

STEP 1 *State the null and alternative hypotheses.*
STEP 2 *Calculate the expected frequencies using the formula*

$$E = np$$

where n denotes the sample size and p denotes the probability for the category given in the null hypothesis.
STEP 3 *Check whether the expected frequencies satisfy Assumptions 1 and 2. If they do not, this procedure should not be used.*
STEP 4 *Decide on the significance level, α.*
STEP 5 *The critical value is χ_α^2 with $df = k - 1$, where k is the number of categories in the distribution.*

[†] Many texts give the rule that all expected frequencies should be at least five. Research by W. G. Cochran, a noted statistician, shows that the "rule of five" is too restrictive.

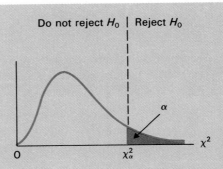

Do not reject H_0 | Reject H_0

α

χ^2

χ_α^2

0

STEP 6 *Compute the value of the test statistic*

$$\chi^2 = \Sigma(O - E)^2/E$$

STEP 7 *If the value of the test statistic falls in the rejection region, then reject H_0; otherwise, do not reject H_0.*

STEP 8 *State the conclusion in words.*

EXAMPLE 11.5   *Illustrates Procedure 11.1*

Let us now return to the hypothesis test of Example 11.4: A researcher wants to know whether the specialty distribution of American physicians currently in practice is different from the 1986 specialty distribution. The 1986 specialty distribution is given in Table 11.1 and is repeated here as Table 11.5.

TABLE 11.5
Specialty distribution of American physicians, 1986

Specialty	Percent	Probability, p
General practice	18.0	0.180
Medical	33.9	0.339
Surgical	27.0	0.270
Other	21.1	0.211

A random sample of 500 American physicians currently in practice yields the frequency distribution shown in Table 11.6.

TABLE 11.6
Observed frequencies

Specialty	Observed frequency O
General practice	80
Medical	162
Surgical	156
Other	102

At the 5% significance level, do the data provide sufficient evidence to conclude that the current specialty distribution of American physicians is different from the 1986 specialty distribution?

SOLUTION   We apply Procedure 11.1.

STEP 1 *State the null and alternative hypotheses.*

The null and alternative hypotheses are

> H_0: The current specialty distribution of American physicians is the same as the 1986 distribution given in Table 11.5.
>
> H_a: The current specialty distribution of American physicians is different from the 1986 distribution.

STEP 2 *Calculate the expected frequencies using the formula*

$$E = np$$

where n denotes the sample size and p denotes the probability for the category given in the null hypothesis.

The calculations are summarized in Table 11.7.

TABLE 11.7
Expected frequencies

Specialty	Probability p	Expected frequency $np = E$
General practice	0.180	$500 \cdot 0.180 = $ 90.0
Medical	0.339	$500 \cdot 0.339 = $ 169.5
Surgical	0.270	$500 \cdot 0.270 = $ 135.0
Other	0.211	$500 \cdot 0.211 = $ 105.5

STEP 3 *Check whether the expected frequencies satisfy Assumptions 1 and 2.*

1. All expected frequencies are at least 1? Yes (see Table 11.7).

2. At most 20% of the expected frequencies are less than 5? Yes, because none of the expected frequencies are less than 5.

STEP 4 *Decide on the significance level, α.*

We are to perform the test at the 5% significance level. Thus, $\alpha = 0.05$.

STEP 5 *The critical value is χ_α^2 with df $= k - 1$, where k is the number of categories in the distribution.*

From Step 4, $\alpha = 0.05$. Also, there are four categories (specialties), so $k = 4$. Referring to Table IV, we find that for df $= k - 1 = 4 - 1 = 3$, $\chi_{0.05}^2 = $ 7.815. See Figure 11.8.

FIGURE 11.8

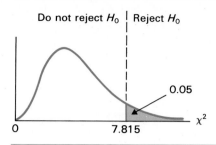

STEP 6 *Compute the value of the test statistic*

$$\chi^2 = \Sigma(O - E)^2/E$$

Using the observed frequencies in Table 11.6 and the expected frequencies in Table 11.7, we can compute the value of the test statistic. This is done in Table 11.8.

TABLE 11.8
Table for
computation of χ^2

Specialty	Observed frequency O	Expected frequency E	Difference $O - E$	Square of difference $(O - E)^2$	$(O - E)^2/E$
General practice	80	90.0	−10.0	100.00	1.111
Medical	162	169.5	−7.5	56.25	0.332
Surgical	156	135.0	21.0	441.00	3.267
Other	102	105.5	−3.5	12.25	0.116
	500	500.0	0		4.826

From the last column of Table 11.8, we see that the value of the test statistic is

$$\chi^2 = \Sigma(O - E)^2/E = \boxed{4.826}$$

STEP 7 *If the value of the test statistic falls in the rejection region, reject H_0; otherwise, do not reject H_0.*

From Step 6, the value of the test statistic is $\chi^2 = 4.826$. Since this does not fall in the rejection region, shown in Figure 11.8, we **do not reject H_0.**

STEP 8 *State the conclusion in words.*

The sample data do not provide sufficient evidence to conclude that the current specialty distribution of American physicians is different from the 1986 specialty distribution.

MTB

Note: In Table 11.8, we calculated the sum of the observed frequencies, the sum of the expected frequencies, and the sum of their differences. Strictly speaking, those sums are not needed. However, they serve as a check for computational errors. The sums of the observed and expected frequencies, ΣO and ΣE, should both equal the sample size, n, which in this case is 500. The sum of the differences, $\Sigma(O - E)$, should equal zero. See Exercise 11.20 for a verification of these facts.

The chi-square goodness-of-fit test provides a method for performing a hypothesis test about the percentage distribution of a population in which each member is classified into one of k categories. If the number of categories is two (i.e., $k = 2$), then the population is a two-category population. In this case, it can be shown that the chi-square goodness-of-fit test is equivalent to the z-test for one population proportion (Procedure 9.5 on page 467). See Exercise 11.19.

Exercises 11.2

___ **11.9** Why do you think the term "goodness-of-fit" is used to describe the type of hypothesis test considered in this section?

___ **11.10** Are the observed frequencies random variables? What about the expected frequencies? Explain your answers.

In each of Exercises 11.11–11.18, apply Procedure 11.1 on page 547 to perform the required hypothesis test.

___ **11.11** As reported by the U.S. Bureau of the Census in the publication *U.S. Census of Population*, the 1983 distribution of the U.S. resident population, by region, is as given here.

Region	Percent
Northeast	21.2
Midwest	25.2
South	34.0
West	19.6

A random sample of 500 current U.S. residents yields the following data:

Region	Frequency
Northeast	97
Midwest	121
South	176
West	106

At the 5% significance level, do the data suggest that the current distribution of the U.S. resident population by region is different from the 1983 distribution?

___ **11.12** According to the Census Bureau publication *Current Population Reports*, the marital-status distribution of the American adult population is as shown in the table below:

Marital status	Percent
Single	21.5
Married	63.9
Widowed	7.7
Divorced	6.9

A random sample of 750 American males, 25–29 years old, yielded the frequency distribution displayed in the following table:

Marital status	Frequency
Single	289
Married	408
Widowed	0
Divorced	53

At the 1% significance level, does it appear that the marital-status distribution of all 25–29-year-old American males is different from that of the American adult population as a whole?

___ **11.13** The table shown at the top of the first column on the next page provides a relative-frequency distribution of the number of years of school completed by United States residents 25 years old and over. [SOURCE: U.S. Bureau of the Census, *U.S. Census of Population*.]

Years of school	Relative frequency
8 or less	0.183
9–11	0.153
12	0.346
13–15	0.157
16 or more	0.161

A random sample of 300 Tennessee residents, 25 years old and over, gave the following statistics:

Years of school	Frequency
8 or less	83
9–11	48
12	95
13–15	36
16 or more	38

Do the data provide sufficient evidence to conclude that the distribution of the number of years of school completed by Tennessee residents is different from the national distribution? Use $\alpha = 0.01$.

___ 11.14   According to the document *Current Housing Reports*, published by the U.S. Bureau of the Census, the primary-heating-fuel distribution for occupied housing units is as follows:

Primary heating fuel	Percent
Natural gas	56.7
Fuel oil, kerosene	14.3
Electricity	16.0
LPG	4.5
Wood	6.7
Other	1.8

A random sample of 250 occupied housing units built after 1974 yields the frequency distribution below.

Primary heating fuel	Frequency
Natural gas	91
Fuel oil, kerosene	16
Electricity	110
LPG	14
Wood	17
Other	2

Do the data provide sufficient evidence to conclude that the primary-heating-fuel distribution for occu-pied housing units built after 1974 differs from that of all occupied housing units? Use $\alpha = 0.05$.

___ 11.15   A gambler thinks that a die may be loaded. That is, he thinks the six numbers may not be equally likely. To test his suspicion, he rolls the die 150 times and obtains the following results:

Number	Frequency
1	23
2	26
3	23
4	21
5	31
6	26

Do the data provide sufficient evidence to conclude that the die is loaded? Perform the hypothesis test at the 0.05 level of significance.

___ 11.16   A roulette wheel contains 18 red numbers, 18 black numbers, and two green numbers. The table below gives the frequency with which the ball landed on each color in 200 trials.

Color	Frequency
Red	88
Black	102
Green	10

At the 5% significance level, do the data suggest that the wheel is out of balance?

___ 11.17   The U.S. Federal Bureau of Investigation (FBI) compiles data on violent crime and reports the information obtained in the publication *Crime in the United States*. According to that publication, a per-centage distribution for the types of violent crime for the entire United States is as shown in the table below.

Violent crime	Percent
Murder	1.6
Forcible rape	6.4
Robbery	40.3
Aggravated assault	51.7

A random sample of 600 violent-crime reports from the state of New Jersey yields the following statistics:

Violent crime	Frequency
Murder	6
Forcible rape	33
Robbery	292
Aggravated assault	269

Do the data provide sufficient evidence to conclude that the distribution of types of violent crime in New Jersey is different from the nation as a whole? Perform the hypothesis test using $\alpha = 0.01$.

___ 11.18 The publication *Current Population Reports,* released by the U.S. Bureau of the Census, contains the following information on the 1983 money income of American households:

Income level	Percent
Under $5,000	9.2
$5,000–$9,999	13.7
$10,000–$14,999	13.0
$15,000–$19,999	12.0
$20,000–$24,999	10.8
$25,000–$34,999	17.0
$35,000–$49,999	14.0
$50,000 and over	10.3

A random sample of 825 household incomes for last year gave the frequency distribution below. The income levels are in 1983 constant dollars; that is, they are adjusted for inflation to 1983 levels.

Income level	Frequency
Under $5,000	70
$5,000–$9,999	117
$10,000–$14,999	92
$15,000–$19,999	111
$20,000–$24,999	101
$25,000–$34,999	140
$35,000–$49,999	111
$50,000 and over	83

Do the data provide sufficient evidence to conclude that last year's income-level distribution for households has changed from the 1983 distribution? Perform the hypothesis test at the 5% level of significance.

═ 11.19 Consider a two-category population with population proportion, p. Suppose that we want to perform a hypothesis test to decide whether p is dif-

ferent from the value p_0. Discuss the method for performing such a hypothesis test using
a) the z-test for one population proportion (Procedure 9.5 on page 467).
b) the chi-square goodness-of-fit test (Procedure 11.1 on page 547).
On page 551, we stated that the hypothesis tests in parts (a) and (b) are *equivalent*. What do you think that means?

═ 11.20 On page 551, we mentioned that the sums of the observed and expected frequencies, ΣO and ΣE, always equal the sample size, n, and that the sum of the differences, $\Sigma(O - E)$, always equals zero.
a) Explain, in words, why $\Sigma O = n$.
b) Show mathematically that $\Sigma E = n$.
c) Prove that $\Sigma(O - E) = 0$.

Testing for normality: Many of the inferential procedures that we have studied require the population(s) under consideration to be normally distributed. For example, Procedure 9.4 on page 458, which provides a method for performing a hypothesis test for a population mean, requires that the population be normally distributed. Sometimes we are reasonably sure, either because of previous experience or theoretical reasons, that the population we are studying is normally distributed. But what if we are not so sure and we want to decide whether the population is normally distributed? One way we can decide is by performing a chi-square goodness-of-fit test. We illustrate the procedure in Exercises 11.21 and 11.22.

═ 11.21 Suppose that we want to decide whether the ages of U.S. residents are normally distributed. We take a random sample of 500 U.S. residents and find that $\bar{x} = 34.93$ years and $s = 21.82$ years. In addition, we obtain the following frequency distribution:

TABLE 11.9

Age	Observed frequency O
Under 15	91
15–under 25	114
25–under 35	84
35–under 45	59
45–under 65	98
65 or older	54

To use this data to test for normality, we must first find the frequencies that would be expected, if the population of ages is, in fact, normally distributed. We proceed as follows: Since the population mean, μ, and the population standard deviation, σ, are unknown, we estimate them by the sample mean, $\bar{x} = 34.93$ years, and the sample standard deviation, $s = 21.82$ years:

$$\mu \approx 34.93 \text{ years} \qquad \sigma \approx 21.82 \text{ years}$$

a) Obtain the probability, p, associated with each age category in Table 11.9, if the population of ages is normally distributed with mean 34.93 years and standard deviation 21.82 years. Fill in the second column of Table 11.10.

TABLE 11.10

Age	Probability p	Expected frequency E
Under 15		
15–under 25		
25–under 35		
35–under 45		
45–under 65		
65 or older		

b) Use the second column of Table 11.10 and the fact that $n = 500$ to fill in the third column.

Now that we have the observed and expected frequencies, we can perform a chi-square goodness-of-fit test by employing Procedure 11.1. However, there is one modification. For six categories ($k = 6$), we would think that df $= 6 - 1 = 5$. But the degrees of freedom are reduced by one for each of the two parameters, μ and σ, that we estimated in order to find the expected frequencies. Thus,

$$\text{df} = \underset{\substack{\uparrow \\ \text{Usual}}}{6 - 1} \; - \; \underset{\substack{\uparrow \\ \text{For} \\ \text{estimating} \\ \mu \text{ by } \bar{x}}}{1} \; - \; \underset{\substack{\uparrow \\ \text{For} \\ \text{estimating} \\ \sigma \text{ by } s}}{1} \; = 3$$

c) Use Procedure 11.1 to perform the test for normality at the 5% significance level.

≡ **11.22** A statistical-process-control (SPC) expert wants to know whether the diameters of bolts produced by a machine are normally distributed. He takes a random sample of 300 bolts and finds that their mean diameter is 10.00 mm and that the sample standard deviation of their diameters is 0.10 mm. Moreover, he obtains the following frequency distribution:

Diameter	Observed frequency O
Under 9.8	8
9.8–under 9.9	42
9.9–under 10.0	112
10.0–under 10.1	97
10.1–under 10.2	38
10.2 or over	3

Can the SPC expert conclude that the diameters of bolts produced by the machine are not normally distributed? Use $\alpha = 0.05$.

11.3   Chi-square independence test

The next chi-square procedure we will study is the **chi-square independence test.** To begin, we need to define what it means for two characteristics of a population to be *statistically independent*. Roughly, it means that knowledge of the category to which a member of the population belongs for one of the characteristics imparts no information about the other characteristic. More precisely, we have:

DEFINITION 11.1   Statistical independence for two characteristics of a population

Two characteristics of a population are called *statistically independent* (or *nonassociated*) if within the categories of one of the characteristics, the distributions of the other characteristic are the same.

If two characteristics of a population are not statistically independent, then we say that they are **statistically dependent** (or **associated**).

Suppose that the characteristics, X and Y, of a population are statistically independent. Specifically, suppose that within the categories of X, the distributions of Y are the same. Then those common distributions for Y will also equal the distribution of Y within the entire population. Furthermore, the reverse must also hold; that is, within the categories of Y, the distributions of X will be the same as each other and as the distribution of X within the entire population.

EXAMPLE 11.6   *Illustrates Definition 11.1*

In Example 4.20, we considered the two characteristics, age and rank, for the population of faculty at Arizona State University. The contingency table in Table 4.8 on page 176 provides a joint frequency distribution for those two characteristics. Use the contingency table to determine whether the characteristics, age and rank, are statistically independent.

SOLUTION   To determine whether age and rank are statistically independent, we will obtain the age distribution within each rank category. If those age distributions are the same, then the characteristics, age and rank, are statistically independent; otherwise, they are statistically dependent.

By dividing each column in Table 4.8 by its column total, we get the relative-frequency distributions of age within the rank categories. These are shown in the columns of Table 11.11.

TABLE 11.11
Age distributions within the rank categories

Rank

Age		Full Professor R_1	Associate Professor R_2	Assistant Professor R_3	Instructor R_4	Total
	Under 30 A_1	0.005	0.008	0.178	0.182	0.058
	30–39 A_2	0.121	0.446	0.509	0.515	0.345
	40–49 A_3	0.363	0.328	0.191	0.182	0.299
	50–59 A_4	0.337	0.178	0.113	0.121	0.217
	60 & over A_5	0.174	0.039	0.009	0.000	0.080
	Total	1.000	1.000	1.000	1.000	1.000

Since the columns in Table 11.11 are not identical, the age distributions within the rank categories are not the same. Hence the characteristics, age and rank, for the population of faculty at Arizona State University are statistically dependent.

Thus, knowledge of a faculty member's rank provides information about the faculty member's age, and vice-versa. For instance, with no knowledge of a faculty member's rank, we can say that there is a 21.7% chance that the faculty member will be in his or her 50s. However, if we know that the faculty member is a full professor, then we can say that there is a 33.7% chance that the faculty member will be in his or her 50s.   ■

We should point out that the concept of statistical independence for two characteristics of a population is closely related to the concept of statistical independence of events, which we studied in Section 4.6. In fact, it can be shown that two characteristics of a population are statistically independent if and only if the events corresponding to the categories of the two characteristics are statistically independent.

For instance, consider the characteristics, age and rank, discussed in Example 11.6. Suppose that a faculty member is selected at random. Let

$$A_1 = \text{event the faculty member selected is under 30}$$

and so on (see the letters labeling the rows and columns of Table 11.11). Then the characteristics, age and rank, are statistically independent if and only if the events $A_1, A_2, \ldots, A_5$ and $R_1, R_2, \ldots, R_4$ are statistically independent; that is, event A_i and event R_j are statistically independent for all i and j.

THE LOGIC BEHIND THE CHI-SQUARE INDEPENDENCE TEST

If we have data for an entire population, then we can always determine for sure whether two characteristics of the population are statistically independent by proceeding as in Example 11.6. However, in most cases, data for an entire population are not available and so it is usually necessary to employ inferential methods in order to decide whether two characteristics are statistically independent.

One of the most commonly used methods for making such decisions is the *chi-square independence test*. The following example introduces and explains the reasoning behind that hypothesis-testing procedure.

EXAMPLE 11.7   *Introduces the chi-square independence test*

A national survey was conducted to obtain information on the alcohol consumption patterns of American adults by marital status. A random sample of 1772 residents, 18 years old and over, yielded the data shown in Table 11.12.[†]

---

† Adapted from: Clark, W. B. and Midanik, L. "Alcohol Use and Alcohol Problems among U.S. Adults: Results of the 1979 National Survey." In: National Institute on Alcohol Abuse and Alcoholism. Alcohol and Health Monograph No 1, Alcohol Consumption and Related Problems. DHHS Pub. No (ADM) 82–1190, 1982.

TABLE 11.12
Contingency table of marital status versus alcohol consumption for 1772 randomly selected adults

Drinks per month

Marital status		Abstain	1–60	Over 60	Total
	Single	67	213	74	354
	Married	411	633	129	1173
	Widowed	85	51	7	143
	Divorced	27	60	15	102
Total		590	957	225	1772

The table shows, for instance, that of the 1772 adults sampled, 1173 are married, 590 abstain, and 411 are married and abstain.

We want to use the sample data to decide whether there is an association between marital status and alcohol consumption patterns. Specifically, we want to perform the hypothesis test

H_0: Marital status and alcohol consumption are statistically independent.

H_a: Marital status and alcohol consumption are statistically dependent.

The idea behind the chi-square independence test is to compare the observed frequencies in Table 11.12 with the frequencies that would be expected if the null hypothesis of statistical independence is true. The test statistic employed to make the comparison is the same as the one used for the goodness-of-fit test, namely,

$$\chi^2 = \Sigma(O - E)^2/E$$

where O represents observed frequency and E represents expected frequency.

We will now develop a formula for computing the expected frequencies. Consider, for instance, the cell of Table 11.12 corresponding to "married and abstain," the cell in the second row and first column of the table. To begin, note that the population proportion of all adults who abstain can be estimated by the sample proportion of the 1772 adults sampled who abstain; that is, by

Number sampled who abstain

$$\frac{590}{1772} = 0.333 \ (33.3\%)$$

Total number sampled

If marital status and alcohol consumption are statistically independent (i.e., if H_0 is true), then the proportion of married adults who abstain will be the same as the proportion of all adults who abstain. Thus, if H_0 is true, the sample proportion, $\frac{590}{1772}$, or 33.3%, will also be an estimate of the population proportion of married adults who abstain.

Now, a total of 1173 of the adults sampled are married and, as we have just seen, if H_0 is true, then approximately $\frac{590}{1772}$ ths, or 33.3%, of married adults abstain. Therefore, if H_0 is true, we would expect about

$$\frac{590}{1772} \cdot 1173 = 390.6$$

of the adults in the survey to be married adults who abstain.

It is useful to rewrite the left-hand side of the above expected-frequency computation in a slightly different way. By using algebra and referring to Table 11.12, we obtain

$$\text{Expected frequency} = \frac{590}{1772} \cdot 1173 = \frac{1173 \cdot 590}{1772} = \frac{(\text{Row total}) \cdot (\text{Column total})}{\text{Sample size}}$$

If we let R denote "row total" and C denote "column total," then we can express this last equation compactly as

$$E = \frac{R \cdot C}{n}$$

where, as usual, E denotes expected frequency and n denotes sample size.

Using this simple formula, we can obtain the expected frequencies for all 12 cells in Table 11.12. We have already done that for the cell in the second row and first column. For the cell in the upper right-hand corner of the table, we get

$$E = \frac{R \cdot C}{n} = \frac{354 \cdot 225}{1772} = 44.9$$

and similar computations give the expected frequencies for the remaining cells. In Table 11.13, we have modified Table 11.12 by placing the expected frequency for each cell beneath the corresponding observed frequency.

TABLE 11.13
Observed and expected frequencies for marital status versus alcohol consumption; expected frequencies are printed below the observed frequencies

Drinks per month

Marital status	Abstain	1–60	Over 60	Total
Single	67 / 117.9	213 / 191.2	74 / 44.9	354
Married	411 / 390.6	633 / 633.5	129 / 148.9	1173
Widowed	85 / 47.6	51 / 77.2	7 / 18.2	143
Divorced	27 / 34.0	60 / 55.1	15 / 13.0	102
Total	590	957	225	1772

Thus, for instance, Table 11.13 shows that the observed frequency of the adults sampled who are single and have over 60 drinks per month is 74; whereas, if marital status and alcohol consumption are statistically independent, then the expected frequency is 44.9.

If the null hypothesis of statistical independence is true, then the observed and expected frequencies should be about the same, resulting in a relatively small value of the test statistic, $\chi^2 = \Sigma(O - E)^2/E$. Consequently, if χ^2 is too large, we will reject the null hypothesis of statistical independence in favor of the alternative hypothesis of statistical dependence.

We will now compute the value of χ^2 for the observed and expected frequencies displayed in Table 11.13:

$$
\begin{aligned}
\chi^2 &= \Sigma(O - E)^2/E \\
&= (67 - 117.9)^2/117.9 + (213 - 191.2)^2/191.2 + (74 - 44.9)^2/44.9 \\
&\quad + (411 - 390.6)^2/390.6 + (633 - 633.5)^2/633.5 + (129 - 148.9)^2/148.9 \\
&\quad + (85 - 47.6)^2/47.6 + (51 - 77.2)^2/77.2 + (7 - 18.2)^2/18.2 \\
&\quad + (27 - 34.0)^2/34.0 + (60 - 55.1)^2/55.1 + (15 - 13.0)^2/13.0 \\
&= 21.952 + 2.489 + 18.776 + 1.070 + 0.000 + 2.670 \\
&\quad + 29.358 + 8.908 + 6.856 + 1.427 + 0.438 + 0.324 \\
&= 94.269 \ ^\dagger
\end{aligned}
$$

Can this value be reasonably attributed to sampling error or is it large enough to indicate that marital status and alcohol consumption are statistically dependent? Before we can answer that question, we must know the probability distribution of the test statistic, $\chi^2 = \Sigma(O - E)^2/E$. ∎

KEY FACT 11.3

Consider a chi-square independence test where the null hypothesis is that two characteristics of a population are statistically independent. Suppose that the sample size is large. Then, if the null hypothesis of statistical independence is true, the random variable

$$\chi^2 = \Sigma(O - E)^2/E$$

has approximately a chi-square distribution. The number of degrees of freedom is

$$df = (r - 1)(c - 1)$$

where r and c are the number of rows and columns in the contingency table, respectively (i.e., r is the number of categories for the characteristic displayed along the side of the contingency table and c is the number of categories for the characteristic displayed along the top of the contingency table).

---

† Although we have displayed the expected frequencies to one decimal place and the chi-square subtotals to three decimal places, the calculations were done using full calculator accuracy.

THE CHI-SQUARE INDEPENDENCE TEST

We will now state a procedure for performing a chi-square independence test. The null hypothesis is that the two characteristics under consideration are statistically independent and the alternative hypothesis is that they are statistically dependent. Note that the hypothesis test is always right-tailed (why?). Also, note carefully the conditions required for using the procedure.

PROCEDURE 11.2

To perform a chi-square independence test.

ASSUMPTIONS

1. All expected frequencies are at least 1.
2. At most 20% of the expected frequencies are less than 5.

STEP 1 *State the null and alternative hypotheses.*

STEP 2 *Calculate the expected frequencies using the formula*

$$E = \frac{R \cdot C}{n}$$

  where R = row total, C = column total, and n = sample size. Place each expected frequency below its corresponding observed frequency in the contingency table.

STEP 3 *Check whether the expected frequencies satisfy Assumptions 1 and 2. If they do not, this procedure should not be used.*

STEP 4 *Decide on the significance level, α.*

STEP 5 *The critical value is χ^2_α with df $= (r - 1)(c - 1)$, where r and c are the number of rows and columns in the contingency table.*

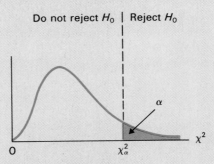

STEP 6 *Compute the value of the test statistic*

$$\chi^2 = \Sigma(O - E)^2/E$$

STEP 7 *If the value of the test statistic falls in the rejection region, then reject H_0; otherwise, do not reject H_0.*

STEP 8 *State the conclusion in words.*

EXAMPLE 11.8 · *Illustrates Procedure 11.2*

Recall that a random sample of 1772 American adults yielded the data on marital status versus alcohol consumption given in Table 11.12. We repeat that contingency table here as Table 11.14.

TABLE 11.14
Contingency table of marital status versus alcohol consumption for 1772 randomly selected adults

Drinks per month

Marital status	Abstain	1–60	Over 60	Total
Single	67	213	74	354
Married	411	633	129	1173
Widowed	85	51	7	143
Divorced	27	60	15	102
Total	590	957	225	1772

Do the data suggest, at the 5% significance level, that marital status and alcohol consumption patterns are statistically dependent?

SOLUTION · We employ Procedure 11.2.

STEP 1 *State the null and alternative hypotheses.*

We want to decide whether the characteristics, marital status and alcohol consumption, are statistically dependent. So, the null and alternative hypotheses are

H_0: Marital status and alcohol consumption are statistically independent.
H_a: Marital status and alcohol consumption are statistically dependent.

STEP 2 *Calculate the expected frequencies using the formula*

$$E = \frac{R \cdot C}{n}$$

where R = row total, C = column total, and n = sample size. Place each expected frequency below its corresponding observed frequency in the contingency table.

The expected frequency for the "single and abstain" cell in the upper left-hand corner of Table 11.14 is

$$E = \frac{R \cdot C}{n} = \frac{354 \cdot 590}{1772} = 117.9$$

Similar computations give the other 11 expected frequencies. See Table 11.15.

Drinks per month

	Abstain	1–60	Over 60	Total
Single	67 117.9	213 191.2	74 44.9	354
Married	411 390.6	633 633.5	129 148.9	1173
Widowed	85 47.6	51 77.2	7 18.2	143
Divorced	27 34.0	60 55.1	15 13.0	102
Total	590	957	225	1772

Marital status

STEP 3 *Check whether the expected frequencies satisfy Assumptions 1 and 2.*

1. All expected frequencies are at least 1? **Yes** (see Table 11.15).
2. At most 20% of the expected frequencies are less than 5? **Yes,** because none of the expected frequencies are less than 5.

STEP 4 *Decide on the significance level, α.*

The test is to be performed at the 5% significance level. Hence, $\alpha = 0.05.$

STEP 5 *The critical value is χ_{α}^2 with $df = (r-1)(c-1)$, where r and c are the number of rows and columns in the contingency table, respectively.*

The number of rows, r, in the contingency table is the number of marital-status categories, which is four; so, $r = 4$. The number of columns, c, in the contingency table is the number of drinks-per-month categories, which is three; so, $c = 3$. Thus,

$$df = (r-1)(c-1) = (4-1)(3-1) = 3 \cdot 2 = 6$$

Since $\alpha = 0.05$, we find from Table IV that the critical value is $\chi_{0.05}^2 = 12.592$. See Figure 11.9.

FIGURE 11.9

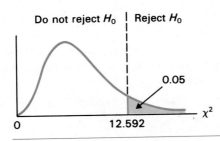

STEP 6 *Compute the value of the test statistic*

$$\chi^2 = \Sigma(O - E)^2/E$$

The observed frequencies (O-values) and expected frequencies (E-values) are displayed in Table 11.15. Using these, we compute the value of the test statistic:

$$\chi^2 = \Sigma(O - E)^2/E$$
$$= (67 - 117.9)^2/117.9 + (213 - 191.2)^2/191.2 + \cdots + (15 - 13.0)^2/13.0$$
$$= 21.952 + 2.489 + \cdots + 0.324 = \boxed{94.269}$$

STEP 7 *If the value of the test statistic falls in the rejection region, reject H_0; otherwise, do not reject H_0.*

From Step 6, the value of the test statistic is $\chi^2 = 94.269$, which falls in the rejection region (see Figure 11.9). Thus, we reject H_0.

STEP 8 *State the conclusion in words.*

MTB

It appears that marital status and alcohol consumption patterns are statistically dependent.

CONCERNING THE ASSUMPTIONS

In Procedure 11.2, we made two assumptions about expected frequencies:

1. All expected frequencies are at least 1.
2. At most 20% of the expected frequencies are less than 5.

What can we do if one or both of these assumptions are violated? Three approaches are possible. We could combine rows or columns in order to increase the expected frequencies in those cells where they are too small; we could eliminate certain rows or columns where the small expected frequencies occur; or, we could increase the sample size. Which, if any, of these modifications is employed depends on the problem under consideration. See Exercises 11.35 and 11.36.

STATISTICAL DEPENDENCE DOES NOT IMPLY CAUSATION

The chi-square independence test is used to decide whether two characteristics of a population are statistically dependent. Specifically, the null hypothesis is that the two characteristics are statistically independent and the alternative hypothesis is that they are statistically dependent. Consequently, if the null hypothesis is rejected, then we can conclude that the two characteristics are statistically dependent. This does not imply a causal relationship between the two characteristics.

For instance, in Example 11.8, we rejected the null hypothesis of statistical independence for marital status and alcohol consumption. This means that knowledge of the marital status of a person provides information about the alcohol consumption patterns of that person, and vice-versa. It does not mean, for example, that being single causes a person to drink more.

DECIDING ON STATISTICAL DEPENDENCE WITH POPULATION DATA

It is important to understand that the chi-square independence test is an inferential procedure and, as such, should be applied only to sample data and not to data for an entire population. How then should we proceed if we have data for an entire population and we want to decide whether two characteristics of the population are statistically dependent?

One way we can proceed is as described in Example 11.6 at the beginning of this section. Another way is to compute the expected frequencies as in the chi-square independence test; that is, using the formula $E = R \cdot C/N$, where N is the size of the population. If the expected and observed frequencies are identical for each cell, then the two characteristics are statistically independent; otherwise, they are statistically dependent. See Exercise 11.39.

USING THE COMPUTER (OPTIONAL)

Procedure 11.2 provides a step-by-step method for carrying out a chi-square independence test. Minitab has a program called **CHISQUARE** that will perform Procedure 11.2 for us. The next example shows how CHISQUARE can be applied to carry out the hypothesis test considered in Example 11.8.

EXAMPLE 11.9   *Illustrates the CHISQUARE command*

A national survey was conducted to obtain information on the alcohol consumption patterns of American adults by marital status. A random sample of 1772 residents, 18 years old and over, yielded the data given in Table 11.14 on page 561. Do the data suggest, at the 5% significance level, that marital status and alcohol consumption patterns are statistically dependent? Use Minitab to carry out the hypothesis test.

SOLUTION   We want to perform the hypothesis test

H_0: Marital status and alcohol consumption are statistically independent.

H_a: Marital status and alcohol consumption are statistically dependent.

at the 5% significance level.

To employ Minitab, we first enter the data from the cells of Table 11.14 into the computer using the READ command, as indicated in Printout 11.1.

PRINTOUT 11.1

```
MTB > READ C1-C3
DATA> 67 213 74
DATA> 411 633 129
DATA> 85 51 7
DATA> 27 60 15
DATA> END
      4 ROWS READ
```

Then we type the command CHISQUARE followed by the storage locations of the sample data; that is, we type

<u>CHISQUARE on C1-C3</u>

Printout 11.2 displays this command and the output resulting from it.

PRINTOUT 11.2
Minitab output
for CHISQUARE

MTB > CHISQUARE on C1-C3

Expected counts are printed below observed counts

	C1	C2	C3	Total
1	67	213	74	354
	117.87	191.18	44.95	
2	411	633	129	1173
	390.56	633.50	148.94	
3	85	51	7	143
	47.61	77.23	18.16	
4	27	60	15	102
	33.96	55.09	12.95	
Total	590	957	225	1772

ChiSq = 21.952 + 2.489 + 18.776 +
          1.070 + 0.000 + 2.670 +
         29.358 + 8.908 + 6.856 +
          1.427 + 0.438 + 0.324 = 94.269
df = 6

The first part of the output provides a table of the observed and expected frequencies. This is Minitab's version of Table 11.15 on page 562. After the table, we find the value of the test statistic, $\chi^2 = \Sigma(O - E)^2/E$, including the cell-by-cell subtotals. Thus, the value of the test statistic is $\chi^2 = 94.269$. The final item given in the output is the number of degrees of freedom, df = 6.

In order to complete the hypothesis test, we need only obtain the critical value and compare it with the value of the test statistic, $\chi^2 = 94.269$. The critical value for a chi-square independence test is χ^2_α with df $= (r - 1)(c - 1)$. Because the hypothesis test is to be performed at the 5% significance level, we have $\alpha = 0.05$. Also, the last line of Printout 11.2 shows that df $= 6$. Consulting Table IV, we find that for df $= 6$, $\chi^2_{0.05} = 12.592$.

Since the value of the test statistic is $\chi^2 = 94.269$, which exceeds the critical value of 12.592, we reject H_0. In other words, it appears that marital status and alcohol consumption patterns are statistically dependent. ∎

Exercises 11.3

In each of Exercises 11.23–11.30, apply Procedure 11.2 on page 560 to perform the required hypothesis test, provided the conditions for the use of that procedure are met.

___ **11.23**   The U.S. Bureau of the Census compiles data on money earnings of persons by educational attainment, sex, and age, and publishes the information in *Current Population Reports*. A study to decide whether annual income is statistically dependent on educational level yielded the sample data shown in the following contingency table:

Years of schooling

Annual income	0–8	9–12	Over 12	Total
Under $10,000	34	36	10	80
$10,000–$24,999	41	72	36	149
$25,000–$39,999	10	78	58	146
$40,000 and over	4	26	45	75
Total	89	212	149	450

Do the data provide sufficient evidence to conclude that annual income and educational level are statistically dependent? Perform the required hypothesis test at the 1% significance level.

___ **11.24**   The Gallup Organization conducts periodic surveys to gauge the support by American adults for regional primary elections. The question asked is: "It has been proposed that four individual primaries be held in different weeks of June during presidential election years. Does this sound like a good idea or a poor idea?" At the top of the next column is a contingency table for responses by political affiliation adapted from the results of a May, 1985, Gallup Poll appearing in the *Arizona Republic*.
a) Fill in the row and column totals.
b) Do the data suggest, at the 5% level of significance, that the feelings of adults on the issue of regional primaries are dependent on political affiliation?

Response

Political affiliation	Good idea	Poor idea	No opinion	Total
Republican	266	266	186	
Democrat	308	250	176	
Independent	28	27	21	
Total				1528

___ **11.25**   The U.S. Bureau of Labor Statistics gathers data on occupations of employed workers by sex and publishes its findings in *Employment and Earnings*. A random sample of 83 employed workers yields the following data:

Sex

Occupation type	Male	Female	Total
Managerial/Professional	14	10	24
Technical sales Administrative	9	16	25
Service	4	6	10
Other	20	4	24
Total	47	36	83

Do the data provide sufficient evidence to conclude that there is an association between the characteristics, sex and occupation type, for employed workers? Use $\alpha = 0.01$.

___ **11.26**   In 1986, more than 62 million Americans suffered injuries, as reported by the National Center for Health Statistics in *Vital and Health Statistics*. More males (34.0 million) were injured than females (28.4 million). Those statistics do not tell us whether males and females tend to be injured in similar circumstances. One set of categories that is commonly used for accident circumstance is "while at work," "home," "motor vehicle," and "other." In order to decide whether there is an association between accident circumstance and gender, a safety official in a large city took a random sample of accident reports. She obtained the following data:

Sex

Circumstance	Male	Female	Total
While at work	18	4	
Home	26	28	
Motor vehicle	4	6	
Other	36	24	
Total			

a) Fill in the row and column totals.
b) Determine the sample size.
c) Perform a hypothesis test at the 5% significance level to decide whether, in this city, accident circumstance and gender are statistically dependent.

___ 11.27 The FBI compiles information on arrests for violent crimes by the type of crime committed and the age of the person arrested. Results are published in *Crime in the United States.* To decide whether there is an association between the type of violent crime committed and the age of the person arrested, 750 arrest records are randomly selected. The sample results are displayed in the following contingency table:

Age

Type of violent crime	18–24	25–44	45+	Total
Murder	11	16	4	
Forcible rape	21	26	4	
Robbery	128	92	6	
Aggravated assault	162	234	46	
Total				750

a) Fill in the row and column totals.
b) Is there evidence that an association exists between the type of violent crime committed and the age of the person arrested? Use $\alpha = 0.01$.

___ 11.28 The National Center for Health Statistics reports information on people with acute medical conditions in the publication *Vital and Health Statistics.* Acute conditions are counted only if they are medically attended or caused at least one day of restricted activity. Suppose that a random sample of 376 acute conditions revealed the data displayed in the contingency table below. Note that we have used the following coding for the columns of the contingency table: A stands for infective and parasitic diseases, B stands for respiratory diseases, C stands for digestive-system diseases, and D stands for injuries.

Type of condition

Family income	A	B	C	D	Total
Under $15,000	4	24	2	8	
$15,000–$19,999	7	36	4	11	
$20,000–$29,999	8	35	4	9	
$30,000–$39,999	12	59	5	17	
$40,000 and over	19	81	6	25	
Total					376

a) Fill in the row and column totals.
b) Do the data provide sufficient evidence to conclude that there is an association between the characteristics, type of acute condition and family income? Perform the appropriate hypothesis test using a significance level of 0.05.

___ 11.29 The publication *Statistics of Income Bulletin,* from the U.S. Internal Revenue Service (IRS), contains data on top wealthholders by marital status. A random sample of 487 top wealthholders yielded the contingency table displayed at the top of the first column on the following page. Do the data provide sufficient evidence to conclude that, for top wealthholders, the characteristics, net worth and marital status, are statistically dependent? Use a significance level of 0.05 for your hypothesis test.

Net worth

Marital status

	Married	Single/ Divorced	Widowed	Total
$100,000– $249,999	227	54	63	344
$250,000– $499,999	60	15	22	97
$500,000– $999,999	20	4	7	31
$1,000,000 or more	10	2	3	15
Total	317	75	95	487

__ 11.30   Data on characteristics of lawyers are published by The American Bar Foundation, Chicago, IL, in *The Lawyer Statistical Report*. The following contingency table cross classifies 307 randomly selected lawyers by status in practice and size of city practicing in:

Status in practice

Size of city

	Less than 250,000	250,000– 499,999	500,000 or more	Total
Government	12	4	14	30
Judicial	8	1	2	11
Private practice	122	31	69	222
Salaried	19	7	18	44
Total	161	43	103	307

Do the data provide sufficient evidence to conclude that the characteristics, size of city and status in practice, are statistically dependent? Use $\alpha = 0.05$.

Exercises 11.31–11.34 are computer exercises.

__ 11.31 (**Computer exercise**) Refer to the contingency table displayed in Exercise 11.23. Suppose the cell data in the columns headed "0–8," "9–12," and "Over 12" are stored, respectively, in columns named 0-8, 9-12, and OVER 12.

a) Which Minitab command and subcommands (if any) should be used to perform the hypothesis test considered in Exercise 11.23?
b) If you have access to Minitab, use it to carry out the hypothesis test.

__ 11.32 (**Computer exercise**) Refer to the contingency table displayed in Exercise 11.24. Suppose that the cell data in the columns headed "Good idea," "Poor idea," and "No opinion" are stored, respectively, in columns named GOODIDEA, POORIDEA, and NOOPIN.

a) Which Minitab command and subcommands (if any) should be used to perform the hypothesis test considered in part (b) of Exercise 11.24?
b) If you have access to Minitab, use it to carry out the hypothesis test.

__ 11.33 (**Computer exercise**) A study was conducted at Arizona State University to decide whether there is an association between grade and study time for intermediate algebra students who complete the course with a grade of C or better. Data were collected for a random sample of 422 students and compiled in a contingency table with categories for the number of hours studied per week across the top and categories for the grade earned along the side. For the number of hours studied, there were four categories: 0–3, 4–6, 7–9, and 10 or more. For the grade earned, there were three categories: A, B, and C. Printout 11.3 on page 571 provides a Minitab printout that was obtained by applying the CHISQUARE command to the data. Determine

a) the number of students in the sample who studied at least 10 hours per week.
b) the number of students in the sample who received a grade of B in the class.
c) the observed number of students in the sample who studied at least 10 hours per week and received a grade of B in the class.
d) the expected number of students in the sample who studied at least 10 hours per week and received a grade of B in the class, if grade and study time are statistically independent.
e) the chi-square subtotal, $(O - E)^2/E$, for the "at least 10 hours per week *and* grade of B" cell.
f) the number of degrees of freedom.
g) the null and alternative hypotheses for a hypothesis test to decide whether there is an association between grade and study time for intermediate algebra students at Arizona State University.

h) the value of the test statistic, χ^2.
i) the conclusion, if the hypothesis test is performed at the 5% significance level. (*Note:* You will need to consult Table IV.)

═ **11.34 (Computer exercise)** The Book Industry Study Group, Inc, performs surveys to obtain information on characteristics of book readers. A *book reader* is defined to be one who read one or more books in the six months prior to the survey; a *non-book reader* is defined to be one who read newspapers or magazines but no books in the six months prior to the survey; and a *nonreader* is defined to be one who did not read a book, newspaper, or magazine in the six months prior to the survey. Printout 11.4 on page 571 was obtained by applying the CHISQUARE command to data from a random sample of people, 16 years old and over. Across the top of the contingency table are the reader-classification categories: book reader, non-book reader, and nonreader. Along the side of the contingency table are the household income categories: less than $15,000, $15,000–$24,999, $25,000–$39,999, and $40,000 or over. Determine

a) the total number of people in the sample.
b) the number of people in the sample with a household income between $25,000 and $39,999.
c) the number of people in the sample who are book readers.
d) the observed number of people in the sample who are book readers with a household income between $25,000 and $39,999.
e) the expected number of people in the sample who are book readers with a household income between $25,000 and $39,999, if the characteristics, household income and reader classification, are statistically independent.
f) the number of degrees of freedom.
g) the null and alternative hypotheses for a hypothesis test to decide whether the characteristics, household income and reader classification, are statistically dependent.
h) the value of the test statistic, χ^2.
i) the conclusion, if the hypothesis test is performed at the 1% significance level. (*Note:* You will need to consult Table IV.)

═ **11.35** In Exercise 11.29, it was not possible to perform the chi-square independence test because the assumptions regarding expected frequencies were not met. As mentioned on page 563, we can try three approaches to remedy the situation: (1) combine rows or

columns, (2) eliminate rows or columns, or (3) increase the sample size.

a) Combine the last two rows of the contingency table in Exercise 11.29 to form a new contingency table.
b) Use the table in part (a) to perform the hypothesis test indicated in Exercise 11.29, if possible.
c) Eliminate the last row of the contingency table in Exercise 11.29 to form a new contingency table.
d) Use the table in part (c) to perform the hypothesis test indicated in Exercise 11.29, if possible.

═ **11.36** In Exercise 11.30, it was not possible to perform the chi-square independence test because the assumptions regarding expected frequencies were not met. As mentioned on page 563, we can try three approaches to remedy the situation: (1) combine rows or columns, (2) eliminate rows or columns, or (3) increase the sample size.

a) Combine the first two rows of the contingency table in Exercise 11.30 to form a new contingency table.
b) Use the table in part (a) to perform the hypothesis test described in Exercise 11.30, if possible.
c) Eliminate the second row of the contingency table in Exercise 11.30 to form a new contingency table.
d) Use the table in part (c) to perform the hypothesis test described in Exercise 11.30, if possible.

═ **11.37** A random sample of 100 members of a union are asked to respond to two questions:

Question 1: Are you happy with your financial situation today?

Question 2: Do you approve of the federal government's economic policies?

The responses are:

	Question 1		
	Yes	No	Total
Yes	22	48	70
No	12	18	30
Total	34	66	100

(Question 2 labels the rows Yes and No)

a) Calculate χ^2 for this data and test the null hypothesis that response to Question 1 is statistically independent of response to Question 2 for members of this union. Use $\alpha = 0.05$.
b) Suppose 200 members of the union are sampled, twice as many as in part (a), and that their responses are in the same proportions as before:

Question 1

	Yes	No	Total
Yes	44	96	140
No	24	36	60
Total	68	132	200

(row label: Question 2)

Calculate χ^2 for this data and compare your answer to the answer from part (a). How would this change affect your hypothesis test?

c) Suppose 1000 members of the union are sampled, 10 times as many as in part (a), and that their responses are still in exactly the same proportions as in part (a):

Question 1

	Yes	No	Total
Yes	220	480	700
No	120	180	300
Total	340	660	1000

(row label: Question 2)

Calculate χ^2 for this data and compare your answer to the answer from part (a). How would this change affect your hypothesis test?

d) What happens to the statistic χ^2 as sample proportions stay exactly the same but sample size increases? Does this indicate that you should be cautious about the results of chi-square tests based on large samples? Justify your answer.

≡ **11.38** Suppose that you are given the following two contingency tables whose cells are in exactly the same proportions:

			Total
	a	b	$a+b$
	c	d	$c+d$
Total	$a+c$	$b+d$	$a+b+c+d$

			Total
	ma	mb	$ma+mb$
	mc	md	$mc+md$
Total	$ma+mc$	$mb+md$	$ma+mb+mc+md$

a) Show that if χ_1^2 is the χ^2-statistic for the first table and χ_2^2 is the χ^2-statistic for the second table, then

$$\chi_2^2 = m\chi_1^2$$

b) Prove that a similar result holds for a contingency table with r rows and c columns.

≡ **11.39** The following contingency table provides a joint frequency distribution for the population of all males on active military duty by classification and race. Data are in thousands. [SOURCE: U.S. Department of Defense, unpublished data.]

Race

	White	Black	Other	Total
Officer	2,401	145	176	2,722
Enlisted	11,085	3,392	1,944	16,421
Total	13,486	3,537	2,120	19,143

(row label: Class)

a) Suppose that a male on active duty is selected at random. Obtain the probability of each of the following events:

$$W = \text{event a white man is selected}$$
$$A = \text{event an officer is selected}$$
$$(W \& A) = \text{event a white officer is selected}$$

b) Determine whether the events "white" and "officer" are statistically independent by employing the special multiplication rule; that is, by checking whether $P(W \& A) = P(W)P(A)$.

c) Determine whether the observed and expected frequencies for the cell in the upper left-hand corner of the contingency table are equal.

d) Prove that the checks done in parts (b) and (c) are equivalent.

e) Are the characteristics, classification and race, statistically independent for males on active military duty? Explain.

PRINTOUT 11.3   Minitab output for Exercise 11.33

Expected counts are printed below observed counts

	C1	C2	C3	C4	Total
1	48	51	34	13	146
	37.02	59.51	37.71	11.76	
2	37	75	43	9	164
	41.58	66.84	42.36	13.21	
3	22	46	32	12	112
	28.40	45.65	28.93	9.02	
Total	107	172	109	34	422

ChiSq = 3.257 + 1.216 + 0.365 + 0.130 +
        0.505 + 0.995 + 0.010 + 1.343 +
        1.441 + 0.003 + 0.326 + 0.982 = 10.574
df = 6

PRINTOUT 11.4   Minitab output for Exercise 11.34

Expected counts are printed below observed counts

	C1	C2	C3	Total
1	173	267	55	495
	247.33	217.88	29.79	
2	168	130	19	317
	158.39	139.53	19.08	
3	160	144	9	313
	156.39	137.77	18.84	
4	213	88	3	304
	151.89	133.81	18.30	
Total	714	629	86	1429

ChiSq = 22.337 + 11.072 + 21.334 +
         0.583 + 0.651 + 0.000 +
         0.083 + 0.281 + 5.137 +
        24.583 + 15.684 + 12.787 = 114.534
df = 6

11.4   Inferences for a population standard deviation

Recall that the standard deviation, σ, of a population is a measure of dispersion or variability of the population values. A population with a great deal of variation will have a large standard deviation, while one with little variation will have a small standard deviation. See Figure 11.10.

FIGURE 11.10

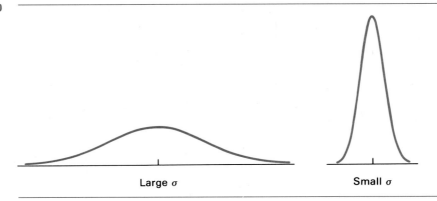

Large σ Small σ

Suppose that we want to obtain information about the standard deviation, σ, of a population. If the population is small, then we can ordinarily determine σ exactly by first taking a census and then computing σ from the population data. In most cases, however, the population will be large and, consequently, taking a census will usually not be feasible. For these situations, we must employ inferential methods to obtain the required information about σ.

In this section, we will learn how to perform hypothesis tests and how to find confidence intervals for a population standard deviation, σ. We will assume that the population under consideration is *normally distributed*.

EXAMPLE 11.10   *Introduces hypothesis tests for a population standard deviation*

A hardware manufacturer produces 10-millimeter bolts. The manufacturer knows that the diameters of the bolts produced vary somewhat from 10 mm and also from each other. But even if he is willing to accept some variation in bolt diameters, he cannot tolerate too much variation. For, if the variation is too large, then too many of the bolts produced will be unusable.

Thus, the manufacturer must make sure that the standard deviation, σ, of the bolt diameters is not unduly large. Since, in this case, it is not possible to obtain the value of σ exactly (why?), inferential methods must be employed.

Let us suppose it has been determined that an acceptable standard deviation for the bolt diameters is one that is less than 0.09 mm.[†] Knowing this, the manu-

---

[†] See Exercise 11.55 for an explanation of how that information could be obtained.

facturer can decide whether or not there is too much variation in the diameters of the bolts being produced by performing the hypothesis test

$$H_0: \sigma = 0.09 \text{ mm (too much variation)}$$
$$H_a: \sigma < 0.09 \text{ mm (not too much variation)}$$

If the null hypothesis can be rejected, then this will provide evidence that the variation in bolt diameters is acceptable.

The basic idea for carrying out the hypothesis test is as follows:

1. Take a random sample of bolts.
2. Compute the standard deviation, s, of the diameters of the bolts sampled.
3. If s is too much smaller than 0.09 mm, reject the null hypothesis in favor of the alternative hypothesis; otherwise, do not reject the null hypothesis.

The manufacturer takes a random sample of 20 bolts and carefully measures their diameters. The results are displayed in Table 11.16.

TABLE 11.16
Diameters, in mm, of 20 randomly selected bolts

10.03	9.89	9.99	9.96	10.10
10.08	9.95	10.00	9.94	10.01
10.05	9.97	10.03	9.98	10.05
10.03	9.99	10.08	10.02	9.98

The sample standard deviation of the bolt diameters in Table 11.16 is

$$s = \sqrt{\frac{n(\Sigma x^2) - (\Sigma x)^2}{n(n-1)}} = \sqrt{\frac{20(2002.6527) - (200.13)^2}{20 \cdot 19}} = 0.052 \text{ mm}$$

Is this value of s too much smaller than 0.09 mm, so that the null hypothesis should be rejected, or can the difference between $s = 0.052$ mm and the null hypothesis value of $\sigma = 0.09$ mm be attributed to sampling error? To answer this question, we need to know the probability distribution of s—the **sampling distribution of the standard deviation.** We will therefore first discuss that sampling distribution and then return to complete the hypothesis test posed in this example. ∎

THE SAMPLING DISTRIBUTION OF THE STANDARD DEVIATION

Recall that in order to perform a hypothesis test for the mean, μ, of a normally distributed population, we use the random variable

$$t = \frac{\bar{x} - \mu_0}{s/\sqrt{n}}$$

as the test statistic and not simply the random variable $\bar{x}$. Similarly, when performing a hypothesis test for the standard deviation, σ, of a normally distributed population, we do not employ the random variable s as the test statistic. Rather, we use the random variable

$$\frac{n-1}{\sigma_0^2} s^2$$

That random variable has a familiar probability distribution.

KEY FACT 11.4   The sampling distribution of the standard deviation[†]

Suppose that a random sample of size n is to be taken from a normally distributed population with standard deviation σ. Then the random variable

$$\chi^2 = \frac{n-1}{\sigma^2} s^2$$

has the chi-square distribution with $n-1$ degrees of freedom. In other words, probabilities for that random variable are equal to areas under the χ^2-curve with df $= n - 1$.

HYPOTHESIS TESTS FOR A POPULATION STANDARD DEVIATION

Now that we know the sampling distribution of the standard deviation, we can state a step-by-step method for performing a hypothesis test for a population standard deviation. The method is presented here as Procedure 11.3.

PROCEDURE 11.3

To perform a hypothesis test for a population standard deviation with null hypothesis H_0: $\sigma = \sigma_0$.

ASSUMPTION

Normal population.

STEP 1 *State the null and alternative hypotheses.*
STEP 2 *Decide on the significance level, α.*
STEP 3 *The critical value(s)*
   a) *for a two-tailed test are $\chi^2_{1-\alpha/2}$ and $\chi^2_{\alpha/2}$,*
   b) *for a left-tailed test is $\chi^2_{1-\alpha}$,*
   c) *for a right-tailed test is χ^2_{α},*
   with df $= n - 1$. Use Table IV to find the critical value(s).

[†] Strictly speaking, the sampling distribution given here is not the sampling distribution of the standard deviation but is the sampling distribution of a function of the standard deviation.

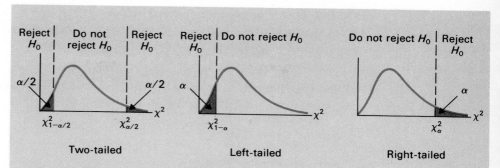

STEP 4 *Compute the value of the test statistic*

$$\chi^2 = \frac{n-1}{\sigma_0^2} s^2$$

STEP 5 *If the value of the test statistic falls in the rejection region, then reject H_0; otherwise, do not reject H_0.*

STEP 6 *State the conclusion in words.*

EXAMPLE 11.11   *Illustrates Procedure 11.3*

We can now complete the hypothesis test proposed in Example 11.10. Recall that a hardware manufacturer needs to decide whether the standard deviation, σ, of bolt diameters is less than 0.09 mm. He randomly samples 20 bolts and measures their diameters. The results are shown in Table 11.17.

TABLE 11.17
Diameters, in mm, of 20 randomly selected bolts

10.03	9.89	9.99	9.96	10.10
10.08	9.95	10.00	9.94	10.01
10.05	9.97	10.03	9.98	10.05
10.03	9.99	10.08	10.02	9.98

At the 5% significance level, do the data provide sufficient evidence to conclude that the standard deviation, σ, of the diameters of all 10-millimeter bolts produced by the manufacturer is less than 0.09 mm? [Assume that the population of diameters of all bolts produced is normally distributed.]

SOLUTION   We apply Procedure 11.3.

STEP 1 *State the null and alternative hypotheses.*

The null and alternative hypotheses are

$$H_0: \sigma = 0.09 \text{ mm (too much variation)}$$
$$H_a: \sigma < 0.09 \text{ mm (not too much variation)}$$

Note that the hypothesis test is left-tailed since there is a less-than sign ($<$) in the alternative hypothesis.

STEP 2 *Decide on the significance level, α.*

The test is to be performed at the 5% level of significance. Thus, $\alpha = 0.05$.

STEP 3 *The critical value for a left-tailed test is $\chi^2_{1-\alpha}$, with df $= n - 1$.*

We have $\alpha = 0.05$. Also, $n = 20$ and so df $= 20 - 1 = 19$. Consulting Table IV, we find that the critical value is $\chi^2_{1-0.05} = \chi^2_{0.95} = \mathbf{10.117}$. See Figure 11.11.

FIGURE 11.11

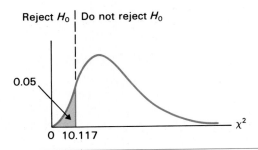

STEP 4 *Compute the value of the test statistic*

$$\chi^2 = \frac{n-1}{\sigma_0^2} s^2$$

First we obtain the sample variance, s^2. From Table 11.17, we find that

$$s^2 = \frac{n(\Sigma x^2) - (\Sigma x)^2}{n(n-1)} = \frac{20(2002.6527) - (200.13)^2}{20 \cdot 19} = 0.0027$$

Therefore, since $n = 20$ and $\sigma_0 = 0.09$, the value of the test statistic is

$$\chi^2 = \frac{n-1}{\sigma_0^2} s^2 = \frac{20-1}{(0.09)^2} \cdot 0.0027 = \mathbf{6.333}$$

STEP 5 *If the value of the test statistic falls in the rejection region, reject H_0; otherwise, do not reject H_0.*

From Step 4, the value of the test statistic is $\chi^2 = 6.333$. This falls in the rejection region, as we see by referring to Figure 11.11. Thus, we **reject H_0.**

STEP 6 *State the conclusion in words.*

The data provide sufficient evidence to conclude that the standard deviation, σ, of the diameters of all 10-millimeter bolts produced by the manufacturer is less than 0.09 mm. So, the variation in bolt diameters is not unduly large. ■

MTB

CONFIDENCE INTERVALS FOR A POPULATION STANDARD DEVIATION

Using Key Fact 11.4 on page 574, we can also obtain the following confidence-interval procedure for a population standard deviation.

PROCEDURE 11.4

To find a confidence interval for a population standard deviation, σ.

ASSUMPTION

Normal population.

STEP 1 *For a confidence level of $1 - \alpha$, use Table IV to find $\chi^2_{1-\alpha/2}$ and $\chi^2_{\alpha/2}$ with df $= n - 1$.*

STEP 2 *The confidence interval for σ is from*

$$\sqrt{\frac{n-1}{\chi^2_{\alpha/2}}} \cdot s \quad to \quad \sqrt{\frac{n-1}{\chi^2_{1-\alpha/2}}} \cdot s$$

where $\chi^2_{1-\alpha/2}$ and $\chi^2_{\alpha/2}$ are found in Step 1, n is the sample size, and s is computed from the actual sample data obtained.

EXAMPLE 11.12   **Illustrates Procedure 11.4**

Refer to Example 11.11. Use the sample data in Table 11.17 on page 575 to determine a 95% confidence interval for the standard deviation, σ, of the diameters of all 10-millimeter bolts produced by the manufacturer.

SOLUTION   We employ Procedure 11.4.

STEP 1 *For a confidence level of $1 - \alpha$, use Table IV to find $\chi^2_{1-\alpha/2}$ and $\chi^2_{\alpha/2}$ with df $= n - 1$.*

We want a 95% confidence interval, so the confidence level is $0.95 = 1 - 0.05$. This means that $\alpha = 0.05$. Also, since $n = 20$, df $= 20 - 1 = 19$. Referring to Table IV, we find that

$$\chi^2_{1-\alpha/2} = \chi^2_{1-0.05/2} = \chi^2_{0.975} = \boxed{8.907}$$

and

$$\chi^2_{\alpha/2} = \chi^2_{0.05/2} = \chi^2_{0.025} = \boxed{32.852}$$

STEP 2 *The confidence interval for σ is from*

$$\sqrt{\frac{n-1}{\chi^2_{\alpha/2}}} \cdot s \quad to \quad \sqrt{\frac{n-1}{\chi^2_{1-\alpha/2}}} \cdot s$$

We have $n = 20$ and, from Step 1, $\chi^2_{\alpha/2} = 32.852$ and $\chi^2_{1-\alpha/2} = 8.907$. Also, we found in Example 11.10 that $s = 0.052$ mm. Consequently, a 95% confidence interval for σ is from

$$\sqrt{\frac{20-1}{32.852}} \cdot 0.052 \quad to \quad \sqrt{\frac{20-1}{8.907}} \cdot 0.052$$

or

$$0.040 \quad \text{to} \quad 0.076$$

In other words, we can be 95% confident that the standard deviation, σ, of the diameters of all 10-millimeter bolts produced by the manufacturer is somewhere between 0.040 mm and 0.076 mm. ■

MTB

Exercises 11.4

___ 11.40   What does σ measure?

In each of Exercises 11.41–11.46, use Procedure 11.3 on page 574 to perform the required hypothesis test. Assume that, in each exercise, the population under consideration is normally distributed.

___ 11.41   Each year, thousands of high-school students bound for college take the Scholastic Aptitude Test (SAT). The test, provided by the Educational Testing Service of Princeton, New Jersey, measures the verbal and mathematical abilities of prospective college students. Student scores are reported on a scale that ranges from a low of 200 to a high of 800. This scale was introduced in 1941. At that time, the standard deviation of scores was 100 points. A random sample of 25 verbal scores for this year yields the following data:

560	405	367	416	540
347	370	629	570	372
396	439	526	610	339
569	314	228	483	353
379	558	432	648	475

At the 5% significance level, do the data suggest that the standard deviation, σ, of this year's verbal scores is different from the 1941 standard deviation of 100? (*Note: $\Sigma x = 11{,}325$ and $\Sigma x^2 = 5{,}425{,}835$.*)

___ 11.42   A company produces cans of stewed tomatoes with an advertised weight of 14 ounces. Recently the company hired a quality-control engineer. The engineer has performed some statistical analyses and is satisfied that, on the average, the cans do contain 14 ounces of stewed tomatoes. However, she must also be concerned about the variability of the weights. Specifically, in order to ensure that the vast majority of the cans have a weight within 10% of the advertised weight, it is required that the standard deviation, σ,

of the weights be less than 0.45 oz. The weights, in ounces, of 10 randomly selected cans are as follows:

13.85	13.95	13.90	13.49	14.17
14.33	14.03	13.48	14.27	14.19

Do the data provide sufficient evidence to conclude that the standard deviation of the weights is less than 0.45 oz? Perform the hypothesis test using $\alpha = 0.05$. (*Note: $\Sigma x = 139.66$ and $\Sigma x^2 = 1951.2932$.*)

___ 11.43   A manufacturer of watches claims that the weekly error made by the watches she produces has a standard deviation of about one second. To test this claim, a random sample of 20 watches are set to the correct time. After one week, the error made by each watch is recorded. The results, in seconds, are:

0.6	2.3	2.0	−2.1	−1.4
−0.5	1.5	−0.3	0.4	0.6
0.4	−2.2	0.7	0.5	−1.3
−2.0	2.6	−0.8	1.0	−0.6

Does it appear that the standard deviation, σ, of the weekly errors exceeds the one-second claim made by the manufacturer? Use $\alpha = 0.01$. (*Note: $\Sigma x = 1.4$ and $\Sigma x^2 = 39.32$.*)

___ 11.44   The designer of a 100-point aptitude test has attempted to construct the test so that the standard deviation of scores is 10 points. For a preliminary documentation, 30 randomly selected people are given the test. Their scores are the following:

83	43	29	37	64	69
79	67	85	61	52	55
70	65	43	64	60	35
76	46	35	43	51	52
61	50	41	87	62	59

Do the data indicate that the standard deviation of all scores on the aptitude test is not 10 points? Use $\alpha = 0.10$. (*Note:* $\Sigma x = 1724$ and $\Sigma x^2 = 106{,}172$.)

___ **11.45** A coffee machine is supposed to dispense six fluid ounces of coffee into a paper cup. In reality, the amounts dispensed vary from cup to cup. However, if the machine is working properly, then the vast majority of cups will contain within 10% of the advertised six fluid ounces. This means that the standard deviation of the amounts dispensed should be less than 0.2 fluid ounces. A random sample of 15 cups provided the following data (in fluid ounces):

6.2	6.0	5.8	5.9	6.3
6.1	6.2	6.0	6.1	6.0
6.1	6.0	6.1	6.0	5.9

At the 5% significance level, do the data provide sufficient evidence to conclude that the standard deviation, σ, of the amounts being dispensed is less than 0.2 fluid ounces?

___ **11.46** Mileage estimates for cars and light-duty trucks are determined and published by the Environmental Protection Agency (EPA). According to the EPA, "...the mileages obtained by most drivers will be within plus or minus 15 percent of the [EPA] estimates ...." The mileage estimate given for one 1990 model is 23 mpg on the highway. If the claim made in quotes by the EPA is true, then the standard deviation of mileages should be about $0.15 \cdot 23/3 = 1.15$ mpg. Suppose that a random sample of 12 cars of this model yields the following highway mileages:

24.1	23.3	22.5	23.2
22.3	21.1	21.4	23.4
23.5	22.8	24.5	24.3

At the 5% significance level, do the data suggest that the standard deviation, σ, of highway mileages for all 1990 cars of this model is different from 1.15 mpg?

In each of Exercises 11.47–11.52, use Procedure 11.4 on page 577 to obtain the required confidence interval.

___ **11.47** Refer to Exercise 11.41. Obtain a 95% confidence interval for the standard deviation, σ, of this year's verbal SAT scores.

___ **11.48** Refer to Exercise 11.42. Find a 90% confidence interval for the standard deviation, σ, of the weights of all cans of stewed tomatoes produced by the company.

___ **11.49** Refer to Exercise 11.43. Obtain a 98% confidence interval for the standard deviation, σ, of the weekly errors of the watches manufactured.

___ **11.50** Refer to Exercise 11.44. Obtain a 90% confidence interval for the standard deviation, σ, of all scores on the aptitude test.

___ **11.51** Refer to Exercise 11.45. Find a 90% confidence interval for the standard deviation, σ, of the amounts of coffee being dispensed.

___ **11.52** Refer to Exercise 11.46. Find a 95% confidence interval for the standard deviation, σ, of highway gas mileages for all 1990 cars of the given model.

___ **11.53** Refer to Exercise 11.45. Why is it important that the standard deviation, σ, of the amounts of coffee being dispensed not be too large?

___ **11.54** Refer to Exercise 11.46. Why is it useful to know the standard deviation of the gas mileages as well as the mean gas mileage?

═══ **11.55** In the bolt-manufacturer problem of Example 11.10, we assumed it had been determined that an acceptable standard deviation, σ, for the bolt diameters is one that is less than 0.09 mm. We will now see how such information might be obtained. Let us suppose the manufacturer has set the tolerance specifications for the 10-millimeter bolts at ±0.3 mm. In other words, a bolt is considered satisfactory if its diameter is between 9.7 and 10.3 mm. Further suppose the manufacturer has decided that less than 0.1% (one out of a thousand) of the bolts produced should be defective.

a) Let x denote the diameter of a randomly selected bolt. Show that the manufacturer's production criteria can be expressed as

$$P(9.7 \leq x \leq 10.3) > 0.999$$

b) Draw a normal-curve picture that illustrates the equation $P(9.7 \leq x \leq 10.3) = 0.999$. Include an x-axis and a z-axis. [Assume $\mu = 10$ mm.]

c) Deduce from your picture in part (b) that the manufacturer's production criteria are equivalent to the condition that

$$\frac{0.3}{\sigma} > z_{0.0005}$$

d) Use part (c) to conclude that the manufacturer's production criteria are equivalent to requiring that the standard deviation of bolt diameters be less than 0.09 mm, that is, $\sigma < 0.09$ mm.

≡ **11.56** The purpose of this exercise is to justify Procedure 11.4 on page 577.

a) Use Key Fact 11.4 on page 574 to show that

$$P\left(\chi_{1-\alpha/2}^2 < \frac{n-1}{\sigma^2} s^2 < \chi_{\alpha/2}^2\right) = 1 - \alpha$$

b) Deduce from part (a) that

$$P\left(\frac{n-1}{\chi_{\alpha/2}^2} \cdot s^2 < \sigma^2 < \frac{n-1}{\chi_{1-\alpha/2}^2} \cdot s^2\right) = 1 - \alpha$$

c) Prove that the result in part (b) implies that once the sample is taken, the interval from

$$\sqrt{\frac{n-1}{\chi_{\alpha/2}^2}} \cdot s \quad \text{to} \quad \sqrt{\frac{n-1}{\chi_{1-\alpha/2}^2}} \cdot s$$

will be a $(1 - \alpha)$-level confidence interval for the population standard deviation, σ. [This last formula is the confidence-interval formula given in Step 2 of Procedure 11.4.]

Chapter review

KEY TERMS   associated, 555
χ_α^2, 541
chi-square (χ^2) curve, 540
chi-square distribution, 540
chi-square procedures, 539
chi-square goodness-of-fit test, 544
chi-square independence test, 554
CHISQUARE,* 564

expected frequencies, 546
nonassociated, 554
observed frequencies, 546
sampling distribution of the standard deviation, 574
skewed, 540
statistically dependent, 555
statistically independent, 554

FORMULAS   In the formulas below,

O = observed frequency
E = expected frequency
k = number of categories in a probability or percentage distribution
p = probability or proportion
n = sample size
R = row total in a contingency table

C = column total in a contingency table
r = number of rows in a contingency table
c = number of columns in a contingency table
σ = population standard deviation
s = sample standard deviation

Expected frequencies for a chi-square goodness-of-fit test, 547

$$E = np$$

Test statistic for a chi-square goodness-of-fit test, 548

$$\chi^2 = \Sigma(O - E)^2/E$$

with df $= k - 1$.

Expected frequencies for a chi-square independence test, 560

$$E = \frac{R \cdot C}{n}$$

Test statistic for a chi-square independence test, 560

$$\chi^2 = \Sigma(O - E)^2/E$$

with df $= (r - 1)(c - 1)$.

Test statistic for H_0: $\sigma = \sigma_0$ (normal population), 575

$$\chi^2 = \frac{n - 1}{\sigma_0^2} s^2$$

with df $= n - 1$.

Confidence interval for σ (normal population), 577

$$\sqrt{\frac{n-1}{\chi_{\alpha/2}^2}} \cdot s \quad \text{to} \quad \sqrt{\frac{n-1}{\chi_{1-\alpha/2}^2}} \cdot s$$

with df $= n - 1$.

YOU SHOULD
BE ABLE TO

1. use and understand the preceding formulas.
2. use the chi-square table, Table IV.
3. explain the reasoning behind the chi-square goodness-of-fit test.
4. perform a chi-square goodness-of-fit test for the percentage distribution of a population or the probability distribution of a random variable.
5. explain the reasoning behind the chi-square independence test.
6. perform a chi-square independence test to decide whether two characteristics of a population are statistically dependent.
7. perform a hypothesis test for the standard deviation, σ, of a normally distributed population.
8. obtain a confidence interval for the standard deviation, σ, of a normally distributed population.
9. use the Minitab commands covered in this chapter.*
10. interpret the output obtained from the application of the Minitab commands discussed in this chapter.*

REVIEW TEST

1. Consider a χ^2-curve with 17 degrees of freedom. Use Table IV to determine
   a) $\chi_{0.99}^2$ b) $\chi_{0.01}^2$
   c) the χ^2-value with area 0.05 to its right.
   d) the χ^2-value with area 0.05 to its left.
   e) the two χ^2-values that divide the area under the curve into a middle 0.95 area and two outside 0.025 areas.

2. The Bureau of the Census and the Department of Housing and Urban Development publish data on characteristics of new, privately-owned, one-family homes in the document *Characteristics of New Housing*. A percentage distribution for the number of bedrooms in homes completed in 1987 is as given in the table shown at the top of the first column on the next page.

No. bedrooms	Percent
2 or less	19
3	58
4 or more	23

A project called *Homestyle 1988*, directed by the Impulse Research Corporation of Santa Monica, California, was designed to provide builders with a profile of the next generation of home buyers in the three southwestern states of Arizona, California, and Nevada. Researchers conducted telephone interviews with 150 randomly selected potential home buyers between the ages of 24 and 35. They obtained the data on preferences for the number of bedrooms as shown in the following frequency distribution:

No. bedrooms	Frequency
2 or less	47
3	93
4 or more	10

Do the data provide sufficient evidence to conclude that the actual preference distribution for the number of bedrooms is different from the distribution for the number of bedrooms in homes completed in 1987? Perform the hypothesis test at the 5% significance level.

3. In a 1985 poll conducted by the Gallup Organization, 1528 randomly selected adults were asked the following question: "The New Jersey Supreme Court recently ruled that all life-sustaining medical treatment may be withheld or withdrawn from terminally ill patients, provided that is what the patients want or would want if they were able to express their wishes. Would you like to see such a ruling in the state in which you live, or not?" In the contingency table displayed at the top of the next column, we have presented a modified version of the sample results which cross-classify response by educational level. Do the data provide sufficient evidence to conclude that response and educational level are statistically dependent? Perform the required hypothesis test using a significance level of 0.01.

Response

	Favor	Oppose	No opinion	Total
College grad	264	17	6	287
Some college	205	26	7	238
High school grad	461	81	34	576
Non high school grad	290	81	56	427
Total	1220	205	103	1528

Educational level (row label)

*4. (**Computer problem**) Refer to the contingency table displayed in Problem 3 (at the top of this column). Suppose the cell data in the columns headed "Favor," "Oppose," and "No opinion" are stored in columns named FAVOR, OPPOSE, and NOOPIN, respectively.
   a) Which Minitab command and subcommands (if any) should be used to perform the hypothesis test considered in Problem 3?
   b) If you have access to Minitab, use it to carry out the hypothesis test.

*5. (**Computer problem**) Printout 11.5 at the bottom of the next page shows the Minitab output obtained by applying the command CHISQUARE to the cell data in the columns of the contingency table in Problem 3. Employ the output in that printout to determine
   a) the number of people sampled who would be opposed to such a ruling in their state.
   b) the number of people sampled who had some college. (*Note:* The row labeled "2" in Printout 11.5 corresponds to the "Some college" category, as we can see from the contingency table in Problem 3.)
   c) the observed number of people sampled who had some college and would be opposed to such a ruling in their state.
   d) the expected number of people sampled who had some college and would be opposed to such a ruling in their state, if response and educational level are statistically independent.

e) the chi-square subtotal, $(O - E)^2/E$, for the "Some college *and* Oppose" cell.

f) the number of degrees of freedom.

g) the value of the test statistic, χ^2.

h) the conclusion, if the hypothesis test concerning statistical independence of response and educational level is performed at the 1% significance level. (*Note:* You will need to consult Table IV.)

6. IQs measured on the Stanford Revision of the *Binet-Simon Intelligence Scale* are supposed to have a standard deviation of 16 points. Suppose that 25 randomly selected persons are given an IQ test and that the results are as displayed in the table at the top of the next column.

91	96	106	116	97
102	96	124	115	121
95	111	105	101	86
88	129	112	82	98
104	118	127	66	102

a) At the 10% significance level, do the data provide sufficient evidence to conclude that the standard deviation of all IQs is not equal to 16? (*Note:* $\Sigma x = 2588$ and $\Sigma x^2 = 273{,}314$.)

b) What assumption must you make about IQs in order to perform the hypothesis test that you did in part (a)?

7. Refer to Problem 6. Determine a 90% confidence interval for the standard deviation of all IQ scores.

PRINTOUT 11.5 Minitab output for Problem 5

Expected counts are printed below observed counts

	FAVOR	OPPOSE	NOOPIN	Total
1	264	17	6	287
	229.15	38.50	19.35	
2	205	26	7	238
	190.03	31.93	16.04	
3	461	81	34	576
	459.90	77.28	38.83	
4	290	81	56	427
	340.93	57.29	28.78	
Total	1220	205	103	1528

ChiSq = 5.300 + 12.010 + 9.207 +
        1.180 + 1.102 + 5.097 +
        0.003 + 0.179 + 0.600 +
        7.608 + 9.815 + 25.735 = 77.837

df = 6

CHAPTER 12

DESCRIPTIVE METHODS IN LINEAR REGRESSION

It is frequently of interest to know whether two or more variables are related and, if so, how they are related. For instance, is there a relationship between college GPA and SAT scores? If these variables are related, how are they related? The president of a large corporation knows there is a tendency for sales to increase as advertising expenditures increase. But how strong is that tendency and how can she predict the approximate sales that will result from various advertising expenditures?

Some commonly used methods for examining the relationship between two or more variables and for making predictions are provided by *linear regression and correlation*. Descriptive methods in linear regression and correlation will be discussed in this chapter. Inferential methods in linear regression and correlation will be discussed in the next chapter.

CHAPTER OUTLINE

12.1 Linear equations with one independent variable Reviews linear equations with one independent variable and the graphs of such equations.

12.2 The regression equation Explains how to determine the best-fitting line to a set of data points and how to use that line for making predictions.

12.3 The coefficient of determination Examines the coefficient of determination, a descriptive measure that indicates how useful the best-fitting line is for making predictions.

12.4 Linear correlation Discusses the linear correlation coefficient, a descriptive measure that gives the strength of the linear relationship between two variables.

12.5 Multiple regression (Optional) Introduces descriptive methods in regression that apply when more than one predictor variable is used.

12.1   Linear equations with one independent variable

In order to study linear regression, we first need to review linear equations with one independent variable. The general form of a **linear equation** with one independent variable is

$$y = b_0 + b_1 x$$

where b_0 and b_1 are fixed numbers, x is the *independent variable,* and y is the *dependent variable.*[†] The graph of a linear equation with one independent variable is a **straight line;** moreover, any nonvertical straight line can be described by such an equation.

Three examples of linear equations with one independent variable are

$$y = 4 + 0.2x, \qquad y = -1.5 - 2x, \qquad \text{and} \qquad y = -3.4 + 1.8x.$$

Figure 12.1 shows the straight-line graphs of these three linear equations.

FIGURE 12.1

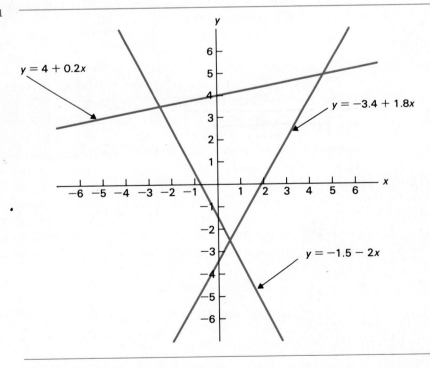

[†] You may be familiar with the form $y = mx + b$ instead of the form $y = b_0 + b_1 x$. In statistics, the latter form is preferred because it allows a smoother transition to the topic of multiple regression in which there is more than one independent variable.

Linear equations with one independent variable occur frequently in applications of mathematics to many different subject areas. These subject areas include the management, life, and social sciences, as well as the physical and mathematical sciences. In Examples 12.1 and 12.2, we will illustrate the use of linear equations in a simple business application.

EXAMPLE 12.1   *Illustrates linear equations*

CJ^2 Business Services does word processing as one of its basic functions. Its rate is $20/hr plus a $25 disk charge. The total cost to a customer depends, of course, on the number of hours it takes to complete the job. Find the equation that expresses the total cost in terms of the number of hours required to complete the job.

SOLUTION   Let

$$y = \text{total cost} = (\textit{dependent variable})$$

and

$$x = \text{number of hours required} \ (\textit{independent variable})$$

Since the rate for word processing is $20/hr, a job that takes x hours will cost $20x$ plus the $25 disk charge. Thus, the total cost, y, of a job that takes x hours is

$$y = 25 + 20x$$

Note that the equation for the total cost of a word-processing job is a *linear equation*. Here $b_0 = 25$ and $b_1 = 20$. Using the equation, we can determine the exact cost for a job once we know the number of hours required. For instance, a job that takes five hours will cost

$$y = 25 + 20 \cdot 5 = \$125$$

and a job that takes 7.5 hours will cost

$$y = 25 + 20 \cdot 7.5 = \$175$$

In Table 12.1, we have displayed these and a few more cost illustrations.

TABLE 12.1

Time (hrs) x	Cost ($) y
5.0	125
7.5	175
15.0	325
20.0	425
22.5	475

As we have already mentioned, a linear equation, such as $y = 25 + 20x$, has a straight-line graph. We can obtain the graph of $y = 25 + 20x$ by plotting the points in Table 12.1 and connecting them with a straight line. This is done in Figure 12.2.

FIGURE 12.2
Graph of
$y = 25 + 20x$
obtained from
points in Table 12.1

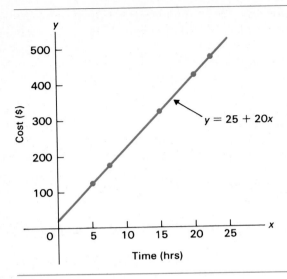

The graph is useful to estimate the costs for various processing times. For instance, a quick glance at the graph shows that a 10-hour job will cost somewhere between $200 and $300. The exact cost is $y = 25 + 20 \cdot 10 = \$225$.

INTERCEPT AND SLOPE

For a linear equation, $y = b_0 + b_1 x$, the numbers b_0 and b_1 have an important geometric interpretation. The number b_0 gives the y-value at which the straight-line graph of the linear equation intersects the y-axis. The number b_1 measures the steepness of the straight line. More precisely, b_1 indicates how much the y-value on the straight line increases (or decreases) when the x-value increases by one unit. See Figure 12.3.

FIGURE 12.3
Graph of
$y = b_0 + b_1 x$

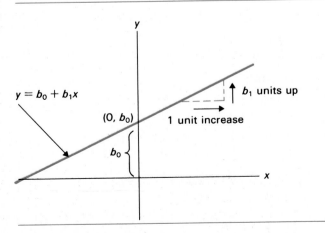

Because of the geometric interpretation of the numbers b_0 and b_1, they are given special names that reflect that interpretation.

DEFINITION 12.1   *y*-intercept and slope

For a linear equation, $y = b_0 + b_1 x$, the number b_0 is called the *y-intercept* and the number b_1 is called the *slope*.

EXAMPLE 12.2   *Illustrates y-intercept and slope*   D ✓

In Example 12.1, we obtained the linear equation that expresses the total cost, y, of a word-processing job in terms of the number of hours, x, required to complete the job. The equation is $y = 25 + 20x$.
a) Find the y-intercept and slope of that linear equation.
b) Interpret the y-intercept and slope in terms of the graph of the equation.
c) Interpret the y-intercept and slope in terms of word-processing costs.

SOLUTION   a) The y-intercept for the equation is $b_0 = 25$ and the slope is $b_1 = 20$.
b) The y-intercept, $b_0 = 25$, gives the y-value at which the straight line, $y = 25 + 20x$, intersects the y-axis. The slope, $b_1 = 20$, indicates that the y-value increases by 20 units for every increase in x of one unit. See Figure 12.4.

FIGURE 12.4
Graph of
$y = 25 + 20x$

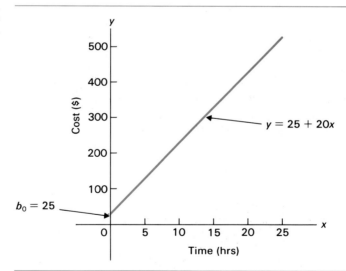

c) In terms of word-processing costs, the y-intercept, $b_0 = 25$, represents the total cost of a job that takes zero hours. In other words, the y-intercept of $25 is a fixed cost that is always there no matter how many hours the job takes. The slope, $b_1 = 20$, represents the fact that the cost per hour is $20. It is the amount that the total cost, y, goes up for every increase of one hour in the time, x, required to complete the job. ■

A straight line is determined by any two distinct points that lie on the line. This means that the straight-line graph of a linear equation, $y = b_0 + b_1x$, can be obtained by first substituting two different x-values into the equation in order to get two distinct points and then connecting those two points with a straight line.

For example, to graph the linear equation, $y = 5 - 3x$, we can use the x-values, $x = 1$ and $x = 3$ (or any other two x-values). The y-values corresponding to those two x-values are $y = 5 - 3 \cdot 1 = 2$ and $y = 5 - 3 \cdot 3 = -4$, respectively. Hence, the graph of the linear equation, $y = 5 - 3x$, is the straight line that passes through the two points $(1, 2)$ and $(3, -4)$. See Figure 12.5.

FIGURE 12.5
Graph of
$y = 5 - 3x$

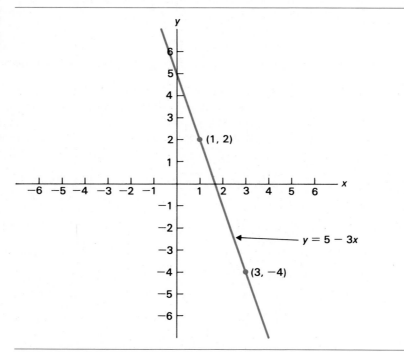

Note that the line in Figure 12.5 slopes downward, the y-values decrease as x increases. This is because the slope of the line is negative: $b_1 = -3 < 0$. Now look back at the line in Figure 12.4 on page 589. It is the graph of the linear equation $y = 25 + 20x$. That line slopes upward, the y-values increase as x increases. This is because the slope of the line is positive: $b_1 = 20 > 0$. In general, we have the following fact:

KEY FACT 12.1   Graphical interpretation of slope

The straight-line graph of the linear equation, $y = b_0 + b_1x$, slopes upward if $b_1 > 0$, slopes downward if $b_1 < 0$, and is horizontal if $b_1 = 0$. See Figure 12.6.

FIGURE 12.6

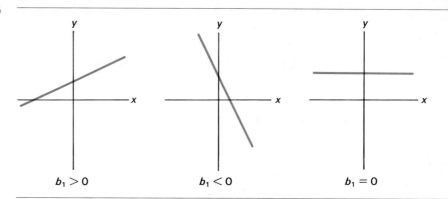

$b_1 > 0$ $b_1 < 0$ $b_1 = 0$

Exercises 12.1

___ **12.1** On January 8, 1990, one of the authors of this book phoned Avis Rent-A-Car to get the <u>rate for renting a midsize car</u> (a Buick Skylark). The quoted rate on that day was $41.88 per day plus $0.33 per mile. For a one-day rental, let x denote the number of miles driven and y denote the total cost.
a) Obtain the equation that expresses y in terms of x.
b) Find b_0 and b_1.
c) Construct a table similar to Table 12.1 on page 587 for the x-values 50, 100, and 250 miles.
d) Draw the graph of the equation in part (a) by plotting the points from part (c) and connecting them with a straight line.
e) Apply the graph from part (d) to visually estimate the cost of driving the car 150 miles. Then calculate that cost exactly using the equation from part (a).

___ **12.2** Encore Air Conditioning charges $36 per hour plus a $30 service charge. Let x denote the number of hours it takes for a given job and y denote the total cost to the customer.
a) Obtain the equation that expresses y in terms of x.
b) Find b_0 and b_1.
c) Construct a table similar to Table 12.1 on page 587 for the x-values 0.5, 1, and 2.25 hours.
d) Draw the graph of the equation in part (a) by plotting the points from part (c) and connecting them with a straight line.
e) Apply the graph from part (d) to visually estimate the cost of a job that takes 1.75 hours. Then calculate that cost exactly using the equation from part (a).

___ **12.3** The most commonly used scales for measuring temperature are the Fahrenheit and Celsius scales. If we let y denote Fahrenheit temperature and x denote Celsius temperature, then we can express the relationship between those two scales with the linear equation $y = 32 + 1.8x$.
a) Determine b_0 and b_1.
b) Find the Fahrenheit temperatures corresponding to the following Celsius temperatures: $-40°$, $0°$, $20°$, and $100°$.
c) Graph the linear equation, $y = 32 + 1.8x$, using the four points found in part (b).
d) Apply the graph obtained in part (c) to visually estimate the Fahrenheit temperature corresponding to a Celsius temperature of $28°$. Then calculate that temperature exactly by employing the linear equation $y = 32 + 1.8x$.

___ **12.4** A ball is thrown straight up in the air with an initial velocity of 64 feet per second. According to the laws of physics, if we let y denote the velocity of the ball after x seconds, then $y = 64 - 32x$. (The constant 32 is the acceleration due to gravity.)
a) Find b_0 and b_1 for this linear equation.
b) Determine the velocity of the ball after one, two, three, and four seconds.
c) Graph the linear equation, $y = 64 - 32x$, using the four points obtained in part (b).
d) Use the graph from part (c) to visually estimate the velocity of the ball after 1.5 seconds. Then calculate that velocity exactly by employing the linear equation $y = 64 - 32x$.

In each of Exercises 12.5–12.8,
a) *determine the y-intercept and slope of the given linear equation.*
b) *explain what the y-intercept and slope represent in terms of the graph of the equation.*
c) *explain what the y-intercept and slope represent in terms relating to the given application.*

___ **12.5** $y = 41.88 + 0.33x$ (from Exercise 12.1)

___ **12.6** $y = 30 + 36x$ (from Exercise 12.2)

___ **12.7** $y = 32 + 1.8x$ (from Exercise 12.3)

___ **12.8** $y = 64 - 32x$ (from Exercise 12.4)

In each of Exercises 12.9–12.18, you will be given a linear equation. For each exercise,
a) *find the y-intercept and slope.*
b) *determine whether the line slopes upward, slopes downward, or is horizontal, without graphing the equation.*
c) *graph the equation using two points.*

___ **12.9** $y = 3 + 4x$

___ **12.10** $y = -1 + 2x$

___ **12.11** $y = 6 - 7x$

___ **12.12** $y = -8 - 4x$

___ **12.13** $y = 0.5x - 2$

___ **12.14** $y = -0.75x - 5$

___ **12.15** $y = 2$

___ **12.16** $y = -3x$

___ **12.17** $y = 1.5x$

___ **12.18** $y = -3$

In each of Exercises 12.19–12.26, you will be given the y-intercept, b_0, and slope, b_1, of a straight line. For each exercise,
a) *determine whether the line slopes upward, slopes downward, or is horizontal, without graphing the equation.*
b) *find the equation of the line.*
c) *graph the equation using two points.*

___ **12.19** $b_0 = 5, b_1 = 2$

___ **12.20** $b_0 = -3, b_1 = 4$

___ **12.21** $b_0 = -2, b_1 = -3$

___ **12.22** $b_0 = 0.4, b_1 = 1$

___ **12.23** $b_0 = 0, b_1 = -0.5$

___ **12.24** $b_0 = -1.5, b_1 = 0$

___ **12.25** $b_0 = 3, b_1 = 0$

___ **12.26** $b_0 = 0, b_1 = 3$

= **12.27**   On page 586, we stated that any nonvertical straight line can be described by an equation of the form $y = b_0 + b_1 x$.
a) Why can't a vertical straight line be expressed in that form?
b) What is the form of the equation of a vertical straight line?
c) Does a vertical straight line have a slope? Explain.

12.2   The regression equation

In Examples 12.1 and 12.2, we discussed the linear equation, $y = 25 + 20x$, which gives the total cost, y, of a word processing job in terms of the time in hours, x, required to complete the job. Given the amount of time required, x, we can use the equation to determine the *exact* cost of the job, y.

Generally speaking, things are not quite as simple as they are in the word processing situation where one variable (cost) can be predicted exactly in terms of another variable (time required). More often than not, we must be content with rough predictions.

For instance, we cannot predict the exact price, y, of a Nissan Z by just knowing its age, x. Indeed, even for a fixed age, say three years old, the price of a Nissan Z varies from car to car. We must be satisfied with making a rough prediction for the price of a three-year-old Nissan Z or with an estimate of the mean price of all three-year-old Nissan Zs.

Table 12.2 displays data on age and price for a sample of Nissan Zs. The data were obtained from the *Asian Import* edition of the *Auto Trader* magazine. Ages are in years; prices are in hundreds of dollars, rounded to the nearest hundred dollars.

TABLE 12.2
Age versus price data for Nissan Zs

Car	Age (yrs) x	Price ($100s) y
1	5	85
2	4	103
3	6	70
4	5	82
5	5	89
6	5	98
7	6	66
8	6	95
9	2	169
10	7	70
11	7	48

It is useful to plot the data so that we can visualize any apparent relationships between age and price. Such a plot is called a **scatter diagram.** The scatter diagram for the data in Table 12.2 is depicted in Figure 12.7.

FIGURE 12.7
Scatter diagram for age versus price data from Table 12.2

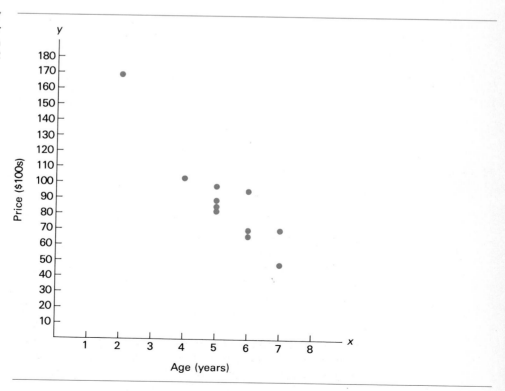

Although it is clear from the scatter diagram that the data points do not lie on a straight line, it appears that they are clustered about a straight line. We would like to fit a straight line to the data points. Then we could use that line to predict the price of a Nissan Z given its age.

Since it is possible to draw many straight lines through the cluster of data points, we need a method to choose the "best" line. The method employed is called the **least-squares criterion.** It is based on an analysis of the errors made in using a straight line to fit the data points.

To introduce the least-squares criterion, we will use a very simple data set. We will return to the Nissan Z data shortly.

EXAMPLE 12.3   *Introduces the least-squares criterion*

Let us consider the problem of fitting a straight line to the four data points given in Table 12.3.

TABLE 12.3

x	y
1	1
1	2
2	2
4	6

The scatter diagram for these data is pictured in Figure 12.8.

FIGURE 12.8

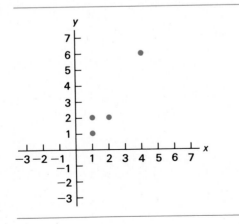

It is possible to fit many straight lines to these data. Figures 12.9(a) and 12.9(b) at the top of the next page show two such lines. We should emphasize that these are only two of many straight lines that could have been chosen.

FIGURE 12.9

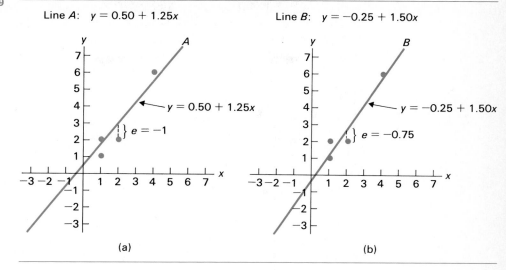

Line A: $y = 0.50 + 1.25x$ Line B: $y = -0.25 + 1.50x$

(a)   (b)

To keep things straight, we use $\hat{y}$ to denote the y-value predicted by a straight line for a value of x. For instance, the y-value predicted by Line A for $x = 2$ is

$$\hat{y} = 0.50 + 1.25 \cdot 2 = 3$$

and the y-value predicted by Line B for $x = 2$ is

$$\hat{y} = -0.25 + 1.50 \cdot 2 = 2.75$$

To obtain a quantitative measure of how well a line fits the data, we first look at the errors, e, made in using the line to predict the y-values of the data points. For instance, as we have just seen, Line A predicts a y-value of $\hat{y} = 3$ when $x = 2$. The actual y-value for $x = 2$ is $y = 2$. Thus, the error made in using Line A to predict the y-value of the data point $(2, 2)$ is

$$e = y - \hat{y} = 2 - 3 = -1$$

See Figure 12.9(a).

The fourth column of Table 12.4(a) gives the errors made for all four data points by Line A. The fourth column of Table 12.4(b) gives that for Line B.

TABLE 12.4

Line A: $y = 0.50 + 1.25x$

x	y	$\hat{y}$	e	e^2
1	1	1.75	−0.75	0.5625
1	2	1.75	0.25	0.0625
2	2	3.00	−1.00	1.0000
4	6	5.50	0.50	0.2500
				1.8750

(a)

Line B: $y = -0.25 + 1.50x$

x	y	$\hat{y}$	e	e^2
1	1	1.25	−0.25	0.0625
1	2	1.25	0.75	0.5625
2	2	2.75	−0.75	0.5625
4	6	5.75	0.25	0.0625
				1.2500

(b)

The rule that decides which line, Line A or Line B, fits the data better is as follows: For each line, compute the sum of the squared errors, Σe^2. This is done in the final columns of Tables 12.4(a) and 12.4(b). The line with the smaller sum of squared errors, in this case Line B, is the one that fits the data better.   ∎

With the previous example in mind, we can now state the least-squares criterion for the straight line that best fits a set of data points.

KEY FACT 12.2   Least-squares criterion

The straight line that best fits a set of data points is the one for which the sum of squared errors is smallest.

The straight line that best fits a set of data points according to the least-squares criterion is called the **regression line** and its equation is called the **regression equation.**

DEFINITION 12.2   Regression line and regression equation

Regression line: The straight line that fits a set of data points the best according to the least-squares criterion.
Regression equation: The equation of the regression line.

The least-squares criterion tells us what property the regression line for a set of data points must have, but it does not tell us how to find that line. In a moment, we will provide formulas for obtaining the equation of the regression line; that is, the regression equation. However, before we do, we need to present some notation that will be used throughout our study of regression and correlation.

DEFINITION 12.3   Notation used in regression and correlation

We define S_{xx}, S_{xy}, and S_{yy} by $S_{xx} = \Sigma(x - \overline{x})^2$, $S_{xy} = \Sigma(x - \overline{x})(y - \overline{y})$, and $S_{yy} = \Sigma(y - \overline{y})^2$. Those three quantities are most easily computed by using the following shortcut formulas:

$$S_{xx} = \Sigma x^2 - (\Sigma x)^2/n$$
$$S_{xy} = \Sigma xy - (\Sigma x)(\Sigma y)/n$$
$$S_{yy} = \Sigma y^2 - (\Sigma y)^2/n$$

Note: It can be shown mathematically that the shortcut formulas for S_{xx}, S_{xy}, and S_{yy} are equivalent to the defining formulas.

We can now give the formulas that permit us to actually determine the regression line for a set of data points. The formulas can be derived using elementary calculus (see Exercise 12.52).

FORMULA 12.1   Regression equation

The equation of the best-fitting line (regression line) to a set of n data points is $\hat{y} = b_0 + b_1 x$, where

$$b_1 = \frac{S_{xy}}{S_{xx}}$$

and

$$b_0 = \frac{1}{n}(\Sigma y - b_1 \Sigma x)$$

EXAMPLE 12.4   *Illustrates Formula 12.1*

Table 12.2 on page 593 displays data on age and price for a sample of 11 Nissan Zs. We repeat that data in the first two columns of Table 12.5.
a) Determine the regression equation for the data; that is, find the equation of the regression line.
b) Graph the regression equation and the data points.
c) Describe the apparent relationship between age and price for Nissan Zs.
d) What does the slope of the regression line represent in terms of the prices for Nissan Zs?
e) Use the regression equation to predict the price of a three-year-old Nissan Z and a four-year-old Nissan Z.

SOLUTION   a) In order to determine the regression equation, we need to compute b_1 and b_0 using Formula 12.1. To that end, it is convenient to construct a table of values for x (age), y (price), xy, x^2, and their sums. This is presented in Table 12.5.

TABLE 12.5
Table for computing the regression equation for the Nissan Z data

Age (yrs) x	Price ($100s) y	xy	x^2
5	85	425	25
4	103	412	16
6	70	420	36
5	82	410	25
5	89	445	25
5	98	490	25
6	66	396	36
6	95	570	36
2	169	338	4
7	70	490	49
7	48	336	49
58	975	4732	326

The slope of the regression line is therefore

$$b_1 = \frac{S_{xy}}{S_{xx}} = \frac{\Sigma xy - (\Sigma x)(\Sigma y)/n}{\Sigma x^2 - (\Sigma x)^2/n} = \frac{4732 - (58)(975)/11}{326 - (58)^2/11} = -20.26$$

and the y-intercept is

$$b_0 = \frac{1}{n}(\Sigma y - b_1 \Sigma x) = \frac{1}{11}(975 - (-20.26) \cdot 58) = 195.47$$

Thus, the regression equation is

$$\hat{y} = 195.47 - 20.26x$$

Note: The usual warnings about rounding apply. When computing the slope, b_1, of the regression line, do not perform any rounding until after the computation is complete. Moreover, when computing the y-intercept, b_0, do not use the rounded value of b_1; instead, keep full calculator accuracy.

b) To graph the regression equation, we need to substitute two different x-values into the regression equation in order to obtain two distinct points. Let us use the x-values, $x = 2$ and $x = 8$. The corresponding y-values are

$$\hat{y} = 195.47 - 20.26 \cdot 2 = 154.95$$

and

$$\hat{y} = 195.47 - 20.26 \cdot 8 = 33.39$$

Consequently, the regression line passes through the two points $(2, 154.95)$ and $(8, 33.39)$. In Figure 12.10, we have plotted these two points using hollow dots. Drawing a straight line through the two hollow dots yields the regression line, the graph of the regression equation.

FIGURE 12.10
Regression line
and data points
for Nissan Z data

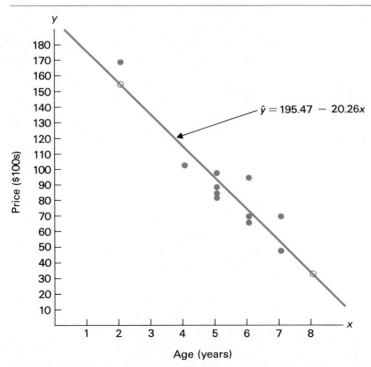

Also included in Figure 12.10 are the data points from Table 12.2. As we know, the regression line in Figure 12.10 is the straight line that best fits the data points according to the least-squares criterion. That is, it is the straight line for which the sum of squared errors is smallest.

c) Here we are to describe the apparent relationship between age and price of Nissan Zs. Since the slope of the regression line is negative, we see that price tends to decrease as age increases—no particular surprise.

d) For this part, we are to interpret the slope of the regression line in terms of the prices for Nissan Zs. To begin, recall that x represents age, in years, and y represents price, in hundreds of dollars. The slope of -20.26, or $-\$2,026$, indicates that Nissan Zs depreciate an estimated $2,026 per year, at least in the two- to seven-year-old range.

e) Finally, we are to use the regression equation, $\hat{y} = 195.47 - 20.26x$, to predict the price of a three-year-old Nissan Z and a four-year-old Nissan Z. For a three-year-old Nissan Z, we have $x = 3$, and so the predicted price is

$$\hat{y} = 195.47 - 20.26 \cdot 3 = 134.69$$

or $13,469. Similarly, the price the regression equation predicts for a four-year-old Nissan Z is

$$\hat{y} = 195.47 - 20.26 \cdot 4 = 114.43$$

MTB

or $11,443. Questions concerning the accuracy and reliability of such predictions will be discussed later. ∎

A WARNING ON THE USE OF LINEAR REGRESSION

The idea behind finding a regression line is based on the assumption that the data points are actually scattered about a straight line.[†] In some cases, data points may be scattered about a curve instead of a straight line, as in Figure 12.11(a). Unfortunately, the formulas for b_0 and b_1 will still work for this data set and fit an inappropriate straight line. Indeed, the data, which really follow a curve, would be fitted by the straight line shown in Figure 12.11(b).

FIGURE 12.11

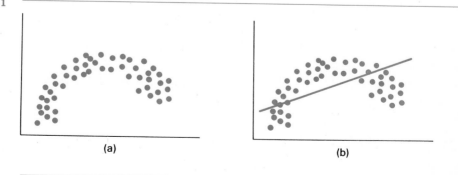

(a)   (b)

[†] We shall discuss this assumption in detail in Section 13.1 and make it more precise.

This procedure is misleading. For instance, it would lead us to predict that y-values in Figure 12.11(a) will keep increasing when they have actually begun to decrease. In summary:

KEY FACT 12.3   Criterion for finding a regression line

Before finding a regression line for a set of data points, draw a scatter diagram. If the data points do not appear to be scattered about a straight line, do not determine a regression line.

There are techniques that allow us to fit curves to data points, as should be done for the data points in Figure 12.11(a). We will discuss those techniques briefly in Section 12.5.

EXTRAPOLATION

It is important to be aware of the danger of *extrapolation* in regression. By **extrapolation,** we mean using the regression equation to make predictions for x-values that are outside the range of the x-values in the sample data.

Here is why such predictions might be unreliable: Employing the regression equation to make predictions is reasonable only when there is a linear relationship between the variables. Although there may be a linear relationship between the variables within the range of the x-values in the sample data, that linear relationship may not necessarily hold true for x-values that lie outside the range of those in the sample data.

The Nissan Z illustration provides an excellent example of where extrapolation can be dangerous. The regression equation is

$$\hat{y} = 195.47 - 20.26x$$

The x-values of the sample points used in computing that regression equation range from $x = 2$ to $x = 7$; that is, from two to seven years old.

Suppose we extrapolate by using the regression equation to predict the price of an 11-year-old Nissan Z. The predicted price is

$$\hat{y} = 195.47 - 20.26 \cdot 11 = -27.39$$

or $-\$2,739$. Clearly, this is ridiculous. In fact, by consulting the classified ads, we find that a more reasonable prediction for the price of an 11-year-old Z is about $\$3,000$. Thus, although the relationship between age and price of Nissan Zs appears to be linear in the range from $x = 2$ to $x = 7$, it is definitely not so in the range from $x = 2$ to $x = 11$. Figure 12.12 summarizes the discussion on extrapolation as it applies to age and price for Nissan Zs.

FIGURE 12.12

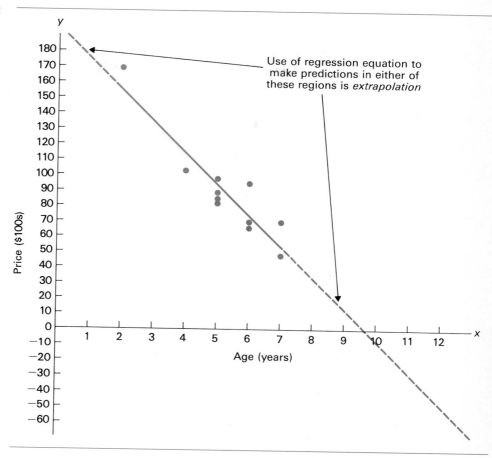

Use of regression equation to make predictions in either of these regions is *extrapolation*

USING THE COMPUTER (OPTIONAL)

As we mentioned earlier, the idea behind finding a regression line is based on the assumption that the data points are actually scattered about a straight line. Consequently, before determining a regression line, we need to look at a scatter diagram of the data. We should then proceed with the regression analysis only if the data points appear to be scattered about a straight line.

Minitab has a command called **PLOT** that can be used to obtain a scatter diagram. We illustrate the use of PLOT in the following example.

EXAMPLE 12.5   *Illustrates the PLOT command*

In Table 12.2, we presented data on age and price of Nissan Zs. We repeat that data in Table 12.6 at the top of the next page. Use Minitab to obtain a scatter diagram of these data.

TABLE 12.6
Age versus price
data for Nissan Zs

Age (yrs) x	Price ($100s) y
5	85
4	103
6	70
5	82
5	89
5	98
6	66
6	95
2	169
7	70
7	48

SOLUTION   We begin by entering the data from Table 12.6 into, say, C1 and C2 using the READ command. Then we name C1 "Age" and C2 "Price" by applying the NAME command. To obtain a scatter diagram of the age-versus-price data, we now type the command PLOT followed by the storage locations of the sample data; that is, we type

PLOT 'PRICE' versus 'AGE'

Printout 12.1 at the top of the next page summarizes the preceding discussion and also shows the resulting output. Note that the variable corresponding to the first storage location typed in the PLOT command, in this case, PRICE, is plotted on the vertical axis and that the variable corresponding to the second storage location typed in the PLOT command, in this case, AGE, is plotted on the horizontal axis.

In general, the data points are plotted using asterisks. However, if two or more data points are the same or very close together, then Minitab prints the number of those points instead of asterisks. For example, the 2 plotted above the 5.0 indicates there are two data points in that vicinity that are either the same or very close together. In this case, they are the points $(5, 85)$ and $(5, 89)$.

The plot in Printout 12.1 is Minitab's version of the scatter diagram that we drew by hand in Figure 12.7 on page 593. From the scatter diagram, it does appear that the data points are scattered about a straight line. Therefore, it is reasonable to find a regression line for that data.   ■

Formula 12.1 on page 597 provides the formulas to determine the regression equation for a set of data points. Minitab has a program that, among other things, will determine the regression equation for us. That program is called **REGRESS**. We will apply REGRESS to obtain the regression equation for the Nissan Z data.

EXAMPLE 12.6   *Illustrates the REGRESS command*

The data on age and price for a sample of Nissan Zs are given in Table 12.6 at the top of this page. Use Minitab to obtain the regression equation for that data.

PRINTOUT 12.1
Minitab output
for PLOT

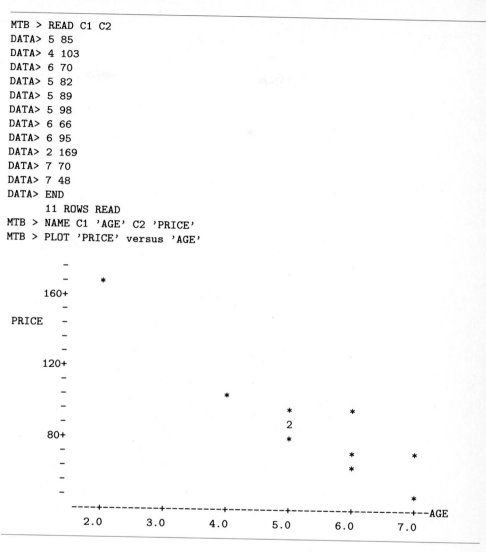

```
MTB > READ C1 C2
DATA> 5  85
DATA> 4  103
DATA> 6  70
DATA> 5  82
DATA> 5  89
DATA> 5  98
DATA> 6  66
DATA> 6  95
DATA> 2  169
DATA> 7  70
DATA> 7  48
DATA> END
        11 ROWS READ
MTB > NAME C1 'AGE' C2 'PRICE'
MTB > PLOT 'PRICE' versus 'AGE'
```

```
           -
           -      *
     160+
           -
PRICE      -
           -
           -
     120+
           -
           -               *
           -                     *       *
           -                     2
      80+                        *
           -                          *        *
           -                          *
           -
           -                                *
        ----+---------+---------+---------+---------+---------+--AGE
           2.0       3.0       4.0       5.0       6.0       7.0
```

SOLUTION Before we have Minitab perform a regression analysis on the data, we need to point out that there are three forms for the output. The command **BRIEF**, followed by an integer between 1 and 3, controls the amount of output for all succeeding REGRESS commands. The larger the integer, the more extensive the output. For this example, we will only need the briefest output, so we type BRIEF 1.

Now we are ready to apply the REGRESS command. We type

REGRESS 'PRICE' on 1 predictor 'AGE'

This command tells Minitab to perform a regression analysis on the age and price data with age as the single independent (predictor) variable. Printout 12.2 displays the above commands and the output that results.

PRINTOUT 12.2
Minitab output
for REGRESS

```
MTB > BRIEF 1
MTB > REGRESS 'PRICE' on 1 predictor 'AGE'

The regression equation is
PRICE = 195 - 20.3 AGE

Predictor        Coef       Stdev      t-ratio        p
Constant       195.47       15.24       12.83      0.000
AGE            -20.261       2.800       -7.24      0.000

s = 12.58      R-sq = 85.3%      R-sq(adj) = 83.7%

Analysis of Variance

SOURCE          DF         SS          MS          F        p
Regression       1      8285.0      8285.0      52.38    0.000
Error            9      1423.5       158.2
Total           10      9708.5
```

The first item in the output gives the required regression equation:

$$PRICE = 195 - 20.3\ AGE$$

In other words, the regression equation is $\hat{y} = 195 - 20.3x$.

After the regression equation, the output provides a table that gives information about the y-intercept, b_0, and slope, b_1, of the regression line. The row labeled Constant provides information on b_0 and the row labeled AGE provides information on b_1. In particular, the entries 195.47 and -20.261 under the column headed Coef are simply the values of b_0 and b_1, respectively. We will discuss the other aspects of the output in Printout 12.2 in future sections. ∎

Exercises 12.2

In each of Exercises 12.28 and 12.29,
a) *graph each linear equation and the data points.*
b) *construct tables for x, y, $\hat{y}$, e, and e^2 similar to Table 12.4 on page 595.*
c) *determine which line fits the set of data points better according to the least-squares criterion.*

___ 12.28   Line A: $y = 1.5 + 0.5x$
Line B: $y = 1.125 + 0.375x$

Data points

x	1	1	5	5
y	1	3	2	4

___ 12.29   Line A: $y = 3 - 0.6x$
Line B: $y = 4 - x$

Data points

x	0	2	2	5	6
y	4	2	0	-2	1

For Exercises 12.30–12.37, be sure to save your worksheets. You will need them in later sections.

___ 12.30   Refer to Exercise 12.28.
a) Find the regression equation for the data points.
b) Graph the regression equation and the data points.

— **12.31** Refer to Exercise 12.29.
a) Find the regression equation for the data points.
b) Graph the regression equation and the data points.

— **12.32** Ten Corvettes, between one and six years of age, were randomly selected from the classified ads of the *Arizona Republic*. The following data were obtained on age and price:

Age (yrs) x	Price ($100s) y
6	125
6	115
6	130
2	260
2	219
5	150
4	190
5	163
1	260
4	160

a) Determine the regression equation for the data.
b) Graph the regression equation and the data points.
c) Describe the apparent relationship between the two variables, age and price, for Corvettes.
d) What does the slope of the regression line represent in terms of Corvette prices?
e) Use the regression equation to predict the price of a two-year-old Corvette; a three-year-old Corvette.

— **12.33** The National Center for Health Statistics publishes data on heights and weights in *Vital and Health Statistics*. A random sample of 11 males, aged 18–24 years, yielded the following data:

Height (inches) x	Weight (lb) y
65	175
67	133
71	185
71	163
66	126
75	198
67	153
70	163
71	159
69	151
69	155

a) Determine the regression equation for the data.
b) Graph the regression equation and the data points.
c) Describe the apparent relationship between height and weight for 18–24-year-old males.
d) What does the slope of the regression line represent in terms of weights of 18–24-year-old males?
e) Use the regression equation to predict the weight of an 18–24-year-old male who is 67 inches tall; 73 inches tall.

— **12.34** Hanna Properties specializes in custom-home resales in the Equestrian Estates, an exclusive subdivision in Phoenix, Arizona. A random sample of nine custom homes, currently listed for sale, provided the following information on size and price. The size data are in hundreds of square feet, rounded to the nearest hundred; the price data are in thousands of dollars, rounded to the nearest thousand.

Size (100 sq ft) x	Price ($1000s) y
26	235
27	249
33	267
29	269
29	295
34	345
30	415
40	475
22	195

a) Determine the regression equation for the data.
b) Graph the regression equation and the data points.
c) Describe the apparent relationship between square-footage and price for custom homes in the Equestrian Estates.
d) What does the slope of the regression line represent in terms of sizes and prices of custom homes in the Equestrian Estates?
e) Use the regression equation to predict the price of a custom home in the Equestrian Estates that has 2600 square feet.

— **12.35** An article read by a physician indicated that the maximum heart rate an individual can reach during intensive exercise decreases with age. The physician decided to do his own study. Ten randomly selected people performed exercise tests and recorded their peak heart rates. The results are shown in the table at the top of the first column on the next page.

| Age | Peak heart rate |
x	y
30	186
38	183
41	171
38	177
29	191
39	177
46	175
41	176
42	171
24	196

a) Determine the regression equation for the data.
b) Graph the regression equation and the data points.
c) Describe the apparent relationship between the two variables, age and peak heart rate.
d) What does the slope of the regression line represent in terms of age and peak heart rate?
e) Use the regression equation to predict the peak heart rate of a person who is 28 years old.

___ **12.36** A calculus instructor asked a random sample of eight students to record their study times per lesson in a beginning calculus course. She then made a table for total study times over two weeks and test scores at the end of the two weeks. Here are the results:

| Study time (hrs) | Grade (percent) |
x	y
10	92
15	81
12	84
20	74
8	85
16	80
14	84
22	80

a) Determine the regression equation for the data.
b) Graph the regression equation and the data points.
c) Describe the apparent relationship between study time and test score. (Does it surprise you?)
d) What does the slope of the regression line represent in terms of study time and test score?
e) Use the regression equation to predict the test score of a student who studies for 15 hours.

___ **12.37** An economist is interested in the relation between the disposable income of a family and the amount of money spent annually on food. For a preliminary study, the economist takes a random sample of eight middle-income families of the same size (father, mother, two children). The results are as follows:

| Disposable income ($1000s) | Food expenditure ($100s) |
x	y
30	55
36	60
27	42
20	40
16	37
24	26
19	39
25	43

a) Determine the regression equation for the data.
b) Graph the regression equation and the data points.
c) Describe the apparent relationship between disposable income and annual food expenditure.
d) What does the slope of the regression line represent in terms of disposable income and annual food expenditure?
e) Use the regression equation to predict the annual food expenditure of a family with a disposable income of $25,000.

___ **12.38** For which of the following sets of data points is it reasonable to determine a regression line?

___ **12.39** For which of the following sets of data points is it reasonable to determine a regression line?

___ **12.40** In Exercise 12.32, you found a regression equation that can be used to predict the price of a Corvette given its age.

a) Should that regression equation be used to predict the price of a four-year-old Corvette? a 10-year-old Corvette? Explain your answers.

b) For which ages is it reasonable to use the regression equation to predict price?

— **12.41** In Exercise 12.33, you found a regression equation relating height and weight of 18–24-year-old males.

a) Should that regression equation be used to predict the weight of an 18–24-year-old male who is 68 inches tall? 60 inches tall? Explain your answers.

b) For which heights is it reasonable to use the regression equation to predict weight?

Exercises 12.42–12.45 are computer exercises.

— **12.42** (**Computer exercise**) Suppose that the data in Exercise 12.36 are stored in columns named TIME and GRADE.

a) Which Minitab command and subcommands (if any) should be used to obtain a scatter diagram for the data?

b) Which Minitab command and subcommands (if any) should be used to obtain the regression equation for the data?

c) If you have access to Minitab, use it to obtain a scatter diagram for the data and to determine the regression equation for the data.

— **12.43** (**Computer exercise**) Suppose that the data in Exercise 12.37 are stored in columns named INCOME and FOODEX.

a) Which Minitab command and subcommands (if any) should be used to obtain a scatter diagram for the data?

b) Which Minitab command and subcommands (if any) should be used to obtain the regression equation for the data?

c) If you have access to Minitab, use it to obtain a scatter diagram for the data and to determine the regression equation for the data.

— **12.44** (**Computer exercise**) The Energy Information Administration publishes data on energy consumption by family income in *Residential Energy Consumption Survey: Consumption and Expenditures*. We applied Minitab's REGRESS command to data on family income and last year's energy consumption from a random sample of 25 families. The income data are in thousands of dollars and the energy-consumption data are in millions of BTU. Printout 12.3 at the top of the next page shows the computer output.

a) Use the printout to obtain the regression equation for the data.

b) Predict last year's energy consumption for a family with an income of $42,000.

— **12.45** (**Computer exercise**) Greene and Touchstone conducted a study on the relationship between the estriol levels of pregnant women and the birth weights of their children. Their findings, entitled "Urinary Tract Estriol: An Index of Placental Function," were published in the *American Journal of Obstetrics and Gynecology*. Printout 12.4 on the next page shows the computer output that results by applying Minitab's REGRESS command to the data obtained by Greene and Touchstone. The estriol levels are in milligrams per 24 hours and the birth weights are in grams.

a) Consult the printout to determine the regression equation for the data.

b) Use the regression equation to predict the birth weight of the child of a pregnant woman with an estriol level of 17 mg/24 hr.

═ **12.46** The negative relation between study time and grade found in Exercise 12.36 has been discovered by many investigators, and has puzzled them. Can you think of a possible explanation for it?

Sample covariance: The *sample covariance*, s_{xy}, of a sample of n data points is defined by

$$(1) \qquad s_{xy} = \frac{\Sigma(x - \bar{x})(y - \bar{y})}{n - 1}$$

═ **12.47** Determine the sample covariance of the data points in Exercise 12.29.

═ **12.48** Determine the sample covariance of the data points in Exercise 12.28.

Obtaining the regression equation using the sample covariance: The sample covariance can be used as an alternate method for determining the slope and y-intercept of the regression line for a set of data points. The formulas are as follows:

$$(2) \qquad b_1 = s_{xy}/s_x^2, \qquad b_0 = \bar{y} - b_1\bar{x}$$

where s_x denotes the sample standard deviation of the x-values.

═ **12.49** Use the equations in Formula (2) to find the regression equation for the data points in Exercise 12.29. Compare your answer to the one that you obtained in part (a) of Exercise 12.31.

PRINTOUT 12.3   Minitab output for Exercise 12.44 (and Exercises 12.64 and 13.18)

```
The regression equation is
CONSUMPT = 82.0 + 0.931 INCOME
```

Predictor	Coef	Stdev	t-ratio	p
Constant	82.036	2.054	39.94	0.000
INCOME	0.93051	0.05727	16.25	0.000

s = 5.375   R-sq = 92.0%   R-sq(adj) = 91.6%

Analysis of Variance

SOURCE	DF	SS	MS	F	p
Regression	1	7626.6	7626.6	264.03	0.000
Error	23	664.4	28.9		
Total	24	8291.0			

PRINTOUT 12.4   Minitab output for Exercise 12.45 (and Exercises 12.65 and 13.19)

```
The regression equation is
WEIGHT = 2152 + 60.8 ESTRIOL
```

Predictor	Coef	Stdev	t-ratio	p
Constant	2152.3	262.0	8.21	0.000
ESTRIOL	60.82	14.68	4.14	0.000

s = 382.1   R-sq = 37.2%   R-sq(adj) = 35.0%

Analysis of Variance

SOURCE	DF	SS	MS	F	p
Regression	1	2505745	2505745	17.16	0.000
Error	29	4234255	146009		
Total	30	6740000			

12.50 Apply the equations in Formula (2) to find the regression equation for the data points in Exercise 12.28. Compare your answer to the one that you obtained in part (a) of Exercise 12.30.

12.51 Prove that the equations in Formula (2) are equivalent to the ones given in Formula 12.1 on page 597.

12.52 In this exercise, we will derive the equations in Formula 12.1 on page 597 for the slope and y-intercept of the regression line. The derivation requires elementary calculus. To begin, recall that according to the least-squares criterion, the regression line is the straight line for which the sum of squared errors is smallest.

a) Show that the regression line is the straight line, $\hat{y} = b_0 + b_1 x$, for which b_0 and b_1 minimize the function

$$f(b_0, b_1) = \Sigma[y - (b_0 + b_1 x)]^2$$

b) Use elementary calculus to determine the values of b_0 and b_1 that minimize $f(b_0, b_1)$ and show that these values give the equations in Formula 12.1. (Hint: Compute the partial derivatives of $f(b_0, b_1)$, set them equal to zero, and solve for b_0 and b_1.)

12.3 The coefficient of determination

In Example 12.4 on page 598, we determined the regression equation

$$\hat{y} = 195.47 - 20.26x$$

for data on age and price of a sample of 11 Nissan Zs. Here x represents age, in years, and $\hat{y}$ predicted price, in hundreds of dollars. We can apply the regression equation to predict the price of a Nissan Z of a given age, x. For instance, we predict that a four-year-old Nissan Z will cost roughly

$$\hat{y} = 195.47 - 20.26 \cdot 4 = 114.43$$

or $11,443. But how valuable are such predictions? Is the regression equation useful for predicting price or could we do just as well by ignoring age?

There are several ways that we can attempt to answer questions on how valuable a regression equation is for making predictions. One method is to measure the reduction in the errors made in prediction by using the regression equation instead of simply predicting the mean of the observed y-values, $\overline{y}$. To illustrate the ideas involved, we return to the Nissan Z data.

EXAMPLE 12.7   *Introduces the coefficient of determination*

The age and price data for a sample of 11 Nissan Zs are repeated in the first two columns of Table 12.7 on the next page. Ages are in years. Prices are in hundreds of dollars, rounded to the nearest hundred.

One way that we can employ this information to predict the price of a Nissan Z is to ignore age and simply use the mean price, $\overline{y}$, of the 11 Nissan Zs sampled. In other words, just use

$$\overline{y} = \frac{\Sigma y}{n} = \frac{975}{11} = 88.64 \ (\$8,864)$$

as the predicted price for a Nissan Z, regardless of its age.

The errors made when the mean price, $\overline{y} = 88.64$, is used as the predicted price for each of the 11 Nissan Zs sampled are portrayed graphically in Figure 12.13 at the top of the next page.

To obtain a quantitative measure of the total error made, we compute the sum of the squared errors. The required computations are shown in Table 12.7.[†] From the final column of Table 12.7, we see that the total squared error is 9708.5, when the mean price, $\overline{y} = 88.64$, is used as the predicted price of each of the 11 Nissan Zs sampled. That total squared error is called the **total sum of squares, SST**. Thus, for the Nissan Z data, the total sum of squares is

$$SST = \Sigma(y - \overline{y})^2 = 9708.5$$

[†] Values in Table 12.7 and all other tables in this section are displayed to various numbers of decimal places, but computations are done using full calculator accuracy.

FIGURE 12.13
Errors made when
the mean price is
used to predict the
observed prices

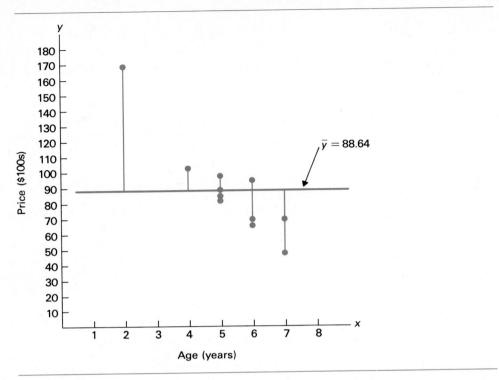

Age (years)

TABLE 12.7
Table for computing
SST for the
Nissan Z data

Age (yrs) x	Price ($100s) y	$y - \bar{y}$	$(y - \bar{y})^2$
5	85	−3.64	13.2
4	103	14.36	206.3
6	70	−18.64	347.3
5	82	−6.64	44.0
5	89	0.36	0.1
5	98	9.36	87.7
6	66	−22.64	512.4
6	95	6.36	40.5
2	169	80.36	6458.3
7	70	−18.64	347.3
7	48	−40.64	1651.3
	975		9708.5

Now, if age (i.e., the regression equation) is useful for predicting price, then we should obtain a reduction in the total squared error by using the regression equation to make the price predictions instead of using the mean price, $\bar{y}$. The errors made when the regression equation, $\hat{y} = 195.47 - 20.26x$, is used to predict the price of each of the 11 Nissan Zs sampled are portrayed graphically in Figure 12.14.

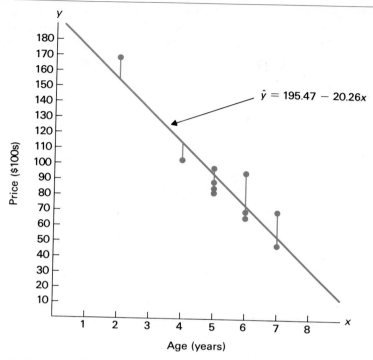

We will employ Table 12.8 to compute the total squared error made when the regression equation is used to predict the price of each of the 11 Nissan Zs sampled. Each predicted price, $\hat{y}$, is obtained by substituting the age of the Nissan Z in question into the regression equation, $\hat{y} = 195.47 - 20.26x$.

Age (yrs) x	Price ($100s) y	$\hat{y}$	$y - \hat{y}$	$(y - \hat{y})^2$
5	85	94.16	−9.16	83.9
4	103	114.42	−11.42	130.5
6	70	73.90	−3.90	15.2
5	82	94.16	−12.16	147.9
5	89	94.16	−5.16	26.6
5	98	94.16	3.84	14.7
6	66	73.90	−7.90	62.4
6	95	73.90	21.10	445.2
2	169	154.95	14.05	197.5
7	70	53.64	16.36	267.7
7	48	53.64	−5.64	31.8
				1423.5

From the final column of Table 12.8, we see that the total squared error is 1423.5, when the regression equation is used to predict the price of each of the 11 Nissan Zs sampled. That total squared error is called the **error sum of squares, SSE.** Thus, for the Nissan Z data, the error sum of squares is

$$SSE = \Sigma(y - \hat{y})^2 = 1423.5$$

Consequently, we have obtained a drastic reduction in the total squared error by using the regression equation to make the price predictions instead of using the mean price, $\bar{y}$. The percentage reduction is

$$\frac{SST - SSE}{SST} = 1 - \frac{SSE}{SST} = 1 - \frac{1423.5}{9708.5} = 0.853$$

or 85.3%. That percentage reduction is called the **coefficient of determination, r^2.** Thus, for the Nissan Z data, the coefficient of determination is

$$r^2 = 1 - \frac{SSE}{SST} = 0.853$$

In any case, by using the regression equation, instead of the mean price, $\bar{y}$, the total squared error for the price predictions of the 11 Nissan Zs sampled has been reduced by 85.3%. This indicates, as we would suspect, that age is extremely useful for predicting price. ■

MTB

The definitions introduced in Example 12.7 are summarized below.

DEFINITION 12.4

Total sum of squares: $SST = \Sigma(y - \bar{y})^2$

Error sum of squares: $SSE = \Sigma(y - \hat{y})^2$

Coefficient of determination: $r^2 = 1 - \frac{SSE}{SST}$

As we discovered in Example 12.7, the coefficient of determination, r^2, is a descriptive measure of the utility of the regression equation for making predictions. Specifically, r^2 gives the percentage reduction obtained in the total squared error by using the regression equation, instead of the sample mean, $\bar{y}$, to predict the observed y-values. Values of r^2 near 0 indicate that the regression equation is not very useful for making predictions, whereas values of r^2 near 1 (100%) indicate that the regression equation is extremely useful for making predictions.

EXPLAINED VARIATION

There is another way to interpret the coefficient of determination, r^2; namely, as the percentage of the variation in the observed y-values that is explained by the regression line. To see why, we return to the Nissan Z data.

EXAMPLE 12.8   *Introduces explained variation*

The scatter diagram for the age and price data of 11 Nissan Zs, shown in Figure 12.7, is reproduced here in Figure 12.15. Also included in Figure 12.15 is the regression line for the data, $\hat{y} = 195.47 - 20.26x$.

FIGURE 12.15
Scatter diagram and regression line for Nissan Z data

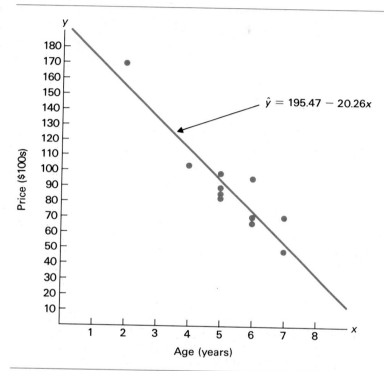

As we see from the scatter diagram in Figure 12.15, there is quite a bit of variation in the prices of the 11 Nissan Zs sampled, ranging from a low of 48 (i.e., $4,800) to a high of 169 (i.e., $16,900). But, as we also see from the regression line in Figure 12.15, much of the variation in the prices of the 11 Nissan Zs sampled is "explained" by age. That is, the regression line predicts a good portion of the type of variation found in the prices.

To describe quantitatively how much of the total variation in the prices sampled is explained by age (i.e., by the regression line), we proceed as follows: First of all, the total sum of squares, *SST*, is used as the measure of total variation of the prices sampled. From the bottom of page 609, we have

$$SST = \Sigma(y - \overline{y})^2 = 9708.5$$

Now let us look at a particular observed price, say, $y = 98$, corresponding to the data point $(5, 98)$. In Figure 12.16, we have drawn a blow-up of a portion of Figure 12.15 showing the data point $(5, 98)$ and no others.

FIGURE 12.16

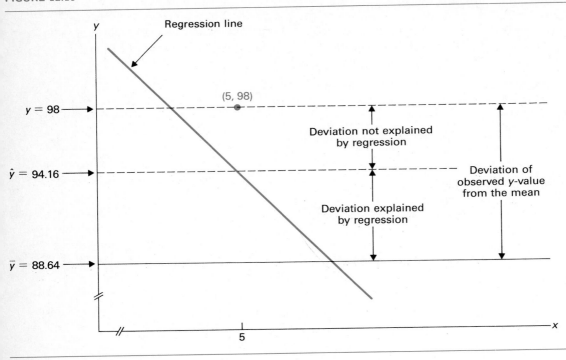

From Figure 12.16, we see that the deviation, $y - \overline{y}$, of a price from the mean price can be decomposed into two parts; namely, the deviation that is explained by the regression line, $\hat{y} - \overline{y}$, and the remaining unexplained deviation, $y - \hat{y}$. Therefore, the total amount of variation (squared deviation) explained by the regression line is $\Sigma(\hat{y} - \overline{y})^2$. This is called the **regression sum of squares, SSR.**

To compute SSR, we need the predicted prices, $\hat{y}$, and the mean of the observed prices, $\overline{y}$. The predicted prices are given in the third column of Table 12.8 and are repeated in the third column of Table 12.9 at the top of the next page. Recalling that $\overline{y} = 88.64$, we obtain the regression sum of squares, SSR, from the final column of Table 12.9. Thus, for the Nissan Z data, the regression sum of squares is

$$SSR = \Sigma(\hat{y} - \overline{y})^2 = 8285.0$$

This is the amount of variation in the prices sampled that is explained by the regression line. Therefore, the percentage of the total variation in the prices sampled that is explained by the regression line is

$$\frac{SSR}{SST} = \frac{8285.0}{9708.5} = 0.853$$

or 85.3%. Consequently, a good deal of the variation in the prices sampled is explained by the regression line. This, in turn, implies that age is quite useful for predicting price.

MTB

TABLE 12.9
Table for computing SSR for the Nissan Z data

Age (yrs) x	Price ($100s) y	$\hat{y}$	$\hat{y} - \bar{y}$	$(\hat{y} - \bar{y})^2$
5	85	94.16	5.53	30.5
4	103	114.42	25.79	665.0
6	70	73.90	−14.74	217.1
5	82	94.16	5.53	30.5
5	89	94.16	5.53	30.5
5	98	94.16	5.53	30.5
6	66	73.90	−14.74	217.1
6	95	73.90	−14.74	217.1
2	169	154.95	66.31	4397.0
7	70	53.64	−35.00	1224.8
7	48	53.64	−35.00	1224.8
				8285.0

In the previous example, we used data on age and price of Nissan Zs to introduce the regression sum of squares. As we observed, that quantity is defined and interpreted as follows:

DEFINITION 12.5   Regression sum of squares

The *regression sum of squares, SSR,* is defined by

$$SSR = \Sigma(\hat{y} - \bar{y})^2$$

and represents the amount of variation in the observed y-values that is explained by the regression.

THE REGRESSION IDENTITY

For the Nissan Z data, we have determined that $SST = 9708.5$, $SSR = 8285.0$, and $SSE = 1423.5$. Since $9708.5 = 8285.0 + 1423.5$, we see that $SST = SSR + SSE$. This equation is always true and is called the **regression identity.** We state that identity formally in the following key fact:

KEY FACT 12.4   Regression identity

The total sum of squares equals the regression sum of squares plus the error sum of squares; that is,

$$SST = SSR + SSE$$

You may have noted that, for the Nissan Z data, the percentage of the total variation in the prices sampled that is explained by the regression,

$$\frac{SSR}{SST} = \frac{8285.0}{9708.5} = 0.853,$$

is equal to the coefficient of determination, r^2:

$$r^2 = 1 - \frac{SSE}{SST} = 1 - \frac{1423.5}{9708.5} = 0.853$$

This is a consequence of the regression identity and, hence, is not only true for the Nissan Z data but holds for any set of data points.

Indeed, because of the regression identity, we have $SSR = SST - SSE$, and so

$$\frac{SSR}{SST} = \frac{SST - SSE}{SST} = 1 - \frac{SSE}{SST} = r^2$$

This shows that the percentage of the total variation in the observed y-values that is explained by the regression equals the coefficient of determination. Thus, we have the following fact:

KEY FACT 12.5   Interpretation of the coefficient of determination
The coefficient of determination, r^2, is defined by

$$r^2 = 1 - \frac{SSE}{SST}$$

and equals the percentage reduction obtained in the total squared error by using the regression equation, instead of the sample mean, $\overline{y}$, to predict the observed y-values. The coefficient of determination can also be computed as

$$r^2 = \frac{SSR}{SST}$$

Thus, it also equals the percentage of the total variation in the observed y-values that is explained by the regression. In any case, r^2 always lies between 0 and 1 and is a descriptive measure of the utility of the regression equation for making predictions. Values of r^2 near 0 indicate that the regression equation is not very useful for making predictions, whereas values of r^2 near 1 indicate that the regression equation is extremely useful for making predictions.

SHORTCUT FORMULAS FOR THE SUMS OF SQUARES

Computing the three sums of squares, SST, SSR, and SSE, using the defining formulas is time consuming and can lead to significant roundoff error unless full accuracy is retained. For those reasons, we usually employ shortcut formulas to compute the sums of squares. These shortcut formulas are given in Formula 12.2. (See Exercise 12.69 for a derivation of the shortcut formulas.)

FORMULA 12.2 Shortcut formulas for the sums of squares

The sums of squares, SST, SSR, and SSE, can be computed by using the following shortcut formulas:

Total sum of squares: $SST = S_{yy}$

Regression sum of squares: $SSR = S_{xy}^2 / S_{xx}$

Error sum of squares: $SSE = S_{yy} - S_{xy}^2 / S_{xx}$

where S_{yy}, S_{xy}, and S_{xx} are given by Definition 12.3 on page 596.

To illustrate Formula 12.2, we will apply the shortcut formulas to compute the three sums of squares for the data on age and price of 11 Nissan Zs. Those sums of squares were obtained previously by using the defining formulas.

EXAMPLE 12.9 *Illustrates Formula 12.2*

The age and price data for a sample of 11 Nissan Zs are repeated in the first two columns of Table 12.10. Use the shortcut formulas in Formula 12.2 to determine the three sums of squares, SST, SSR, and SSE.

SOLUTION

To apply the shortcut formulas, we will need a table of values for x (age), y (price), xy, x^2, y^2, and their sums. This is presented in Table 12.10.

TABLE 12.10
Table for computing the three sums of squares for the Nissan Z data using the shortcut formulas

Age (yrs) x	Price ($100s) y	xy	x^2	y^2
5	85	425	25	7,225
4	103	412	16	10,609
6	70	420	36	4,900
5	82	410	25	6,724
5	89	445	25	7,921
5	98	490	25	9,604
6	66	396	36	4,356
6	95	570	36	9,025
2	169	338	4	28,561
7	70	490	49	4,900
7	48	336	49	2,304
58	975	4732	326	96,129

Using the last row of Table 12.10 and Formula 12.2, we can now obtain the three sums of squares for the Nissan Z data. The total sum of squares equals

$$SST = S_{yy} = \Sigma y^2 - (\Sigma y)^2/n = 96{,}129 - (975)^2/11 = \boxed{9708.5}$$

The regression sum of squares equals

$$SSR = \frac{S_{xy}^2}{S_{xx}} = \frac{[\Sigma xy - (\Sigma x)(\Sigma y)/n]^2}{\Sigma x^2 - (\Sigma x)^2/n} = \frac{[4732 - (58)(975)/11]^2}{326 - (58)^2/11} = \boxed{8285.0}$$

And, using the previous two results, we see that the error sum of squares equals

$$SSE = S_{yy} - \frac{S_{xy}^2}{S_{xx}} = 9708.5 - 8285.0 = \boxed{1423.5}$$

The values for the three sums of squares that we just obtained by applying the shortcut formulas are, of course, the same as the values that we found earlier by using the defining formulas. However, when the shortcut formulas are employed, the computations are much simpler and less subject to roundoff error. ■

USING THE COMPUTER (OPTIONAL)

In Section 12.2, we learned that Minitab's REGRESS command can be used to obtain the regression equation for a set of data points. The output resulting from the application of that command contains more than the regression equation. In particular, it also provides the coefficient of determination, r^2, and the three sums of squares, SST, SSR, and SSE. Let us return once again to the Nissan Z data.

EXAMPLE 12.10   *Illustrates the REGRESS command*

Printout 12.2 on page 604 shows the output obtained by applying the REGRESS command to the data on age and price for a sample of 11 Nissan Zs. We repeat that printout here as Printout 12.5.

PRINTOUT 12.5
Minitab output for REGRESS

```
MTB > BRIEF 1
MTB > REGRESS 'PRICE' on 1 predictor 'AGE'

The regression equation is
PRICE = 195 - 20.3 AGE

Predictor       Coef      Stdev    t-ratio       p
Constant      195.47      15.24      12.83   0.000
AGE          -20.261      2.800      -7.24   0.000

s = 12.58      R-sq = 85.3%      R-sq(adj) = 83.7%

Analysis of Variance

SOURCE       DF        SS        MS       F       p
Regression    1    8285.0    8285.0   52.38   0.000
Error         9    1423.5     158.2
Total        10    9708.5
```

Use the printout to find
a) the coefficient of determination, r^2.
b) the three sums of squares, SST, SSR, and SSE.

SOLUTION

a) The coefficient of determination, r^2, is displayed as the second entry in the sixth line of the output in Printout 12.5: R-sq = 85.3%. In other words, $r^2 = 0.853$.

b) To obtain the three sums of squares, we use the table in the output entitled Analysis of Variance. Specifically, the values of the three sums of squares can be found in the column headed SS. The entries in the column headed SOURCE identify the sums of squares. Hence, the first entry in the SS column is the regression sum of squares, SSR, the second is the error sum of squares, SSE, and the third is the total sum of squares, SST. So, we see that $SSR = 8285.0$, $SSE = 1423.5$, and $SST = 9708.5$. ∎

Exercises 12.3

— **12.53** In this section we introduced a descriptive measure of the utility of the regression equation for making predictions.

a) What term and symbol are used for that descriptive measure?

b) Provide two different interpretations of that descriptive measure.

In Exercises 12.54 and 12.55, we have repeated the data sets from Exercises 12.28 and 12.29, respectively, of Section 12.2. We have also provided the regression equations for those data sets which were found in Exercises 12.30 and 12.31, respectively. For each exercise,

a) *compute SST, SSR, and SSE using the defining formulas.*

b) *verify the regression identity, $SST = SSR + SSE$.*

c) *compute the coefficient of determination by using both the definition, $r^2 = 1 - SSE/SST$, and the formula $r^2 = SSR/SST$.*

d) *determine the percentage reduction obtained in the total squared error by using the regression equation to predict the observed y-values instead of simply using the mean, $\bar{y}$, of the observed y-values.*

e) *determine the percentage of variation in the observed y-values that is explained by the regression.*

f) *state how useful the regression equation appears to be for making predictions. [The answer given here is somewhat subjective.]*

— **12.54** The data from Exercise 12.28:

Data points

x	1	1	5	5
y	1	3	2	4

Regression equation is $\hat{y} = 1.75 + 0.25x$.

— **12.55** The data from Exercise 12.29:

Data points

x	0	2	2	5	6
y	4	2	0	−2	1

Regression equation is $\hat{y} = 2.875 - 0.625x$.

For each of Exercises 12.56–12.61,

a) *compute SST, SSR, and SSE using the shortcut formulas.*

b) *compute the coefficient of determination, r^2.*

c) *determine the percentage of variation in the observed y-values that is explained by the regression and interpret your result in words.*

d) *state how useful the regression equation appears to be for making predictions.*

— **12.56** The age and price data for Corvettes from Exercise 12.32 of Section 12.2:

Age (yrs) x	Price ($100s) y
6	125
6	115
6	130
2	260
2	219
5	150
4	190
5	163
1	260
4	160

___ **12.57**   The height and weight data for males aged 18–24 years from Exercise 12.33 of Section 12.2:

Height (inches) x	Weight (lb) y
65	175
67	133
71	185
71	163
66	126
75	198
67	153
70	163
71	159
69	151
69	155

___ **12.58**   The size and price data for custom homes from Exercise 12.34 of Section 12.2:

Size (100 sq ft) x	Price ($1000s) y
26	235
27	249
33	267
29	269
29	295
34	345
30	415
40	475
22	195

___ **12.59**   The data on age and peak-heart-rate from Exercise 12.35 of Section 12.2:

Age x	Peak heart rate y
30	186
38	183
41	171
38	177
29	191
39	177
46	175
41	176
42	171
24	196

___ **12.60**   The study-time and test-score data from Exercise 12.36 of Section 12.2:

Study time (hrs) x	Grade (percent) y
10	92
15	81
12	84
20	74
8	85
16	80
14	84
22	80

___ **12.61**   The disposable-income and annual-food-expenditure data from Exercise 12.37 of Section 12.2:

Disposable income ($1000s) x	Food expenditure ($100s) y
30	55
36	60
27	42
20	40
16	37
24	26
19	39
25	43

Exercises 12.62–12.65 are computer exercises.

___ **12.62 (Computer exercise)**   Suppose that the data in Exercise 12.60 are stored in columns named TIME and GRADE.
a) Which Minitab command and subcommands (if any) should be used to obtain the coefficient of determination, r^2, and the three sums of squares, SST, SSR, and SSE, for the data?
b) If you have access to Minitab, use it to determine the coefficient of determination and the three sums of squares for the data.

___ **12.63 (Computer exercise)**   Suppose that the data in Exercise 12.61 are stored in columns named INCOME and FOODEX.
a) Which Minitab command and subcommands (if any) should be used to obtain the coefficient of determination, r^2, and the three sums of squares, SST, SSR, and SSE, for the data?

b) If you have access to Minitab, use it to determine the coefficient of determination and the three sums of squares for the data.

___ **12.64 (Computer exercise)** The Energy Information Administration publishes data on energy consumption by family income in *Residential Energy Consumption Survey: Consumption and Expenditures*. We applied Minitab's REGRESS command to data on family income and last year's energy consumption from a random sample of 25 families. The income data are in thousands of dollars and the energy-consumption data are in millions of BTU. Printout 12.3 on page 608 shows the computer output. Use the printout to find
a) the coefficient of determination for the data.
b) the regression sum of squares, the error sum of squares, and the total sum of squares.
c) the percentage of variation in the sample of energy consumptions that is explained by family income.

___ **12.65 (Computer exercise)** Greene and Touchstone conducted a study on the relationship between the estriol levels of pregnant women and the birth weights of their children. Their findings, entitled "Urinary Tract Estriol: An Index of Placental Function," were published in the *American Journal of Obstetrics and Gynecology*. Printout 12.4 on page 608 shows the computer output that results by applying Minitab's REGRESS command to the data obtained by Greene and Touchstone. The estriol levels are in milligrams per 24 hours and the birth weights are in grams. Using the printout, determine
a) the coefficient of determination for the data.
b) the regression sum of squares, the error sum of squares, and the total sum of squares.
c) the percentage of variation in the birth weights that is explained by estriol level.

= **12.66** Suppose that $r^2 = 1$ for a data set. What can you say about
a) *SSE*?
b) *SSR*?
c) the utility of the regression equation for making predictions?
d) Repeat parts (a)–(c) if $r^2 = 0$.

= **12.67** This exercise shows that the coefficient of determination always lies between 0 and 1.
a) Explain why the quantities *SST*, *SSR*, and *SSE* are nonnegative.
b) Verify that $SSE \le SST$. (*Hint:* Use part (a) and the regression identity, $SST = SSR + SSE$.)
c) Prove that $0 \le r^2 \le 1$.

≡ **12.68** This exercise provides a proof of the regression identity, $SST = SSR + SSE$. It uses some of the results obtained in the course of deriving the formulas for b_0 and b_1 in Exercise 12.52 of Section 12.2.
a) Show that

$$\Sigma(y-\overline{y})^2 = \Sigma(y-\hat{y})^2 + \Sigma(\hat{y}-\overline{y})^2 + 2\Sigma(y-\hat{y})(\hat{y}-\overline{y})$$

(*Hint:* Write $(y - \overline{y})^2$ as $[(y - \hat{y}) + (\hat{y} - \overline{y})]^2$ and apply the binomial formula.)
b) Verify that

$$\Sigma(y-\hat{y})(\hat{y}-\overline{y}) = \Sigma\hat{y}(y-\hat{y}) - \overline{y}\Sigma(y-\hat{y})$$

c) In Exercise 12.52, we found that

$$\Sigma(y-\hat{y}) = 0 \qquad \text{and} \qquad \Sigma x(y-\hat{y}) = 0$$

Use these results and part (b) to show that

$$\Sigma(y-\hat{y})(\hat{y}-\overline{y}) = 0$$

d) Deduce the regression identity from the results obtained in parts (a) and (c).

≡ **12.69** In this exercise, we will derive the shortcut formulas, given in Formula 12.2 on page 617, for the three sums of squares.
a) Obtain the shortcut formula, $SST = S_{yy}$, for the total sum of squares by showing that

$$\Sigma(y-\overline{y})^2 = \Sigma y^2 - (\Sigma y)^2/n$$

(*Hint:* Expand the square on the left of the above equation and then apply properties of summation.)
b) Show that

$$\Sigma(\hat{y}-\overline{y})^2 = b_1^2\Sigma(x-\overline{x})^2$$

(*Hint:* Substitute $b_0 + b_1x$ for $\hat{y}$ and then use the formula for b_0 given in Formula 12.1 on page 597.)
c) Apply the results of part (b) and the formula for b_1 in Formula 12.1 to obtain the shortcut formula, $SSR = S_{xy}^2/S_{xx}$, for the regression sum of squares.
d) Use the regression identity and the shortcut formulas for *SST* and *SSR* to obtain the shortcut formula, $SSE = S_{yy} - S_{xy}^2/S_{xx}$, for the error sum of squares.

12.4   Linear correlation

We often hear statements pertaining to the correlation or lack of correlation between two variables: "There is a positive correlation between advertising expenditures and sales" or "IQ and alcohol consumption are uncorrelated." In this section, we will explain the meaning of such statements.

There are several statistics that can be employed to measure the correlation between two variables. Probably the most commonly used one is the **linear correlation coefficient, *r*,** also called the **Pearson product moment correlation coefficient.** The linear correlation coefficient is a descriptive measure of the strength of the linear (straight-line) relationship between two variables. Formula 12.3 provides a formula for computing the linear correlation coefficient.

FORMULA 12.3   Linear correlation coefficient[†]

The *linear correlation coefficient, r,* of n data points can be computed by using the formula

$$r = \frac{S_{xy}}{\sqrt{S_{xx}S_{yy}}}$$

where S_{xx}, S_{xy}, and S_{yy} are given in Definition 12.3 on page 596.

The linear correlation coefficient, r, always lies between -1 and 1. Values of r close to -1 or 1 indicate a strong linear relationship between the variables and that the variable, x, is a good linear predictor of the variable, y (i.e., the regression equation is quite useful for making predictions). On the other hand, values of r near 0 indicate a weak linear relationship between the variables and that the variable, x, is not too useful as a linear predictor of the variable, y (i.e., the regression equation is not very valuable for making predictions).

Positive values of r suggest that the variables are **positively linearly correlated,** meaning that y tends to increase linearly as x increases, with the tendency being greater the closer that r is to 1. Negative values of r suggest that the variables are **negatively linearly correlated,** meaning that y tends to decrease linearly as x increases, with the tendency being greater the closer that r is to -1. The sign of r is the same as the sign of the slope of the regression line.

Graphically speaking, we can summarize the discussion in the previous two paragraphs as follows (refer to Figure 12.17 at the top of the next page): If the linear correlation coefficient, r, is close to ± 1, then the data points are clustered closely about the regression line. If r is farther from ± 1, then the data points are more widely scattered about the regression line. And if r is near 0, then the slope of the regression line is also near 0, indicating that there is probably no linear relationship between the variables.

---

† This is actually the shortcut formula for r. The defining formula and its equivalence to Formula 12.3 are discussed in the exercises. See, in particular, Exercise 12.87.

FIGURE 12.17

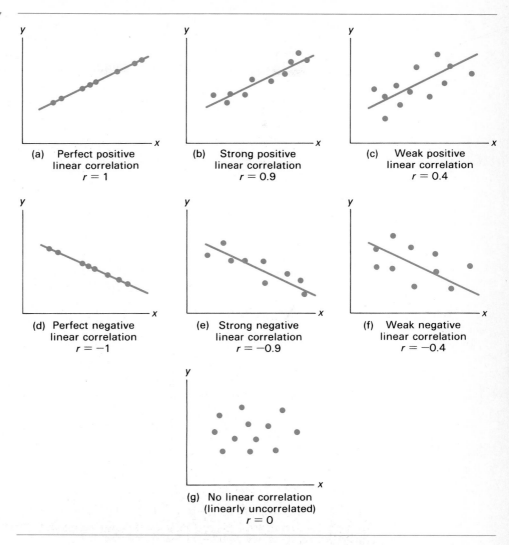

(a) Perfect positive
linear correlation
$r = 1$

(b) Strong positive
linear correlation
$r = 0.9$

(c) Weak positive
linear correlation
$r = 0.4$

(d) Perfect negative
linear correlation
$r = -1$

(e) Strong negative
linear correlation
$r = -0.9$

(f) Weak negative
linear correlation
$r = -0.4$

(g) No linear correlation
(linearly uncorrelated)
$r = 0$

We will now illustrate how to compute and interpret the linear correlation coefficient of a set of data points. To do this, we return to the data on age and price for a sample of Nissan Zs.

EXAMPLE 12.11   *Illustrates the linear correlation coefficient*

The age and price data for a sample of 11 Nissan Zs are repeated in the first two columns of Table 12.11 on the next page. Ages are in years. Prices are in hundreds of dollars, rounded to the nearest hundred.

a) Compute the linear correlation coefficient, r, of the data.
b) Interpret the value of r obtained in part (a) in terms of the linear relationship between the variables, age and price, of Nissan Zs.
c) Discuss the graphical implications of the value of r.

SOLUTION a) We see from Formula 12.3 that in order to compute the linear correlation coefficient, r, we need a table of values for x, y, xy, x^2, y^2, and their sums. This is presented in Table 12.11.

TABLE 12.11

Age (yrs) x	Price ($100s) y	xy	x^2	y^2
5	85	425	25	7,225
4	103	412	16	10,609
6	70	420	36	4,900
5	82	410	25	6,724
5	89	445	25	7,921
5	98	490	25	9,604
6	66	396	36	4,356
6	95	570	36	9,025
2	169	338	4	28,561
7	70	490	49	4,900
7	48	336	49	2,304
58	975	4732	326	96,129

Applying Formula 12.3, we obtain

$$r = \frac{S_{xy}}{\sqrt{S_{xx}S_{yy}}} = \frac{\Sigma xy - (\Sigma x)(\Sigma y)/n}{\sqrt{[\Sigma x^2 - (\Sigma x)^2/n][\Sigma y^2 - (\Sigma y)^2/n]}}$$

$$= \frac{4732 - (58)(975)/11}{\sqrt{[326 - (58)^2/11][96,129 - (975)^2/11]}} = -0.924$$

b) The linear correlation coefficient of $r = -0.924$ suggests that there is a strong negative linear correlation between age and price of Nissan Zs. In particular then, it indicates that as age increases, there is a strong tendency for price to decrease, which is not surprising. It also implies that the regression equation, $\hat{y} = 195.47 - 20.26x$, is quite useful for making predictions.

c) Since the correlation coefficient, $r = -0.924$, is quite close to -1, the data points should be clustered rather closely about the regression line. Figure 12.15 on page 613 shows that this is indeed the case. ∎

MTB

RELATIONSHIP BETWEEN THE CORRELATION COEFFICIENT AND THE COEFFICIENT OF DETERMINATION

In Section 12.3, we discussed the coefficient of determination, r^2. That descriptive measure has two interpretations: It is the percentage reduction obtained in the total squared error by using the regression equation, instead of $\bar{y}$, to predict the observed y-values; and it is the percentage of the total variation in the observed y-values that is explained by the regression line. In any case, the coefficient of determination is a descriptive measure of the utility of the regression equation for making predictions.

Now we have introduced the linear correlation coefficient, r, as a descriptive measure of the strength of the linear relationship between two variables. We would expect the strength of the linear relationship to also give an indication of the usefulness of the regression equation for making predictions. That is, we would expect a connection between the linear correlation coefficient and the coefficient of determination. As a matter of fact, the connection is precisely the one suggested by the notation: r = linear correlation coefficient, r^2 = coefficient of determination. In other words, we have the following key fact (see Exercise 12.88 for a verification).

KEY FACT 12.6

The coefficient of determination is the square of the linear correlation coefficient.

In Example 12.11 on page 624, we found that the linear correlation coefficient for the data on age and price of a sample of 11 Nissan Zs is $r = -0.924$. From this and Key Fact 12.6, we can easily obtain the coefficient of determination:

$$r^2 = (-0.924)^2 = 0.854$$

This, of course, is the same value (except for roundoff error) that we found for r^2 in Example 12.7 on page 612 by using the defining formula, $r^2 = 1 - SSE/SST$. In general then, we can compute the coefficient of determination for a set of data points by using the defining formula, $r^2 = 1 - SSE/SST$, or by first obtaining the linear correlation coefficient and then squaring the result.

A WARNING ON THE USE OF THE LINEAR CORRELATION COEFFICIENT

As we mentioned in Section 12.2, an assumption for finding the regression line for a set of data points is that the data points are actually scattered about a straight line. That same assumption applies to the use of the linear correlation coefficient. In other words, the linear correlation coefficient, r, is used to describe the strength of the *linear* relationship between two variables. It should be employed as a descriptive measure only when a scatter diagram indicates that the data points are scattered about a straight line.

CORRELATION IS NOT CAUSATION

Two variables may have a high correlation without being causally related. For example, Table 12.12 on the next page displays data on total parimutuel turnover (money wagered) at American race tracks and on college enrollment for five randomly selected years between 1970 and 1982. [SOURCE: National Association of State Racing Commissioners and U.S. National Center for Education Statistics.]

TABLE 12.12

Parimutuel turnover ($millions) x	College enrollment (thousands) y
5,977	8,581
7,862	11,185
10,029	11,260
11,677	12,372
11,888	12,426

The linear correlation coefficient of the data points in Table 12.12 is $r = 0.931$, suggesting that there is a strong positive linear correlation between parimutuel wagering and college enrollment. But this does not mean that there is a causal relationship between the two variables, such as that when people go to racetracks, they are somehow inspired to go to college. On the contrary, we can only infer that the two variables have a strong tendency to increase (or decrease) simultaneously and that total parimutuel turnover is a good predictor of college enrollment.

It may happen that two variables are strongly correlated because they are both associated with a third variable. For example, a study was done that showed that teachers' salaries and the dollar amount of liquor sales are positively linearly correlated. A possible explanation for this curious fact might be that both of the variables, teachers' salaries and liquor sales, are tied to other variables, such as the rate of inflation, that pull them along together.

USING THE COMPUTER (OPTIONAL)

Minitab has a program called **CORRELATION** that will compute the linear correlation coefficient, r, of a set of data points. The next example shows how to apply the CORRELATION command.

EXAMPLE 12.12   *Illustrates the CORRELATION command*

The data on age and price for a sample of 11 Nissan Zs are displayed in the first two columns of Table 12.11 on page 624. Use Minitab to determine the linear correlation coefficient of the data.

SOLUTION   Recall that we have previously stored the age and price data in columns named AGE and PRICE. To have Minitab compute the linear correlation coefficient of the data, we type the command CORRELATION followed by the storage locations of the sample data; that is, we type CORRELATION of 'AGE' and 'PRICE'. See Printout 12.6.

PRINTOUT 12.6
Minitab output for
CORRELATION

```
MTB > CORRELATION of 'AGE' and 'PRICE'

Correlation of AGE and PRICE = -0.924
```

From the printout, we conclude that the linear correlation coefficient for the age and price data is -0.924; that is, $r = -0.924$. Evidently, there is a strong negative linear correlation between age and price of Nissan Zs. ∎

Exercises 12.4

In Exercises 12.70–12.77, we have repeated the data from exercises in Section 12.2. For each exercise,
a) compute the linear correlation coefficient, r.
b) interpret the value of r in terms of the linear relationship between the two variables in question.
c) discuss the graphical interpretation of the value of r and check that it is consistent with the graph drawn in the corresponding exercise in Section 12.2.
d) square the value of r and compare the result with the value of the coefficient of determination found in the corresponding exercise in Section 12.3.

___ **12.70** The data for Exercise 12.30:

Data points

x	1	1	5	5
y	1	3	2	4

___ **12.71** The data for Exercise 12.31:

Data points

x	0	2	2	5	6
y	4	2	0	-2	1

___ **12.72** The age and price data for a sample of 10 Corvettes (from Exercise 12.32 of Section 12.2):

Age (yrs) x	Price ($100s) y
6	125
6	115
6	130
2	260
2	219
5	150
4	190
5	163
1	260
4	160

___ **12.73** The height and weight data for 11 randomly selected males aged 18–24 years (from Exercise 12.33 of Section 12.2):

Height (inches) x	Weight (lb) y
65	175
67	133
71	185
71	163
66	126
75	198
67	153
70	163
71	159
69	151
69	155

___ **12.74** The size and price data for a random sample of nine custom homes in the Equestrian Estates (from Exercise 12.34 of Section 12.2):

Size (100 sq ft) x	Price ($1000s) y
26	235
27	249
33	267
29	269
29	295
34	345
30	415
40	475
22	195

___ **12.75** The data on age and peak-heart-rate for a random sample of 10 people (from Exercise 12.35 of Section 12.2):

Age	Peak heart rate
x	y
30	186
38	183
41	171
38	177
29	191
39	177
46	175
41	176
42	171
24	196

Height (inches)	Exam score
x	y
71	87
68	96
71	66
65	71
66	71
68	55
68	83
64	67
62	86
65	60

___ **12.76** The study-time and test-score data for a random sample of eight students in a beginning calculus course (from Exercise 12.36 of Section 12.2):

Study time (hrs)	Grade (percent)
x	y
10	92
15	81
12	84
20	74
8	85
16	80
14	84
22	80

___ **12.77** The disposable-income and annual-food-expenditure data for eight randomly selected middle-income families (from Exercise 12.37 of Section 12.2):

Disposable income ($1000s)	Food expenditure ($100s)
x	y
30	55
36	60
27	42
20	40
16	37
24	26
19	39
25	43

___ **12.78** We took a random sample of 10 students from an introductory statistics class and obtained the following data on height and final-exam score:

a) What sort of value of r would you expect to find for these data? Explain your answer.
b) Compute r.

___ **12.79**   Consider the following set of data points:

x	-3	-2	-1	0	1	2	3
y	9	4	1	0	1	4	9

a) Compute the linear correlation coefficient, r.
b) Can you conclude from your result in part (a) that the variables, x and y, are unrelated? Explain.
c) Draw a scatter diagram for the data.
d) Is it appropriate to use the linear correlation coefficient as a descriptive measure for the data? Why?
e) Show that the data are related by the equation $y = x^2$ and graph that equation along with the data points.

___ **12.80**   Determine whether r is positive, negative, or zero for each of the following data sets:

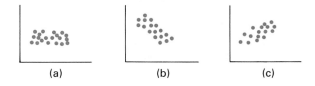

    (a)   (b)   (c)

Exercises 12.81 and 12.82 are computer exercises.

___ **12.81** **(Computer exercise)** Suppose that the data in Exercise 12.77 are stored in columns named INCOME and FOODEX.
a) Which Minitab command and subcommands (if any) should be used to obtain the linear correlation coefficient of the data?

b) If you have access to Minitab, use it to determine the linear correlation coefficient.

___ **12.82 (Computer exercise)** Suppose that the data in Exercise 12.76 are stored in columns named TIME and GRADE.
a) Which Minitab command and subcommands (if any) should be used to obtain the linear correlation coefficient of the data?
b) If you have access to Minitab, use it to determine the linear correlation coefficient.

Defining formula for the linear correlation coefficient: In Exercises 12.47–12.51 of Section 12.2, we examined the concept of the sample covariance. Recall that the *sample covariance*, s_{xy}, of a sample of n data points is defined by

$$(3) \qquad s_{xy} = \frac{\Sigma(x - \bar{x})(y - \bar{y})}{n - 1}$$

The defining formula for the **linear correlation coefficient** is

$$(4) \qquad r = \frac{s_{xy}}{s_x s_y}$$

where s_x and s_y are the sample standard deviations of the x-values and y-values, respectively.

═ **12.83** Use Formula (4) to compute r for the set of data points in Exercise 12.71 and compare your answer with the value obtained for r in that exercise.

═ **12.84** Use Formula (4) to compute r for the set of data points in Exercise 12.70 and compare your answer with the value obtained for r in that exercise.

═ **12.85** In this exercise, we will discuss the interpretation of the sample covariance.
a) Consider the following data set:

x	1	2	3	4	5	6
y	1	3	4	6	6	7

For these data, $\bar{x} = 3.5$ and $\bar{y} = 4.5$. At the top of the next column, we have drawn a coordinate system with a second set of axes passing through the point $(3.5, 4.5)$. Construct a scatter diagram for the data points on the coordinate system provided.

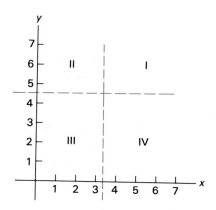

b) The dashed lines divide the above coordinate system into four regions, which we have labeled I, II, III, and IV. For a point, (x, y), in region I, $x - \bar{x}$ and $y - \bar{y}$ are both positive, so $(x - \bar{x})(y - \bar{y})$ is also positive. Fill in the remainder of the following table by using similar reasoning.

Region	Sign of $(x - \bar{x})(y - \bar{y})$
I	+
II	
III	
IV	

c) Without performing any calculations, determine whether the sample covariance of the data points in part (a) is positive or negative. (*Hint:* Use the graph from part (a) and the table from part (b).)
d) Use a similar graphing procedure to decide whether the sample covariance is positive or negative for the following data points:

x	1	2	3	4	5	6
y	7	6	6	4	3	1

e) Below are two data sets, one scattered about a straight line with positive slope and the other scattered about a straight line with negative slope.

Complete the following statements:
  (i) For data scattered about a straight line with positive slope, the covariance is _____ .
  (ii) For data scattered about a straight line with negative slope, the covariance is _____ .

≡ **12.86**   Exercise 12.67 of Section 12.3 shows that the coefficient of determination always lies between 0 and 1. Use that result and Key Fact 12.6 on page 625 to deduce that the linear correlation coefficient always lies between -1 and 1; that is, $-1 \le r \le 1$.

≡ **12.87**   Show that, for the linear correlation coefficient, the defining formula, Formula (4), and the shortcut formula, Formula 12.3, are equivalent. *(Hint: Use the defining formulas for S_{xy}, S_{xx}, and S_{yy} given in Definition 12.3 on page 596.)*

≡ **12.88**   This exercise verifies Key Fact 12.6. Prove that the square of the correlation coefficient is equal to the coefficient of determination. *(Hint: Rewrite the defining formula for the coefficient of determination in terms of the shortcut formulas for SSE and SST.)*

12.5   Multiple regression (Optional)

Up to this point, we have considered linear regression analysis with one predictor (independent) variable. That type of linear regression analysis is called **simple linear regression.** Now we will study **multiple regression,** which applies when there is more than one predictor variable.

In simple linear regression, where there is only one predictor variable, x, the regression equation is of the form

$$\hat{y} = b_0 + b_1 x$$

For multiple regression with two predictor variables, x_1 and x_2, the regression equation is of the form

$$\hat{y} = b_0 + b_1 x_1 + b_2 x_2$$

And, in general, for multiple regression with k predictor variables, $x_1, x_2, \ldots, x_k$, the regression equation is of the form

$$\hat{y} = b_0 + b_1 x_1 + \cdots + b_k x_k$$

Basically, multiple regression uses the same principles and techniques as those in simple linear regression: The regression equation is obtained by employing the least squares criterion; the three sums of squares, *SST, SSR,* and *SSE,* and the coefficient of determination are defined in the same way; and so forth.

We should point out that, in multiple regression, the **coefficient of determination** is denoted by R^2 instead of r^2. However, the interpretation is the same. In particular, the coefficient of determination, R^2, equals the percentage of the total variation in the observed y-values that is explained by the regression. Thus, R^2 is a descriptive measure of the utility of the regression equation for making predictions. Values of R^2 near 0 indicate that the regression equation is not very useful for making predictions, whereas values of R^2 near 1 indicate that the regression equation is extremely useful for making predictions.

In Formula 12.1 on page 597, we presented formulas that can be used to obtain the regression equation in simple linear regression. Similar formulas exist for multiple regression but they are much more complicated, especially if there are a large number of predictor variables. Because of this and other computational difficulties, multiple regression is almost always done by computer. So, in this section, we will present computer output as a means for examining multiple regression.

EXAMPLE 12.13   *Illustrates multiple regression*

The data on age and price for a sample of 11 Nissan Zs are repeated in the second and fourth columns of Table 12.13. Ages are in years. Prices are in hundreds of dollars, rounded to the nearest hundred. Those data have been used to illustrate simple linear regression throughout this chapter.

In Example 12.7 on page 612, we found that the coefficient of determination for the age and price data is $r^2 = 0.853$. This means that 85.3% of the variation in the price data is explained by age. Perhaps by using some additional predictor variables we could explain more of the variation in the price data and, thereby, obtain a regression equation that is a better predictor of price.

For instance, besides "age," we might include "number of miles driven" as a predictor variable. The third column of Table 12.13 gives the number of miles, in thousands, that each of the 11 Nissan Zs has been driven.

TABLE 12.13
Data on age, miles driven, and price for Nissan Zs

Car	Age (yrs) x_1	Miles (thous) x_2	Price ($100s) y
1	5	57	85
2	4	40	103
3	6	77	70
4	5	60	82
5	5	49	89
6	5	47	98
7	6	58	66
8	6	39	95
9	2	8	169
10	7	69	70
11	7	89	48

We employed a statistical computer program to perform a multiple regression analysis on the data in Table 12.13 with the variables, age and miles, as predictor variables. The results are displayed in Printout 12.7 at the top of the next page.

a) Use the output in Printout 12.7 to obtain the regression equation for price in terms of age and miles.

b) Apply the regression equation to predict the price of a Nissan Z that is five years old and has been driven 52,000 miles.

c) Use the output in Printout 12.7 to find the coefficient of determination, R^2.

d) Which regression equation better explains the variation in the price data: the multiple regression equation, using both age and miles as predictor variables; or the simple linear regression equation, using only age as a predictor variable?

PRINTOUT 12.7
Computer output for
multiple regression
of Nissan Z data
in Table 12.13

```
The regression equation is
PRICE = 183 - 9.50 AGE - 0.821 MILES

Predictor         Coef        Stdev     t-ratio        p
Constant        183.04        11.35       16.13    0.000
AGE             -9.504        3.874       -2.45    0.040
MILES          -0.8215       0.2552       -3.22    0.012

s = 8.805         R-sq = 93.6%      R-sq(adj) = 92.0%


Analysis of "     ınce

SOURCE        DF          SS          MS         F        p
Regression     2       9088.3      4544.2     58.61    0.000
Error          8        620.2        77.5
Total         10       9708.5
```

SOLUTION   a) The regression equation is given in the second line of the computer output in Printout 12.7: PRICE = 183 - 9.50 AGE - 0.821 MILES. In other words, the regression equation is

$$\hat{y} = 183 - 9.50x_1 - 0.821x_2$$

where x_1 denotes age, in years, x_2 denotes miles driven, in thousands, and $\hat{y}$ denotes predicted price, in hundreds of dollars.

b) For a five-year-old Nissan Z with 52,000 miles, we have $x_1 = 5$ and $x_2 = 52$. The predicted price for such a car is

$$\hat{y} = 183 - 9.50 \cdot 5 - 0.821 \cdot 52 = 92.81$$

or $9,281.

c) The coefficient of determination is displayed as the second entry in the seventh line of the computer output in Printout 12.7: R-sq = 93.6%. That is, the coefficient of determination is $R^2 = 0.936$.

d) From part (c), we conclude that 93.6% of the variation in prices of the 11 Nissan Zs sampled is explained by age and miles driven. On the other hand, as we have seen, only 85.3% of the variation in prices of the 11 Nissan Zs sampled is explained by age alone. Thus, the multiple regression equation provides a much better explanation of the variation in the price data than the simple linear regression equation. As a consequence, we would expect to be able to make better price predictions by using both age and miles driven instead of age alone. ■

MTB

CURVILINEAR REGRESSION

We pointed out in Section 12.2 (page 599) that finding a regression line for a set of data points is based on the assumption that the data points are scattered about a straight line. If the data points are not scattered about a straight line, but follow a curve, then a regression line should not be determined. Instead, a curve should be fit to the data points.

Multiple regression can be used to fit curves to a set of data points. This is called **curvilinear regression.** We will illustrate the use of curvilinear regression in Exercise 12.95.

USING THE COMPUTER (OPTIONAL)

In Section 12.2, we learned how to use Minitab's REGRESS command to perform a regression analysis with one predictor variable. For instance, suppose that the data on age and price for a sample of 11 Nissan Zs are stored in columns named AGE and PRICE. Then to perform a regression analysis for the price of a Nissan Z with age as the predictor variable, we type

REGRESS 'PRICE' on 1 predictor 'AGE'

The 1 signifies that there is one predictor variable being used for the regression.

To perform a regression analysis with two or more predictor variables, we proceed in a similar manner. Example 12.14 provides the details.

EXAMPLE 12.14   *Illustrates the REGRESS command*

Table 12.13 on page 631 gives data on age, miles driven, and price for a sample of 11 Nissan Zs. Explain how to obtain the Minitab output shown in Printout 12.7 on page 632.

SOLUTION   To begin, we enter the age, miles, and price data from Table 12.13 into, say, C3, C4, and C5, using the READ command. Then we name C3 "Age," C4 "Miles," and C5 "Price" by employing the NAME command. See Printout 12.8.

PRINTOUT 12.8

```
MTB > READ C3 C4 C5
DATA> 5 57 85
DATA> 4 40 103
DATA> 6 77 70
DATA> 5 60 82
DATA> 5 49 89
DATA> 5 47 98
DATA> 6 58 66
DATA> 6 39 95
DATA> 2 8 169
DATA> 7 69 70
DATA> 7 89 48
DATA> END
      11 ROWS READ
MTB > NAME C3 'AGE' C4 'MILES' C5 'PRICE'
```

Finally, we apply the REGRESS command with *two* predictor variables, age and miles driven. In other words, we type

REGRESS 'PRICE' on 2 predictors 'AGE' and 'MILES'

The result of this command is the computer output shown in Printout 12.7.   ■

Exercises 12.5

___ 12.89 A manufacturer of household appliances wants to analyze the relationship between total sales and the three primary means of advertising. The first three columns of the following table provide the expenditures on advertising, by type, for each of 10 randomly selected sales periods. The fourth column contains the total sales. All data are in millions of dollars.

Television x_1	Magazines x_2	Radio x_3	Sales y
8.3	4.4	6.1	361.1
6.3	4.2	4.9	344.0
9.9	5.9	6.3	377.9
9.4	3.3	6.1	371.5
10.4	2.7	5.2	365.4
9.0	3.5	5.1	364.5
9.2	4.1	6.0	372.9
10.6	4.8	6.4	379.4
9.3	4.2	5.5	362.6
10.5	6.0	5.9	387.5

We used a computer package to perform a multiple regression analysis on the data with the variables, television, magazine, and radio advertising expenditures, as predictor variables. The computer output is shown in Printout 12.9 at the top of the next page.
a) Use the computer output to obtain the regression equation for sales in terms of television, magazine, and radio advertising expenditures.
b) Apply the regression equation to predict total sales if the amounts spent on television, magazine, and radio advertising are \$9.5, \$4.3, and \$5.2 million, respectively.
c) Obtain and interpret the coefficient of determination, R^2.

___ 12.90   The data on age and price for 10 Corvettes from Exercise 12.32 of Section 12.2 are repeated in the first and third columns of the following table. [SOURCE: *Arizona Republic.*] In Exercise 12.56 of Section 12.3, we found that the coefficient of determination for the age and price data is $r^2 = 0.937$. This means that 93.7% of the variation in the price data is explained by age. The second column of the table below gives the number of miles, in thousands, that each of the 10 Corvettes has been driven.

Age (yrs) x_1	Miles (thous) x_2	Price (\$100s) y
6	36	125
6	36	115
6	36	130
2	22	260
2	5	219
5	31	150
4	22	190
5	39	163
1	9	260
4	27	160

We used a computer package to perform a multiple regression analysis on the data with the variables, age and miles, as predictor variables. The computer output is shown in Printout 12.10 on the next page.
a) Use the output to obtain the regression equation for price in terms of age and miles.
b) Apply the regression equation to predict the price of a Corvette that is four years old and has been driven 28,000 miles.
c) Use the computer output to find the coefficient of determination, R^2.
d) Which regression equation better explains the variation in the price data: the multiple regression equation, using both age and miles as predictor variables; or the simple linear regression equation, using only age as a predictor variable?

PRINTOUT 12.9   Computer output for Exercise 12.89

```
The regression equation is
SALES = 266 + 6.73 TV + 3.26 MAG + 4.51 RADIO

Predictor        Coef       Stdev      t-ratio         p
Constant       266.23       16.34       16.29       0.000
TV              6.727        1.344        5.01       0.002
MAG             3.257        1.642        1.98       0.095
RADIO           4.507        3.703        1.22       0.269

s = 4.418        R-sq = 91.1%      R-sq(adj) = 86.6%

Analysis of Variance

SOURCE         DF          SS          MS         F         p
Regression      3       1194.53      398.18     20.40     0.002
Error           6        117.11       19.52
Total           9       1311.64
```

PRINTOUT 12.10   Computer output for Exercise 12.90

```
The regression equation is
PRICE = 287 - 37.4 AGE + 1.64 MILES

Predictor        Coef       Stdev      t-ratio         p
Constant       287.362      9.943       28.90       0.000
AGE            -37.375      5.174       -7.22       0.000
MILES           1.6378      0.8116       2.02       0.083

s = 12.11        R-sq = 96.0%      R-sq(adj) = 94.9%

Analysis of Variance

SOURCE         DF          SS          MS         F         p
Regression      2        24655       12328      84.07     0.000
Error           7         1026         147
Total           9        25682
```

___ **12.91** Colleges and universities often require prospective students to take college-entrance exams. The theory is that those examinations are helpful for predicting success in college. We used the *Focus* system at Arizona State University to obtain data on high-school GPA (cumulative grade-point average), SAT math score, SAT verbal score, and sophomore GPA (cumulative grade-point average through two years of college). This was done for a random sample of 47 students. Printout 12.11 at the top of the next page provides the output obtained by applying a statistical computer package to perform a regression analysis on the data. The regression uses high-school GPA, SAT math score, and SAT verbal score as predictor variables for sophomore GPA.
a) Determine the regression equation.
b) Apply the regression equation to predict the sophomore GPA of an Arizona State University student with a high-school GPA of 3.20, an SAT math score of 650, and an SAT verbal score of 500.
c) Find the coefficient of determination.
d) How useful do high-school GPA and SAT scores appear to be for predicting sophomore GPA at Arizona State University?

___ **12.92** Hanna Properties specializes in custom-home resales in the Equestrian Estates, an exclusive subdivision in Phoenix, Arizona. A random sample of 33 properties was selected and data on the following variables were obtained: square-footage, number of bedrooms, number of bathrooms, number of days on the market, and selling price (in thousands of dollars). Then a computer package was used to perform a regression analysis for selling price in terms of the other four variables. The resulting computer output is displayed in Printout 12.12 on the next page.
a) Use the output to obtain the regression equation.
b) Apply the regression equation to find the predicted selling price for a home in the Equestrian Estates that has 3200 square feet, four bedrooms, three bathrooms, and has been on the market for 60 days.
c) Find and interpret the coefficient of determination.

Exercises 12.93 and 12.94 are computer exercises.

___ **12.93 (Computer exercise)** Suppose that the data in Exercise 12.89 are stored in columns named TV, MAG, RADIO, and SALES.
a) Which Minitab command should be used to obtain Printout 12.9 on page 635?
b) If you have access to Minitab, use it to obtain Printout 12.9.

___ **12.94 (Computer exercise)** Suppose that the data in Exercise 12.90 are stored in columns named AGE, MILES, and PRICE.
a) Which Minitab command should be used to obtain Printout 12.10 on page 635?
b) If you have access to Minitab, use it to obtain Printout 12.10.

=== **12.95 Curvilinear regression:** This exercise illustrates how to fit a curve to a set of data points using multiple regression. We stated earlier that although the relationship between age and price of Nissan Zs appears to be linear in the age-range from two to seven years, it is definitely not so in the age-range from two to 11 years. From the *Auto Trader* magazine, we obtained the following data on age and price for a sample of 31 Nissan Zs (ages are in years, prices are in hundreds of dollars). Below the table is a scatter diagram for the data.

Age x	Price y	Age x	Price y	Age x	Price y	Age x	Price y
5	85	4	103	10	25	3	135
6	70	4	100	5	82	9	44
4	90	6	75	10	35	9	36
2	150	3	140	5	89	11	33
5	98	6	66	6	95	1	180
6	95	2	169	4	65	5	80
3	129	6	60	6	82	5	105
4	115	8	50	9	42		

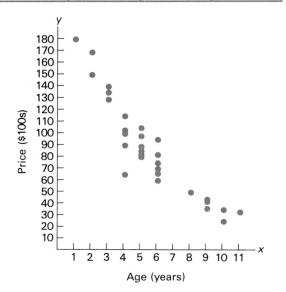

PRINTOUT 12.11   Computer output for Exercise 12.91

The regression equation is
SOPHGPA = 0.648 + 0.562 HSGPA +0.000267 SATMATH +0.000223 SATVERBL

Predictor	Coef	Stdev	t-ratio	p
Constant	0.6479	0.4294	1.51	0.139
HSGPA	0.5616	0.1521	3.69	0.001
SATMATH	0.0002665	0.0009694	0.27	0.785
SATVERBL	0.0002234	0.0009959	0.22	0.824

s = 0.4860   R-sq = 34.9%   R-sq(adj) = 30.3%

Analysis of Variance

SOURCE	DF	SS	MS	F	p
Regression	3	5.4431	1.8144	7.68	0.000
Error	43	10.1585	0.2362		
Total	46	15.6015			

PRINTOUT 12.12   Computer output for Exercise 12.92

The regression equation is
SELL\$ = - 212 + 0.0754 SQFT + 20.9 BEDROOMS + 62.1 BATHS - 0.0899 DAYS

Predictor	Coef	Stdev	t-ratio	p
Constant	-211.95	73.05	-2.90	0.007
SQFT	0.07542	0.02345	3.22	0.003
BEDROOMS	20.94	19.99	1.05	0.304
BATHS	62.11	21.62	2.87	0.008
DAYS	-0.08986	0.07624	-1.18	0.248

s = 39.48   R-sq = 84.0%   R-sq(adj) = 81.7%

Analysis of Variance

SOURCE	DF	SS	MS	F	p
Regression	4	229575	57394	36.83	0.000
Error	28	43633	1558		
Total	32	273209			

As you can see from the scatter diagram, the data points are not clustered about a straight line but instead follow a curve. This indicates that we should not determine a regression line but that we should try to fit a curve to the data. From the curvature of the scatter diagram, it appears that a parabola might be an appropriate curve to fit to the data points. To fit a parabola to the data points, we need a regression equation of the form

$$\hat{y} = b_0 + b_1 x + b_2 x^2$$

If we let $x_1 = x$ and $x_2 = x^2$, then the previous equation becomes

$$\hat{y} = b_0 + b_1 x_1 + b_2 x_2$$

which is a multiple regression equation with two predictor variables, age and age^2 (square of the age). We used a computer package to perform a multiple regression analysis for price with age and age^2 as predictor variables. Printout 12.13 shows the computer output.

a) Use the computer output to obtain the regression equation; that is, the equation of the parabola that best fits the data points in the scatter diagram.
b) Graph the parabola that you obtained in part (a) on the scatter diagram (bottom of page 636).
c) Apply the regression equation to predict the price of a nine-year-old Nissan Z.
d) What percentage of the total variation in the prices is explained by age and age^2 (i.e., by the parabola)?

═ **12.96 (Computer exercise)** Suppose that the data in Exercise 12.95 are stored in columns named AGE and PRICE. Further suppose that the squares of the age data are stored in a column named AGESQ.
a) Which Minitab command and subcommands (if any) should be used to obtain Printout 12.13 shown below?
b) If you have access to Minitab, use it to obtain Printout 12.13. (*Note:* To store the squares of the age data, first name a column AGESQ and then type the command LET 'AGESQ'='AGE'**2.)

PRINTOUT 12.13   Computer output for Exercise 12.95

```
The regression equation is
PRICE = 209 - 30.8 AGE + 1.33 AGESQ
```

Predictor	Coef	Stdev	t-ratio	p
Constant	209.44	11.48	18.24	0.000
AGE	-30.776	4.056	-7.59	0.000
AGESQ	1.3297	0.3219	4.13	0.000

```
s = 12.81      R-sq = 90.3%      R-sq(adj) = 89.6%
```

Analysis of Variance

SOURCE	DF	SS	MS	F	p
Regression	2	42895	21448	130.70	0.000
Error	28	4595	164		
Total	30	47490			

Chapter review

KEY TERMS	BRIEF,* 603	CORRELATION,* 626
	coefficient of determination (r^2), 612	curvilinear regression,* 633
	coefficient of determination (R^2),* 630	error sum of squares (SSE), 612

FORMULAS

In the formulas below,

n = sample size (number of data points)
SST = total sum of squares
SSR = regression sum of squares

SSE = error sum of squares
r^2 = coefficient of determination
r = linear correlation coefficient

S_{xx}, S_{xy}, and S_{yy}, 596

$$S_{xx} = \Sigma(x - \overline{x})^2 = \Sigma x^2 - (\Sigma x)^2/n$$
$$S_{xy} = \Sigma(x - \overline{x})(y - \overline{y}) = \Sigma xy - (\Sigma x)(\Sigma y)/n$$
$$S_{yy} = \Sigma(y - \overline{y})^2 = \Sigma y^2 - (\Sigma y)^2/n$$

Regression equation, 597

$$\hat{y} = b_0 + b_1 x$$

where

$$b_1 = \frac{S_{xy}}{S_{xx}}, \qquad b_0 = \frac{1}{n}(\Sigma y - b_1 \Sigma x)$$

Total sum of squares, 612, 617

$$SST = \Sigma(y - \overline{y})^2 = S_{yy}$$

Regression sum of squares, 615, 617

$$SSR = \Sigma(\hat{y} - \overline{y})^2 = S_{xy}^2/S_{xx}$$

Error sum of squares, 612, 617

$$SSE = \Sigma(y - \hat{y})^2 = S_{yy} - S_{xy}^2/S_{xx}$$

Regression identity, 615

$$SST = SSR + SSE$$

Coefficient of determination, 612, 616

$$r^2 = 1 - \frac{SSE}{SST} = \frac{SSR}{SST}$$

Linear correlation coefficient, 622

$$r = \frac{S_{xy}}{\sqrt{S_{xx}S_{yy}}}$$

Multiple regression equation,* 630

$$\hat{y} = b_0 + b_1 x_1 + \cdots + b_k x_k$$

(k = number of predictor variables)

YOU SHOULD
BE ABLE TO

1. use and understand the preceding formulas.
2. apply the concepts related to linear equations with one independent variable.
3. explain the least-squares criterion.
4. obtain and graph the regression equation for a set of data points, interpret the slope of the regression line, and use the regression equation to make predictions.
5. calculate and interpret the three sums of squares, SST, SSE, and SSR, and the coefficient of determination, r^2.
6. determine and interpret the linear correlation coefficient, r.
7. obtain the regression equation and coefficient of determination for a multiple regression from computer output.*
8. apply a multiple regression equation to make predictions.*
9. use the Minitab commands covered in this chapter.*
10. interpret the output obtained from the application of the Minitab commands discussed in this chapter.*

REVIEW TEST

1. A small company has purchased a microcomputer system for $7200 and plans to depreciate the value of the equipment by $1200 per year for six years. Let x denote the age of the equipment, in years, and y denote the value of the equipment, in hundreds of dollars.

a) Determine the equation that expresses y in terms of x.

b) Find the y-intercept, b_0, and slope, b_1, for the linear equation in part (a).

c) Without graphing the equation in part (a), decide whether the line slopes upward, slopes downward, or is horizontal.

d) Find the value of the computer equipment after two years; after five years.

e) Obtain the graph of the equation in part (a) by plotting the points from part (d) and connecting them with a straight line.

f) Use the graph from part (e) to visually estimate the value of the equipment after four years. Then calculate that value exactly by employing the equation from part (a).

2. The director of a large mathematics course hires upper-division science students to grade papers. On each grading day, he records the number of papers graded and the total amount of money paid to the graders. The table shown at the top of the first column on the next page provides data for 12 randomly selected days from last semester:

No. papers (100s)	Cost ($)
x	y
16	234
16	220
18	258
22	298
19	273
16	227
18	246
15	210
19	265
17	250
15	223
18	251

a) Draw a scatter diagram of the data.

b) Is it reasonable to find a regression line for the data? Explain.

c) Determine the regression equation for the data and draw its graph on the scatter diagram that you drew in part (a).

d) Describe the apparent relationship that exists between the two variables, number of papers graded and cost of grading.

e) What does the slope of the regression line represent in terms of paper-grading costs?

f) Use the regression equation to predict the cost of grading 1600 papers.

3. Refer to Problem 2.

a) Compute SST, SSR, and SSE using the shortcut formulas.

b) Calculate the coefficient of determination using the defining formula.

c) Determine the percentage reduction obtained in the total squared error by using the regression equation, instead of the sample mean, $\bar{y}$, to predict the observed costs.

d) Obtain the percentage of the variation in the observed costs that is explained by the number of papers graded (i.e., by the regression line).

e) State how useful the regression equation appears to be for making predictions.

*4. **(Computer problem)** Suppose that the data in Problem 2 are stored in columns named PAPERS and COST.

a) Which Minitab command and subcommands (if any) should be used to obtain a scatter diagram for the data?

b) Which Minitab command and subcommands (if any) should be used to obtain the regression equation for the data?

c) Which Minitab command and subcommands (if any) should be used to obtain the coefficient of determination, r^2, and the three sums of squares, SST, SSR, and SSE, for the data?

d) If you have access to Minitab, use it to obtain a scatter diagram for the data and to determine the regression equation for the data, the coefficient of determination for the data, and the three sums of squares for the data.

*5. **(Computer problem)** Printout 12.14 on the next page displays the Minitab output obtained by applying the REGRESS command to the data in Problem 2. Employ the output to determine

a) the regression equation.

b) the coefficient of determination.

c) the regression sum of squares, the error sum of squares, and the total sum of squares.

6. Refer to Problem 2.

a) Compute the linear correlation coefficient, r.

b) Interpret the value of r in terms of the linear relationship between the number of papers graded and the cost of grading.

c) Discuss the graphical implications of the value of the linear correlation coefficient, r.

d) Use the value of the linear correlation coefficient that you computed in part (a) to obtain the coefficient of determination.

*7. **(Computer problem)** Suppose that the data in Problem 2 are stored in columns named PAPERS and COST.

a) Which Minitab command and subcommands (if any) should be used to obtain the linear correlation coefficient of the data?

b) If you have access to Minitab, use it to determine the linear correlation coefficient.

*8. The U.S. Bureau of the Census collects data on income by educational attainment, sex, and age. Results are published in *Current Population Reports*. From a random sample of 75 males between the ages of 25 and 50, all of whom have at least a ninth-grade education, data were collected on age, number of years of school completed, and annual income. Then a statistical computer program was used to perform a multiple regression analysis for annual income with the variables, age and number of years of school completed, as predictor variables. The resulting computer output is displayed below in Printout 12.15.

a) Use the computer output to obtain the regression equation for annual income in terms of age and number of years of school completed.

b) Apply the regression equation to predict the annual income of a male who is 32 years old and has completed four years of college (i.e., 16 years of school).

c) Obtain and interpret the coefficient of determination, R^2.

*9. **(Computer problem)** Refer to Problem 8. Assume that the data on age, number of years of school completed, and annual income are stored, respectively, in columns named AGE, EDUC, and INCOME. Which Minitab command and subcommands (if any) should be used to obtain the output shown in Printout 12.15?

PRINTOUT 12.14 Minitab output for Problem 5

The regression equation is
COST = 35.8 + 12.1 PAPERS

Predictor	Coef	Stdev	t-ratio	p
Constant	35.80	17.06	2.10	0.062
PAPERS	12.0835	0.9738	12.41	0.000

s = 6.526   R-sq = 93.9%   R-sq(adj) = 93.3%

Analysis of Variance

SOURCE	DF	SS	MS	F	p
Regression	1	6558.3	6558.3	153.97	0.000
Error	10	425.9	42.6		
Total	11	6984.3			

PRINTOUT 12.15 Computer output for Problem 8

The regression equation is
INCOME = - 40.9 + 0.772 AGE + 3.11 EDUC

Predictor	Coef	Stdev	t-ratio	p
Constant	-40.855	5.511	-7.41	0.000
AGE	0.7719	0.1165	6.62	0.000
EDUC	3.1052	0.2250	13.80	0.000

s = 7.886   R-sq = 76.6%   R-sq(adj) = 75.9%

Analysis of Variance

SOURCE	DF	SS	MS	F	p
Regression	2	14657.2	7328.6	117.84	0.000
Error	72	4477.6	62.2		
Total	74	19134.7			

CHAPTER 13

INFERENTIAL METHODS IN LINEAR REGRESSION

In the previous chapter we examined descriptive methods in linear regression and correlation. We discovered how to determine the regression equation for a set of data points and how to use the regression equation to make predictions. We also learned how to compute and interpret the coefficient of determination and the linear correlation coefficient for a set of data points.

Now we will consider some inferential methods in linear regression and correlation. For example, we will see how the regression equation can be used to obtain a confidence interval for the mean price of all Nissan Zs of any particular age; and how the linear correlation coefficient, r, can be used to decide whether there is a negative correlation between the variables, age and price, for Nissan Zs.

CHAPTER OUTLINE

13.1 The regression model Describes the conditions that must be met by two variables in order to use inferential methods in linear regression and correlation.

13.2 Inferences for the slope of the population regression line Explains how to perform hypothesis tests and how to obtain confidence intervals for the slope of the population regression line.

13.3 Estimation and prediction Shows how to find a confidence interval for the mean of the population of y-values corresponding to a particular x-value and how to obtain a prediction interval for a population y-value corresponding to a particular x-value.

13.4 Inferences in correlation Discusses how to perform a hypothesis test to decide whether two variables are linearly correlated.

13.5 Multiple regression (Optional) Introduces the regression model for multiple regression and explains how the computer can be used to perform inferences in multiple regression.

13.1 The regression model

To perform statistical inferences in linear regression and correlation, it is necessary that the variables under consideration satisfy certain conditions. We will discuss those conditions in this section.

To begin, let us return to the Nissan Z illustration. In Table 13.1 we have reproduced the data on age and price for a sample of 11 Nissan Zs. Ages are in years. Prices are in hundreds of dollars, rounded to the nearest hundred.

TABLE 13.1
Age versus price
data for Nissan Zs

Age (yrs) x	Price ($100s) y
5	85
4	103
6	70
5	82
5	89
5	98
6	66
6	95
2	169
7	70
7	48

In Section 12.2 (page 598), we found that the regression equation for these data is $\hat{y} = 195.47 - 20.26x$. We can use the regression equation to predict the price of a Nissan Z given its age. However, we cannot expect such predictions to be completely accurate since prices vary even for Nissan Zs of the same age. For example, in the sample data of Table 13.1, there are four five-year-old Nissan Zs. Their prices are $8500, $8200, $8900, and $9800. [The predicted price for a five-year-old Nissan Z is $\hat{y} = 195.47 - 20.26 \cdot 5 = 94.17$, or $9417.]

This variation in price for Nissan Zs of the same age should be anticipated. Cars of the same age have different mileages, interior conditions, and paint quality. These and other random factors result in variations in price for Nissan Zs of the same age. Therefore, to each age there corresponds an entire population of prices, namely, the prices of all Nissan Zs of that age. There is a population of prices for two-year-old Nissan Zs, another population of prices for three-year-old Nissan Zs, and so on. In other words, to each x-value (age) there corresponds a population of y-values (prices).

With the preceding discussion in mind, we now state the conditions required for using inferential methods in regression analysis.

KEY FACT 13.1   Assumptions for regression inferences

1. *Population regression line:* There is a straight line, $y = \beta_0 + \beta_1 x$, such that for each x-value, the mean of the corresponding population of y-values

lies on that straight line. We refer to the straight line as the *population regression line* and to its equation as the *population regression equation.*

2. *Equal standard deviations:* The standard deviation, σ, of the population of y-values corresponding to a particular x-value is the same, regardless of the x-value.

3. *Normality:* For each x-value, the corresponding population of y-values is normally distributed.

In other words, Assumptions 1, 2, and 3 require that there exist constants, β_0, β_1, and σ, such that for each x-value, the corresponding population of y-values is normally distributed with mean $\beta_0 + \beta_1 x$ and standard deviation σ. These assumptions are often referred to as the *regression model.*

Note: β_0, β_1, and σ are generally unknown and hence must be estimated from sample data. Point estimates for the y-intercept, β_0, and slope, β_1, of the population regression line are provided, respectively, by the y-intercept, b_0, and slope, b_1, of a sample regression line. A point estimate for σ will be discussed momentarily.

EXAMPLE 13.1   *Illustrates the assumptions for regression inferences*

Consider again the variables, age and price, for Nissan Zs.
a) Discuss what it would mean for the assumptions for regression inferences to be satisfied by those two variables.
b) Display the assumptions graphically.

SOLUTION   a) For the assumptions for regression inferences to be satisfied, it would mean there are constants, β_0, β_1, and σ, such that for each age, x, the prices of all Nissan Zs of that age are normally distributed with mean $\beta_0 + \beta_1 x$ and standard deviation σ. Thus, it would mean that the prices of all two-year-old Nissan Zs ($x = 2$) are normally distributed with mean $\beta_0 + \beta_1 \cdot 2$ and standard deviation σ; the prices of all three-year-old Nissan Zs ($x = 3$) are normally distributed with mean $\beta_0 + \beta_1 \cdot 3$ and standard deviation σ; and so on.

b) To display the assumptions for regression inferences graphically, let us first consider Assumption 1. This assumption requires that for each age, x, the mean price of all Nissan Zs of that age lies on the straight line $y = \beta_0 + \beta_1 x$. See Figure 13.1 on the next page.

The population regression line is usually not known and one of the main reasons for determining a sample regression line is to estimate the population regression line. Of course, a sample regression line, in this case $\hat{y} = 195.47 - 20.26x$, ordinarily will not be the same as the population regression line, just as a sample mean, $\bar{x}$, generally will not equal the population mean, μ. We picture the situation in Figure 13.2 on the next page.

The solid line in Figure 13.2 is the population regression line. The dashed line is a sample regression line, which is the best approximation that can be made to the population regression line by using the sample data in Table 13.1 on page 644. We should emphasize that a different sample of Nissan Zs would yield a different sample regression line.

FIGURE 13.1
Population
regression line

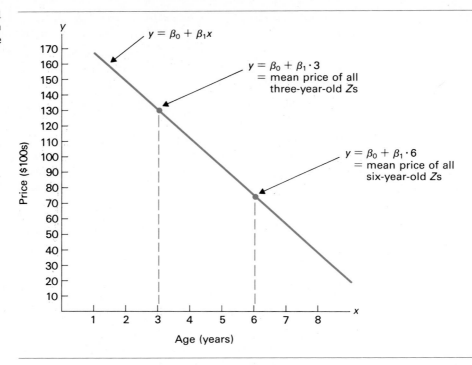

FIGURE 13.2
Population regression
line and sample
regression line

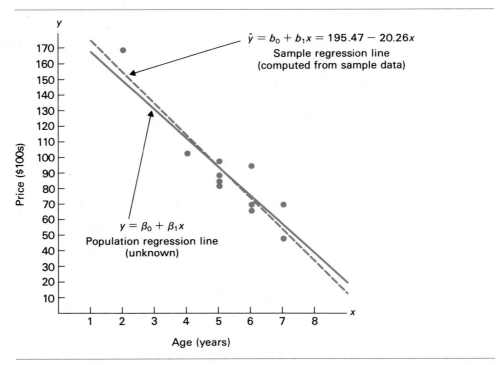

Next we will present a graph that depicts Assumptions 2 and 3. Those assumptions require that the price distributions for the various ages of Nissan Zs are all normally distributed with the same standard deviation, σ. Figure 13.3 shows this for the price distributions of two-year-old, five-year-old, and seven-year-old Nissan Zs.

Note that the shapes of the three normal curves in Figure 13.3 are identical. This reflects the fact that the shape of a normal distribution is determined by its standard deviation and, under Assumption 2, the standard deviations of the price distributions are the same for all ages of Nissan Zs.

FIGURE 13.3
Price distributions for two-year-old, five-year-old, and seven-year-old Nissan Zs under Assumptions 2 and 3

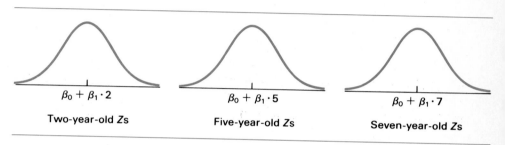

Two-year-old Zs   Five-year-old Zs   Seven-year-old Zs

We can graphically portray all three assumptions for regression inferences, as they pertain to the Nissan Z illustration, by combining Figures 13.1 and 13.3 into a three-dimensional graph. This is pictured in Figure 13.4.

FIGURE 13.4   Graphical portrayal of assumptions for regression inferences for Nissan Z illustration

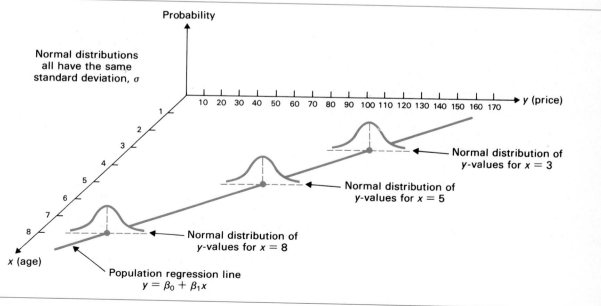

Finally, we should point out that Figure 13.4 depicts the various price distributions for Nissan Zs under the presumption that the assumptions for regression inferences hold; in other words, under the presumption that Assumptions 1, 2, and 3 are true for the variables, age, x, and price, y, of Nissan Zs. Whether that is actually the case remains to be seen. ∎

THE STANDARD ERROR OF THE ESTIMATE

Suppose that we are considering two variables, x and y, for which the assumptions for regression inferences are met; that is, the variables, x and y, satisfy Assumptions 1, 2 and 3 on pages 644–645. Then, in particular, the populations of y-values corresponding to the various x-values all have the same standard deviation, σ.

As we mentioned earlier, the common population standard deviation, σ, is usually unknown and must be estimated from sample data. The statistic used to obtain a point estimate for σ is called the **standard error of the estimate** or the **residual standard deviation** and is defined as follows:

DEFINITION 13.1   Standard error of the estimate

The *standard error of the estimate*, s_e, is defined by

$$s_e = \sqrt{\frac{SSE}{n-2}}$$

where $SSE = \Sigma(y - \hat{y})^2 = S_{yy} - S_{xy}^2/S_{xx}$.

Recall that SSE is the *error sum of squares* and represents the total squared error made when the regression equation is used to predict the observed y-values. Thus, roughly speaking, the standard error of the estimate, s_e, indicates how far the observed y-values are from the predicted y-values, on the average.

EXAMPLE 13.2   *Illustrates Definition 13.1*

Refer to the age and price data for a sample of 11 Nissan Zs given in Table 13.1 on page 644.
a) Compute the standard error of the estimate, s_e.
b) Interpret the result from part (a).

SOLUTION   a) The error sum of squares, *SSE*, was computed earlier in Example 12.9 on page 618. We found that $SSE = 1423.5$. Consequently, the standard error of the estimate is

$$s_e = \sqrt{\frac{SSE}{n-2}} = \sqrt{\frac{1423.5}{11-2}} = 12.58$$

b) Presuming that the variables age, x, and price, y, for Nissan Zs satisfy Assumptions 1–3, the standard error of the estimate, $s_e = 12.58$, or \$1258, provides an estimate for the (common) population standard deviation, σ, of prices for all Nissan Zs of any particular age. ∎

MTB

USING THE COMPUTER (OPTIONAL)

In Section 12.2, we learned that Minitab's REGRESS command can be used to obtain the (sample) regression equation for a set of data points. The output resulting from the application of that command displays the value of the standard error of the estimate, s_e, as well as the regression equation and several other statistics.

EXAMPLE 13.3   *Illustrates the REGRESS command*

We have repeated in Printout 13.1 the output obtained by applying the REGRESS command to the data on age and price for a sample of 11 Nissan Zs given in Table 13.1 on page 644.

PRINTOUT 13.1
Minitab output
for REGRESS

```
MTB > BRIEF 1
MTB > REGRESS 'PRICE' on 1 predictor 'AGE'

The regression equation is
PRICE = 195 - 20.3 AGE

Predictor         Coef        Stdev      t-ratio        p
Constant        195.47        15.24        12.83      0.000
AGE             -20.261        2.800        -7.24      0.000

s = 12.58        R-sq = 85.3%       R-sq(adj) = 83.7%

Analysis of Variance

SOURCE        DF          SS          MS         F         p
Regression     1        8285.0      8285.0     52.38     0.000
Error          9        1423.5       158.2
Total         10        9708.5
```

Use the output in Printout 13.1 to determine the standard error of the estimate for the sample of 11 Nissan Zs.

SOLUTION   The standard error of the estimate, s_e, is the first entry in the sixth line of the output: s = 12.58 (Minitab uses s instead of s_e to denote the standard error of the estimate). Thus, for the Nissan Z data, $s_e = 12.58$. ∎

Exercises 13.1

___ **13.1** State the three assumptions for regression inferences.

In Exercises 13.2–13.7, we have repeated the information from Exercises 12.32–12.37 of Section 12.2. For each exercise, discuss what it would mean for the assumptions for regression inferences to be satisfied by the variables under consideration.

___ **13.2** Ten Corvettes, between one and six years of age, were randomly selected from the classified ads of the *Arizona Republic*. The following data were obtained on age and price:

Age (yrs) x	Price ($100s) y
6	125
6	115
6	130
2	260
2	219
5	150
4	190
5	163
1	260
4	160

___ **13.3** The National Center for Health Statistics publishes data on heights and weights in *Vital and Health Statistics*. A random sample of 11 males, aged 18–24 years, yielded the following data:

Height (inches) x	Weight (lb) y
65	175
67	133
71	185
71	163
66	126
75	198
67	153
70	163
71	159
69	151
69	155

___ **13.4** Hanna Properties, Incorporated, specializes in custom-home resales in the Equestrian Estates, an exclusive subdivision in Phoenix, Arizona. A random sample of nine custom homes, currently listed for sale in that subdivision, provided the following information on size and price:

Size (100 sq ft) x	Price ($1000s) y
26	235
27	249
33	267
29	269
29	295
34	345
30	415
40	475
22	195

___ **13.5** A research article read by a physician indicated that the maximum heart rate an individual can reach during intensive exercise decreases with age. The physician decided to do his own study. Ten randomly selected people performed exercise tests and recorded their peak heart rates. The results are shown in the table below:

Age x	Peak heart rate y
30	186
38	183
41	171
38	177
29	191
39	177
46	175
41	176
42	171
24	196

___ **13.6** A calculus instructor asked a random sample of eight students to record their study times per lesson in a beginning calculus course. She then made a table for total study times over two weeks and test scores at the end of the two weeks. The results are displayed in the table shown at the top of the first column on the next page.

| Study time (hrs) | Grade (percent) |
x	y
10	92
15	81
12	84
20	74
8	85
16	80
14	84
22	80

__ **13.7**   An economist is interested in the relation between the disposable income of a family and the amount of money spent annually on food. For a preliminary study, the economist takes a random sample of eight middle-income families of the same size (father, mother, two children). The results are as follows:

| Disposable income ($1000s) | Food expenditure ($100s) |
x	y
30	55
36	60
27	42
20	40
16	37
24	26
19	39
25	43

For each of Exercises 13.8–13.13,
a) compute the standard error of the estimate, s_e, for the specified data.
b) interpret the result from part (a).

__ **13.8**   The age and price data for Corvettes from Exercise 13.2.

__ **13.9**   The height and weight data for males aged 18–24 years from Exercise 13.3.

__ **13.10**   The size and price data for custom homes from Exercise 13.4.

__ **13.11**   The data on age and peak-heart-rate from Exercise 13.5.

__ **13.12**   The study-time and grade data from Exercise 13.6.

__ **13.13**   The disposable-income and annual-food-expenditure data from Exercise 13.7.

In Exercises 13.14 and 13.15, compute the standard error of the estimate, s_e. Perform these computations from scratch; that is, do not use results obtained from previous exercises.

__ **13.14**

Data points

x	1	1	5	5
y	1	3	2	4

__ **13.15**

Data points

x	0	2	2	5	6
y	4	2	0	−2	1

Exercises 13.16–13.19 are computer exercises.

__ **13.16** (**Computer exercise**)   Suppose that the data in Exercise 13.6 are stored in columns named TIME and GRADE.
a) Which Minitab command and subcommands (if any) should be used to obtain the standard error of the estimate, s_e, for the data?
b) If you have access to Minitab, use it to determine the standard error of the estimate.

__ **13.17** (**Computer exercise**)   Suppose that the data in Exercise 13.7 are stored in columns named INCOME and FOODEX.
a) Which Minitab command and subcommands (if any) should be used to obtain the standard error of the estimate, s_e, for the data?
b) If you have access to Minitab, use it to determine the standard error of the estimate.

__ **13.18** (**Computer exercise**)   The Energy Information Administration publishes data on energy consumption by family income in *Residential Energy Consumption Survey: Consumption and Expenditures*. We applied Minitab's REGRESS command to data on family income and last year's energy consumption from a random sample of 25 families. The income data are in thousands of dollars and the energy-consumption data are in millions of BTU. Printout 12.3 on page 608 displays the computer output. Employ the computer output to determine the standard error of the estimate for the data.

__ **13.19** (**Computer exercise**) Greene and Touchstone conducted a study on the relationship between the estriol levels of pregnant women and the birth weights of their children. Their findings, entitled "Urinary Tract Estriol: An Index of Placental Function," were published in the *American Journal of Obstetrics and Gynecology*. Printout 12.4 on page 608 shows the computer output that results by applying Minitab's REGRESS command to the data obtained by Greene and Touchstone. The estriol levels are in milligrams per 24 hours and the birth weights are in grams. Use the printout to find the standard error of the estimate.

13.2   Inferences for the slope of the population regression line

In this and the next section, we will examine several statistical-inference procedures that are used in regression analysis. Those inferential techniques presume that the assumptions for regression inferences, Assumptions 1–3 on pages 644–645, are satisfied by the variables under consideration.

The first inferential methods that we will study are those concerning the slope, β_1, of the population regression line. We begin with hypothesis tests.

HYPOTHESIS TESTS FOR THE SLOPE OF THE POPULATION REGRESSION LINE

Suppose that the variables, x and y, satisfy the assumptions for regression inferences (Assumptions 1–3). Then for each x-value, the corresponding population of y-values is normally distributed with mean $\beta_0 + \beta_1 x$ and standard deviation σ.

Of particular importance, is whether the slope, β_1, of the population regression line is zero. If $\beta_1 = 0$, then for each x-value, the corresponding population of y-values is normally distributed with mean β_0 $(= \beta_0 + 0 \cdot x)$ and standard deviation σ; and neither of these two parameters involves x. So we see that if $\beta_1 = 0$, then the value of x provides no information about the distribution of the population of y-values. This implies that there is no linear relationship between the variables, x and y, and, consequently, that the variable, x, is useless as a predictor of the variable, y.

Thus, we can decide whether the variables, x and y, are linearly related and, hence, whether the variable, x, is useful as a predictor of the variable, y, by performing the hypothesis test

$$H_0: \beta_1 = 0 \ (x \text{ is not useful for predicting } y)$$
$$H_a: \beta_1 \neq 0 \ (x \text{ is useful for predicting } y)$$

Remark: We should point out that, although x alone may not be useful for predicting y, it may be useful in conjunction with another variable or variables. Thus, in this section, when we say that x is not useful for predicting y, we really mean that the regression equation with x as the predictor variable is not useful for predicting y. Similarly, although x alone may be useful for predicting y, it may not be useful in conjunction with another variable or variables. Thus, in this section, when we say that x is useful for predicting y, we really mean that the regression equation with x as the predictor variable is useful for predicting y.

To perform a hypothesis test for the slope, β_1, of the population regression line, we will employ the statistic b_1, the slope of the sample regression line. Using Assumptions 1–3, we can determine the probability distribution of the random variable b_1—the **sampling distribution of the slope of the regression line.**

KEY FACT 13.2 **The sampling distribution of the slope of the regression line**

Suppose that the variables, x and y, satisfy the assumptions for regression inferences. Then the random variable b_1 is normally distributed and has mean $\mu_{b_1} = \beta_1$ and standard deviation $\sigma_{b_1} = \sigma/\sqrt{S_{xx}}$. Thus, the standardized random variable

$$z = \frac{b_1 - \beta_1}{\sigma/\sqrt{S_{xx}}}$$

has the standard normal distribution.

The standardized random variable in Key Fact 13.2 cannot be used as a basis for obtaining the required test statistic since the common population standard deviation, σ, is generally unknown. We therefore replace σ by its sample estimate, s_e, the standard error of the estimate. This yields the random variable

$$\frac{b_1 - \beta_1}{s_e/\sqrt{S_{xx}}}$$

which, as you might suspect, has a t-distribution.

KEY FACT 13.3

Suppose that the variables, x and y, satisfy the assumptions for regression inferences. Then the random variable

$$t = \frac{b_1 - \beta_1}{s_e/\sqrt{S_{xx}}}$$

has the t-distribution with df $= n - 2$.

In view of Key Fact 13.3, we see that for a hypothesis test with null hypothesis H_0: $\beta_1 = 0$, we can use the random variable

$$t = \frac{b_1}{s_e/\sqrt{S_{xx}}}$$

as the test statistic and obtain the critical values from the t-table, Table III. Specifically, we have the following procedure:

PROCEDURE 13.1

To perform a hypothesis test to decide whether the slope of a population regression line is not zero and, hence, whether x is useful as a predictor of y.

ASSUMPTIONS

The Assumptions 1–3 for regression inferences.

STEP 1 *State the null and alternative hypotheses.*
STEP 2 *Decide on the significance level, α.*
STEP 3 *The critical values are $\pm t_{\alpha/2}$, with $df = n - 2$. Use Table III to find the critical values.*

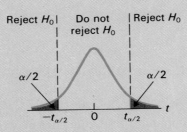

STEP 4 *Compute the value of the test statistic*

$$t = \frac{b_1}{s_e/\sqrt{S_{xx}}}$$

STEP 5 *If the value of the test statistic falls in the rejection region, then reject H_0; otherwise, do not reject H_0.*
STEP 6 *State the conclusion in words.*

EXAMPLE 13.4   Illustrates Procedure 13.1

The data on age and price for a sample of 11 Nissan Zs are repeated in Table 13.2. Ages are in years; prices are in hundreds of dollars, rounded to the nearest hundred.

TABLE 13.2
Age versus price data for Nissan Zs

Age (yrs) x	Price ($100s) y
5	85
4	103
6	70
5	82
5	89
5	98
6	66
6	95
2	169
7	70
7	48

Presuming that the variables, age, x, and price, y, satisfy the assumptions for regression inferences, do the data provide sufficient evidence to conclude that age is useful as a predictor of price for Nissan Zs? Perform the appropriate hypothesis test at the 5% significance level.

SOLUTION   We apply Procedure 13.1.

STEP 1 *State the null and alternative hypotheses.*

Let β_1 denote the slope of the population regression line that relates price to age for Nissan Zs. Then the null and alternative hypotheses are

$$H_0: \beta_1 = 0 \text{ (age is not useful for predicting price)}$$
$$H_a: \beta_1 \neq 0 \text{ (age is useful for predicting price)}$$

STEP 2 *Decide on the significance level, α.*

We are to perform the hypothesis test at the 5% significance level; consequently, $\alpha = 0.05$.

STEP 3 *The critical values are $\pm t_{\alpha/2}$, with df $= n - 2$.*

From Step 2, $\alpha = 0.05$. Also, from Table 13.2, we see that $n = 11$ and so df $= n - 2 = 11 - 2 = 9$. Using Table III, we find that the critical values are $\pm t_{\alpha/2} = \pm t_{0.05/2} = \pm t_{0.025} = \pm 2.262$. See Figure 13.5.

FIGURE 13.5

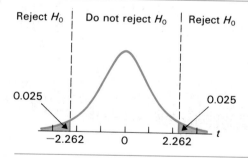

STEP 4 *Compute the value of the test statistic*

$$t = \frac{b_1}{s_e / \sqrt{S_{xx}}}$$

In Example 12.4 on page 597, we found that $b_1 = -20.26$, $\Sigma x^2 = 326$, and $\Sigma x = 58$. Also, in Example 13.2 on page 648, we determined that $s_e = 12.58$.

Therefore, since $n = 11$, the value of the test statistic is

$$t = \frac{b_1}{s_e/\sqrt{S_{xx}}} = \frac{b_1}{s_e/\sqrt{\Sigma x^2 - (\Sigma x)^2/n}} = \frac{-20.26}{12.58/\sqrt{326 - (58)^2/11}} = -7.235$$

STEP 5 *If the value of the test statistic falls in the rejection region, reject H_0; otherwise, do not reject H_0.*

The value of the test statistic, found in Step 4, is $t = -7.235$. Since this falls in the rejection region, we **reject H_0**.

STEP 6 *State the conclusion in words.*

Evidently, the slope of the population regression line is not zero and, hence, age is useful as a predictor of price for Nissan Zs.   ■

MTB

Note: Procedure 13.1, which is based on the statistic b_1, is used for performing a hypothesis test to decide whether the slope of the population regression line is not zero or, equivalently, to decide whether the regression equation is useful for making predictions. In Section 12.3, we introduced the coefficient of determination, r^2, as a descriptive measure of how useful the regression equation is for making predictions. This suggests that we should also be able to use the statistic r^2 as a basis for performing a hypothesis test to decide whether the regression equation is useful for making predictions and, indeed, that is the case. However, we will not cover the hypothesis test based on r^2 since it is equivalent to the hypothesis test based on b_1.

CONFIDENCE INTERVALS FOR THE SLOPE OF THE POPULATION REGRESSION LINE

Recall that the slope of a straight line gives the amount of change in y resulting from an increase in x by one unit. Also recall that the population regression line, whose slope is β_1, gives the means of the populations of y-values corresponding to the various x-values. Thus, β_1 represents the change in the mean of the population of y-values for every increase in the value of x by one unit.

For example, consider the variables, age, x, and price, y, of Nissan Zs. In this case, β_1 is the amount, in hundreds of dollars, that the mean price decreases for every increase in age by one year. In other words, β_1 gives the mean yearly depreciation of Nissan Zs.

Consequently, we see that it is worthwhile to obtain an estimate of the slope, β_1, of the population regression line. As we know, a point estimate for β_1 is provided by b_1. To determine a confidence-interval estimate for β_1, we apply Key Fact 13.3 on page 653 to get the following procedure:

PROCEDURE 13.2

To find a confidence interval for the slope of a population regression line.

ASSUMPTIONS

The Assumptions 1–3 for regression inferences.

STEP 1 *For a confidence level of* $1 - \alpha$, *use Table III to find* $t_{\alpha/2}$ *with* $df = n - 2$.

STEP 2 *The endpoints of the confidence interval for* β_1 *are*

$$b_1 \pm t_{\alpha/2} \cdot \frac{s_e}{\sqrt{S_{xx}}}$$

EXAMPLE 13.5 **Illustrates Procedure 13.2**

Use the data in Table 13.2 on page 654 to obtain a 95% confidence interval for the slope, β_1, of the population regression line that relates price to age for Nissan Zs.

SOLUTION We apply Procedure 13.2.

STEP 1 *For a confidence level of* $1 - \alpha$, *use Table III to find* $t_{\alpha/2}$ *with* $df = n - 2$.

For a 95% confidence interval, $\alpha = 0.05$. Since $n = 11$, $df = 11 - 2 = 9$. Using Table III, we find that $t_{\alpha/2} = t_{0.05/2} = t_{0.025} = 2.262.$

STEP 2 *The endpoints of the confidence interval for* β_1 *are*

$$b_1 \pm t_{\alpha/2} \cdot \frac{s_e}{\sqrt{S_{xx}}}$$

From Example 12.4, $b_1 = -20.26$, $\Sigma x^2 = 326$, and $\Sigma x = 58$. Also, from Example 13.2, $s_e = 12.58$. Hence, the endpoints of the confidence interval for β_1 are

$$-20.26 \pm 2.262 \cdot \frac{12.58}{\sqrt{326 - (58)^2/11}}$$

or

$$-20.26 \pm 6.33$$

Consequently, a 95% confidence interval for β_1 is from

$$-26.59 \quad \text{to} \quad -13.93$$

We can be 95% confident that the slope, β_1, of the population regression line is somewhere between -26.59 and -13.93. In other words, we can be 95% confident that the yearly decrease in mean price for Nissan Zs is somewhere between \$1393 and \$2659.

MTB

USING THE COMPUTER (OPTIONAL)

Procedure 13.1 on page 654 provides a step-by-step method for performing a hypothesis test to decide whether the slope, β_1, of a population regression line is not zero. Alternatively, we can use Minitab to carry out such a hypothesis test. In fact, the output obtained by applying Minitab's REGRESS command to a set of data points contains all the information that we need. To illustrate, we return again to the Nissan Z example.

EXAMPLE 13.6   *Illustrates the REGRESS command*

In Section 12.2 we applied the REGRESS command to the data on age and price for a sample of 11 Nissan Zs shown in Table 13.2 on page 654. The resulting computer output is repeated here in Printout 13.2. Use that output to perform the hypothesis test considered in Example 13.4.

PRINTOUT 13.2
Minitab output
for REGRESS

```
MTB > BRIEF 1
MTB > REGRESS 'PRICE' on 1 predictor 'AGE'

The regression equation is
PRICE = 195 - 20.3 AGE

Predictor        Coef        Stdev     t-ratio        p
Constant       195.47        15.24       12.83    0.000
AGE            -20.261        2.800       -7.24    0.000

s = 12.58      R-sq = 85.3%      R-sq(adj) = 83.7%


Analysis of Variance

SOURCE         DF          SS          MS        F        p
Regression      1        8285.0      8285.0    52.38    0.000
Error           9        1423.5       158.2
Total          10        9708.5
```

SOLUTION   Let β_1 denote the slope of the population regression line that relates price to age for Nissan Zs. The problem in Example 13.4 is to perform the hypothesis test

$$H_0: \beta_1 = 0 \text{ (age is not useful for predicting price)}$$
$$H_a: \beta_1 \neq 0 \text{ (age is useful for predicting price)}$$

at the 5% significance level.

Look at the fifth line of the computer output in Printout 13.2, the line labeled AGE. The second entry in that line, which is under the column headed **Coef**, gives the slope, b_1, of the sample regression line; hence, $b_1 = -20.261$. The third entry in that line, which is under the column headed **Stdev**, shows the estimated standard deviation of b_1, $s_e/\sqrt{S_{xx}}$. So, $s_e/\sqrt{S_{xx}} = 2.800$. The fourth entry in that line, which is under the column headed t-ratio, displays the value of the test statistic

$$t = \frac{b_1}{s_e/\sqrt{S_{xx}}}$$

Thus, we see that $t = -7.24$.

The final entry of the line labeled AGE, under the column headed p, gives the P-value for the hypothesis test. This is really the only quantity that we need in order to decide whether the null hypothesis should be rejected. Recall that the P-value is equal to the smallest significance level at which the null hypothesis can be rejected. Therefore, if the P-value is less than or equal to the specified significance level, then we reject H_0; otherwise, we do not reject H_0.

From Printout 13.2, we see that the P-value is 0.000 (to three decimal places). Since this is less than the specified significance level of $\alpha = 0.05$, we reject H_0. In other words, the data provide sufficient evidence to conclude that the slope of the population regression line is not zero and, hence, that age is useful as a predictor of price for Nissan Zs. ■

Exercises 13.2

In Exercises 13.20–13.25, we have repeated the information from Exercises 12.32–12.37 of Section 12.2. Presuming that the assumptions for regression inferences are met, apply Procedure 13.1 on page 654 to perform each of the required hypothesis tests. Note: You previously obtained the sample regression equations in Exercises 12.32–12.37 and the standard errors of the estimate in Exercises 13.8–13.13.

— **13.20** Ten Corvettes, between one and six years of age, were randomly selected from the classified ads of the *Arizona Republic*. The following data were obtained on age and price:

Age (yrs)	Price ($100s)
x	y
6	125
6	115
6	130
2	260
2	219
5	150
4	190
5	163
1	260
4	160

At the 10% significance level, do the data provide sufficient evidence to conclude that the slope of the population regression line is not zero and, hence, that age is useful as a predictor of price for Corvettes?

— **13.21** The National Center for Health Statistics publishes data on heights and weights in *Vital and Health Statistics*. A random sample of 11 males, aged 18–24 years, yielded the following data:

Height (inches)	Weight (lb)
x	y
65	175
67	133
71	185
71	163
66	126
75	198
67	153
70	163
71	159
69	151
69	155

At the 10% significance level, do the data provide sufficient evidence to conclude that the slope of the population regression line is not zero and, hence, that height is useful as a predictor of weight for 18–24-year-old males?

__ 13.22　Hanna Properties specializes in custom-home resales in the Equestrian Estates, an exclusive subdivision in Phoenix, Arizona. A random sample of nine custom homes, currently listed for sale, provided the following information on size and price:

Size (100 sq ft) x	Price ($1000s) y
26	235
27	249
33	267
29	269
29	295
34	345
30	415
40	475
22	195

Do the data suggest that size is useful as a predictor of price for custom homes in the Equestrian Estates? Perform the required hypothesis test at the 0.01 level of significance.

__ 13.23　An article read by a physician indicated that the maximum heart rate an individual can reach during intensive exercise decreases with age. The physician decided to do his own study. Ten randomly selected people performed exercise tests and recorded their peak heart rates. The results are shown in the table below:

Age x	Peak heart rate y
30	186
38	183
41	171
38	177
29	191
39	177
46	175
41	176
42	171
24	196

Does it appear that age is useful as a predictor of peak heart rate? Perform the required hypothesis test at the 0.05 level of significance.

__ 13.24　A calculus instructor asked a random sample of eight students to record their study times per lesson in a beginning calculus course. She then made a table for total study times over two weeks and test scores at the end of the two weeks. The table that the calculus instructor obtained is as follows:

Study time (hrs) x	Grade (percent) y
10	92
15	81
12	84
20	74
8	85
16	80
14	84
22	80

Do the data provide sufficient evidence to conclude that study time is useful as a predictor of test score in beginning calculus courses? Perform the hypothesis test using a significance level of $\alpha = 0.05$.

__ 13.25　An economist is interested in the relation between the disposable income of a family and the amount of money spent annually on food. For a preliminary study, the economist takes a random sample of eight middle-income families of the same size (father, mother, two children). The resulting data are displayed in the table below.

Disposable income ($1000s) x	Food expenditure ($100s) y
30	55
36	60
27	42
20	40
16	37
24	26
19	39
25	43

Do the data provide sufficient evidence to conclude that disposable income is useful as a predictor of annual food expenditure for middle-income families with a father, mother, and two children? Use a significance level of $\alpha = 0.01$.

In each of Exercises 13.26–13.31, apply Procedure 13.2 on page 657 to obtain the required confidence interval.

___ **13.26**   Refer to Exercise 13.20.
a) Determine a 90% confidence interval for the slope, β_1, of the population regression line that relates price to age for Corvettes.
b) Interpret your result from part (a) in words.

___ **13.27**   Refer to Exercise 13.21.
a) Obtain a 90% confidence interval for the slope, β_1, of the population regression line that relates weight to height for males aged 18–24.
b) Interpret your result from part (a) in words.

___ **13.28**   Refer to Exercise 13.22.
a) Find a 99% confidence interval for the slope of the population regression line that relates price to size for custom homes in the Equestrian Estates.
b) Interpret your result from part (a) in words.

___ **13.29**   Refer to Exercise 13.23.
a) Find a 95% confidence interval for the slope of the population regression line that relates peak heart rate to age.
b) Interpret your result from part (a) in words.

___ **13.30**   Refer to Exercise 13.24.
a) Obtain a 95% confidence interval for the slope, β_1, of the population regression line that relates grade to study time in beginning calculus courses.
b) Interpret your result from part (a) in words.

___ **13.31**   Refer to Exercise 13.25.
a) Determine a 99% confidence interval for the slope, β_1, of the population regression line that relates annual food expenditure to disposable income for middle-income families with a father, mother, and two children.
b) Interpret your result from part (a) in words.

Exercises 13.32–13.35 are computer exercises.

___ **13.32 (Computer exercise)**   Suppose that the data in Exercise 13.24 are stored in columns named TIME and GRADE.
a) Which Minitab command and subcommands (if any) should be used to perform the hypothesis test considered in that exercise?
b) If you have access to Minitab, use it to carry out the hypothesis test.

___ **13.33 (Computer exercise)**   Suppose that the data in Exercise 13.25 are stored in columns named INCOME and FOODEX.

a) Which Minitab command and subcommands (if any) should be used to perform the hypothesis test considered in that exercise?
b) If you have access to Minitab, use it to carry out the hypothesis test.

___ **13.34 (Computer exercise)**   The Energy Information Administration publishes data on energy consumption by family income in *Residential Energy Consumption Survey: Consumption and Expenditures*. We applied Minitab's REGRESS command to data on family income and last year's energy consumption from a random sample of 25 families. The income data are in thousands of dollars and the energy-consumption data are in millions of BTU. Printout 13.3 on the next page shows the computer output. Using the printout,
a) find the slope, b_1, of the sample regression line.
b) determine the estimated standard deviation of the random variable b_1.
c) obtain the value of the test statistic, t, for a hypothesis test to decide whether the slope, β_1, of the population regression line is not zero.
d) determine the P-value for the hypothesis test referred to in part (c).
e) decide whether family income is useful for predicting last year's energy consumption. Use $\alpha = 0.01$.
f) obtain a 99% confidence interval for the slope, β_1, of the population regression line and interpret your result in words. (*Note:* You will need to use Table III here to determine $t_{\alpha/2}$, but everything else that is required to obtain the confidence interval can be found in the printout.)

___ **13.35 (Computer exercise)**   Greene and Touchstone conducted a study on the relationship between the estriol levels of pregnant women and the birth weights of their children. Their findings, entitled "Urinary Tract Estriol: An Index of Placental Function," were published in the *American Journal of Obstetrics and Gynecology*. Printout 13.4 on the next page shows the computer output that results by applying Minitab's REGRESS command to the 31 pairs of data obtained by Greene and Touchstone. The estriol levels are in milligrams per 24 hours and the birth weights are in grams. Using the printout,
a) find the slope, b_1, of the sample regression line.
b) determine the estimated standard deviation of the random variable b_1.
c) obtain the value of the test statistic, t, for a hypothesis test to decide whether the slope, β_1, of the population regression line is not zero.

PRINTOUT 13.3   Minitab output for Exercise 13.34

```
The regression equation is
CONSUMPT = 82.0 + 0.931 INCOME
```

Predictor	Coef	Stdev	t-ratio	p
Constant	82.036	2.054	39.94	0.000
INCOME	0.93051	0.05727	16.25	0.000

```
s = 5.375       R-sq = 92.0%      R-sq(adj) = 91.6%
```

Analysis of Variance

SOURCE	DF	SS	MS	F	p
Regression	1	7626.6	7626.6	264.03	0.000
Error	23	664.4	28.9		
Total	24	8291.0			

PRINTOUT 13.4   Minitab output for Exercise 13.35

```
The regression equation is
WEIGHT = 2152 + 60.8 ESTRIOL
```

Predictor	Coef	Stdev	t-ratio	p
Constant	2152.3	262.0	8.21	0.000
ESTRIOL	60.82	14.68	4.14	0.000

```
s = 382.1       R-sq = 37.2%      R-sq(adj) = 35.0%
```

Analysis of Variance

SOURCE	DF	SS	MS	F	p
Regression	1	2505745	2505745	17.16	0.000
Error	29	4234255	146009		
Total	30	6740000			

d) determine the P-value for the hypothesis test referred to in part (c).

e) decide, at the 5% significance level, whether estriol level is useful for predicting birth weights.

f) obtain a 95% confidence interval for the slope, β_1, of the population regression line and interpret your result in words. (*Note:* You will need to use Table III here to determine $t_{\alpha/2}$, but everything else that is required to obtain the confidence interval can be found in the printout.)

13.36   This exercise justifies Procedure 13.2 on page 657. Suppose that the variables, x and y, satisfy

the assumptions for regression inferences (Assumptions 1, 2, and 3 on pages 644–645).

a) Use Key Fact 13.3 on page 653 to show that

$$P\left(-t_{\alpha/2} < \frac{b_1 - \beta_1}{s_e/\sqrt{S_{xx}}} < t_{\alpha/2}\right) = 1 - \alpha$$

b) Use part (a) to show that the probability is $1 - \alpha$ that the interval from

$$b_1 - t_{\alpha/2} \cdot \frac{s_e}{\sqrt{S_{xx}}} \quad \text{to} \quad b_1 + t_{\alpha/2} \cdot \frac{s_e}{\sqrt{S_{xx}}}$$

will contain β_1.

c) Deduce from part (b) that once the sample is taken, the interval with endpoints

$$b_1 \pm t_{\alpha/2} \cdot \frac{s_e}{\sqrt{S_{xx}}}$$

will be a $(1 - \alpha)$-level confidence interval for β_1.

≡ 13.37 On page 656, we mentioned that the coefficient of determination, r^2, can also be used as a basis for performing a hypothesis test to decide whether the regression equation is useful for making predictions.

The test statistic, based on r^2, for performing such a hypothesis test is

$$F = \frac{(n-2)r^2}{1-r^2}$$

Show that the test statistic, F, is the square of the test statistic, t, that is used in Procedure 13.1. (*Hint:* Express both statistics in terms of the quantities S_{xx}, S_{xy}, and S_{yy}.)

13.3 Estimation and prediction

In this section, we will learn how the (sample) regression equation can be used to make two important types of inferences. One is to estimate the mean of the population of y-values corresponding to a particular x-value. The other is to predict an individual y-value corresponding to a particular x-value.

We will employ the Nissan Z example to illustrate the pertinent ideas. In doing so, we will presume that the variables, age and price, for Nissan Zs satisfy the assumptions for regression inferences, Assumptions 1–3 on pages 644–645.

EXAMPLE 13.7   *Illustrates the estimation of means in regression*

The data on age and price for a sample of 11 Nissan Zs are repeated in Table 13.3. Ages are in years; prices are in hundreds of dollars, rounded to the nearest hundred.

TABLE 13.3
Age versus price
data for Nissan Zs

Age (yrs) x	Price ($100s) y
5	85
4	103
6	70
5	82
5	89
5	98
6	66
6	95
2	169
7	70
7	48

Use the sample data to obtain an estimate for the mean price of, say, all three-year-old Nissan Zs.

SOLUTION   By Assumption 1 of the assumptions for regression inferences, the population regression line gives the mean prices for the various ages of Nissan Zs. Thus, the mean price of all three-year-old Nissan Zs is exactly equal to $\beta_0 + \beta_1 \cdot 3$. Since β_0 and β_1 are unknown, we will estimate the mean price of all three-year-old Nissan Zs, $\beta_0 + \beta_1 \cdot 3$, by the corresponding value, $b_0 + b_1 \cdot 3$, on the sample regression line.

Recall that the sample regression line for the data in Table 13.3 is $\hat{y} = 195.47 - 20.26x$. Thus, our estimate for the mean price of all three-year-old Nissan Zs is

$$\hat{y} = 195.47 - 20.26 \cdot 3 = 134.69$$

or $13,469. [Note that the estimate for the mean price of all three-year-old Nissan Zs is the same as the predicted price of a randomly selected three-year-old Nissan Z. Both are obtained by substituting $x = 3$ into the sample regression equation.]   ∎

The estimate of $13,469 for the mean price of all three-year-old Nissan Zs is a point estimate. As we know, it would be more informative if we had some idea of the accuracy of that point estimate. In other words, it would be better to provide a confidence-interval estimate for the mean price of all three-year-old Nissan Zs. We will now see how such confidence-interval estimates can be obtained.

CONFIDENCE INTERVALS FOR MEANS IN REGRESSION

To develop a confidence-interval procedure for means in regression, we must first identify the probability distribution of the random variable $\hat{y}$. This is given in Key Fact 13.4.

KEY FACT 13.4

Suppose that the variables, x and y, satisfy the assumptions for regression inferences. Let x_p denote a particular value of the predictor variable, x, and let $\hat{y}_p = b_0 + b_1 x_p$. Then the random variable $\hat{y}_p$ is normally distributed and has mean $\mu_{\hat{y}_p} = \beta_0 + \beta_1 x_p$ and standard deviation

$$\sigma_{\hat{y}_p} = \sigma \sqrt{\frac{1}{n} + \frac{(x_p - \Sigma x/n)^2}{S_{xx}}}$$

Key Fact 13.4 implies that the standardized random variable

$$z = \frac{\hat{y}_p - (\beta_0 + \beta_1 x_p)}{\sigma \sqrt{\dfrac{1}{n} + \dfrac{(x_p - \Sigma x/n)^2}{S_{xx}}}}$$

has the standard normal distribution. However, since σ is usually unknown, we cannot employ that random variable to find a confidence-interval formula. Thus, we replace σ by its estimate, s_e, the standard error of the estimate. The resulting random variable has a t-distribution:

KEY FACT 13.5

Suppose that the variables, x and y, satisfy the assumptions for regression inferences. Let x_p denote a particular value of the predictor variable, x, and let $\hat{y}_p = b_0 + b_1 x_p$. Then the random variable

$$t = \frac{\hat{y}_p - (\beta_0 + \beta_1 x_p)}{s_e \sqrt{\dfrac{1}{n} + \dfrac{(x_p - \Sigma x/n)^2}{S_{xx}}}}$$

has the t-distribution with df $= n - 2$.

If we now recall that $\beta_0 + \beta_1 x_p$ is the mean of the population of y-values corresponding to x_p, then we can use Key Fact 13.5 to derive the following confidence-interval procedure for means in regression:

PROCEDURE 13.3

To find a confidence interval for the mean of the population of y-values corresponding to a particular x-value, x_p.

ASSUMPTIONS

The Assumptions 1–3 for regression inferences.

STEP 1 *For a confidence level of $1 - \alpha$, use Table III to find $t_{\alpha/2}$ with df $= n - 2$.*

STEP 2 *Compute the point estimate,*

$$\hat{y}_p = b_0 + b_1 x_p,$$

for the mean of the population of y-values corresponding to x_p.

STEP 3 *The endpoints of the confidence interval for the mean are*

$$\hat{y}_p \pm t_{\alpha/2} \cdot s_e \sqrt{\frac{1}{n} + \frac{(x_p - \Sigma x/n)^2}{S_{xx}}}$$

EXAMPLE 13.8 *Illustrates Procedure 13.3*

Use the sample data in Table 13.3 on page 663 to obtain a 95% confidence interval for the mean price of all three-year-old Nissan Zs.

SOLUTION We apply Procedure 13.3.

STEP 1 *For a confidence level of $1 - \alpha$, use Table III to find $t_{\alpha/2}$ with df $= n - 2$.*

We want a 95% confidence interval, which means $\alpha = 0.05$. Since $n = 11$, df $= 11 - 2 = 9$. Consulting Table III, we find that $t_{\alpha/2} = t_{0.05/2} = t_{0.025} = 2.262$.

STEP 2 *Compute the point estimate,*

$$\hat{y}_p = b_0 + b_1 x_p,$$

for the mean of the population of y-values corresponding to x_p.

From Example 12.4, the sample regression equation for the data in Table 13.3 is $\hat{y} = 195.47 - 20.26x$. Here we want $x = x_p = 3$ (three-year-old Nissan Zs). So,

$$\hat{y}_p = 195.47 - 20.26 \cdot 3 = \boxed{134.69}$$

STEP 3 *The endpoints of the confidence interval for the mean are*

$$\hat{y}_p \pm t_{\alpha/2} \cdot s_e \sqrt{\frac{1}{n} + \frac{(x_p - \Sigma x/n)^2}{S_{xx}}}$$

In Example 12.4 we found that $\Sigma x = 58$ and $\Sigma x^2 = 326$; and in Example 13.2 we determined that $s_e = 12.58$. Also, from Step 1, $t_{\alpha/2} = 2.262$ and, from Step 2, $\hat{y}_p = 134.69$. So, the endpoints of the confidence interval for the mean are

$$134.69 \pm 2.262 \cdot 12.58 \sqrt{\frac{1}{11} + \frac{(3 - 58/11)^2}{326 - (58)^2/11}}$$

or

$$134.69 \pm 16.76$$

Thus, a 95% confidence interval for the mean is from

$$\boxed{117.93 \quad \text{to} \quad 151.45}$$

We can be 95% confident that the mean price of all three-year-old Nissan Zs is somewhere between $11,793 and $15,145. ■

MTB

PREDICTION INTERVALS

One of the main reasons for determining the regression equation for a sample of data points is to use it for making predictions. The regression equation for the Nissan Z data in Table 13.3 is $\hat{y} = 195.47 - 20.26x$. Thus, for example, the predicted price for a three-year-old Nissan Z is

$$\hat{y} = 195.47 - 20.26 \cdot 3 = 134.69$$

or $13,469. However, since the prices of such cars vary, it makes more sense to find a **prediction interval** for the price of a three-year-old Nissan Z than to give a single predicted value.

Prediction intervals are similar to confidence intervals. The term "confidence" is usually reserved for interval estimates of a parameter, such as the mean price of all three-year-old Nissan Zs. The term "prediction" is used for interval estimates of random variables, such as the price of a randomly selected three-year-old Nissan Z.

The procedure for obtaining prediction intervals in regression is quite similar to that for obtaining confidence intervals. The prediction-interval procedure is based on the following fact:

KEY FACT 13.6

Suppose that the variables, x and y, satisfy the assumptions for regression inferences. Let x_p denote a particular value of the predictor variable, x, and y_p denote the value of a randomly selected member from the population of y-values corresponding to x_p. Then the random variable $y_p - \hat{y}_p$ is normally distributed and has mean $\mu_{y_p - \hat{y}_p} = 0$ and standard deviation

$$\sigma_{y_p - \hat{y}_p} = \sigma \sqrt{1 + \frac{1}{n} + \frac{(x_p - \Sigma x/n)^2}{S_{xx}}}$$

If we standardize the random variable $y_p - \hat{y}_p$, then the resulting random variable has the standard normal distribution. However, since the standardized random variable contains the unknown parameter σ, we cannot use it to develop a prediction-interval formula. Consequently, we replace the parameter σ by its estimate, s_e, the standard error of the estimate. The random variable so obtained has a t-distribution:

KEY FACT 13.7

Suppose that the variables, x and y, satisfy the assumptions for regression inferences. Let x_p denote a particular value of the predictor variable, x, and y_p denote the value of a randomly selected member from the population of y-values corresponding to x_p. Then the random variable

$$t = \frac{y_p - \hat{y}_p}{s_e \sqrt{1 + \frac{1}{n} + \frac{(x_p - \Sigma x/n)^2}{S_{xx}}}}$$

has the t-distribution with df $= n - 2$.

Using Key Fact 13.7, we can derive the following prediction-interval procedure for a population y-value corresponding to a particular x-value:

PROCEDURE 13.4

To find a prediction interval for a population y-value corresponding to a particular x-value, x_p.

ASSUMPTIONS

The Assumptions 1–3 for regression inferences.

STEP 1 *For a prediction level of $1 - \alpha$, use Table III to find $t_{\alpha/2}$ with df $= n - 2$.*

STEP 2 *Compute the predicted y-value,*

$$\hat{y}_p = b_0 + b_1 x_p$$

STEP 3 *The endpoints of the prediction interval for the y-value are*

$$\hat{y}_p \pm t_{\alpha/2} \cdot s_e \sqrt{1 + \frac{1}{n} + \frac{(x_p - \Sigma x/n)^2}{S_{xx}}}$$

EXAMPLE 13.9   *Illustrates Procedure 13.4*

Using the sample data in Table 13.3 on page 663, obtain a 95% prediction interval for the price of a randomly selected three-year-old Nissan Z.

SOLUTION   We employ the step-by-step method given in Procedure 13.4.

STEP 1 *For a prediction level of $1 - \alpha$, use Table III to find $t_{\alpha/2}$ with df $= n - 2$.*

We want a 95% prediction interval; therefore $\alpha = 0.05$. Also, since $n = 11$, df $= 11 - 2 = 9$. Consulting Table III, we find that $t_{\alpha/2} = t_{0.05/2} = t_{0.025} = $ **2.262.**

STEP 2 *Compute the predicted y-value,*

$$\hat{y}_p = b_0 + b_1 x_p$$

The sample regression equation for the data in Table 13.3 is $\hat{y} = 195.47 - 20.26x$. Thus, the predicted y-value for a three-year-old Nissan Z is

$$\hat{y}_p = 195.47 - 20.26 \cdot 3 = \textbf{134.69}$$

STEP 3 *The endpoints of the prediction interval for the y-value are*

$$\hat{y}_p \pm t_{\alpha/2} \cdot s_e \sqrt{1 + \frac{1}{n} + \frac{(x_p - \Sigma x/n)^2}{S_{xx}}}$$

From Example 12.4, $\Sigma x = 58$ and $\Sigma x^2 = 326$; and from Example 13.2, $s_e = 12.58$. Also $n = 11$, $t_{\alpha/2} = 2.262$, $x_p = 3$, and $\hat{y}_p = 134.69$. Consequently, the endpoints of the prediction interval are

$$134.69 \pm 2.262 \cdot 12.58 \sqrt{1 + \frac{1}{11} + \frac{(3 - 58/11)^2}{326 - (58)^2/11}}$$

or

$$134.69 \pm 33.02$$

Thus, a 95% prediction interval is from

$$\boxed{101.67 \quad \text{to} \quad 167.71}$$

MTB

We can be 95% certain that the price of a randomly selected three-year-old Nissan Z will be somewhere between $10,167 and $16,771. ∎

We have just seen that a 95% prediction interval for the price of a randomly selected three-year-old Nissan Z is from $10,167 to $16,771. In Example 13.8, we found that a 95% confidence interval for the mean price of all three-year-old Nissan Zs is from $11,793 to $15,145. We picture both of these intervals in Figure 13.6.

FIGURE 13.6
Prediction and confidence intervals for three-year-old Nissan Zs

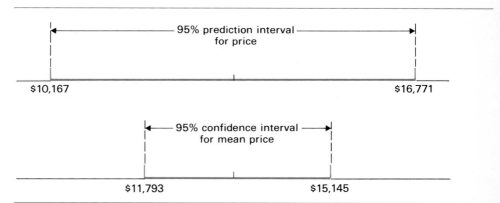

Note that the prediction interval is more extensive than the confidence interval. This is to be expected for the following reason: The error in the estimate of the mean price of all three-year-old Nissan Zs is due only to the fact that the population regression line is being estimated by a sample regression line. On the other hand, the error in the prediction of the price of a randomly selected three-year-old Nissan Z is due to the aforementioned error in estimating the mean price plus the variation in prices for three-year-old Nissan Zs.

USING THE COMPUTER (OPTIONAL)

We can apply Minitab to obtain both confidence intervals and prediction intervals in regression. As previously, we use the REGRESS command, but for estimation and prediction, we also employ the **PREDICT** subcommand.

EXAMPLE 13.10   *Illustrates the REGRESS; PREDICT command*

The data on age and price for a sample of 11 Nissan Zs are displayed in Table 13.3 on page 663. Apply Minitab to that data to simultaneously obtain a 95% confidence interval for the mean price of all three-year-old Nissan Zs and a 95% prediction interval for the price of a randomly selected three-year-old Nissan Z.

SOLUTION   Recall that we have previously stored the age and price data from Table 13.3 in columns named AGE and PRICE. To obtain the required confidence and prediction intervals, we first employ the REGRESS command as described earlier, except this time we type a semicolon at the end of the command in preparation for a subcommand. Thus, we type

REGRESS 'PRICE' on 1 predictor 'AGE';

On the next line we type the subcommand <u>PREDICT 3.</u>. The 3 indicates that we are considering three-year-old Nissan Zs. Printout 13.5 shows these commands and the resulting output.

PRINTOUT 13.5
Minitab output
for REGRESS;
PREDICT

```
MTB > BRIEF 1
MTB > REGRESS 'PRICE' on 1 predictor 'AGE';
SUBC> PREDICT 3.

The regression equation is
PRICE = 195 - 20.3 AGE

Predictor        Coef        Stdev      t-ratio          p
Constant       195.47        15.24        12.83      0.000
AGE           -20.261        2.800        -7.24      0.000

s = 12.58        R-sq = 85.3%      R-sq(adj) = 83.7%

Analysis of Variance

SOURCE         DF           SS          MS          F          p
Regression      1       8285.0      8285.0      52.38      0.000
Error           9       1423.5       158.2
Total          10       9708.5


     Fit   Stdev.Fit        95% C.I.           95% P.I.
   134.68        7.41   ( 117.92, 151.44)  ( 101.66, 167.71)
```

The first part of the output in Printout 13.5 is the same as that given by the REGRESS command without the PREDICT subcommand. It is only the last two lines of the output that are new; and it is those two lines that contain the information about the confidence and prediction intervals.

So, let us concentrate on the last two lines of the output in Printout 13.5. The first item, headed `Fit`, gives the point estimate, $\hat{y}_p = 134.68$ ($\$13{,}468$), for the mean price of all three-year-old Nissan Zs or for the price of a randomly selected three-year-old Nissan Z. The second item, headed `Stdev.Fit`, displays the estimated standard deviation of the random variable $\hat{y}_p = b_0 + b_1 x_p$, which is

$$s_e \sqrt{\frac{1}{n} + \frac{(x_p - \Sigma x/n)^2}{S_{xx}}}$$

Here, of course, $x_p = 3$.

The third item, headed `95% C.I.`, provides the required confidence interval. Hence, a 95% confidence interval for the mean price of all three-year-old Nissan Zs is from 117.92 to 151.44. We can be 95% confident that the mean price of all three-year-old Nissan Zs is somewhere between $\$11{,}792$ and $\$15{,}144$.

The final item, headed `95% P.I.`, gives the prediction interval. So, a 95% prediction interval for the price of a randomly selected three-year-old Nissan Z is from 101.66 to 167.71. We can be 95% certain that the price of a randomly selected three-year-old Nissan Z will be somewhere between $\$10{,}166$ and $\$16{,}771$. ∎

You may have noted that, in solving Example 13.10, we did not specify either the confidence level or the prediction level. The reason is that Minitab automatically uses levels of 95%, and those were the levels that we wanted. If levels other than 95% are required, then we must proceed somewhat differently. See the *Minitab Supplement* for details.

Exercises 13.3

In Exercises 13.38–13.43, we have repeated the information from Exercises 12.32–12.37 of Section 12.2. Presuming that the assumptions for regression inferences are met, determine the required confidence and prediction intervals. Note: You previously obtained the sample regression equations in Exercises 12.32–12.37 and the standard errors of the estimate in Exercises 13.8–13.13.

___ **13.38** Ten Corvettes, between one and six years of age, were randomly selected from the classified ads of the *Arizona Republic*. Data on age and price for those 10 Corvettes are displayed in the table shown at the top of the next column.

a) Obtain a point estimate for the mean price of all four-year-old Corvettes.

b) Determine a 90% confidence interval for the mean price of all four-year-old Corvettes.

Age (yrs) x	Price ($100s) y
6	125
6	115
6	130
2	260
2	219
5	150
4	190
5	163
1	260
4	160

c) Find the predicted price of a randomly selected four-year-old Corvette.

d) Determine a 90% prediction interval for the price of a randomly selected four-year-old Corvette.

e) Draw graphs similar to the ones in Figure 13.6 on page 669 showing both the 90% confidence interval from part (b) and the 90% prediction interval from part (d).

f) Why is the prediction interval more extensive than the confidence interval?

____ 13.39   The National Center for Health Statistics compiles data on heights and weights, by age and sex, in the publication *Vital and Health Statistics*. A random sample of 11 males, aged 18–24 years, yielded the following data:

Height (inches) x	Weight (lb) y
65	175
67	133
71	185
71	163
66	126
75	198
67	153
70	163
71	159
69	151
69	155

a) Obtain a point estimate for the mean weight of all 18–24-year-old males who are 70 inches tall.

b) Find a 90% confidence interval for the mean weight of all 18–24-year-old males who are 70 inches tall.

c) Find the predicted weight of a randomly selected 18–24-year-old male who is 70 inches tall.

d) Determine a 90% prediction interval for the weight of a randomly selected 18–24-year-old male who is 70 inches tall.

e) Draw graphs similar to the ones in Figure 13.6 on page 669 showing both the 90% confidence interval from part (b) and the 90% prediction interval from part (d).

f) Why is the prediction interval more extensive than the confidence interval?

____ 13.40   Hanna Properties specializes in custom-home resales in the Equestrian Estates, an exclusive subdivision in Phoenix, Arizona. A random sample of nine custom homes, currently listed for sale, provided the information on size and price shown in the table at the top of the next column.

Size (100 sq ft) x	Price ($1000s) y
26	235
27	249
33	267
29	269
29	295
34	345
30	415
40	475
22	195

a) Obtain a point estimate for the mean price of all 2800-square-foot Equestrian-Estate homes.

b) Find a 99% confidence interval for the mean price of all 2800-square-foot Equestrian-Estate homes.

c) Determine the predicted price of a randomly selected 2800-square-foot Equestrian-Estate home.

d) Find a 99% prediction interval for the price of a randomly selected 2800-square-foot Equestrian-Estate home.

____ 13.41   An article read by a physician indicated that the maximum heart rate an individual can reach during intensive exercise decreases with age. The physician decided to do his own study. Ten randomly selected people performed exercise tests and recorded their peak heart rates. The results are as follows:

Age x	Peak heart rate y
30	186
38	183
41	171
38	177
29	191
39	177
46	175
41	176
42	171
24	196

a) Obtain a point estimate for the mean peak heart rate of all 40-year-olds.

b) Find a 95% confidence interval for the mean peak heart rate of all 40-year-olds.

c) Determine the predicted peak heart rate of a randomly selected 40-year-old.

d) Find a 95% prediction interval for the peak heart rate of a randomly selected 40-year-old.

— 13.42 A calculus instructor asked a random sample of eight students to record their study times per lesson in a beginning calculus course. She then made a table for total study times over two weeks and test scores at the end of the two weeks. The table that the calculus instructor obtained is as follows:

Study time (hrs)	Grade (percent)
x	y
10	92
15	81
12	84
20	74
8	85
16	80
14	84
22	80

a) Obtain a 95% confidence interval for the mean test score of all beginning calculus students who study for 15 hours.
b) Obtain a 95% prediction interval for the test score of a randomly selected beginning calculus student who studies for 15 hours.

— 13.43 An economist is interested in the relation between the disposable income of a family and the amount of money spent annually on food. For a preliminary study, the economist takes a random sample of eight middle-income families of the same size (father, mother, two children). The resulting data are displayed in the table below.

Disposable income ($1000s)	Food expenditure ($100s)
x	y
30	55
36	60
27	42
20	40
16	37
24	26
19	39
25	43

a) Determine a 99% confidence interval for the mean annual food expenditure of all middle-income families consisting of a father, mother, and two children that have a disposable income of $25,000.

b) Find a 99% prediction interval for the annual food expenditure of a randomly selected middle-income family consisting of a father, mother, and two children that has a disposable income of $25,000.

Exercises 13.44–13.47 are computer exercises.

— 13.44 (Computer exercise) Suppose that the data in Exercise 13.42 are stored in columns named TIME and GRADE.
a) Which Minitab command and subcommands (if any) should be used to determine the confidence interval considered in part (a) of Exercise 13.42?
b) Which Minitab command and subcommands (if any) should be used to determine the prediction interval considered in part (b) of Exercise 13.42?
c) If you have access to Minitab, use it to obtain both the confidence and prediction intervals.

— 13.45 (Computer exercise) Suppose that the data in Exercise 13.43 are stored in columns named INCOME and FOODEX.
a) Which Minitab command and subcommands (if any) should be used to determine a 95% confidence interval for the mean annual food expenditure of all families consisting of a father, mother, and two children that have a disposable income of $25,000?
b) Which Minitab command and subcommands (if any) should be used to determine a 95% prediction interval for the annual food expenditure of a randomly selected family consisting of a father, mother, and two children that has a disposable income of $25,000?
c) If you have access to Minitab, use it to obtain both the confidence and prediction intervals.
d) In Exercise 13.43, we asked you to obtain 99% confidence and prediction intervals, whereas, in this exercise, we asked you to obtain 95% confidence and prediction intervals. Why didn't we ask you to obtain 99% confidence and prediction intervals in this exercise also?

— 13.46 (Computer exercise) The Energy Information Administration publishes data on energy consumption by family income in *Residential Energy Consumption Survey: Consumption and Expenditures*. We applied Minitab's REGRESS command to data on family income and last year's energy consumption from a random sample of 25 families. The income data are in thousands of dollars and the energy-consumption data are in millions of BTU. Printout 13.6 at the top of the following page shows the computer output.

PRINTOUT 13.6   Minitab output for Exercise 13.46

```
The regression equation is
CONSUMPT = 82.0 + 0.931 INCOME

Predictor        Coef        Stdev     t-ratio        p
Constant       82.036        2.054       39.94    0.000
INCOME        0.93051      0.05727       16.25    0.000

s = 5.375      R-sq = 92.0%      R-sq(adj) = 91.6%

Analysis of Variance

SOURCE        DF          SS          MS          F          p
Regression     1      7626.6      7626.6     264.02    0.000
Error         23       664.4        28.9
Total         24      8291.0


    Fit   Stdev.Fit        95% C.I.            95% P.I.
 119.26        1.20   ( 116.77, 121.75)   ( 107.86, 130.65)
```

PRINTOUT 13.7   Minitab output for Exercise 13.47

```
The regression equation is
WEIGHT = 2152 + 60.8 ESTRIOL

Predictor        Coef        Stdev     t-ratio        p
Constant       2152.3        262.0        8.21    0.000
ESTRIOL         60.82        14.68        4.14    0.000

s = 382.1      R-sq = 37.2%      R-sq(adj) = 35.0%

Analysis of Variance

SOURCE        DF          SS          MS          F          p
Regression     1     2505745     2505745      17.16    0.000
Error         29     4234255      146009
Total         30     6740000


    Fit   Stdev.Fit        95% C.I.            95% P.I.
 3064.6        76.0   ( 2909.1, 3220.1)   ( 2267.6, 3861.6)
```

a) What subcommand was used to obtain the output shown in the last two lines of Printout 13.6? *(Note: The confidence and prediction intervals are for an income level of $40,000.)*

b) Determine a point estimate for last year's mean energy consumption of all families with an annual income of $40,000.

c) Find a 95% confidence interval for last year's mean energy consumption of all families with an annual income of $40,000.

d) Obtain a 95% prediction interval for last year's energy consumption by a randomly selected family with an annual income of $40,000.

— 13.47 **(Computer exercise)** Greene and Touchstone conducted a study on the relationship between the estriol levels of pregnant women and the birth weights of their children. Their findings, entitled "Urinary Tract Estriol: An Index of Placental Function," were published in the *American Journal of Obstetrics and Gynecology.* Printout 13.7 on the previous page shows the computer output that results by applying Minitab's REGRESS command to the 31 pairs of data obtained by Greene and Touchstone. The estriol levels are in milligrams per 24 hours and the birth weights are in grams.

a) What subcommand was used to obtain the output shown in the last two lines of Printout 13.7? *(Note: The confidence and prediction intervals are for an estriol level of 15 mg/24 hours.)*

b) Obtain a point estimate for the mean birth weight of all babies whose mothers have an estriol level of 15 mg/24 hours.

c) Determine a 95% confidence interval for the mean birth weight of all babies whose mothers have an estriol level of 15 mg/24 hours.

d) Find a 95% prediction interval for the birth weight of a randomly selected baby whose mother has an estriol level of 15 mg/24 hours.

13.4 Inferences in correlation

Frequently we want to decide whether two variables, x and y, are linearly correlated, that is, whether there is a linear relationship between the two variables. As we learned in Section 13.2, we can make that decision by performing a hypothesis test for the slope, β_1, of the population regression line.

Alternatively, we can perform a hypothesis test for the **population linear correlation coefficient, ρ** (rho). The population linear correlation coefficient, ρ, measures the linear correlation between the population of all data points in the same way that the sample linear correlation coefficient, r, measures the linear correlation between a sample of data points. Thus, it is ρ that actually describes the strength of the linear relationship between two variables; r is just an estimate of ρ.

The population linear correlation coefficient, ρ, lies between -1 and 1. Values of ρ near -1 or 1 indicate a strong linear relationship between the variables, whereas values of ρ near 0 indicate a weak linear relationship between the variables. If $\rho > 0$, the variables are **positively linearly correlated,** meaning that y tends to increase linearly as x increases, with the tendency being greater the closer that ρ is to 1. If $\rho < 0$, the variables are **negatively linearly correlated,** meaning that y tends to decrease linearly as x increases, with the tendency being greater the closer that ρ is to -1. If $\rho = 0$, the variables are **linearly uncorrelated,** meaning that there is no linear relationship between the variables.

As you might suspect, we can use the sample linear correlation coefficient, r, as a basis for performing a hypothesis test concerning a population linear correlation coefficient, ρ. For a test with null hypothesis H_0: $\rho = 0$ (i.e., the variables are linearly uncorrelated), we will employ the following fact:

KEY FACT 13.8

Suppose that the variables, x and y, satisfy the assumptions for regression infer-ences. Then, if $\rho = 0$, the random variable

$$t = \frac{r}{\sqrt{\dfrac{1 - r^2}{n - 2}}}$$

has the t-distribution with df $= n - 2$.

In view of Key Fact 13.8, we see that for a hypothesis test with null hypothesis H_0: $\rho = 0$, we can use the random variable

$$t = \frac{r}{\sqrt{\dfrac{1 - r^2}{n - 2}}}$$

as the test statistic and obtain the critical values from the t-table, Table III. Specif-ically, we have the following procedure:

PROCEDURE 13.5

To perform a hypothesis test for a population linear correlation coefficient with null hypothesis H_0: $\rho = 0$.

ASSUMPTIONS

The Assumptions 1–3 for regression inferences.

STEP 1 *State the null and alternative hypotheses.*
STEP 2 *Decide on the significance level, α.*
STEP 3 *The critical value(s)*
      a) *for a two-tailed test are $\pm t_{\alpha/2}$,*
      b) *for a left-tailed test is $-t_{\alpha}$,*
      c) *for a right-tailed test is t_{α},*
      with df $= n - 2$. Use Table III to find the critical value(s).

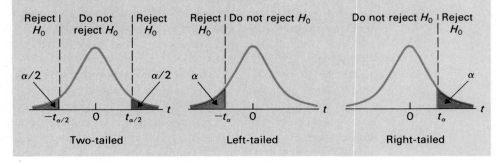

STEP 4 *Compute the value of the test statistic*

$$t = \frac{r}{\sqrt{\dfrac{1 - r^2}{n - 2}}} .$$

STEP 5 *If the value of the test statistic falls in the rejection region, then reject H_0; otherwise, do not reject H_0.*

STEP 6 *State the conclusion in words.*

EXAMPLE 13.11   **Illustrates Procedure 13.5**

Consider once more the age and price data for a sample of 11 Nissan Zs, which we repeat here in Table 13.4. Ages are in years; prices are in hundreds of dollars, rounded to the nearest hundred.

TABLE 13.4
Age versus price
data for Nissan Zs

Age (yrs) x	Price ($100s) y
5	85
4	103
6	70
5	82
5	89
5	98
6	66
6	95
2	169
7	70
7	48

Presuming that the variables, age, x, and price, y, satisfy the assumptions for regression inferences, do the data provide sufficient evidence to conclude that age and price for Nissan Zs are negatively linearly correlated? Use $\alpha = 0.05$.

SOLUTION   We apply Procedure 13.5.

STEP 1 *State the null and alternative hypotheses.*

Let ρ denote the population linear correlation coefficient for the variables, age and price, of Nissan Zs. Then the null and alternative hypotheses are

H_0: $\rho = 0$ (age and price are linearly uncorrelated)
H_a: $\rho < 0$ (age and price are negatively linearly correlated)

Note that the hypothesis test is left-tailed since there is a less-than sign ($<$) in the alternative hypothesis.

STEP 2 *Decide on the significance level, α.*

We are to use $\alpha = 0.05.$

STEP 3 *The critical value for a left-tailed test is $-t_\alpha$, with $df = n - 2$.*

We have $n = 11$, so that df $= 9$. Also, $\alpha = 0.05$. Consulting Table III, we find that for df $= 9$, $t_{0.05} = 1.833$. Thus, the critical value is $-t_{0.05} = -1.833.$ See Figure 13.7.

FIGURE 13.7

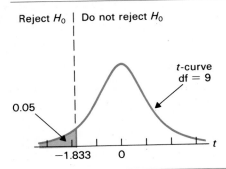

STEP 4 *Compute the value of the test statistic*

$$t = \frac{r}{\sqrt{\dfrac{1 - r^2}{n - 2}}}$$

We have already computed the sample linear correlation coefficient, r, for the age and price data shown in Table 13.4. This was done in Example 12.11 on page 624 where we found that $r = -0.924$. Therefore, the value of the test statistic is

$$t = \frac{r}{\sqrt{\dfrac{1 - r^2}{n - 2}}} = \frac{-0.924}{\sqrt{\dfrac{1 - (-0.924)^2}{11 - 2}}} = -7.249$$

STEP 5 *If the value of the test statistic falls in the rejection region, reject H_0; otherwise, do not reject H_0.*

The value of the test statistic, found in Step 4, is $t = -7.249$. A glance at Figure 13.7 shows that this falls in the rejection region. Hence, we reject H_0.

STEP 6 *State the conclusion in words.*

Evidently, the variables, age and price, for Nissan Zs are negatively linearly correlated. In other words, the price of a Nissan Z tends to decrease linearly as its age increases (at least for Nissan Zs between two and seven years old). ∎

MTB

Exercises 13.4

In Exercises 13.48–13.53, we have repeated the information from Exercises 12.32–12.37 of Section 12.2. Presuming that the assumptions for regression inferences are met, apply Procedure 13.5 on page 676 to perform each of the required hypothesis tests. Note: You previously obtained the sample linear correlation coefficients in Exercises 12.72–12.77 of Section 12.4.

___ **13.48** Ten Corvettes, between one and six years of age, were randomly selected from the classified ads of the *Arizona Republic*. The following data were obtained on age and price:

Age (yrs) x	Price ($100s) y
6	125
6	115
6	130
2	260
2	219
5	150
4	190
5	163
1	260
4	160

Do the data provide sufficient evidence to conclude that the variables, age and price, are negatively linearly correlated for Corvettes? Use $\alpha = 0.05$.

___ **13.49** The National Center for Health Statistics publishes data on heights and weights in *Vital and Health Statistics*. A random sample of 11 males, aged 18–24 years, yielded the following data:

Height (inches) x	Weight (lb) y
65	175
67	133
71	185
71	163
66	126
75	198
67	153
70	163
71	159
69	151
69	155

Do the data provide sufficient evidence to conclude that the variables, height and weight, are positively linearly correlated for 18–24-year-old males? Perform the hypothesis test at the 5% significance level.

___ **13.50** Hanna Properties specializes in custom-home resales in the Equestrian Estates, an exclusive subdivision in Phoenix, Arizona. A random sample of nine custom homes, currently listed for sale, provided the following information on size and price:

Size (100 sq ft) x	Price ($1000s) y
26	235
27	249
33	267
29	269
29	295
34	345
30	415
40	475
22	195

At the 0.5% significance level, do the data indicate that, for custom homes in the Equestrian Estates, size and price are positively linearly correlated?

___ **13.51** An article read by a physician indicated that the maximum heart rate an individual can reach during intensive exercise decreases with age. The physician decided to do his own study. Ten randomly selected people performed exercise tests and recorded their peak heart rates. The results are shown in the table below:

Age x	Peak heart rate y
30	186
38	183
41	171
38	177
29	191
39	177
46	175
41	176
42	171
24	196

Do the data provide sufficient evidence to conclude that age and peak heart rate are negatively linearly correlated? Use $\alpha = 0.025$.

___ **13.52** A calculus instructor asked a random sample of eight students to record their study times per lesson in a beginning calculus course. She then made a table for total study times over two weeks and test scores at the end of the two weeks. The table that the calculus instructor obtained is as follows:

| Study time (hrs) | Grade (percent) |
x	y
10	92
15	81
12	84
20	74
8	85
16	80
14	84
22	80

Do the data provide sufficient evidence to conclude that, in beginning calculus courses, study time and test score are linearly correlated? Perform the hypothesis test using a significance level of $\alpha = 0.05$.

___ **13.53** An economist is interested in the relation between the disposable income of a family and the amount of money spent annually on food. For a preliminary study, the economist takes a random sample of eight middle-income families of the same size (father, mother, two children). The resulting data are displayed in the table below.

| Disposable income ($1000s) | Food expenditure ($100s) |
x	y
30	55
36	60
27	42
20	40
16	37
24	26
19	39
25	43

At the 1% significance level, do the data indicate that family disposable income and annual food expenditure are linearly correlated for middle-income families with a father, mother, and two children?

___ **13.54** A random sample of 10 students in an introductory statistics class provided the following data on height and final-exam score:

Height (inches) x	Exam score y
71	87
68	96
71	66
65	71
66	71
68	55
68	83
64	67
62	86
65	60

Do the data provide sufficient evidence to conclude that, for students in introductory statistics courses, height and final-exam score are linearly correlated? Perform the hypothesis test using $\alpha = 0.05$.

___ **13.55** Is the population linear correlation coefficient, ρ, a random variable? What about the sample linear correlation coefficient, r? Explain your answers.

≡ **13.56** Procedure 13.1 on page 654 employs the random variable

$$t = \frac{b_1}{s_e/\sqrt{S_{xx}}}$$

as the test statistic for a hypothesis test with null hypothesis H_0: $\beta_1 = 0$. Show that this test statistic is equal to the one used in Procedure 13.5, that is, to

$$t = \frac{r}{\sqrt{\dfrac{1 - r^2}{n - 2}}}$$

(Hint: Express both statistics in terms of the quantities S_{xx}, S_{xy}, and S_{yy}.) This result proves that the hypothesis tests

$$H_0: \beta_1 = 0$$
$$H_a: \beta_1 \neq 0$$

and

$$H_0: \rho = 0$$
$$H_a: \rho \neq 0$$

are equivalent.

13.5   Multiple regression (Optional)

So far, we have considered inferential methods in regression analysis only for simple linear regression; that is, when dealing with only one predictor variable. Now we will study the corresponding methods in multiple regression, where there is more than one predictor variable.

THE MULTIPLE REGRESSION MODEL

As in simple linear regression, to perform statistical inferences in multiple regression, it is necessary that the variables under consideration satisfy certain conditions. For multiple regression with k predictor variables, those conditions are as follows:

KEY FACT 13.9   Assumptions for multiple regression inferences

1. *Population regression equation:* For each set of values, $x_1, x_2, \ldots, x_k$, of the predictor variables, the mean of the corresponding population of y-values is $\beta_0 + \beta_1 x_1 + \cdots + \beta_k x_k$. The equation

$$y = \beta_0 + \beta_1 x_1 + \cdots + \beta_k x_k$$

is called the *population regression equation*.

2. *Equal standard deviations:* The standard deviation, σ, of the population of y-values corresponding to a particular set of values, $x_1, x_2, \ldots, x_k$, of the predictor variables is the same, regardless of $x_1, x_2, \ldots, x_k$.

3. *Normality:* For each set of values, $x_1, x_2, \ldots, x_k$, of the predictor variables, the corresponding population of y-values is normally distributed.

So, Assumptions 1, 2, and 3 require that there exist constants, $\beta_0, \beta_1, \ldots, \beta_k$, and σ, such that for each set of values, $x_1, x_2, \ldots, x_k$, of the predictor variables, the corresponding population of y-values is normally distributed with mean $\beta_0 + \beta_1 x_1 + \cdots + \beta_k x_k$ and standard deviation σ. These assumptions are often referred to as the *multiple regression model*.

EXAMPLE 13.12   *Illustrates the assumptions for multiple regression inferences*

Consider the variables, age, miles driven, and price, for Nissan Zs, where age is in years, miles driven is in thousands, and price is in hundreds of dollars. Discuss what it would mean for the assumptions for multiple regression inferences to be satisfied with age and miles driven as predictor variables for price.

SOLUTION   Note that here we have $k = 2$ since there are two predictor variables, age and miles driven. For the assumptions for multiple regression inferences to be satisfied, it would mean the following: There exist constants, $\beta_0, \beta_1, \beta_2$, and σ, such that for each age, x_1, and number of miles driven, x_2, the prices of all Nissan Zs of that age which have been driven that number of miles are normally distributed with mean $\beta_0 + \beta_1 x_1 + \beta_2 x_2$ and standard deviation σ.

Thus, it would mean that the prices of all Nissan Zs that are two years old ($x_1 = 2$) and have been driven 15,000 miles ($x_2 = 15$) are normally distributed with mean $\beta_0 + \beta_1 \cdot 2 + \beta_2 \cdot 15$ and standard deviation σ; the prices of all Nissan Zs that are three years old ($x_1 = 3$) and have been driven 32,000 miles ($x_2 = 32$) are normally distributed with mean $\beta_0 + \beta_1 \cdot 3 + \beta_2 \cdot 32$ and standard deviation σ; and so on.   ■

The interpretation of the coefficients, β_1, β_2, $\ldots$, β_k, is similar to that in simple linear regression. Specifically, the coefficient, β_j, of the predictor variable, x_j, represents the change in the mean of the population of y-values for every increase in x_j by one unit, with all other predictor variables held fixed.

For instance, in the Nissan Z illustration of Example 13.12, the population regression equation is $y = \beta_0 + \beta_1 x_1 + \beta_2 x_2$, where x_1 denotes age, in years, x_2 denotes miles driven, in thousands, and y denotes price, in hundreds of dollars. So, for any fixed number of miles driven, β_1 is the amount, in hundreds of dollars, that the mean price changes for every increase in age by one year. Similarly, for any fixed age, β_2 is the amount, in hundreds of dollars, that the mean price changes for every increase in the number of miles driven by one thousand.

When we determine a sample regression equation, $\hat{y} = b_0 + b_1 x_1 + \cdots + b_k x_k$, we obtain the best estimate, based on the sample data, of the (unknown) population regression equation, $y = \beta_0 + \beta_1 x_1 + \cdots + \beta_k x_k$. In fact, for each j, b_j is the best estimate of β_j.

THE STANDARD ERROR OF THE ESTIMATE

Assumption 2 for inferences in multiple regression requires that the standard deviations of the various populations of y-values be the same, regardless of the values of the predictor variables. The statistic used to estimate the common population standard deviation, σ, is called the **standard error of the estimate** and is defined as follows:

DEFINITION 13.2   Standard error of the estimate

For multiple regression with k predictor variables, the *standard error of the estimate*, s_e, is defined by

$$s_e = \sqrt{\frac{SSE}{n - (k + 1)}}$$

where $SSE = \Sigma(y - \hat{y})^2$.

Note that the definition of the standard error of the estimate in multiple regression is consistent with that in simple linear regression. Indeed, in simple linear regression, we have $k = 1$ (one predictor variable) and so

$$s_e = \sqrt{\frac{SSE}{n - (k + 1)}} = \sqrt{\frac{SSE}{n - (1 + 1)}} = \sqrt{\frac{SSE}{n - 2}}$$

which is the definition of s_e given in Definition 13.1 on page 648.

As we mentioned in Section 12.5, because of computational complexity, multiple regression is almost always done by computer. So, in this section, we will present computer output as a means for examining multiple regression.

EXAMPLE 13.13 *Illustrates the multiple regression model*

In Example 12.13 we presented data on age, miles driven, and price for a sample of 11 Nissan Zs. Those data are repeated here in Table 13.5.

TABLE 13.5
Data on age, miles driven, and price for Nissan Zs

Car	Age (yrs) x_1	Miles (thous) x_2	Price (\$100s) y
1	5	57	85
2	4	40	103
3	6	77	70
4	5	60	82
5	5	49	89
6	5	47	98
7	6	58	66
8	6	39	95
9	2	8	169
10	7	69	70
11	7	89	48

We employed a statistical computer program to perform a multiple regression analysis on the data in Table 13.5 with the variables, age and miles driven, as predictor variables of price. The results are displayed in Printout 12.7 and are repeated in Printout 13.8 at the top of the next page.

Presuming that the variables, age, miles driven, and price, satisfy the assumptions for multiple regression inferences, use Printout 13.8 to
a) obtain and interpret the sample regression equation.
b) find and interpret the standard error of the estimate.

SOLUTION a) The sample regression equation is given in the second line of the computer output in Printout 13.8: PRICE = 183 - 9.50 AGE - 0.821 MILES. In other words, the sample regression equation is

$$\hat{y} = 183 - 9.50x_1 - 0.821x_2$$

where x_1 denotes age, in years, x_2 denotes miles driven, in thousands, and $\hat{y}$ denotes predicted price, in hundreds of dollars. Based on the sample data in Table 13.5, this equation is the best estimate of the unknown population regression equation, $y = \beta_0 + \beta_1x_1 + \beta_2x_2$.

PRINTOUT 13.8
Computer output for
multiple regression
of Nissan Z data
in Table 13.5

```
The regression equation is
PRICE = 183 - 9.50 AGE - 0.821 MILES

Predictor        Coef       Stdev     t-ratio        p
Constant       183.04       11.35       16.13    0.000
AGE            -9.504        3.874      -2.45     0.040
MILES         -0.8215       0.2552      -3.22     0.012

s = 8.805        R-sq = 93.6%      R-sq(adj) = 92.0%

Analysis of Variance

SOURCE        DF         SS          MS         F         p
Regression     2      9088.3      4544.2     58.61    0.000
Error          8       620.2        77.5
Total         10      9708.5
```

MTB

b) The standard error of the estimate, s_e, is the first entry in the seventh line of the output: **s = 8.805** (Minitab uses s instead of s_e to denote the standard error of the estimate). Thus, $s_e = 8.805$. So, based on the sample data in Table 13.5, the best estimate for the (common) population standard deviation, σ, of prices for all Nissan Zs of any particular age and number of miles driven is $880.5. ∎

INFERENCES CONCERNING THE UTILITY OF THE REGRESSION

Suppose that the variables, x_1, x_2, ..., x_k, and y, satisfy the assumptions for multiple regression inferences. A first question is whether the regression as a whole is useful for making predictions; that is, whether the variables, x_1, x_2, ..., x_k, taken together are useful for predicting y. The answer to that question depends on the parameters β_1, β_2, ..., β_k, the coefficients of x_1, x_2, ..., x_k.

If β_1, β_2, ..., β_k are all zero, then for each set of values, x_1, x_2, ..., x_k, of the predictor variables, the corresponding population of y-values is normally distributed with mean β_0 ($= \beta_0 + 0 \cdot x_1 + 0 \cdot x_2 + \cdots + 0 \cdot x_k$) and standard deviation σ; and neither of those two parameters involves the predictor variables. So we see that if $\beta_1 = \beta_2 = \cdots = \beta_k = 0$, then the predictor variables provide no information about the distribution of the population of y-values. This implies that the predictor variables taken together are useless for predicting y.

Consequently, we can decide on the overall utility of the regression—whether the variables, x_1, x_2, ..., x_k, taken together are useful for predicting y—by performing the hypothesis test

$$H_0: \beta_1 = \beta_2 = \cdots = \beta_k = 0$$

$$H_a: \text{At least one of the } \beta\text{s is not zero.}$$

Rejection of the null hypothesis indicates that the regression is useful for making predictions. Nonrejection of the null hypothesis indicates that the regression may not be useful for making predictions.

Now we must identify a test statistic that can be used to perform the above hypothesis test. As we learned in Section 12.5, the coefficient of determination, R^2, is a descriptive measure of the utility of the regression equation for making predictions. So it seems reasonable that we could use R^2, or some expression involving R^2, as a test statistic. This is indeed the case. In fact, the test statistic is

$$F = \frac{R^2/k}{(1 - R^2)/[n - (k + 1)]}$$

The random variable F has what is called an F-distribution. We will study that distribution in detail in Chapter 14.

EXAMPLE 13.14

Illustrates inferences concerning the utility of the regression

Consider again the data shown in Table 13.5 on age, miles driven, and price for a sample of 11 Nissan Zs. At the 5% significance level, do the data provide sufficient evidence to conclude that the regression, with age and miles as predictor variables, is useful for making price predictions? Employ the computer output in Printout 13.8 on page 684 to perform the required hypothesis test.

SOLUTION   We want to perform the hypothesis test

H_0: $\beta_1 = \beta_2 = 0$

H_a: At least one of β_1 and β_2 is not zero.

with $\alpha = 0.05$. As we just learned, the test statistic for this hypothesis test is

$$F = \frac{R^2/k}{(1 - R^2)/[n - (k + 1)]}$$

(Here $k = 2$ and $n = 11$.) Look at the tenth line of the computer output in Printout 13.8, the line labeled **Regression**. The fifth entry in that line, which is under the column headed **F**, displays the value of the test statistic, F. Thus, $F = 58.61$.

The final entry of the line labeled **Regression,** under the column headed **p,** gives the smallest significance level (to three decimal places) at which the null hypothesis can be rejected. So we see that the null hypothesis can be rejected at any significance level that is greater than 0.000.

Since the designated significance level for this hypothesis test is 0.05, which exceeds 0.000, we **reject H_0.** In other words, the data provide sufficient evidence to conclude that the regression, with age and miles driven as predictor variables, is useful for making price predictions.

MTB

INFERENCES CONCERNING THE UTILITY OF A PARTICULAR PREDICTOR VARIABLE

To decide whether a particular predictor variable, say x_j, is useful as a predictor of the variable y, we proceed as we did in simple linear regression. Namely, we perform the hypothesis test

$$H_0: \beta_j = 0 \ (x_j \text{ is not useful for predicting } y)$$
$$H_a: \beta_j \neq 0 \ (x_j \text{ is useful for predicting } y)$$

Rejection of the null hypothesis indicates that the variable x_j is useful as a predictor of the variable y. Nonrejection of the null hypothesis indicates that the variable x_j may not be useful as a predictor of the variable y and that it may be worthwhile to do a regression analysis with the variable x_j omitted.

The test statistic for this hypothesis test is essentially the same as the one used in simple linear regression:

$$t = \frac{b_j}{s_{b_j}}$$

where s_{b_j} is the estimated standard deviation of the random variable b_j. However, in multiple regression, df $= n - (k + 1)$ and the formula for s_{b_j} is much more complicated than in simple linear regression.

We should convey a warning similar to the one in Section 13.2 about interpreting the conclusion of a hypothesis test for a β-parameter: Although x_j may not be useful for predicting y in conjunction with the other predictor variables under consideration, it may be useful when employed as the only predictor variable or with some other collection of predictor variables. Thus, in this section, when we say that x_j is not useful for predicting y, we really mean that in the regression with $x_1, x_2, \ldots, x_k$ as the predictor variables, x_j is not useful for predicting y.

Likewise, although x_j may be useful for predicting y in conjunction with the other predictor variables under consideration, it may not be useful when employed as the only predictor variable or with some other collection of predictor variables. Thus, in this section, when we say that x_j is useful for predicting y, we really mean that in the regression with $x_1, x_2, \ldots, x_k$ as the predictor variables, x_j is useful for predicting y.

EXAMPLE 13.15   *Illustrates inferences concerning the utility of a particular predictor variable*

Use the data in Table 13.5 to decide whether, in conjunction with age, number of miles driven is useful as a predictor of price for Nissan Zs. Apply Printout 13.8 on page 684 to perform the required hypothesis test at the 5% significance level.

SOLUTION   Here we want to perform the hypothesis test

$$H_0: \beta_2 = 0 \text{ (miles driven is not useful for predicting price)}$$
$$H_a: \beta_2 \neq 0 \text{ (miles driven is useful for predicting price)}$$

with $\alpha = 0.05$.

Look at the sixth line of the computer output in Printout 13.8, the line labeled MILES. The second entry in that line, which is under the column headed Coef, gives the coefficient, b_2, of the sample regression line; hence, $b_2 = -0.8215$. The third entry in that line, which is under the column headed Stdev, shows the estimated standard deviation of b_2, that is, s_{b_2}. So, $s_{b_2} = 0.2552$. The fourth entry in that line, which is under the column headed t-ratio, displays the value of the test statistic

$$t = \frac{b_2}{s_{b_2}}$$

Thus, we see that $t = -3.22$.

The final entry of the line labeled MILES, under the column headed p, gives the smallest significance level at which the null hypothesis can be rejected. So, the null hypothesis can be rejected at any significance level that is 0.012 or greater. Since the designated significance level for the hypothesis test is 0.05, which exceeds 0.012, we reject H_0. The data provide sufficient evidence to conclude that number of miles driven is useful as a predictor of price or, more precisely, that in conjunction with age, number of miles driven is useful as a predictor of price. ■

MTB

CONFIDENCE INTERVALS FOR MEANS

To determine a point estimate or confidence-interval estimate for the mean of the population of y-values corresponding to particular values, x_{1p}, x_{2p}, ..., x_{kp}, of the predictor variables, we proceed as in simple linear regression. A point estimate for the mean is obtained by substituting the predictor-variable values into the sample regression equation:

$$\hat{y}_p = b_0 + b_1 x_{1p} + b_2 x_{2p} + \cdots + b_k x_{kp}$$

And, the endpoints of a $(1 - \alpha)$-level confidence interval for the mean are found by applying the formula

$$\hat{y}_p \pm t_{\alpha/2} \cdot s_{\hat{y}_p}$$

where $t_{\alpha/2}$ is computed for the t-curve with df $= n - (k+1)$ and $s_{\hat{y}_p}$ is the estimated standard deviation of the random variable $\hat{y}_p$.

In practice, a computer is almost always used to implement these formulas and obtain the required point estimate or confidence interval. Here is an example.

EXAMPLE 13.16   *Illustrates point estimates and confidence intervals for means*

Printout 13.9, at the top of the next page, displays the computer output that results from applying a statistical computer program to the Nissan Z data shown in Table 13.5 on page 683. The computer output is the same as that in Printout 13.8 except for the last two lines. Those two lines, which provide information on price for Nissan Zs that are five years old and have been driven 52,000 miles, were obtained by using a special command.

PRINTOUT 13.9
Computer output for
multiple regression
of Nissan Z data
in Table 13.5

```
The regression equation is
PRICE = 183 - 9.50 AGE - 0.821 MILES

Predictor          Coef         Stdev      t-ratio          p
Constant         183.04         11.35        16.13      0.000
AGE               -9.504         3.874        -2.45      0.040
MILES            -0.8215        0.2552        -3.22      0.012

s = 8.805        R-sq = 93.6%        R-sq(adj) = 92.0%

Analysis of Variance

SOURCE         DF          SS           MS          F          p
Regression      2       9088.3       4544.2      58.61      0.000
Error           8        620.2         77.5
Total          10       9708.5


      Fit   Stdev.Fit          95% C.I.          95% P.I.
    92.80        2.74     ( 86.47,  99.12)   ( 71.53, 114.07)
```

Employ Printout 13.9 to

a) determine a point estimate for the mean price of all Nissan Zs that are five years old and have been driven 52,000 miles.

b) obtain a 95% confidence interval for the mean price of all Nissan Zs that are five years old and have been driven 52,000 miles.

SOLUTION   Both problems can be solved by referring to the last two lines of Printout 13.9.

a) The required point estimate for the mean is the first item in the last line of the printout, the item headed Fit. So, based on the sample data, we estimate that the mean price of all Nissan Zs that are five years old and have been driven 52,000 miles is about 92.80, or $9280.

b) Here we want to obtain a 95% confidence interval for the mean price of all Nissan Zs that are five years old and have been driven 52,000 miles. This is displayed as the third item in the last line of the printout, the item headed 95% C.I. Hence, the required confidence interval is from 86.47 to 99.12. We can be 95% confident that the mean price of all Nissan Zs that are five years old and have been driven 52,000 miles is somewhere between $8647 and $9912.   ■

MTB

PREDICTION INTERVALS

To determine the predicted value or a prediction interval for a randomly selected member of the population of y-values corresponding to particular values, x_{1p}, x_{2p}, ..., x_{kp}, of the predictor variables, we again proceed as in simple linear regression. Like a point estimate for the mean, the predicted y-value is obtained by substituting the predictor-variable values into the sample regression equation:

$$\hat{y}_p = b_0 + b_1 x_{1p} + b_2 x_{2p} + \cdots + b_k x_{kp}$$

And, the endpoints of a $(1 - \alpha)$-level prediction interval are found by applying the formula

$$\hat{y}_p \pm t_{\alpha/2} \cdot s_{y_p - \hat{y}_p}$$

where $t_{\alpha/2}$ is computed for the t-curve with df $= n - (k + 1)$ and $s_{y_p - \hat{y}_p}$ is the estimated standard deviation of the random variable $y_p - \hat{y}_p$.

EXAMPLE 13.17

Illustrates predicted values and prediction intervals

Use Printout 13.9 on page 688 to
a) find the predicted price of a randomly selected Nissan Z that is five years old and has been driven 52,000 miles.
b) obtain a 95% prediction interval for the price of a randomly selected Nissan Z that is five years old and has been driven 52,000 miles.

SOLUTION   The required information can be found in the last two lines of Printout 13.9.
a) As we know, the predicted price is the same as the point estimate for the mean price, which we found in part (a) of Example 13.16. Thus, based on the sample data, the predicted price of a randomly selected Nissan Z that is five years old and has been driven 52,000 miles is 92.80, or $9280.
b) Here we want to obtain a 95% prediction interval for the price of a randomly selected Nissan Z that is five years old and has been driven 52,000 miles. This is given as the final item in the last line of Printout 13.9, the item headed 95% P.I. Hence, the required prediction interval is from 71.53 to 114.07. We can be 95% certain that the price of a randomly selected Nissan Z that is five years old and has been driven 52,000 miles will be somewhere between $7153 and $11,407.

MTB

■

USING THE COMPUTER (OPTIONAL)

In Section 13.3, we learned how Minitab's REGRESS command and PREDICT subcommand can be used to obtain both confidence intervals and prediction intervals in simple linear regression. As the next example shows, the procedure used to obtain confidence intervals and prediction intervals in multiple regression is essentially identical to that in simple linear regression.

EXAMPLE 13.18

Illustrates the REGRESS; PREDICT command

Table 13.5 on page 683 gives data on age, miles driven, and price for a sample of 11 Nissan Zs. Explain how to obtain the Minitab output in Printout 13.9 on page 688 that, in addition to the usual regression output, provides the information shown in the last two lines about Nissan Zs that are five years old and have been driven 52,000 miles.

SOLUTION   Recall that we have previously stored the age, miles, and price data from Table 13.5 in columns named AGE, MILES, and PRICE, respectively. To obtain the computer output in Printout 13.9, we first employ the REGRESS command as described in Example 12.14 on pages 633–634, except this time we type a semicolon at the end of the command in preparation for a subcommand. Thus, we type

REGRESS 'PRICE' on 2 predictors 'AGE' and 'MILES';

On the next line we type the subcommand

PREDICT 5 52.

The 5 52 indicates that we want price information about Nissan Zs that are five years old and have been driven 52,000 miles. The result of applying these commands is the computer output shown in Printout 13.9 on page 688. ∎

Exercises 13.5

___ **13.57** A manufacturer of household appliances wants to analyze the relationship between total sales and the three primary means of advertising. The first three columns of the following table provide the expenditures on advertising, by type, for each of 10 randomly selected sales periods. The fourth column contains the total sales. All data are in millions of dollars.

Television x_1	Magazines x_2	Radio x_3	Sales y
8.3	4.4	6.1	361.1
6.3	4.2	4.9	344.0
9.9	5.9	6.3	377.9
9.4	3.3	6.1	371.5
10.4	2.7	5.2	365.4
9.0	3.5	5.1	364.5
9.2	4.1	6.0	372.9
10.6	4.8	6.4	379.4
9.3	4.2	5.5	362.6
10.5	6.0	5.9	387.5

Explain what it would mean for the assumptions for multiple regression inferences to be satisfied with tele-

vision, magazine, and radio advertising expenditures as predictor variables for sales.

___ **13.58** Ten Corvettes, between one and six years of age, were randomly selected from the classified ads of the *Arizona Republic*. The following data were obtained on age, miles driven, and price:

Age (yrs) x_1	Miles (thous) x_2	Price ($100s) y
6	36	125
6	36	115
6	36	130
2	22	260
2	5	219
5	31	150
4	22	190
5	39	163
1	9	260
4	27	160

Explain what it would mean for the assumptions for multiple regression inferences to be satisfied with age and miles driven as predictor variables for price.

__ **13.59** Refer to Exercise 13.57. We used a statistical computer package to perform a regression analysis on the data with the variables, television, magazine, and radio advertising expenditures, as predictor variables for sales. Printout 13.10, at the top of the next page, displays the resulting computer output. The last two lines of the output give information about sales when the amounts spent on television, magazine, and radio advertising are $9.5, $4.3, and $5.2 million, respectively. Presuming that the variables under consideration satisfy the assumptions for multiple regression inferences, employ the computer output in Printout 13.10 to solve the following problems:

a) Find and interpret the sample regression equation.
b) Obtain and interpret the standard error of the estimate, s_e.
c) At the 5% significance level, do the data provide sufficient evidence to conclude that the regression, with television, magazine, and radio advertising expenditures as predictor variables, is useful for predicting sales?
d) At the 5% significance level, do the data provide sufficient evidence to conclude that television advertising expenditure is useful as a predictor of sales? Be precise in your conclusion.
e) Repeat part (d) for radio advertising expenditure.
f) Determine a point estimate for mean sales when the amounts spent on television, magazine, and radio advertising are $9.5, $4.3, and $5.2 million.
g) Find a 95% confidence interval for mean sales when the amounts spent on television, magazine, and radio advertising are $9.5, $4.3, and $5.2 million.
h) Determine the predicted sales if the amounts spent on television, magazine, and radio advertising are $9.5, $4.3, and $5.2 million.
i) Determine a 95% prediction interval for sales if the amounts spent on television, magazine, and radio advertising are $9.5, $4.3, and $5.2 million.

__ **13.60** Refer to Exercise 13.58. We used a statistical computer package to perform a multiple regression analysis on the data with the variables, age and miles driven, as predictor variables for price. Printout 13.11 on the next page shows the computer output obtained. The last two lines of the output provide information about price for Corvettes that are four years old and have been driven 28,000 miles. Presuming that the variables under consideration satisfy the assumptions for multiple regression inferences, employ the computer output displayed in Printout 13.11 to solve the following problems:

a) Find and interpret the sample regression equation.
b) Obtain and interpret the standard error of the estimate, s_e.
c) At the 5% significance level, do the data provide sufficient evidence to conclude that the regression, with age and miles driven as predictor variables, is useful for predicting price?
d) At the 5% significance level, do the data provide sufficient evidence to conclude that age is useful as a predictor of price? Be precise in your conclusion.
e) Repeat part (d) for the predictor variable, miles driven. What if the hypothesis test is performed at the 10% significance level?
f) Determine a point estimate for the mean price of all Corvettes that are four years old and have been driven 28,000 miles.
g) Obtain a 95% confidence interval for the mean price of all Corvettes that are four years old and have been driven 28,000 miles.
h) Determine the predicted price of a randomly selected Corvette that is four years old and has been driven 28,000 miles.
i) Find a 95% prediction interval for the price of a randomly selected Corvette that is four years old and has been driven 28,000 miles.

__ **13.61** Colleges and universities often require prospective students to take college-entrance exams. The theory is that those examinations are helpful for predicting success in college. We used the *Focus* system at Arizona State University to obtain data on high-school GPA (cumulative grade-point average), SAT math score, SAT verbal score, and sophomore GPA (cumulative grade-point average through two years of college). This was done for a random sample of 47 students. Printout 13.12, at the top of page 694, provides the output obtained by applying a statistical computer package to perform a regression analysis on the data. The regression uses high-school GPA, SAT math score, and SAT verbal score as predictor variables for sophomore GPA. In the last two lines of Printout 13.12, you will find information about sophomore GPA for students who had a high-school GPA of 3.20, an SAT math score of 650, and an SAT verbal score of 500. Solve the following problems:

a) At the 1% significance level, do the data provide sufficient evidence to conclude that the regression, with high-school GPA, SAT math score, and SAT verbal score as predictor variables, is useful for predicting sophomore GPA of Arizona State University students?

PRINTOUT 13.10   Computer output for Exercise 13.59

```
The regression equation is
SALES = 266 + 6.73 TV + 3.26 MAG + 4.51 RADIO

Predictor        Coef       Stdev     t-ratio         p
Constant       266.23       16.34       16.29     0.000
TV              6.727        1.344        5.01     0.002
MAG             3.257        1.642        1.98     0.095
RADIO           4.507        3.703        1.22     0.269

s = 4.418       R-sq = 91.1%      R-sq(adj) = 86.6%

Analysis of Variance

SOURCE          DF          SS         MS          F          p
Regression      3       1194.53     398.18      20.40      0.002
Error           6        117.11      19.52
Total           9       1311.64


    Fit  Stdev.Fit        95% C.I.            95% P.I.
 367.58       2.59   ( 361.24, 373.92)   ( 355.05, 380.12)
```

PRINTOUT 13.11   Computer output for Exercise 13.60

```
The regression equation is
PRICE = 287 - 37.4 AGE + 1.64 MILES

Predictor        Coef       Stdev     t-ratio         p
Constant       287.362       9.943      28.90     0.000
AGE            -37.375       5.174      -7.22     0.000
MILES            1.6378      0.8116      2.02     0.083

s = 12.11       R-sq = 96.0%      R-sq(adj) = 94.9%

Analysis of Variance

SOURCE          DF          SS         MS          F          p
Regression      2        24655      12328      84.07      0.000
Error           7         1026        147
Total           9        25682


    Fit  Stdev.Fit        95% C.I.            95% P.I.
 183.72       4.26   ( 173.65, 193.79)   ( 153.36, 214.08)
```

b) At the 1% significance level, do the data provide sufficient evidence to conclude that the variable, high-school GPA, is useful as a predictor of sophomore GPA for Arizona State University students? Be precise in your conclusion.

c) Do the data suggest that either SAT math score or SAT verbal score is useful as a predictor of sophomore GPA for Arizona State University students?

d) Do you think that another regression analysis is called for? If so, which predictor variable(s) would you include? Explain.

e) Obtain a 95% confidence interval for the mean sophomore GPA of all Arizona State University students who had a high-school GPA of 3.20, an SAT math score of 650, and an SAT verbal score of 500.

f) Find a 95% prediction interval for the sophomore GPA of a randomly selected student who plans to attend Arizona State University, if that student had a high-school GPA of 3.20, an SAT math score of 650, and an SAT verbal score of 500.

g) What assumptions are you making in answering parts (a)–(f)?

— 13.62   Hanna Properties specializes in custom-home resales in the Equestrian Estates, an exclusive subdivision in Phoenix, Arizona. A random sample of 33 properties was selected and data on the following variables were obtained: square-footage, number of bedrooms, number of bathrooms, number of days on the market, and selling price (in thousands of dollars). Then a regression analysis was performed for selling price in terms of the other four variables. The resulting computer output is displayed in Printout 13.13 on the next page. In the last two lines of the output you will find information on selling price for homes in the Equestrian Estates that have 3200 square feet, four bedrooms, three bathrooms, and remain on the market for 60 days. Solve the following problems:

a) Do the data provide sufficient evidence to conclude that the regression, with square-footage, number of bedrooms, number of bathrooms, and number of days on the market as predictor variables, is useful for predicting selling price of homes in the Equestrian Estates? Use $\alpha = 0.01$.

b) Do the data provide sufficient evidence to conclude that square-footage is useful as a predictor of selling price for homes in the Equestrian Estates? Use $\alpha = 0.01$. Be precise in your conclusion.

c) Repeat part (b) for the variable, number of bedrooms. Explain why number of bedrooms might

not be useful for predicting selling price of homes in the Equestrian Estates.

d) Suppose you decide to run another regression analysis on the data with fewer predictor variables. Which predictor variables would you include?

e) Obtain a 95% confidence interval for the mean selling price of all homes in the Equestrian Estates that have 3200 square feet, four bedrooms, three bathrooms, and remain on the market for 60 days.

f) Determine a 95% prediction interval for the selling price of a home in the Equestrian Estates that has 3200 square feet, four bedrooms, three bathrooms, and remains on the market for 60 days.

g) What assumptions are you making in answering parts (a)–(f)?

Exercises 13.63 and 13.64 are computer exercises.

— 13.63   (Computer exercise)   Suppose that the data in Exercise 13.57 are stored in columns named TV, MAG, RADIO, and SALES.

a) Which Minitab command and subcommands (if any) should be used to obtain the computer output shown in Printout 13.10? *(Note: The confidence and prediction intervals are for television, magazine, and radio advertising expenditures of $9.5, $4.3, and $5.2 million, respectively.)*

b) If you have access to Minitab, use it to obtain Printout 13.10.

— 13.64   (Computer exercise)   Suppose that the data in Exercise 13.58 are stored in columns named AGE, MILES, and PRICE.

a) Which Minitab command and subcommands (if any) should be used to obtain the computer output shown in Printout 13.11? *(Note: The confidence and prediction intervals are for Corvettes that are four years old and have been driven 28,000 miles.)*

b) If you have access to Minitab, use it to obtain Printout 13.11.

Individual confidence intervals for the β parameters: A $(1 - \alpha)$-level confidence interval for β_j has endpoints

$$(1) \qquad b_j \pm t_{\alpha/2} \cdot s_{b_j}$$

where $t_{\alpha/2}$ is computed for the t-curve with $n - (k + 1)$ degrees of freedom and s_{b_j} is the estimated standard deviation of the random variable b_j. Here, n denotes the sample size and k the number of predictor variables. We will apply this confidence-interval formula in Exercises 13.65 and 13.66.

PRINTOUT 13.12   Computer output for Exercise 13.61

```
The regression equation is
SOPHGPA = 0.648 + 0.562 HSGPA +0.000267 SATMATH +0.000223 SATVERBL

Predictor        Coef        Stdev      t-ratio         p
Constant       0.6479      0.4294         1.51      0.139
HSGPA          0.5616      0.1521         3.69      0.001
SATMATH     0.0002665   0.0009694         0.27      0.785
SATVERBL    0.0002234   0.0009959         0.22      0.824

s = 0.4860      R-sq = 34.9%     R-sq(adj) = 30.3%

Analysis of Variance

SOURCE         DF          SS          MS          F          p
Regression      3      5.4431      1.8144       7.68      0.000
Error          43     10.1585      0.2362
Total          46     15.6015


    Fit  Stdev.Fit        95% C.I.           95% P.I.
 2.7299     0.1181   ( 2.4916, 2.9682)   ( 1.7209, 3.7389)
```

PRINTOUT 13.13   Computer output for Exercise 13.62

```
The regression equation is
SELL$ = - 212 + 0.0754 SQFT + 20.9 BEDROOMS + 62.1 BATHS - 0.0899 DAYS

Predictor        Coef        Stdev      t-ratio         p
Constant      -211.95       73.05        -2.90      0.007
SQFT          0.07542     0.02345         3.22      0.003
BEDROOMS        20.94       19.99         1.05      0.304
BATHS           62.11       21.62         2.87      0.008
DAYS         -0.08986     0.07624        -1.18      0.248

s = 39.48      R-sq = 84.0%     R-sq(adj) = 81.7%

Analysis of Variance

SOURCE         DF          SS          MS          F          p
Regression      4      229575       57394      36.83      0.000
Error          28       43633        1558
Total          32      273209


    Fit  Stdev.Fit        95% C.I.           95% P.I.
 294.12      8.28   ( 277.15, 311.08)   ( 211.48, 376.76)
```

$=$ **13.65**   Refer to Exercise 13.57 and Printout 13.10.

a) Use Formula (1), at the bottom of page 693, to find a 95% confidence interval for the coefficient, β_1, of the predictor variable x_1 (television advertising expenditure).

b) Interpret your result from part (a) in words.

c) Repeat part (a) for the coefficient, β_3, of the predictor variable x_3 (radio advertising expenditure).

$=$ **13.66**   Refer to Exercise 13.58 and Printout 13.11.

a) Apply Formula (1), at the bottom of page 693, to determine a 95% confidence interval for the coefficient, β_1, of the predictor variable x_1 (age).

b) Interpret your result from part (a) in words.

c) Repeat part (a) for the coefficient, β_2, of the predictor variable x_2 (miles driven).

$=$ **13.67**   **Curvilinear regression:** Refer to Exercise 12.95 on page 636. In that exercise, we used a statistical computer program to perform a curvilinear regression analysis for the price of Nissan Zs with age and age^2 as the predictor variables. Printout 12.13, which shows the computer output, is repeated below as Printout 13.14.

a) Explain what it would mean for the assumptions for multiple regression inferences to be satisfied with age and age^2 as predictor variables for price.

b) Obtain the standard error of the estimate, s_e.

c) At the 1% significance level, do the data provide

sufficient evidence to conclude that the regression, with age and age^2 as predictor variables, is useful for predicting price?

d) At the 1% significance level, do the data provide sufficient evidence to conclude that age is useful as a predictor of price? Be precise in your conclusion.

e) Repeat part (d) for the predictor variable, age^2.

f) Determine a point estimate for the mean price of all nine-year-old Nissan Zs using age and age^2 as the predictor variables.

g) Find the predicted price of a randomly selected nine-year-old Nissan Z using age and age^2 as the predictor variables.

$=$ **13.68**   **(Computer exercise)**   Assume that the age and price data in Exercise 12.95 on page 636 are stored in columns named AGE and PRICE; and that the squares of the age data are stored in a column named AGESQ. Suppose now that you want to apply Minitab to obtain 95% confidence and prediction intervals for prices of nine-year-old Nissan Zs using age and age^2 as predictor variables.

a) Which Minitab command and subcommands (if any) should be used?

b) If you have access to Minitab, use it to obtain the required confidence and prediction intervals. *(Note:* To store the squares of the age data, first name a column AGESQ and then type the command LET 'AGESQ'='AGE'**2.)*

PRINTOUT 13.14   Computer output for Exercise 13.67

```
The regression equation is
PRICE = 209 - 30.8 AGE + 1.33 AGESQ

Predictor        Coef        Stdev       t-ratio         p
Constant        209.44       11.48        18.24       0.000
AGE            -30.776        4.056        -7.59       0.000
AGESQ            1.3297       0.3219        4.13       0.000

s = 12.81       R-sq = 90.3%      R-sq(adj) = 89.6%

Analysis of Variance

SOURCE          DF          SS          MS          F           p
Regression       2        42895       21448      130.70      0.000
Error           28         4595         164
Total           30        47490
```

Chapter review

FORMULAS   In the formulas below,

$b_0 = y$-intercept of sample regression line
$b_1 = $ slope of sample regression line
$n = $ sample size
$SSE = $ error sum of squares
$\rho = $ population linear correlation
    coefficient

$r = $ (sample) linear correlation coefficient
$\beta_0 = y$-intercept of population regression
    line
$\beta_1 = $ slope of population regression line
$s_e = $ standard error of the estimate
$\hat{y}_p = b_0 + b_1 x_p$

Note: You may also wish to refer to the formulas given in the Chapter Review in Chapter 12 on pages 639–640.

Population regression line (equation), 644

$$y = \beta_0 + \beta_1 x$$

Standard error of the estimate, 648

$$s_e = \sqrt{\frac{SSE}{n-2}}$$

Test statistic for H_0: $\beta_1 = 0$, 654

$$t = \frac{b_1}{s_e / \sqrt{S_{xx}}}$$

with df $= n - 2$.

Confidence interval for β_1, 657

$$b_1 \pm t_{\alpha/2} \cdot \frac{s_e}{\sqrt{S_{xx}}}$$

(df $= n - 2$)

Confidence interval for the mean of the population of y-values corresponding to x_p, 665

$$\hat{y}_p \pm t_{\alpha/2} \cdot s_e \sqrt{\frac{1}{n} + \frac{(x_p - \Sigma x/n)^2}{S_{xx}}}$$

(df $= n - 2$)

Prediction interval for a population y-value corresponding to x_p, 668

$$\hat{y}_p \pm t_{\alpha/2} \cdot s_e \sqrt{1 + \frac{1}{n} + \frac{(x_p - \Sigma x/n)^2}{S_{xx}}}$$

(df $= n - 2$)

Test statistic for H_0: $\rho = 0$, 677

$$t = \frac{r}{\sqrt{\dfrac{1 - r^2}{n - 2}}}$$

with df $= n - 2$.

Population regression equation in multiple regression,* 681

$$y = \beta_0 + \beta_1 x_1 + \cdots + \beta_k x_k$$

(k = number of predictor variables)

YOU SHOULD
BE ABLE TO

1. use and understand the preceding formulas.
2. state the assumptions for regression inferences.
3. determine the standard error of the estimate.
4. perform a hypothesis test to decide whether the slope of the population regression line is not zero and, hence, whether the regression is useful for making predictions.
5. obtain a confidence interval for β_1.
6. find a confidence interval for the mean of the population of y-values corresponding to a particular x-value.
7. determine a prediction interval for a population y-value corresponding to a particular x-value.
8. perform a hypothesis test for a population linear correlation coefficient with null hypothesis H_0: $\rho = 0$.
9. state the assumptions for multiple regression inferences.*
10. determine the standard error of the estimate from computer output.*
11. use computer output to decide on the overall utility of a multiple regression.*
12. use computer output to decide on the utility of a particular predictor variable in a multiple regression.*
13. obtain from computer output, confidence and prediction intervals corresponding to particular values, $x_{1p}, x_{2p}, \ldots, x_{kp}$, of the predictor variables.*
14. use the Minitab commands covered in this chapter.*
15. interpret the output obtained from the application of the Minitab commands discussed in this chapter.*

REVIEW TEST

1. The director of a large mathematics course hires upper-division science students to grade papers. On each grading day, he records the number of papers graded and the total amount of money paid to the graders. The following table provides data for 12 randomly selected days from last semester:

No. papers (100s) x	Cost ($) y
16	234
16	220
18	258
22	298
19	273
16	227
18	246
15	210
19	265
17	250
15	223
18	251

Discuss what it would mean for the assumptions for regression inferences to be satisfied by the variables, number of papers graded and cost.

In Problems 2–7, presume that the variables, number of papers graded, x, and cost, y, satisfy the assumptions for regression inferences.

2. Refer to Problem 1.
   a) Determine the regression equation for the data.
   b) Compute the standard error of the estimate.
   c) Interpret the result from part (b).

3. Refer to Problems 1 and 2.
   a) Do the data provide sufficient evidence to conclude that the number of papers graded is useful as a predictor of cost? Use $\alpha = 0.05$.
   b) Obtain a 95% confidence interval for the slope, β_1, of the population regression line that relates cost to number of papers graded. Interpret your result in words.

4. Refer to Problems 1 and 2.
   a) Find a point estimate for the mean cost of grading 1600 papers.
   b) Determine a 95% confidence interval for the mean cost of grading 1600 papers.
   c) Find the predicted cost of grading 1600 papers.

   d) Obtain a 95% prediction interval for the cost of grading 1600 papers.
   e) Explain precisely why the prediction interval in part (d) is more extensive than the confidence interval in part (b).

*5. **(Computer problem)** Suppose that the data in Problem 1 are stored in columns named PAPERS and COST.
   a) Which Minitab command and subcommands (if any) should be used to determine the standard error of the estimate?
   b) Which Minitab command and subcommands (if any) should be used to carry out the hypothesis test in Problem 3(a)?
   c) Which Minitab command and subcommands (if any) should be used to determine the confidence interval in Problem 4(b) and the prediction interval in Problem 4(d)?
   d) If you have access to Minitab, use it to perform the hypothesis test in Problem 3(a), to obtain the confidence interval in Problem 4(b), and to determine the prediction interval in Problem 4(d).

*6. **(Computer problem)** Printout 13.15, at the top of the next page, gives the computer output that results from applying Minitab's REGRESS command to the data shown in Problem 1. Use the computer output to
   a) determine the sample regression equation.
   b) obtain the standard error of the estimate.
   c) find the slope, b_1, of the sample regression line.
   d) determine the estimated standard deviation of the random variable b_1.
   e) obtain the value of the test statistic, t, for a hypothesis test to decide whether the slope, β_1, of the population regression line is not zero.
   f) determine the P-value for the hypothesis test referred to in part (e).
   g) decide whether the number of papers graded is useful for predicting cost. Take $\alpha = 0.05$.
   h) obtain a 95% confidence interval for the slope, β_1, of the population regression line and interpret your result in words. *(Note: You will need to use Table III to find $t_{\alpha/2}$, but everything else that is required to obtain the confidence interval can be found in the printout.)*

(continued at the top of page 700)

PRINTOUT 13.15   Minitab output for Problem 6

```
The regression equation is
COST = 35.8 + 12.1 PAPERS
```

Predictor	Coef	Stdev	t-ratio	p
Constant	35.80	17.06	2.10	0.062
PAPERS	12.0835	0.9738	12.41	0.000

s = 6.526   R-sq = 93.9%   R-sq(adj) = 93.3%

Analysis of Variance

SOURCE	DF	SS	MS	F	p
Regression	1	6558.3	6558.3	153.97	0.000
Error	10	425.9	42.6		
Total	11	6984.3			

Fit	Stdev.Fit	95% C.I.	95% P.I.
229.13	2.34	(223.93, 234.34)	(213.68, 244.58)

PRINTOUT 13.16   Computer output for Problems 8 and 9

```
The regression equation is
INCOME = - 40.9 + 0.772 AGE + 3.11 EDUC
```

Predictor	Coef	Stdev	t-ratio	p
Constant	-40.855	5.511	-7.41	0.000
AGE	0.7719	0.1165	6.62	0.000
EDUC	3.1052	0.2250	13.80	0.000

s = 7.886   R-sq = 76.6%   R-sq(adj) = 75.9%

Analysis of Variance

SOURCE	DF	SS	MS	F	p
Regression	2	14657.2	7328.6	117.84	0.000
Error	72	4477.6	62.2		
Total	74	19134.7			

Fit	Stdev.Fit	95% C.I.	95% P.I.
33.528	1.096	(31.342, 35.714)	(17.653, 49.403)

i) identify the subcommand used to obtain the output shown in the last two lines of Printout 13.15. *(Note:* The confidence and prediction intervals are for 1600 papers.*)*

j) find a point estimate for the mean cost of grading 1600 papers.

k) determine a 95% confidence interval for the mean cost of grading 1600 papers.

l) find the predicted cost of grading 1600 papers.

m) obtain a 95% prediction interval for the cost of grading 1600 papers.

7. Refer to Problem 1. At the 2.5% significance level, do the data provide sufficient evidence to conclude that the variables, number of papers graded and cost, are positively linearly correlated?

*8. The U.S. Bureau of the Census collects data on income by educational attainment, sex, and age. Results are published in *Current Population Reports.* From a random sample of 75 males between the ages of 25 and 50, all of whom have at least a ninth-grade education, data were collected on age, number of years of school completed, and annual income. Then a statistical computer program was used to perform a multiple regression analysis for annual income with the variables, age and number of years of school completed, as predictor variables. The resulting computer output is displayed in Printout 13.16 on the previous page. In the last two lines of the output you will find information on annual income of males who are 32 years old and have completed exactly four years of college (i.e., 16 years of school). Use Printout 13.16 to solve the following problems:

a) Determine and interpret the sample regression equation.

b) Obtain and interpret the standard error of the estimate, s_e.

c) At the 5% significance level, do the data provide sufficient evidence to conclude that the

regression, with age and number of years of school completed as predictor variables, is useful for predicting annual income for males?

d) At the 5% significance level, do the data provide sufficient evidence to conclude that age is useful as a predictor of annual income for males? Be precise in your conclusion.

e) Repeat part (d) for the predictor variable, number of years of school completed.

f) What assumptions are you making in answering parts (a)–(e)?

*9. Refer to Problem 8 and to the computer output in Printout 13.16 on page 699.

a) Find a point estimate for the mean annual income of all males who are 32 years old and have completed exactly four years of college.

b) Obtain a 95% confidence interval for the mean annual income of all males who are 32 years old and have completed exactly four years of college.

c) Determine the predicted annual income of a randomly selected male who is 32 years old and has completed exactly four years of college.

d) Find a 95% prediction interval for the annual income of a randomly selected male who is 32 years old and has completed exactly four years of college.

e) What assumptions are you making in answering parts (b) and (d)?

*10. **(Computer problem)** Suppose that the data in Problem 8 are stored in columns named AGE, EDUC, and INCOME.

a) Which Minitab command and subcommands (if any) should be used to obtain the computer output in Printout 13.16 on page 699 if the last two lines of that printout are omitted?

b) Which Minitab command and subcommands (if any) should be used to obtain the computer output in Printout 13.16 if the last two lines are to be included?

CHAPTER 14

ANALYSIS OF VARIANCE (ANOVA)

In Chapter 10 we studied inferential methods for comparing the means of two populations. Now we will study **analysis of variance,** or **ANOVA,** which provides methods for comparing the means of more than two populations. There are several different ANOVA procedures, just as there are several different procedures for comparing the means of two populations.

We will examine two types of ANOVA procedures. The first, called a *one-way analysis of variance,* is the generalization of the pooled-*t* procedure (Procedure 10.3) to more than two populations. The second, called a *two-way analysis of variance,* is similar to the paired-difference procedure (Procedure 10.7) but can be applied to more than two populations.

CHAPTER OUTLINE

14.1   The *F*-distribution

The analysis-of-variance procedures utilize a class of continuous probability distributions called **F-distributions,** named in honor of Ronald Fisher (1890–1962). We will study *F*-distributions now so that in later sections we can concentrate on the development and application of analysis of variance.

Probabilities for a random variable that has an *F*-distribution are equal to areas under a curve that, not surprisingly, is called an **F-curve.** Recall that a *t*-distribution or a chi-square distribution depends on the number of degrees of freedom, df. An *F*-distribution also depends on the number of degrees of freedom, but there are two numbers of degrees of freedom instead of one. Figure 14.1 depicts two different *F*-curves. One has df = $(10, 2)$ and the other has df = $(9, 50)$.

FIGURE 14.1
Two different
F-curves

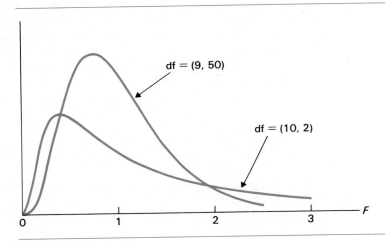

The first number of degrees of freedom for an *F*-curve is called the **degrees of freedom for the numerator;** and the second number of degrees of freedom is called the **degrees of freedom for the denominator.** (The reason for this terminology will become clear in Section 14.2.) Thus, for the *F*-curve in Figure 14.1 with df = $(10, 2)$, we have

$$\text{df} \; = \; (10, 2)$$

degrees of freedom   degrees of freedom
for numerator   for denominator

Some of the fundamental properties of *F*-curves are as follows:

KEY FACT 14.1   Basic properties of *F*-curves

PROPERTY 1   The total area under an *F*-curve is equal to 1.

PROPERTY 2   An *F*-curve starts at 0 on the horizontal axis and extends indefinitely to the right, approaching the horizontal axis as it does so.

PROPERTY 3   An *F*-curve is not symmetrical, but is skewed to the right; that is, it climbs to its high point rapidly and comes back to the horizontal axis more slowly.

USING THE *F*-TABLES

Areas under *F*-curves have been compiled and put into tables. These tables provide areas that are likely to be used as significance levels, such as 0.01 and 0.05. For *F*-curves, there are entire tables corresponding to each area. This is because critical values are needed for each different combination of degrees of freedom for the numerator and degrees of freedom for the denominator.

As you might expect, the symbol F_α is used to denote the *F*-value with area α to its right. In this book, tables for $F_{0.01}$ and $F_{0.05}$ are presented. These are given as Tables V and VI, respectively, in the appendix.

Let us now consider Table VI in the appendix, which gives values for $F_{0.05}$. The values of $F_{0.05}$ are displayed inside the table; and the degrees of freedom for the numerator and denominator on the top and sides, respectively. We will illustrate the use of Table VI in Example 14.1.

EXAMPLE 14.1   *Illustrates how to find the F-value for a specified area*

For an *F*-curve with df $= (4, 12)$, find $F_{0.05}$. That is, find the *F*-value with area 0.05 to its right. See Figure 14.2.

FIGURE 14.2

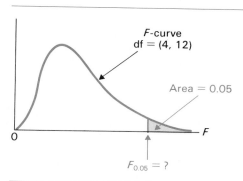

SOLUTION   To obtain the *F*-value in question, we use Table VI. In this case, the degrees of freedom for the numerator is four and the degrees of freedom for the denominator is 12. Thus, we first go down the outside columns to the row labeled "12." Then we go across that row until we are under the column headed "4." The number in the body of the table there, 3.26, is the required *F*-value. That is, for an *F*-curve with df $= (4, 12)$, the *F*-value with area 0.05 to its right is 3.26: $F_{0.05} = 3.26$.   ∎

MTB

Exercises 14.1

___ **14.1**   An F-curve has df $= (12, 7)$. What is the number of degrees of freedom for the
a) numerator?   b) denominator?

___ **14.2**   An F-curve has df $= (8, 19)$. What is the number of degrees of freedom for the
a) denominator?   b) numerator?

In Exercises 14.3–14.8, use Tables V and VI to determine the required F-values. Illustrate your work with graphs similar to Figure 14.2 on page 703.

___ **14.3**   For an F-curve with df $= (24, 40)$, find the F-value with
a) area 0.05 to its right.   b) area 0.01 to its right.

___ **14.4**   For an F-curve with df $= (12, 5)$, find the F-value with
a) area 0.01 to its right.   b) area 0.05 to its right.

___ **14.5**   For an F-curve with df $= (20, 21)$, find
a) $F_{0.01}$.   b) $F_{0.05}$.

___ **14.6**   For an F-curve with df $= (6, 10)$, find
a) $F_{0.05}$.   b) $F_{0.01}$.

___ **14.7**   An F-curve has df $= (30, 15)$. Determine the F-value with
a) area 0.95 to its left.   b) area 0.99 to its left.

___ **14.8**   An F-curve has df $= (9, 8)$. Determine the F-value with
a) area 0.99 to its left.   b) area 0.95 to its left.

$=$ **14.9**   Refer to Table VI in the appendix. Because of space restrictions, the numbers of degrees of freedom are not consecutive. For instance, the degrees of

freedom for the numerator goes from 24 to 30. So, if you had only Table VI to work with and you needed to find $F_{0.05}$ for df $= (25, 20)$, how would you do it?

Using Tables V and VI to obtain left-tailed critical values: For a given F-curve, Tables V and VI give the F-values, $F_{0.01}$ and $F_{0.05}$, with areas 0.01 and 0.05 to their right, respectively. But what if we want to obtain the F-value, $F_{0.99}$, with area 0.01 to its left or the F-value, $F_{0.95}$, with area 0.05 to its left? Consider, for instance, an F-curve with df $= (60, 8)$. To determine the F-value, $F_{0.95}$, with area 0.05 to its left, we use the following fact: The value $F_{0.95}$ for df $= (60, 8)$ is the reciprocal of the value $F_{1-0.95} = F_{0.05}$ for df $= (8, 60)$. (*Note:* We switched the degrees of freedom.) From Table VI, we see that for df $= (8, 60)$, $F_{0.05} = 2.10$. So, for df $= (60, 8)$, $F_{0.95} = 1/2.10 = 0.48$. We summarize this discussion below in Figure 14.3.

In general, we have the following fact: For an F-curve with df $= (\nu_1, \nu_2)$, the F-value, $F_{1-\alpha}$, with area α to its left is equal to the reciprocal, $1/F_\alpha$, of the F-value with area α to its right for an F-curve with df $= (\nu_2, \nu_1)$. We will use this fact in Exercises 14.10 and 14.11.

$=$ **14.10**   For an F-curve with df $= (10, 24)$, determine the F-value with
a) area 0.05 to its left.   b) area 0.01 to its left.
Illustrate your work with a pair of graphs.

$=$ **14.11**   For an F-curve with df $= (30, 7)$, find
a) $F_{0.99}$.   b) $F_{0.95}$.
Illustrate your work with a pair of graphs.

FIGURE 14.3

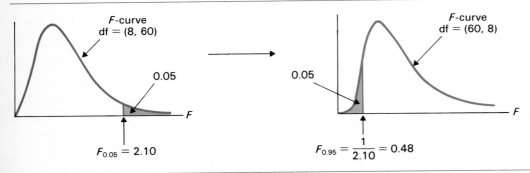

14.2   One-way analysis of variance

We have already mentioned that analysis of variance (ANOVA) is an inferential procedure that is used to compare the means of several populations. The reason for the word "variance" in "analysis of variance" is that the procedure for comparing the means involves analyzing the variation in the sample data.

In this section we will examine **one-way analysis of variance**. It is called *one-way* analysis of variance because each piece of data is classified in one way; namely, according to the population from which it was sampled. One-way analysis of variance is the generalization to more than two populations of the pooled-t procedure, Procedure 10.3 on page 489. As in the pooled-t procedure, we make the following assumptions:

KEY FACT 14.2   Assumptions for one-way ANOVA

1. *Independent samples:* The samples taken from the populations under consideration are independent of one another.

2. *Normal populations:* The populations under consideration are normally distributed.

3. *Equal standard deviations:* The standard deviations of the populations under consideration are equal.

Assumption 1, the independent-samples assumption, is often referred to as the assumption of a **completely randomized design.**

The following example presents the essential ideas behind one-way analysis of variance. It also introduces some of the pertinent terminology and notation.

EXAMPLE 14.2   *Introduces one-way ANOVA*

The U.S. Energy Information Administration gathers data on residential energy consumption and expenditures. Results are published in *Residential Energy Consumption Survey: Consumption and Expenditures*. A researcher wants to know whether there is a difference in mean annual energy consumptions among households in the four regions of the United States. Let μ_1, μ_2, μ_3, and μ_4 denote last year's mean energy consumptions for households in the Northeast, Midwest, South, and West, respectively. Then the hypotheses to be tested are:

$$H_0\colon \mu_1 = \mu_2 = \mu_3 = \mu_4 \text{ (mean energy consumptions are all equal)}$$
$$H_a\colon \text{Not all the means are equal.}$$

The basic strategy for carrying out the hypothesis test is as follows:

1. Take independent random samples of last year's energy consumptions for households in the four regions.

2. Compute the means, $\overline{x}_1$, $\overline{x}_2$, $\overline{x}_3$, and $\overline{x}_4$, of the four samples.

3. Reject the null hypothesis if the sample means differ by too much; otherwise, do not reject the null hypothesis.

This process is depicted in Figure 14.4.

FIGURE 14.4

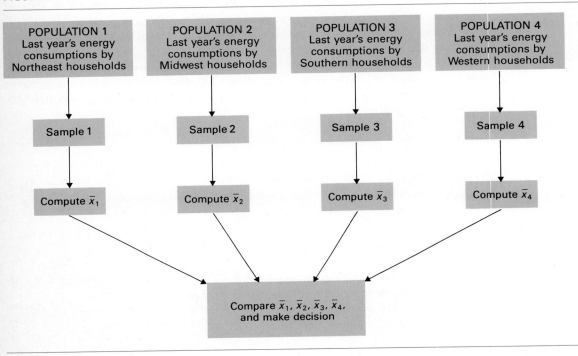

Steps 1 and 2 entail collecting the sample data and computing the sample means. Suppose that the results of those steps are as displayed in Table 14.1 (data are given to the nearest 10 million BTU).

TABLE 14.1
Samples and their means of last year's energy consumptions for households in the four U.S. regions

Northeast	Midwest	South	West
15	17	11	10
10	12	7	12
13	18	9	8
14	13	13	7
13	15		9
	12		
13.0	14.5	10.0	9.2	← *Means*

Step 3 involves comparing the four sample means shown at the bottom of Table 14.1. Specifically, we must decide whether the variation among the four sample means can be reasonably attributed to sampling error or whether that variation is large enough to indicate that the population means are not all the same.

In hypothesis tests for two population means, we measure the variation among the two sample means by calculating their difference, $\overline{x}_1 - \overline{x}_2$. When more than two populations are involved, as in the problem at hand, we cannot measure the variation among the sample means by simply taking a difference. However, we can measure that variation by computing the standard deviation or variance of the sample means or, for that matter, by computing any descriptive statistic that measures the variation among the sample means.

In one-way analysis of variance, we measure the variation among the sample means by a weighted average of their squared deviations about the mean, $\overline{x}$, of all the sample data. That measure of variation is called the **treatment mean square, MSTR,** and is defined by

$$MSTR = \frac{SSTR}{k - 1}$$

where k denotes the number of populations being sampled (in this case, $k = 4$) and

$$SSTR = n_1(\overline{x}_1 - \overline{x})^2 + n_2(\overline{x}_2 - \overline{x})^2 + \cdots + n_k(\overline{x}_k - \overline{x})^2$$

The quantity $SSTR$ is called the **treatment sum of squares.**

$MSTR$ is similar to the sample variance of the sample means. In fact, if the sample sizes are all the same, then $MSTR$ is equal to that common sample size times the sample variance of the sample means (see Exercise 14.45).

If the null hypothesis of equal population means is true, then we would expect the sample means to be roughly equal, resulting in a small value for $MSTR$. In other words, if $MSTR$ is too large, then this gives us evidence that the null hypothesis of equal population means is false.

Let us determine $MSTR$ for the sample data in Table 14.1. We have $k = 4$ and, referring to Table 14.1, we see that $n_1 = 5$, $n_2 = 6$, $n_3 = 4$, $n_4 = 5$, and $\overline{x}_1 = 13.0$, $\overline{x}_2 = 14.5$, $\overline{x}_3 = 10.0$, $\overline{x}_4 = 9.2$. To obtain the overall mean, $\overline{x}$, we need to divide the sum of all the data in Table 14.1 by the total number of pieces of data:

$$\overline{x} = \frac{\Sigma x}{n} = \frac{15 + 10 + 13 + \cdots + 7 + 9}{20} = \frac{238}{20} = 11.9$$

Therefore,

$$\begin{aligned}
SSTR &= n_1(\overline{x}_1 - \overline{x})^2 + n_2(\overline{x}_2 - \overline{x})^2 + n_3(\overline{x}_3 - \overline{x})^2 + n_4(\overline{x}_4 - \overline{x})^2 \\
&= 5(13.0 - 11.9)^2 + 6(14.5 - 11.9)^2 + 4(10.0 - 11.9)^2 + 5(9.2 - 11.9)^2 \\
&= 97.5
\end{aligned}$$

and so

$$MSTR = \frac{SSTR}{k - 1} = \frac{97.5}{4 - 1} = \boxed{32.5}$$

This is our measure of variation among the four sample means shown at the bottom of Table 14.1.

The question now, of course, is whether the value, 32.5, for $MSTR$ is large enough to conclude that the null hypothesis of equal population means is false. To decide, we compare $MSTR$ to a measure of variation within the samples. This latter measure is simply the pooled estimate of the common population variance, σ^2. It is called the **error mean square, MSE,** and is defined by

$$MSE = \frac{SSE}{n - k}$$

where k denotes the number of populations under consideration, n denotes the total number of pieces of sample data, and

$$SSE = (n_1 - 1)s_1^2 + (n_2 - 1)s_2^2 + \cdots + (n_k - 1)s_k^2$$

The quantity SSE is called the **error sum of squares.**[†]

For the sample data in Table 14.1, we have $k = 4$, $n_1 = 5$, $n_2 = 6$, $n_3 = 4$, $n_4 = 5$, and $n = 20$. Computing the sample variance for each of the four data sets in Table 14.1, we find that $s_1^2 = 3.5$, $s_2^2 = 6.7$, $s_3^2 = 6.\overline{6}$, and $s_4^2 = 3.7$. Consequently,

$$
\begin{aligned}
SSE &= (n_1 - 1)s_1^2 + (n_2 - 1)s_2^2 + (n_3 - 1)s_3^2 + (n_4 - 1)s_4^2 \\
&= (5 - 1) \cdot 3.5 + (6 - 1) \cdot 6.7 + (4 - 1) \cdot 6.\overline{6} + (5 - 1) \cdot 3.7 \\
&= 82.3
\end{aligned}
$$

and so

$$MSE = \frac{SSE}{n - k} = \frac{82.3}{20 - 4} = \boxed{5.144}$$

This is our measure of variation within the samples.

As we said, we will decide whether the variation among the sample means, $MSTR$, is large enough to conclude that the null hypothesis of equal population means is false by comparing it to the variation within the samples, MSE. The idea is the following: MSE is an estimate of the common variance of the populations under consideration. So, if $MSTR$ is large relative to MSE, then we can infer that the variation among the sample means is due to a difference among the population means and not to the variation within the populations. Thus, we use the random variable

$$F = \frac{MSTR}{MSE}$$

as the test statistic. Large values of F indicate that $MSTR$ is large relative to MSE and, hence, that the null hypothesis of equal population means should be rejected.

[†] The terms **treatment** and **error** arose from the fact that many ANOVA techniques were first developed to analyze agricultural experiments. In any case, the treatments refer to the different populations, whereas the errors pertain to the variation within the populations.

For the energy-consumption data in Table 14.1, we have seen that $MSTR = 32.5$ and $MSE = 5.144$. Thus, the value of the F-statistic is

$$F = \frac{MSTR}{MSE} = \frac{32.5}{5.144} = \mathbf{6.32}$$

Is this value of F large enough to conclude that the null hypothesis of equal population means is false? To answer that question, we need to know the probability distribution of F. We will discuss that and then return to complete the hypothesis test considered in this example. ∎

First we summarize the definitions presented in the previous example and then we give the probability distribution of the random variable F.

DEFINITION 14.1 Sums of squares and mean squares for one-way ANOVA

Treatment sum of squares, SSTR:

$$SSTR = n_1(\overline{x}_1 - \overline{x})^2 + n_2(\overline{x}_2 - \overline{x})^2 + \cdots + n_k(\overline{x}_k - \overline{x})^2$$

Treatment mean square, MSTR:

$$MSTR = \frac{SSTR}{k-1}$$

Error sum of squares, SSE:

$$SSE = (n_1 - 1)s_1^2 + (n_2 - 1)s_2^2 + \cdots + (n_k - 1)s_k^2$$

Error mean square, MSE:

$$MSE = \frac{SSE}{n-k}$$

where,

k = number of populations being sampled

$n = n_1 + n_2 + \cdots + n_k$ = total number of pieces of data

$\overline{x}$ = mean of all n pieces of data

and, for $j = 1, 2, \ldots, k,$

n_j = size of sample from Population j

$\overline{x}_j$ = mean of sample from Population j

s_j^2 = variance of sample from Population j

KEY FACT 14.3   Test statistic for one-way ANOVA

Suppose that independent random samples of sizes $n_1, n_2, \ldots, n_k$ are to be taken from k normally distributed populations with means $\mu_1, \mu_2, \ldots, \mu_k$, respectively. Further suppose that the standard deviations of the k populations are equal. Then, if $\mu_1 = \mu_2 = \cdots = \mu_k$, the random variable

$$F = \frac{MSTR}{MSE}$$

has the F-distribution with df $= (k - 1, n - k)$, where n denotes the total number of pieces of data.

We have now studied all of the elements necessary to construct a procedure for performing a one-way analysis of variance. However, it will be helpful to consider two additional concepts before presenting that procedure.

ONE-WAY ANOVA IDENTITY

To begin, we will define another sum of squares. This sum of squares gives a measure of total variation among all the sample data. It is called the **total sum of squares, SST,** and is defined as follows:

DEFINITION 14.2   Total sum of squares

The *total sum of squares* is defined by

$$SST = \Sigma(x - \overline{x})^2$$

where the sum extends over all n pieces of sample data.

If we divide SST by $n - 1$, then we get the sample variance of all the data. So, SST really is a measure of total variation.

For the energy-consumption data in Table 14.1, we have $\overline{x} = 11.9$ and so

$$SST = \Sigma(x - \overline{x})^2 = (15 - 11.9)^2 + (10 - 11.9)^2 + \cdots + (9 - 11.9)^2$$
$$= 9.61 + 3.61 + \cdots + 8.41$$
$$= 179.8$$

Previously we found that $SSTR = 97.5$ and $SSE = 82.3$ for the energy-consumption data. Since $179.8 = 97.5 + 82.3$, we see that $SST = SSTR + SSE$. This equation is always true and is called the **one-way ANOVA identity.**

KEY FACT 14.4   One-way ANOVA identity

The total sum of squares equals the treatment sum of squares plus the error sum of squares; that is,

$$SST = SSTR + SSE$$

The one-way ANOVA identity shows that we can partition the total variation in the data into a component representing variation among the sample means and a component representing variation within the samples. We can picture this partitioning as in Figure 14.5.

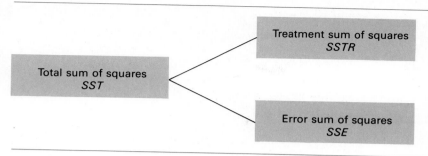

ONE-WAY ANOVA TABLES

Next, we will discuss the **one-way ANOVA table.** This table is useful to organize and summarize the quantities required for performing a one-way analysis of variance. The general format of a one-way ANOVA table is shown in Table 14.2.

Source	df	SS	MS = SS/df	F-statistic
Treatment	$k - 1$	$SSTR$	$MSTR = \dfrac{SSTR}{k - 1}$	$F = \dfrac{MSTR}{MSE}$
Error	$n - k$	SSE	$MSE = \dfrac{SSE}{n - k}$	
Total	$n - 1$	SST		

For the energy-consumption data in Table 14.1, we have already computed all of the quantities that appear in the one-way ANOVA table. Table 14.3 displays the one-way ANOVA table for that data.

Source	df	SS	MS = SS/df	F-statistic
Treatment	3	97.5	32.500	6.32
Error	16	82.3	5.144	
Total	19	179.8		

THE ONE-WAY ANOVA PROCEDURE

We now present a step-by-step method that can be used to perform a one-way analysis of variance. The procedure utilizes shortcut formulas to compute the sums of squares since those formulas are easier to work with and reduce the possibility of roundoff error. Note that the hypothesis test is always right-tailed since the null hypothesis is rejected only when the test statistic, F, is too large.

PROCEDURE 14.1

To perform a one-way ANOVA for k population means.

ASSUMPTIONS

1. Independent samples.
2. Normal populations.
3. Equal population standard deviations.

STEP 1 *State the null and alternative hypotheses.*

STEP 2 *Decide on the significance level, α.*

STEP 3 *The critical value is F_α, with $df = (k-1, n-k)$, where n is the total number of pieces of data.*

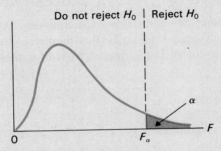

STEP 4 *Calculate the sums of squares using the shortcut formulas*

$$SST = \Sigma x^2 - \frac{(\Sigma x)^2}{n}$$

$$SSTR = \left(\frac{T_1^2}{n_1} + \frac{T_2^2}{n_2} + \cdots + \frac{T_k^2}{n_k} \right) - \frac{(\Sigma x)^2}{n}$$

$$SSE = SST - SSTR$$

where, for $j = 1, 2, \ldots, k,$

$n_j =$ size of sample from Population j,

$T_j =$ sum of sample data from Population j.

STEP 5 *Construct the one-way ANOVA table:*

Source	df	SS	MS = SS/df	F-statistic
Treatment	$k-1$	SSTR	$MSTR = \dfrac{SSTR}{k-1}$	$F = \dfrac{MSTR}{MSE}$
Error	$n-k$	SSE	$MSE = \dfrac{SSE}{n-k}$	
Total	$n-1$	SST		

> **STEP 6** *If the value of the F-statistic falls in the rejection region, then re-ject H_0; otherwise, do not reject H_0.*
> **STEP 7** *State the conclusion in words.*

Note: In Step 4 of Procedure 14.1, we need to compute Σx, the sum of all the data. Generally, it is easiest to do that by summing the T_js, that is, by using the formula $\Sigma x = T_1 + T_2 + \cdots + T_k$.

EXAMPLE 14.3   Illustrates Procedure 14.1

Recall that independent random samples of households in the four U.S. regions yielded the data shown in Table 14.4 on last year's energy consumptions. The data are given to the nearest 10 million BTU. At the bottom of Table 14.4, we have also recorded the sum of the data for each sample.

TABLE 14.4
Samples and their sums of last year's energy consumptions for households in the four U.S. regions

Northeast	Midwest	South	West
15	17	11	10
10	12	7	12
13	18	9	8
14	13	13	7
13	15		9
	12		
65	87	40	46

At the 5% significance level, do the data provide sufficient evidence to conclude that there is a difference in last year's mean energy consumptions among households in the four U.S. regions? Assume that last year's energy consumptions in the four regions are normally distributed and have equal standard deviations.

SOLUTION   We apply Procedure 14.1.

STEP 1 *State the null and alternative hypotheses.*

Let μ_1, μ_2, μ_3, and μ_4 denote last year's mean energy consumptions for households in the Northeast, Midwest, South, and West, respectively. Then the null and alternative hypotheses are

> H_0: $\mu_1 = \mu_2 = \mu_3 = \mu_4$ (mean energy consumptions are equal)
> H_a: Not all the means are equal.

STEP 2 *Decide on the significance level, α.*

We are to perform the test at the 5% significance level; thus, $\alpha = 0.05$.

STEP 3 *The critical value is F_α, with* $df = (k - 1, n - k)$.

From Step 2, $\alpha = 0.05$. Also, as we see from Table 14.4, the number of populations under consideration is four $(k = 4)$ and the total number of pieces of data is 20 $(n = 20)$. Hence, df $= (k - 1, n - k) = (4 - 1, 20 - 4) = (3, 16)$. Consulting Table VI, we find that the critical value is $F_\alpha = F_{0.05} = 3.24$. See Figure 14.6.

FIGURE 14.6

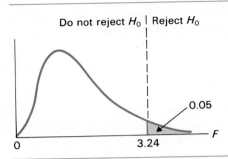

Do not reject H_0 | Reject H_0

0.05

F

0   3.24

STEP 4 *Calculate the sums of squares using the shortcut formulas.*

Referring to Table 14.4, we observe that

$$k = 4$$

$$n_1 = 5 \qquad n_2 = 6 \qquad n_3 = 4 \qquad n_4 = 5$$
$$T_1 = 65 \qquad T_2 = 87 \qquad T_3 = 40 \qquad T_4 = 46$$

and

$$n = 5 + 6 + 4 + 5 = 20$$
$$\Sigma x = 65 + 87 + 40 + 46 = 238$$

Summing the squares of all the data in Table 14.4 gives

$$\Sigma x^2 = 15^2 + 10^2 + 13^2 + \cdots + 7^2 + 9^2 = 3012$$

Consequently,

$$SST = \Sigma x^2 - \frac{(\Sigma x)^2}{n} = 3012 - \frac{(238)^2}{20}$$
$$= 3012 - 2832.2 = 179.8$$

and

$$SSTR = \left(\frac{T_1^2}{n_1} + \frac{T_2^2}{n_2} + \cdots + \frac{T_k^2}{n_k} \right) - \frac{(\Sigma x)^2}{n}$$

$$= \left(\frac{65^2}{5} + \frac{87^2}{6} + \frac{40^2}{4} + \frac{46^2}{5} \right) - \frac{(238)^2}{20}$$

$$= 2929.7 - 2832.2 = 97.5$$

and

$$SSE = SST - SSTR = 179.8 - 97.5 = \boxed{82.3}$$

STEP 5 *Construct the one-way ANOVA table:*

The one-way ANOVA table for the energy-consumption data was constructed earlier in Table 14.3 on page 711.

STEP 6 *If the value of the F-statistic falls in the rejection region, then reject H_0; otherwise, do not reject H_0.*

As we see from Table 14.3, the value of the F-statistic is $F = 6.32$. A glance at Figure 14.6 shows that this value falls in the rejection region. Thus, we **reject H_0**.

STEP 7 *State the conclusion in words.*

The data provide sufficient evidence to conclude that last year's mean energy consumptions for households in the four U.S. regions are not all equal. That is, at least two of the regions have different mean energy consumptions for last year. ■

MTB

CONCERNING THE ASSUMPTIONS FOR ONE-WAY ANOVA

The three assumptions for one-way ANOVA, which are stated in Key Fact 14.2 on page 705, are (1) independent samples, (2) normal populations, and (3) equal standard deviations. Here are some qualitative remarks concerning those assumptions.

Assumption 1 on independent samples is absolutely essential to the one-way ANOVA procedure. Assumption 2 on normality is not too critical as long as the populations are not too far from being normally distributed. Assumption 3 on equal standard deviations is also not that important provided that the sample sizes are roughly the same.

MULTIPLE COMPARISONS

Suppose that we perform a one-way ANOVA test and reject the null hypothesis. Then we can conclude that the means of the populations under consideration are not all the same. But how can we decide on such things as which means are different, which mean is largest, and so forth?

There are methods, called **multiple comparisons,** that can be used to answer these and other related questions. We will not cover those methods here but instead refer the reader to any of the more advanced books on inferential statistics.

USING THE COMPUTER (OPTIONAL)

Procedure 14.1 on pages 712–713 provides a step-by-step method for performing a one-way ANOVA test. Alternatively, we can apply Minitab to carry out such a hypothesis test. The appropriate command is called **AOVONEWAY.** Example 14.4 explains how AOVONEWAY is used.

EXAMPLE 14.4   *Illustrates the AOVONEWAY command*

Refer to Example 14.3 on page 713. Use Minitab to perform the hypothesis test considered in that example.

SOLUTION   Let μ_1, μ_2, μ_3, and μ_4 denote last year's mean energy consumptions for households in the Northeast, Midwest, South, and West, respectively. Then we want to perform the hypothesis test

H_0: $\mu_1 = \mu_2 = \mu_3 = \mu_4$ (mean energy consumptions are equal)

H_a: Not all the means are equal.

at the 5% significance level.

To employ Minitab, we first enter the four data sets in Table 14.4 on page 713 into, say, C1, C2, C3, and C4 using the SET command. We also name C1 "Nrtheast," C2 "Midwest," C3 "South," and C4 "West" by applying the NAME command. Next we type the command AOVONEWAY followed by the storage locations of the sample data. That is, we type

AOVONEWAY for 'NRTHEAST' 'MIDWEST' 'SOUTH' 'WEST'

This command instructs Minitab to perform a one-way analysis of variance on the data stored in the four designated columns. Printout 14.1, at the top of the next page, displays the above commands along with the resulting output.

The first part of the computer output gives a one-way ANOVA table. This is Minitab's version of the one-way ANOVA table in Table 14.3 on page 711. Note that Minitab uses the terminology "Factor" instead of "Treatment."

To the right of the one-way ANOVA table in Printout 14.1, under the column headed **p**, is the *P*-value. As we know, this is really the only item that is required to make the decision concerning the hypothesis test. Since the *P*-value is 0.005, which is less than the specified significance level of $\alpha = 0.05$, we reject H_0. In other words, the data provide sufficient evidence to conclude that last year's mean energy consumptions for households in the four U.S. regions are not all the same.   ■

We will now examine some of the other items in the output shown in Printout 14.1. As we see, the output contains more than just the one-way ANOVA table and the *P*-value. Below the ANOVA table is another table that gives the sample sizes, sample means, and sample standard deviations of the four samples. Below that table we find the item POOLED STDEV = 2.268. This is the pooled estimate of the common population standard deviation, σ, of the four populations.

Finally, the lower right side of the output gives individual 95% confidence intervals for the means, μ_1, μ_2, μ_3, and μ_4, of the four populations under consideration. The formula used to obtain those confidence intervals is presented in the exercises (see page 721).

PRINTOUT 14.1
Minitab output
for AOVONEWAY

```
MTB > SET C1
DATA> 15 10 13 14 13
DATA> END
MTB > SET C2
DATA> 17 12 18 13 15 12
DATA> END
MTB > SET C3
DATA> 11 7 9 13
DATA> END
MTB > SET C4
DATA> 10 12 8 7 9
DATA> END
MTB > NAME C1 'NRTHEAST' C2 'MIDWEST' C3 'SOUTH' C4 'WEST'
MTB > AOVONEWAY for 'NRTHEAST' 'MIDWEST' 'SOUTH' 'WEST'
```

```
ANALYSIS OF VARIANCE
SOURCE      DF        SS        MS        F        P
FACTOR       3     97.50     32.50     6.32    0.005
ERROR       16     82.30      5.14
TOTAL       19    179.80
```

```
                                    INDIVIDUAL 95 PCT CI'S FOR MEAN
                                    BASED ON POOLED STDEV
LEVEL        N      MEAN     STDEV  -------+---------+---------+---------
NRTHEAST     5    13.000     1.871                 (------*-------)
MIDWEST      6    14.500     2.588                      (-----*------)
SOUTH        4    10.000     2.582      (-------*-------)
WEST         5     9.200     1.924  (-------*------)
                                    -------+---------+---------+---------
POOLED STDEV =      2.268              9.0       12.0      15.0
```

Exercises 14.2

___ 14.12   State the three assumptions required for one-way ANOVA.

___ 14.13   One-way ANOVA is a procedure for comparing the means of several populations. It is the generalization of what procedure for comparing the means of two populations?

___ 14.14   One of the assumptions required for one-way ANOVA is that the samples taken are independent of one another. There is another name for that assumption. What is it?

___ 14.15   If we define $s = \sqrt{MSE}$, then of which parameter is s an estimate?

___ 14.16   Identify the quantity used as a measure of the total variation among all the sample data.

___ 14.17   In each of parts (a)–(c) below, we have given the notation for one of the three sums of squares defined in this section. For each sum of squares, state its name, its defining formula, and the source of variation that it represents.
a) SSE b) $SSTR$ c) SST

___ 14.18   True or False: Given any two of the three sums of squares, SST, $SSTR$, and SSE, the remaining one can be determined. Explain your answer.

___ 14.19   State the one-way ANOVA identity and interpret its meaning with regard to partitioning the total variation in the data.

___ 14.20   Explain the reason for the word "variance" in the phrase "analysis of variance."

In each of Exercises 14.21–14.24,
a) compute SSTR, SSE, and SST using the defining formulas.
b) verify that the one-way ANOVA identity holds.
c) determine MSTR and MSE.
d) obtain the one-way ANOVA table for the data.

___ 14.21   The times required by three workers to perform an assembly-line task were recorded on five randomly selected occasions by a quality-control engineer. Here are the times, to the nearest minute.

Hank	Joseph	Susan
8	8	10
10	9	9
9	9	10
11	8	11
10	10	9

___ 14.22   A pig farmer wants to test three different diets designed to maximize weight gain. The farmer randomly selects 12 pigs and divides them randomly into three groups of four pigs each. Each group is given one of the diets. The weight gains, in pounds, after a three week period are shown below.

Diet A	Diet B	Diet C
10.5	11.5	9.8
10.9	10.9	10.5
10.7	11.8	10.4
10.3	11.8	10.5

___ 14.23   The College Entrance Examination Board, New York, NY, publishes information on Scholastic Aptitude Test (SAT) scores in National College-Bound Senior. SAT scores for randomly selected students from each of four different high-school rank categories are displayed in the following table:

Top tenth	Second tenth	Second fifth	Third fifth
528	514	649	372
586	457	506	440
680	521	556	495
718	370	413	321
	532	470	424
			330

___ 14.24   The U.S. Bureau of the Census collects data on monthly rents of newly completed apartments by region. Data are published in Current Housing Reports. Suppose that independent random samples of monthly rents for newly completed apartments in the four U.S. regions yield the following sample data:

Northeast	Midwest	South	West
470	408	428	379
363	386	167	366
413	337	398	444
646	452	573	280
709	359		623
	344		

In Exercises 14.25 and 14.26 we have presented two partially completed one-way ANOVA tables. Fill in the missing entries in each table.

___ 14.25

Source	df	SS	MS = SS/df	F-statistic
Treatment	2		21.652	
Error		84.400		
Total	14			

___ 14.26

Source	df	SS	MS = SS/df	F-statistic
Treatment		2.124	0.708	0.75
Error	20			
Total				

In each of Exercises 14.27–14.32, presume that the populations under consideration satisfy the assumptions for one-way ANOVA. For each exercise, use Procedure 14.1 on pages 712–713 to perform the appropriate hypothesis test.

___ 14.27   A consumer-advocacy group wants to compare four different brands of flashlight batteries. Each brand is to be tested five times. Twenty flashlights are randomly selected and divided into four groups of five flashlights each. Then each group of flashlights uses a different brand of batteries. The lifetimes of the batteries to the nearest hour are as follows:

Brand A	Brand B	Brand C	Brand D
42	28	24	20
30	31	36	32
39	31	28	38
28	32	28	28
29	27	33	25

At the 5% significance level, does there appear to be a difference in the mean lifetimes of the four brands of batteries?

___ 14.28 The general manager of a large chain of convenience stores wants to try three different advertising policies. The three policies are:

Policy 1: No advertising
Policy 2: Advertise in neighborhood with circulars
Policy 3: Use circulars and advertise in newspapers

Eighteen stores are randomly selected and divided at random into three groups of six stores. Each group uses one of the three policies. Following the implementation of the policies, sales figures are obtained for each of the stores during a one month period. The figures are displayed, in thousands of dollars, in the following table:

Policy 1	Policy 2	Policy 3
22	21	29
20	25	24
21	25	31
21	20	32
24	22	26
22	26	27

Do the data provide evidence of a difference in mean monthly sales among the three policies? Perform the required hypothesis test at the 1% significance level.

___ 14.29 The U.S. Bureau of Labor Statistics gathers data on hourly earnings of nonsupervisory workers in nonfarm U.S. jobs, by industry. Results are published in *Employment and Earnings*. The following data, in dollars per hour, were obtained from random samples of workers in three different industries:

Wholesale trade	Finance, Insurance, and Real Estate	Services
9.78	9.68	9.16
11.86	9.32	9.61
7.60	8.67	3.67
8.05	4.42	
	10.30	

Do the data provide sufficient evidence to conclude that a difference exists in the mean hourly earnings for nonsupervisory workers in the three industries? Perform the hypothesis test with $\alpha = 0.05$.

___ 14.30 Manufacturers of golf balls seem to always be saying that their ball goes the farthest. A writer for a sports magazine decides to conduct an impartial test. She randomly selects 20 golf professionals and then randomly assigns four golfers to each of the five brands. Finally, each golfer drives the assigned brand of ball. Here are the results, in yards.

Brand 1	Brand 2	Brand 3	Brand 4	Brand 5
279	284	270	281	281
276	277	262	271	293
281	284	277	269	276
274	288	286	275	292

Do the data provide sufficient evidence to conclude that a difference exists in mean driving distances for the five brands of golf balls? Use $\alpha = 0.05$.

___ 14.31 The U.S. Bureau of Prisons compiles data on the time served by prisoners released from federal institutions for the first time. Data are published in the document *Statistical Report*. Independent random samples of released prisoners for five different offense categories yielded the following information on time served, in months:

	Counter-feiting	Drug laws	Firearms	Forgery	Fraud
n_j	15	17	12	10	11
T_j	218	313	218	156	127

$$\Sigma x^2 = 17,769$$

At the 1% significance level, do the data indicate that a difference exists in the mean times served by prisoners in the five offense groups?

___ **14.32**   Data are collected by the Northwestern University Placement Center at Evanston, Illinois, on annual starting salaries of college graduates, by major. Findings are reported in *The Northwestern Lindquist-Endicott Report*. Independent random samples of college graduates in marketing, statistics, economics, and computer science provided the following information on annual starting salaries. The salary data are given to the nearest thousand dollars.

	Marketing	Statistics	Economics	CS
n_j	35	25	30	34
T_j	683	600	622	870

$$\Sigma x^2 = 64{,}239$$

Do the data imply that a difference exists in the mean annual starting salaries among the four majors? Take $\alpha = 0.05$.

___ **14.33**   The null and alternative hypotheses for a one-way ANOVA test are

$$H_0: \mu_1 = \mu_2 = \cdots = \mu_k$$
$$H_a: \text{Not all means are equal.}$$

Suppose that, in reality, the null hypothesis is false. Does this mean that no two of the populations have the same mean? If not, what does it mean?

___ **14.34**   In Section 10.2, we learned how to perform a pooled-t test to decide whether two normally distributed populations with equal standard deviations have different means. This procedure could be used in place of one-way ANOVA for comparing the means of several populations by testing two population means at a time. For instance, in Exercise 14.27, we could compare the brands of batteries two at a time: first compare Brand A and Brand B, second compare Brand A and Brand C, and so forth. Why isn't that a desirable procedure?

Exercises 14.35–14.38 are computer exercises.

___ **14.35**   **(Computer exercise)**   Suppose that the data in Exercise 14.29 are stored in columns named WHTRADE, FIRE, and SERVICES.
a) Which Minitab command and subcommands (if any) should be used to carry out the hypothesis test considered in that exercise?
b) If you have access to Minitab, use it to perform the hypothesis test.

___ **14.36**   **(Computer exercise)**   Assume the data in Exercise 14.30 are stored in columns named BRAND1, BRAND2, BRAND3, BRAND4, and BRAND5.
a) Which Minitab command and subcommands (if any) should be used to carry out the hypothesis test considered in that exercise?
b) If you have access to Minitab, use it to perform the hypothesis test.

___ **14.37**   **(Computer exercise)**   The U.S. Bureau of the Census collects data on income by educational attainment, sex, and age. Results are published in the document *Current Population Reports*. Independent random samples were taken of women from three categories of educational attainment: elementary school, secondary school, and college (four-year degree). Then Minitab was applied to perform a one-way analysis of variance on the annual incomes of the women sampled. Printout 14.2 at the top of the next page displays the resulting computer output (data are in thousands of dollars). Determine
a) the three sums of squares, *SSTR, SSE,* and *SST*.
b) the treatment mean square, *MSTR*, and the error mean square, *MSE*.
c) the value of the test statistic, *F*.
d) the null and alternative hypotheses.
e) the *P*-value for the hypothesis test.
f) the conclusion if the hypothesis test is performed at the 1% significance level.
g) the sample size, sample mean, and sample standard deviation of each of the three samples.
h) a 95% confidence interval for the mean annual income of all women whose educational attainment is at the secondary level.

___ **14.38**   **(Computer exercise)**   The Motor Vehicle Manufacturers Association of the United States conducts surveys on the costs of owning and operating a motor vehicle. Data are published in *Motor Vehicle Facts and Figures* and include costs for gas and oil, tires, maintenance, insurance, license and registration, and depreciation. Independent random samples of owners of large, intermediate, and compact cars were taken to obtain information on annual insurance premiums. Then Minitab's AOVONEWAY command was applied to the resulting data. Printout 14.3 on the next page displays the computer output generated by Minitab. Determine
a) the three sums of squares, *SSTR, SSE,* and *SST*.
b) the treatment mean square, *MSTR*, and the error mean square, *MSE*.

PRINTOUT 14.2   Minitab output for Exercise 14.37

```
ANALYSIS OF VARIANCE
SOURCE      DF        SS        MS        F         p
FACTOR       2      6545.7    3272.9     69.65     0.000
ERROR      154      7236.0      47.0
TOTAL      156     13781.7
```

				INDIVIDUAL 95 PCT CI'S FOR MEAN
				BASED ON POOLED STDEV
LEVEL	N	MEAN	STDEV	------+---------+---------+---------+
ELEMENTA	40	11.099	6.720	(--*---)
HIGHSCHO	62	16.587	7.324	(--*--)
COLLEGE	55	27.190	6.388	(--*--)

```
                             ------+---------+---------+---------+
POOLED STDEV =    6.855       12.0      18.0      24.0      30.0
```

PRINTOUT 14.3   Minitab output for Exercise 14.38

```
ANALYSIS OF VARIANCE
SOURCE      DF        SS        MS        F         p
FACTOR       2       36763     18382     0.55      0.581
ERROR       74     2483155     33556
TOTAL       76     2519918
```

				INDIVIDUAL 95 PCT CI'S FOR MEAN
				BASED ON POOLED STDEV
LEVEL	N	MEAN	STDEV	----------+---------+---------+------
LARGE	20	902.4	152.5	(------------*------------)
INTERMED	30	851.7	169.2	(----------*----------)
COMPACT	27	890.5	215.8	(---------*----------)

```
                             ----------+---------+---------+------
POOLED STDEV =    183.2            840       900       960
```

---

c) the value of the test statistic, F.

d) the null and alternative hypotheses for the hypothesis test.

e) the P-value for the hypothesis test.

f) the conclusion if the hypothesis test is performed at the 5% significance level.

g) the sample size, sample mean, and sample standard deviation of each of the three samples.

h) a 95% confidence interval for the mean annual insurance premium of all compact-car owners.

Confidence intervals for means and differences between means in one-way ANOVA: Suppose independent random samples of sizes $n_1, n_2, \ldots, n_k$ are to be taken from k normally distributed populations with means $\mu_1, \mu_2, \ldots, \mu_k$, respectively. Further sup-

pose the standard deviations of the k populations are equal. Let $s = \sqrt{MSE}$. Then:

- A $(1 - \alpha)$-level confidence interval for any particular population mean, say μ_i, has endpoints

$$\bar{x}_i \pm t_{\alpha/2} \cdot \frac{s}{\sqrt{n_i}}$$

where df $= n - k$.

- A $(1 - \alpha)$-level confidence interval for the difference between any two particular population means, say μ_i and μ_j, has endpoints

$$(\bar{x}_i - \bar{x}_j) \pm t_{\alpha/2} \cdot s\sqrt{(1/n_i) + (1/n_j)}$$

where df $= n - k$.

We will apply these confidence-interval formulas in Exercises 14.39 and 14.40.

═ 14.39 Refer to Exercise 14.23.
a) Obtain a 90% confidence interval for the mean SAT score, μ_3, of all students ranked in the second fifth of their high school class.
b) Obtain a 90% confidence interval for the difference, $\mu_1 - \mu_4$, between the mean SAT scores of students ranked in the top tenth and third fifth of their high school class.
c) What assumptions are you making in parts (a) and (b)?

═ 14.40 Refer to Exercise 14.24.
a) Determine a 99% confidence interval for the mean monthly rent, μ_2, of newly completed apartments in the Midwest.
b) Find a 99% confidence interval for the difference, $\mu_1 - \mu_3$, between the mean monthly rents of newly completed apartments in the Northeast and South.
c) What assumptions are you making in parts (a) and (b)?

═ 14.41 Refer to Exercise 14.39. Suppose that you have obtained a 90% confidence interval for each of the two differences, $\mu_1 - \mu_2$ and $\mu_1 - \mu_3$. Can you be 90% confident of both results simultaneously? In other words, can you be 90% confident that both differences are contained in their corresponding confidence intervals? Explain your answer.

═ 14.42 Show that for two populations ($k = 2$), $MSE = s_p^2$, where s_p^2 is the pooled variance defined in Section 10.2 on page 488. Conclude that $\sqrt{MSE}$ is the pooled sample standard deviation, s_p.

═ 14.43 Consider two normally distributed populations with equal standard deviations and with means μ_1 and μ_2. Suppose that we want to perform a hypothesis test to decide whether $\mu_1 \neq \mu_2$. If independent samples are used, identify two hypothesis-testing procedures that can be employed to carry out the test.

═ 14.44 Recall that $\bar{x}$ is the mean of all n pieces of sample data.
a) Show that $\bar{x}$ is a weighted average of the k sample means, weighted according to sample size. That is,

$$\bar{x} = \frac{n_1 \bar{x}_1 + n_2 \bar{x}_2 + \cdots + n_k \bar{x}_k}{n_1 + n_2 + \cdots + n_k}$$

b) Prove that if the sample sizes are all equal, then $\bar{x}$ is just the mean of the sample means.

═ 14.45 Suppose the sample sizes, $n_1, n_2, \ldots, n_k$, are all equal, say to m. Show that, under those circumstances, $MSTR$ is equal to m times the sample variance of the k sample means, $\bar{x}_1, \bar{x}_2, \ldots, \bar{x}_k$. (*Hint:* Use part (b) of Exercise 14.44.)

═ 14.46 In this exercise we will derive the shortcut formulas, given on page 712, for the three sums of squares, SST, $SSTR$, and SSE.
a) Show that

$$\Sigma(x - \bar{x})^2 = \Sigma x^2 - \frac{(\Sigma x)^2}{n}$$

where the sums extend over all the data.
b) Use part (a) and the defining formula for SST, given in Definition 14.2 on page 710, to conclude that the shortcut formula

$$SST = \Sigma x^2 - \frac{(\Sigma x)^2}{n}$$

is valid.
c) Show that for each j, the equation

$$n_j \left(\frac{T_j}{n_j} - \frac{\Sigma x}{n} \right)^2 = \frac{T_j^2}{n_j} - \frac{2\Sigma x}{n} T_j + \frac{(\Sigma x)^2}{n^2} n_j$$

holds true.
d) Use part (c) to show that

$$\Sigma n_j \left(\frac{T_j}{n_j} - \frac{\Sigma x}{n} \right)^2 = \Sigma \frac{T_j^2}{n_j} - \frac{(\Sigma x)^2}{n}$$

(*Hint:* Recall that $\Sigma T_j = \Sigma x$ and $\Sigma n_j = n$.)
e) Conclude from part (d) and the defining formula for $SSTR$, given in Definition 14.1 on page 709, that the shortcut formula

$$SSTR = \Sigma \frac{T_j^2}{n_j} - \frac{(\Sigma x)^2}{n}$$

is valid.
f) Why does $SSE = SST - SSTR$?

14.3   Two-way analysis of variance

In Section 14.2, we discussed *one-way analysis of variance*. This is the type of analysis of variance that is used to compare the means of two or more populations for a *completely randomized design* (independent random samples from the populations). In this respect, we should emphasize that the type of *design* refers to the plan for selecting the sample data, whereas the type of *analysis of variance* refers to the method for analyzing the sample data.

Sometimes there is so much variation within the populations under consideration that the use of a completely randomized design will fail to detect a difference among the population means when one exists. This is because it is not possible to decide whether the variation among the sample means is due to a difference among the population means or whether it is due to the variation within the populations.

If a large portion of the variation within the populations is due to one extraneous variable, then it is often appropriate to use a *randomized block design* instead of a completely randomized design. In a **randomized block design,** the extraneous source of variation is isolated and removed so that it is easier to detect differences among the population means when such differences exist. The type of analysis of variance that is used to compare the means of two or more populations for a randomized block design is called **two-way analysis of variance.**[†]

Although the randomized block design is not always appropriate or feasible, it offers a viable alternative to the completely randomized design in the presence of a single extraneous source of variability. Consider the following example:

EXAMPLE 14.5   *Introduces the randomized block design*

Suppose that we want to compare the times it takes three analgesics, Brand A, Brand B, and Brand C, to relieve a headache. Let μ_1, μ_2, and μ_3 denote, respectively, the mean times it takes Brand A, Brand B, and Brand C to relieve a headache. Then the hypotheses to be tested are

$$H_0: \mu_1 = \mu_2 = \mu_3 \text{ (mean relief times are equal)}$$

$$H_a: \text{Not all three means are equal.}$$

Suppose that each brand is to be administered six times.
a) Explain how a completely randomized design can be used.
b) Explain how a randomized block design can be used.
c) Which experimental design is probably better?

SOLUTION   a)   To use a completely randomized design, we can proceed as follows: First we randomly and independently select three groups of six people each (a total of 18 people). Next we have the first group of six people take Brand A, the second group of six people take Brand B, and the third group of six people take

---

[†] Two-way analysis of variance is used for other experimental designs besides the randomized block design.

Brand C. And then, as described in Section 14.2, we employ Procedure 14.1 to perform a one-way analysis of variance on the resulting data.

Table 14.5 shows the arrangement of the sample data for the completely randomized design. Here x_{11} denotes the time until headache relief for the first person who takes Brand A, x_{21} denotes the time until headache relief for the second person who takes Brand A, ..., x_{63} denotes the time until headache relief for the sixth person who takes Brand C. Note, for instance, that in this case, x_{11}, x_{12}, and x_{13} represent times for three different people.

TABLE 14.5
Arrangement of sample data for the completely randomized design to compare the three analgesics

Brand A	Brand B	Brand C
x_{11}	x_{12}	x_{13}
x_{21}	x_{22}	x_{23}
x_{31}	x_{32}	x_{33}
x_{41}	x_{42}	x_{43}
x_{51}	x_{52}	x_{53}
x_{61}	x_{62}	x_{63}

b) To use a randomized block design, we can proceed as follows: First we select a single group of six people (a total of six people). Next we have each person take each of the three analgesics on different occasions. And then, as will be described in this section, we employ Procedure 14.2 to perform a two-way analysis of variance on the resulting data.

In the randomized block design, the six people are called the **blocks,** as opposed to the three analgesics, which are the **treatments.** Table 14.6 shows the arrangement of the sample data for the randomized block design. Here x_{11} denotes the time until headache relief when Person 1 takes Brand A, x_{21} denotes the time until headache relief when Person 2 takes Brand A, ..., x_{63} denotes the time until headache relief when Person 6 takes Brand C. Note, for instance, that in this case, x_{11}, x_{12}, and x_{13} represent times for the same person (block).

TABLE 14.6
Arrangement of sample data for the randomized block design to compare the three analgesics

Analgesic (Treatment)

Person (Block)	Brand A	Brand B	Brand C
1	x_{11}	x_{12}	x_{13}
2	x_{21}	x_{22}	x_{23}
3	x_{31}	x_{32}	x_{33}
4	x_{41}	x_{42}	x_{43}
5	x_{51}	x_{52}	x_{53}
6	x_{61}	x_{62}	x_{63}

We can now see why the analysis of variance for a randomized block design is called *two-way* analysis of variance: Each piece of data is classified in two ways, according to treatment and block.

c) The randomized block design is probably better than the completely randomized design. This is because by blocking we can isolate an extraneous source of variation in the time it takes an analgesic to relieve a headache, namely, the variation due to differences in people. Once that variation is isolated, it can then be removed. This, in turn, will make it easier to detect differences among the mean relief times of the three analgesics, if such differences exist.   ∎

In a randomized block design, the treatments should be randomly assigned within each block. For instance, in the analgesics illustration, the three analgesics (treatments) should be taken by each of the six persons (blocks) in a random order. Thus the adjective "randomized" in "randomized block design."

Now that we have discussed the randomized block design, we need to address the details for using two-way analysis of variance to carry out the hypothesis test. As we said, the basic idea of a randomized block design is to isolate the variation among the blocks and remove it from the variation within the samples for the treatments. Once this is done, we decide whether the null hypothesis should be rejected in precisely the same way as in a completely randomized design. That is, we compare a measure of variation among the sample means, *MSTR*, to a measure of variation within the samples, *MSE*, using the *F*-statistic,

$$F = \frac{MSTR}{MSE}$$

But for a randomized block design, *MSE* is computed differently since it represents the variation within the samples for the treatments with the variation due to the blocks removed.

TWO-WAY ANOVA IDENTITY

To develop the two-way ANOVA procedure for a randomized block design, we begin by partitioning the total variation among all the sample data, the **total sum of squares, *SST*,** into three components (instead of the two-component partitioning done in a one-way ANOVA for a completely randomized design). The first component represents the variation among the sample means for the treatments—it is the **treatment sum of squares, *SSTR*,** defined exactly as in one-way ANOVA.

The second component represents the variation among the sample means for the blocks—it is the **block sum of squares, *SSB*.** This sum of squares is defined in the same way as the treatment sum of squares except that it measures the variation among the sample means for the blocks instead of the variation among the sample means for the treatments.

The third and final component in the partitioning of the total sum of squares is the **error sum of squares, *SSE*.** This represents the variation within the samples (for the treatments) with the variation due to the blocks removed. It is important to note that the error sum of squares, *SSE,* for a two-way ANOVA is not the same as the error sum of squares, *SSE,* for a one-way ANOVA, even though the same notation is employed.

Key Fact 14.5 summarizes the three-component partitioning of the total sum of squares that is used in a two-way analysis of variance for a randomized block design. We will call this partitioning the **two-way ANOVA identity.**

KEY FACT 14.5   Two-way ANOVA identity

The total sum of squares equals the treatment sum of squares plus the block sum of squares plus the error sum of squares; that is,

$$SST = SSTR + SSB + SSE$$

The two-way ANOVA identity shows that we can partition the total variation in the data into a component representing variation among the sample means for the treatments, a component representing variation among the sample means for the blocks, and a component representing variation within the samples for the treatments with the variation due to the blocks removed. We can picture this partitioning as in Figure 14.7.

FIGURE 14.7
Partitioning of the total sum of squares for a randomized block design

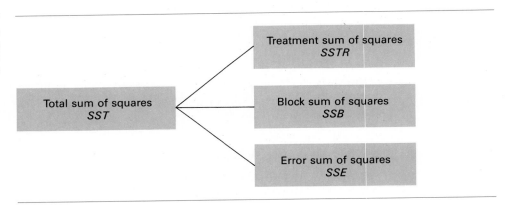

FORMULAS FOR THE SUMS OF SQUARES

Instead of presenting the defining formulas for the four sums of squares shown in Figure 14.7, we will give the shortcut formulas. The shortcut formulas are easier to use and reduce the possibility of roundoff error.

For a randomized block design, we denote the number of blocks by the letter m and the number of treatments (populations) by the letter k. We can depict the general format of a randomized block design as shown in Table 14.7 at the top of the next page.

TABLE 14.7
General format
of a randomized
block design

Treatment

Here now are the shortcut formulas to compute the four sums of squares when performing a two-way analysis of variance for a randomized block design.

FORMULA 14.1   Shortcut formulas for sums of squares in two-way ANOVA

Total sum of squares, SST:

$$SST = \Sigma x^2 - \frac{(\Sigma x)^2}{n}$$

Treatment sum of squares, $SSTR$:

$$SSTR = \frac{1}{m}\left(T_1^2 + T_2^2 + \cdots + T_k^2\right) - \frac{(\Sigma x)^2}{n}$$

Block sum of squares, SSB:

$$SSB = \frac{1}{k}\left(B_1^2 + B_2^2 + \cdots + B_m^2\right) - \frac{(\Sigma x)^2}{n}$$

Error sum of squares, SSE:

$$SSE = SST - SSTR - SSB$$

where,

k = number of treatments

m = number of blocks

$n = km$ = total number of pieces of data

and, for $i = 1, 2, \ldots, m$, and $j = 1, 2, \ldots, k$,

T_j = sum of sample data for Treatment j

B_i = sum of sample data for Block i

EXAMPLE 14.6   *Illustrates Formula 14.1*

Three analgesics are to be compared for speed in relieving a headache. A randomized block design is to be used. Six people are selected, with each person taking each of the three analgesics on different occasions. The times, in minutes, until headache relief are displayed in Table 14.8. Data are given to the nearest minute.

TABLE 14.8
Times until
headache relief

Analgesic (Treatment)

		Brand A	Brand B	Brand C	B_i
Person (Block)	1	18	22	25	65
	2	14	23	13	50
	3	21	19	23	63
	4	20	29	24	73
	5	12	19	13	44
	6	13	20	20	53
	T_j	98	132	118	348

Obtain and interpret the four sums of squares, *SST*, *SSTR*, *SSB*, and *SSE*.

SOLUTION   We apply the shortcut formulas given in Formula 14.1. Note that we have already computed the treatment and block sums and have displayed them in Table 14.8. From the table we see that

$$k = 3 \qquad m = 6$$
$$n = 3 \cdot 6 = 18$$
$$T_1 = 98 \qquad T_2 = 132 \qquad T_3 = 118$$
$$B_1 = 65 \qquad B_2 = 50 \qquad B_3 = 63 \qquad B_4 = 73 \qquad B_5 = 44 \qquad B_6 = 53$$
$$\Sigma x = 348$$

Also, summing the squares of the data in Table 14.8, we find that

$$\Sigma x^2 = 18^2 + 14^2 + 21^2 + \cdots + 13^2 + 20^2 = 7118$$

Now we apply Formula 14.1 to obtain the four sums of squares. First we compute the total sum of squares, *SST*:

$$SST = \Sigma x^2 - \frac{(\Sigma x)^2}{n} = 7118 - \frac{(348)^2}{18} = 7118 - 6728 = \mathbf{390}$$

and this represents the total variation among all the relief times.

Next we compute the treatment sum of squares, *SSTR*:

$$SSTR = \frac{1}{m}\left(T_1^2 + T_2^2 + T_3^2\right) - \frac{(\Sigma x)^2}{n} = \frac{1}{6}\left(98^2 + 132^2 + 118^2\right) - \frac{(348)^2}{18}$$

$$= 6825.333 - 6728 = \boxed{97.333}$$

and this represents the variation among the sample mean relief times for the treatments (analgesics).

Then we compute the block sum of squares, *SSB*:

$$SSB = \frac{1}{k}\left(B_1^2 + B_2^2 + B_3^2 + B_4^2 + B_5^2 + B_6^2\right) - \frac{(\Sigma x)^2}{n}$$

$$= \frac{1}{3}\left(65^2 + 50^2 + 63^2 + 73^2 + 44^2 + 53^2\right) - \frac{(348)^2}{18}$$

$$= 6922.667 - 6728 = \boxed{194.667}$$

and this represents the variation among the sample mean relief times for the blocks (people).

Finally, we compute the error sum of squares, *SSE*:

$$SSE = SST - SSTR - SSB = 390 - 97.333 - 194.667 = \boxed{98}$$

and this represents the variation among the relief times within the samples for the treatments (analgesics) with the variation due to the blocks (people) removed. ∎

MEAN SQUARES AND THE *F*-STATISTIC

Next we discuss *mean squares* and the *F-statistic*. To each of the sums of squares, *SSTR*, *SSB*, and *SSE*, there corresponds a mean square. Essentially, a **mean square** is the average variation for any particular source of variation. More precisely, we have the following definition:

DEFINITION 14.3   Mean squares for two-way ANOVA

Treatment mean square, MSTR:

$$MSTR = \frac{SSTR}{k-1}$$

Block mean square, MSB:

$$MSB = \frac{SSB}{m-1}$$

Error mean square, MSE:

$$MSE = \frac{SSE}{n-k-m+1}$$

where,

$$k = \text{number of treatments}$$
$$m = \text{number of blocks}$$
$$n = km = \text{total number of pieces of data}$$

As we mentioned earlier, the test statistic for a two-way analysis of variance for a randomized block design is

$$F = \frac{MSTR}{MSE}$$

To obtain the critical value for a two-way ANOVA, we need to know the probability distribution of F when the null hypothesis of equal population means is true. That probability distribution is identified in Key Fact 14.6.

KEY FACT 14.6   Test statistic for two-way ANOVA

Suppose that a randomized block design with m blocks is to be employed to compare the means, $\mu_1, \mu_2, \ldots, \mu_k$, of k populations (treatments). Further suppose that the km populations of treatment-block combinations are all normally distributed and have equal standard deviations. Then, if $\mu_1 = \mu_2 = \cdots = \mu_k$, the random variable

$$F = \frac{MSTR}{MSE}$$

has the F-distribution with df $= (k-1, n-k-m+1)$, where $n = km$ is the total number of pieces of data.

TWO-WAY ANOVA TABLES

Next we will discuss the **two-way ANOVA table.** This table is useful for organizing and summarizing the quantities required to perform a two-way analysis of variance for a randomized block design. It is the analogue of the one-way ANOVA table used in a one-way analysis of variance for a completely randomized design. The general format of a two-way ANOVA table is shown in Table 14.9.

TABLE 14.9
ANOVA table format for a randomized block design

Source	df	SS	MS = SS/df	F-statistic
Treatment	$k-1$	$SSTR$	$MSTR = \dfrac{SSTR}{k-1}$	$F = \dfrac{MSTR}{MSE}$
Block	$m-1$	SSB	$MSB = \dfrac{SSB}{m-1}$	
Error	$n-k-m+1$	SSE	$MSE = \dfrac{SSE}{n-k-m+1}$	
Total	$n-1$	SST		

Note that the test statistic, F, for a two-way analysis of variance appears in the final column of the two-way ANOVA table.

EXAMPLE 14.7 *Illustrates the two-way ANOVA table*

Three analgesics are to be compared for speed in relieving a headache. A randomized block design is to be used. Six people are selected, with each person taking each of the three analgesics on different occasions. The times, in minutes, until headache relief are displayed in Table 14.8 on page 728. Data are given to the nearest minute.
a) Determine *MSTR*, *MSB*, *MSE*, and *F*.
b) Construct the two-way ANOVA table for the data.

SOLUTION a) The sums of squares, *SSTR*, *SSB*, and *SSE*, were computed earlier in Example 14.6 (pages 728–729). Using those results and Definition 14.3, we find that

$$MSTR = \frac{SSTR}{k-1} = \frac{97.333}{3-1} = 48.667$$

$$MSB = \frac{SSB}{m-1} = \frac{194.667}{6-1} = 38.933$$

and

$$MSE = \frac{SSE}{n-k-m+1} = \frac{98}{18-3-6+1} = 9.800$$

From *MSTR* and *MSE*, we get the *F*-statistic:

$$F = \frac{MSTR}{MSE} = \frac{48.667}{9.8} = 4.97$$

b) The two-way ANOVA table for the data is displayed in Table 14.10.

TABLE 14.10
Two-way ANOVA
table for the
analgesics illustration

Source	df	SS	MS = SS/df	F-statistic
Treatment	2	97.333	48.667	4.97
Block	5	194.667	38.933	
Error	10	98.000	9.800	
Total	17	390.000		

THE TWO-WAY ANOVA PROCEDURE

We can now state a step-by-step method that can be used to perform a two-way analysis of variance for a randomized block design. Note that the hypothesis test is always right-tailed since the null hypothesis is rejected only when the test statistic, F, is too large.

PROCEDURE 14.2
To perform a two-way ANOVA for k population means.

ASSUMPTIONS

1. Blocked samples.
2. Normal populations.
3. Equal population standard deviations.[†]

STEP 1 *State the null and alternative hypotheses.*
STEP 2 *Decide on the significance level, α.*
STEP 3 *The critical value is F_α, with df $= (k-1, n-k-m+1)$, where m is the number of blocks and n is the total number of pieces of data.*

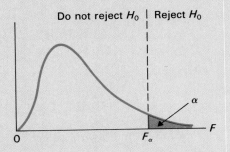

STEP 4 *Calculate the sums of squares using the shortcut formulas*

$$SST = \Sigma x^2 - \frac{(\Sigma x)^2}{n}$$

$$SSTR = \frac{1}{m}\left(T_1^2 + T_2^2 + \cdots + T_k^2\right) - \frac{(\Sigma x)^2}{n}$$

$$SSB = \frac{1}{k}\left(B_1^2 + B_2^2 + \cdots + B_m^2\right) - \frac{(\Sigma x)^2}{n}$$

$$SSE = SST - SSTR - SSB$$

[†] More precisely, Assumptions 1, 2 and 3 require that independent random samples (of size 1) be taken from the populations of treatment-block combinations and that all of those populations be normally distributed with equal standard deviations. We are also tacitly assuming that there is no *interaction* between blocks and treatments. Roughly speaking this means that the treatments behave consistently across the blocks and vice versa.

STEP 5 *Construct the two-way ANOVA table:*

Source	df	SS	MS = SS/df	F-statistic
Treatment	$k - 1$	SSTR	$MSTR = \dfrac{SSTR}{k - 1}$	$F = \dfrac{MSTR}{MSE}$
Block	$m - 1$	SSB	$MSB = \dfrac{SSB}{m - 1}$	
Error	$n - k - m + 1$	SSE	$MSE = \dfrac{SSE}{n - k - m + 1}$	
Total	$n - 1$	SST		

STEP 6 *If the value of the F-statistic falls in the rejection region, then reject H_0; otherwise, do not reject H_0.*

STEP 7 *State the conclusion in words.*

EXAMPLE 14.8   *Illustrates Procedure 14.2*

Three analgesics are to be compared for speed in relieving a headache. A randomized block design is to be employed. Six people are selected, with each person taking each of the three analgesics on different occasions and in a randomly chosen order. The times until headache relief are shown in Table 14.11. Data are given to the nearest minute.

TABLE 14.11
Times until headache relief

Analgesic (Treatment)

Person (Block)	Brand A	Brand B	Brand C
1	18	22	25
2	14	23	13
3	21	19	23
4	20	29	24
5	12	19	13
6	13	20	20

At the 5% significance level, do the data provide sufficient evidence to conclude that a difference exists among the mean times it takes the three analgesics to relieve a headache? Assume that for any given person and any given analgesic, the time until headache relief is normally distributed; and that the standard deviations of the times are equal for all person-analgesic combinations.

SOLUTION   We apply Procedure 14.2.

STEP 1 *State the null and alternative hypotheses.*

Let μ_1, μ_2, and μ_3 denote the mean relief times for Brand A, Brand B, and Brand C, respectively. Then the null and alternative hypotheses are

$$H_0: \mu_1 = \mu_2 = \mu_3 \text{ (mean relief times are equal)}$$
$$H_a: \text{Not all three means are equal.}$$

STEP 2 *Decide on the significance level, α.*

We are to perform the test at the 5% level of significance; so, $\alpha = 0.05$.

STEP 3 *The critical value is F_α, with df $= (k - 1, n - k - m + 1)$.*

From Table 14.11, we see that $k = 3$, $m = 6$, and $n = 3 \cdot 6 = 18$. Therefore, df $= (k - 1, n - k - m + 1) = (3 - 1, 18 - 3 - 6 + 1) = (2, 10)$. Consulting Table VI, we find that the critical value is $F_\alpha = F_{0.05} = 4.10$. See Figure 14.8.

FIGURE 14.8

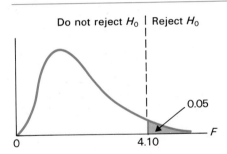

STEP 4 *Calculate the sums of squares using the shortcut formulas.*

We have already completed this step in Example 14.6 on pages 728–729.

STEP 5 *Construct the two-way ANOVA table:*

This was done in Example 14.7 (see Table 14.10 on page 731). Note in particular that the value of the F-statistic is $F = 4.97$.

STEP 6 *If the value of the F-statistic falls in the rejection region, then reject H_0; otherwise, do not reject H_0.*

As we noted in Step 5, the value of the F-statistic is $F = 4.97$. Referring to Figure 14.8, we see that this falls in the rejection region; hence, we reject H_0.

STEP 7 *State the conclusion in words.*

MTB   The data provide sufficient evidence to conclude that there is a difference among the mean relief times for the three analgesics.

OTHER TYPES OF ANOVA

In this chapter we have examined two types of analysis-of-variance procedures: one-way ANOVA for a completely randomized design and two-way ANOVA for a randomized block design. Both of those procedures are used to analyze the effect of one factor on some response variable. Such procedures are called **one-factor procedures.**

For instance, in Example 14.3 of Section 14.2, we analyzed the effect of the four U.S. regions (the treatments) on annual energy consumption (the response variable); and in Example 14.8 of this section, we analyzed the effect of three analgesics (the treatments) on the time until headache relief (the response variable). The factor in the first example is "region" and in the second example is "analgesic."

There are analysis-of-variance procedures that are used to analyze the effect of two or more factors on a response variable. For example, in the case where the response variable is annual energy consumption, we might want to simultaneously analyze the effect of region and income level. Then there would be two factors—region and income level—and we would need to use a **two-factor procedure** to perform the analysis of variance.

Time does not permit us to cover these and many other ANOVA procedures. However, it is useful to note that the essential concept in all ANOVA procedures is the partitioning of the total sum of squares into several components representing different sources of variation.

USING THE COMPUTER (OPTIONAL)

Minitab has a program that can be used to perform a two-way analysis of variance for a randomized block design. The appropriate command is called **TWOWAY.** We will illustrate the use of TWOWAY by employing it to perform the hypothesis test for the three analgesics.

EXAMPLE 14.9 *Illustrates the TWOWAY command*

Refer to Example 14.8 on page 733. Apply Minitab to carry out the hypothesis test considered in that example.

SOLUTION Let μ_1, μ_2, and μ_3 denote the mean relief times for Brand A, Brand B, and Brand C, respectively. Then we want to perform the hypothesis test

$$H_0: \mu_1 = \mu_2 = \mu_3 \text{ (mean relief times are equal)}$$
$$H_a: \text{Not all three means are equal.}$$

at the 5% significance level.

To employ Minitab, we must first enter the sample data into the computer. For a randomized block design, all of the sample data go into one column, say, C5. Then we use two more columns, say, C6 and C7, to identify the treatment and block for each piece of sample data. For instance, by referring to Table 14.11 on page 733, we see that it took 29 minutes for Brand B (Treatment 2) to relieve the headache of Person 4. Thus, for that piece of sample data, we would enter 29 into C5, 2 into C6, and 4 into C7. We can use the READ command to enter all of the data in this way. See Printout 14.4. Note that, in Printout 14.4, we have also named C5 "Times," C6 "Treatmnt," and C7 "Block."

PRINTOUT 14.4

```
MTB > READ C5-C7
DATA> 18 1 1
DATA> 14 1 2
DATA> 21 1 3
DATA> 20 1 4
DATA> 12 1 5
DATA> 13 1 6
DATA> 22 2 1
DATA> 23 2 2
DATA> 19 2 3
DATA> 29 2 4
DATA> 19 2 5
DATA> 20 2 6
DATA> 25 3 1
DATA> 13 3 2
DATA> 23 3 3
DATA> 24 3 4
DATA> 13 3 5
DATA> 20 3 6
DATA> END
      12 ROWS READ
MTB > NAME C5 'TIMES' C6 'TREATMNT' C7 'BLOCK'
```

Next we type the command TWOWAY followed by the storage locations of the sample data, treatment data, and block data. That is, we type

TWOWAY on 'TIMES' 'TREATMNT' 'BLOCK'

Printout 14.5 displays the above command and the output that results.

PRINTOUT 14.5
Minitab output
for TWOWAY

```
MTB > TWOWAY on 'TIMES' 'TREATMNT' 'BLOCK'

ANALYSIS OF VARIANCE   TIMES

SOURCE         DF        SS         MS
TREATMNT        2     97.33      48.67
BLOCK           5    194.67      38.93
ERROR          10     98.00       9.80
TOTAL          17    390.00
```

The output of the TWOWAY command is an ANOVA table similar to the two-way ANOVA table in Table 14.10 on page 731. The main difference between the two tables is that Minitab does not provide the value of the F-statistic, $F = MSTR/MSE$.

Since Minitab also does not give the P-value for the hypothesis test, we must do three things ourselves in order to carry out the two-way analysis of variance using Printout 14.5: We need to obtain the critical value for the hypothesis test, compute the value of the F-statistic, and compare those two values.

The degrees of freedom for the numerator and denominator are displayed as the first and third entries of the column headed DF in Printout 14.5. Hence, df $= (2, 10)$. Recalling that $\alpha = 0.05$ and consulting Table VI, we find that the critical value is $F_\alpha = F_{0.05} = 4.10$.

Next we compute the value of the F-statistic. The treatment mean square, $MSTR$, and error mean square, MSE, are the first and third entries of the column headed MS in Printout 14.5. Thus, we see that

$$F = \frac{MSTR}{MSE} = \frac{48.67}{9.80} = 4.97$$

Since the value of the F-statistic, 4.97, exceeds the critical value, 4.10, we reject H_0. In other words, the data provide sufficient evidence to conclude that a difference exists among the mean relief times for the three analgesics. ∎

Exercises 14.3

___ 14.47 Which type of analysis of variance is used for a randomized block design? a completely randomized design?

___ 14.48 To what does the type of design refer? the type of analysis of variance?

___ 14.49 What is the purpose of blocking in a randomized block design?

___ 14.50 Below we have given the notation that is used for the four sums of squares in a two-way analysis of variance for a randomized block design. For each sum of squares, state its name, the source of variation it represents, and its shortcut formula.
a) SSB b) SSE c) $SSTR$ d) SST

___ 14.51 State the two-way ANOVA identity and interpret its meaning with regard to partitioning the total variation in the data.

___ 14.52 For each part below, answer true or false.
a) The total sum of squares, SST, for a randomized block design is defined in the same way as for a completely randomized design.

b) The treatment sum of squares, $SSTR$, for a randomized block design is defined in the same way as for a completely randomized design.

c) The error sum of squares, SSE, for a randomized block design is defined in the same way as for a completely randomized design.

In Exercises 14.53–14.56, we have presented partially completed two-way ANOVA tables for a randomized block design. Fill in the missing entries in each table.

___ 14.53

Source	df	SS	MS = SS/df	F-statistic
Treatment	2		21.667	
Block		55.060		
Error	8			
Total	14	127.733		

__ 14.54

Source	df	SS	MS = SS/df	F-statistic
Treatment		248.625		4.25
Block	5		63.300	
Error			19.500	
Total	23			

__ 14.55

Source	df	SS	MS = SS/df	F-statistic
Treatment		250.488		9.80
Block	4	15.364		
Error			4.260	
Total				

(Hint: n = km.)

__ 14.56

Source	df	SS	MS = SS/df	F-statistic
Treatment	4		98.732	
Block				
Error		1875.968		
Total	44	4223.553		

(Hint: n = km.)

In each of Exercises 14.57 and 14.58, a randomized block design is to be employed to compare the means of several populations. For each exercise,
a) identify the treatments.
b) identify the blocks.
c) construct the two-way ANOVA table for the data.

__ 14.57   An agronomist has three new fertilizers to test for use with wheat farming. She selects one wheat farm from each of five wheat-producing states. On each farm, three one-acre plots with similar soil and water conditions are selected and the three fertilizers are randomly assigned to the three plots. The wheat yields, in bushels, are shown in the following table:

Fertilizer brand

	A	B	C
1	72	81	75
2	40	39	30
3	73	83	70
4	60	62	52
5	39	35	30

Farm (rows 1–5)

__ 14.58   Four brands of gasoline are to be compared for gas mileage. For the test, six cars are selected. Each car is driven a distance of 300 miles over a specified route with each of the four gasolines. The resulting mileages, to the nearest mile per gallon, are as follows:

Gasoline brand

	A	B	C	D
1	19	17	15	19
2	16	15	13	17
3	21	20	19	22
4	27	27	24	28
5	27	25	24	26
6	16	15	13	15

Car (rows 1–6)

For each of Exercises 14.59–14.64, use Procedure 14.2 on page 732 to perform the required hypothesis test.

__ 14.59   Refer to Exercise 14.57. At the 5% significance level, do the data provide sufficient evidence to conclude that a difference exists among the mean wheat yields of the three fertilizers?

__ 14.60   Refer to Exercise 14.58. At the 1% significance level, do the data imply that there is a difference in mean mileages among the four brands of gasoline?

__ 14.61   The U.S. Department of Agriculture conducts surveys to estimate weekly food costs for families, by region and family type. Results are published in News. Suppose that we want to compare the mean weekly food costs among the four U.S. regions using a randomized block design, with blocking done by family type, and that we obtain the following data. Here A = couple, 20–50 years old; B = couple, 51 years old or over; C = four-person family with preschool children; D = four-person family with school children.

Region

		Northeast	Midwest	South	West
Family type	A	65	69	40	57
	B	84	52	54	50
	C	108	92	84	108
	D	118	104	108	110

Do the data provide sufficient evidence to conclude that there is a difference in mean weekly food costs among the four U.S. regions? Use $\alpha = 0.01$.

___ 14.62 A real estate broker in the greater Phoenix area wants to compare the resale prices for tract homes in three suburban locations—Chandler, Gilbert, and Mesa. The realtor uses a randomized block design and blocks by square footage. She obtains the following data, in thousands of dollars, from the *Arizona Regional Multiple Listing Service (ARMLS):*

Suburb

		Chandler	Gilbert	Mesa
Square footage	Under 1000	34	52	48
	1000–1499	72	47	70
	1500–1799	82	83	84
	1800–1999	95	99	110
	2000 & over	121	116	133

Do the data provide evidence of a difference among the mean resale prices of tract homes in the three suburbs? Use $\alpha = 0.05$.

___ 14.63 In Exercise 14.27 of Section 14.2, a completely randomized design was used for comparing the mean lifetimes of four brands of flashlight batteries. The data did not provide sufficient evidence to conclude that a difference exists among the mean lifetimes of the four brands. Now a randomized block design is

employed. Specifically, five flashlights are selected and each flashlight is tested with all four brands of batteries, where the order in which a given flashlight uses the four brands is randomly determined. The lifetimes of the batteries to the nearest hour are as follows:

Battery brand

		A	B	C	D
Flashlight	1	23	20	20	19
	2	38	36	33	31
	3	45	40	37	34
	4	34	32	27	29
	5	28	24	28	26

a) At the 5% significance level, does there appear to be a difference in the mean lifetimes of the four brands of batteries?
b) Explain why a difference in the means was detected using a randomized block design but was not detected using a completely randomized design.

___ 14.64 In Exercise 14.30 of Section 14.2, a writer for a sports magazine wanted to compare the mean driving distances of five brands of golf balls. She used a completely randomized design: 20 golf professionals were randomly selected, randomly divided into five groups of four golfers each, and then each group drove one of the brands of golf balls. The data did not provide sufficient evidence to conclude that a difference exists among the mean driving distances of the five brands. Now a randomized block design is employed. Four golf professionals are selected and each golfer drives all five brands of balls in a random order. Here are the results, in yards.

Golf-ball brand

		1	2	3	4	5
Golfer	1	265	277	257	266	265
	2	271	282	275	277	287
	3	296	295	282	285	299
	4	283	287	281	278	284

a) Do the data provide sufficient evidence to conclude that a difference exists in mean driving distances for the five brands of golf balls? Use $\alpha = 0.05$.

b) Explain why a difference in the means was detected using a randomized block design but was not detected using a completely randomized design.

Exercises 14.65–14.68 are computer exercises.

___ **14.65 (Computer exercise)** Suppose that all of the weekly-food-cost data in Exercise 14.61 are stored in a column named FOODCOST and that the corresponding treatment and block data are stored in columns named TREATMNT and BLOCK.

a) Which Minitab command and subcommands (if any) should be used to carry out the hypothesis test considered in that exercise?

b) If you have access to Minitab, use it to perform the hypothesis test.

___ **14.66 (Computer exercise)** Suppose that all of the data on resale home prices in Exercise 14.62 are stored in a column named PRICES and that the corresponding treatment and block data are stored in columns named TREATMNT and BLOCK.

a) Which Minitab command and subcommands (if any) should be used to carry out the hypothesis test considered in that exercise?

b) If you have access to Minitab, use it to perform the hypothesis test.

___ **14.67 (Computer exercise)** The College Placement Council, Inc., of Bethlehem, PA, conducts surveys to obtain information on starting monthly salary offers to candidates for degrees by field of study. Results are published in *Salary Survey, A Study of Beginning Offers*. The mean starting-monthly-salary offers are to be compared for four engineering fields—civil, chemical, electrical, and mechanical. A randomized block design is used with blocking by level of degree—bachelor's, master's, and doctor's. One salary offer is randomly selected from each of the 12 field-degree population combinations. The following computer output was obtained by applying Minitab's TWOWAY command to the sample data:

ANALYSIS OF VARIANCE SALARIES

SOURCE	DF	SS	MS
TREATMNT	3	596452	198817
BLOCK	2	2488982	1244491
ERROR	6	124905	20818
TOTAL	11	3210339	

Determine

a) the sums of squares, *SST*, *SSTR*, *SSB*, and *SSE*.

b) the mean squares, *MSTR*, *MSB*, and *MSE*.

c) the null and alternative hypotheses for the hypothesis test.

d) the value of the test statistic, F.

e) the critical value if the hypothesis test is performed at the 5% significance level. (*Note:* You will need to refer to Table VI.)

f) the conclusion if the hypothesis test is performed at the 5% significance level.

___ **14.68 (Computer exercise)** In Exercise 14.38 of Section 14.2, a completely randomized design was used in comparing the mean annual insurance premiums for owners of large, intermediate, and compact cars. The data did not provide sufficient evidence to conclude that a difference exists among the three mean annual insurance premiums. As another attempt to detect a difference among the three means, we might employ a randomized block design. Insurance companies classify drivers as *good, average,* or *poor* depending on the number of tickets received in the last three years. So, suppose that we block by driver classification and select independently and randomly one annual insurance premium from each of the nine populations of car-size and driver-classification combinations. Further suppose that the computer output obtained by applying Minitab's TWOWAY command to the resulting data is as follows:

ANALYSIS OF VARIANCE PREMIUMS

SOURCE	DF	SS	MS
TREATMNT	2	46812	23406
BLOCK	2	222442	111221
ERROR	4	168378	42094
TOTAL	8	437631	

Determine

a) the sums of squares, *SST*, *SSTR*, *SSB*, and *SSE*.

b) the mean squares, *MSTR*, *MSB*, and *MSE*.

c) the null and alternative hypotheses for the hypothesis test.

d) the value of the test statistic, F.

e) the critical value if the hypothesis test is performed at the 5% significance level. (*Note:* You will need to refer to Table VI.)

f) the conclusion if the hypothesis test is performed at the 5% significance level.

Confidence intervals for the difference between two means in two-way ANOVA: Suppose that a randomized block design with m blocks is to be employed to compare the means, $\mu_1, \mu_2, \ldots, \mu_k$, of k populations (treatments). Further suppose that the km populations of treatment-block combinations are all normally distributed and have equal standard deviations. Let $s = \sqrt{MSE}$. Then a $(1-\alpha)$-level confidence interval for the difference between any two particular population (treatment) means, say, μ_i and μ_j, has endpoints

$$(\bar{x}_i - \bar{x}_j) \pm t_{\alpha/2} \cdot s\sqrt{2/m}$$

where df $= n - k - m + 1$. We will apply this confidence-interval formula in Exercises 14.69 and 14.70.

$=$ **14.69** Refer to Exercise 14.57.
a) Find a 95% confidence interval for the difference, $\mu_1 - \mu_2$, between the mean wheat yields of Brand A and Brand B fertilizers.
b) Find a 95% confidence interval for the difference, $\mu_2 - \mu_3$, between the mean wheat yields of Brand B and Brand C fertilizers.

$=$ **14.70** Refer to Exercise 14.58.
a) Determine a 99% confidence interval for the difference, $\mu_1 - \mu_2$, between the mean mileages of Brand A and Brand B gasolines.

b) Determine a 99% confidence interval for the difference, $\mu_3 - \mu_4$, between the mean mileages of Brand C and Brand D gasolines.

$=$ **14.71** This exercise shows what can happen when one-way ANOVA, which is intended for use with a completely randomized design, is applied to perform an analysis of variance for a randomized block design. In Example 14.8 on pages 733–734, we applied Procedure 14.2 to perform a two-way ANOVA to decide whether a difference exists among the mean times it takes three analgesics to relieve a headache. Specifically, if we let μ_1, μ_2, and μ_3 denote the mean relief times for Brand A, Brand B, and Brand C, respectively, then the hypothesis test is

H_0: $\mu_1 = \mu_2 = \mu_3$ (mean relief times are equal)
H_a: Not all three means are equal.

a) Apply Procedure 14.1 on page 712 (the one-way ANOVA procedure) to the sample data displayed in Table 14.11 on page 733 to perform the hypothesis test. Use $\alpha = 0.05$.
b) Why is it inappropriate to perform the hypothesis test in the way that you did in part (a)?
c) Compare your result in part (a) to the one obtained in Example 14.8.

Chapter review

FORMULAS

In the formulas below,

k = number of populations (treatments)
n = total number of pieces of data
n_j = sample size for Population j
    (in one-way ANOVA)
T_j = sum of sample data for
    Population (Treatment) j
SST = total sum of squares
$SSTR$ = treatment sum of squares

SSE = error sum of squares
$MSTR$ = treatment mean square
MSE = error mean square
m = number of blocks
B_i = sum of sample data for Block i
SSB = block sum of squares
MSB = block mean square

One-way ANOVA identity, 710

$$SST = SSTR + SSE$$

Shortcut formulas for sums of squares in one-way ANOVA, 712

$$SST = \Sigma x^2 - \frac{(\Sigma x)^2}{n}$$

$$SSTR = \left(\frac{T_1^2}{n_1} + \frac{T_2^2}{n_2} + \cdots + \frac{T_k^2}{n_k} \right) - \frac{(\Sigma x)^2}{n}$$

$$SSE = SST - SSTR$$

Mean squares in one-way ANOVA, 709

$$MSTR = \frac{SSTR}{k-1}, \qquad MSE = \frac{SSE}{n-k}$$

Test statistic for one-way ANOVA, 710

$$F = \frac{MSTR}{MSE}$$

with df $= (k-1, n-k)$.

Two-way ANOVA identity, 726

$$SST = SSTR + SSB + SSE$$

Shortcut formulas for sums of squares in two-way ANOVA, 727

$$SST = \Sigma x^2 - \frac{(\Sigma x)^2}{n}$$

$$SSTR = \frac{1}{m} \left(T_1^2 + T_2^2 + \cdots + T_k^2 \right) - \frac{(\Sigma x)^2}{n}$$

$$SSB = \frac{1}{k} \left(B_1^2 + B_2^2 + \cdots + B_m^2 \right) - \frac{(\Sigma x)^2}{n}$$

$$SSE = SST - SSTR - SSB$$

Mean squares in two-way ANOVA, 729

$$MSTR = \frac{SSTR}{k-1}, \qquad MSB = \frac{SSB}{m-1}, \qquad MSE = \frac{SSE}{n-k-m+1}$$

Test statistic for two-way ANOVA, 730

$$F = \frac{MSTR}{MSE}$$

with df $= (k-1, n-k-m+1)$.

YOU SHOULD BE ABLE TO

1. use and understand the preceding formulas.
2. use the F-tables, Tables V and VI.
3. explain the essential ideas behind a one-way analysis of variance for a completely randomized design.
4. compute the sums of squares for a one-way ANOVA using the shortcut formulas.
5. compute the mean squares and the F-statistic for a one-way ANOVA.
6. construct a one-way ANOVA table.
7. perform a one-way ANOVA for k population means.
8. explain the essential ideas behind a two-way analysis of variance for a randomized block design.
9. compute the sums of squares for a two-way ANOVA using the shortcut formulas.
10. compute the mean squares and the F-statistic for a two-way ANOVA.
11. construct a two-way ANOVA table.
12. perform a two-way ANOVA for k population means.
13. use the Minitab commands covered in this chapter.*
14. interpret the output obtained from the application of the Minitab commands discussed in this chapter.*

REVIEW TEST

1. Consider an F-curve with df $= (24, 5)$.
   a) Obtain the number of degrees of freedom for the numerator.
   b) Obtain the number of degrees of freedom for the denominator.
   c) Find $F_{0.05}$.
   d) Determine the F-value for which the area under the curve to its right is 0.01.
   e) Find the F-value with area 0.95 to its left.

2. Identify a statistic that measures the variation among the sample means in a one-way ANOVA for a completely randomized design.

3. Identify a statistic that measures the variation within the samples in a one-way ANOVA for a completely randomized design.

4. The FBI compiles data on the value of losses due to various types of robberies and publishes those data in *Population-at-Risk Rates and Crime Indicators*. Independent random samples of reports for three types of robberies—highway, gas station, and convenience store—gave the following data, in dollars, on value of losses.

Highway	Gas station	Convenience store
411	314	575
320	356	442
496	379	458
410	424	475
429	365	376
	532	548

a) Compute the sample mean and sample variance of each of the three data sets.
b) Determine *MSTR* and *MSE* using the defining formulas.
c) What is *MSTR* measuring?
d) What is *MSE* measuring?

5. Refer to Problem 4.
   a) Suppose that we want to perform a one-way ANOVA to compare the mean losses for the three types of robberies. What assumptions are necessary? How crucial are those assumptions?
   b) At the 5% significance level, do the data provide evidence of a difference in mean losses among the three types of robberies?

6. For a one-way ANOVA:
   a) List and interpret the three sums of squares.
   b) State the one-way ANOVA identity and interpret its meaning with regard to partitioning the total variation in the data.

7. A cereal company wants to test the possible market effect of four different designs for its boxes. Independent random samples of five markets each are selected. Each design is tried in one of the five-market groups. The number of cases sold during the test period are given below.

Design *A*	Design *B*	Design *C*	Design *D*
41	51	44	58
51	65	58	37
52	35	37	24
50	66	75	54
43	57	54	65

Apply the one-way ANOVA procedure to decide, at the 1% significance level, whether a difference exists in mean sales among the four designs.

*8. **(Computer problem)** Suppose that the data shown in Problem 7 are stored in columns named DESIGN A, DESIGN B, DESIGN C, and DESIGN D.
   a) Which Minitab command and subcommands (if any) should be used to carry out the hypothesis test in that problem?
   b) If you have access to Minitab, use it to perform the hypothesis test.

*9. **(Computer problem)** In Problem 7, we considered a hypothesis test to decide whether a difference exists in mean sales among four designs for cereal boxes. Printout 14.6, on the next page, shows the output obtained by applying Minitab's AOVONEWAY command to the data displayed in Problem 7. Employ that computer output to determine
   a) the sums of squares, *SSTR, SSE,* and *SST*.
   b) the treatment mean square, *MSTR,* and the error mean square, *MSE*.
   c) the value of the test statistic, *F*.
   d) the *P*-value for the hypothesis test.
   e) the conclusion if the hypothesis test is performed at the 1% significance level.
   f) the sample mean, sample standard deviation, and sample size for each of the four samples.

10. Explain why it is sometimes preferable to employ a randomized block design instead of a completely randomized design when comparing the means of several populations.

11. For a two-way ANOVA:
    a) List and interpret the four sums of squares.
    b) State the two-way ANOVA identity and interpret its meaning with regard to partitioning the total variation in the data.

12. The Energy Information Administration collects data on household energy expenditures by main fuel type and region. Results are published in *Residential Energy Consumption Survey: Consumption and Expenditures*. A researcher wants to compare the mean energy expenditures for households using natural gas, electricity, and fuel oil as their main heating fuels. He decides to employ a randomized block design, blocking by geographical region. The following 12 energy expenditures, in dollars, are obtained for last year:

Region	Type: Gas	Electric	Fuel oil
Northeast	1808	1767	1947
Midwest	1388	1441	1695
South	1418	1157	1392
West	1074	971	1499

a) What assumptions should be satisfied in order to perform a two-way analysis of variance for this randomized block design?

b) Presuming the required assumptions are met, do the data provide sufficient evidence to conclude that a difference exists among last year's mean energy expenditures for households using the three types of fuel? Take $\alpha = 0.05$.

*13. **(Computer problem)** Suppose that all of the energy-expenditure data in Problem 12 are stored in a column named ENERGY\$ and that the corresponding treatment and block data are stored in columns named TREATMNT and BLOCK.
a) Which Minitab command and subcommands (if any) should be used to carry out the hypothesis test in that problem?
b) If you have access to Minitab, use it to perform the hypothesis test.

*14. **(Computer problem)** In Problem 12, a hypothesis test was considered to decide whether a difference exists among last year's mean energy expenditures for households using natural gas, electricity, and fuel oil as their main heat-

ing fuels. The following computer output was obtained by applying Minitab's TWOWAY command to the data displayed in Problem 12:

ANALYSIS OF VARIANCE   ENERGY\$

SOURCE	DF	SS	MS
TREATMNT	2	189228	94614
BLOCK	3	731341	243780
ERROR	6	80444	13407
TOTAL	11	1001013	

Determine
a) the four sums of squares, SST, $SSTR$, SSB, and SSE.
b) the mean squares, $MSTR$, MSB, and MSE.
c) the value of the test statistic, F.
d) the critical value if the hypothesis test is performed at the 5% significance level. *(Note: You will need to refer to Table VI.)*
e) the conclusion if the hypothesis test is performed at the 5% significance level.

PRINTOUT 14.6   Minitab output for Problem 9

ANALYSIS OF VARIANCE

SOURCE	DF	SS	MS	F	p
FACTOR	3	228	76	0.45	0.721
ERROR	16	2708	169		
TOTAL	19	2937			

```
                                   INDIVIDUAL 95 PCT CI'S FOR MEAN
                                   BASED ON POOLED STDEV
LEVEL      N    MEAN    STDEV   -----+---------+---------+---------+-
DESIGN A   5   47.40    5.03    (-----------*------------)
DESIGN B   5   54.80   12.66        (------------*-----------)
DESIGN C   5   53.60   14.54       (------------*-----------)
DESIGN D   5   47.60   16.74    (------------*-----------)
                                   -----+---------+---------+---------+-
POOLED STDEV =   13.01             40        50        60        70
```

CHAPTER 15

NONPARAMETRIC STATISTICS

Many of the inferential methods we have studied so far require that certain conditions be met by the population(s) under consideration. For instance, the pooled-t procedure requires that the populations being sampled are normally distributed and have equal standard deviations.

However, there are many situations where these kinds of assumptions are not met or where it is simply not known whether they are met. Consequently, statistical procedures have been developed that require few or no assumptions about the population(s) under investigation. Those procedures are referred to as *distribution-free methods* or, more commonly, as *nonparametric methods*. In this chapter we will examine some widely used nonparametric methods.

CHAPTER OUTLINE

15.1 What is nonparametric statistics? Introduces nonparametric statistics and explains the basic difference between nonparametric and parametric statistics.

15.2 The sign test Discusses the sign test, a nonparametric test that can be used to make inferences about the median of a population.

15.3 The Wilcoxon signed-rank test Examines the Wilcoxon signed-rank test, a nonparametric procedure to perform a hypothesis test for the median or mean of a population having a symmetric distribution.

15.4 The Mann-Whitney test Introduces the Mann-Whitney test. This nonparametric procedure can be employed to compare the medians or means of two populations whose distributions have the same shape.

15.5 Rank correlation Discusses the rank correlation coefficient, a nonparametric alternative to the linear correlation coefficient.

15.6 Parametric and nonparametric procedures: a comparison Summarizes several parametric procedures and their corresponding nonparametric alternatives; and provides a strategy for choosing between parametric and nonparametric methods.

15.1   What is nonparametric statistics?

In previous chapters, we have examined several statistical procedures that are designed explicitly for making inferences about normally distributed populations. For instance, the *t*-tests, regression-analysis procedures, and analysis-of-variance procedures are all derived for use with normal populations. However, those methods still work reasonably well if the populations being sampled are not normally distributed, as long as the deviation from normality is not severe. In other words, they are robust to moderate deviations of the normality assumption.

But what if the normality assumption is badly violated? Populations whose distributions are extremely non-normal are by no means rare. For instance, consider the 1987 age distribution, shown in Figure 15.1, for United States residents between 15 and 65 years old. [SOURCE: U.S. Bureau of the Census, *Current Population Reports*.]

FIGURE 15.1
1987 age distribution
for U.S. residents
between 15 and 65
years of age

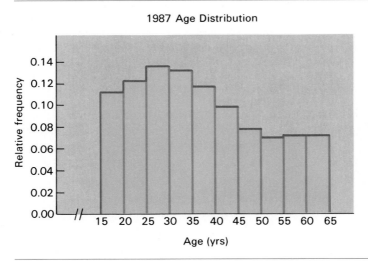

As we can see, the population of ages depicted in Figure 15.1 is far from being normally distributed. Hence, for example, it would be inappropriate to use a *t*-test to make inferences about the mean of this population.

DISTRIBUTION-FREE AND NONPARAMETRIC METHODS

Inferential methods have been developed that do not depend on the distribution of the population(s) being sampled. Such methods are aptly called **distribution-free methods.** In this chapter, we will study some of the more important distribution-free methods.

Closely related to distribution-free methods are *nonparametric methods*. Recall that descriptive measures for a population, such as μ and σ, are called parameters.

Inferential methods concerned with questions about parameters are called **parametric methods;** those that are not are called **nonparametric methods.** For instance, analysis of variance is a parametric method; it compares the means of several populations. On the other hand, the chi-square independence test is a nonparametric method; it addresses whether two characteristics of a population are statistically dependent and has nothing at all to do with parameters.

Many distribution-free methods are actually parametric since they deal with such parameters as the population mean or population median. However, it has become common practice to refer to most methods that can be used without assuming normality as nonparametric. Books and articles entitled "nonparametric statistics" typically cover both nonparametric and distribution-free methods.

Nonparametric methods have advantages beyond the distribution-free aspects. For example, they often entail fewer or simpler computations than parametric methods and do not necessarily require that the data be metric. Since nonparametric methods are generally simpler than parametric ones and are usually based on fewer assumptions, we might wonder why parametric methods are used at all. The reason is that parametric methods tend to give more accurate results when the requirements for their use are met. Thus, in situations where both a parametric and nonparametric method are appropriate, the parametric method should be employed.

Exercises 15.1

__ **15.1** Explain what *distribution-free* means in the context of distribution-free methods.

__ **15.2** Strictly speaking, define what is meant by a *nonparametric method.*

15.2 The sign test

The first nonparametric test that we will discuss is called the **sign test.** This test is used to make inferences about the median of a population. Recall that the median of a population divides the bottom 50% of the population data from the top 50%. We will use the Greek letter η (eta) to denote a **population median.**

The sign test is based on the binomial distribution, which we discussed in Section 5.4. We will illustrate the reasoning behind the sign test in the next example.

EXAMPLE 15.1 *Introduces the sign test*

An economic analyst in a northeastern city is given an estimate that his city's median family income equals $32,000. The city is actually a middle-class suburb of a large metropolitan area and the economist feels that the figure of $32,000 might well be too low. To check the figure, he selects a random sample of 12 family incomes. The results are displayed in Table 15.1 at the top of the next page. Do the data provide sufficient evidence to conclude that the median family income in the city exceeds $32,000? Use the sign test to carry out the hypothesis test.

TABLE 15.1
Sample data for
family incomes

$57,968	39,041	61,502
33,625	29,671	43,430
56,157	48,011	65,003
28,820	37,093	38,310

SOLUTION Let η denote the median family income in the city. Then we want to perform the hypothesis test

$$H_0\!: \eta = \$32{,}000 \text{ (median income does not exceed \$32,000)}$$
$$H_a\!: \eta > \$32{,}000 \text{ (median income does exceed \$32,000)}$$

The logic behind the sign test is this: If the null hypothesis is true (i.e., the median family income equals $32,000), then half of the families have incomes below $32,000 and half have incomes above $32,000. Thus, in a sample of 12 families, we would expect roughly six to have incomes below $32,000 and six to have incomes above $32,000. If too many more than six of the 12 incomes are greater than $32,000, then this provides evidence that the null hypothesis is false and hence that the median family income actually exceeds $32,000.

From Table 15.1, we see that 10 out of the 12 family incomes sampled are greater than $32,000. This seems to indicate that we should reject H_0 in favor of H_a. However, as we are now well aware, it is certainly possible that H_0 is true and that the large percentage of families in the sample who have incomes above $32,000 is due to chance (i.e., sampling error).

Thus, we need to decide whether observing 10 or more incomes above $32,000 can be reasonably attributed to sampling error or whether it indicates that the median income exceeds $32,000. To make that decision, we need to know how likely it would be to obtain such a sample if the median family income in the city is in fact $32,000. That likelihood can determined by applying the binomial probability formula, as we will now see.

The process of sampling 12 family incomes can be regarded as a sequence of Bernoulli trials. Each trial consists of the random selection of a family income and so the number of trials is $n = 12$. Suppose that we consider a family income above $32,000 a "success." If the median family income is in fact $32,000, then half of the incomes are below $32,000 and half are above. Therefore, the probability is 0.5 that a randomly selected income will exceed $32,000; that is, the success probability is $p = 0.5$. Consequently, if the median family income is $32,000, then the total number of incomes in the sample that are above $32,000 has the binomial distribution with parameters $n = 12$ and $p = 0.5$.

According to Table I, when $n = 12$ and $p = 0.5$, the probability of obtaining 10 or more successes is equal to

$$0.016 + 0.003 + 0.000 = 0.019$$

In other words, if the median family income is $32,000, then in a sample of 12 incomes, the probability that 10 or more will be above $32,000 is only 0.019, less than a 2% chance. So, the fact that 10 of the 12 incomes sampled turned out to be above $32,000 does not seem to be attributable to sampling error. Rather it provides sufficient evidence to conclude that the median family income in the city exceeds $32,000.

∎

In the previous example, we discovered an important relationship between the binomial distribution and sampling from a population with median η. That relationship is presented formally in Key Fact 15.1.

KEY FACT 15.1

Suppose that a random sample of size n is to be taken from a population with median η. Let x denote the number of pieces of sample data that exceed η. Then x has the binomial distribution with parameters n and 0.5.

Before we present a procedure for performing a sign test, we need to comment on the specification of the significance level and the determination of the rejection region. The test statistic, x, for a sign test has a binomial distribution. Since a binomial distribution is discrete, it is generally not possible to perform a sign test at a significance level exactly equal to the one specified. Usually, we must be satisfied with a significance level close to the one desired.†

For instance, consider the hypothesis test of Example 15.1:

$$H_0\text{: } \eta = \$32{,}000 \text{ (median income does not exceed }\$32{,}000)$$

$$H_a\text{: } \eta > \$32{,}000 \text{ (median income does exceed }\$32{,}000)$$

Note that the hypothesis test is right-tailed since there is a greater-than sign ($>$) in the alternative hypothesis.

Suppose we want to perform the hypothesis test at the 5% significance level, using a sample size of $n = 12$. Let x denote the number of incomes sampled that are above $32,000. Then by Key Fact 15.1, if H_0 is true, the random variable x has the binomial distribution with parameters $n = 12$ and $p = 0.5$. That probability distribution can be found in Table I. We display it here as Table 15.2.

TABLE 15.2
Binomial distribution with parameters $n = 12$ and $p = 0.5$

x	$P(x)$	x	$P(x)$
0	0.000	7	0.193
1	0.003	8	0.121
2	0.016	9	0.054
3	0.054	10	0.016
4	0.121	11	0.003
5	0.193	12	0.000
6	0.226		

Since the hypothesis test is right-tailed, the null hypothesis will be rejected only when x is too large. Using Table 15.2, we can list some of the possible rejection regions and significance levels. This is done in Table 15.3 at the top of the next page.

---

† If we employ the P-value approach to hypothesis testing, then this problem does not arise. See the optional material on computer usage at the end of this section.

TABLE 15.3

Rejection region	Significance level
$x \geq 12$	0.000
$x \geq 11$	0.003
$x \geq 10$	0.019
$x \geq 9$	0.073
$x \geq 8$	0.194

For example, by referring to Table 15.2, we see that the significance level associated with the rejection region $x \geq 9$ is

$$\alpha = P(x \geq 9) = P(9) + P(10) + P(11) + P(12)$$
$$= 0.054 + 0.016 + 0.003 + 0.000 = 0.073$$

We picture this rejection region and significance level in Figure 15.2.

FIGURE 15.2

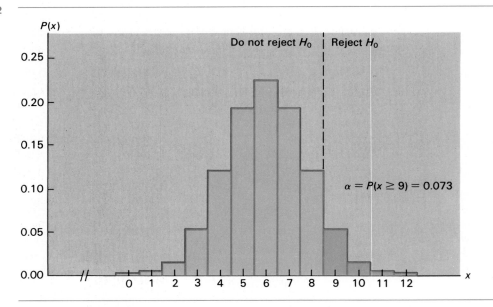

In any case, Table 15.3 reveals that we cannot use a 5% (0.05) significance level. The nearest we can come to that is either 1.9% (0.019) or 7.3% (0.073), with 7.3% actually being the closest to 5%.

THE SIGN TEST FOR A POPULATION MEDIAN

Procedure 15.1 below provides a step-by-step method to perform a sign test for a population median, η. To be perfectly correct in applying the sign test, we need to assume that the population under consideration has a continuous distribution. However, in practice, we can ignore that restriction and so we have not listed any assumptions for the use of Procedure 15.1.

PROCEDURE 15.1

To perform a sign test for a population median with null hypothesis H_0: $\eta = \eta_0$.

STEP 1 *State the null and alternative hypotheses.*

STEP 2 *Decide on a significance level and use the binomial distribution with parameters n and 0.5 to obtain a rejection region with an associated significance level, α, as close as possible to the one required.*

STEP 3 *Assign a "+" sign to each piece of data larger than η_0 and a "−" sign to each piece smaller than η_0.*

STEP 4 *Determine the value of the test statistic*

$$x = \text{number of "+" signs}$$

STEP 5 *If the value of the test statistic falls in the rejection region, then reject H_0; otherwise, do not reject H_0.*

STEP 6 *State the conclusion in words.*

EXAMPLE 15.2   *Illustrates Procedure 15.1*

As a first application of Procedure 15.1, let us reconsider the hypothesis-testing situation of Example 15.1. A random sample of 12 families in a northeastern city yielded the income data shown in Table 15.4.

TABLE 15.4
Sample data for
family incomes

$57,968	39,041	61,502
33,625	29,671	43,430
56,157	48,011	65,003
28,820	37,093	38,310

Do the data provide sufficient evidence to conclude that the median family income in the city exceeds $32,000? Use a significance level as close to 5% as possible.

SOLUTION   We apply Procedure 15.1.

STEP 1 *State the null and alternative hypotheses.*

Let η denote the median family income in the city. Then the null and alternative hypotheses are

$$H_0: \eta = \$32,000 \text{ (median income does not exceed \$32,000)}$$
$$H_a: \eta > \$32,000 \text{ (median income does exceed \$32,000)}$$

Note that the hypothesis test is right-tailed since there is a greater-than sign ($>$) in the alternative hypothesis.

STEP 2 *Decide on a significance level and use the binomial distribution with parameters n and 0.5 to obtain a rejection region with an associated significance level, α, as close as possible to the one required.*

We are to use a significance level as close to 0.05 as possible. The information necessary to find the appropriate rejection region has already been obtained and is summarized in Table 15.3 on page 752. From the table, we see that the rejection region giving the significance level closest to 0.05 is $x \geq 9$, and its associated significance level is $\alpha = 0.073$. See Figure 15.3.

FIGURE 15.3

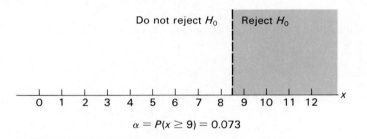

$$\alpha = P(x \geq 9) = 0.073$$

STEP 3 *Assign a "+" sign to each piece of data larger than η_0 and a "−" sign to each piece smaller than η_0.*

We have $\eta_0 = \$32{,}000$. Thus, we assign a "+" sign to each income in Table 15.4 that is larger than \$32,000 and a "−" sign to each one that is smaller than \$32,000. This is done in Table 15.5.

TABLE 15.5

\$57,968 (+)	39,041 (+)	61,502 (+)
33,625 (+)	29,671 (−)	43,430 (+)
55,157 (+)	48,011 (+)	65,003 (+)
28,820 (−)	37,093 (+)	38,310 (+)

STEP 4 *Determine the value of the test statistic*

$$x = number\ of\ \text{"+"}\ signs$$

From Table 15.5, we see that the number of "+" signs is 10. Thus, $x = 10$.

STEP 5 *If the value of the test statistic falls in the rejection region, reject H_0; otherwise, do not reject H_0.*

The value of the test statistic, found in Step 4, is $x = 10$. A glance at Figure 15.3 shows that this falls in the rejection region. Therefore, we reject H_0.

STEP 6 *State the conclusion in words.*

The data provide sufficient evidence to conclude that the median family income in the city exceeds \$32,000. ∎

EXAMPLE 15.3   *Illustrates Procedure 15.1*

The median number of years of teaching experience for elementary-school teachers was 14 years in 1986, as reported by the National Education Association in *Status of the American Public School Teacher.* A random sample of nine elementary-school teachers taken this year yielded the data on years of teaching experience displayed in Table 15.6.

TABLE 15.6
Years of teaching
experience

32	16	8
9	21	2
18	15	22

Do the data indicate that this year's median number of years of teaching experience for elementary-school teachers has changed from the 1986 median? Use a significance level as close to 5% as possible.

SOLUTION   We apply Procedure 15.1.

STEP 1 *State the null and alternative hypotheses.*

Let η denote this year's median number of years of teaching experience for elementary-school teachers. Then the null and alternative hypotheses are

$$H_0: \eta = 14 \text{ years (median is the same as in 1986)}$$
$$H_a: \eta \neq 14 \text{ years (median is not the same as in 1986)}$$

Note that the hypothesis test is two-tailed since there is a not-equal sign ($\neq$) in the alternative hypothesis.

STEP 2 *Decide on a significance level and use the binomial distribution with parameters n and 0.5 to obtain a rejection region with an associated significance level, α, as close as possible to the one required.*

We want a significance level as close to 0.05 as possible. The sample size is nine and so we must use the binomial distribution with parameters $n = 9$ and $p = 0.5$. That probability distribution can be found in Table I and we display it here as Table 15.7.

TABLE 15.7
Binomial distribution
with parameters
$n = 9$ and $p = 0.5$

x	$P(x)$	
0	0.002	} *0.020*
1	0.018	
2	0.070	
3	0.164	
4	0.246	
5	0.246	
6	0.164	
7	0.070	
8	0.018	} *0.020*
9	0.002	

Since the test is two-tailed and the significance level is to be as close to 0.05 as possible, we must find a two-sided rejection region with each tail having probability as close to 0.025 as possible. Table 15.7 shows that the best we can do is to use the rejection region consisting of those x-values for which either $x \leq 1$ or $x \geq 8$. The associated significance level for that rejection region is

$$\alpha = P(x \leq 1 \quad \text{or} \quad x \geq 8) = P(x \leq 1) + P(x \geq 8)$$
$$= [P(0) + P(1)] + [P(8) + P(9)] = 0.02 + 0.02$$

or $\alpha = 0.04$. See Figure 15.4.

FIGURE 15.4

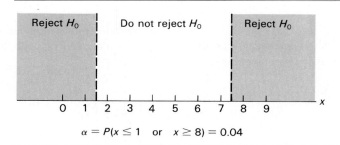

$$\alpha = P(x \leq 1 \quad \text{or} \quad x \geq 8) = 0.04$$

STEP 3 *Assign a "+" sign to each piece of data larger than η_0 and a "−" sign to each piece smaller than η_0.*

Here $\eta_0 = 14$ years. Thus, we assign a "+" sign to each data value in Table 15.6 that is larger than 14 and a "−" sign to each data value that is smaller than 14. This is done in Table 15.8.

TABLE 15.8

32 (+)	16 (+)	8 (−)
9 (−)	21 (+)	2 (−)
18 (+)	15 (+)	22 (+)

STEP 4 *Determine the value of the test statistic*

$$x = \text{number of "+" signs}$$

The number of "+" signs in Table 15.8 is 6. Therefore, $x = 6$.

STEP 5 *If the value of the test statistic falls in the rejection region, reject H_0; otherwise, do not reject H_0.*

From Step 4, the value of the test statistic is $x = 6$ which, as we see from Figure 15.4, does not fall in the rejection region. Thus, we do not reject H_0.

STEP 6 *State the conclusion in words.*

The data do not provide sufficient evidence to conclude that the median number of years of teaching experience for elementary-school teachers is different now than it was in 1986.

MTB

DATA VALUES EQUAL TO THE NULL HYPOTHESIS MEDIAN

In applying the sign test, we assign a "+" sign to data values larger than η_0 and a "−" sign to data values smaller than η_0. Occasionally, we observe data values equal to η_0. When such a data value appears, it is customary to remove it from the data set and reduce the sample size by one.

USING THE COMPUTER (OPTIONAL)

Procedure 15.1 on page 753 provides a step-by-step method to perform a sign test for a population median, η. Alternatively, we can use Minitab to carry out a sign test by employing the **STEST** command. Example 15.4 explains how to apply STEST.

EXAMPLE 15.4   *Illustrates the STEST command*

Refer to Example 15.2 on pages 753–754. Use Minitab to perform the hypothesis test considered in that example.

SOLUTION   Let η denote the median family income in the city. The problem is to use the sign test to perform the hypothesis test

$$H_0: \eta = \$32{,}000 \text{ (median income does not exceed \$32,000)}$$
$$H_a: \eta > \$32{,}000 \text{ (median income does exceed \$32,000)}$$

at the 5% significance level. Note that the hypothesis test is right-tailed since there is a greater-than sign ($>$) in the alternative hypothesis.

To apply STEST, we first enter the sample data from Table 15.4 on page 753 into C1 using the SET command and name C1 "Incomes" using the NAME command. Then we type the command STEST, followed by the null hypothesis and the storage location of the sample data. That is, we type

STEST of eta=32000, data in 'INCOMES';

Because the test is right-tailed, we also type the subcommand **ALTERNATIVE=1**.. Printout 15.1 displays the above commands and the resulting output.

PRINTOUT 15.1
Minitab output
for STEST

```
MTB > SET C1
DATA> 57968 39041 61502 33625 29671 43430
DATA> 56157 48011 65003 28820 37093 38310
DATA> END
MTB > NAME C1 'INCOMES'
MTB > STEST of eta=32000, data in 'INCOMES';
SUBC> ALTERNATIVE=1.

SIGN TEST OF MEDIAN = 32000 VERSUS  G.T.  32000
```

	N	BELOW	EQUAL	ABOVE	P-VALUE	MEDIAN
INCOMES	12	2	0	10	0.0193	41235

The first line of the output shows the null and alternative hypotheses for the hypothesis test: SIGN TEST OF MEDIAN = 32000 VERSUS G.T. 32000. Then, on the next line, we find the sample size; the number of data values below, equal, and above the value of 32,000 given for the median in the null hypothesis; the P-value; and the sample median.

We see, in particular, that the P-value equals 0.0193. Since this is less than the designated significance level of $\alpha = 0.05$, we reject H_0. In other words, the data provide sufficient evidence to conclude that the median family income in the city exceeds \$32,000. ∎

Exercises 15.2

__ **15.3** Upon what probability distribution is the sign test based?

__ **15.4** Explain why a sign test always uses a binomial distribution with success probability $p = 0.5$.

In each of Exercises 15.5–15.10, use Procedure 15.1 on page 753 to perform the required hypothesis test.

__ **15.5** The median age of recipients of science and engineering doctoral degrees was 32.8 years in 1983, as reported by the National Science Foundation, Division of Science Resources Studies, in *Survey of Earned Doctorates*. A random sample of this year's recipients gave the following data on ages:

37	28	36	33
37	43	41	28
24	44	27	24

Do the data provide sufficient evidence to conclude that the median age of recipients of science and engineering doctoral degrees has increased since 1983? Use a significance level as close to 5% as possible.

__ **15.6** The U.S. Bureau of Justice Statistics reports in *Profile of Jail Inmates* that the median educational attainment of jail inmates was 10.2 years in 1978. Ten current inmates were randomly selected and found to have the following educational attainments, in years:

9	11	8	8	9
13	11	6	5	9

Can we conclude that the median educational attainment of current jail inmates has changed from that

in 1978? Use a significance level as close to 0.10 as possible.

__ **15.7** The 1986 median net earnings of all office-based American MDs was \$112.8 thousand. [SOURCE: Medical Economics Company, Oradell, NJ, *Medical Economics*.] Twenty randomly selected internists reported the following earnings for that same year (data in thousands of dollars):

84.4	88.9	109.7	128.9
97.2	112.5	96.1	98.6
71.7	96.9	91.2	119.5
98.5	105.0	110.3	104.0
84.1	97.1	84.5	91.3

At a significance level as close to 0.05 as possible, can you conclude that the 1986 median net earnings of office-based internists was less than that of all office-based MDs?

__ **15.8** A personnel officer for an electronics company has designed a qualifying test in mathematics for prospective employees. She wants the median test score to be about 70 but suspects the test is too hard. Fifteen randomly selected prospective employees are given the test. Their scores are shown below.

57	78	60	92	61
68	55	70	66	63
85	56	43	59	54

Do the scores provide sufficient evidence to conclude that the median score on the exam by all prospective employees will be less than 70? Choose a significance level as close to 0.05 as possible. (*Note:* There is a data value equal to η_0.)

___ **15.9** According to *Food Cost Review*, published by the U.S. Department of Agriculture, the median retail price for oranges in 1983 was 38.5 cents per pound. Recently, a random sample of 15 markets reported the following prices for oranges in cents per pound:

43.0	40.0	42.6	40.2	37.5
44.1	45.2	41.8	35.6	34.6
37.9	44.2	44.5	38.2	42.4

Do the data provide sufficient evidence to conclude that the median retail price for oranges now is different from the 1983 median of 38.5 cents per pound? Use a significance level as close as possible to 0.05.

___ **15.10** The U.S. Bureau of the Census reports in *Current Housing Reports* that the median number of rooms for housing units was 5.2 in 1985. A random sample taken this year of 15 housing units revealed that six had five rooms or less and nine had six rooms or more. Do the data provide sufficient evidence to conclude that the median number of rooms in housing units today is greater than in 1985? Use a significance level as close to 0.05 as possible.

Exercises 15.11–15.14 are computer exercises.

___ **15.11** (**Computer exercise**) Suppose that the sample data in Exercise 15.7 are stored in a column named EARNINGS.
a) Which Minitab command and subcommands (if any) should be used to carry out the hypothesis test considered in that exercise?
b) If you have access to Minitab, use it to perform the hypothesis test.

___ **15.12** (**Computer exercise**) Suppose the data in Exercise 15.6 are stored in a column named EDUC.
a) Which Minitab command and subcommands (if any) should be used to carry out the hypothesis test considered in that exercise?
b) If you have access to Minitab, use it to perform the hypothesis test.

___ **15.13** (**Computer exercise**) The National Center for Health Statistics publishes data on the duration of marriages in *Vital Statistics of the United States*. In 1985, the median duration of a marriage was 6.8 years. A random sample is taken from last year's divorce certificates and the marriage durations are recorded. Then Minitab's STEST is applied. The resulting computer output is shown in Printout 15.2 at the top of the next page. Use the computer output to determine

a) the null and alternative hypotheses for the hypothesis test.
b) the subcommand used, if any.
c) the number of divorce certificates sampled.
d) the number of certificates in the sample for which the marriages lasted less than 6.8 years; exactly 6.8 years; more than 6.8 years.
e) the sample median marriage duration.
f) the *P*-value of the hypothesis test.
g) the smallest significance level at which the null hypothesis can be rejected.
h) the conclusion if the hypothesis test is performed at the 5% significance level.

___ **15.14** (**Computer exercise**) The Census Bureau estimates that the *U.S. Census Form* takes the average household 14 minutes to complete. A random sample of households is obtained and the time that each household takes to complete the form is recorded. Printout 15.3, on the next page, shows the computer output resulting from applying Minitab's STEST command to the completion-time data. Using the computer output, determine
a) the null and alternative hypotheses for the hypothesis test.
b) the subcommand used, if any.
c) the number of households sampled.
d) the number of households in the sample that took less than 14 minutes to complete the form; greater than 14 minutes to complete the form.
e) the sample median completion time.
f) the *P*-value of the hypothesis test.
g) the smallest significance level at which the null hypothesis can be rejected.
h) the conclusion if the hypothesis test is performed at the 10% significance level.

Large-sample sign test using the normal approximation: To perform a sign test by employing Procedure 15.1, we need to use binomial tables. Since, by necessity, tables of binomial probabilities are limited, we must apply a different method if the sample size is out of the range of our binomial tables. One possibility is to use a computer and another is to employ a normal approximation. In Exercises 15.15–15.18, we will develop and apply this latter approach.

=== **15.15** Let x denote the number of "+" signs; that is, the number of data values that exceed η_0.
a) Show that if the null hypothesis, $H_0: \eta = \eta_0$, is true, then the mean and standard deviation of the

PRINTOUT 15.2   Minitab output for Exercise 15.13

```
SIGN TEST OF MEDIAN = 6.800 VERSUS  L.T.   6.800

            N   BELOW   EQUAL   ABOVE   P-VALUE     MEDIAN
YEARS      50    27       0      23     0.3359      6.555
```

PRINTOUT 15.3   Minitab output for Exercise 15.14

```
SIGN TEST OF MEDIAN = 14.00 VERSUS  N.E.   14.00

            N   BELOW   EQUAL   ABOVE   P-VALUE     MEDIAN
MINUTES    36    13       0      23     0.1325      15.31
```

random variable x are

$$\mu_x = n/2 \quad \text{and} \quad \sigma_x = \sqrt{n}/2$$

(Hint: Refer to Formulas 5.3 and 5.4 on page 275.)
b) Apply part (a) to explain why the random variable

$$(1) \qquad\qquad z = \frac{2x - n}{\sqrt{n}}$$

has approximately the standard normal distribution for large n.

= **15.16**   Formulate a large-sample procedure for a sign test that employs the test statistic displayed in Equation (1).

= **15.17**   The median age of United States residents was 32.1 years in 1987, as reported by the Bureau of the Census in *Current Population Reports*. In a random sample of 250 current residents, 144 were older than 32.1 years and 106 were younger than 32.1 years.

Employ the procedure that you formulated in Exercise 15.16 to decide whether the median age of today's United States residents has increased over the 1987 median age of 32.1 years. Perform the hypothesis test using a significance level of 0.05.

= **15.18**   Arizona State University recommends that students study two hours outside class for each hour in class. A professor decides to conduct a survey to see whether, on the average, students are following the university's recommendation. From a random sample of 250 students, the professor discovers that 114 of the students study less than two hours per class hour and 136 study more than two hours per class hour. Do the data provide sufficient evidence to conclude that the median study time per class hour for all ASU students is different from the recommended two hours? Perform the required hypothesis test at the 5% significance level using the procedure that you formulated in Exercise 15.16.

15.3   The Wilcoxon signed-rank test

One of the main advantages of the sign test is that the assumptions for its use are minimal. But the sign test considers only whether each data value is above ($+$) or below ($-$) the value, η_0, given for the median in the null hypothesis. It does not take into account *how much* each data value is above or below η_0. Thus, a large amount of pertinent information is ignored.

In this section we will discuss the **Wilcoxon signed-rank test.** Like the sign test, the Wilcoxon signed-rank test is a nonparametric test that can be used to make inferences about the median of a population. However, the Wilcoxon signed-rank test does not disregard as much information as the sign test and, consequently, is more powerful.

To employ the Wilcoxon signed-rank test, though, we need to make an assumption that is not required for the sign test; namely, that the population under consideration has a **symmetric distribution.** This means that the portion of the distribution to the left of the median is the mirror image of the portion to the right of the median. For example, normal and t distributions are symmetric, whereas F and χ^2 distributions are not. Since, for a population with a symmetric distribution, the mean and the median are identical, the Wilcoxon signed-rank test can also be used to make inferences about the mean of a population.

We will now present an example to introduce the basic reasoning behind the Wilcoxon signed-rank test. Following the example, we will give a step-by-step procedure for performing such a test.

EXAMPLE 15.5   *Introduces the Wilcoxon signed-rank test*

In January, 1984, the U.S. Department of Agriculture estimated that a typical American family of four with an intermediate budget would spend about $92 per week for food. A consumer researcher in Kansas suspected that the median weekly cost was less in her state. She took a random sample of 10 Kansas families of four with intermediate budgets. Their weekly food costs were as given in Table 15.9.

TABLE 15.9
Weekly food costs

$78	104	84	70	96
73	87	85	76	94

Do the data provide sufficient evidence to conclude that, in 1984, the median weekly food cost for Kansas families of four with intermediate budgets was less than the national median of $92? Apply the Wilcoxon signed-rank test. [Assume that the 1984 distribution of weekly food costs for all Kansas families of four with intermediate budgets is symmetric.]

SOLUTION   Let η denote the 1984 median weekly food cost for all Kansas families of four with intermediate budgets. Then we want to perform the hypothesis test

$$H_0: \eta = \$92 \text{ (median weekly food cost was not less than \$92)}$$
$$H_a: \eta < \$92 \text{ (median weekly food cost was less than \$92)}$$

To apply the Wilcoxon signed-rank test, we first rank each observation (piece of sample data) according to distance and direction from the null hypothesis median, $\eta_0 = \$92$. The steps for doing this are depicted in Table 15.10.

TABLE 15.10

Cost ($) x	Difference D = x − 92	|D|	Rank of |D|	Signed rank R
78	−14	14	7	−7
73	−19	19	9	−9
104	12	12	6	6
87	−5	5	3	−3
84	−8	8	5	−5
85	−7	7	4	−4
70	−22	22	10	−10
76	−16	16	8	−8
96	4	4	2	2
94	2	2	1	1

STEP 1 *Subtract η_0 from x.*

STEP 2 *Make each difference positive by taking absolute values.*

STEP 3 *Rank the absolute differences in order from smallest (1) to largest (10).*

STEP 4 *Give each rank the same sign as the sign in Column 2.*

The absolute differences, $|D|$, displayed in the third column of Table 15.10, reveal how far each observation is from \$92; the ranks of those absolute differences, displayed in the fourth column, show which observations are closer to \$92 and which are further away; and the signed ranks, R, displayed in the last column, indicate additionally whether an observation is greater than \$92 (+) or less than \$92 (−). Figure 15.5 graphically illustrates this information for the second and third rows of Table 15.10.

FIGURE 15.5

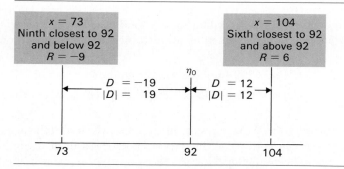

The reasoning behind the Wilcoxon signed-rank test is as follows: If the median weekly food cost is in fact $92, then, because the distribution of weekly food costs is symmetric, we would expect the sum of the positive ranks and the sum of the negative ranks to be roughly the same magnitude. Since the sample size here is 10, the sum of all the ranks must be $1 + 2 + \cdots + 10 = 55$; and half of 55 is 27.5. So, if the median weekly food cost is $92, we would expect the sum of the positive ranks (and the sum of the negative ranks) to be about 27.5. To put it another way, if the sum of the positive ranks is too much smaller than 27.5, then we would take this as evidence that the null hypothesis, $\eta = \$92$, is false and conclude that the median weekly food cost is less than $92.

From the last column of Table 15.10, we see that the sum of the positive ranks, denoted by W, is equal to

$$W = 6 + 2 + 1 = 9$$

This is quite a bit smaller than 27.5, the value that we would expect if the median is in fact $92. Thus, it appears that the median is actually less than $92. But, as we know, appearances can be deceiving. To be objective, we need to have a table of critical values for W, such as Table VII in the appendix. We will discuss that table and then return to complete the hypothesis test. ∎

USING THE WILCOXON SIGNED-RANK TABLE

Table VII provides critical values for a Wilcoxon signed-rank test. The first column of the table gives the sample size; the second and third columns give the significance levels for a one-tailed and two-tailed test, respectively; and the fourth and fifth columns give the left-hand and right-hand critical values, respectively. We should emphasize that *a critical value from Table VII is to be included in the rejection region.*

For instance, suppose that the sample size is $n = 10$ and that we want to perform a left-tailed test with a significance level as close to 0.05 as possible. To determine the critical value and the associated significance level, we first go down the sample-size column of Table VII to 10. Then we look for a significance level for a one-tailed test that is as close to 0.05 as possible, which in this case is $\alpha = 0.053$. Finally, we go across the row containing 0.053 to the W_ℓ column, where we find the corresponding left-hand critical value, 11. Thus, the critical value for a left-tailed test with $n = 10$ and $\alpha = 0.053$ is 11. In other words, the null hypothesis is rejected if $W \leq 11$ and is not rejected if $W > 11$.

THE WILCOXON SIGNED-RANK TEST FOR A POPULATION MEDIAN OR A POPULATION MEAN

We now present a step-by-step procedure for performing a Wilcoxon signed-rank test. As we mentioned earlier, that test can be used to make inferences for either a population median or a population mean. We will state the procedure in terms of a population median. To employ the procedure for a population mean, simply replace η by μ and η_0 by μ_0.

PROCEDURE 15.2

To perform a Wilcoxon signed-rank test for a population median with null hypothesis H_0: $\eta = \eta_0$.

ASSUMPTION

Symmetric population.[†]

STEP 1 *State the null and alternative hypotheses.*

STEP 2 *Decide on a significance level and use Table VII to find a significance level, α, as close as possible to the one required.*

STEP 3 *The critical value(s)*
  a) *for a two-tailed test are W_ℓ and W_r.*
  b) *for a left-tailed test is W_ℓ.*
  c) *for a right-tailed test is W_r.*
  Use Table VII to find the critical value(s).

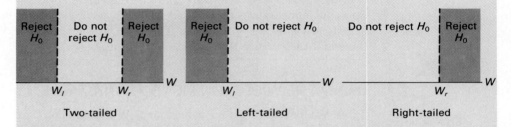

Two-tailed · Left-tailed · Right-tailed

STEP 4 *Construct a work table of the form:*

Data value x	Difference $D = x - \eta_0$	$\lvert D \rvert$	Rank of $\lvert D \rvert$	Signed rank R
.	.	.	.	.
.	.	.	.	.
.	.	.	.	.

STEP 5 *Compute the value of the test statistic*

$$W = \text{sum of the positive ranks}$$

STEP 6 *If the value of the test statistic falls in the rejection region, then reject H_0; otherwise, do not reject H_0.*

STEP 7 *State the conclusion in words.*

[†] For brevity, we will sometimes use the phrase **symmetric population** to indicate that a population has a symmetric distribution.

EXAMPLE 15.6   *Illustrates Procedure 15.2*

As a first application of Procedure 15.2, let us complete the hypothesis test of Example 15.5. A random sample of 10 Kansas families of four with intermediate budgets yielded the data on 1984 weekly food costs shown in Table 15.11.

TABLE 15.11
Weekly food costs

$78	104	84	70	96
73	87	85	76	94

Do the data provide sufficient evidence to conclude that, in 1984, the median weekly food cost for Kansas families of four with intermediate budgets was less than the national median of $92? Apply the Wilcoxon signed-rank test with a significance level as close to 0.05 as possible.

SOLUTION   We apply Procedure 15.2.

STEP 1 *State the null and alternative hypotheses.*

Let η denote the 1984 median weekly food cost for all Kansas families of four with intermediate budgets. Then the null and alternative hypotheses are

H_0: $\eta = \$92$ (median weekly food cost was not less than $92)
H_a: $\eta < \$92$ (median weekly food cost was less than $92)

Note that the hypothesis test is left-tailed since there is a less-than sign ($<$) in the alternative hypothesis.

STEP 2 *Decide on a significance level and use Table VII to find a significance level, α, as close as possible to the one required.*

The test is to be performed at a significance level as close to 0.05 as possible. Since $n = 10$ and the test is one-tailed, we see from Table VII that the closest possible significance level is $\alpha = 0.053$.

STEP 3 *The critical value for a left-tailed test is W_ℓ.*

Referring again to Table VII with $n = 10$ and $\alpha = 0.053$, we find that the critical value is $W_\ell = 11$. See Figure 15.6.

FIGURE 15.6

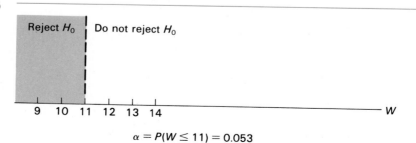

$\alpha = P(W \leq 11) = 0.053$

STEP 4 *Construct a work table.*

We have already done this in Table 15.10 on page 762.

STEP 5 *Compute the value of the test statistic*

$$W = sum\ of\ the\ positive\ ranks$$

The last column of Table 15.10 shows that the sum of the positive ranks equals

$$W = 6 + 2 + 1 = 9$$

STEP 6 *If the value of the test statistic falls in the rejection region, reject H_0; otherwise, do not reject H_0.*

The value of the test statistic is $W = 9$, as found in Step 5. This falls in the rejection region, as we observe from Figure 15.6. Thus, we reject H_0.

STEP 7 *State the conclusion in words.*

The data provide sufficient evidence to conclude that, in 1984, the median weekly food cost for Kansas families of four with intermediate budgets was less than the national median of $92. ∎

Before we illustrate the Wilcoxon signed-rank test further, we need to discuss a couple of fine points. First, as in the sign test, if a data value is equal to the value, η_0, given for the median in the null hypothesis, then that data value should be removed and the sample size reduced by one.

Second, we need to address the method for assigning ranks when there are ties among the absolute differences, $|D|$. The method is the following: *If two or more absolute differences are tied, each is assigned the mean of the ranks they would have had if there were no ties.* For example, if two absolute differences are tied for second place, each is assigned rank $(2+3)/2 = 2.5$ and rank 4 is assigned to the next largest absolute difference (which really is fourth!). If three absolute differences are tied for fifth place, each is assigned rank $(5+6+7)/3 = 6$ and rank 8 is assigned to the next largest absolute difference.

EXAMPLE 15.7   *Illustrates Procedure 15.2*

A highway official in California wants to know whether the average speed by drivers on a certain section of highway differs from the legal limit of 55 mph. Twelve randomly selected cars are monitored for speed. The speeds, in miles per hour, are:

TABLE 15.12
Highway speeds (mph)

65	58	54	50
48	60	54	64
57	61	55	56

Can we conclude that the median speed by drivers on the section of highway is different from 55 mph? Employ the Wilcoxon signed-rank test with a significance level as close to 0.05 as possible. [Assume that the speeds have a symmetric distribution.]

SOLUTION   We apply Procedure 15.2.

STEP 1 *State the null and alternative hypotheses.*

Let η denote the median speed by drivers on the section of highway. Then the null and alternative hypotheses are

$$H_0: \eta = 55 \text{ mph (median speed is 55 mph)}$$
$$H_a: \eta \neq 55 \text{ mph (median speed is not 55 mph)}$$

Note that the hypothesis test is two-tailed since there is a not-equal sign ($\neq$) in the alternative hypothesis.

STEP 2 *Decide on a significance level and use Table VII to find a significance level, α, as close as possible to the one required.*

To begin, note that there is a data value in the sample that equals the value, 55, given for the median in the null hypothesis; namely, the third entry in the third row of Table 15.12. As we said, that data value must be deleted from the sample. This reduces the sample size from 12 to 11. Now we look in Table VII for a significance level as close as possible to 0.05 for a two-tailed test with $n = 11$. From the table, we see that the best we can do is $\alpha = 0.054.$

STEP 3 *The critical values for a two-tailed test are W_ℓ and W_r.*

Consulting Table VII, we find that the critical values for a two-tailed test with $n = 11$ and $\alpha = 0.054$ are $W_\ell = 11$ and $W_r = 55.$ See Figure 15.7.

FIGURE 15.7

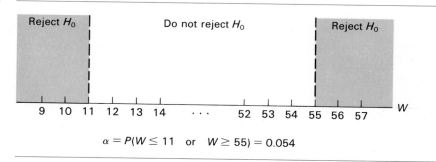

$\alpha = P(W \leq 11 \quad \text{or} \quad W \geq 55) = 0.054$

STEP 4 *Construct a work table.*

Referring to Table 15.12 and recalling that the data value 55 has been deleted, we obtain the work table shown in Table 15.13 at the top of the next page. Note that the three 1s in the $|D|$-column are all tied for smallest. Hence, each is assigned the rank $(1 + 2 + 3)/3 = 2$. Also, the two 5s in the $|D|$-column are tied for sixth smallest, so each is assigned the rank $(6 + 7)/2 = 6.5$.

TABLE 15.13

Speed (mph) x	Difference D = x − 55	|D|	Rank of |D|	Signed rank R
65	10	10	11	11
48	−7	7	9	−9
57	2	2	4	4
58	3	3	5	5
60	5	5	6.5	6.5
61	6	6	8	8
54	−1	1	2	−2
54	−1	1	2	−2
50	−5	5	6.5	−6.5
64	9	9	10	10
56	1	1	2	2

STEP 5 *Compute the value of the test statistic*

$$W = \text{sum of the positive ranks}$$

From the final column of Table 15.13, we see that the sum of the positive ranks is equal to

$$W = 11 + 4 + 5 + 6.5 + 8 + 10 + 2 = \textbf{46.5}$$

STEP 6 *If the value of the test statistic falls in the rejection region, reject H_0; otherwise, do not reject H_0.*

The value of the test statistic, $W = 46.5$, does not fall in the rejection region pictured in Figure 15.7. Thus, we **do not reject H_0.**

STEP 7 *State the conclusion in words.*

MTB

The data do not provide sufficient evidence to conclude that the median speed differs from 55 mph. ∎

COMPARISON OF THE WILCOXON SIGNED-RANK TEST AND THE SIGN TEST

Both the sign test and the Wilcoxon signed-rank test can be used to perform a hypothesis test for the median, η, of a population. The sign test has the advantage of being essentially assumption free, whereas the Wilcoxon signed-rank test requires that the population have a symmetric distribution.

On the other hand, the Wilcoxon signed-rank test utilizes more of the available information provided by the sample data than does the sign test. Consequently, if the population being sampled has a symmetric distribution, then the Wilcoxon signed-rank test is preferable to the sign test. In summary, then, we have the following key fact:

KEY FACT 15.2 Wilcoxon signed-rank test versus the sign test

Suppose a hypothesis test is to be performed for the median, η, of a population. When deciding between the sign test and the Wilcoxon signed-rank test, employ the following rule: If you are reasonably sure that the population has a symmetric distribution, use the Wilcoxon signed-rank test; otherwise use the sign test.

COMPARISON OF THE WILCOXON SIGNED-RANK TEST AND THE t-TEST

In Chapter 9, we learned how to perform a t-test for the mean of a normally distributed population (Procedure 9.4, pages 458–459). Since a normal population necessarily has a symmetric distribution, we can also use the Wilcoxon signed-rank test (Procedure 15.2, page 764) to perform such a hypothesis test.

So now the question is: If we want to perform a hypothesis test for the mean of a population and we know that the population is normally distributed, which test should we use—the t-test or the Wilcoxon signed-rank test? As you might expect, the t-test is better, because it is designed expressly for normal populations. Surprisingly, however, it is not much better.[†]

On the other hand, if the population being sampled has a symmetric distribution but is not normal, the Wilcoxon signed-rank test is usually better and is often much better. In summary, then, we have the following fact:

KEY FACT 15.3 Wilcoxon signed-rank test versus the t-test

Suppose a hypothesis test is to be performed for the mean, μ, of a population. When deciding between the t-test and the Wilcoxon signed-rank test, employ the following rule: If you are sure that the population is normally distributed, use the t-test. However, if you are not sure that the population is normally distributed but are reasonably sure that it has a symmetric distribution, use the Wilcoxon signed-rank test.

USING THE COMPUTER (OPTIONAL)

Procedure 15.2 on page 764 provides a step-by-step method to perform a Wilcoxon signed-rank test for a population median, η, or a population mean, μ. Minitab has a program that will carry out Procedure 15.2 for us. The appropriate command is **WTEST.** The next example shows how WTEST is applied.

EXAMPLE 15.8   *Illustrates the WTEST command*

Refer to Example 15.6 on pages 765–766. Use Minitab to perform the hypothesis test considered in that example.

[†] Roughly speaking, by "better" we mean less chance of making an error using the same size sample. See Section 16.3 for details.

SOLUTION   Let η denote the 1984 median weekly food cost for all Kansas families of four with intermediate budgets. Then the problem is to use the Wilcoxon signed-rank test to perform the hypothesis test

$$H_0: \eta = \$92 \text{ (median weekly food cost was not less than \$92)}$$

$$H_a: \eta < \$92 \text{ (median weekly food cost was less than \$92)}$$

at the 5% significance level. Note that the hypothesis test is left-tailed since there is a less-than sign ($<$) in the alternative hypothesis.

To apply WTEST, we begin by entering the sample data from Table 15.11 into C2 and naming C2 "Costs." Next we type the command WTEST, followed by the null hypothesis and the storage location of the sample data. That is, we type

WTEST of eta=92, data in 'COSTS';

Since the test is left-tailed, we also type the subcommand ALTERNATIVE=-1.. Printout 15.4 summarizes the above discussion and also shows the resulting output.

PRINTOUT 15.4
Minitab output
for WTEST

```
MTB > SET C2
DATA> 78 104 84 70 96
DATA> 73 87 85 76 94
DATA> END
MTB > NAME C2 'COSTS'
MTB > WTEST of eta=92, data in 'COSTS';
SUBC> ALTERNATIVE=-1.

TEST OF MEDIAN = 92.00 VERSUS MEDIAN L.T. 92.00

                  N FOR   WILCOXON          ESTIMATED
           N      TEST    STATISTIC  P-VALUE   MEDIAN
COSTS      10     10          9.0     0.033    84.50
```

The first line of the output displays the null and alternative hypotheses for the hypothesis test: TEST OF MEDIAN = 92.00 VERSUS MEDIAN L.T. 92.00. Below that we find several entries. The first entry, headed N, gives the sample size, which in this case is 10.

The next entry, headed N FOR TEST, shows the number of data values in the sample that are not equal to the null hypothesis median of 92. Recall that for a Wilcoxon signed-rank test, if a data value is equal to the null hypothesis median, then that data value should be removed and the sample size reduced by one. Since here the sample size is 10 and none of the data values in the sample are equal to 92, the sample size for the test (N FOR TEST) is also 10.

The third entry, headed WILCOXON STATISTIC, provides the value of the test statistic

$$W = \text{sum of the positive ranks}$$

So, we see that $W = 9$.

The last two entries, headed P-VALUE and ESTIMATED MEDIAN, respectively, give the P-value for the hypothesis test and a point estimate for the population median. Since the P-value of 0.033 is less than the designated significance level of $\alpha = 0.05$, we reject H_0. In other words, the data provide sufficient evidence to conclude that, in 1984, the median weekly food cost for Kansas families of four with intermediate budgets was less than the national median of $92. ∎

Exercises 15.3

__ **15.19** We said on page 766 that if a data value is equal to the value, η_0, given for the median in the null hypothesis, then that data value should be removed and the sample size reduced by one. Why do you think that is done?

__ **15.20** What population assumption must be met for use of the Wilcoxon signed-rank test?

__ **15.21** Suppose you want to perform a hypothesis test for the median, η, of a population. For each of the following types of populations, decide whether you would use the t-test, the sign test, or the Wilcoxon signed-rank test.
a) Symmetric, non-normal population.
b) Normal population.
c) Non-symmetric population.

__ **15.22** Suppose you want to perform a hypothesis test for the mean, μ, of a population. For each of the following types of populations, decide whether you would use the t-test, the sign test, or the Wilcoxon signed-rank test.
a) Symmetric, non-normal population.
b) Normal population.

__ **15.23** Can the sign test be used to perform a hypothesis test for a population mean? Explain.

__ **15.24** When performing a hypothesis test for the median of a population, what makes the Wilcoxon signed-rank test preferable to the sign test? What makes the sign test preferable to the Wilcoxon signed-rank test?

In each of Exercises 15.25–15.28, use Procedure 15.2 on page 764 to perform the required hypothesis test. For each exercise, assume that the population under consideration has a symmetric distribution.

__ **15.25** The median age of United States residents was 32.1 years in 1987, as reported by the Bureau of the Census in *Current Population Reports*. A random

sample taken this year of 10 United States residents yielded the following ages:

38	58	10	53	32
41	45	35	7	22

Do the data provide sufficient evidence to conclude that the median age of today's United States residents has increased over the 1987 median age of 32.1 years? Use a significance level as close to 0.05 as possible.

__ **15.26** A commuter in New York City is told that the median commuting time to work from his neighborhood is 29 minutes. The commuter thinks that figure is too low. He decides to obtain the commuting times of 12 randomly selected commuters in his neighborhood. The times, in minutes, are as follows:

42	27	54	61
38	36	48	14
39	43	21	26

At a significance level as close to 0.05 as possible, do the data support the commuter's contention?

__ **15.27** A chemist working for a pharmaceutical company has developed a new antacid tablet that she feels will relieve pain more quickly than the company's present tablet. Experience indicates that the present tablet requires an average of 12 minutes to take effect. The chemist records the following times, in minutes, for relief with the new tablet:

10.9	11.4	12.0	8.8	4.4
15.0	7.1	10.1	9.8	14.8
14.2	9.2	9.2	6.6	8.0

At a significance level as close to 5% as possible, does it appear that the new antacid tablet works faster?

___ **15.28** The National Center for Health Statistics reports in *Vital Statistics of the United States* that the median birth weight of American babies was 7.4 lb in 1986. A random sample of this year's births provided the following weights, in pounds:

8.6	7.4	5.3	13.8	7.8	5.7	9.2
8.8	8.2	9.2	5.6	6.0	11.6	7.2

Can we conclude that this year's median birth weight differs from that in 1986? Use a significance level as close to 5% as possible.

___ **15.29** According to *Food Cost Review,* published by the U.S. Department of Agriculture, the mean retail price for oranges in 1983 was 38.5 cents per pound. Recently, a random sample of 15 markets reported the following prices for oranges in cents per pound:

43.0	40.0	42.6	40.2	37.5
44.1	45.2	41.8	35.6	34.6
37.9	44.2	44.5	38.2	42.4

a) Do the data indicate that the mean retail price for oranges now is different from the 1983 mean of 38.5 cents per pound? Perform a Wilcoxon signed-rank test at a significance level as close as possible to 0.05.
b) The hypothesis test considered in part (a) was done previously in Exercise 9.81 on page 463 using a *t*-test. The assumption in that exercise is that the retail prices for oranges are normally distributed. Under that assumption, why is it permissible to perform a Wilcoxon signed-rank test for the mean retail price of oranges?

___ **15.30** Atlas Fishing Line, Inc., manufactures a 10-lb test line. Twelve randomly selected spools are subjected to tensile-strength tests. The results are:

9.8	10.2	9.8	9.4
9.7	9.7	10.1	10.1
9.8	9.6	9.1	9.7

a) Use the data to decide whether Atlas Fishing Line's 10-lb test line is not up to specifications. Perform a Wilcoxon signed-rank test for the mean tensile strength, μ, at a significance level as close to 0.05 as possible.
b) The hypothesis test of part (a) was done previously in Exercise 9.82 on page 463 using a *t*-test. The assumption in that exercise is that tensile strengths

are normally distributed. Under that assumption, why is it permissible to perform a Wilcoxon signed-rank test for the mean tensile strength?

___ **15.31** A manufacturer of liquid soap produces a bottle with an advertised content of 310 ml. Sixteen bottles are randomly selected and found to have the following contents:

297	318	306	300
311	303	291	298
322	307	312	300
315	296	309	311

Assume the contents are normally distributed and let μ denote the mean content of all bottles produced. In order to determine whether the mean content is less than advertised, perform the hypothesis test

H_0: $\mu = 310$ ml (mean is not less than advertised)

H_a: $\mu < 310$ ml (mean is less than advertised)

a) Use the *t*-test, Procedure 9.4 on pages 458–459, with $\alpha = 0.05$.
b) Use the Wilcoxon signed-rank test with a significance level as close as possible to 0.05.
c) Assuming that the mean content is, in fact, less than 310 ml, how do you explain the discrepancy between the two tests?

___ **15.32** Refer to Exercise 15.6 on page 758, where a random sample of 10 current jail inmates gave the following data on educational attainment, in years:

9	11	8	8	9
13	11	6	5	9

Assume that the population of all such educational attainments has a symmetric distribution.
a) Apply the Wilcoxon signed-rank test to the above data to decide whether the median educational attainment of current jail inmates has changed from the 1978 median of 10.2 years. Select a significance level as close to 0.10 as possible.
b) Presuming that the median has changed, how do you explain the discrepancy between the results of the Wilcoxon signed-rank test in part (a) and the sign test in Exercise 15.6?

Exercises 15.33–15.36 are computer exercises.

___ **15.33** (**Computer exercise**) Suppose the data in Exercise 15.25 are stored in a column named AGES.

a) Which Minitab command and subcommands (if any) should be used to carry out the hypothesis test considered in that exercise?

b) If you have access to Minitab, use it to perform the hypothesis test.

___ **15.34** (**Computer exercise**) Suppose the data in Exercise 15.26 are stored in a column named TIMES.

a) Which Minitab command and subcommands (if any) should be used to carry out the hypothesis test considered in that exercise?

b) If you have access to Minitab, use it to perform the hypothesis test.

___ **15.35** (**Computer exercise**) The National Center for Health Statistics publishes data on the duration of marriages in the document *Vital Statistics of the United States*. In 1985, the median duration of a marriage was 6.8 years. A random sample is taken from last year's divorce certificates and the marriage durations are recorded. Then Minitab's WTEST is applied. The resulting computer output is displayed in Printout 15.5 at the top of the following page. Using the computer output, determine

a) the null and alternative hypotheses for the hypothesis test.

b) the subcommand used, if any.

c) the number of divorce certificates sampled.

d) the number of certificates in the sample for which the marriages lasted exactly 6.8 years.

e) a point estimate for the population median marriage duration.

f) the P-value of the hypothesis test.

g) the smallest significance level at which the null hypothesis can be rejected.

h) the conclusion if the hypothesis test is performed at the 5% significance level.

In Exercise 15.13 on page 759, we employed Minitab's STEST to carry out the same hypothesis test.

i) Compare the results of STEST and WTEST.

j) Which test procedure do you think is more appropriate here? Explain your answer.

___ **15.36** (**Computer exercise**) The Census Bureau estimates that the *U.S. Census Form* takes the average household 14 minutes to complete. A random sample of households is obtained and the time that it takes each household to complete the form is recorded. Printout 15.6 on the next page shows the output resulting from applying Minitab's WTEST command to the completion-time data. Use the computer output to determine

a) the null and alternative hypotheses for the hypothesis test.

b) the subcommand used, if any.

c) the number of households sampled.

d) the number of households in the sample that took exactly 14 minutes to complete the form.

e) the estimated population median completion time.

f) the P-value of the hypothesis test.

g) the smallest significance level at which the null hypothesis can be rejected.

h) the conclusion if the hypothesis test is performed at the 10% significance level.

In Exercise 15.14 on page 759, we employed Minitab's STEST to carry out the same hypothesis test.

i) Compare the results of STEST and WTEST.

j) Which test procedure do you think is more appropriate here? Explain your answer.

Large-sample Wilcoxon signed-rank test using the normal approximation: The table of critical values for the Wilcoxon signed-rank test, Table VII, stops at $n = 20$. For larger samples, a normal approximation can be used. In fact, the normal approximation works well even for sample sizes as small as 10. Specifically, we have the following result:

Suppose a random sample of size $n \geq 10$ is to be taken from a symmetric population with median η_0. Then the random variable W is approximately normally distributed with mean $\mu_W = n(n + 1)/4$ and standard deviation $\sigma_W = \sqrt{n(n + 1)(2n + 1)/24}$. In particular then, the standardized random variable

$$(2) \qquad z = \frac{W - n(n + 1)/4}{\sqrt{n(n + 1)(2n + 1)/24}}$$

has approximately the standard normal distribution.

In Exercises 15.37–15.39, we will develop and apply a large-sample procedure for a Wilcoxon signed-rank test based on the previous fact.

═ **15.37** Formulate a large-sample procedure for a Wilcoxon signed-rank test that uses the test statistic in Equation (2).

═ **15.38** Refer to Exercise 15.28.

a) Use your procedure from Exercise 15.37 to perform the hypothesis test of Exercise 15.28 at the 0.048 level of significance.

b) Compare your result in part (a) to the one obtained in Exercise 15.28, where the normal approximation was not used.

PRINTOUT 15.5   Minitab output for Exercise 15.35

TEST OF MEDIAN = 6.800 VERSUS MEDIAN L.T. 6.800

	N	N FOR TEST	WILCOXON STATISTIC	P-VALUE	ESTIMATED MEDIAN
YEARS	50	50	696.0	0.716	7.231

PRINTOUT 15.6   Minitab output for Exercise 15.36

TEST OF MEDIAN = 14.00 VERSUS MEDIAN N.E. 14.00

	N	N FOR TEST	WILCOXON STATISTIC	P-VALUE	ESTIMATED MEDIAN
MINUTES	36	36	462.0	0.044	15.05

15.39   Refer to Exercise 15.27.

a) Use your procedure from Exercise 15.37 to perform the hypothesis test of Exercise 15.27 at the 5.2% significance level.

b) Compare your result in part (a) to the one obtained in Exercise 15.27, where the normal approximation was not used.

15.40   In this exercise, we will find the probability distribution of the random variable W when $n = 3$ and $n = 4$. This will enable you to see how the critical values for the Wilcoxon signed-rank test are derived.

a) The rows of the table shown at the top of the next column give all possible signs for the signed ranks in a Wilcoxon signed-rank test with $n = 3$. For example, the first row covers the possibility that all three data values are greater than η_0 and thus have the sign +. There is an empty column for values of W. Fill it in. (*Hint:* The first entry is 6 and the last is 0.)

b) If the null hypothesis, H_0: $\eta = \eta_0$, is true, what is the probability that a sample will match any particular row of the table? (*Hint:* The answer is the same for each row.)

Rank			
1	2	3	W
+	+	+	
+	+	−	
+	−	+	
+	−	−	
−	+	+	
−	+	−	
−	−	+	
−	−	−	

c) Use the answer from part (b) to find the probability distribution of the random variable W when $n = 3$.

d) Draw a histogram for the probability distribution of W when $n = 3$.

e) Use your histogram from part (d) to obtain the critical value for a left-tailed Wilcoxon signed-rank test with a sample size of $n = 3$ and a significance level of $\alpha = 0.125$.

f) Compare your critical value from part (e) with the critical value in Table VII.

g) Repeat parts (a)–(f) for a Wilcoxon signed-rank test with $n = 4$.

15.4   The Mann-Whitney test

In the previous section we discussed the Wilcoxon signed-rank test. That nonparametric test is used to make statistical inferences for the median or mean of one population. Now we will examine a nonparametric test for comparing the medians or means of two populations using independent samples. The test is usually called the **Mann-Whitney test** (or **Wilcoxon rank-sum test**). We introduce the Mann-Whitney test in Example 15.9.

EXAMPLE 15.9   *Introduces the Mann-Whitney test*

A nationwide shipping firm purchased a new computer system to keep track of the present status of all its current shipments, pickups, and deliveries. The system was linked to computer terminals in all regional offices where office personnel could type in requests for information on the location of shipments and get answers immediately on display screens.

   The company had to set up a training program for use of the computer terminals and decided to hire a technical writer to write a short self-study manual for that purpose. The manual was designed so that a person could read it and be ready to use the computer terminal in two hours.

   It was discovered that, in practice, the manual took very little time for some employees and quite a bit of time for others. Someone suggested that the reason for this might be that some employees had previous experience with computers and others did not. To test this suggestion, independent samples of employees with and without computer experience were randomly selected. The times, in minutes, required for the employees to complete the manual are displayed in Table 15.14.

TABLE 15.14
Times, in minutes, required for completion of the computer manual

Without experience	With experience
139	142
118	109
164	130
151	107
182	155
140	88
134	95
	104

At the 5% significance level, do the data provide sufficient evidence to conclude that the median completion time for all employees without computer experience exceeds the median completion time for all employees with computer experience?

SOLUTION   Let η_1 and η_2 denote the median completion times for all employees without computer experience and with computer experience, respectively. Then the null and alternative hypotheses are

H_0: $\eta_1 = \eta_2$ (median time for inexperienced employees is not greater)

H_a: $\eta_1 > \eta_2$ (median time for inexperienced employees is greater)

   To apply the Mann-Whitney test, we first rank all the data from both samples combined. The result of doing that is depicted in Table 15.15.

TABLE 15.15
Result of ranking
the combined data
from Table 15.14

Without experience	Overall rank	With experience	Overall rank
139	9	142	11
118	6	109	5
164	14	130	7
151	12	107	4
182	15	155	13
140	10	88	1
134	8	95	2
		104	3

The table shows, for example, that the first employee without computer experience had the ninth shortest completion time among all 15 employees in the sample.

The idea behind the Mann-Whitney test is a simple one: If the sum of the ranks for the sample of employees without experience is too large, then we take this as evidence that the null hypothesis is false and conclude that the median completion time for all employees without experience exceeds that for all employees with experience. From Table 15.15, we see that the sum of the ranks for the sample of employees without experience, denoted by M, is

$$M = 9 + 6 + 14 + 12 + 15 + 10 + 8 = 74$$

Of course, to decide whether this value of M is large enough to warrant rejection of the null hypothesis, we need a table of critical values for the random variable M. We will discuss such a table, Table IX in the appendix, and then return to complete the hypothesis test considered here.   ■

USING THE MANN-WHITNEY TABLE

The test statistic, M, for a Mann-Whitney test is the sum of the ranks associated with the smaller sample size. For instance, in Example 15.9, the smaller sample size is the one for the employees without computer experience (see Table 15.14). Thus, in that example, the test statistic, M, is the sum of the ranks for the sample of employees without computer experience.

It is convenient to arrange things so that the sample size for Population 1 is less than or equal to that for Population 2. This can always be accomplished by interchanging the roles of the populations, if necessary. Therefore, *we will assume that the populations have been designated so that the sample size for Population 1 is less than or equal to the sample size for Population 2; that is, $n_1 \leq n_2$.* Under that assumption, the test statistic, M, for a Mann-Whitney test is the sum of the ranks for the sample from Population 1.

Tables VIII and IX in the appendix provide critical values for a Mann-Whitney test. Table VIII supplies the critical values for a one-tailed test with $\alpha = 0.025$ or a two-tailed test with $\alpha = 0.05$. Table IX supplies the critical values for a one-tailed test with $\alpha = 0.05$ or a two-tailed test with $\alpha = 0.10$. [Actually these are only approximate significance levels, but they are considered close enough in practice.]

It should be noted that *a critical value from Table VIII or Table IX is to be included in the rejection region.*

Let us illustrate the use of Table IX. Suppose, for instance, that the sample sizes are $n_1 = 7$ and $n_2 = 8$, and that we want to obtain the critical value for a right-tailed test with $\alpha = 0.05$. First, we locate the column labeled 7 ($n_1 = 7$) along the top of the table. Then we go down that column until we are in the row labeled 8 ($n_2 = 8$) along the side of the table. There we find two numbers, 41 and 71. Those numbers are the critical values for a left-tailed and right-tailed test, respectively. Thus, the critical value for a right-tailed Mann-Whitney test with $n_1 = 7$, $n_2 = 8$, and $\alpha = 0.05$ is 71. In other words, the null hypothesis is rejected if $M \geq 71$ and is not rejected if $M < 71$.

THE MANN-WHITNEY TEST FOR TWO POPULATION MEDIANS OR TWO POPULATION MEANS

We now present a step-by-step procedure for performing a Mann-Whitney test. That test can be used to compare two population medians or two population means. We will state the procedure in terms of population medians. To employ the procedure for population means, simply replace η_1 by μ_1 and η_2 by μ_2.

PROCEDURE 15.3

To perform a Mann-Whitney test for two population medians with null hypothesis H_0: $\eta_1 = \eta_2$.[†]

ASSUMPTIONS

1. Independent samples.
2. Populations have the same shape.
3. $n_1 \leq n_2$.

STEP 1 *State the null and alternative hypotheses.*
STEP 2 *Decide on the significance level, α.*
STEP 3 *The critical value(s)*
    a) *for a two-tailed test are M_ℓ and M_r.*
    b) *for a left-tailed test is M_ℓ.*
    c) *for a right-tailed test is M_r.*

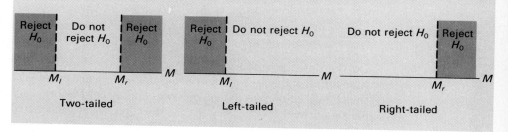

[†] In some texts the null hypothesis is "the two populations have the same distribution." We have stated that hypothesis in terms of medians to maintain consistency with previous sections.

STEP 4 *Construct a work table of the form:*

Sample from Population 1	Overall rank	Sample from Population 2	Overall rank
.	.	.	.
.	.	.	.
.	.	.	.

STEP 5 *Compute the value of the test statistic*

$M = $ *sum of the ranks for sample data from Population 1*

STEP 6 *If the value of the test statistic falls in the rejection region, then reject H_0; otherwise, do not reject H_0.*

STEP 7 *State the conclusion in words.*

EXAMPLE 15.10   *Illustrates Procedure 15.3*

We will now apply Procedure 15.3 to perform the hypothesis test considered in Example 15.9. Samples of employees with and without computer experience were timed to see how long it would take them to complete a self-study manual for the use of computer terminals. The times, in minutes, are repeated in Table 15.16.

TABLE 15.16
Times, in minutes, required for completion of the computer manual

Without experience	With experience
139	142
118	109
164	130
151	107
182	155
140	88
134	95
	104

Do the data provide sufficient evidence to conclude that the median completion time for all employees without computer experience exceeds the median completion time for all employees with computer experience? Perform the hypothesis test at the 5% significance level. [Assume that the completion-time distributions for employees with and without experience have the same shape.]

SOLUTION   We apply Procedure 15.3.

STEP 1 *State the null and alternative hypotheses.*

Let η_1 and η_2 denote the median completion times for all employees without and with computer experience, respectively. Then the null and alternative hypotheses for the hypothesis test are

H_0: $\eta_1 = \eta_2$ (median time for inexperienced employees is not greater)
H_a: $\eta_1 > \eta_2$ (median time for inexperienced employees is greater)

Note that the hypothesis test is right-tailed since there is a greater-than sign ($>$) in the alternative hypothesis.

STEP 2 *Decide on the significance level, α.*

We are to perform the hypothesis test at the 5% significance level; therefore, $\alpha = 0.05$.

STEP 3 *The critical value for a right-tailed test is M_r.*

We have $n_1 = 7$, $n_2 = 8$, and $\alpha = 0.05$. Since the hypothesis test is right-tailed, we consult Table IX to obtain the critical value. We find that the critical value is $M_r = 71$. See Figure 15.8.

FIGURE 15.8

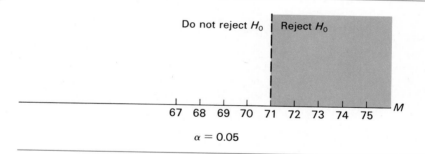

$\alpha = 0.05$

STEP 4 *Construct a work table.*

We have already done this in Table 15.15 on page 776.

STEP 5 *Compute the value of the test statistic*

$$M = \text{sum of the ranks for sample data from Population 1}$$

Referring to the second column of Table 15.15, we find that

$$M = 9 + 6 + 14 + 12 + 15 + 10 + 8 = 74$$

STEP 6 *If the value of the test statistic falls in the rejection region, reject H_0; otherwise, do not reject H_0.*

From Step 5, the value of the test statistic is $M = 74$. Figure 15.8 shows that this falls in the rejection region. Thus, we reject H_0.

STEP 7 *State the conclusion in words.*

The data provide sufficient evidence to conclude that the median completion time for all employees without computer experience exceeds the median completion time for all employees with computer experience.

MTB

When there are ties in the sample data, ranks are assigned in the same way as in the Wilcoxon signed-rank test. Namely, *if two or more data values are tied, each is assigned the mean of the ranks they would have had if there were no ties.* The next example gives a further illustration of the Mann-Whitney test and also provides a case in which ties occur.

EXAMPLE 15.11   *Illustrates Procedure 15.3*

The U.S. National Center for Health Statistics collects information on the ages of Americans at the time of their first marriage. Data are published in *Vital Statistics of the United States.* Independent random samples of 10 married males and 10 married females yielded the ages at first marriage shown in Table 15.17.

TABLE 15.17
Ages at first marriage

Male	Female
27	20
23	19
19	28
36	23
33	23
22	27
27	22
21	34
22	31
41	23

Can we conclude that the median ages at the time of first marriage are different for men and women? Perform a Mann-Whitney test at the 5% significance level. [Assume that the distribution of ages at first marriage for males has the same shape as the distribution of ages at first marriage for females.]

SOLUTION   We apply Procedure 15.3.

STEP 1 *State the null and alternative hypotheses.*

Let η_1 and η_2 denote the median ages at first marriage for males and females, respectively. Then the null and alternative hypotheses are

$$H_0: \eta_1 = \eta_2 \text{ (median ages are the same)}$$
$$H_a: \eta_1 \neq \eta_2 \text{ (median ages are different)}$$

Note that the hypothesis test is two-tailed since there is a not-equal sign ($\neq$) in the alternative hypothesis.

STEP 2 *Decide on the significance level, α.*

The test is to be performed at the 5% significance level; so, $\alpha = 0.05.$

STEP 3 *The critical values for a two-tailed test are M_ℓ and M_r.*

We have $n_1 = 10$ and $n_2 = 10$. Since the hypothesis test is two-tailed and $\alpha = 0.05$, we use Table VIII to obtain the critical values. From that table, we find that $M_\ell = 79$ and $M_r = 131$. See Figure 15.9.

FIGURE 15.9

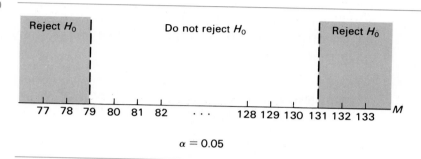

$$\alpha = 0.05$$

STEP 4 *Construct a work table.*

From Table 15.17, we obtain the rankings given in Table 15.18.

TABLE 15.18

Male	Overall rank	Female	Overall rank
27	13	20	3
23	9.5	19	1.5
19	1.5	28	15
36	19	23	9.5
33	17	23	9.5
22	6	27	13
27	13	22	6
21	4	34	18
22	6	31	16
41	20	23	9.5

Note that there are several ties in the data. For instance, there are two 19s tied for smallest; so each is assigned the rank $(1+2)/2 = 1.5$. As another instance, there are three 22s tied for fifth smallest; hence each is assigned the rank $(5 + 6 + 7)/3 = 6$.

STEP 5 *Compute the value of the test statistic*

$$M = \text{sum of the ranks for sample data from Population 1}$$

The second column of Table 15.18 shows that the value of the test statistic is

$$M = 13 + 9.5 + 1.5 + 19 + 17 + 6 + 13 + 4 + 6 + 20 = \boxed{109}$$

STEP 6 *If the value of the test statistic falls in the rejection region, reject H_0; otherwise, do not reject H_0.*

From Step 5, $M = 109$. Referring to Figure 15.9, we see that this value does not fall in the rejection region. Consequently, we do not reject H_0.

STEP 7 *State the conclusion in words.*

The data do not provide sufficient evidence to conclude that there is a difference in the median ages at first marriage for males and females.   ■

COMPARISON OF THE MANN-WHITNEY TEST AND THE POOLED-t TEST

As we know, the Mann-Whitney test can be used to compare the medians or means of two populations whose distributions have the same shape. Recall that the standard deviation of a normal distribution determines its shape. So, two normal distributions with equal standard deviations have the same shape. Consequently, the Mann-Whitney test can be used to compare the means of two normally distributed populations with equal standard deviations.

In Chapter 10 we studied a parametric test for comparing the means of two normally distributed populations with equal standard deviations; namely, the pooled-t test (Procedure 10.3, pages 489–490). Since the pooled-t test is designed specifically for normal populations, we would expect it to be better than the Mann-Whitney test when the populations being sampled are indeed normal. This is true, although the Mann-Whitney test is almost as good as the pooled-t test in the normal case.

On the other hand, if the populations being sampled are not normally distributed, then the Mann-Whitney test is generally better than the pooled-t test, and is frequently much better. Thus, we present the following recommendations:

KEY FACT 15.4   **Mann-Whitney test versus the pooled-t test**

Suppose that a hypothesis test is to be performed to compare the means of two populations whose distributions have the same shape. If you are sure that the populations are normally distributed, use the pooled-t test; otherwise, use the Mann-Whitney test.

USING THE COMPUTER (OPTIONAL)

Procedure 15.3 on pages 777–778 gives a method for performing a Mann-Whitney test to compare the medians of two populations whose distributions have the same shape. The corresponding Minitab program is called **MANNWHITNEY**. Example 15.12 shows how MANNWHITNEY is applied.

EXAMPLE 15.12   *Illustrates the MANNWHITNEY command*

Refer to Example 15.11 on page 780. Use Minitab to perform the hypothesis test considered in that example.

SOLUTION Let η_1 and η_2 denote the median ages at first marriage for males and females, respectively. Then the problem is to use the Mann-Whitney test to perform the hypothesis test

$$H_0: \eta_1 = \eta_2 \text{ (median ages are the same)}$$
$$H_a: \eta_1 \neq \eta_2 \text{ (median ages are different)}$$

at the 5% significance level. Note that the hypothesis test is two-tailed since there is a not-equal sign ($\neq$) in the alternative hypothesis.

To apply MANNWHITNEY, we first enter the sample data from Table 15.17 into C3 and C4 using the SET command and name C3 "Male" and C4 "Female" using the NAME command. Then we type the command MANNWHITNEY, followed by the storage locations of the sample data. That is, we type

MANNWHITNEY for 'MALE' vs 'FEMALE'

No subcommand is required since the test is two-tailed. See Printout 15.7.

PRINTOUT 15.7
Minitab output for
MANNWHITNEY

```
MTB > SET C3
DATA> 27 23 19 36 33 22 27 21 22 41
DATA> END
MTB > SET C4
DATA> 20 19 28 23 23 27 22 34 31 23
DATA> END
MTB > NAME C3 'MALE' C4 'FEMALE'
MTB > MANNWHITNEY for 'MALE' vs 'FEMALE'

Mann-Whitney Confidence Interval and Test

MALE       N =  10     Median =      25.00
FEMALE     N =  10     Median =      23.00
Point estimate for ETA1-ETA2 is       0.50
95.5 pct c.i. for ETA1-ETA2 is (-4.00,8.00)
W = 109.0
Test of ETA1 = ETA2  vs.  ETA1 n.e. ETA2 is significant at 0.7913
The test is significant at 0.7899 (adjusted for ties)

Cannot reject at alpha = 0.05
```

As we see from Printout 15.7, the output first gives the sample sizes and the sample medians for the two samples. Then it provides a point estimate and confidence interval for the difference, $\eta_1 - \eta_2$, between the population medians. The confidence level is chosen to be as close to 95% as possible.

Next we find the value of the test statistic,

$$M = \text{sum of the ranks for sample data from Population 1,}$$

for a Mann-Whitney test (Minitab uses W instead of M). Consequently, we see that $M = 109$.

Below the value of the test statistic is a statement of the null and alternative hypotheses and an approximate P-value:

```
Test of ETA1 = ETA2 vs. ETA1 n.e. ETA2 is significant at 0.7913
```

Minitab employs a normal approximation with a continuity correction factor to obtain the approximate P-value. If there are ties in the data, as there are in this example, another approximate P-value is printed:

The test is significant at 0.7899 (adjusted for ties)

This second approximation is usually closer to the actual P-value than the first approximation.

Since the (approximate) P-value is 0.7899, which exceeds the designated significance level of $\alpha = 0.05$, we do not reject H_0. In other words, the data do not provide sufficient evidence to conclude that there is a difference in the median ages at first marriage for males and females.   ■

Exercises 15.4

In each of Exercises 15.41–15.44, use Procedure 15.3 on pages 777–778 to perform the required hypothesis test. For each exercise, assume that the distributions of the two populations have the same shape.

__ **15.41**   A college chemistry teacher is concerned about the detrimental effects of poor math background on his students. He randomly selects 15 students and divides them according to mathematics background. Their semester averages turn out to be the following:

Fewer than two years of high-school algebra	Two or more years of high-school algebra
58	84
81	67
74	65
61	75
64	74
43	92
	83
	52
	81

Do the data provide sufficient evidence to conclude that students with fewer than two years of high-school algebra have a lower median semester average in this teacher's chemistry courses than do students with two or more years of high-school algebra? Perform the appropriate hypothesis test at the 5% significance level.

__ **15.42**   The lifetimes up to major breakdown for two brands of power lawn mowers are to be compared. Independent random samples yield the following data on lifetimes, in years, for the two brands:

Brand 1	Brand 2
2.3	1.9
3.7	3.8
5.9	6.4
6.8	5.6
3.5	4.9

Do the data indicate that there is a difference in median lifetimes between the two brands of power lawn mowers? Perform the required hypothesis test using a significance level of 0.05.

__ 15.43 The U.S. National Center for Education Statistics surveys college and university libraries to obtain information on the number of volumes held. Independent random samples of public and private colleges and universities yield the following data on number of volumes held, in thousands:

Public	Private
79	289
41	953
516	23
15	64
24	117
411	650
265	

Can we conclude that the median number of volumes held by public colleges and universities is less than that held by private colleges and universities? Perform the required hypothesis test at the 5% significance level. (*Note:* The sample size for the public colleges and universities is larger than the sample size for the private colleges and universities.)

__ 15.44 The U.S. Bureau of Labor Statistics gathers data on weekly earnings of full-time wage and salary workers, by sex. Results are published in *Employment and Earnings.* Independent random samples of male and female workers gave the following data on weekly earnings, in dollars:

Male	Female
191	298
812	327
442	334
366	400
475	280
389	487
520	206
728	310
	366
	485

Do the data provide sufficient evidence to conclude that the median weekly earnings of male full-time wage and salary workers exceeds the median weekly earnings of female full-time wage and salary workers? Perform the required hypothesis test using a significance level of 5%.

__ 15.45 A highway official wants to compare two brands of paint used for striping roads. Ten stripes of each paint are run across the highway. The number of months that each stripe lasts is given below.

Brand *A*		Brand *B*	
35.6	36.1	37.2	36.4
37.0	35.8	39.7	37.5
34.9	34.9	37.2	40.5
36.0	38.8	38.8	38.2
36.6	36.5	37.7	36.6

a) Based on the sample data, does there appear to be a difference in mean lasting times between the two paints? Perform a Mann-Whitney test using a significance level of $\alpha = 0.05$.
b) The hypothesis test in part (a) was done previously in Exercise 10.17 on page 496 using the pooled-t procedure. The assumption in that exercise is that the lasting times for both brands of paint are normally distributed and have equal standard deviations. Under those assumptions, why is it permissible to perform a Mann-Whitney test for comparing the means?

__ 15.46 In a packing plant, a machine packs cartons with jars. A salesperson claims the machine she sells will pack faster. To test that claim, the time it takes each machine to pack 10 cartons is recorded. The results, in seconds, are as follows:

New machine		Present machine	
42.0	41.0	42.7	43.6
41.3	41.8	43.8	43.3
42.4	42.8	42.5	43.5
43.2	42.3	43.1	41.7
41.8	42.7	44.0	44.1

a) Do the data suggest that, on the average, the new machine packs faster? Perform a Mann-Whitney test at the 5% significance level.
b) The hypothesis test in part (a) was performed in Exercise 10.18 on page 496 using the pooled-t procedure. The assumption in that exercise is that the packing times for both machines are normally distributed and have equal standard deviations. Under those assumptions, why is it permissible to perform a Mann-Whitney test to compare the means?

Exercises 15.47–15.50 are computer exercises.

___ **15.47 (Computer exercise)** Suppose the data in Exercise 15.43 are stored in columns named PUBLIC and PRIVATE.

a) Which Minitab command and subcommands (if any) should be used to carry out the hypothesis test considered in that exercise?

b) If you have access to Minitab, use it to perform the hypothesis test.

___ **15.48 (Computer exercise)** Suppose the data in Exercise 15.44 are stored in columns named MALE and FEMALE.

a) Which Minitab command and subcommands (if any) should be used to carry out the hypothesis test considered in that exercise?

b) If you have access to Minitab, use it to perform the hypothesis test.

___ **15.49 (Computer exercise)** The research and development (R&D) department of a light-bulb manufacturing company claims to have developed a new bulb that, on the average, will outlast the bulb currently produced. To try to justify the claim, R&D takes independent random samples of the current bulb and the new bulb. After the lifetimes of the bulbs sampled are obtained, Minitab's MANNWHITNEY command is applied to get the output shown in Printout 15.8 at the top of the next page. Determine

a) the null and alternative hypotheses for R&D's hypothesis test.

b) the command and subcommand(s) used.

c) the sample median lifetime for both the current bulb and the new bulb.

d) the number of current bulbs sampled and the number of new bulbs sampled.

e) the value of the test statistic.

f) the approximate P-value of the hypothesis test.

g) the smallest significance level at which the null hypothesis can be rejected.

h) the conclusion if the hypothesis test is performed at the 1% significance level.

i) a 95.5% confidence interval for the difference between the median lifetimes of the current bulb and the new bulb.

j) the assumptions made by R&D in using the Minitab program MANNWHITNEY.

___ **15.50 (Computer exercise)** A psychology professor, teaching at a large university in the northeast, wants to know whether there is a difference between the average IQs of male and female students in attendance. She randomly selects 20 female and 20 male students and has them take IQ tests. Following that, she applies Minitab's MANNWHITNEY command to the resulting data and obtains the output shown in Printout 15.9 on the next page. Determine

a) the null and alternative hypotheses for the professor's hypothesis test.

b) the command and subcommand(s) used.

c) the median IQ of the females sampled and the median IQ of the males sampled.

d) the value of the test statistic.

e) the approximate P-value of the hypothesis test.

f) the smallest significance level at which the null hypothesis can be rejected.

g) the conclusion if the hypothesis test is performed at the 5% significance level.

h) a 95% confidence interval for the difference between the median IQs of female and male students at the university.

i) the assumptions made by the professor in order to use MANNWHITNEY.

Large-sample Mann-Whitney test using the normal approximation: The tables of critical values for the Mann-Whitney test, Tables VIII and IX, stop at $n_1 = 10$ and $n_2 = 10$. For larger samples, a normal approximation can be used. In fact, the normal approximation works well even for sample sizes as small as 10. Specifically, we have the following result:

> Suppose that independent random samples of sizes n_1 and n_2, both at least 10, are to be taken from two populations with medians η_1 and η_2, respectively. Further suppose that the distributions of the two populations have the same shape. Then, if $\eta_1 = \eta_2$, the random variable M is approximately normally distributed with mean $\mu_M = n_1(n_1 + n_2 + 1)/2$ and standard deviation $\sigma_M = \sqrt{n_1 n_2 (n_1 + n_2 + 1)/12}$. Consequently, the standardized random variable
>
> $$(3) \qquad z = \frac{M - n_1(n_1 + n_2 + 1)/2}{\sqrt{n_1 n_2 (n_1 + n_2 + 1)/12}}$$
>
> has approximately the standard normal distribution.

In Exercises 15.51–15.53, we will develop and apply a large-sample procedure for a Mann-Whitney test based on the previous fact.

═ **15.51** Formulate a large-sample procedure for a Mann-Whitney test that uses the test statistic displayed in Equation (3).

PRINTOUT 15.8   Minitab output for Exercise 15.49

Mann-Whitney Confidence Interval and Test

```
CURRENT     N =  20      Median =        1033.5
NEW         N =  10      Median =        1105.0
Point estimate for ETA1-ETA2 is      -72.0
95.5 pct c.i. for ETA1-ETA2 is (-109.0,-25.0)
W = 242.5
Test of ETA1 = ETA2  vs.  ETA1 l.t. ETA2 is significant at 0.0016
The test is significant at 0.0016 (adjusted for ties)
```

PRINTOUT 15.9   Minitab output for Exercise 15.50

Mann-Whitney Confidence Interval and Test

```
FEMALE     N =  20      Median =        121.00
MALE       N =  20      Median =        119.50
Point estimate for ETA1-ETA2 is       2.00
95.0 pct c.i. for ETA1-ETA2 is (-5.00,9.00)
W = 426.0
Test of ETA1 = ETA2  vs.  ETA1 n.e. ETA2 is significant at 0.6750
The test is significant at 0.6747 (adjusted for ties)

Cannot reject at alpha = 0.05
```

═ **15.52**   Refer to Exercise 15.46.

a) Use your procedure from Exercise 15.51 to perform the hypothesis test of Exercise 15.46 at the 0.05 level of significance.

b) Compare your result in part (a) to the one obtained in Exercise 15.46, where the normal approximation was not used.

═ **15.53**   Refer to Exercise 15.45.

a) Use your procedure from Exercise 15.51 to perform the hypothesis test of Exercise 15.45 at the 5% significance level.

b) Compare your result in part (a) to the one obtained in Exercise 15.45, where the normal approximation was not used.

≡ **15.54**   In this exercise, we will obtain the probability distribution of the random variable M, when the sample sizes are $n_1 = 3$ and $n_2 = 3$. This will enable you to see how the critical values for the Mann-Whitney test are derived. We can display all possible ranks for the data by constructing the following table

in which the letter A stands for a member from Population 1 and the letter B stands for a member from Population 2.

Rank						
1	2	3	4	5	6	M
A	A	A	B	B	B	6
A	A	B	A	B	B	7
A	A	B	B	A	B	8
.	.	.	.	.	.	.
.	.	.	.	.	.	.
.	.	.	.	.	.	.
B	B	A	B	A	A	14
B	B	B	A	A	A	15

a) Complete the table. (*Hint:* There are 20 rows.)

b) Use your result from part (a) to find the probability distribution of the random variable M if $\eta_1 = \eta_2$. (*Hint:* Each row of the above table is equally likely.)

c) If $\eta_1 = \eta_2$, draw a histogram for the probability distribution of M when $n_1 = 3$ and $n_2 = 3$.

d) Apply your result from part (b) to obtain the entries in Table IX for M_ℓ and M_r when $n_1 = 3$ and $n_2 = 3$.

≡ **15.55 Wilcoxon signed-rank test for paired differences:** The Mann-Whitney test is a nonparametric test designed to compare the medians or means of two populations having the same shape using *independent samples*. We can also derive a nonparametric test for comparing the medians or means of two populations having the same shape using *paired samples*. Consider, for instance, the situation of Example 10.10 on pages 512–513. A major oil company has developed a new gasoline additive that is supposed to increase mileage. To test that hypothesis, 10 cars are randomly selected. Each car sampled is driven both with and without the additive. The resulting gas mileages, in miles per gallon, are displayed in the second and third columns of the following table:

Car	With additive x_1	Without additive x_2	Paired difference $d = x_1 - x_2$
1	25.7	24.9	
2	20.0	18.8	
3	28.4	27.7	
4	13.7	13.0	
5	18.8	17.8	
6	12.5	11.3	
7	28.4	27.8	
8	8.1	8.2	
9	23.1	23.1	
10	10.4	9.9	

a) Fill in the difference column in the above table.

b) Apply Procedure 15.2 on page 764, the Wilcoxon signed-rank test, to the difference data in order to decide whether, on the average, the gasoline additive improves gas mileage. Perform the hypothesis test at the 5% significance level.

15.5 Rank correlation

In Chapters 12 and 13, we discussed the *linear correlation coefficient, r*. Recall that r is a descriptive measure of the strength of the linear relationship between two variables, x and y. Now we will study the *rank correlation coefficient* which is a nonparametric alternative to the linear correlation coefficient.

The **rank correlation coefficient, r_s,** of a sample of data points is defined to be the linear correlation of the ranks of those data points. Since the rank correlation coefficient was developed by Charles Spearman (1863–1945), it is also called **Spearman's rank correlation coefficient.** The subscript s in r_s is for Spearman.

The rank correlation coefficient, r_s, has properties similar to that of the linear correlation coefficient, r. Specifically, r_s is always between -1 and 1. Positive values of r_s suggest that the variables are **positively correlated,** meaning that y tends to increase as x increases, with the tendency being greater the closer that r_s is to 1. Negative values of r_s suggest that the variables are **negatively correlated,** meaning that y tends to decrease as x increases, with the tendency being greater the closer that r_s is to -1. Values of r_s close to zero suggest that the variables, x and y, are **uncorrelated.**

Recall that the linear correlation coefficient, r, is used to describe the strength of the *linear* relationship between two variables and, as such, should be employed as a descriptive measure only when a scatter diagram indicates that the data points are scattered about a straight line. Thus, for instance, r can be used to measure the strength of a positive linear relationship like the one shown in Figure 15.10(a), but should not be used to measure the strength of a positive nonlinear relationship like the one shown in Figure 15.10(b).

On the other hand, the rank correlation coefficient, r_s, can be used to describe not only the strength of a positive or negative linear relationship between two variables, but also the strength of a positive or negative nonlinear relationship between two variables. Hence, for example, r_s can be used to measure the strength of both the positive linear relationship shown in Figure 15.10(a) and the positive nonlinear relationship shown in Figure 15.10(b).

FIGURE 15.10

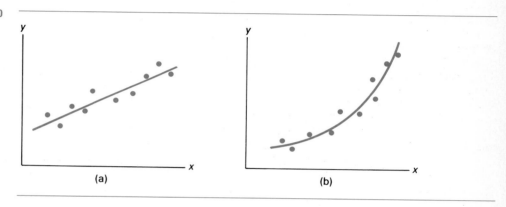

(a)   (b)

We will now formally present the definition of the rank correlation coefficient, r_s, of a set of data points. Following that definition, we will consider an illustrative example.

DEFINITION 15.1   Rank correlation coefficient

The *rank correlation coefficient*, r_s, of n data points is defined to be the linear correlation coefficient, r, of the ranks of those data points. It can be computed using the formula

$$r_s = \frac{S_{uv}}{\sqrt{S_{uu}S_{vv}}}$$

where the u-values represent the ranks of the x-values and the v-values represent the ranks of the y-values; and where $S_{uv} = \Sigma uv - (\Sigma u)(\Sigma v)/n$, $S_{uu} = \Sigma u^2 - (\Sigma u)^2/n$, and $S_{vv} = \Sigma v^2 - (\Sigma v)^2/n$.

EXAMPLE 15.13   **Illustrates Definition 15.1**

A differential equations instructor wants to know whether there is a correlation between students' midterm averages and their final-exam scores. The instructor takes a random sample of nine students from previous differential equations courses and obtains the data displayed in Table 15.19.

TABLE 15.19
Midterm averages
and final-exam scores

Midterm average x	Final-exam score y
60	52
96	97
74	48
86	80
92	86
90	95
72	49
67	71
77	73

The table shows, for instance, that the first student selected had a midterm average of 60 and a final-exam score of 52.

a) Determine the rank correlation coefficient, r_s, of the data.

b) Interpret the value of r_s obtained in part (a) in terms of the relationship between midterm average and final-exam score in differential equations.

SOLUTION   a) According to Definition 15.1, the rank correlation coefficient, r_s, of the data is the linear correlation coefficient, r, of the ranks of the data. Consequently, we begin by ranking the midterm averages among themselves and the final-exam scores among themselves. The result of doing those rankings is displayed below in Table 15.20.

TABLE 15.20
Result of ranking the
x-values and y-values
in Table 15.19
among themselves

Midterm average x	Rank u	Final-exam score y	Rank v
60	1	52	3
96	9	97	9
74	4	48	1
86	6	80	6
92	8	86	7
90	7	95	8
72	3	49	2
67	2	71	4
77	5	73	5

Next we compute the linear correlation coefficient, r, of the ranks; that is, of the (u, v) pairs from the second and fourth columns of Table 15.20. This is accomplished in the usual manner, as described in Section 12.4, by first constructing a table and then applying the formula for the linear correlation coefficient, Formula 12.3 on page 622.

TABLE 15.21
Table for computing the linear correlation coefficient of the ranks

u	v	uv	u^2	v^2
1	3	3	1	9
9	9	81	81	81
4	1	4	16	1
6	6	36	36	36
8	7	56	64	49
7	8	56	49	64
3	2	6	9	4
2	4	8	4	16
5	5	25	25	25
45	45	275	285	285

From the final row of Table 15.21, we see that the linear correlation coefficient, r, of the ranks is

$$r = \frac{S_{uv}}{\sqrt{S_{uu}S_{vv}}} = \frac{\Sigma uv - (\Sigma u)(\Sigma v)/n}{\sqrt{[\Sigma u^2 - (\Sigma u)^2/n][\Sigma v^2 - (\Sigma v)^2/n]}}$$

$$= \frac{275 - (45)(45)/9}{\sqrt{[285 - (45)^2/9][285 - (45)^2/9]}} = 0.833$$

Since the rank correlation coefficient, r_s, of the data in Table 15.19 is, by definition, the linear correlation coefficient, r, of the ranks, we see that $r_s = 0.833$.

b) The rank correlation coefficient of 0.833, obtained in part (a), suggests that there is a strong positive correlation between midterm average and final-exam score in differential equations courses. ∎

MTB

A SHORTCUT FORMULA FOR THE RANK CORRELATION COEFFICIENT

We see from Example 15.13 that the calculation of the rank correlation coefficient, r_s, is quite time consuming. Fortunately, there is an alternative way to compute r_s that is somewhat quicker and easier. We present this as Formula 15.1.

FORMULA 15.1 Shortcut formula for r_s

The rank correlation coefficient of n data points can be computed using the shortcut formula

$$r_s = 1 - \frac{6\Sigma d^2}{n(n^2 - 1)}$$

where the d-values denote the paired differences of the ranks; i.e., $d = u - v$.

Note: Strictly speaking, the shortcut formula applies only when there are no ties in the data. However, in practice, the shortcut formula provides a reasonable approximation to the value of r_s as long as the number of ties is not large relative to the sample size. See Exercises 15.73–15.76.

EXAMPLE 15.14   *Illustrates Formula 15.1*

We will illustrate the use of the shortcut formula for r_s by applying it to the data considered in Example 15.13. The data on midterm averages and final-exam scores are repeated in the first and third columns of Table 15.22. Use the shortcut formula, Formula 15.1, to compute the rank correlation coefficient, r_s.

SOLUTION   To use the shortcut formula, we first rank the midterm averages among themselves and the final-exam scores among themselves. See the first four columns of Table 15.22.

TABLE 15.22
Table for computing r_s using the shortcut formula

Midterm average x	Rank u	Final-exam score y	Rank v	Difference of ranks $d = u - v$	d^2
60	1	52	3	-2	4
96	9	97	9	0	0
74	4	48	1	3	9
86	6	80	6	0	0
92	8	86	7	1	1
90	7	95	8	-1	1
72	3	49	2	1	1
67	2	71	4	-2	4
75	5	73	5	0	0
					20

Next we compute the paired differences, $d = u - v$, of the ranks (fifth column) and the squares, d^2, of those paired differences (sixth column). Finally, we apply Formula 15.1. We have $n = 9$ and, from the last column of Table 15.22, $\Sigma d^2 = 20$. Consequently,

$$r_s = 1 - \frac{6 \Sigma d^2}{n(n^2 - 1)} = 1 - \frac{6 \cdot 20}{9(9^2 - 1)} = \boxed{0.833}$$

Note that this agrees with the value of r_s obtained in Example 15.13, where the defining formula was used. However, the calculation of r_s using the shortcut formula is considerably easier.   ■

INFERENCES IN RANK CORRELATION

Up to this point we have employed the rank correlation coefficient, r_s, as a descriptive measure of the correlation between two variables, x and y. For example, the rank correlation coefficient of the data on midterm averages and final-exam scores is $r_s = 0.833$. This seems to indicate that the variables, midterm average, x, and final-exam score, y, are positively correlated.

However, the data constitute only a sample from the population of midterm-average and final-exam-score pairs for all differential equations students. Thus, a hypothesis test must be performed in order to decide whether the value, 0.833, for the rank correlation coefficient is large enough to conclude that the variables, midterm average and final-exam score, are positively correlated. We will call such a hypothesis test a **rank-correlation test.**

The null hypothesis for a rank-correlation test is that the two variables are uncorrelated. To carry out the hypothesis test, we use r_s as the test statistic and employ Table X in the appendix to obtain the critical value(s). Specifically, we have the following procedure:

PROCEDURE 15.4

To perform a rank-correlation test with null hypothesis
H_0: **The variables are uncorrelated.**

STEP 1 *State the null and alternative hypotheses.*
STEP 2 *Decide on the significance level, α.*
STEP 3 *The critical value(s)*
    a) *for a two-tailed test are $\pm r_{s,\alpha/2}$.*
    b) *for a left-tailed test is $-r_{s,\alpha}$.*
    c) *for a right-tailed test is $r_{s,\alpha}$.*
    Use Table X to find the critical value(s).

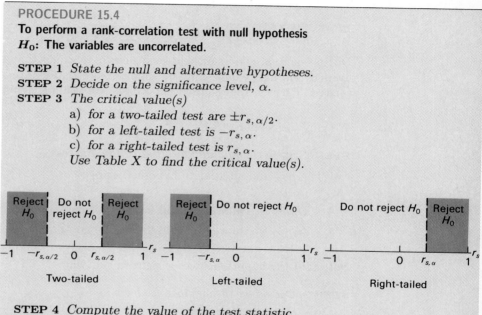

Two-tailed   Left-tailed   Right-tailed

STEP 4 *Compute the value of the test statistic*

$$r_s = 1 - \frac{6\Sigma d^2}{n(n^2 - 1)}$$

STEP 5 *If the value of the test statistic falls in the rejection region, then reject H_0; otherwise, do not reject H_0.*
STEP 6 *State the conclusion in words.*

EXAMPLE 15.15   *Illustrates Procedure 15.4*

As a first application of Procedure 15.4, we return to the data on midterm averages and final-exam scores from Example 15.13. A random sample of nine students who have previously taken differential equations yields the data on midterm averages and final-exam scores given in Table 15.19. We repeat that table here as Table 15.23.

TABLE 15.23
Midterm averages
and final-exam scores

Midterm average x	Final-exam score y
60	52
96	97
74	48
86	80
92	86
90	95
72	49
67	71
77	73

Do the data provide sufficient evidence to conclude that the variables, midterm average and final-exam score, for differential equations students are positively correlated? Perform the required hypothesis test using a significance level of 0.05.

SOLUTION   We apply Procedure 15.4.

STEP 1 *State the null and alternative hypotheses.*

The null and alternative hypotheses are

> H_0: Midterm average and final-exam score are uncorrelated.
> H_a: Midterm average and final-exam score are positively correlated.

Note that the hypothesis test is right-tailed (why?).

STEP 2 *Decide on the significance level, α.*

We are to use $\alpha = 0.05.$

STEP 3 *The critical value for a right-tailed test is $r_{s,\alpha}$.*

We have $n = 9$ and $\alpha = 0.05$. Consulting Table X, we find that the critical value is $r_{s,0.05} = 0.600$. See Figure 15.11.

FIGURE 15.11

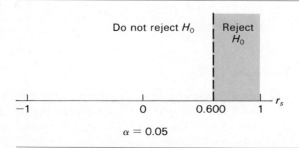

$\alpha = 0.05$

STEP 4 *Compute the value of the test statistic*

$$r_s = 1 - \frac{6\Sigma d^2}{n(n^2 - 1)}$$

We have already computed r_s in Example 15.14 on page 792 where we found that $r_s = 0.833.$

STEP 5 *If the value of the test statistic falls in the rejection region, reject H_0; otherwise, do not reject H_0.*

The value of the test statistic, found in Step 4, is $r_s = 0.833$. This value falls in the rejection region, as we can see by referring to Figure 15.11. Consequently, we reject H_0.

STEP 6 *State the conclusion in words.*

The sample data in Table 15.23 provide sufficient evidence to conclude that the variables, midterm average and final-exam score, for differential equations students are positively correlated. In other words, we can conclude that differential equations students with high midterm averages tend to have higher final-exam scores than those with lower midterm averages. ∎

DEALING WITH TIES

Frequently we encounter ties in the data. There may be ties among the x-values or ties among the y-values. In such situations, ranks are assigned in the manner discussed previously: *If two or more data values are tied, then each is assigned the mean of the ranks they would have had if there were no ties.*

COMPARISON OF THE RANK-CORRELATION TEST AND THE LINEAR-CORRELATION TEST

In Chapter 13, we studied a parametric test for correlation; namely, the linear-correlation test, Procedure 13.5 on pages 676–677. The requirements for the use of that test are the Assumptions 1–3 for regression inferences given on pages 644–645. If those assumptions are met, then, as we would expect, the linear-correlation test is better than the rank-correlation test. What is somewhat surprising is that the rank-correlation test works almost as well.

On the other hand, if one or more of the assumptions for regression inferences does not strictly hold, then the rank-correlation test is generally better than the linear-correlation test, and is oftentimes considerably better. In summary, we have the following key fact:

KEY FACT 15.5   Rank-correlation test versus the linear-correlation test

Suppose a hypothesis test is to be performed for the correlation between two variables. If you are sure that the assumptions for regression inferences are satisfied, use the linear-correlation test; otherwise, use the rank-correlation test.

Exercises 15.5

___ **15.56** True or False: The rank correlation coefficient, r_s, can be used to measure the strength of a positive or negative nonlinear relationship between two variables.

___ **15.57** True or False: A negative value of the rank correlation coefficient suggests that the variable, y, tends to decrease linearly as the variable, x, increases.

___ **15.58** For each of the following scatter diagrams, decide whether the rank correlation coefficient, r_s, is positive, negative, or zero:

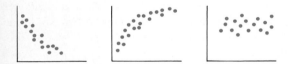

___ **15.59** For each of the following scatter diagrams, decide whether the rank correlation coefficient, r_s, is positive, negative, or zero:

___ **15.60** Colleges and universities often require prospective students to take college entrance examinations. The theory is that scores on those examinations are positively correlated with success in college. In the table at the top of the next column are the mathematics percentile scores on a college entrance examination and the freshman GPAs for a random sample of 10 students attending a southeastern university.

a) Compute the rank correlation coefficient using the defining formula, Definition 15.1 on page 789.
b) Compute the rank correlation coefficient using the shortcut formula, Formula 15.1 on page 791, and compare your answer to the one found in part (a).
c) Interpret the value of r_s in terms of the relationship between math percentile score and freshman GPA.

Math percentile score x	Freshman GPA y
85	3.23
77	2.95
98	3.70
86	3.85
66	2.37
55	2.75
75	2.73
64	2.16
71	2.59
82	2.78

___ **15.61** A social psychologist is studying the association between alcohol consumption patterns and IQ for people who drink alcoholic beverages. A random sample of nine people who drink, at least occasionally, yields the following data:

IQ x	Drinks per month y
121	15
122	6
112	55
108	23
94	38
98	30
120	73
109	25
105	10

a) Compute the rank correlation coefficient using the defining formula, Definition 15.1 on page 789.
b) Compute the rank correlation coefficient using the shortcut formula, Formula 15.1 on page 791, and compare your answer to the one found in part (a).
c) Interpret the value of r_s in terms of the relationship between the variables, IQ and number of drinks per month, for people who drink alcoholic beverages.

In Exercises 15.62 and 15.63, compute the rank correlation coefficient of the data using the shortcut formula. Interpret your answers.

__ 15.62 The following table is reprinted by permission from the October issue of *Science 85*. Copyright © 1985 by the American Association for the Advancement of Science.

DO YOUR WORRIES MATCH THOSE OF THE EXPERTS?

Experts and lay people were asked to rank the risk of dying in any year from various activities and technologies. The experts' ranking closely matches known fatality statistics.

PUBLIC		EXPERTS
1	Nuclear power	20
2	Motor vehicles	1
3	Handguns	4
4	Smoking	2
5	Motorcycles	6
6	Alcoholic beverages	3
7	General (private) aviation	12
8	Police work	17
9	Pesticides	8
10	Surgery	5
11	Fire fighting	18
12	Large construction	13
13	Hunting	23
14	Spray cans	26
15	Mountain climbing	29
16	Bicycles	15
17	Commercial aviation	16
18	Electric power (nonnuclear)	9
19	Swimming	10
20	Contraceptives	11
21	Skiing	30
22	X rays	7
23	High school and college football	27
24	Railroads	19
25	Food preservatives	14
26	Food coloring	21
27	Power mowers	28
28	Prescription antibiotics	24
29	Home appliances	22
30	Vaccinations	25

__ 15.63 Two wine connoisseurs are asked to rank eight different wines from 1 (most favorable) to 8 (least favorable). The results are as follows:

Wine	Judge I	Judge II
A	5	6
B	7	4
C	1	2
D	3	5
E	2	1
F	6	7
G	4	3
H	8	8

In each of Exercises 15.64–15.69, use Procedure 15.4 on page 793 to perform the required hypothesis test.

__ 15.64 Refer to Exercise 15.60. At the 5% significance level, do the data provide sufficient evidence to conclude that the variables, math percentile score and freshman GPA, are positively correlated?

__ 15.65 Refer to Exercise 15.61. Do the data provide sufficient evidence to conclude that there is a correlation between the variables, IQ and number of drinks per month, for people who drink alcoholic beverages? Use $\alpha = 0.05$.

__ 15.66 A random sample of 11 nations yielded the following data on area and population. [SOURCE: *Reader's Digest 1985 Almanac and Yearbook.*]

Country	Area (sq. mi.)	Population
India	1,269,346	754,225,000
San Marino	24	23,200
Transkei	16,070	3,345,670
Britain	94,515	57,621,900
Syria	71,498	10,314,100
Jamaica	4,244	2,408,300
Lesotho	11,720	1,492,400
South Yemen	128,560	2,259,790
Malaysia	127,317	15,498,600
Netherlands	15,770	14,466,000
Argentina	1,068,301	30,570,500

Do the data provide sufficient evidence to conclude that there is a correlation between the variables, area and population, for nations of the world? Perform the hypothesis test at the 1% significance level.

__ 15.67 Refer to Exercise 15.63. Do the data provide sufficient evidence to conclude that there is a positive correlation between the tastes of the two wine connoisseurs? Take $\alpha = 0.05$.

__ **15.68** The following data on horsepower and gas mileage were obtained from a sample of 12 makes of automobiles. [SOURCE: *Consumer Reports.*]

Horsepower x	MPG y
155	16.9
68	30.0
95	27.5
97	27.4
125	17.0
115	21.6
110	18.6
120	18.1
68	34.1
80	27.4
70	34.2
78	30.5

Do the data provide sufficient evidence to conclude that the variables, horsepower and gas mileage, are negatively correlated? Use $\alpha = 0.05$.

__ **15.69** Below are data on birth rates and literacy rates for a random sample of 10 nations of the world. The birth rates are per 1000 population and the literacy rates represent the percent of the population over 15 years old that can read and write. [SOURCE: *Reader's Digest 1985 Almanac and Yearbook.*]

Country	Birth rate	Literacy rate
Mauritania	50.2	17%
Saudi Arabia	45.9	25%
Tanzania	46.3	66%
Uruguay	18.6	94%
Antigua	16.5	88%
Switzerland	11.6	99%
Taiwan	26.0	90%
Bahrain	34.4	40%
Liechtenstein	14.3	100%
Kirbata	21.9	90%

At the 1% significance level, do the data provide sufficient evidence to conclude that the variables, birth rate and literacy rate, are negatively correlated?

Large-sample rank-correlation test using the normal approximation: The table of critical values for the rank-correlation test, Table X, stops at $n = 20$. For larger samples, a normal approximation

can be used. In fact, the normal approximation works well even for sample sizes as small as 10. Specifically, we have the following result:

Suppose a random sample of n data points is to be taken from two uncorrelated variables, x and y. Then, for $n \geq 10$, the random variable r_s is approximately normally distributed with mean $\mu_{r_s} = 0$ and standard deviation $\sigma_{r_s} = 1/\sqrt{n-1}$. Thus, the standardized random variable

$$(4) \qquad z = \frac{r_s}{1/\sqrt{n-1}} = \sqrt{n-1}\, r_s$$

has approximately the standard normal distribution.

In Exercises 15.70–15.72, we will develop and apply a large-sample procedure for a rank-correlation test based on the previous fact.

== **15.70** Formulate a large-sample procedure for a rank-correlation test that uses the test statistic displayed in Equation (4).

== **15.71** Refer to Exercise 15.69.
a) Use your procedure from Exercise 15.70 to perform the hypothesis test of Exercise 15.69 with $\alpha = 0.01$.
b) Compare your result in part (a) to the one obtained in Exercise 15.69, where the normal approximation was not used.

== **15.72** Refer to Exercise 15.66.
a) Use your procedure from Exercise 15.70 to perform the hypothesis test of Exercise 15.66 at the 1% significance level.
b) Compare your result in part (a) to the one obtained in Exercise 15.66, where the normal approximation was not used.

We noted on page 791 that although strictly speaking the shortcut formula for r_s applies only when there are no ties in the data, it works reasonably well in practice provided that the number of ties is not large relative to the sample size. In each of Exercises 15.73–15.76, we have presented data sets in which ties occur. For each exercise,
a) *compute r_s by applying the defining formula given in Definition 15.1 on page 789.*
b) *compute r_s by applying the shortcut formula given in Formula 15.1 on page 791.*
c) *remark on the discrepancy between the approximate value of r_s obtained in part (b) and the actual value of r_s obtained in part (a).*

≡ **15.73**   The data on birth rates and literacy rates from Exercise 15.69.

≡ **15.74**   The data on horsepower and gas mileage from Exercise 15.68.

≡ **15.75**   The data on heights and weights for a sample of 18–24-year-old males from Exercise 12.33:

Height (inches)	Weight (lb)
x	y
65	175
67	133
71	185
71	163
66	126
75	198
67	153
70	163
71	159
69	151
69	155

≡ **15.76**   The data on age and price for a sample of Corvettes from Exercise 12.32:

Age (yrs)	Price ($100s)
x	y
6	125
6	115
6	130
2	260
2	219
5	150
4	190
5	163
1	260
4	160

≡ **15.77**   In this exercise, we will derive the shortcut formula, Formula 15.1, for the rank correlation coefficient, r_s. Suppose then that we have n data points, (x_1, y_1), (x_2, y_2), $\ldots$, (x_n, y_n), whose corresponding rank pairs are (u_1, v_1), (u_2, v_2), $\ldots$, (u_n, v_n); that is, the ranks of the x-values, x_1, x_2, $\ldots$, x_n, are u_1, u_2, $\ldots$, u_n, respectively, and the ranks of the y-values, y_1, y_2, $\ldots$, y_n, are v_1, v_2, $\ldots$, v_n, respectively. Further suppose that there are neither ties among the x-values nor ties among the y-values. In deriving Formula 15.1, we will assume as known the following mathematical formulas:

(5) $$1 + 2 + \cdots + n = n(n+1)/2$$

(6) $$1^2 + 2^2 + \cdots + n^2 = n(n+1)(2n+1)/6$$

a) Show that for any two numbers, a and b,

$$ab = \left[a^2 + b^2 - (a-b)^2\right]/2$$

b) Use Formulas (5) and (6) to show that for the ranks, u_1, u_2, $\ldots$, u_n,

$$\Sigma u = n(n+1)/2$$

and

$$\Sigma u^2 = n(n+1)(2n+1)/6$$

and similarly for the ranks, v_1, v_2, $\ldots$, v_n.

c) Use parts (a) and (b) to show that

$$\Sigma uv = n(n+1)(2n+1)/6 - \Sigma d^2/2$$

where $d = u - v$.

d) Use the defining formula for r_s, Definition 15.1 on page 789, and the results of parts (b) and (c) to obtain the shortcut formula for r_s.

15.6   Parametric and nonparametric procedures: a comparison

In the previous sections of this chapter we have discussed a few of the many nonparametric tests. The Wilcoxon signed-rank test of Section 15.3 is the non-parametric alternative to the t-test of Section 9.5; the Mann-Whitney test of Section 15.4 is the nonparametric alternative to the pooled-t test of Section 10.2; and the rank-correlation test of Section 15.5 is the nonparametric alternative to the linear-correlation test of Section 13.4.

There are also nonparametric tests corresponding to the paired-t test of Section 10.4, to the one-way ANOVA test of Section 14.2, and to many other parametric procedures that we have studied. For reference purposes, we present in Table 15.24 some of the more important parametric tests and their nonparametric alternatives.

TABLE 15.24

Type	Parametric test	Nonparametric test
One mean	t-test (Procedure 9.4)	Wilcoxon signed-rank test (Procedure 15.2)
Two means using independent samples	Pooled-t test (Procedure 10.3)	Mann-Whitney test (Procedure 15.3)
Two means using paired samples	Paired-t test (Procedure 10.7)	Wilcoxon signed-rank test for paired differences[†]
Correlation	Linear-correlation test (Procedure 13.5)	Rank-correlation test (Procedure 15.4)
Several means using independent samples	One-way ANOVA (Procedure 14.1)	Kruskal-Wallis test[‡]
Several means using blocked samples	Two-way ANOVA (Procedure 14.2)	Friedman F_r test[‡]

[†] Discussed in Exercise 15.55 on page 788.   [‡] Not discussed in this book.

We should emphasize that Table 15.24 by no means provides a complete list of parametric and nonparametric tests. Its purpose is simply to convey that many parametric tests have corresponding nonparametric alternatives.

CHOOSING BETWEEN PARAMETRIC AND NONPARAMETRIC METHODS

As we mentioned earlier, one important consideration in deciding between a parametric and nonparametric procedure is *distribution type*. Parametric procedures are preferable to nonparametric procedures when the populations being sampled are normally distributed, although the nonparametric procedures usually work almost as well. On the other hand, nonparametric procedures are preferable to parametric procedures when the populations being sampled are not normally distributed.

Another important consideration in deciding between a parametric and nonparametric procedure is *data type*. Parametric procedures are designed for use with metric data and should generally not be used with other types of data. Nonparametric procedures, however, apply to data that are either ordinal or metric.

In summary, we present the following guidelines for choosing between parametric and nonparametric procedures:

KEY FACT 15.6 Choosing between parametric and nonparametric methods

1. If you are sure that the populations are normally distributed, use a parametric procedure.
2. If you are unsure that the populations are normally distributed or if you have reason to suspect that they are not, use a nonparametric procedure.
3. If the population data are not metric, use a nonparametric procedure.

FURTHER CONSIDERATIONS

You may have noted that, although we are favorably inclined toward nonparametric statistics, we have devoted the majority of the book to parametric statistics. There is an explanation for this: The normal distribution played (and continues to play) a fundamental role in the development of inferential statistics. As a consequence, you will probably encounter more parametric than nonparametric procedures in your studies and research.

One important reason for learning statistics is to attain the skills required for understanding and interpreting reports that contain statistical analyses. Many of you will never conduct a real statistical analysis but almost all of you will read analyses that employ such things as a z-score, a linear-correlation test, or a confidence interval for a mean. A knowledge of parametric statistics is therefore required.

Chapter review

KEY TERMS

binomial distribution, 751
distribution-free methods, 748
Mann-Whitney test, 775
MANNWHITNEY,* 782
negatively correlated, 788
nonparametric methods, 749
parametric methods, 749
population median (η), 749
positively correlated, 788

rank correlation coefficient (r_s), 789
rank-correlation test, 793
sign test, 749
STEST,* 757
symmetric distribution, 761
uncorrelated, 788
Wilcoxon rank-sum test, 775
Wilcoxon signed-rank test, 761
WTEST,* 769

FORMULAS

In the formulas below,

η = population median
μ = population mean
n = sample size

η_1, η_2 = population medians
μ_1, μ_2 = population means
n_1, n_2 = sample sizes

Test statistic for a sign test with H_0: $\eta = \eta_0$, 753

$$x = \text{number of "+" signs}$$

where a "+" sign indicates that a data value is larger than η_0.

Test statistic for a Wilcoxon signed-rank test with H_0: $\eta = \eta_0$ or H_0: $\mu = \mu_0$ (symmetric population), 764

$$W = \text{sum of the positive ranks}$$

Test statistic for a Mann-Whitney test with H_0: $\eta_1 = \eta_2$ or H_0: $\mu_1 = \mu_2$ (independent samples, populations have same shape, and $n_1 \leq n_2$), 778

$$M = \text{sum of the ranks for sample data from Population 1}$$

Rank correlation coefficient, 789

$$r_s = \frac{S_{uv}}{\sqrt{S_{uu}S_{vv}}}$$

where the u-values represent the ranks of the x-values and the v-values represent the ranks of the y-values; and where $S_{uv} = \Sigma uv - (\Sigma u)(\Sigma v)/n$, $S_{uu} = \Sigma u^2 - (\Sigma u)^2/n$, and $S_{vv} = \Sigma v^2 - (\Sigma v)^2/n$.

Shortcut formula for r_s, 791

$$r_s = 1 - \frac{6\,\Sigma d^2}{n(n^2 - 1)}$$

where the d-values denote the paired differences of the ranks; i.e., $d = u - v$.

YOU SHOULD BE ABLE TO

1. use and understand the preceding formulas.
2. explain the difference between parametric and nonparametric statistics.
3. perform a sign test for the median of a population.
4. perform a Wilcoxon signed-rank test for the median or mean of a population with a symmetric distribution.
5. perform a Mann-Whitney test to compare the medians or means of two populations whose distributions have the same shape using independent samples.
6. compute and interpret the rank correlation coefficient.
7. perform a rank-correlation test.
8. decide whether it is better to use a parametric or nonparametric procedure in any given situation.
9. use the Minitab commands covered in this chapter.*
10. interpret the output obtained from the application of the Minitab commands discussed in this chapter.*

REVIEW TEST

1. In 1980, the median monthly cost for housing to American homeowners was $365, as reported in *1980 Census of Housing* by the U.S. Bureau of the Census. During that same year, a random sample of 10 homeowners in Indiana reported the following monthly costs:

$346	199	488	292	368
287	191	220	383	278

Use the sign test, with a significance level as close to 0.05 as possible, to decide whether the 1980 median monthly cost for housing to Indiana homeowners was less than the national median of $365.

*2. **(Computer problem)** Suppose that the data in Problem 1 have been stored in a column named HOMECOST.
   a) Which Minitab command and subcommands (if any) should be used to carry out the hypothesis test in that problem?
   b) If you have access to Minitab, use it to perform the hypothesis test.

*3. **(Computer problem)** The output obtained by applying Minitab's STEST command to the data in Problem 1 is displayed in Printout 15.10 on page 805. From the output, determine
   a) the null and alternative hypotheses for the hypothesis test.
   b) the subcommand used, if any.
   c) the number of homeowners sampled.
   d) the number of homeowners sampled for whom the monthly cost for housing was less than $365; exactly $365; more than $365.
   e) the median monthly housing cost of the homeowners sampled.
   f) the *P*-value of the hypothesis test.
   g) the smallest significance level at which the null hypothesis can be rejected.
   h) the conclusion if the hypothesis test is performed at the 5% significance level.

4. Each year, manufacturers perform mileage tests on new car models and submit the results to the Environmental Protection Agency. The EPA then tests the vehicles to determine whether the manufacturers are correct. In 1989, one company reported that a particular model equipped with a four-speed manual transmission averaged 29 miles per gallon (mpg) on the highway. Let us suppose that the EPA tested 15 of the cars and obtained the gas mileages given below.

27.3	31.2	29.4	31.6	28.6
30.9	29.7	28.5	27.8	27.3
25.9	28.8	28.9	27.8	27.6

   a) At a significance level as close to 5% as possible, what decision would you make regarding the company's report on the average gas mileage of the car? Perform a Wilcoxon signed-rank test for the mean gas mileage, μ.
   b) In performing the hypothesis test of part (a), what assumption are you making about the distribution of the gas mileages?
   c) In Problem 8 of the Review Test for Chapter 9, we performed the hypothesis test in part (a) using a *t*-test. The assumption in that problem is that the gas mileages are normally distributed. Under that assumption, why is it permissible to perform a Wilcoxon signed-rank test for the mean gas mileage?

*5. **(Computer problem)** Suppose that the data in Problem 4 have been stored in a column named MILEAGES.
   a) Which Minitab command and subcommands (if any) should be used to carry out the hypothesis test in that problem?
   b) If you have access to Minitab, use it to perform the hypothesis test.

*6. **(Computer problem)** We have displayed in Printout 15.11 on page 805 the computer output obtained by applying Minitab's WTEST command to the data in Problem 4. Employ the output to determine
   a) the null and alternative hypotheses for the hypothesis test.
   b) the subcommand used, if any.
   c) the number of cars tested.
   d) the number of cars tested whose gas mileage is exactly 29 mpg.
   e) the *P*-value of the hypothesis test.
   f) the smallest significance level at which the null hypothesis can be rejected.
   g) the conclusion if the hypothesis test is performed at the 5% significance level.

7. Refer to Problem 4. Assume that the gas mileages are normally distributed.
   a) Would it be acceptable to use the sign test to perform the hypothesis test for the mean gas mileage? Explain.
   b) Of the t-test, Wilcoxon signed-rank test, and sign test, which is preferred for a hypothesis test concerning the mean gas mileage? Why?

8. The U.S. Immigration and Naturalization Service collects information on various characteristics of naturalized aliens. Data are published in *Statistical Yearbook*. Independent random samples of this year's naturalized males and females yielded the following data on age, in years:

Male	Female
73	33
25	34
31	35
47	10
33	13
39	63
20	43
47	33
19	20
	30

   a) Do the data provide sufficient evidence to conclude that the median age of this year's naturalized alien males exceeds the median age of this year's naturalized alien females? Apply the Mann-Whitney test using a significance level of $\alpha = 0.05$.
   b) In performing the hypothesis test of part (a), what assumption are you making about the age distributions of this year's naturalized alien males and females?
   c) If the age distributions of this year's naturalized alien males and females are both normal, would it be better to use the pooled-t test instead of the Mann-Whitney test? Why?

*9. (**Computer problem**) Suppose that the data in Problem 8 are stored in columns named MALE and FEMALE.
   a) Which Minitab command and subcommands (if any) should be used to carry out the hypothesis test in that problem?
   b) If you have access to Minitab, use it to perform the hypothesis test.

*10. (**Computer problem**) The Minitab command, MANNWHITNEY, was applied to the age data, displayed in Problem 8, obtained from samples of this year's naturalized males and females. The resulting output is shown in Printout 15.12 on page 805. Use the computer output to determine
   a) the null and alternative hypotheses for the hypothesis test.
   b) the subcommands used, if any.
   c) the median age of the males sampled and the median age of the females sampled.
   d) the number of males sampled and the number of females sampled.
   e) the value of the test statistic.
   f) the approximate P-value.
   g) the smallest significance level at which the null hypothesis can be rejected.
   h) the conclusion if the hypothesis test is performed at the 5% significance level.
   i) a point estimate for the difference, $\eta_1 - \eta_2$, between the median age of this year's naturalized alien males and the median age of this year's naturalized alien females.
   j) a 95.5% confidence interval for the difference, $\eta_1 - \eta_2$, between the median age of this year's naturalized alien males and the median age of this year's naturalized alien females.
   k) the assumptions made in using the Minitab program MANNWHITNEY.

11. Suppose that independent random samples are taken from two populations in order to compare their means. For each of the following situations, state which hypothesis-testing procedure would be the best one to employ:
   a) The populations are normally distributed and have the same shape.
   b) The populations are normally distributed but may not have the same shape.
   c) The populations have the same shape but may not be normally distributed.
   d) The populations may neither be normally distributed nor have the same shape.

12. The FBI collects data on crime rates, by type, and publishes the results in the document *Crime in the United States*. A random sample of 10 U.S. cities gave the data on murder rates and burglary rates shown in the table at the top of the first column on the next page. The rates are offenses known to the police per 100,000 population.

Murder x	Burglary y	Murder x	Burglary y
25.5	2398	16.0	1870
5.7	912	11.0	1768
17.7	1688	10.6	2381
11.8	2162	19.4	2108
8.2	1083	24.4	2410

Determine the rank correlation coefficient of the data by employing
a) the defining formula.
b) the shortcut formula.

13. Refer to Problem 12. At the 5% significance level, do the data provide sufficient evidence to conclude that the variables, murder rate and burglary rate, for U.S. cities are positively correlated?

PRINTOUT 15.10 Minitab output for Problem 3

SIGN TEST OF MEDIAN = 365.0 VERSUS L.T. 365.0

	N	BELOW	EQUAL	ABOVE	P-VALUE	MEDIAN
HOMECOST	10	7	0	3	0.1719	289.5

PRINTOUT 15.11 Minitab output for Problem 6

TEST OF MEDIAN = 29.00 VERSUS MEDIAN N.E. 29.00

	N	N FOR TEST	WILCOXON STATISTIC	P-VALUE	ESTIMATED MEDIAN
MILEAGES	15	15	48.5	0.532	28.65

PRINTOUT 15.12 Minitab output for Problem 10

Mann-Whitney Confidence Interval and Test

```
MALE      N =   9     Median =      33.00
FEMALE    N =  10     Median =      33.00
Point estimate for ETA1-ETA2 is      5.00
95.5 pct c.i. for ETA1-ETA2 is (-10.99,19.00)
W = 97.5
Test of ETA1 = ETA2  vs.  ETA1 g.t. ETA2 is significant at 0.2838
The test is significant at 0.2833 (adjusted for ties)

Cannot reject at alpha = 0.05
```

CHAPTER 16

CONDUCTING A STUDY

Since Chapter 7, we have been concentrating on inferential methods that employ data obtained from random samples. In the majority of the problems, the sample data were given to us; we did not have to think about how to obtain the random sample. Furthermore, most of the problems were designed to illustrate a specific procedure. For instance, if a problem appeared in Chapter 11, Chi-Square Procedures, then we knew in advance that we would very likely use a chi-square procedure to solve that problem.

For an introductory course in statistics, it is necessary to keep things that simple. However, in reality, a more comprehensive approach is required. For example, when researchers plan and conduct studies, they must not only be able to perform a variety of statistical procedures, but they must also know how to obtain the sample data and which statistical procedure to use. In this chapter, we will consider the key aspects of planning and conducting a complete statistical study.

CHAPTER OUTLINE

16.1 Is a study necessary? Explains how it is sometimes possible to avoid the effort and expense of a study by surveying the available literature.

16.2 Sampling Presents some of the most commonly used sampling techniques that are employed by researchers.

16.3 Operating characteristic curves; power Provides a detailed examination of the types of errors that can be made in statistical tests of hypotheses and explains methods for evaluating the overall effectiveness of a hypothesis-testing procedure.

16.4 Design of experiments Discusses the fundamental ideas of experimental design and presents a few of the most widely used experimental designs.

16.5 The study: putting it all together Examines the entire process of conducting a statistical study, from the preliminary stages of a literature search to the final conclusions and decisions.

16.1   Is a study necessary?

Throughout this book we have seen examples of people conducting their own studies: A consumer group wants information about the gas mileage of a particular make of car, so it performs mileage tests on a sample of such cars and carries out a statistical analysis on the resulting data; or a teacher wants to know about the comparative merits of two teaching methods, so she tests those methods on two randomly selected samples of students.

This reflects a healthy attitude—if you don't know about something, collect data and find out about it. However, there is always the possibility that the study under consideration has already been done. Repeating the study would then be simply a waste of time, energy, and money. Therefore, before a complicated research effort is planned and conducted, a literature search should be made. This does not require going through all the books in the library. There are information-collection agencies that specialize in finding studies on specific topics in specific areas.

For example, the Educational Resources Information Center (ERIC) assembles educational studies; publications entitled *Psychological Abstracts* gather the results of studies in psychology; the National Library of Medicine compiles lists of medical studies and makes them accessible to research centers and universities. A considerable amount of information can also be found in publications by governmental agencies such as the Bureau of the Census and the Environmental Protection Agency. Data are published on income, age, energy consumption, and hundreds of other variables. Many of the examples and exercises in this book are based on information obtained from the Census Bureau's *Statistical Abstract of the United States.*

It is not the purpose of this book to explain how to search through journal articles, abstracts, or census data. In most cities and on most campuses, services are available that will perform those tasks. The important point is this: *It is often possible to avoid the effort and expense of a study if someone else has already done that study and published the results.*

16.2   Sampling

Once we have decided that the information required is not already available from a previous study, we must proceed to plan a study of our own. One of the first steps in planning a study is to choose a procedure for obtaining the sample. Many different sampling techniques are used by researchers. In this section, we will examine a few of the most important ones. We begin by discussing simple random sampling, a sampling procedure that we studied earlier in Chapter 7.

SIMPLE RANDOM SAMPLING

Recall that a sampling procedure is called **simple random sampling,** or more briefly, **random sampling,** if each possible sample of a given size is equally likely to be the one selected. A sample obtained by a simple random sampling procedure is called a **(simple) random sample.** The following example illustrates one way of implementing simple random sampling.

EXAMPLE 16.1   *Illustrates simple random sampling*

Student questionnaires, known as "teacher evaluations," gained widespread use in the late 1960s and early 1970s. Generally, student evaluations of teaching are not done at final exam time. It is more common for professors to hand out evaluation forms a week or so before the final.

There are, however, several problems with that practice. On some days, only about 60% of the students registered for a class are in attendance. Moreover, because many of those that are in attendance have other classes to prepare for, they will often fill out their teacher evaluation forms in a hurry in order to leave class early. It may well be better, therefore, to select a sample of students from the class and interview them individually. This is exactly the kind of situation in which a simple random sample should be obtained.

In the fall of 1980, Professor H wanted to sample the attitudes of the students taking college algebra at his school. He decided to interview 15 of the 728 students enrolled in the course. Since Professor H had a registration list on which the 728 students were numbered 1–728, he could obtain a simple random sample of 15 students by randomly selecting 15 numbers between 1 and 728. To do this, he used a **table of random numbers.** Such tables are constructed so that all the digits in them are chosen at random. The random-number table employed by Professor H is given as Table XI in the appendix. For reference purposes, we repeat it on the next page as Table 16.1.

To select 15 random numbers between 1 and 728, we first pick a random starting point, say, by closing our eyes and putting our finger down on Table 16.1. Then beginning with the three digits under our finger, we go down the table and record the numbers as we go. Since we want numbers between 1 and 728 only, we discard the number 000 and numbers between 729 and 999. To avoid repetition, we also eliminate numbers that have occurred previously. If not enough numbers have been found by the time we reach the bottom of the table, we move over to the next column of three digit numbers and go up.

Using this procedure, Professor H obtained 069 as a starting point (circled in Table 16.1). Reading down from 069 to the bottom of Table 16.1 and then up the next column of three digit numbers, he found the 15 random numbers displayed in Figure 16.1 on the next page and in the following table:

069	303	458	652	178
386	097	009	694	578
539	628	036	024	404

Thus, Professor H interviewed the 15 students whose numbers on the registration list were the ones given in the previous table.   ∎

TABLE 16.1
Random numbers

Line number	Column number									
	00–09		10–19		20–29		30–39		40–49	
00	15544	80712	97742	21500	97081	42451	50623	56071	28882	28739
01	01011	21285	04729	39986	73150	31548	30168	76189	56996	19210
02	47435	53308	40718	29050	74858	64517	93573	51058	68501	42723
03	91312	75137	86274	59834	69844	19853	06917	17413	44474	86530
04	12775	08768	80791	16298	22934	09630	98862	39746	64623	32768
05	31466	43761	94872	92230	52367	13205	38634	55882	77518	36252
06	09300	43847	40881	51243	97810	18903	53914	31688	06220	40422
07	73582	13810	57784	72454	68997	72229	30340	08844	53924	89630
08	11092	81392	58189	22697	41063	09451	09789	00637	06450	85990
09	93322	98567	00116	35605	66790	52965	62877	21740	56476	49296
10	80134	12484	67089	08674	70753	90959	45842	59844	45214	36505
11	97888	31797	95037	84400	76041	96668	75920	68482	56855	97417
12	92612	27082	59459	69380	98654	20407	88151	56263	27126	63797
13	72744	45586	43279	44218	83638	05422	00995	70217	78925	39097
14	96256	70653	45285	26293	78305	80252	03625	40159	68760	84716
15	07851	47452	66742	83331	54701	06573	98169	37499	67756	68301
16	25594	41552	96475	56151	02089	33748	65289	89956	89559	33687
17	65358	15155	59374	80940	03411	94656	69440	47156	77115	99463
18	09402	31008	53424	21928	02198	61201	02457	87214	59750	51330
19	97424	90765	01634	37328	41243	33564	17884	94747	93650	77668

FIGURE 16.1
Procedure used by
Professor H to obtain
15 random numbers
between 1 and 728
from Table 16.1

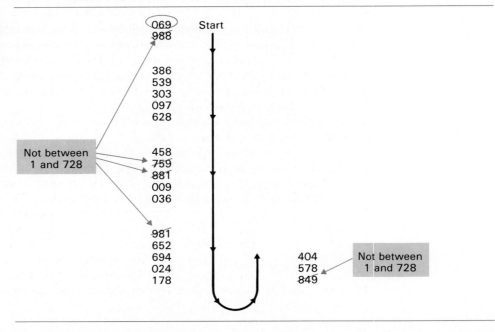

Note: Many calculators and most computers have **random-number generators.** A random-number generator makes it possible to automatically obtain a list of random numbers within any specified range. Random-number tables, such as Table 16.1, were used much more frequently before calculators with random-number generators became available at low prices. If you have a calculator with a random-number generator, you will find it easier to use than a table of random numbers.

SAMPLES OF CONVENIENCE

Sometimes it is not practical to obtain a simple random sample and other times it is not even possible. In such cases, an alternate sampling procedure must be chosen. One possibility is to employ a method that gives a **sample of convenience.**

EXAMPLE 16.2   *Illustrates samples of convenience*

A teacher at a state university wanted to compare two methods of teaching college algebra. Her plan was to use one method on one randomly selected group of students and the other method on another randomly selected group of students.

The university had 10,000 students who had never taken college algebra and so the ideal procedure for obtaining the two groups of students would have been to select them at random from those 10,000 students. Unfortunately, the university did not require anyone to take college algebra and most of the 10,000 students would not. In fact, only 728 students registered for the algebra course.

The next obvious procedure would be to randomly select the two groups from the 728 students that did register for college algebra. However, this would require that all of the students within each of the two groups selected be placed in the same class. For instance, students selected for one of the teaching methods might be told that they are in a class starting at 8:40 a.m. and students selected for the other teaching method might be told that they are in a class starting at 12:40 p.m. But students at large universities cannot be scheduled like this—some students work, some are in car pools, etc. One further difficulty was, frankly, the other teachers. Of the six other teachers scheduled to teach college algebra, not one was willing to take part in the comparison study.

So the teacher who wanted to conduct the study wound up doing the following: She randomly divided her class, which was at 9:40 a.m., into two groups, Group I and Group II. Then she found a teacher from a different course who was willing to teach one of the two groups using one of the two teaching methods. The entire process can be pictured as in Figure 16.2 at the top of the next page.

Samples of this kind are called *samples of convenience.* When samples of convenience are used, it is a good idea to make some common-sense checks to ensure that the samples are not biased. In this particular case, the teacher gave both Groups I and II a standard placement test. She found that there was no significant difference between the test scores of the two groups and that the scores were similar to those of other algebra classes. The teacher decided that it was reasonable to go ahead with her comparative study of teaching methods using Groups I and II as samples for comparison.   ∎

FIGURE 16.2

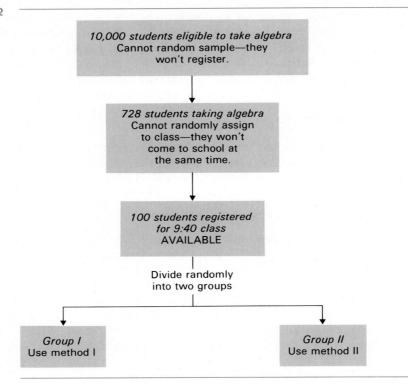

There is a danger in samples of convenience. We can never be sure that they are truly random. For example, at one school we know of, it has been observed that classes in remedial algebra scheduled at 9:40 a.m. always have better average grades than those scheduled at 1:40 p.m. In particular then, a sample of convenience consisting of students taking remedial algebra at 9:40 a.m. is not representative of all remedial algebra students. Samples of convenience should be examined with great care.

CLUSTER SAMPLING

Although it is sometimes not possible to obtain a simple random sample, we can often do better than a sample of convenience. Consider, for instance, the situation described in the next example.

EXAMPLE 16.3   *Illustrates cluster sampling*

The city council of a southwestern city of 100,000 people was under pressure from citizens' groups to install bike paths. The members of the council wanted to be sure that they had the support of a majority of the taxpayers and decided to poll the homeowners in the city.

Their first attempt at surveying opinion was a questionnaire mailed out with the city's 18,000 homeowner water bills. Unfortunately, this did not work very well. Only 3500 questionnaires were returned and a good number of those had comments written on them indicating that they came from avid bicyclists (pro) or people strongly resenting bicyclists (con). The questionnaire generally had not been returned by the average voter and the city council realized that.

The city had an employee in the planning department with some sample-survey experience. The council called her in and asked her to do a survey. She was given two assistants to help interview a representative sample of voters and was instructed to report back in 10 days.

The planner thought about taking a simple random sample of 300 voters—100 interviews for herself and for each of the two assistants. However, using a simple random sample created some time problems. The city was so spread out that an interviewer with a list of 100 voters randomly scattered around the city would have to drive an average of 18 minutes from one interview to the next. This would require approximately 30 hours of driving time for each interviewer and could delay completion of the report. Obviously, simple random sampling would not do.

To save time, the planner decided to use a sampling method called **cluster sampling.** The residential portion of the city was divided into 947 blocks, each containing approximately 20 houses (see Figure 16.3).

FIGURE 16.3
A typical block of homes

The planner numbered the blocks on the city map from 1 to 947 and then used a table of random numbers to randomly select 15 of the 947 blocks. Each of the three interviewers was then assigned five of the 15 blocks selected. This method gave each interviewer roughly 100 homes to visit but saved a great deal of travel time. Indeed, an interviewer could work on a block for nearly a full day without having to drive to another neighborhood. The report was finished on time. ■

The method used by the planner in the previous example is called *cluster sampling* because each interviewer concentrates on clusters of voters. Though this method saves time and money, it can have drawbacks. For instance, let us look at a simplified small town, as depicted in Figure 16.4 at the top of the next page.

FIGURE 16.4

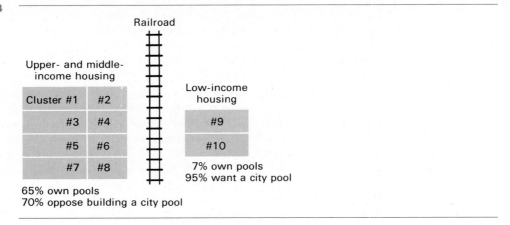

The town council is thinking about building a town swimming pool. A planner for the town needs to sample voter sentiment on using public funds to build the pool. Many upper-income and middle-income homeowners will probably say "No" because they own pools or can use a neighbor's. Many low-income voters will probably say "Yes" because they generally do not have access to pools.

If the planner decides to use cluster sampling and interview the voters of, say, three randomly selected clusters, then there is a good chance that no low-income voters will be interviewed. [This is because 46.7% of all possible three-cluster samples will contain neither of the low-income clusters, #9 and #10.] And, if no low-income voters are interviewed, the results of the survey will be misleading.

Suppose, for instance, that the planner selects clusters #3, #5, and #8. Then his survey will show that only about 30% of the voters want a pool. But that is not true. More than 40% of the voters want a pool. The planner will have left out the clusters that most strongly support the town swimming pool.

In this hypothetical example, the town is so small that common sense indicates that a cluster sample may very well not be representative. However, in situations where there are hundreds of clusters, the problems are more difficult to detect.

STRATIFIED SAMPLING

Another sampling method, known as **stratified sampling,** is frequently more reliable than cluster sampling. With a stratified-sampling procedure, the population is first subdivided into subpopulations, called **strata,** and then sampling is done from each of the strata.

For example, in the town-swimming-pool illustration, voters would be divided into three strata: high-income, middle-income, and low-income voters. A simple random sample would then be taken from each of the three strata. This stratified-sampling procedure would ensure that no income group is missed and would improve the precision of the statistical estimates. Moreover, it would make it possible to estimate the separate opinions of each of the three strata. For instance, the opinions of the high-income voters could be estimated since they would be represented in a separate sample.

In stratified sampling, the strata are often sampled in proportion to their percentages. For example, suppose that the strata consist of the three income groups (high, middle, and low) in a city and that the percentages of those groups are, respectively, 20%, 70%, and 10%. Then, for a sample size of, say, 500, the number of high-income, middle-income, and low-income individuals sampled would be, respectively, 100 (20% of 500), 350 (70% of 500), and 50 (10% of 500).

SOME WORDS OF WARNING

The methods that we have discussed in the previous chapters of this book are designed expressly for use with simple random samples and can sometimes be used with samples of convenience. However, different methods are required when dealing with stratified samples or cluster samples.

The whole question of how to sample is a complex one to which entire books are devoted. A highly readable introduction to sampling can be found in the book *An Introduction to Survey Research and Data Analysis,* by Herbert Weisberg and Bruce Bowen (San Francisco: W.H. Freeman & Co., 1977).

USING THE COMPUTER (OPTIONAL)

Minitab can be used to obtain a simple random sample from a population. The appropriate command is called **SAMPLE.** The next example shows how the SAMPLE command is applied.

EXAMPLE 16.4   *Illustrates the SAMPLE command*

Recall that in Example 16.1, Professor H wanted to interview a simple random sample of 15 students from a class of 728 college-algebra students. Since the students were numbered 1–728 on the registration list, he obtained the simple random sample by selecting 15 numbers between 1 and 728 from a table of random numbers. Explain how Minitab could have been used to obtain the simple random sample.

SOLUTION   We will store the numbers, 1–728, corresponding to the population of students in the class, into C1 and we will store the simple random sample obtained by Minitab into C2. So, we first name C1 "Class" and C2 "srs" (simple random sample) using the NAME command.

To store the numbers, 1–728, into C1, we employ the SET command. Fortunately, however, we do not have to type all 728 numbers because, in Minitab, a colon between two numbers stands for "through." In other words, typing 1:728 in response to the **DATA>** prompt, stores the numbers, 1, 2, ..., 728. See Printout 16.1 at the top of the following page.

PRINTOUT 16.1

```
MTB > NAME C1 'CLASS' C2 'SRS'
MTB > SET C1
DATA> 1:728
DATA> END
```

To obtain the simple random sample, we type the command SAMPLE followed by the sample size, the storage location of the population data, and the storage location for the sample data. That is, we type

SAMPLE 15 from 'CLASS' put into 'SRS'

The simple random sample of 15 numbers (students) is now stored in SRS. We can, of course, observe that sample by typing PRINT SRS. Printout 16.2 summarizes the discussion in this paragraph and also displays the resulting simple random sample of 15 numbers. ■

PRINTOUT 16.2

```
MTB > SAMPLE 15 from 'CLASS' put into 'SRS'
MTB > PRINT 'SRS'

SRS
    442    364     58    176     47    486     46    258    672     62    114
    428    471    165    254
```

Exercises 16.2

In each of Exercises 16.1–16.4, use Table XI to obtain the required list of random numbers.

___ **16.1** The owner of a business with 685 employees wants to select 25 of them at random for extensive interviewing. Construct a list of 25 random numbers between 1 and 685 that can be used in obtaining the required simple random sample.

___ **16.2** A university committee on parking has been formed to gauge the sentiment of the people using the university's parking facilities. Each person that uses the facilities has a parking sticker with a number. This year the numbers range between 1 and 8493. Make a list of 30 numbers that can be employed to obtain a simple random sample of 30 people who use the parking facilities.

___ **16.3** Each year *Fortune Magazine* publishes an article entitled "The International 500" which provides a ranking by sales of the top 500 firms outside the United States. Suppose that you want to examine various characteristics of successful firms. Further suppose that for your study, you decide to take a simple random sample of 10 firms from *Fortune's* list of "The International 500." Determine 10 numbers that you can use to obtain your sample.

___ **16.4** In the game of Keno, there are 80 balls, numbered 1–80, and 20 of the 80 balls are randomly selected. Simulate one game of Keno by obtaining 20 random numbers between 1 and 80.

___ **16.5** Students in the dormitories of a certain university in the state of New York live in clusters of four double rooms, called *suites*. There are 48 suites, with eight students per suite.
a) Describe a cluster-sampling procedure for obtaining a sample of 24 dormitory residents.
b) Students typically choose friends from their classes as suitemates. With that in mind, do you think

that cluster sampling is a good procedure for obtaining a representative sample of dormitory residents? Explain your answer.

c) The university housing office has separate lists of dormitory residents by class level. Using those lists, the following frequency distribution for the class level of dormitory residents was obtained:

Class	Frequency
Freshman	128
Sophomore	112
Junior	96
Senior	48

Use the frequency distribution to design a procedure for obtaining a stratified sample of 24 dormitory residents.

__ 16.6 Discuss the advantages and disadvantages of samples of convenience. Illustrate your discussion with a specific example.

Exercises 16.7 and 16.8 are computer exercises.

__ 16.7 (**Computer exercise**) Consider again the situation of Exercise 16.3. Suppose that the numbers, 1–500, are stored in a column named INTER500 and that a simple random sample of 10 of those numbers is to be stored in a column named SRS.

a) Which Minitab command and subcommands (if any) should be used to obtain the required simple random sample?

b) If you have access to Minitab, use it to obtain the simple random sample.

__ 16.8 (**Computer exercise**) Consider again the situation of Exercise 16.4. Suppose that the numbers, 1–80, are stored in a column named KENO and that a simple random sample of 20 of those numbers is to be stored in a column named SRS.

a) Which Minitab command and subcommands (if any) should be used to obtain the required simple random sample?

b) If you have access to Minitab, use it to obtain the simple random sample.

16.3 Operating characteristic curves; power

As we learned in Section 9.2, hypothesis tests do not always yield correct conclusions; they have built-in margins of error. Another important part of planning a study is to take an advance look at the types of errors that can be made and the effects those errors might have. The risks involved in committing the errors can then be assessed and, in turn, that assessment can be used in the selection of the decision criterion.

TYPE I AND TYPE II ERRORS

Two types of errors are associated with hypothesis tests. One is a **Type I error:** Rejecting a true null hypothesis. The other is a **Type II error:** Not rejecting a false null hypothesis. See Table 16.2.

TABLE 16.2
The four possible outcomes for a hypothesis test

	H_0 is: True	H_0 is: False
Do not reject H_0	Correct decision	Type II error
Reject H_0	Type I error	Correct decision

Decision:

In this section, we will use a gas-mileage study to illustrate Type I and Type II errors and other related concepts.

EXAMPLE 16.5   *Illustrates Type I and Type II errors*

The manufacturer of a new model car, called the Orion, claims that a typical car gets 26 mpg (miles per gallon). An independent consumer group is somewhat skeptical of that claim and thinks that the mean gas mileage of all Orions may very well be less than 26 mpg. The consumer group plans to perform the hypothesis test

$$H_0: \mu = 26 \text{ mpg (manufacturer's claim)}$$
$$H_a: \mu < 26 \text{ mpg (consumer group's conjecture)}$$

where μ is the mean gas mileage of all Orions.
a) What is the meaning of a Type I error and a Type II error?
b) Discuss the implication of a Type I error from both the manufacturer's point of view and the consumer group's point of view.
c) Discuss the implication of a Type II error from both the manufacturer's point of view and the consumer group's point of view.

SOLUTION   a) A Type I error is the mistake of rejecting a true null hypothesis. In this case that would be rejecting the manufacturer's claim that the mean gas mileage of all Orions is 26 mpg, when in fact the mean gas mileage is 26 mpg. A Type II error is the mistake of not rejecting a false null hypothesis. Here that would be not rejecting the manufacturer's claim that the mean gas mileage of all Orions is 26 mpg, when in fact the mean gas mileage is less than 26 mpg.
b) A Type I error would cause the consumer group to announce that the manufacturer's claim is incorrect when, in actuality, the claim is correct. Undoubtedly, this would upset the manufacturer because of the resulting bad publicity and other detrimental effects on sales. On the other hand, although the consumer group is primarily concerned with protecting the public from false claims, it certainly does not want to make a Type I error either (such an error might result in a lawsuit from the manufacturer or loss of the consumer group's credibility).
c) A Type II error would result in the consumer group not announcing that the manufacturer's claim is incorrect when, in actuality, it is incorrect. This might or might not disturb the manufacturer but it is something that the consumer group definitely wants to avoid. If the mean gas mileage is less than 26 mpg, the consumer group would like to detect that fact.   ■

PROBABILITIES OF TYPE I AND TYPE II ERRORS

The probability of making a Type I error is the probability of rejecting a true null hypothesis. As we learned in Section 9.2, that probability is called the **significance level, α,** of the hypothesis test.

DEFINITION 16.1   Probability of a Type I error

The *probability of a Type I error* is the probability of rejecting a true null hypothesis. That probability is called the *significance level*, α, of the hypothesis test. In symbols:

$$\alpha = P(\text{Type I error})$$

Thus, for instance, if the hypothesis test for the gas-mileage illustration is performed at the 5% significance level, then the probability of a Type I error will be $\alpha = 0.05$. There will be only a 5% chance of rejecting the manufacturer's claim, if that claim is in fact true.

The probability of making a Type II error is the probability of not rejecting a false null hypothesis. We use the Greek letter β to denote the probability of a Type II error.

DEFINITION 16.2   Probability of a Type II error

The *probability of a Type II error* is the probability of not rejecting a false null hypothesis and is denoted by β. In symbols:

$$\beta = P(\text{Type II error})$$

The probability, β, of a Type II error depends on the decision criterion which, in turn, depends on the sample size and the significance level. Let us return to the gas-mileage example in order to illustrate the determination of β.

EXAMPLE 16.6   **Illustrates Type II error probabilities**

Recall that a consumer group wants to perform the hypothesis test

$$H_0: \mu = 26 \text{ mpg (manufacturer's claim)}$$
$$H_a: \mu < 26 \text{ mpg (consumer group's conjecture)}$$

where μ is the mean gas mileage of all Orions.

Suppose that, for the hypothesis test, the consumer group decides to use a significance level of 0.05 and a sample size of 30. Find the probability, β, of a Type II error if the true mean gas mileage, μ, is
a) 25.8 mpg.
b) 25.0 mpg.
[Assume that the standard deviation of the gas mileages for all Orions is 1.5 mpg.]

SOLUTION   To begin, note that we are dealing with a large-sample hypothesis test for a population mean and that the hypothesis test is left-tailed. Referring to Procedure 9.1 on page 438, we see that the test statistic is

(1) $$z = \frac{\bar{x} - \mu_0}{\sigma/\sqrt{n}}$$

and the critical value is $-z_\alpha = -z_{0.05} = -1.645$. Thus, the decision criterion for the hypothesis test is: If $z \le -1.645$, reject H_0; otherwise, do not reject H_0.

It is somewhat simpler to compute Type II error probabilities if the decision criterion is expressed in terms of $\bar{x}$ instead of z. To do that, we first solve for $\bar{x}$ in Equation (1). The result is Equation (2):

$$(2) \qquad\qquad\qquad \bar{x} = \mu_0 + z \cdot \frac{\sigma}{\sqrt{n}}$$

Next we determine the value of $\bar{x}$ corresponding to $z = -1.645$. Since $\mu_0 = 26$, $\sigma = 1.5$, and $n = 30$, we see that when $z = -1.645$,

$$\bar{x} = 26 - 1.645 \cdot \frac{1.5}{\sqrt{30}} = 25.6$$

Therefore, the decision criterion can be expressed in terms of $\bar{x}$ as follows: If $\bar{x} \le 25.6$ mpg, reject H_0; otherwise, do not reject H_0. See Figure 16.5.

FIGURE 16.5
Graphical display of decision criterion for the gas-mileage illustration

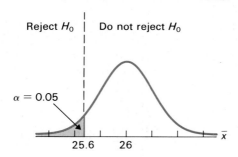

a) For this part, we want to determine the probability, β, of a Type II error if the true mean gas mileage of all Orions is 25.8 mpg; that is, we need to obtain the probability of not rejecting the null hypothesis if $\mu = 25.8$ mpg. According to the decision criterion, pictured in Figure 16.5, the null hypothesis is not rejected if the sample mean gas mileage, $\bar{x}$, of the 30 Orions tested exceeds 25.6 mpg.

Consequently, we need to determine $P(\bar{x} > 25.6)$ given that the true mean is $\mu = 25.8$ mpg. Since the sample size is large ($n = 30$), the random variable $\bar{x}$ is approximately normally distributed and has mean, $\mu_{\bar{x}} = \mu = 25.8$, and standard deviation, $\sigma_{\bar{x}} = \sigma/\sqrt{n} = 1.5/\sqrt{30} = 0.27$. So, $P(\bar{x} > 25.6)$ is equal to the area under the normal curve with parameters 25.8 and 0.27 that lies to the right of 25.6. That area is obtained in the usual manner, as explained in Section 6.2. See Figure 16.6 at the top of the next page. *Note:* The curve that we have drawn in Figure 16.6 is not the curve based on the null hypothesis value of μ, which is 26 mpg, but rather is the curve based on the true value of μ, which in this case is assumed to be 25.8 mpg.

FIGURE 16.6
Determination of
the area under
the normal curve
with parameters
$\mu_{\bar{x}} = 25.8$ and
$\sigma_{\bar{x}} = 0.27$ that lies
to the right of 25.6

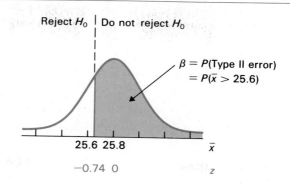

Reject H_0 | Do not reject H_0

$\beta = P(\text{Type II error})$
$\quad = P(\bar{x} > 25.6)$

25.6 25.8

-0.74 0

$\bar{x}$

z

z-score computation: Area between 0 and z:

$\bar{x} = 25.6 \longrightarrow z = \dfrac{25.6 - 25.8}{0.27} = -0.74 \qquad 0.2704$

Shaded area $= 0.2704 + 0.5000 = 0.7704$

Thus, as we see from Figure 16.6, if the true mean gas mileage of all Orions is 25.8 mpg, then the probability of making a Type II error is $\beta = 0.7704$. In other words, there is about a 77% chance that the consumer group will fail to reject the manufacturer's claim that the mean gas mileage of all Orions is 26 mpg when, in fact, the true mean is 25.8 mpg. Although this is a rather high chance of error, we probably would not expect the hypothesis test to detect such a small difference in mean gas mileage (25.8 mpg as opposed to 26 mpg).

b) For this part, we want to determine the probability, β, of a Type II error if the true mean gas mileage is 25.0 mpg. As in part (a), this means we need to obtain $P(\bar{x} > 25.6)$, but this time assuming that $\mu = 25.0$ mpg. See Figure 16.7.

FIGURE 16.7
Determination of
the area under
the normal curve
with parameters
$\mu_{\bar{x}} = 25.0$ and
$\sigma_{\bar{x}} = 0.27$ that lies
to the right of 25.6

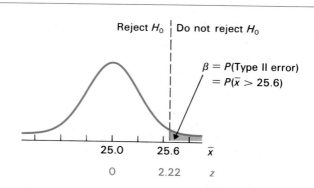

Reject H_0 | Do not reject H_0

$\beta = P(\text{Type II error})$
$\quad = P(\bar{x} > 25.6)$

25.0   25.6

0   2.22

$\bar{x}$

z

z-score computation: Area between 0 and z:

$\bar{x} = 25.6 \longrightarrow z = \dfrac{25.6 - 25.0}{0.27} = 2.22 \qquad 0.4868$

Shaded area $= 0.5000 - 0.4868 = 0.0132$

Hence, as we see from Figure 16.7, if the true mean gas mileage of all Orions is 25.0 mpg, then the probability of making a Type II error is $\beta = 0.0132$. In other words, there is only about a 1.3% chance that the consumer group will fail to reject the manufacturer's claim that the mean gas mileage of all Orions is 26 mpg when, in fact, the true mean is 25.0 mpg. ∎

By combining figures such as Figures 16.6 and 16.7, we can gain a better understanding of Type II error probabilities. In Figure 16.8, we combined those two figures with two others. The Type II error probabilities for the two additional values of μ were obtained using the same techniques as the ones used in Example 16.6.

FIGURE 16.8
Type II error probabilities for $\mu = 25.8$, 25.6, 25.3, and 25.0

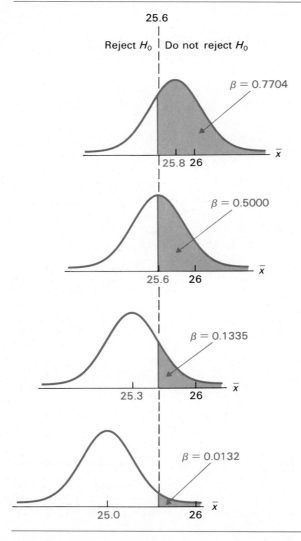

Figure 16.8 makes it clear that the further the true value of μ is from the null hypothesis value of 26 mpg, the smaller the probability, β, of making a Type II error. This is hardly surprising. We would expect it to be more likely for a false null hypothesis to be detected when the true value of μ is far from the null hypothesis value than when it is close.

OPERATING CHARACTERISTIC CURVES

Since, in reality, the true value of μ will be unknown, it is helpful to construct a table of Type II error probabilities for various values of μ. For the gas-mileage illustration, we have already computed β when the true mean is 25.8 mpg and when the true mean is 25.0 mpg. Similar calculations yield the β-values given in Table 16.3.

TABLE 16.3
Selected Type II error probabilities for the gas-mileage illustration

True mean μ	P(Type II error) β	True mean μ	P(Type II error) β
25.9	0.8665	25.3	0.1335
25.8	0.7704	25.2	0.0694
25.7	0.6443	25.1	0.0322
25.6	0.5000	25.0	0.0132
25.5	0.3557	24.9	0.0048
25.4	0.2296	24.8	0.0015

Table 16.3 can be used as an aid in evaluating the overall effectiveness of the hypothesis test. We can also employ the table to obtain a visual display of that effectiveness. This is accomplished by plotting points of β versus μ and then connecting the points with a smooth curve. See Figure 16.9.

FIGURE 16.9
Operating characteristic curve for the gas-mileage illustration

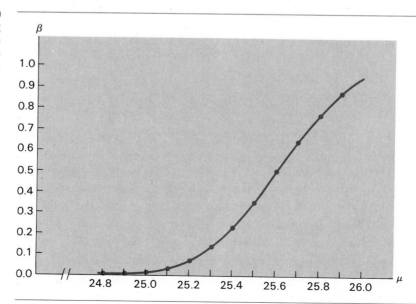

The curve in Figure 16.9—a graph of the Type II error probability, β, versus the true value of μ—is called an **operating characteristic curve** or, more briefly, an **OC curve.** For a given hypothesis test, the operating characteristic curve depends on both the sample size and the significance level. In other words, if either the sample size or the significance level is changed, then the operating characteristic curve will also change.

POWER AND POWER CURVES

Closely related to the operating characteristic curve is the *power curve*. To discuss power curves, we first need to define the concept of *power*. The **power** of a hypothesis test is the probability of not making a Type II error. By the complementation rule, Formula 4.2 on page 167, the probability of not making a Type II error is equal to 1 minus the probability of making a Type II error. Therefore, we have the following definition:

DEFINITION 16.3   Power

The *power* of a hypothesis test is defined to be the probability of not making a Type II error; that is, the probability of rejecting a false null hypothesis. In symbols:

$$\text{Power} = 1 - P(\text{Type II error}) = 1 - \beta$$

Once we know the probability, β, of a Type II error, it is a simple matter to obtain the power by subtracting β from 1. For the gas-mileage illustration, the power of the hypothesis test for various values of μ is presented in Table 16.4. That table is easily constructed by referring to Table 16.3 on page 823, which provides selected Type II error probabilities.

TABLE 16.4
Selected powers for the gas-mileage illustration

True mean μ	Power $1 - \beta$	True mean μ	Power $1 - \beta$
25.9	0.1335	25.3	0.8665
25.8	0.2296	25.2	0.9306
25.7	0.3557	25.1	0.9678
25.6	0.5000	25.0	0.9868
25.5	0.6443	24.9	0.9952
25.4	0.7704	24.8	0.9985

The **power curve** is a graph of the power, $1 - \beta$, versus the true value of μ. Using Table 16.4, we obtained the power curve for the gas-mileage illustration. See Figure 16.10 at the top of the next page.

FIGURE 16.10
Power curve for
the gas-mileage
illustration

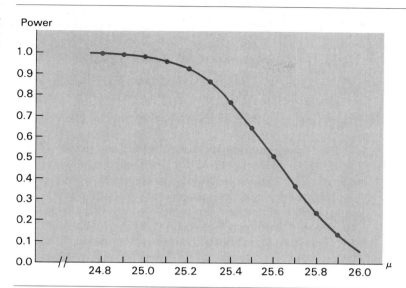

Usually there is no need to plot both an operating characteristic curve and a power curve since they provide essentially the same information. We have discussed both curves here since they are both used extensively by researchers.

SAMPLE SIZE CONSIDERATIONS

Ideally, we would like both Type I and Type II errors to have small probabilities. For then the chances of making an incorrect decision would be small regardless of which hypothesis is true.

Since the probability of a Type I error is the significance level, α, we can control the size of the Type I error probability by selecting the appropriate significance level. For instance, if we want a small Type I error probability, then we simply choose a small value for the significance level, α. However, as we learned in Section 9.2, we must keep the following fact in mind:

KEY FACT 16.1

For a fixed sample size, the smaller the Type I error probability, α, of rejecting a true null hypothesis, the larger the Type II error probability, β, of not rejecting a false null hypothesis; and vice versa.

Nonetheless, there is a way that we can make both Type I and Type II error probabilities small: We can specify a small significance level, α, which makes the Type I error probability small, and we can use a large sample size, which makes the Type II error probabilities small. Consider Example 16.7.

EXAMPLE 16.7   *Illustrates the effect of sample size on the Type II error probabilities*

Refer once again to the gas-mileage illustration of Example 16.5. A consumer group wants to perform the hypothesis test

$$H_0\colon \mu = 26 \text{ mpg (manufacturer's claim)}$$
$$H_a\colon \mu < 26 \text{ mpg (consumer group's conjecture)}$$

where μ is the mean gas mileage of all Orions.

In Table 16.3 on page 823, we presented a table of Type II error probabilities when $\alpha = 0.05$ and $n = 30$; and in Figure 16.9 on page 823, we drew the corresponding operating characteristic curve. Now suppose that the significance level is kept at 0.05 but that the sample size is increased from 30 to 100.
a) Construct a table for the Type II error probabilities similar to Table 16.3.
b) Use the table from part (a) to draw the operating characteristic curve.
c) Compare the Type II error probabilities for the sample sizes, $n = 30$ and $n = 100$.

SOLUTION   As before, we first express the decision criterion in terms of $\bar{x}$. The critical value for a left-tailed test with $\alpha = 0.05$ is $-z_{0.05} = -1.645$. Referring to Equation (2) on page 820 and noting that $\mu_0 = 26$, $\sigma = 1.5$, and $n = 100$, we see that when $z = -1.645$,

$$\bar{x} = 26 - 1.645 \cdot \frac{1.5}{\sqrt{100}} = 25.75$$

Therefore, the decision criterion can be expressed in terms of $\bar{x}$ as follows: If $\bar{x} \le 25.75$ mpg, reject H_0; otherwise, do not reject H_0. See Figure 16.11.

FIGURE 16.11
Graphical display of
decision criterion for
the gas-mileage
illustration
when $n = 100$

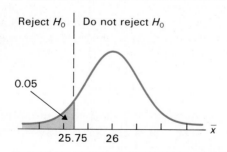

a) Now that the decision criterion has been expressed in terms of the sample mean gas mileage, $\bar{x}$, Type II error probabilities can be obtained by using the same techniques as in Example 16.6. We computed several Type II error probabilities and have displayed them in Table 16.5 at the top of the next page.

TABLE 16.5
Selected Type II error probabilities for the gas-mileage illustration when $n = 100$

True mean μ	P(Type II error) β	True mean μ	P(Type II error) β
25.9	0.8413	25.3	0.0013
25.8	0.6293	25.2	0.0001
25.7	0.3707	25.1	0.0000[†]
25.6	0.1587	25.0	0.0000
25.5	0.0475	24.9	0.0000
25.4	0.0099	24.8	0.0000

† For $\mu \leq 25.1$, the β probabilities are zero to four decimal places.

b) Using Table 16.5, we can now draw the operating characteristic curve for the gas-mileage illustration when $n = 100$. This is shown in Figure 16.12. For comparison purposes, we have also reproduced from Figure 16.9 the operating characteristic curve for the sample size $n = 30$.

FIGURE 16.12
Operating characteristic curves for the gas-mileage illustration when $n = 30$ and $n = 100$

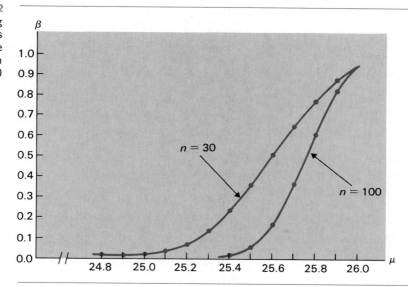

c) By comparing Tables 16.3 and 16.5, we see that each Type II error probability is smaller when $n = 100$ than when $n = 30$. Figure 16.12 displays that fact visually. ∎

In Example 16.7, we found that by increasing the sample size while keeping the significance level the same, we can reduce the Type II error probabilities. This is true in general, as indicated by the following key fact:

KEY FACT 16.2

Increasing the sample size for a hypothesis test, without changing the significance level, decreases the Type II error probabilities. In other words, for a fixed significance level, increasing the sample size increases the power.

Thus, we see that by employing a sufficiently large sample size, we can obtain a hypothesis test with as much power as we want. However, in practice, one needs to keep in mind that larger sample sizes tend to increase the cost of a study. Consequently, the planner of a study must balance, among other things, the cost of a large sample against the cost of possible errors.

As we have seen, power is useful for evaluating the overall effectiveness of a hypothesis-testing procedure. In addition, power can be used to compare different hypothesis-testing procedures. For example, a researcher might decide between a Wilcoxon signed-rank test and a t-test on the basis of which test is more powerful for the situation under consideration.

Exercises 16.3

__ **16.9** Define the following terms:
a) Type I error.   b) Type II error.

In each of Exercises 16.10–16.15, you will be given a hypothesis-testing situation. Explain what each of the following would mean:
a) A Type I error.
b) A Type II error.

__ **16.10** As reported by the U.S. National Center for Health Statistics in *Vital Health Statistics,* the average hospital stay was 7.1 days in 1982. A researcher thinks that this year's average will be less. The researcher wants to test the hypotheses

$$H_0: \mu = 7.1 \text{ days}$$
$$H_a: \mu < 7.1 \text{ days}$$

where μ is the mean hospital stay for this year.

__ **16.11** A consumer group is concerned that a soft-drink bottler who sells "one-liter" bottles of soda may be shortchanging the public. The bottler claims that the mean content, μ, of the bottles of soda is 1000 ml and the consumer group thinks that μ may very well be less than 1000 ml. The consumer group plans to take a random sample of bottles of soda in order to perform the hypothesis test

$$H_0: \mu = 1000 \text{ ml}$$
$$H_a: \mu < 1000 \text{ ml}$$

__ **16.12** A few years ago, the owner of a menswear store decided to advertise in local papers in an attempt to improve sales. From past records he knows that, without advertising, average weekly sales had been $1700. Of course, the owner does not wish to continue spending money on advertising if sales have not increased. Let μ denote the mean weekly sales with the advertising. By testing the hypotheses

$$H_0: \mu = \$1700$$
$$H_a: \mu > \$1700$$

the owner can decide whether to continue advertising.

__ **16.13** As reported by the Motor Vehicle Manufacturers Association in *Motor Vehicle Facts and Figures,* the average age of the trucks in use in 1983 was 8.1 years. The trend over the past 10 years has been an increase in the average age. Say that you want to decide whether this year's average age for trucks in use exceeds the 1983 mean of 8.1 years. That is, suppose you want to perform the hypothesis test

$$H_0: \mu = 8.1 \text{ years}$$
$$H_a: \mu > 8.1 \text{ years}$$

where μ is the mean age of trucks in use this year.

__ **16.14** The general partner of a limited partnership firm has told a potential investor that the mean monthly rent for three-bedroom homes in the area is \$525. To check this claim, the investor obtains the monthly rents for a random sample of three-bedroom homes in order to test the hypotheses

$$H_0: \mu = \$525$$
$$H_a: \mu \neq \$525$$

where μ is the mean monthly rent for three-bedroom homes in the area.

___ 16.15   The U.S. Department of Agriculture reports in *Food Consumption, Prices, and Expenditures* that the mean annual consumption of beef per person was 106.5 lb in 1983. To decide whether last year's mean differs from the 1983 figure, 50 people are to be randomly selected. Their mean beef consumption, $\bar{x}$, for last year will be used to test the hypotheses

$$H_0: \mu = 106.5 \text{ lb}$$
$$H_a: \mu \neq 106.5 \text{ lb}$$

where μ is last year's mean beef consumption.

In each of Exercises 16.16–16.21, we have given (i) a value for σ, (ii) a significance level, (iii) a sample size, and (iv) some values of μ for the corresponding problems in Exercises 16.10–16.15. For each exercise,
a) *express the decision criterion for the hypothesis test in terms of $\bar{x}$.*
b) *determine the probability of a Type I error.*
c) *determine the probability of a Type II error for each of the given values of μ and construct a table similar to Table 16.3 on page 823.*
d) *use the table that you obtained in part (c) to draw the operating characteristic curve.*
e) *use the table that you obtained in part (c) to construct a table of powers similar to Table 16.4 on page 824.*
f) *use the table that you obtained in part (e) to draw the power curve.*

___ 16.16   Refer to Exercise 16.10.
(i) $\sigma = 7.7$ (ii) $\alpha = 0.10$ (iii) $n = 40$
(iv) $\mu = 1, 2, 3, 4, 5, 6, 7$

___ 16.17   Refer to Exercise 16.11.
(i) $\sigma = 33.3$ (ii) $\alpha = 0.05$ (iii) $n = 30$
(iv) $\mu = 970, 975, 980, 985, 990, 992.5, 995, 997.5$

___ 16.18   Refer to Exercise 16.12.
(i) $\sigma = 250$ (ii) $\alpha = 0.05$ (iii) $n = 32$
(iv) $\mu = 1725, 1750, 1775, 1800, 1825, 1850, 1875, 1900$

___ 16.19   Refer to Exercise 16.13.
(i) $\sigma = 5.2$ (ii) $\alpha = 0.10$ (iii) $n = 37$
(iv) $\mu = 8.5, 9.0, 9.5, 10.0, 10.5, 11.0$

___ 16.20   Refer to Exercise 16.14 and note that the test is two-tailed.
(i) $\sigma = 118.73$ (ii) $\alpha = 0.10$ (iii) $n = 35$
(iv) $\mu = 440, 460, 480, 500, 520, 530, 550, 570, 590, 610$

___ 16.21   Refer to Exercise 16.15 and note that the test is two-tailed.
(i) $\sigma = 16.7$ (ii) $\alpha = 0.05$ (iii) $n = 50$
(iv) $\mu = 96, 98, 100, 102, 104, 106, 108, 110, 112, 114, 116, 118$

___ 16.22   Repeat parts (a)–(d) of Exercise 16.16 using a sample size of 80 instead of 40. Compare your operating characteristic curves for the two sample sizes and explain the principle being illustrated.

___ 16.23   Repeat parts (a)–(d) of Exercise 16.17 using a sample size of 100 instead of 30. Compare your operating characteristic curves for the two sample sizes and explain the principle being illustrated.

___ 16.24   Suppose that you must choose between two hypothesis-testing procedures, say Procedure 1 and Procedure 2. Further suppose that, for the same sample size, Procedure 1 has less power than Procedure 2. Which procedure would you choose? Explain.

16.4   Design of experiments

In conducting a study, the plan for selecting the sample data plays a crucial role. That plan is called the **experimental design** for the study. The general principle of experimental design is to ensure that all of the pertinent variables are taken into account. These include not only the variables of interest but also extraneous variables that influence the experimental results. By taking the extraneous variables into consideration, we can increase the chances of making a correct decision.

We have already discussed several experimental designs. Two important ones were considered in Chapter 14 (analysis of variance). They are the **completely randomized design,** used to analyze a single factor with no blocking, and the **randomized block design,** used to analyze a single factor with blocking of one extraneous variable. There are, however, many other experimental designs. For example, there is the **Latin square design,** which is used to analyze a single factor with blocking of two extraneous variables, and the **factorial design,** which is used to simultaneously analyze two or more factors.

Entire books and courses are devoted to the design of experiments. In this text, we have studied only the simplest experimental designs. Nonetheless, those simple experimental designs illustrate a fundamental point: *When planning a study, it is important to think about other variables that may affect the variables of interest and to choose an experimental design that takes those other variables into consideration.*

16.5   The study: putting it all together

In this chapter, we have discussed some of the key aspects of planning and conducting a study, but we have discussed those aspects one by one. It is now time for us to examine the entire process. We will do that for a real study from the field of education carried out at the United States Air Force Academy.

EXAMPLE 16.8   *Illustrates the aspects of a complete study*

In the fall of 1977, the mathematics department at the United States Air Force Academy sponsored a study to compare the traditional method of teaching calculus with a new individualized instructional method. The study was designed and supervised by Major Samuel Thompson who has a doctorate in Mathematics Education. Since the final report by Major Thompson exceeds 160 pages, we can only present the highlights of the study. However, those highlights will aptly illustrate the planning that goes into a well-designed study.

1. *Literature search:* After carefully surveying a large number of articles on individualized instruction, Major Thompson found several studies that compared traditional and individualized instruction. He analyzed those studies before coming to the conclusion that none of them answered the specific question that the Air Force Academy was asking: "Is the individualized instruction method better for our calculus course?"

2. *Other preparation:* To ensure that the comparison was fair to the new (individualized instruction) method, the Academy ran a pilot study in which the new method was tried the semester before the actual experiment was conducted. In

that way, the experimenters gained some experience with the new method and solved some of the more obvious instructional difficulties. Academy personnel also attended conferences on educational innovation to become more familiar with individualized instruction and to make sure that the new method was implemented correctly.

3. *Population and sample:* The population of interest in the study was the collection of all students who would ever take calculus at the Air Force Academy. The sample consisted of the 800 students who took calculus in the fall of 1977.

4. *Experimental design:* A number of extraneous variables can affect student performance in a comparison of teaching methods. Differences can arise because of student aptitude, instructor abilities, time of day for the class, and so forth. The experimenters considered all these variables in advance and set up an elaborate design that had students of all aptitudes taking calculus using both instructional methods at all class hours. Even the teaching assignments were balanced between the two methods; each method was assigned the same mix of experienced and inexperienced instructors.

5. *Variables for comparison:* Many variables are important to students and faculty alike. The two instructional methods were compared using
   a) a common final exam given to students taught by both methods.
   b) records of student study time to determine whether one method produced time savings.
   c) measurements of instructor work time with the two methods.
   d) failure rates for the two methods.
   e) a follow-up study on the students' success in their next mathematics course.
   f) a study of student and faculty attitudes toward the two methods.
   Although these were not the only variables studied, they were the main ones. The experimenters decided beforehand how they would collect data for each of the variables and planned their mathematical analyses considering such important factors as the powers of the various statistical tests that could be used.

6. *Analysis:* As supervisor of the study, Major Thompson did not teach a section of calculus. Instead, he spent the entire semester observing, collecting data, and ensuring that the experiment was conducted properly. Major Thompson also took an additional semester to perform an intensive analysis of the data according to previously made plans. Other researchers on the Academy faculty also conducted independent analyses of the data.

7. *Results:* The two methods produced no significant differences in student performance on the common final examination. However, the students taught by the new (individualized instruction) method were found, on the average, to have spent less total study time in calculus and to have higher grades in the other classes they took that semester. Moreover, the failure rate for the new method was less than half that for the traditional method.

8. *Decision:* Although the results of the study suggested that the new method of instruction had several advantages over the traditional method, both students and faculty indicated that they did not like the new method. It was decided not to force students and faculty to use a method that they did not care for. Consequently, the Air Force Academy chose to continue teaching calculus using the traditional method of instruction.   ∎

The entire process, from asking the question "Is the individualized instruction method better for our calculus course?" to the decision not to implement it, took more than two years. It also took hundreds of hours of careful planning and analysis.

Not all statistical studies are this elaborate. But the experiment supervised by Major Thompson illustrates an important fact about real-life uses of statistics: They are generally part of a decision-making process that requires a considerable amount of time and effort. Real applications of statistics are rarely as simple as the 10 minute problems in which one takes given data, calculates a predetermined test statistic, and draws a conclusion.

Chapter review

KEY TERMS

cluster sampling, 813
completely randomized design, 830
experimental design, 829
factorial design, 830
Latin square design, 830
operating characteristic curve, 824
power, 824
power curve, 824
probability of Type I error (α), 819
probability of Type II error (β), 819
random sample, 809
random sampling, 808

randomized block design, 830
random-number generator, 811
SAMPLE,* 815
sample of convenience, 811
significance level (α), 819
simple random sample, 809
simple random sampling, 808
strata, 814
stratified sampling, 814
table of random numbers, 809
Type I error, 817
Type II error, 817

YOU SHOULD
BE ABLE TO

1. describe some of the more commonly used sampling techniques.
2. use a table of random numbers to obtain a simple random sample.
3. explain the meaning of a Type I error and a Type II error and discuss the implications of those errors.
4. compute Type II error probabilities.
5. draw an operating characteristic curve.
6. calculate the power of a hypothesis test.
7. draw a power curve.
8. explain the general principle of experimental design.
9. identify the key aspects of planning and conducting a statistical study.
10. use the Minitab commands covered in this chapter.*

REVIEW TEST

1. What is the first thing that should be done when a statistical study is being contemplated? Explain your answer.

2. In a small town, there are 7246 registered voters. The mayor wants to send a detailed questionnaire to a simple random sample of 50 voters.
   a) Explain how Table XI can be employed to obtain the sample.
   b) Starting at the four digit number in line number 14 and column numbers 16–19 of Table XI, read down the column, up the next, etc., to find 50 numbers that can be used to identify the voters who will receive the questionnaire.

*3. **(Computer problem)** Refer to Problem 2. Suppose that the numbers, 1–7246, are stored in a column named VOTERS and that a simple random sample of 50 of those numbers is to be stored in a column named SRS.
   a) Which Minitab command and subcommands (if any) should be used to obtain the required simple random sample?
   b) If you have access to Minitab, use it to obtain the simple random sample.

4. The faculty at a university consists of 820 members. A new president has just been appointed. The president wants to get an idea of what the faculty considers the most important issues. She does not have the time to interview all of the faculty members and so decides to select a stratified sample of 40 faculty members, using rank for stratification. There are 205 full professors, 328 associate professors, 246 assistant professors, and 41 instructors.
   a) How many of each rank should be selected for the interviewing?
   b) Use Table XI to obtain a stratified sample of 40 faculty. Explain your procedure in detail.

5. The Food and Nutrition Board of the National Academy of Sciences states that the recommended daily allowance (RDA) of iron is 18 mg for adult females under the age of 51. A nutritional researcher thinks that, on the average, such females get less than the RDA. In an attempt to verify her conjecture, the researcher plans to obtain the 24-hour iron intakes for 45 randomly selected adult females under the age of 51. Using that data, she will then perform the hypothesis test

$$H_0: \mu = 18 \text{ mg}$$
$$H_a: \mu < 18 \text{ mg}$$

where μ is the mean daily intake of iron of all adult females under the age of 51.
   a) Explain the meaning of a Type I error and a Type II error.
   b) Discuss the implications of those errors.

6. Refer to Problem 5. Suppose that the nutritionist chooses a significance level of $\alpha = 0.01$ for her hypothesis test.
   a) Express the decision criterion for the hypothesis test in terms of $\bar{x}$. [Assume $\sigma = 3.08$ mg.]
   b) Determine the probability of a Type I error.
   c) Determine the probability of a Type II error in case the true value of μ is 15.75; 16.00; 16.25; 16.50; 16.75; 17.00; 17.25; 17.50; 17.75. Construct a table similar to Table 16.3 on page 823.
   d) Use the table from part (c) to draw the operating characteristic curve. Explain what this curve portrays.
   e) Use the table from part (c) to obtain a power table similar to Table 16.4 on page 824.
   f) Use the table from part (e) to draw the power curve. Explain what this curve portrays.

7. Refer to Problems 5 and 6. Repeat parts (a)–(d) of Problem 6 using a sample size of 80 instead of 45. Compare your operating characteristic curves for the two sample sizes and indicate the principle being illustrated.

8. Discuss the general principle of experimental design. That is, explain the role of the experimental design in a statistical study.

9. A corporate farm grows corn on three farms, two in Indiana and one in Iowa. The company has been using the same fertilizer for the past three years but is now considering two new fertilizers. This means that the company needs to perform a statistical analysis to decide which fertilizer produces the best yield. The company has set aside one large field on each farm for testing purposes. Since there are three fertilizers and three farms, one possible approach is to use one fertilizer on each farm:

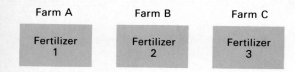

Farm A	Farm B	Farm C
Fertilizer 1	Fertilizer 2	Fertilizer 3

a) Explain why that approach is not a good one.

b) What kind of experimental design would be preferable?

10. An industrial psychologist, hired as a consultant by a toy company, has recommended that a motivational procedure be instituted to improve the output of assembly-line workers. The company is not convinced that the motivational procedure is worth the expense involved and wants to be shown proof that it would increase output over present personnel management procedures. The assembly-line workers work on three eight-hour shifts and company records indicate that there are differences in output, absenteeism, and morale across those shifts. Outline the steps that the psychologist should take in order to substantiate the case for the new motivational procedure. [These steps should cover the essential points discussed in this chapter for planning and conducting a study.]

APPENDIX A

STATISTICAL TABLES

CONTENTS

TABLE I

Binomial probabilities:

$$\binom{n}{x} p^x (1 - p)^{n-x}$$

							p					
n	x	0.1	0.2	0.25	0.3	0.4	0.5	0.6	0.7	0.75	0.8	0.9
1	0	0.900	0.800	0.750	0.700	0.600	0.500	0.400	0.300	0.250	0.200	0.100
	1	0.100	0.200	0.250	0.300	0.400	0.500	0.600	0.700	0.750	0.800	0.900
2	0	0.810	0.640	0.563	0.490	0.360	0.250	0.160	0.090	0.063	0.040	0.010
	1	0.180	0.320	0.375	0.420	0.480	0.500	0.480	0.420	0.375	0.320	0.180
	2	0.010	0.040	0.063	0.090	0.160	0.250	0.360	0.490	0.563	0.640	0.810
3	0	0.729	0.512	0.422	0.343	0.216	0.125	0.064	0.027	0.016	0.008	0.001
	1	0.243	0.384	0.422	0.441	0.432	0.375	0.288	0.189	0.141	0.096	0.027
	2	0.027	0.096	0.141	0.189	0.288	0.375	0.432	0.441	0.422	0.384	0.243
	3	0.001	0.008	0.016	0.027	0.064	0.125	0.216	0.343	0.422	0.512	0.729
4	0	0.656	0.410	0.316	0.240	0.130	0.063	0.026	0.008	0.004	0.002	0.000
	1	0.292	0.410	0.422	0.412	0.346	0.250	0.154	0.076	0.047	0.026	0.004
	2	0.049	0.154	0.211	0.265	0.346	0.375	0.346	0.265	0.211	0.154	0.049
	3	0.004	0.026	0.047	0.076	0.154	0.250	0.346	0.412	0.422	0.410	0.292
	4	0.000	0.002	0.004	0.008	0.026	0.063	0.130	0.240	0.316	0.410	0.656
5	0	0.590	0.328	0.237	0.168	0.078	0.031	0.010	0.002	0.001	0.000	0.000
	1	0.328	0.410	0.396	0.360	0.259	0.156	0.077	0.028	0.015	0.006	0.000
	2	0.073	0.205	0.264	0.309	0.346	0.312	0.230	0.132	0.088	0.051	0.008
	3	0.008	0.051	0.088	0.132	0.230	0.312	0.346	0.309	0.264	0.205	0.073
	4	0.000	0.006	0.015	0.028	0.077	0.156	0.259	0.360	0.396	0.410	0.328
	5	0.000	0.000	0.001	0.002	0.010	0.031	0.078	0.168	0.237	0.328	0.590
6	0	0.531	0.262	0.178	0.118	0.047	0.016	0.004	0.001	0.000	0.000	0.000
	1	0.354	0.393	0.356	0.303	0.187	0.094	0.037	0.010	0.004	0.002	0.000
	2	0.098	0.246	0.297	0.324	0.311	0.234	0.138	0.060	0.033	0.015	0.001
	3	0.015	0.082	0.132	0.185	0.276	0.313	0.276	0.185	0.132	0.082	0.015
	4	0.001	0.015	0.033	0.060	0.138	0.234	0.311	0.324	0.297	0.246	0.098
	5	0.000	0.002	0.004	0.010	0.037	0.094	0.187	0.303	0.356	0.393	0.354
	6	0.000	0.000	0.000	0.001	0.004	0.016	0.047	0.118	0.178	0.262	0.531
7	0	0.478	0.210	0.133	0.082	0.028	0.008	0.002	0.000	0.000	0.000	0.000
	1	0.372	0.367	0.311	0.247	0.131	0.055	0.017	0.004	0.001	0.000	0.000
	2	0.124	0.275	0.311	0.318	0.261	0.164	0.077	0.025	0.012	0.004	0.000
	3	0.023	0.115	0.173	0.227	0.290	0.273	0.194	0.097	0.058	0.029	0.003
	4	0.003	0.029	0.058	0.097	0.194	0.273	0.290	0.227	0.173	0.115	0.023
	5	0.000	0.004	0.012	0.025	0.077	0.164	0.261	0.318	0.311	0.275	0.124
	6	0.000	0.000	0.001	0.004	0.017	0.055	0.131	0.247	0.311	0.367	0.372
	7	0.000	0.000	0.000	0.000	0.002	0.008	0.028	0.082	0.133	0.210	0.478

TABLE I (continued)
Binomial probabilities:

$$\binom{n}{x} p^x (1-p)^{n-x}$$

							p						
n	x	0.1	0.2	0.25	0.3	0.4	0.5	0.6	0.7	0.75	0.8	0.9
8	0	0.430	0.168	0.100	0.058	0.017	0.004	0.001	0.000	0.000	0.000	0.000
	1	0.383	0.336	0.267	0.198	0.090	0.031	0.008	0.001	0.000	0.000	0.000
	2	0.149	0.294	0.311	0.296	0.209	0.109	0.041	0.010	0.004	0.001	0.000
	3	0.033	0.147	0.208	0.254	0.279	0.219	0.124	0.047	0.023	0.009	0.000
	4	0.005	0.046	0.087	0.136	0.232	0.273	0.232	0.136	0.087	0.046	0.005
	5	0.000	0.009	0.023	0.047	0.124	0.219	0.279	0.254	0.208	0.147	0.033
	6	0.000	0.001	0.004	0.010	0.041	0.109	0.209	0.296	0.311	0.294	0.149
	7	0.000	0.000	0.000	0.001	0.008	0.031	0.090	0.198	0.267	0.336	0.383
	8	0.000	0.000	0.000	0.000	0.001	0.004	0.017	0.058	0.100	0.168	0.430
9	0	0.387	0.134	0.075	0.040	0.010	0.002	0.000	0.000	0.000	0.000	0.000
	1	0.387	0.302	0.225	0.156	0.060	0.018	0.004	0.000	0.000	0.000	0.000
	2	0.172	0.302	0.300	0.267	0.161	0.070	0.021	0.004	0.001	0.000	0.000
	3	0.045	0.176	0.234	0.267	0.251	0.164	0.074	0.021	0.009	0.003	0.000
	4	0.007	0.066	0.117	0.172	0.251	0.246	0.167	0.074	0.039	0.017	0.001
	5	0.001	0.017	0.039	0.074	0.167	0.246	0.251	0.172	0.117	0.066	0.007
	6	0.000	0.003	0.009	0.021	0.074	0.164	0.251	0.267	0.234	0.176	0.045
	7	0.000	0.000	0.001	0.004	0.021	0.070	0.161	0.267	0.300	0.302	0.172
	8	0.000	0.000	0.000	0.000	0.004	0.018	0.060	0.156	0.225	0.302	0.387
	9	0.000	0.000	0.000	0.000	0.000	0.002	0.010	0.040	0.075	0.134	0.387
10	0	0.349	0.107	0.056	0.028	0.006	0.001	0.000	0.000	0.000	0.000	0.000
	1	0.387	0.268	0.188	0.121	0.040	0.010	0.002	0.000	0.000	0.000	0.000
	2	0.194	0.302	0.282	0.233	0.121	0.044	0.011	0.001	0.000	0.000	0.000
	3	0.057	0.201	0.250	0.267	0.215	0.117	0.042	0.009	0.003	0.001	0.000
	4	0.011	0.088	0.146	0.200	0.251	0.205	0.111	0.037	0.016	0.006	0.000
	5	0.001	0.026	0.058	0.103	0.201	0.246	0.201	0.103	0.058	0.026	0.001
	6	0.000	0.006	0.016	0.037	0.111	0.205	0.251	0.200	0.146	0.088	0.011
	7	0.000	0.001	0.003	0.009	0.042	0.117	0.215	0.267	0.250	0.201	0.057
	8	0.000	0.000	0.000	0.001	0.011	0.044	0.121	0.233	0.282	0.302	0.194
	9	0.000	0.000	0.000	0.000	0.002	0.010	0.040	0.121	0.188	0.268	0.387
	10	0.000	0.000	0.000	0.000	0.000	0.001	0.006	0.028	0.056	0.107	0.349
11	0	0.314	0.086	0.042	0.020	0.004	0.000	0.000	0.000	0.000	0.000	0.000
	1	0.384	0.236	0.155	0.093	0.027	0.005	0.001	0.000	0.000	0.000	0.000
	2	0.213	0.295	0.258	0.200	0.089	0.027	0.005	0.001	0.000	0.000	0.000
	3	0.071	0.221	0.258	0.257	0.177	0.081	0.023	0.004	0.001	0.000	0.000
	4	0.016	0.111	0.172	0.220	0.236	0.161	0.070	0.017	0.006	0.002	0.000
	5	0.002	0.039	0.080	0.132	0.221	0.226	0.147	0.057	0.027	0.010	0.000
	6	0.000	0.010	0.027	0.057	0.147	0.226	0.221	0.132	0.080	0.039	0.002
	7	0.000	0.002	0.006	0.017	0.070	0.161	0.236	0.220	0.172	0.111	0.016
	8	0.000	0.000	0.001	0.004	0.023	0.081	0.177	0.257	0.258	0.221	0.071
	9	0.000	0.000	0.000	0.001	0.005	0.027	0.089	0.200	0.258	0.295	0.213
	10	0.000	0.000	0.000	0.000	0.001	0.005	0.027	0.093	0.155	0.236	0.384
	11	0.000	0.000	0.000	0.000	0.000	0.000	0.004	0.020	0.042	0.086	0.314

(Continued)

TABLE I (continued)
Binomial probabilities:

$$\binom{n}{x} p^x (1 - p)^{n-x}$$

							p						
n	x	0.1	0.2	0.25	0.3	0.4	0.5	0.6	0.7	0.75	0.8	0.9
12	0	0.282	0.069	0.032	0.014	0.002	0.000	0.000	0.000	0.000	0.000	0.000
	1	0.377	0.206	0.127	0.071	0.017	0.003	0.000	0.000	0.000	0.000	0.000
	2	0.230	0.283	0.232	0.168	0.064	0.016	0.002	0.000	0.000	0.000	0.000
	3	0.085	0.236	0.258	0.240	0.142	0.054	0.012	0.001	0.000	0.000	0.000
	4	0.021	0.133	0.194	0.231	0.213	0.121	0.042	0.008	0.002	0.001	0.000
	5	0.004	0.053	0.103	0.158	0.227	0.193	0.101	0.029	0.011	0.003	0.000
	6	0.000	0.016	0.040	0.079	0.177	0.226	0.177	0.079	0.040	0.016	0.000
	7	0.000	0.003	0.011	0.029	0.101	0.193	0.227	0.158	0.103	0.053	0.004
	8	0.000	0.001	0.002	0.008	0.042	0.121	0.213	0.231	0.194	0.133	0.021
	9	0.000	0.000	0.000	0.001	0.012	0.054	0.142	0.240	0.258	0.236	0.085
	10	0.000	0.000	0.000	0.000	0.002	0.016	0.064	0.168	0.232	0.283	0.230
	11	0.000	0.000	0.000	0.000	0.000	0.003	0.017	0.071	0.127	0.206	0.377
	12	0.000	0.000	0.000	0.000	0.000	0.000	0.002	0.014	0.032	0.069	0.282
13	0	0.254	0.055	0.024	0.010	0.001	0.000	0.000	0.000	0.000	0.000	0.000
	1	0.367	0.179	0.103	0.054	0.011	0.002	0.000	0.000	0.000	0.000	0.000
	2	0.245	0.268	0.206	0.139	0.045	0.010	0.001	0.000	0.000	0.000	0.000
	3	0.100	0.246	0.252	0.218	0.111	0.035	0.006	0.001	0.000	0.000	0.000
	4	0.028	0.154	0.210	0.234	0.184	0.087	0.024	0.003	0.001	0.000	0.000
	5	0.006	0.069	0.126	0.180	0.221	0.157	0.066	0.014	0.005	0.001	0.000
	6	0.001	0.023	0.056	0.103	0.197	0.209	0.131	0.044	0.019	0.006	0.000
	7	0.000	0.006	0.019	0.044	0.131	0.209	0.197	0.103	0.056	0.023	0.001
	8	0.000	0.001	0.005	0.014	0.066	0.157	0.221	0.180	0.126	0.069	0.006
	9	0.000	0.000	0.001	0.003	0.024	0.087	0.184	0.234	0.210	0.154	0.028
	10	0.000	0.000	0.000	0.001	0.006	0.035	0.111	0.218	0.252	0.246	0.100
	11	0.000	0.000	0.000	0.000	0.001	0.010	0.045	0.139	0.206	0.268	0.245
	12	0.000	0.000	0.000	0.000	0.000	0.002	0.011	0.054	0.103	0.179	0.367
	13	0.000	0.000	0.000	0.000	0.000	0.000	0.001	0.010	0.024	0.055	0.254
14	0	0.229	0.044	0.018	0.007	0.001	0.000	0.000	0.000	0.000	0.000	0.000
	1	0.356	0.154	0.083	0.041	0.007	0.001	0.000	0.000	0.000	0.000	0.000
	2	0.257	0.250	0.180	0.113	0.032	0.006	0.001	0.000	0.000	0.000	0.000
	3	0.114	0.250	0.240	0.194	0.085	0.022	0.003	0.000	0.000	0.000	0.000
	4	0.035	0.172	0.220	0.229	0.155	0.061	0.014	0.001	0.000	0.000	0.000
	5	0.008	0.086	0.147	0.196	0.207	0.122	0.041	0.007	0.002	0.000	0.000
	6	0.001	0.032	0.073	0.126	0.207	0.183	0.092	0.023	0.008	0.002	0.000
	7	0.000	0.009	0.028	0.062	0.157	0.209	0.157	0.062	0.028	0.009	0.000
	8	0.000	0.002	0.008	0.023	0.092	0.183	0.207	0.126	0.073	0.032	0.001
	9	0.000	0.000	0.002	0.007	0.041	0.122	0.207	0.196	0.147	0.086	0.008
	10	0.000	0.000	0.000	0.001	0.014	0.061	0.155	0.229	0.220	0.172	0.035
	11	0.000	0.000	0.000	0.000	0.003	0.022	0.085	0.194	0.240	0.250	0.114
	12	0.000	0.000	0.000	0.000	0.001	0.006	0.032	0.113	0.180	0.250	0.257
	13	0.000	0.000	0.000	0.000	0.000	0.001	0.007	0.041	0.083	0.154	0.356
	14	0.000	0.000	0.000	0.000	0.000	0.000	0.001	0.007	0.018	0.044	0.229

TABLE I (continued)
Binomial probabilities:

$$\binom{n}{x} p^x (1-p)^{n-x}$$

n	x	0.1	0.2	0.25	0.3	0.4	0.5	0.6	0.7	0.75	0.8	0.9
15	0	0.206	0.035	0.013	0.005	0.000	0.000	0.000	0.000	0.000	0.000	0.000
	1	0.343	0.132	0.067	0.031	0.005	0.000	0.000	0.000	0.000	0.000	0.000
	2	0.267	0.231	0.156	0.092	0.022	0.003	0.000	0.000	0.000	0.000	0.000
	3	0.129	0.250	0.225	0.170	0.063	0.014	0.002	0.000	0.000	0.000	0.000
	4	0.043	0.188	0.225	0.219	0.127	0.042	0.007	0.001	0.000	0.000	0.000
	5	0.010	0.103	0.165	0.206	0.186	0.092	0.024	0.003	0.001	0.000	0.000
	6	0.002	0.043	0.092	0.147	0.207	0.153	0.061	0.012	0.003	0.001	0.000
	7	0.000	0.014	0.039	0.081	0.177	0.196	0.118	0.035	0.013	0.003	0.000
	8	0.000	0.003	0.013	0.035	0.118	0.196	0.177	0.081	0.039	0.014	0.000
	9	0.000	0.001	0.003	0.012	0.061	0.153	0.207	0.147	0.092	0.043	0.002
	10	0.000	0.000	0.001	0.003	0.024	0.092	0.186	0.206	0.165	0.103	0.010
	11	0.000	0.000	0.000	0.001	0.007	0.042	0.127	0.219	0.225	0.188	0.043
	12	0.000	0.000	0.000	0.000	0.002	0.014	0.063	0.170	0.225	0.250	0.129
	13	0.000	0.000	0.000	0.000	0.000	0.003	0.022	0.092	0.156	0.231	0.267
	14	0.000	0.000	0.000	0.000	0.000	0.000	0.005	0.031	0.067	0.132	0.343
	15	0.000	0.000	0.000	0.000	0.000	0.000	0.000	0.005	0.013	0.035	0.206
20	0	0.122	0.012	0.003	0.001	0.000	0.000	0.000	0.000	0.000	0.000	0.000
	1	0.270	0.058	0.021	0.007	0.000	0.000	0.000	0.000	0.000	0.000	0.000
	2	0.285	0.137	0.067	0.028	0.003	0.000	0.000	0.000	0.000	0.000	0.000
	3	0.190	0.205	0.134	0.072	0.012	0.001	0.000	0.000	0.000	0.000	0.000
	4	0.090	0.218	0.190	0.130	0.035	0.005	0.000	0.000	0.000	0.000	0.000
	5	0.032	0.175	0.202	0.179	0.075	0.015	0.001	0.000	0.000	0.000	0.000
	6	0.009	0.109	0.169	0.192	0.124	0.037	0.005	0.000	0.000	0.000	0.000
	7	0.002	0.055	0.112	0.164	0.166	0.074	0.015	0.001	0.000	0.000	0.000
	8	0.000	0.022	0.061	0.114	0.180	0.120	0.035	0.004	0.001	0.000	0.000
	9	0.000	0.007	0.027	0.065	0.160	0.160	0.071	0.012	0.003	0.000	0.000
	10	0.000	0.002	0.010	0.031	0.117	0.176	0.117	0.031	0.010	0.002	0.000
	11	0.000	0.000	0.003	0.012	0.071	0.160	0.160	0.065	0.027	0.007	0.000
	12	0.000	0.000	0.001	0.004	0.035	0.120	0.180	0.114	0.061	0.022	0.000
	13	0.000	0.000	0.000	0.001	0.015	0.074	0.166	0.164	0.112	0.055	0.002
	14	0.000	0.000	0.000	0.000	0.005	0.037	0.124	0.192	0.169	0.109	0.009
	15	0.000	0.000	0.000	0.000	0.001	0.015	0.075	0.179	0.202	0.175	0.032
	16	0.000	0.000	0.000	0.000	0.000	0.005	0.035	0.130	0.190	0.218	0.090
	17	0.000	0.000	0.000	0.000	0.000	0.001	0.012	0.072	0.134	0.205	0.190
	18	0.000	0.000	0.000	0.000	0.000	0.000	0.003	0.028	0.067	0.137	0.285
	19	0.000	0.000	0.000	0.000	0.000	0.000	0.000	0.007	0.021	0.058	0.270
	20	0.000	0.000	0.000	0.000	0.000	0.000	0.000	0.001	0.003	0.012	0.122

(Continued)

TABLE I (continued)
Binomial probabilities:

$$\binom{n}{x} p^x (1-p)^{n-x}$$

n	x	0.1	0.2	0.25	0.3	0.4	0.5	0.6	0.7	0.75	0.8	0.9
25	0	0.072	0.004	0.001	0.000	0.000	0.000	0.000	0.000	0.000	0.000	0.000
	1	0.199	0.024	0.006	0.001	0.000	0.000	0.000	0.000	0.000	0.000	0.000
	2	0.266	0.071	0.025	0.007	0.000	0.000	0.000	0.000	0.000	0.000	0.000
	3	0.226	0.136	0.064	0.024	0.002	0.000	0.000	0.000	0.000	0.000	0.000
	4	0.138	0.187	0.118	0.057	0.007	0.000	0.000	0.000	0.000	0.000	0.000
	5	0.065	0.196	0.165	0.103	0.020	0.002	0.000	0.000	0.000	0.000	0.000
	6	0.024	0.163	0.183	0.147	0.044	0.005	0.000	0.000	0.000	0.000	0.000
	7	0.007	0.111	0.165	0.171	0.080	0.014	0.001	0.000	0.000	0.000	0.000
	8	0.002	0.062	0.124	0.165	0.120	0.032	0.003	0.000	0.000	0.000	0.000
	9	0.000	0.029	0.078	0.134	0.151	0.061	0.009	0.000	0.000	0.000	0.000
	10	0.000	0.012	0.042	0.092	0.161	0.097	0.021	0.001	0.000	0.000	0.000
	11	0.000	0.004	0.019	0.054	0.147	0.133	0.043	0.004	0.001	0.000	0.000
	12	0.000	0.001	0.007	0.027	0.114	0.155	0.076	0.011	0.002	0.000	0.000
	13	0.000	0.000	0.002	0.011	0.076	0.155	0.114	0.027	0.007	0.001	0.000
	14	0.000	0.000	0.001	0.004	0.043	0.133	0.147	0.054	0.019	0.004	0.000
	15	0.000	0.000	0.000	0.001	0.021	0.097	0.161	0.092	0.042	0.012	0.000
	16	0.000	0.000	0.000	0.000	0.009	0.061	0.151	0.134	0.078	0.029	0.000
	17	0.000	0.000	0.000	0.000	0.003	0.032	0.120	0.165	0.124	0.062	0.002
	18	0.000	0.000	0.000	0.000	0.001	0.014	0.080	0.171	0.165	0.111	0.007
	19	0.000	0.000	0.000	0.000	0.000	0.005	0.044	0.147	0.183	0.163	0.024
	20	0.000	0.000	0.000	0.000	0.000	0.002	0.020	0.103	0.165	0.196	0.065
	21	0.000	0.000	0.000	0.000	0.000	0.000	0.007	0.057	0.118	0.187	0.138
	22	0.000	0.000	0.000	0.000	0.000	0.000	0.002	0.024	0.064	0.136	0.226
	23	0.000	0.000	0.000	0.000	0.000	0.000	0.000	0.007	0.025	0.071	0.266
	24	0.000	0.000	0.000	0.000	0.000	0.000	0.000	0.001	0.006	0.024	0.199
	25	0.000	0.000	0.000	0.000	0.000	0.000	0.000	0.000	0.001	0.004	0.072

TABLE II
Areas under the standard normal curve

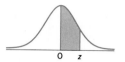

z	0.00	0.01	0.02	0.03	0.04	0.05	0.06	0.07	0.08	0.09
				Second decimal place in z						
0.0	0.0000	0.0040	0.0080	0.0120	0.0160	0.0199	0.0239	0.0279	0.0319	0.0359
0.1	0.0398	0.0438	0.0478	0.0517	0.0557	0.0596	0.0636	0.0675	0.0714	0.0753
0.2	0.0793	0.0832	0.0871	0.0910	0.0948	0.0987	0.1026	0.1064	0.1103	0.1141
0.3	0.1179	0.1217	0.1255	0.1293	0.1331	0.1368	0.1406	0.1443	0.1480	0.1517
0.4	0.1554	0.1591	0.1628	0.1664	0.1700	0.1736	0.1772	0.1808	0.1844	0.1879
0.5	0.1915	0.1950	0.1985	0.2019	0.2054	0.2088	0.2123	0.2157	0.2190	0.2224
0.6	0.2257	0.2291	0.2324	0.2357	0.2389	0.2422	0.2454	0.2486	0.2517	0.2549
0.7	0.2580	0.2611	0.2642	0.2673	0.2704	0.2734	0.2764	0.2794	0.2823	0.2852
0.8	0.2881	0.2910	0.2939	0.2967	0.2995	0.3023	0.3051	0.3078	0.3106	0.3133
0.9	0.3159	0.3186	0.3212	0.3238	0.3264	0.3289	0.3315	0.3340	0.3365	0.3389
1.0	0.3413	0.3438	0.3461	0.3485	0.3508	0.3531	0.3554	0.3577	0.3599	0.3621
1.1	0.3643	0.3665	0.3686	0.3708	0.3729	0.3749	0.3770	0.3790	0.3810	0.3830
1.2	0.3849	0.3869	0.3888	0.3907	0.3925	0.3944	0.3962	0.3980	0.3997	0.4015
1.3	0.4032	0.4049	0.4066	0.4082	0.4099	0.4115	0.4131	0.4147	0.4162	0.4177
1.4	0.4192	0.4207	0.4222	0.4236	0.4251	0.4265	0.4279	0.4292	0.4306	0.4319
1.5	0.4332	0.4345	0.4357	0.4370	0.4382	0.4394	0.4406	0.4418	0.4429	0.4441
1.6	0.4452	0.4463	0.4474	0.4484	0.4495	0.4505	0.4515	0.4525	0.4535	0.4545
1.7	0.4554	0.4564	0.4573	0.4582	0.4591	0.4599	0.4608	0.4616	0.4625	0.4633
1.8	0.4641	0.4649	0.4656	0.4664	0.4671	0.4678	0.4686	0.4693	0.4699	0.4706
1.9	0.4713	0.4719	0.4726	0.4732	0.4738	0.4744	0.4750	0.4756	0.4761	0.4767
2.0	0.4772	0.4778	0.4783	0.4788	0.4793	0.4798	0.4803	0.4808	0.4812	0.4817
2.1	0.4821	0.4826	0.4830	0.4834	0.4838	0.4842	0.4846	0.4850	0.4854	0.4857
2.2	0.4861	0.4864	0.4868	0.4871	0.4875	0.4878	0.4881	0.4884	0.4887	0.4890
2.3	0.4893	0.4896	0.4898	0.4901	0.4904	0.4906	0.4909	0.4911	0.4913	0.4916
2.4	0.4918	0.4920	0.4922	0.4925	0.4927	0.4929	0.4931	0.4932	0.4934	0.4936
2.5	0.4938	0.4940	0.4941	0.4943	0.4945	0.4946	0.4948	0.4949	0.4951	0.4952
2.6	0.4953	0.4955	0.4956	0.4957	0.4959	0.4960	0.4961	0.4962	0.4963	0.4964
2.7	0.4965	0.4966	0.4967	0.4968	0.4969	0.4970	0.4971	0.4972	0.4973	0.4974
2.8	0.4974	0.4975	0.4976	0.4977	0.4977	0.4978	0.4979	0.4979	0.4980	0.4981
2.9	0.4981	0.4982	0.4982	0.4983	0.4984	0.4984	0.4985	0.4985	0.4986	0.4986
3.0	0.4987	0.4987	0.4987	0.4988	0.4988	0.4989	0.4989	0.4989	0.4990	0.4990
3.1	0.4990	0.4991	0.4991	0.4991	0.4992	0.4992	0.4992	0.4992	0.4993	0.4993
3.2	0.4993	0.4993	0.4994	0.4994	0.4994	0.4994	0.4994	0.4995	0.4995	0.4995
3.3	0.4995	0.4995	0.4995	0.4996	0.4996	0.4996	0.4996	0.4996	0.4996	0.4997
3.4	0.4997	0.4997	0.4997	0.4997	0.4997	0.4997	0.4997	0.4997	0.4997	0.4998
3.5	0.4998	0.4998	0.4998	0.4998	0.4998	0.4998	0.4998	0.4998	0.4998	0.4998
3.6	0.4998	0.4998	0.4999	0.4999	0.4999	0.4999	0.4999	0.4999	0.4999	0.4999
3.7	0.4999	0.4999	0.4999	0.4999	0.4999	0.4999	0.4999	0.4999	0.4999	0.4999
3.8	0.4999	0.4999	0.4999	0.4999	0.4999	0.4999	0.4999	0.4999	0.4999	0.4999
3.9	0.5000[†]									

† For $z \geq 3.90$, the areas are 0.5000 to four decimal places.

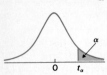

df	$t_{0.10}$	$t_{0.05}$	$t_{0.025}$	$t_{0.01}$	$t_{0.005}$	df
1	3.078	6.314	12.706	31.821	63.657	1
2	1.886	2.920	4.303	6.965	9.925	2
3	1.638	2.353	3.182	4.541	5.841	3
4	1.533	2.132	2.776	3.747	4.604	4
5	1.476	2.015	2.571	3.365	4.032	5
6	1.440	1.943	2.447	3.143	3.707	6
7	1.415	1.895	2.365	2.998	3.499	7
8	1.397	1.860	2.306	2.896	3.355	8
9	1.383	1.833	2.262	2.821	3.250	9
10	1.372	1.812	2.228	2.764	3.169	10
11	1.363	1.796	2.201	2.718	3.106	11
12	1.356	1.782	2.179	2.681	3.055	12
13	1.350	1.771	2.160	2.650	3.012	13
14	1.345	1.761	2.145	2.624	2.977	14
15	1.341	1.753	2.131	2.602	2.947	15
16	1.337	1.746	2.120	2.583	2.921	16
17	1.333	1.740	2.110	2.567	2.898	17
18	1.330	1.734	2.101	2.552	2.878	18
19	1.328	1.729	2.093	2.539	2.861	19
20	1.325	1.725	2.086	2.528	2.845	20
21	1.323	1.721	2.080	2.518	2.831	21
22	1.321	1.717	2.074	2.508	2.819	22
23	1.319	1.714	2.069	2.500	2.807	23
24	1.318	1.711	2.064	2.492	2.797	24
25	1.316	1.708	2.060	2.485	2.787	25
26	1.315	1.706	2.056	2.479	2.779	26
27	1.314	1.703	2.052	2.473	2.771	27
28	1.313	1.701	2.048	2.467	2.763	28
29	1.311	1.699	2.045	2.462	2.756	29
∞	1.282	1.645	1.960	2.326	2.576	∞

TABLE IV
Values of χ^2_α

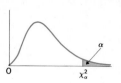

df	$\chi^2_{0.995}$	$\chi^2_{0.99}$	$\chi^2_{0.975}$	$\chi^2_{0.95}$	$\chi^2_{0.05}$	$\chi^2_{0.025}$	$\chi^2_{0.01}$	$\chi^2_{0.005}$	df
1	0.000	0.000	0.001	0.004	3.841	5.024	6.635	7.879	1
2	0.010	0.020	0.051	0.103	5.991	7.378	9.210	10.597	2
3	0.072	0.115	0.216	0.352	7.815	9.348	11.345	12.838	3
4	0.207	0.297	0.484	0.711	9.488	11.143	13.277	14.860	4
5	0.412	0.554	0.831	1.145	11.070	12.832	15.086	16.750	5
6	0.676	0.872	1.237	1.635	12.592	14.449	16.812	18.548	6
7	0.989	1.239	1.690	2.167	14.067	16.013	18.475	20.278	7
8	1.344	1.646	2.180	2.733	15.507	17.535	20.090	21.955	8
9	1.735	2.088	2.700	3.325	16.919	19.023	21.666	23.589	9
10	2.156	2.558	3.247	3.940	18.307	20.483	23.209	25.188	10
11	2.603	3.053	3.816	4.575	19.675	21.920	24.725	26.757	11
12	3.074	3.571	4.404	5.226	21.026	23.337	26.217	28.300	12
13	3.565	4.107	5.009	5.892	22.362	24.736	27.688	29.819	13
14	4.075	4.660	5.629	6.571	23.685	26.119	29.141	31.319	14
15	4.601	5.229	6.262	7.261	24.996	27.488	30.578	32.801	15
16	5.142	5.812	6.908	7.962	26.296	28.845	32.000	34.267	16
17	5.697	6.408	7.564	8.672	27.587	30.191	33.409	35.718	17
18	6.265	7.015	8.231	9.390	28.869	31.526	34.805	37.156	18
19	6.844	7.633	8.907	10.117	30.144	32.852	36.191	38.582	19
20	7.434	8.260	9.591	10.851	31.410	34.170	37.566	39.997	20
21	8.034	8.897	10.283	11.591	32.671	35.479	38.932	41.401	21
22	8.643	9.542	10.982	12.338	33.924	36.781	40.289	42.796	22
23	9.260	10.196	11.689	13.091	35.172	38.076	41.638	44.181	23
24	9.886	10.856	12.401	13.848	36.415	39.364	42.980	45.558	24
25	10.520	11.524	13.120	14.611	37.652	40.646	44.314	46.928	25
26	11.160	12.198	13.844	15.379	38.885	41.923	45.642	48.290	26
27	11.808	12.879	14.573	16.151	40.113	43.194	46.963	49.645	27
28	12.461	13.565	15.308	16.928	41.337	44.461	48.278	50.993	28
29	13.121	14.256	16.047	17.708	42.557	45.722	49.588	52.336	29
30	13.787	14.953	16.791	18.493	43.773	46.979	50.892	53.672	30

TABLE V
Values of $F_{0.01}$

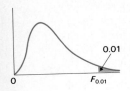

				df for numerator					
	1	*2*	*3*	*4*	*5*	*6*	*7*	*8*	*9*
1	4052	4999.5	5403	5625	5764	5859	5928	5981	6022
2	98.50	99.00	99.17	99.25	99.30	99.33	99.36	99.37	99.39
3	34.12	30.82	29.46	28.71	28.24	27.91	27.67	27.49	27.35
4	21.20	18.00	16.69	15.98	15.52	15.21	14.98	14.80	14.66
5	16.26	13.27	12.06	11.39	10.97	10.67	10.46	10.29	10.16
6	13.75	10.92	9.78	9.15	8.75	8.47	8.26	8.10	7.98
7	12.25	9.55	8.45	7.85	7.46	7.19	6.99	6.84	6.72
8	11.26	8.65	7.59	7.01	6.63	6.37	6.18	6.03	5.91
9	10.56	8.02	6.99	6.42	6.06	5.80	5.61	5.47	5.35
10	10.04	7.56	6.55	5.99	5.64	5.39	5.20	5.06	4.94
11	9.65	7.21	6.22	5.67	5.32	5.07	4.89	4.74	4.63
12	9.33	6.93	5.95	5.41	5.06	4.82	4.64	4.50	4.39
13	9.07	6.70	5.74	5.21	4.86	4.62	4.44	4.30	4.19
14	8.86	6.51	5.56	5.04	4.69	4.46	4.28	4.14	4.03
15	8.68	6.36	5.42	4.89	4.56	4.32	4.14	4.00	3.89
16	8.53	6.23	5.29	4.77	4.44	4.20	4.03	3.89	3.78
17	8.40	6.11	5.18	4.67	4.34	4.10	3.93	3.79	3.68
18	8.29	6.01	5.09	4.58	4.25	4.01	3.84	3.71	3.60
19	8.18	5.93	5.01	4.50	4.17	3.94	3.77	3.63	3.52
20	8.10	5.85	4.94	4.43	4.10	3.87	3.70	3.56	3.46
21	8.02	5.78	4.87	4.37	4.04	3.81	3.64	3.51	3.40
22	7.95	5.72	4.82	4.31	3.99	3.76	3.59	3.45	3.35
23	7.88	5.66	4.76	4.26	3.94	3.71	3.54	3.41	3.30
24	7.82	5.61	4.72	4.22	3.90	3.67	3.50	3.36	3.26
25	7.77	5.57	4.68	4.18	3.85	3.63	3.46	3.32	3.22
26	7.72	5.53	4.64	4.14	3.82	3.59	3.42	3.29	3.18
27	7.68	5.49	4.60	4.11	3.78	3.56	3.39	3.26	3.15
28	7.64	5.45	4.57	4.07	3.75	3.53	3.36	3.23	3.12
29	7.60	5.42	4.54	4.04	3.73	3.50	3.33	3.20	3.09
30	7.56	5.39	4.51	4.02	3.70	3.47	3.30	3.17	3.07
40	7.31	5.18	4.31	3.83	3.51	3.29	3.12	2.99	2.89
60	7.08	4.98	4.13	3.65	3.34	3.12	2.95	2.82	2.72
120	6.85	4.79	3.95	3.48	3.17	2.96	2.79	2.66	2.56
∞	6.63	4.61	3.78	3.32	3.02	2.80	2.64	2.51	2.41

df for denominator

Adapted from D.B. Owen, *Handbook of Statistical Tables*. Courtesy of the Atomic Energy Commission. Reading, MA: Addison-Wesley, 1962.

TABLE V (continued)
Values of $F_{0.01}$

				df for numerator							
	10	*12*	*15*	*20*	*24*	*30*	*40*	*60*	*120*	∞	
	6056	6106	6157	6209	6235	6261	6287	6313	6339	6366	*1*
	99.40	99.42	99.43	99.45	99.46	99.47	99.47	99.48	99.49	99.50	*2*
	27.23	27.05	26.87	26.69	26.60	26.50	26.41	26.32	26.22	26.13	*3*
	14.55	14.37	14.20	14.02	13.93	13.84	13.75	13.65	13.56	13.46	*4*
	10.05	9.89	9.72	9.55	9.47	9.38	9.29	9.20	9.11	9.02	*5*
	7.87	7.72	7.56	7.40	7.31	7.23	7.14	7.06	6.97	6.88	*6*
	6.62	6.47	6.31	6.16	6.07	5.99	5.91	5.82	5.74	5.65	*7*
	5.81	5.67	5.52	5.36	5.28	5.20	5.12	5.03	4.95	4.86	*8*
	5.26	5.11	4.96	4.81	4.73	4.65	4.57	4.48	4.40	4.31	*9*
	4.85	4.71	4.56	4.41	4.33	4.25	4.17	4.08	4.00	3.91	*10*
	4.54	4.40	4.25	4.10	4.02	3.94	3.86	3.78	3.69	3.60	*11*
	4.30	4.16	4.01	3.86	3.78	3.70	3.62	3.54	3.45	3.36	*12*
	4.10	3.96	3.82	3.66	3.59	3.51	3.43	3.34	3.25	3.17	*13*
	3.94	3.80	3.66	3.51	3.43	3.35	3.27	3.18	3.09	3.00	*14*
	3.80	3.67	3.52	3.37	3.29	3.21	3.13	3.05	2.96	2.87	*15*
	3.69	3.55	3.41	3.26	3.18	3.10	3.02	2.93	2.84	2.75	*16*
	3.59	3.46	3.31	3.16	3.08	3.00	2.92	2.83	2.75	2.65	*17*
	3.51	3.37	3.23	3.08	3.00	2.92	2.84	2.75	2.66	2.57	*18*
	3.43	3.30	3.15	3.00	2.92	2.84	2.76	2.67	2.58	2.49	*19*
	3.37	3.23	3.09	2.94	2.86	2.78	2.69	2.61	2.52	2.42	*20*
	3.31	3.17	3.03	2.88	2.80	2.72	2.64	2.55	2.46	2.36	*21*
	3.26	3.12	2.98	2.83	2.75	2.67	2.58	2.50	2.40	2.31	*22*
	3.21	3.07	2.93	2.78	2.70	2.62	2.54	2.45	2.35	2.26	*23*
	3.17	3.03	2.89	2.74	2.66	2.58	2.49	2.40	2.31	2.21	*24*
	3.13	2.99	2.85	2.70	2.62	2.54	2.45	2.36	2.27	2.17	*25*
	3.09	2.96	2.81	2.66	2.58	2.50	2.42	2.33	2.23	2.13	*26*
	3.06	2.93	2.78	2.63	2.55	2.47	2.38	2.29	2.20	2.10	*27*
	3.03	2.90	2.75	2.60	2.52	2.44	2.35	2.26	2.17	2.06	*28*
	3.00	2.87	2.73	2.57	2.49	2.41	2.33	2.23	2.14	2.03	*29*
	2.98	2.84	2.70	2.55	2.47	2.39	2.30	2.21	2.11	2.01	*30*
	2.80	2.66	2.52	2.37	2.29	2.20	2.11	2.02	1.92	1.80	*40*
	2.63	2.50	2.35	2.20	2.12	2.03	1.94	1.84	1.73	1.60	*60*
	2.47	2.34	2.19	2.03	1.95	1.86	1.76	1.66	1.53	1.38	*120*
	2.32	2.18	2.04	1.88	1.79	1.70	1.59	1.47	1.32	1.00	∞

df for denominator

TABLE VI
Values of $F_{0.05}$

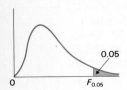

				df for numerator					
	1	*2*	*3*	*4*	*5*	*6*	*7*	*8*	*9*
1	161.4	199.5	215.7	224.6	230.2	234.0	236.8	238.9	240.5
2	18.51	19.00	19.16	19.25	19.30	19.33	19.35	19.37	19.38
3	10.13	9.55	9.28	9.12	9.01	8.94	8.89	8.85	8.81
4	7.71	6.94	6.59	6.39	6.26	6.16	6.09	6.04	6.00
5	6.61	5.79	5.41	5.19	5.05	4.95	4.88	4.82	4.77
6	5.99	5.14	4.76	4.53	4.39	4.28	4.21	4.15	4.10
7	5.59	4.74	4.35	4.12	3.97	3.87	3.79	3.73	3.68
8	5.32	4.46	4.07	3.84	3.69	3.58	3.50	3.44	3.39
9	5.12	4.26	3.86	3.63	3.48	3.37	3.29	3.23	3.18
10	4.96	4.10	3.71	3.48	3.33	3.22	3.14	3.07	3.02
11	4.84	3.98	3.59	3.36	3.20	3.09	3.01	2.95	2.90
12	4.75	3.89	3.49	3.26	3.11	3.00	2.91	2.85	2.80
13	4.67	3.81	3.41	3.18	3.03	2.92	2.83	2.77	2.71
14	4.60	3.74	3.34	3.11	2.96	2.85	2.76	2.70	2.65
15	4.54	3.68	3.29	3.06	2.90	2.79	2.71	2.64	2.59
16	4.49	3.63	3.24	3.01	2.85	2.74	2.66	2.59	2.54
17	4.45	3.59	3.20	2.96	2.81	2.70	2.61	2.55	2.49
18	4.41	3.55	3.16	2.93	2.77	2.66	2.58	2.51	2.46
19	4.38	3.52	3.13	2.90	2.74	2.63	2.54	2.48	2.42
20	4.35	3.49	3.10	2.87	2.71	2.60	2.51	2.45	2.39
21	4.32	3.47	3.07	2.84	2.68	2.57	2.49	2.42	2.37
22	4.30	3.44	3.05	2.82	2.66	2.55	2.46	2.40	2.34
23	4.28	3.42	3.03	2.80	2.64	2.53	2.44	2.37	2.32
24	4.26	3.40	3.01	2.78	2.62	2.51	2.42	2.36	2.30
25	4.24	3.39	2.99	2.76	2.60	2.49	2.40	2.34	2.28
26	4.23	3.37	2.98	2.74	2.59	2.47	2.39	2.32	2.27
27	4.21	3.35	2.96	2.73	2.57	2.46	2.37	2.31	2.25
28	4.20	3.34	2.95	2.71	2.56	2.45	2.36	2.29	2.24
29	4.18	3.33	2.93	2.70	2.55	2.43	2.35	2.28	2.22
30	4.17	3.32	2.92	2.69	2.53	2.42	2.33	2.27	2.21
40	4.08	3.23	2.84	2.61	2.45	2.34	2.25	2.18	2.12
60	4.00	3.15	2.76	2.53	2.37	2.25	2.17	2.10	2.04
120	3.92	3.07	2.68	2.45	2.29	2.17	2.09	2.02	1.96
∞	3.84	3.00	2.60	2.37	2.21	2.10	2.01	1.94	1.88

df for denominator (left axis label)

Adapted from D.B. Owen, *Handbook of Statistical Tables.* Courtesy of the Atomic Energy Commission. Reading, MA: Addison-Wesley, 1962.

TABLE VI (continued)
Values of $F_{0.05}$

df for numerator

10	12	15	20	24	30	40	60	120	∞	
241.9	243.9	245.9	248.0	249.1	250.1	251.1	252.2	253.3	254.3	1
19.40	19.41	19.43	19.45	19.45	19.46	19.47	19.48	19.49	19.50	2
8.79	8.74	8.70	8.66	8.64	8.62	8.59	8.57	8.55	8.53	3
5.96	5.91	5.86	5.80	5.77	5.75	5.72	5.69	5.66	5.63	4
4.74	4.68	4.62	4.56	4.53	4.50	4.46	4.43	4.40	4.36	5
4.06	4.00	3.94	3.87	3.84	3.81	3.77	3.74	3.70	3.67	6
3.64	3.57	3.51	3.44	3.41	3.38	3.34	3.30	3.27	3.23	7
3.35	3.28	3.22	3.15	3.12	3.08	3.04	3.01	2.97	2.93	8
3.14	3.07	3.01	2.94	2.90	2.86	2.83	2.79	2.75	2.71	9
2.98	2.91	2.85	2.77	2.74	2.70	2.66	2.62	2.58	2.54	10
2.85	2.79	2.72	2.65	2.61	2.57	2.53	2.49	2.45	2.40	11
2.75	2.69	2.62	2.54	2.51	2.47	2.43	2.38	2.34	2.30	12
2.67	2.60	2.53	2.46	2.42	2.38	2.34	2.30	2.25	2.21	13
2.60	2.53	2.46	2.39	2.35	2.31	2.27	2.22	2.18	2.13	14
2.54	2.48	2.40	2.33	2.29	2.25	2.20	2.16	2.11	2.07	15
2.49	2.42	2.35	2.28	2.24	2.19	2.15	2.11	2.06	2.01	16
2.45	2.38	2.31	2.23	2.19	2.15	2.10	2.06	2.01	1.96	17
2.41	2.34	2.27	2.19	2.15	2.11	2.06	2.02	1.97	1.92	18
2.38	2.31	2.23	2.16	2.11	2.07	2.03	1.98	1.93	1.88	19
2.35	2.28	2.20	2.12	2.08	2.04	1.99	1.95	1.90	1.84	20
2.32	2.25	2.18	2.10	2.05	2.01	1.96	1.92	1.87	1.81	21
2.30	2.23	2.15	2.07	2.03	1.98	1.94	1.89	1.84	1.78	22
2.27	2.20	2.13	2.05	2.01	1.96	1.91	1.86	1.81	1.76	23
2.25	2.18	2.11	2.03	1.98	1.94	1.89	1.84	1.79	1.73	24
2.24	2.16	2.09	2.01	1.96	1.92	1.87	1.82	1.77	1.71	25
2.22	2.15	2.07	1.99	1.95	1.90	1.85	1.80	1.75	1.69	26
2.20	2.13	2.06	1.97	1.93	1.88	1.84	1.79	1.73	1.67	27
2.19	2.12	2.04	1.96	1.91	1.87	1.82	1.77	1.71	1.65	28
2.18	2.10	2.03	1.94	1.90	1.85	1.81	1.75	1.70	1.64	29
2.16	2.09	2.01	1.93	1.89	1.84	1.79	1.74	1.68	1.62	30
2.08	2.00	1.92	1.84	1.79	1.74	1.69	1.64	1.58	1.51	40
1.99	1.92	1.84	1.75	1.70	1.65	1.59	1.53	1.47	1.39	60
1.91	1.83	1.75	1.66	1.61	1.55	1.50	1.43	1.35	1.25	120
1.83	1.75	1.67	1.57	1.52	1.46	1.39	1.32	1.22	1.00	∞

df for denominator

Sample size n	Significance level, α One-tailed	Two-tailed	Critical value W_l	W_r
3	0.125	0.250	0	6
4	0.063	0.125	0	10
	0.125	0.250	1	9
5	0.031	0.063	0	15
	0.063	0.125	1	14
	0.094	0.188	2	13
6	0.016	0.031	0	21
	0.031	0.063	1	20
	0.047	0.094	2	19
	0.109	0.219	4	17
7	0.008	0.016	0	28
	0.023	0.047	2	26
	0.055	0.109	4	24
	0.109	0.219	6	22
8	0.004	0.008	0	36
	0.012	0.023	2	34
	0.027	0.055	4	32
	0.055	0.109	6	30
	0.098	0.195	8	28
9	0.006	0.012	2	43
	0.010	0.020	3	42
	0.027	0.055	6	39
	0.049	0.098	8	37
	0.102	0.203	11	34
10	0.005	0.010	3	52
	0.010	0.020	5	50
	0.024	0.049	8	47
	0.053	0.105	11	44
	0.097	0.193	14	41
11	0.005	0.010	5	61
	0.009	0.019	7	59
	0.027	0.054	11	55
	0.051	0.102	14	52
	0.103	0.206	18	48
12	0.005	0.009	7	71
	0.010	0.021	10	68
	0.026	0.052	14	64
	0.046	0.092	17	61
	0.102	0.204	22	56

TABLE VII (continued)
Critical values and
significance levels
for a Wilcoxon
signed-rank test

| Sample size | Significance level, α | | Critical value | |
n	One-tailed	Two-tailed	W_l	W_r
13	0.005	0.010	10	81
	0.011	0.021	13	78
	0.024	0.048	17	74
	0.047	0.094	21	70
	0.095	0.191	26	65
14	0.005	0.011	13	92
	0.010	0.020	16	89
	0.025	0.049	21	84
	0.052	0.104	26	79
	0.097	0.194	31	74
15	0.005	0.010	16	104
	0.011	0.022	20	100
	0.024	0.048	25	95
	0.047	0.095	30	90
	0.104	0.208	37	83
16	0.005	0.009	19	117
	0.011	0.021	24	112
	0.025	0.051	30	106
	0.052	0.105	36	100
	0.096	0.193	42	94
17	0.005	0.009	23	130
	0.010	0.020	28	125
	0.025	0.051	35	118
	0.049	0.098	41	112
	0.103	0.207	49	104
18	0.005	0.010	28	143
	0.010	0.021	33	138
	0.024	0.048	40	131
	0.049	0.099	47	124
	0.098	0.196	55	116
19	0.005	0.009	32	158
	0.010	0.020	38	152
	0.025	0.049	46	144
	0.052	0.104	54	136
	0.098	0.196	62	128
20	0.005	0.009	37	173
	0.010	0.019	43	167
	0.024	0.048	52	158
	0.049	0.097	60	150
	0.101	0.202	70	140

TABLE VIII

Critical values for a one-tailed Mann-Whitney test with $\alpha = 0.025$ or a two-tailed Mann-Whitney test with $\alpha = 0.05$

n_2 \ n_1	3		4		5		6		7		8		9		10	
	M_l	M_r	M_l	M_r	M_l	M_r	M_l	M_r	M_l	M_r	M_l	M_r	M_l	M_r	M_l	M_r
3	—	—														
4	6	18	11	25												
5	6	21	12	28	18	37										
6	7	23	12	32	19	41	26	52								
7	7	26	13	35	20	45	28	56	37	68						
8	8	28	14	38	21	49	29	61	39	73	49	87				
9	8	31	15	41	22	53	31	65	41	78	51	93	63	108		
10	9	33	16	44	24	56	32	70	43	83	54	98	66	114	79	131

TABLE IX

Critical values for a one-tailed Mann-Whitney test with $\alpha = 0.05$ or a two-tailed Mann-Whitney test with $\alpha = 0.10$

n_2 \ n_1	3		4		5		6		7		8		9		10	
	M_l	M_r	M_l	M_r	M_l	M_r	M_l	M_r	M_l	M_r	M_l	M_r	M_l	M_r	M_l	M_r
3	6	15														
4	7	17	12	24												
5	7	20	13	27	19	36										
6	8	22	14	30	20	40	28	50								
7	9	24	15	33	22	43	30	54	39	66						
8	9	27	16	36	24	46	32	58	41	71	52	84				
9	10	29	17	39	25	50	33	63	43	76	54	90	66	105		
10	11	31	18	42	26	54	35	67	46	80	57	95	69	111	83	127

TABLE X
Values of $r_{s,\alpha}$

n	$r_{s,\,0.05}$	$r_{s,\,0.025}$	$r_{s,\,0.01}$	$r_{s,\,0.005}$	n
5	0.900	—	—	—	5
6	0.829	0.886	0.943	—	6
7	0.714	0.786	0.893	—	7
8	0.643	0.738	0.833	0.881	8
9	0.600	0.683	0.783	0.833	9
10	0.564	0.648	0.745	0.794	10
11	0.523	0.623	0.736	0.818	11
12	0.497	0.591	0.703	0.780	12
13	0.475	0.566	0.673	0.745	13
14	0.457	0.545	0.646	0.716	14
15	0.441	0.525	0.623	0.689	15
16	0.425	0.507	0.601	0.666	16
17	0.412	0.490	0.582	0.645	17
18	0.399	0.476	0.564	0.625	18
19	0.388	0.462	0.549	0.608	19
20	0.377	0.450	0.534	0.591	20

TABLE XI
Random numbers

Line number	Column number									
	00–09		10–19		20–29		30–39		40–49	
00	15544	80712	97742	21500	97081	42451	50623	56071	28882	28739
01	01011	21285	04729	39986	73150	31548	30168	76189	56996	19210
02	47435	53308	40718	29050	74858	64517	93573	51058	68501	42723
03	91312	75137	86274	59834	69844	19853	06917	17413	44474	86530
04	12775	08768	80791	16298	22934	09630	98862	39746	64623	32768
05	31466	43761	94872	92230	52367	13205	38634	55882	77518	36252
06	09300	43847	40881	51243	97810	18903	53914	31688	06220	40422
07	73582	13810	57784	72454	68997	72229	30340	08844	53924	89630
08	11092	81392	58189	22697	41063	09451	09789	00637	06450	85990
09	93322	98567	00116	35605	66790	52965	62877	21740	56476	49296
10	80134	12484	67089	08674	70753	90959	45842	59844	45214	36505
11	97888	31797	95037	84400	76041	96668	75920	68482	56855	97417
12	92612	27082	59459	69380	98654	20407	88151	56263	27126	63797
13	72744	45586	43279	44218	83638	05422	00995	70217	78925	39097
14	96256	70653	45285	26293	78305	80252	03625	40159	68760	84716
15	07851	47452	66742	83331	54701	06573	98169	37499	67756	68301
16	25594	41552	96475	56151	02089	33748	65289	89956	89559	33687
17	65358	15155	59374	80940	03411	94656	69440	47156	77115	99463
18	09402	31008	53424	21928	02198	61201	02457	87214	59750	51330
19	97424	90765	01634	37328	41243	33564	17884	94747	93650	77668

APPENDIX B

ANSWERS TO SELECTED EXERCISES

Note: Most of the numerical answers presented here were obtained using a computer. If you obtain an answer by hand and perform some intermediate rounding, then your answer may differ somewhat.

CHAPTER 1

Exercises 1.1

— 1.1

a) The *population* is the collection of all individuals, items, or data under consideration in a statistical study.

b) A *sample* is that part of the population from which information is collected.

Exercises 1.2

— 1.3 inferential

— 1.5 inferential

— 1.7 descriptive

— 1.9 descriptive

— 1.11

a) inferential   b) descriptive   c) descriptive
d) inferential   e) inferential

Exercises 1.4

— 1.13

a) ```
MTB > SET C1
DATA> 130 55 45 64 155 66 60 80 102 62
DATA> 58 101 75 111 151 139 81 55 66 90
DATA> 97 77 51 67 125 50 136 55 83 91
DATA> 54 86 100 78 93 113 111 104 96 113
DATA> 96 87 129 109 69 94 99 97 83 97
DATA> END
```

b) `MTB > NAME C1 'ENERGY'`
c) `MTB > PRINT 'ENERGY' (or PRINT C1)`

— 1.15

a) ```
MTB > READ C3 C4
DATA> 65 175
DATA> 67 133
DATA> 71 185
DATA> 71 163
DATA> 66 126
DATA> 75 198
DATA> 67 153
DATA> 70 163
DATA> 71 159
DATA> 69 151
DATA> 69 155
DATA> END
```

b) `MTB > NAME C3 'HEIGHT' C4 'WEIGHT'`
c) `MTB > PRINT 'HEIGHT' 'WEIGHT'`
   `(or PRINT C3 C4)`

REVIEW TEST FOR CHAPTER 1

1. Answers will vary.

2. In conducting an inferential study, information will be obtained from a sample of the population. Generally, that information must be organized and summarized in a clear and effective way prior to the application of inferential methods. Thus, almost any inferential study will involve aspects of descriptive statistics.

3. descriptive

4. inferential

5. inferential

6. descriptive

7. inferential

8. a) MTB > SET C1
    DATA> 48 41 57 83 41 55 59
    DATA> 61 38 48 79 75 77 7
    DATA> 54 23 47 56 79 68 61
    DATA> 64 45 53 82 68 38 70
    DATA> 10 60 83 76 21 65 47
    DATA> END
  b) MTB > NAME C1 'AGES'
  c) MTB > PRINT 'AGES' (or PRINT C1)

9. a) MTB > READ C2 C3
    DATA> 85 3.23
    DATA> 77 2.95
    DATA> 98 3.70
    DATA> 86 3.85
    DATA> 66 2.37
    DATA> 55 2.75
    DATA> 75 2.73
    DATA> 64 2.16
    DATA> 71 2.59
    DATA> 82 2.78
    DATA> END
  b) MTB > NAME C2 'MATHPSC' C3 'FRESHGPA'
  c) MTB > PRINT 'MATHPSC' 'FRESHGPA'
    (or PRINT C2 C3)

CHAPTER 2

Exercises 2.1

__ **2.1**   It serves as an aid in the choice of the correct statistical method.

__ **2.3**   metric

__ **2.5**   ordinal

__ **2.7**

a) metric   b) ordinal   c) frequency
d) qualitative

__ **2.9**

a) ordinal
b) metric (could also be considered frequency data)

__ **2.11**   qualitative

__ **2.13**   height, weight, and age

Exercises 2.2

__ **2.15**   No, data must be numerical for class limits and class marks to make sense.

__ **2.17**

a) The frequency of a class is the number of data values in the class, whereas the relative frequency of a class is the ratio of the class frequency to the total number of pieces of data.
b) The percentage of a class is 100 times the relative frequency of the class. Equivalently, the relative frequency of a class is the percentage of the class expressed as a decimal.

__ **2.19**

Content (ml)	Frequency	Relative frequency	Class mark
910–929	1	0.033	919.5
930–949	1	0.033	939.5
950–969	3	0.100	959.5
970–989	9	0.300	979.5
990–1009	7	0.233	999.5
1010–1029	6	0.200	1019.5
1030–1049	2	0.067	1039.5
1050–1069	1	0.033	1059.5
	30	0.999	

__ **2.21**

Consumption (mil. BTU)	Frequency	Relative frequency	Class mark
40–49	1	0.02	44.5
50–59	7	0.14	54.5
60–69	7	0.14	64.5
70–79	3	0.06	74.5
80–89	6	0.12	84.5
90–99	10	0.20	94.5
100–109	5	0.10	104.5
110–119	4	0.08	114.5
120–129	2	0.04	124.5
130–139	3	0.06	134.5
140–149	0	0.00	144.5
150–159	2	0.04	154.5
	50	1.00	

__ 2.23

Number of cars sold	Frequency	Relative frequency
0	7	0.135
1	15	0.288
2	12	0.231
3	9	0.173
4	5	0.096
5	3	0.058
6	1	0.019
	52	1.000

__ 2.25

Number of days missed	Frequency	Relative frequency
0	4	0.050
1	2	0.025
2	14	0.175
3	10	0.125
4	16	0.200
5	18	0.225
6	10	0.125
7	6	0.075
	80	1.000

__ 2.27

Starting salary ($thousands)	Frequency	Relative frequency	Class mark
16–under 17	3	0.086	16.5
17–under 18	3	0.086	17.5
18–under 19	5	0.143	18.5
19–under 20	9	0.257	19.5
20–under 21	9	0.257	20.5
21–under 22	4	0.114	21.5
22–under 23	1	0.029	22.5
23–under 24	1	0.029	23.5
	35	1.001	

__ 2.29

Sales (millions)	Frequency	Relative frequency	Class mark
0–under 1	1	0.042	0.5
1–under 2	3	0.125	1.5
2–under 3	10	0.417	2.5
3–under 4	4	0.167	3.5
4–under 5	1	0.042	4.5
5–under 6	1	0.042	5.5
6–under 7	2	0.083	6.5
7–under 8	1	0.042	7.5
8–under 9	1	0.042	8.5
	24	1.002	

__ 2.31

Champion	Frequency	Relative frequency
Oklahoma	2	0.077
Oklahoma State	4	0.154
Iowa State	7	0.269
Michigan State	1	0.038
Iowa	11	0.423
Arizona State	1	0.038
	26	0.999

Exercises 2.3

__ 2.35   A frequency histogram displays the class frequencies on the vertical axis, whereas a relative-frequency histogram displays the class relative frequencies on the vertical axis.

___ 2.37

a)

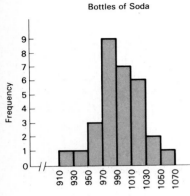

Bottles of Soda

b)

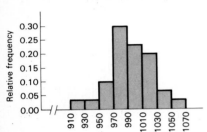

Bottles of Soda

___ 2.39

a)

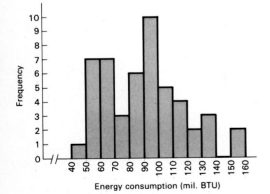

Energy Consumption for Southern Households

b)

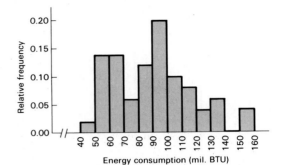

Energy Consumption for Southern Households

___ 2.41

a)

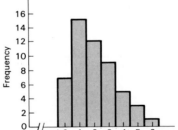

Car Sales

b)

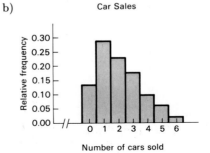

Car Sales

___ 2.43

a)

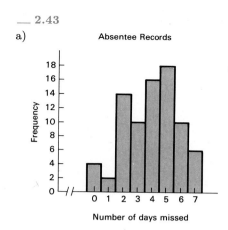

b)

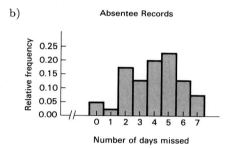

___ 2.45

a)

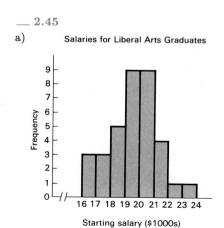

b)

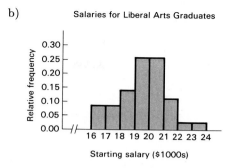

___ 2.47

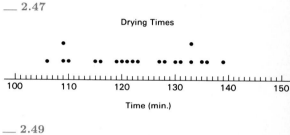

___ 2.49

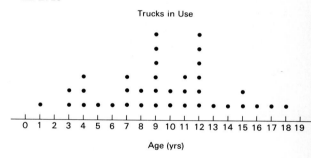

___ 2.51

a)

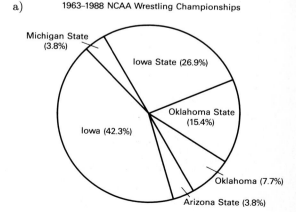

b)

1963–1988 NCAA Wrestling Championships

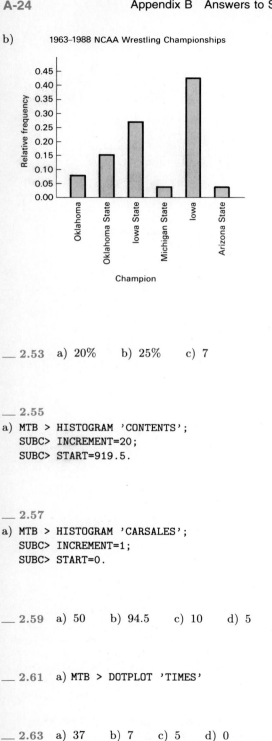

___ 2.53   a) 20%   b) 25%   c) 7

___ 2.55

a)
```
MTB > HISTOGRAM 'CONTENTS';
SUBC> INCREMENT=20;
SUBC> START=919.5.
```

___ 2.57

a)
```
MTB > HISTOGRAM 'CARSALES';
SUBC> INCREMENT=1;
SUBC> START=0.
```

___ 2.59   a) 50   b) 94.5   c) 10   d) 5

___ 2.61   a) MTB > DOTPLOT 'TIMES'

___ 2.63   a) 37   b) 7   c) 5   d) 0

Exercises 2.4

___ 2.71

a)
91	4
92	
93	
94	6
95	9 7
96	4
97	7 5 4 7
98	6 9 4 8 7
99	0 6 1 9 5 7
100	1
101	4 8 0 7
102	5 8
103	0 1
104	
105	
106	0

b)
91	4
92	
93	
94	6
95	7 9
96	4
97	4 5 7 7
98	4 6 7 8 9
99	0 1 5 6 7 9
100	1
101	0 4 7 8
102	5 8
103	0 1
104	
105	
106	0

___ 2.73

a)
4	5
5	8 4 5 1 0 5 5
6	4 7 9 6 0 6 2
7	7 5 8
8	6 7 1 0 3 3
9	7 6 3 4 9 7 6 0 1 7
10	1 0 9 4 2
11	1 3 1 3
12	9 5
13	0 9 6
14	
15	5 1

b)
```
 4 | 5
 5 | 0 1 4 5 5 5 8
 6 | 0 2 4 6 6 7 9
 7 | 5 7 8
 8 | 0 1 3 3 6 7
 9 | 0 1 3 4 6 6 7 7 7 9
10 | 0 1 2 4 9
11 | 1 1 3 3
12 | 5 9
13 | 0 6 9
14 |
15 | 1 5
```

___ 2.75

a)
```
2 | 6 7 3
3 | 9 9 1 5 3 9 3 1 9
4 | 3 8 8 2 2 8 7 4 7 5 3 4 9 5 0 1 4
5 | 5 7 5 6 2 8 1 5
6 | 2 8 1 5 3 6 0 9
7 | 3 0 1 4
8 | 2
```

b)
```
2 | 3
2 | 6 7
3 | 1 3 3 1
3 | 9 9 5 9 9
4 | 3 2 2 4 3 4 0 1 4
4 | 8 8 8 7 7 5 9 5
5 | 2 1
5 | 5 7 5 6 8 5
6 | 2 1 3 0
6 | 8 5 6 9
7 | 3 0 1 4
7 |
8 | 2
8 |
```

___ 2.77

a)
```
2 |
2 | 9 9
3 | 3 1 3 2 3 4 3
3 | 6 9 5 9 5 7 9 9 5 7 8
4 | 0 0 1 2 0 4 4 0 3 2 1 2 4 1 1 4 4
4 | 8 8 5 9 5 6 5 5 5 7 9 8 5 9 5 9 9 7
5 | 1 1 3 1 2 1
5 | 8 7 5 7 9 5 7
6 |
6 | 9
7 | 0 3
7 |
```

b)
```
2 |
2 |
2 |
2 |
2 | 9 9
3 | 1
3 | 3 3 2 3 3
3 | 5 5 4 5
3 | 6 7 7
3 | 9 9 9 9 8
4 | 0 0 1 0 0 1 1 1
4 | 2 3 2 2
4 | 5 5 5 4 4 5 5 4 5 4 4
4 | 6 7 7
4 | 8 8 9 9 8 9 9 9
5 | 1 1 1 1
5 | 3 2
5 | 5 5
5 | 7 7 7
5 | 8 9
6 |
6 |
6 |
6 |
6 | 9
7 | 0
7 | 3
7 |
7 |
7 |
```

___ 2.79

a)   (i) MTB > STEM-AND-LEAF 'CONTENTS';
      SUBC> INCREMENT=10.
  (ii) MTB > STEM-AND-LEAF 'CONTENTS';
      SUBC> INCREMENT=5.
  (iii) MTB > STEM-AND-LEAF 'CONTENTS';
      SUBC> INCREMENT=2.

b) *Note:* In (iii) of part (a), Minitab will give the error message: * INCREMENT specified was too small. This is because the number of lines required to construct a stem-and-leaf diagram for this data set with five lines per stem exceeds what Minitab will accept. However, the command and subcommand stated in (iii) of part (a) are, nonetheless, correct.

___ 2.81

a) 20   b) 9.89 mm   c) 5   d) 5   e) 5
f) 9.89, 9.94, 9.95, 9.96, 9.97, 9.98, 9.98, 9.99, 9.99

Exercises 2.5

___ **2.83**

c) They give the misleading impression that the district average is much greater relative to the national average than it actually is.

___ **2.85**

a) It is a truncated graph.

b)

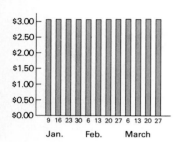

Money Supply
(weekly average of M2 in trillions)

c)

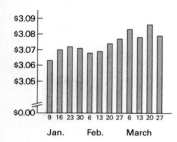

Money Supply
(weekly average of M2 in trillions)

2. a)

Age at inauguration	Frequency	Relative frequency	Class mark
40–44	2	0.050	42
45–49	5	0.125	47
50–54	12	0.300	52
55–59	12	0.300	57
60–64	6	0.150	62
65–69	3	0.075	67
	40	1.000	

b)

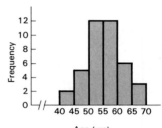

Ages at Inauguration
for First 40 U.S. Presidents

Data from *The World Almanac, 1989*

3.

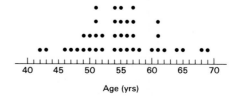

Ages at Inauguration
for First 40 U.S. Presidents

Data from *The World Almanac, 1989*

REVIEW TEST FOR CHAPTER 2

1. a) ordinal   b) metric   c) qualitative
   d) frequency

4. a) **4** | 2 3 6 7 8 9 9
   5 | 0 0 1 1 1 1 2 2 4 4 4 4 5 5 5 5 6 6 6 7 7 7 7 8
   6 | 0 1 1 1 2 4 5 8 9

b) **4** | 2 3
   4 | 6 7 8 9 9
   5 | 0 0 1 1 1 1 2 2 4 4 4 4
   5 | 5 5 5 5 6 6 6 7 7 7 7 8
   6 | 0 1 1 1 2 4
   6 | 5 8 9

c) The second one (that is, the one with two lines per stem).

b)
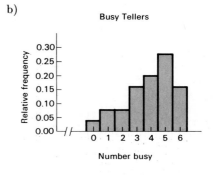

5. a)
```
MTB > HISTOGRAM 'AGES';
SUBC> INCREMENT=5;
SUBC> START=42.
```
b)
```
MTB > DOTPLOT 'AGES'
```
c)
```
MTB > STEM-AND-LEAF 'AGES';
SUBC> INCREMENT=5.
```

6. a) 22   b) 27   c) 3
  d) 5, 12
  e) Times

7. a) 2   b) 2   c) 4
  d) 48 minutes
  e) 30, 31, 31, 37

8. a)

Number busy	Frequency	Relative frequency
0	1	0.04
1	2	0.08
2	2	0.08
3	4	0.16
4	5	0.20
5	7	0.28
6	4	0.16
	25	1.00

9. a)
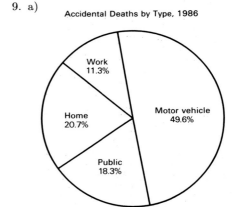

b)
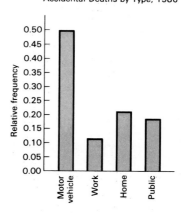

10. a)

High	Freq.	Relative frequency	Class mark
200–under 400	2	0.056	300
400–under 600	5	0.139	500
600–under 800	5	0.139	700
800–under 1000	13	0.361	900
1000–under 1200	6	0.167	1100
1200–under 1400	2	0.056	1300
1400–under 1600	1	0.028	1500
1600–under 1800	0	0.000	1700
1800–under 2000	1	0.028	1900
2000–under 2200	0	0.000	2100
2200–under 2400	0	0.000	2300
2400–under 2600	0	0.000	2500
2600–under 2800	1	0.028	2700
	36	1.002	

b)

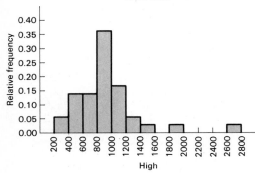

Dow Jones Highs
1952–1987

Data from *The World Almanac, 1989*

11. b) The percentage of women in the labor force for 1985 is about two and two-thirds times that for 1970.
c) About one and one-half.
d) Because it is a truncated graph.
e) Start the graph at zero instead of 30.

CHAPTER 3

Exercises 3.1

____ 3.1   To indicate where the center or most typical value of a data set lies.

____ 3.3   mean = 33.5 years, median = 34.5 years, modes = 24, 28, 37 years

____ 3.5   mean = 40.79¢/lb, median = 41.80¢/lb, no mode

____ 3.7   mean = 306.0 ml, median = 306.5 ml, modes = 300, 311 ml

____ 3.9   mean = 193.0 thous, median = 79.0 thous, no mode

____ 3.11   The *median,* because, unlike the mean, it is not affected strongly by the relatively few homes that have an extremely large or small square footage.

____ 3.13
a) Iowa
b) No, neither the mean nor the median can be used as a measure of central tendency for qualitative data.

____ 3.15   a) 13.9   b) 4   c) median

____ 3.17
a) MTB > MEAN 'AGES'
b) MTB > MEDIAN 'AGES'

Exercises 3.2

____ 3.27   a) 46   b) 4   c) 11.5

____ 3.29   a) 331,143 mi   b) 8   c) 41,392.9 mi

____ 3.31   a) $31,061   b) 18   c) $1,725.6

____ 3.33
a) $\bar{x} = \$97.5$
b)

x	x^2	$x - \bar{x}$	$(x - \bar{x})^2$
75	5,625	−22.5	506.25
98	9,604	0.5	0.25
130	16,900	32.5	1056.25
63	3,969	−34.5	1190.25
112	12,544	14.5	210.25
107	11,449	9.5	90.25
585	60,091	0	3053.50

____ 3.35
a) $\bar{x} = 105.1$

b)

x	x^2	$x - \bar{x}$	$(x - \bar{x})^2$
106	11,236	0.9	0.81
94	8,836	−11.1	123.21
118	13,924	12.9	166.41
109	11,881	3.9	15.21
118	13,924	12.9	166.41
95	9,025	−10.1	102.01
99	9,801	−6.1	37.21
97	9,409	−8.1	65.61
109	11,881	3.9	15.21
106	11,236	0.9	0.81
1051	111,153	0	692.90

c) In the amount of variation.
d) The data set for Brand A.
e) Brand A: $s = 0.50$; Brand B: $s = 1.62$
f) Yes, because the data for Brand A, which has less variation than the data for Brand B, also has a smaller standard deviation.

___ **3.53** a) 16.1   b) 16.1

___ **3.55**

a) MTB > MAX 'PRICES'
   MTB > MIN 'PRICES'
b) *Note:* In response to the MAX command, Minitab will print the highest price; and in response to the MIN command, Minitab will print the lowest price. To obtain the range, subtract the lowest price from the highest price.

___ **3.57** a) MTB > STDEV 'PRICES'

Exercises 3.3

___ **3.41**   To indicate the amount of variation in a data set.

___ **3.43**   a) 20 yr   b) 7.2 yr   c) 7.2 yr

___ **3.45**   a) 10.6¢/lb   b) 3.38¢/lb   c) 3.38¢/lb

___ **3.47**   a) 31 ml   b) 8.7 ml   c) 8.7 ml

___ **3.49**

a) 501 thous   b) 205.5 thous   c) 205.5 thous

___ **3.51**

a) Brand A: $\bar{x} = 9.80$; Brand B: $\bar{x} = 9.80$
b) Brand A: median = 9.7; Brand B: median = 9.7

Exercises 3.4

___ **3.63**

a) Data Set 4
b) Data Set 3: $\bar{x} = 83$, $s = 7.8$
   Data Set 4: $\bar{x} = 83$, $s = 22.3$
c) See Figures A.1 and A.2 at the bottom of the page.
e) Yes. In fact, all of the data in each data set lies within three standard deviations to either side of the mean.

FIGURE A.1   Data Set 3. $\bar{x} = 83$, $s = 7.8$

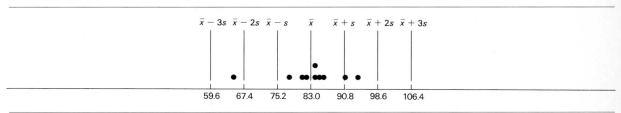

FIGURE A.2   Data Set 4. $\bar{x} = 83$, $s = 22.3$

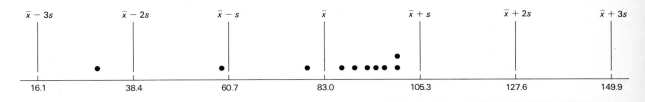

___ **3.65**

a) For any data set, at least 36% of the data lies within 1.25 standard deviations to either side of its mean.

b) For any data set, at least 92% of the data lies within 3.5 standard deviations to either side of its mean.

c) For any data set, at least 96% of the data lies within five standard deviations to either side of its mean.

___ **3.67**

a) At least 75%; at least 89%.

b) 90%; 100%

c) 90%; 100%

d) Chebychev's rule does not necessarily give precise estimates for the percentage of data that lies within a specified number of standard deviations to either side of the mean.

___ **3.69**

a)

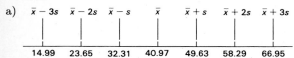

b) At least 225 of the 300 lodging establishments have room rates between $23.65 and $58.29.

c) At least 267 of the 300 lodging establishments have room rates between $14.99 and $66.95.

___ **3.71**

a)

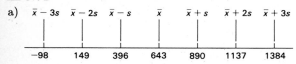

b) At least 188 of the 250 households have vegetable gardens that are between 149 and 1137 sq ft.

c) At least 223 of the 250 households have vegetable gardens that are at most 1384 sq ft.

___ **3.73**

a)

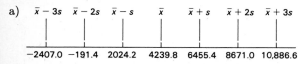

b) At least 23 of the 30 taxable returns have an income tax of at most $8671.0.

c) At least 27 of the 30 taxable returns have an income tax of at most $10,886.6.

___ **3.75**

a) 3.01, −2.42, −0.46. A room rate of $67 is 3.01 standard deviations above the mean; a room rate of $20 is 2.42 standard deviations below the mean; and a room rate of $37 is 0.46 standard deviations below the mean.

b) A room rate of $67 is greater than at least 89.0% ($= 1 − 1/3.01^2$) of the room rates in the sample; a room rate of $20 is less than at least 82.9% ($= 1 − 1/2.42^2$) of the room rates in the sample; and a room rate of $37 is near the mean and below it.

___ **3.77**

a) 4, −1.75, −0.10. A vegetable garden whose size is 1631 square feet is four standard deviations above the mean; a vegetable garden whose size is 211 square feet is 1.75 standard deviations below the mean; and a vegetable garden whose size is 618 square feet is 0.10 standard deviations below the mean.

b) A vegetable garden whose size is 1631 square feet is larger than at least 93.8% ($= 1 − 1/4^2$) of the vegetable gardens in the sample; a vegetable garden whose size is 211 square feet is smaller than at least 67.3% ($= 1 − 1/1.75^2$) of the vegetable gardens in the sample; and a vegetable garden whose size is 618 square feet is near the mean size and below it.

___ **3.79**   A sale price of $95,500 has a z-score of $(95{,}500 − 110{,}258)/5237 = −2.82$, relative to the sample of 25 sale prices obtained by the realtor. By Property 3 of Chebychev's rule, this implies that a sale price of $95,500 is less than at least 87.4% ($= 1 − 1/2.82^2$) of the 25 sale prices obtained by the realtor. It does appear that the home you are contemplating buying is a good deal.

Exercises 3.5

___ **3.85**   a) $\bar{x} = 3.2$ persons   b) $s = 1.3$ persons

___ **3.87**   a) $\bar{x} = 21.6$ years   b) $s = 3.8$ years

___ **3.89**

a) $\bar{x} = 39.5$ bu/acre, $s = 1.9$ bu/acre

b)

Yield	Frequency
36	1
37	3
38	5
39	8
40	5
41	3
42	1
43	4

c) $\bar{x} = 39.5$ bu/acre, $s = 1.9$ bu/acre

___ **3.91** $\bar{x} \approx 105.0$, $s \approx 16.4$

___ **3.93** $\bar{x} \approx 62.8$ minutes, $s \approx 10.7$ minutes

___ **3.95**
a) $\bar{x} = 89.7$ mil BTU, $s = 27.3$ mil BTU
b) $\bar{x} \approx 89.9$ mil BTU, $s \approx 27.3$ mil BTU

___ **3.97** Because the class mark of each class provides only a typical value for the data values in the class and not the data values themselves.

Exercises 3.6

___ **3.101** $Q_1 = 68.5$, $Q_2 = 83.0$, $Q_3 = 89.5$. Thus, 25% of the exam scores are below 68.5, 25% are between 68.5 and 83.0, 25% are between 83.0 and 89.5, and 25% are above 89.5.

___ **3.103** $Q_1 = \$1105.5$, $Q_2 = \$1231.0$, $Q_3 = \$1386.0$. Consequently, 25% of the energy expenditures are less than $1105.5, 25% are between $1105.5 and $1231.0, 25% are between $1231.0 and $1386.0, and 25% are greater than $1386.0.

___ **3.105** The deciles, $D_1, D_2, \ldots, D_9$, are, respectively, 51.0, 65.5, 72.5, 78.5, 83.0, 85.5, 88.5, 90.0, and 96.0. Thus, 10% of the exam scores are below 51.0, 10% are between 51.0 and 65.5, and so forth.

___ **3.107**

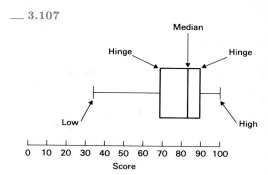

From the box-and-whisker diagram, we see that there is less variation in the top two quarters of the exam-score data than in the bottom two, and that the first quarter has the greatest variation of all.

___ **3.109**

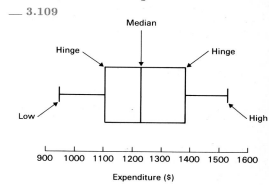

From the box-and-whisker diagram, we see that there is somewhat more variation in the first and third quarters of the energy-expenditure data than in the second and fourth quarters.

___ **3.111**
a) MTB > BOXPLOT 'SCORES'
b) MTB > LVALS 'SCORES'

___ **3.113**
a) The least variation occurs in the third quarter of the beef-consumption data, the next least in the second quarter, the next least in the fourth quarter, and the greatest in the first quarter. Also, $\min \approx 48$, $\max \approx 148$, $Q_1 \approx 100$, $Q_2 \approx 114$, $Q_3 \approx 126$.
b) $\min = 48$, $\max = 148$, $Q_1 = 100$, $Q_2 = 113.5$, $Q_3 = 126$.

Exercises 3.7

___ **3.119** To describe the entire population.

— 3.121

a) $\bar{x} = 75.0$ in   b) $s = 6.2$ in   c) $\mu = 75.0$ in

d) $\sigma = 5.6$ in

e) Both means are computed in the same way: Sum the data and then divide by the total number of pieces of data.

f) The sample standard deviation, s, and the population standard deviation, σ, are computed differently: In the defining formula for s, we divide by one less than the total number of pieces of data, whereas in the defining formula for σ, we divide by the total number of pieces of data.

— 3.123   a) $\mu = \$46.3$ b) $\sigma = \$17.8$

— 3.125   a) $\mu = \$103.8$ mil   b) $\sigma = \$114.4$ mil

— 3.127   It is a parameter, being a descriptive measure for a population.

— 3.129   a) $\mu = 2$ b) $\sigma = 1.57$

— 3.131

a) $\mu \approx 13.0$ years   b) $\sigma \approx 6.2$ years

c) Because the class mark of each class provides only a typical value for the data values in the class and not the data values themselves.

— 3.133

a) 313.0, 366.2

b) 299.7, 379.5

c) −3.50, 2.14, −0.12

d) A gestation period of 293 days is shorter than at least 91.8% (= $1 - 1/3.50^2$) of all gestation periods for the Morgan horse; a gestation period of 368 days is longer than at least 78.2% (= $1 - 1/2.14^2$) of all gestation periods for the Morgan horse; and a gestation period of 338 days is near the mean gestation period for the Morgan horse and below it.

— 3.135

a) $z = -3.13$ (3.13 standard deviations below the mean)

b) By Property 3 of Chebychev's rule, a gas mileage of 21.4 mpg is less than at least 89.8% (= $1 - 1/3.13^2$) of the gas mileages for all cars of this model. It does appear that your car is getting unusually low gas mileage.

_____

REVIEW TEST FOR CHAPTER 3

1. a) 11.3 kg; 16.1 kg

   b) 12.0 kg; 17.0 kg

   c) 17 kg; 12, 16, 19, 23 kg

2. the median

3. the mode

4. a) $\bar{x} = 4.0$ minutes

   b) Range = 14 minutes

   c) $s = 4.1$ minutes

5. a) See Figure A.3 at the bottom of the page.

   b) 75%   c) 91.7%

   d) Its generality—Chebychev's rule holds for any data set.

6. a) $\bar{x} - 3s$ $\bar{x} - 2s$ $\bar{x} - s$ $\bar{x}$ $\bar{x} + s$ $\bar{x} + 2s$ $\bar{x} + 3s$

18.3	31.7	45.1	58.5	71.9	85.3	98.7

   b) 31.7, 85.3   c) 89

7. a) 0.04, −2.05, 1.53. An age of 59 years is 0.04 standard deviations above the mean; an age of 31 years is 2.05 standard deviations below the mean; and an age of 79 years is 1.53 standard deviations above the mean.

   b) An age of 59 years is near the mean and above it; an age of 31 years is less than at least 76.2% (= $1 - 1/2.05^2$) of the ages in the sample; and an age of 79 years is greater than at least 57.3% (= $1 - 1/1.53^2$) of the ages in the sample.

FIGURE A.3

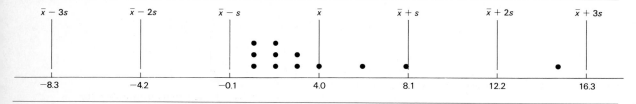

8. a) $Q_1 = 48.0$ yr, $Q_2 = 59.5$ yr, $Q_3 = 68.5$ yr

b)

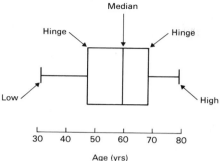

9. a) MTB > MEAN 'AGES'
   b) MTB > MEDIAN 'AGES'
   c) MTB > MAX 'AGES'
      MTB > MIN 'AGES'
      Note: Alternatively, we can use the LVALS command; i.e., MTB > LVALS 'AGES'.
   d) MTB > STDEV 'AGES'

10. a) MTB > BOXPLOT 'AGES'
    b) MTB > LVALS 'AGES'

11. a) The least variation occurs in the third quarter of the age data, the next least in the fourth quarter, the next least in the second quarter, and the greatest in the first quarter. Also, min ≈ 31, max ≈ 79, $Q_1 \approx 48$, $Q_2 \approx 60$, $Q_3 \approx 69$.
    b) min $= 31$, max $= 79$, $Q_1 = 48$, $Q_2 = 59.5$, $Q_3 = 68.5$.

12. a) $\bar{x} = 71.7$ b) $s = 2.5$

13. a) $\mu = 17.43$ thous   b) $\sigma = 10.22$ thous
    c) 1.76, -0.08. The enrollment at UCLA is 1.76 standard deviations above the mean and the enrollment at UCSD is 0.08 standard deviations below the mean.

14. a) $\mu \approx \$76.8$ thous   b) $\sigma \approx \$15.5$ thous
    c) Because the class mark of each class provides only a typical value for the data values in the class and not the data values themselves.

15. a) A sample mean. It is the mean price per gallon for the sample of 10,000 gasoline service stations.
    b) $\bar{x}$
    c) A statistic. It is a descriptive measure for a sample.

CHAPTER 4

Exercises 4.1

___ **4.1**
a) 0.125   b) 0.250   c) 0.750   d) 0   e) 1

___ **4.3**
a) 0.291   b) 0.288   c) 0.652   d) 0.192

___ **4.5**
a) 0.139   b) 0.500   c) 0.222   d) 0.111

___ **4.7**   0.020

___ **4.9**   a) 0.705   b) 0.899   c) 0.194

___ **4.11**   (b) and (d)

___ **4.13**   The event in part (e) is certain; the event in part (d) is impossible.

Exercises 4.2

___ **4.19**

$A =$ ▪ ▪ ▪

$B =$ ▪ ▪ ▪

$C =$ ▪ ▪

$D =$ ▪

___ **4.21**
$A = \{HHTT, HTHT, HTTH, THHT, THTH, TTHH\}$
$B = \{TTHH, TTHT, TTTH, TTTT\}$
$C = \{HHHH, HHHT, HHTH, HHTT, HTHH, HTHT, HTTH, HTTT\}$
$D = \{HHHH, TTTT\}$

___ **4.23**
a) $(not\ A) =$ ▪ ▪ ▪

   The event the die comes up odd.

b) $(A\ \&\ B) =$ ▪ ▪

   The event the die comes up four or six.

c) $(B \text{ or } C) =$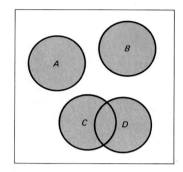

The event the die does *not* come up three.

___ **4.25**

a) $(\text{not } B) = \{$HHHH, HHHT, HHTH, HHTT,
HTHH, HTHT, HTTH, HTTT,
THHH, THHT, THTH, THTT$\}$
The event that at least one of the first two tosses is heads.

b) $(A \& B) = \{$TTHH$\}$
The event that the first two tosses are tails and the last two are heads.

c) $(C \text{ or } D) = \{$HHHH, HHHT, HHTH, HHTT,
HTHH, HTHT, HTTH, HTTT,
TTTT$\}$
The event that the first toss is a head or all four tosses are tails.

___ **4.27**

a) $(\text{not } A)$ is the event that the employee selected missed at least four days. There are 50 employees that missed at least four days.

b) $(A \& B)$ is the event that the employee selected missed between one and three days, inclusive. There are 26 employees that missed between one and three days, inclusive.

c) $(C \text{ or } D)$ is the event that the employee selected missed at least four days. There are 50 employees that missed at least four days. [*Note:* From part (a), we see that $(\text{not } A) = (C \text{ or } D)$.]

___ **4.29**

a) $(\text{not } C)$ is the event that the person selected is 45 years old or older. There are 28,116 thousand such people.

b) $(\text{not } B)$ is the event that the person selected is either under 20 or over 54. There are 18,799 thousand such people.

c) $(B \& C)$ is the event that the person selected is between 20 and 44, inclusive. There are 63,774 thousand such people.

d) $(A \text{ or } D)$ is the event that the person selected is either under 20 or over 54. There are 18,799 thousand such people. [*Note:* From part (b), we see that $(\text{not } B) = (A \text{ or } D)$.]

___ **4.31**

a) No.   b) Yes.   c) No.
d) Yes, events B, C, and D. No.

___ **4.33**

a) mutually exclusive

b) not mutually exclusive
c) mutually exclusive
d) not mutually exclusive
e) not mutually exclusive

___ **4.35**

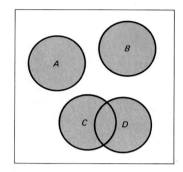

Exercises 4.3

___ **4.37**   0.2; $P(E) = 0.2$

___ **4.39**

a) 0.740
b) $S = (A \text{ or } B \text{ or } C)$
c) 0.07, 0.28, 0.39
d) 0.740

___ **4.41**

a) 0.816   b) 0.054   c) 0.335
d) 81.6% of the businesses in the United States had receipts under \$100,000; 5.4% had receipts of at least \$500,000; and 33.5% had receipts between \$25,000 and \$499,999.

___ **4.43**   a) 0.930   b) 0.940
Note: You should have used the complementation rule to obtain these two probabilities.

___ **4.45**   a) 0.971   b) 0.279
Note: You should have used the complementation rule to obtain these two probabilities.

___ **4.47**

a) 0.167, 0.056, 0.028, 0.056, 0.028, 0.139, 0.167
b) 0.223   c) 0.112
d) (i) 0.278   (ii) 0.278

___ **4.49**

a) 0.526, 0.070, 0.042
b) 0.554; 55.4% of U.S. adults are either female or divorced (or both).
c) 0.474

___ **4.51**   0.267

Exercises 4.4

___ 4.55

a) 8   b) 3274   c) 863   d) 1471   e) 502

___ 4.57

a) 12   b) 108,674.4 thous   c) 44,670.4 thous
d) 10,564.6 thous   e) 686.0 thous

___ 4.59

a) The missing entries in the second, third, and
   fourth rows are, respectively, 157, 247, and 30.
b) 15   c) 690 thous   d) 313 thous
e) 128 thous   f) 1026 thous

___ 4.61

a) The institution selected is private; the institution
   selected is in the South; the institution selected is
   a public school in the West.
b) 0.551; 0.316; 0.096. 55.1% of institutions of higher
   education are private; 31.6% are in the South;
   and 9.6% are public schools in the West.
c)

Type

	Public T_1	Private T_2	$P(R_i)$
Northeast R_1	0.081	0.170	0.251
Midwest R_2	0.110	0.154	0.264
South R_3	0.163	0.153	0.316
West R_4	0.096	0.074	0.170
$P(T_j)$	0.449	0.551	1.000

Region

___ 4.63

a) The person selected is employed; the person selected has completed between 13 and 15 years of
   school; the person selected is employed and has
   completed between 13 and 15 years of school.
b) 0.903; 0.180; 0.168
c) (i) 0.915   (ii) 0.915

d)

Employment status

	Employed E_1	Unemployed E_2	$P(S_i)$
Less than 8 S_1	0.033	0.006	0.038
8 S_2	0.030	0.005	0.035
9–11 S_3	0.117	0.026	0.143
12 S_4	0.369	0.042	0.411
13–15 S_5	0.168	0.012	0.180
16 or more S_6	0.186	0.006	0.192
$P(E_j)$	0.903	0.097	1.000

Years of school completed

___ 4.65

a) (i) A_3 (ii) T_2 (iii) (T_1 & A_5)
b) (i) 0.241   (ii) 0.288   (iii) 0.015
c)

Tenure of operator

	Full owner T_1	Part owner T_2	Tenant T_3	Total
Under 50 A_1	21.5	3.0	3.4	27.9
50–179 A_2	22.7	6.3	3.8	32.9
180–499 A_3	10.6	10.0	3.5	24.1
500–999 A_4	2.3	5.2	1.2	8.7
1000+ A_5	1.5	4.3	0.7	6.5
Total	58.6	28.8	12.6	100.0

Acreage

Exercises 4.5

__ **4.69**
a) 0.077   b) 0.333   c) 0.077   d) 0
e) 0.231   f) 1   g) 0.231   h) 0.167

__ **4.71**
a) 0.125   b) 0.132   c) 0.342
d) 12.5% of the employees at Cudahey Masonry, Inc., missed exactly three days of work; 13.2% of those who missed at least one day of work, missed exactly three days; and 34.2% of those who missed at least one day of work, missed at most three days.

__ **4.73**
a) 0.251   b) 0.308   c) 0.676
d) 25.1% of all institutions of higher education are in the Northeast; 30.8% of all private institutions of higher education are in the Northeast; 67.6% of all institutions of higher education in the Northeast are private schools.

__ **4.75**
a) 0.126   b) 0.035   c) 0.278   d) 0.278

__ **4.77**
a) 0.187   b) 0.103   c) 0.551   d) 0.248
e) 18.7% of the members of the 98th Congress are senators; 10.3% of the members of the 98th Congress are Republican senators; 55.1% of the senators in the 98th Congress are Republicans; and 24.8% of the Republicans in the 98th Congress are senators.

__ **4.79**   33.0%

Exercises 4.6

__ **4.83**   0.008; 0.8% of all families are farm families making at least $25,000 per year.

__ **4.85**
a) 0.1   b) 0.111   c) 0.011   d) 0.222

__ **4.87**
a) $\frac{23}{50} \cdot \frac{27}{49} = 0.253$ b) $\frac{23}{50} \cdot \frac{22}{49} = 0.207$
c) Let $R1$ denote the event that the first governor selected is a Republican, $R2$ denote the event that the second governor selected is a Republican, $D1$ denote the event that the first governor selected is a Democrat, and $D2$ denote the event that the second governor selected is a Democrat. See Figure A.4 at the bottom of the page.
d) 0.493

__ **4.89**
a) 0.151   b) 0.050
c) No, because $P(C_1 \mid S_2) \neq P(C_1)$.
d) No, because $P(S_1) = 0.580$ and $P(S_1 \mid C_2) = 0.458$, so that $P(S_1 \mid C_2) \neq P(S_1)$.

FIGURE A.4

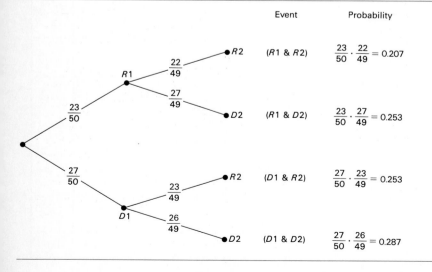

	Event	Probability
R2	(R1 & R2)	$\frac{23}{50} \cdot \frac{22}{49} = 0.207$
D2	(R1 & D2)	$\frac{23}{50} \cdot \frac{27}{49} = 0.253$
R2	(D1 & R2)	$\frac{27}{50} \cdot \frac{23}{49} = 0.253$
D2	(D1 & D2)	$\frac{27}{50} \cdot \frac{26}{49} = 0.287$

__ **4.91**

a) 0.5, 0.5, 0.375   b) 0.5
c) Yes, because $P(B \mid A) = P(B)$.
d) 0.25
e) No, because $P(C \mid A) \neq P(C)$.

__ **4.93**

a) 0.584, 0.187, 0.084
b) No, because $P(P_1 \& C_2) \neq P(P_1)P(C_2)$
   $(0.084 \neq 0.584 \cdot 0.187)$

__ **4.95**   a) 0.006   b) 0.005

__ **4.97**   a) 0.083   b) 0.5

__ **4.99**

a) 0.928   b) 0.072
c) There was a 7.2% chance that at least one "criticality 1" item would fail. In other words, on the
   average, at least one "criticality 1" item will fail
   in 7.2 out of every 100 such missions.

__ **4.101**

a) 0.105   b) 0.105   c) 0.032   d) 0.420

Exercises 4.7

__ **4.111**   a) 48.4%   b) 49.5%

__ **4.113**   a) 1.2%   b) 22.9%

__ **4.115**

a) 0.287   b) 0.332   c) 0.371
d) 28.7% of the people participating in the survey
   said that drug abuse is the nation's top problem;
   33.2% were teen-agers; and 37.1% of those who
   said that drug abuse is the nation's top problem
   were teen-agers.

__ **4.117**

a) 93.4% of those having the disease will test positive and 96.8% of those not having the disease
   will test negative.
b) 0.055
c) Only 5.5% of those testing positive actually have
   the disease.

Exercises 4.8

__ **4.123**

a) 900   b) 9,000,000   c) 8,100,000,000

__ **4.125**   1,021,440

__ **4.127**

a) 24   b) 32,760   c) 30   d) 1   e) 40,320

__ **4.129**   657,720

__ **4.131**   a) 3,628,800   b) $\dfrac{1}{3{,}628{,}800}$

__ **4.133**

a) 4   b) 1365   c) 15   d) 1   e) 1

__ **4.135**

a) 75,287,520   b) 67,800,320   c) 0.901

__ **4.137**   8,145,060

__ **4.139**   a) 0.243   b) 0.972   c) 0.271

__ **4.141**   a) 0.0000002   b) 0.002   c) 0.998

REVIEW TEST FOR CHAPTER 4

1. a) 0.361   b) 0.194
   c)

Adjusted gross income	No. of returns	Event	Prob.
Under $10,000	34,081	A	0.361
$10,000–$19,999	24,842	B	0.263
$20,000–$29,999	16,425	C	0.174
$30,000–$39,999	9,863	D	0.104
$40,000–$49,999	4,717	E	0.050
$50,000–$99,999	3,759	F	0.040
$100,000 & over	740	G	0.008
	94,427		

2. a) (not J) is the event that the return selected
      shows an adjusted gross income of at least
      $100,000. There are 740 thousand such re-
      turns.
   b) ($H \& I$) is the event that the return selected
      shows an adjusted gross income of between
      $20,000 and $49,999. There are 31,005 thou-
      sand such returns.
   c) (H or K) is the event that the return selected
      shows an adjusted gross income of at least
      $20,000. There are 35,504 thousand such re-
      turns.
   d) ($H \& K$) is the event that the return selected
      shows an adjusted gross income of between
      $50,000 and $99,999. There are 3759 thou-
      sand such returns.

3. a) not mutually exclusive
   b) mutually exclusive

c) mutually exclusive
d) not mutually exclusive

4. a) 0.368, 0.952, 0.992, 0.048
   b) $H = (C$ or D or E or $F)$
      $I = (A$ or B or C or D or $E)$
      $J = (A$ or B or C or D or E or $F)$
      $K = (F$ or $G)$
   c) 0.368, 0.952, 0.992, 0.048

5. a) 0.008, 0.328, 0.376, 0.040
   b) 0.992   c) 0.376

6. a) 6   b) 13,615 thous   c) 48,778 thous
   d) 2562 thous

7. a) (i) the student selected is in college
      (ii) the student selected attends a public
           school
      (iii) the student selected attends a public
            college
   b) 0.216; 0.866; 0.171. 21.6% of students at-
      tend college, 86.6% attend public schools, and
      17.1% attend public colleges.
   c)

	Type		
	Public T_1	Private T_2	$P(L_i)$
Elementary L_1	0.478	0.064	0.542
High school L_2	0.217	0.025	0.242
College L_3	0.171	0.045	0.216
$P(T_j)$	0.866	0.134	1.000

   d) (i) 0.911   (ii) 0.911

8. a) 0.197. 19.7% of students attending public
      schools are in college.
   b) 0.197

9. a) 0.134, 0.103
   b) No, because $P(T_2 | L_2) \neq P(T_2)$. We see that
      10.3% of high school students attend private
      schools, whereas 13.4% of all students attend
      private schools.

c) No, because both events can occur; namely,
   if the student selected is any one of the 1400
   thousand students who attend a private high
   school.
d) $P(L_1) = 0.542$, $P(L_1 | T_1) = 0.553$. Since
   $P(L_1 | T_1) \neq P(L_1)$, the event that a student
   is in elementary school is *not* independent
   of the event that a student attends public
   school.

10. a) 0.023   b) 0.309
    c) Let $A1$, $P1$, and $S1$ denote, respectively, the
       events that the first student selected received
       a master of arts, a master of public adminis-
       tration, and a master of science; and let $A2$,
       $P2$, and $S2$ denote, respectively, the events
       that the second student selected received a
       master of arts, a master of public administra-
       tion, and a master of science. See Figure A.5
       on the next page.
    d) 0.451

11. a) 0.004   b) 0.012   c) 0.047

12. a) 0.45   b) 0.507   c) 0.685   d) 0.608
    e) 45% of the women surveyed answered 'no' to
       the question; 50.7% of the people surveyed
       answered 'no' to the question; 68.5% of the
       people surveyed were women; 60.8% of the
       people who answered 'no' to the question
       were women.
    f) The probabilities in parts (b) and (c) are
       prior; those in parts (a) and (d) are posterior.

13. a) 66   b) 1320   c) 28; 336

14. a) 635,013,559,600
    b) 0.213   c) 0.00045   d) 0.032

CHAPTER 5

Exercises 5.1

—— 5.1

a) 1, 2, 3, 4, 5, 6, and 7
b) $\{x = 5\}$
c) 0.073. 7.3% of U.S. households consist of exactly
   five people.
d) 0.175

FIGURE A.5

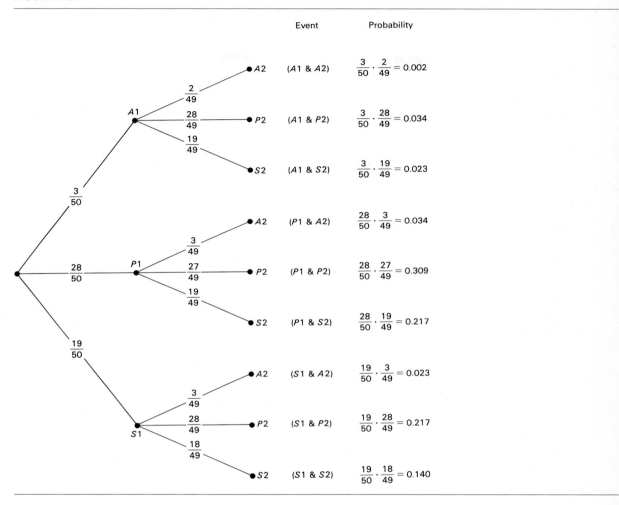

e)

Number of persons x	Probability $P(x)$
1	0.232
2	0.317
3	0.175
4	0.154
5	0.073
6	0.030
7	0.019

___ 5.3

a) 2, 3, 4, 5, 6, 7, 8, 9, 10, 11, and 12
b) $\{y = 7\}$
c) $\frac{1}{6}$ d) $\frac{1}{18}$
e)

Sum of dice y	Probability $P(y)$
2	1/36
3	1/18
4	1/12
5	1/9
6	5/36
7	1/6
8	5/36
9	1/9
10	1/12
11	1/18
12	1/36

___ 5.5

a) $\{x = 4\}$ b) $\{x \geq 2\}$ c) $\{x < 5\}$
d) $\{2 \leq x < 5\}$ e) 0.21   f) 0.92
g) 0.52   h) 0.44

___ 5.7

a) $\{y = 2\}$ b) $\{y \leq 4\}$ c) $\{y > 1\}$
d) $\{2 \leq y \leq 4\}$ e) 0.134   f) 0.923
g) 0.415   h) 0.338   i) 0.338

___ 5.9   0.975

Exercises 5.2

___ 5.15   a) 2.7   b) 1.5   c) 1.5
___ 5.17   a) 7   b) 2.4   c) 2.4
___ 5.19   a) 4.1   b) 1.6   c) 1.6
___ 5.21   a) 1.5   b) 1.7   c) 1.7
___ 5.23

a) $P(x = 1) = \frac{18}{38} = 0.474$, $P(x = -1) = \frac{20}{38} = 0.526$

b) -0.052 c) 5.2¢   d) $5.20, $52   e) No.
___ 5.25

a) $\mu_w = 0.25$, $\sigma_w = 0.536$ b) 0.25   c) 62.5

Exercises 5.3

___ 5.29   5040; 40,320; 362,880
___ 5.31   a) 10   b) 35   c) 120   d) 792
___ 5.33   a) 10   b) 1   c) 1   d) 126
___ 5.35

a) $p = 0.487$

b)

Outcome	Probability
sss	$(0.487)(0.487)(0.487) = 0.116$
ssf	$(0.487)(0.487)(0.513) = 0.122$
sfs	$(0.487)(0.513)(0.487) = 0.122$
sff	$(0.487)(0.513)(0.513) = 0.128$
fss	$(0.513)(0.487)(0.487) = 0.122$
fsf	$(0.513)(0.487)(0.513) = 0.128$
ffs	$(0.513)(0.513)(0.487) = 0.128$
fff	$(0.513)(0.513)(0.513) = 0.135$

c) See Figure A.6 at the top of the next page.
d) ssf, sfs, fss
e) Each has probability 0.122. Because each probability is obtained by multiplying two success probabilities of 0.487 and one failure probability of 0.513.

___ 5.37

a) $p = 0.2$

b)

Outcome	Probability
ssss	$(0.2)(0.2)(0.2)(0.2) = 0.0016$
sssf	$(0.2)(0.2)(0.2)(0.8) = 0.0064$
ssfs	$(0.2)(0.2)(0.8)(0.2) = 0.0064$
ssff	$(0.2)(0.2)(0.8)(0.8) = 0.0256$
sfss	$(0.2)(0.8)(0.2)(0.2) = 0.0064$
sfsf	$(0.2)(0.8)(0.2)(0.8) = 0.0256$
sffs	$(0.2)(0.8)(0.8)(0.2) = 0.0256$
sfff	$(0.2)(0.8)(0.8)(0.8) = 0.1024$
fsss	$(0.8)(0.2)(0.2)(0.2) = 0.0064$
fssf	$(0.8)(0.2)(0.2)(0.8) = 0.0256$
fsfs	$(0.8)(0.2)(0.8)(0.2) = 0.0256$
fsff	$(0.8)(0.2)(0.8)(0.8) = 0.1024$
ffss	$(0.8)(0.8)(0.2)(0.2) = 0.0256$
ffsf	$(0.8)(0.8)(0.2)(0.8) = 0.1024$
fffs	$(0.8)(0.8)(0.8)(0.2) = 0.1024$
ffff	$(0.8)(0.8)(0.8)(0.8) = 0.4096$

FIGURE A.6

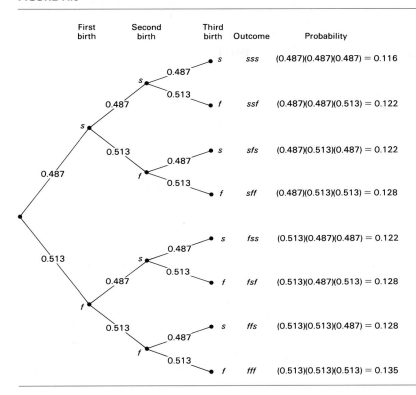

c) See Figure A.7 at the top of the next page.
d) *sssf, ssfs, sfss, fsss*
e) Each has probability 0.0064. Because each probability is obtained by multiplying three success probabilities of 0.2 and one failure probability of 0.8.

___ **5.39**

a) $p = 0.232$

b)

Outcome	Probability
ssss	$(0.232)(0.232)(0.232)(0.232) = 0.003$
sssf	$(0.232)(0.232)(0.232)(0.768) = 0.010$
ssfs	$(0.232)(0.232)(0.768)(0.232) = 0.010$
ssff	$(0.232)(0.232)(0.768)(0.768) = 0.032$
sfss	$(0.232)(0.768)(0.232)(0.232) = 0.010$

(continued in the next column)

Outcome	Probability
sfsf	$(0.232)(0.768)(0.232)(0.768) = 0.032$
sffs	$(0.232)(0.768)(0.768)(0.232) = 0.032$
sfff	$(0.232)(0.768)(0.768)(0.768) = 0.105$
fsss	$(0.768)(0.232)(0.232)(0.232) = 0.010$
fssf	$(0.768)(0.232)(0.232)(0.768) = 0.032$
fsfs	$(0.768)(0.232)(0.768)(0.232) = 0.032$
fsff	$(0.768)(0.232)(0.768)(0.768) = 0.105$
ffss	$(0.768)(0.768)(0.232)(0.232) = 0.032$
ffsf	$(0.768)(0.768)(0.232)(0.768) = 0.105$
fffs	$(0.768)(0.768)(0.768)(0.232) = 0.105$
ffff	$(0.768)(0.768)(0.768)(0.768) = 0.348$

c) *ssff, sfsf, sffs, fssf, fsfs, ffss*
d) Each has probability 0.032. Because each probability is obtained by multiplying two success probabilities of 0.232 and two failure probabilities of 0.768.

FIGURE A.7

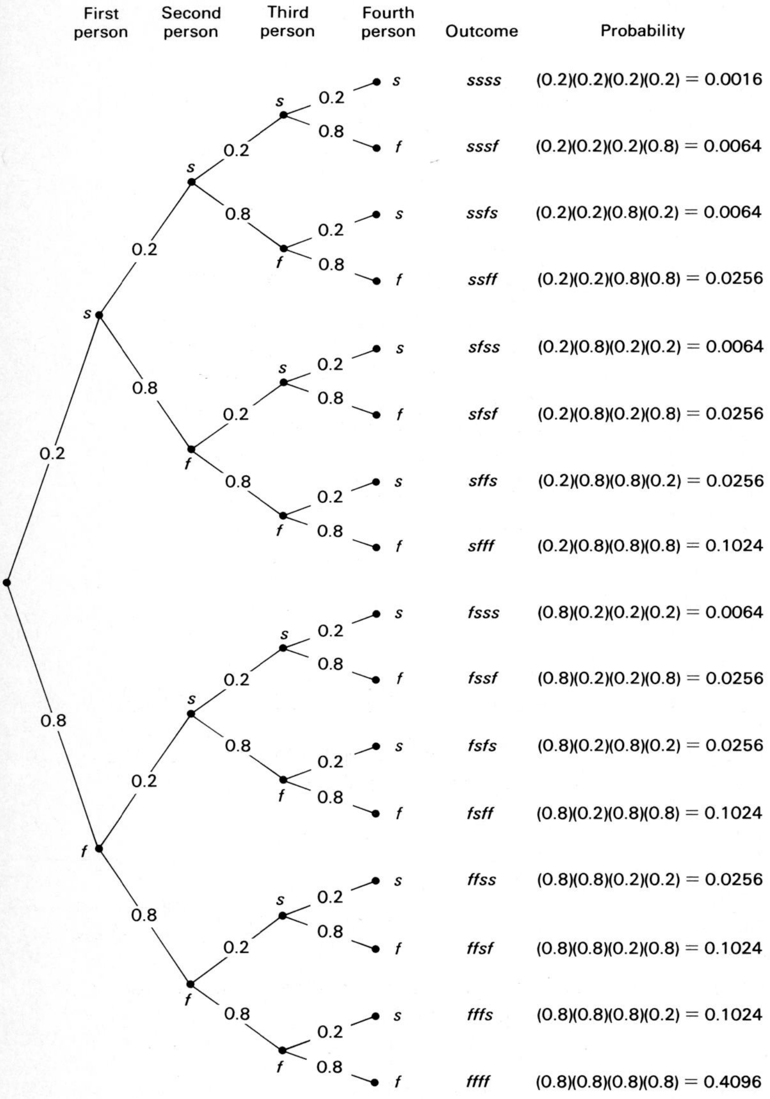

First person	Second person	Third person	Fourth person	Outcome	Probability
			s	ssss	(0.2)(0.2)(0.2)(0.2) = 0.0016
		s 0.2	f	sssf	(0.2)(0.2)(0.2)(0.8) = 0.0064
		0.8	s	ssfs	(0.2)(0.2)(0.8)(0.2) = 0.0064
	s 0.2	f 0.2	f	ssff	(0.2)(0.2)(0.8)(0.8) = 0.0256
	0.8	s 0.2	s	sfss	(0.2)(0.8)(0.2)(0.2) = 0.0064
		0.8	f	sfsf	(0.2)(0.8)(0.2)(0.8) = 0.0256
s 0.2	f 0.2	0.8	s	sffs	(0.2)(0.8)(0.8)(0.2) = 0.0256
		f 0.2	f	sfff	(0.2)(0.8)(0.8)(0.8) = 0.1024
		s 0.2	s	fsss	(0.8)(0.2)(0.2)(0.2) = 0.0064
	s 0.2	0.8	f	fssf	(0.8)(0.2)(0.2)(0.8) = 0.0256
0.8		0.8	s	fsfs	(0.8)(0.2)(0.8)(0.2) = 0.0256
	0.2	f 0.2	f	fsff	(0.8)(0.2)(0.8)(0.8) = 0.1024
f	0.8	s 0.2	s	ffss	(0.8)(0.8)(0.2)(0.2) = 0.0256
		0.8	f	ffsf	(0.8)(0.8)(0.2)(0.8) = 0.1024
	f 0.2	0.8	s	fffs	(0.8)(0.8)(0.8)(0.2) = 0.1024
		f 0.2	f	ffff	(0.8)(0.8)(0.8)(0.8) = 0.4096

Exercises 5.4

__ 5.45

a) $4 \cdot 0.0064 = 0.0256$

b) $\binom{4}{3}(0.2)^3(0.8)^1 = 0.0256$

__ 5.47

a) 0.384   b) 0.519   c) 0.865   d) 0.749

e)

Number of girls x	Probability $P(x)$
0	0.135
1	0.384
2	0.365
3	0.116

__ 5.49

a) 0.169   b) 0.957   c) 0.609   d) 0.609

e)

Number having a color TV x	Probability $P(x)$
0	0.000
1	0.000
2	0.005
3	0.038
4	0.169
5	0.398
6	0.391

__ 5.51

a) 0.177   b) 0.302   c) 0.996
d) 0.501   e) 0.125

__ 5.53

a) 0.302   b) 0.738   c) 1
d) 0.846   e) 0.739

__ 5.55

a) 0.267   b) 0.816   c) 0.451   d) 0.739

__ 5.57

a) Part (a):
   MTB > PDF 4;
   SUBC> BINOMIAL with n=6 and p=0.855.
   Part (b):
   MTB > PDF;
   SUBC> BINOMIAL with n=6 and p=0.855.
   Note: This command and subcommand will yield
   the probability distribution. We must then add
   $P(4)$, $P(5)$, and $P(6)$ to obtain the required
   probability. Alternatively (and more simply), we
   can use:
   MTB > CDF 3;
   SUBC> BINOMIAL with n=6 and p=0.855.
   and subtract the result from 1. Why?
   Part (c):
   MTB > CDF 5;
   SUBC> BINOMIAL with n=6 and p=0.855.
   Part (d):
   MTB > PDF;
   SUBC> BINOMIAL with n=6 and p=0.855.
   Part (e):
   MTB > PDF;
   SUBC> BINOMIAL with n=6 and p=0.855.

__ 5.59

a) 7   b) 0.34   c) 0.2610   d) 0.7411
e) 0.9508   f) 0.1837

__ 5.61

a) 0.3874   b) 0.6513   c) 0.7361
d) 0.2639   e) 0.6385

Exercises 5.5

__ 5.69

a)

x	$P(x)$	$xP(x)$	x^2	$x^2P(x)$
0	0.4096	0.0000	0	0.0000
1	0.4096	0.4096	1	0.4096
2	0.1536	0.3072	4	0.6144
3	0.0256	0.0768	9	0.2304
4	0.0016	0.0064	16	0.0256
		0.8000		1.2800

From the table, we see that

$$\mu_x = \Sigma xP(x) = 0.8$$

and

$$\sigma_x = \sqrt{\Sigma x^2 P(x) - \mu_x^2} = \sqrt{1.28 - (0.8)^2} = 0.8$$

Note: It is only a coincidence that $\mu_x = \sigma_x$.
b) Using the special formulas, we get

$$\mu_x = np = 4 \cdot 0.2 = 0.8$$

and

$$\sigma_x = \sqrt{np(1-p)} = \sqrt{4 \cdot 0.2 \cdot 0.8} = 0.8$$

c) Much less work is required when the special formulas are employed.

__ 5.71 $\mu_x = 5.1$, $\sigma_x = 0.9$

__ 5.73 $\mu_x = 4.4$, $\sigma_x = 1.7$

__ 5.75 $\mu_x = 7.2$, $\sigma_x = 1.2$

__ 5.77 $\mu_x = 1.5$, $\sigma_x = 1.2$

__ 5.79 $\mu_x = 12.5$, $\sigma_x = 3.1$

REVIEW TEST FOR CHAPTER 5

1. a) 1, 2, 3, and 4
   b) $\{x = 3\}$

c) 0.253. 25.3% of the undergraduates at this university are juniors.

d) 0.211

e)

Class level x	Probability $P(x)$
1	0.195
2	0.211
3	0.253
4	0.341

2. a) $\{y = 4\}$ b) $\{y \geq 4\}$
   c) $\{2 \leq y \leq 4\}$ d) $\{y \geq 1\}$
   e) 0.174   f) 0.322   g) 0.646   h) 0.948

3. a) 2.8   b) 2.8   c) 1.5   d) 1.5

4. 1, 6, 24, 5040

5. a) 56   b) 56   c) 1
   d) 45   e) 91,390   f) 1

6. a) $p = 0.493$

b)

Outcome	Probability
sss	$(0.493)(0.493)(0.493) = 0.120$
ssf	$(0.493)(0.493)(0.507) = 0.123$
sfs	$(0.493)(0.507)(0.493) = 0.123$
sff	$(0.493)(0.507)(0.507) = 0.127$
fss	$(0.507)(0.493)(0.493) = 0.123$
fsf	$(0.507)(0.493)(0.507) = 0.127$
ffs	$(0.507)(0.507)(0.493) = 0.127$
fff	$(0.507)(0.507)(0.507) = 0.130$

c) See Figure A.8 at the bottom of the page.

d) ssf, sfs, fss

e) Each has probability 0.123. Because each probability is obtained by multiplying two success probabilities of 0.493 and one failure probability of 0.507.

FIGURE A.8

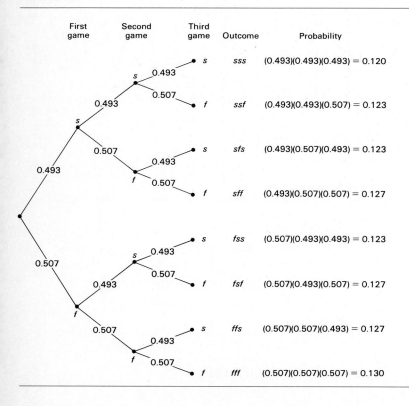

7. a) 0.410   b) 0.590   c) 0.819

d)

Number literate x	Probability $P(x)$
0	0.002
1	0.026
2	0.154
3	0.410
4	0.410

8. a) 0.410   b) 0.592   c) 0.820

   Note: The discrepancy between the answers to parts (b) and (c) given here and those given in Problem 7 are due to roundoff error.

   d) See answer to Problem 7(d).

9. a) Part (a):

   MTB > PDF 3;
   SUBC> BINOMIAL with n=4 and p=0.8.

   Part (b):
   MTB > CDF 3;
   SUBC> BINOMIAL with n=4 and p=0.8.

   Part (c):
   MTB > PDF;
   SUBC> BINOMIAL with n=4 and p=0.8.

   Part (d):
   MTB > PDF;
   SUBC> BINOMIAL with n=4 and p=0.8.

10. a) 5   b) 0.65   c) 0.0488
    d) 0.5663   e) 0.0541   f) 0.9947

11. a) 8   b) 0.57   c) 0.4762
    d) 0.7765   e) 0.2527

12. a) 0.2613   b) 0.8413   c) 0.1586
    d) 0.8225

13. $\mu_x = 3.2$, $\sigma_x = 0.8$

14. Because the trials are not independent and the success probability varies from trial to trial. Because the sample size is small relative to the size of the population.

CHAPTER 6

Exercises 6.1

— 6.1

a) 0.3413   b) 0.4772   c) 0.4987
d) 0.3997   e) 0.4495   f) 0.4750

— 6.3

a) 0.1772   b) 0.4830   c) 0.3643
d) 0.4999   e) 0.0199   f) 0.4591

— 6.5

a) 0.0505   b) 0.0250   c) 0.0099   d) 0.0049

— 6.7

a) 0.1359   b) 0.0215   c) 0.1498   d) 0.0919

— 6.9

a) 0.9625   b) 0.8264   c) 0.9918   d) 0.9990

— 6.11

a) 0.8185   b) 0.9759   c) 0.6789   d) 0.7638

— 6.13

a) 0.1587   b) 0.0228   c) 0.1003   d) 0.0505

— 6.15

a) 0.9162   b) 0.0645   c) 0.7975

— 6.17

a) 0.7994   b) 0.8990   c) 0.0500   d) 0.0198

— 6.21   0.44

— 6.23   a) 1.88   b) 2.575

— 6.25   −1.96

— 6.27   0.67

— 6.29   −1.645

— 6.31   ±1.645

— 6.33

$z_{0.10}$	$z_{0.05}$	$z_{0.025}$	$z_{0.01}$	$z_{0.005}$
1.28	1.645	1.96	2.33	2.575

Exercises 6.2

— 6.35   The one with parameters $\mu = 1$ and $\sigma = 2$.

— 6.37   True, because the parameter μ affects only where the normal curve is centered. The shape of the normal curve is determined by the parameter σ.

— 6.39

a)

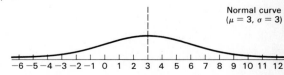

Normal curve
($\mu = 3$, $\sigma = 3$)

b)

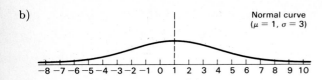

Normal curve
($\mu = 1$, $\sigma = 3$)

c)

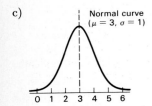

Normal curve
($\mu = 3$, $\sigma = 1$)

__ 6.41	a) 0.6554	b) 0.1587	c) 0.5403
__ 6.43	a) 0.9591	b) 0.1359	c) 0.0668
__ 6.45	a) 0.9104	b) 0.1056	
__ 6.47	a) 0.9115	b) 0.2417	
__ 6.49	a) 0.2033	b) 0.0493	
__ 6.51	a) 77.29	b) 71.44	c) 70.08, 77.92
__ 6.53	a) 337.50	b) 332.50	
__ 6.55	36.14, 40.90, 45.66		

Exercises 6.3

__ 6.63

a) Exact percentage = 11.70% (0.1170)
   Normal curve area = 12.23% (0.1223)
b) Exact percentage = 39.83% (0.3983)
   Normal curve area = 39.14% (0.3914)

__ 6.65   a) 69.43%   b) 97.26%
__ 6.67   a) 13.61%   b) 3.67%
__ 6.69   a) 2   b) −1.75   c) 0
__ 6.71

a) 68.26%   b) 95.44%   c) 99.74%
d) 86.64%

__ 6.73

a) $78.20, $112.60   b) $61.00, $129.80
c) $43.80, $147.00   d) See Figure A.9.

__ 6.75   a) 95   b) 89.9

__ 6.77   a) 2.575   b) 1.28

__ 6.79

a) $Q_1 = \$521.01$, $Q_2 = \$586.00$, $Q_3 = \$650.99$
b) $P_{15} = \$485.12$ c) $P_{98} = \$784.85$

__ 6.81

a) $Q_1 = \$83.88$, $Q_2 = \$95.40$, $Q_3 = \$106.92$
b) $P_{30} = \$86.46$ c) $P_{85} = \$113.29$

Exercises 6.4

__ 6.87

a) 0.2389   b) 0.2095
c) The probability is 0.2389 that a randomly selected secondary school teacher makes less than
   $25 thousand per year. The probability is 0.2095
   that a randomly selected secondary school teacher makes between $32 thousand and $37 thousand per year.

__ 6.89

a) 0.9633   b) 0.9926
c) 96.33% of the batteries last longer than 20 hours.
   99.26% of the batteries last between 15 and
   45 hours.

__ 6.91

a) 0.0594   b) 0.2699
c) The probability is 0.0594 that the time of a randomly selected finisher exceeds 75 minutes. The
   probability is 0.2699 that the time of a randomly
   selected finisher is either less than 50 minutes or
   greater than 70 minutes.

FIGURE A.9   Weekly food cost

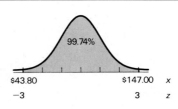

(a)	(b)	(c)

__ **6.93** The probability is 0.6826 that a randomly selected bolt will have a diameter between 9.9 and 10.1 mm; in other words, about 68.26% of the bolts produced have diameters between 9.9 and 10.1 mm. The probability is 0.9544 that a randomly selected bolt will have a diameter between 9.8 and 10.2 mm; in other words, about 95.44% of the bolts produced have diameters between 9.8 and 10.2 mm. The probability is 0.9974 that a randomly selected bolt will have a diameter between 9.7 and 10.3 mm; in other words, about 99.74% of the bolts produced have diameters between 9.7 and 10.3 mm.

__ **6.95** The probability is 0.6826 that a randomly selected flashlight battery of this brand will last between 24.4 and 35.6 hours; in other words, about 68.26% of the batteries will last between 24.4 and 35.6 hours. The probability is 0.9544 that a randomly selected flashlight battery of this brand will last between 18.8 and 41.2 hours; in other words, about 95.44% of the batteries will last between 18.8 and 41.2 hours. The probability is 0.9974 that a randomly selected flashlight battery of this brand will last between 13.2 and 46.8 hours; in other words, about 99.74% of the batteries will last between 13.2 and 46.8 hours.

__ **6.97**

a) $z = (x - 16.3)/17.9$ b) 0.22, 2.70, -0.68
c) No.

__ **6.99**

a) $z = (x - 61)/9$
b) The number of standard deviations that a randomly selected finisher's time is away from the mean of 61 minutes.
c) The standard normal distribution.

__ **6.101**

a) 1.96   b) 2.575
c) $P(\mu_x - 1.96\sigma_x < x < \mu_x + 1.96\sigma_x) = 0.95$
   $P(\mu_x - 2.575\sigma_x < x < \mu_x + 2.575\sigma_x) = 0.99$

Exercises 6.5

__ **6.107**

a) (i) 0.4512   (ii) 0.8907
b) (i) 0.4544   (ii) 0.8858

__ **6.109** The normal curve with parameters $\mu = 15$ and $\sigma = 2.74$.

__ **6.111**   a) 0.0398   b) 0.2835   c) 0.0099

__ **6.113**   a) 0.0263   b) 0.8242   c) 0.9991
__ **6.115**   a) 0.0233   b) 0.1599   c) 0.0516

REVIEW TEST FOR CHAPTER 6

1. It is often appropriate to use the normal distribution as the distribution of a population or random variable; and the normal distribution is frequently employed in inferential statistics.

2. a) 0.4932   b) 0.4678   c) 0.2709
   d) 0.0013   e) 0.1305   f) 0.2749
   g) 0.9441   h) 0.9099   i) 0.9104
   j) 0.8426

3. a) 1.96, 1.645, 2.33, 2.575
   b) 1.28   c) -0.52 d) 0.44   e) -1.04
   f) ± 2.575

4. a)

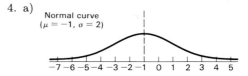

Normal curve ($\mu = -1$, $\sigma = 2$)

   b)

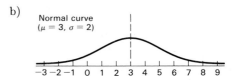

Normal curve ($\mu = 3$, $\sigma = 2$)

   c)

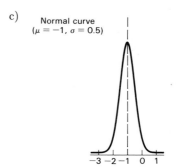

Normal curve ($\mu = -1$, $\sigma = 0.5$)

5. a) the second curve
   b) the first and second curves
   c) the first and third curves
   d) the third curve
   e) the fourth curve

6. a) 0.1125   b) 0.9671   c) 0.0465

7. a) 3.90   b) −5.11   c) 1.48   d) −5.90
   e) −5.11, 3.11

8. a) 82.76%   b) 89.44%   c) 0.62%

9. a) $Q_1 = 433$, $Q_2 = 500$, $Q_3 = 567$. Thus,
   25% of GRE scores are below 433, 25% are
   between 433 and 500, 25% are between 500
   and 567, and 25% are above 567.
   b) $P_{99} = 733$. Thus, 99% of GRE scores are
   below 733 and 1% are above 733.

10. a) 1.45   b) −1.80   c) 0

11. a) 400, 600   b) 300, 700   c) 200, 800

12. a) 99.4   b) 1.645

13. a) $\mu_x = 339.6$, $\sigma_x = 13.3$
    b) 0.1650   c) 0.0110
    d) The probability is 0.1650 that the gestation
    period of a randomly selected Morgan horse
    will be between 320 and 330 days. The prob-
    ability is 0.0110 that the gestation period of
    a randomly selected Morgan horse will exceed
    370 days.

14. a) 326.3, 352.9   b) 313.0, 366.2
    c) 299.7, 379.5

15. a) $z = (x - 339.6)/13.3$
    b) The number of standard deviations that the
    gestation period of a randomly selected Mor-
    gan horse is away from the mean gestation
    period.
    c) −1.10, 2.14, 0
    d) the standard normal distribution
    e) No. Yes.

16. 2.33

17. a) 0.0076   b) 0.9505   c) 0.9988

____ 7.3

a)

Officials selected	Sample obtained
G, L, S	70, 40, 37
G, L, A	70, 40, 55
G, L, T	70, 40, 50
G, S, A	70, 37, 55
G, S, T	70, 37, 50
G, A, T	70, 55, 50
L, S, A	40, 37, 55
L, S, T	40, 37, 50
L, A, T	40, 55, 50
S, A, T	37, 55, 50

b) $\frac{1}{10}$, $\frac{1}{10}$, $\frac{1}{10}$

____ 7.5

a)

Sample
AR, FS, BD, LP
AR, FS, BD, JS
AR, FS, BD, RT
AR, FS, LP, JS
AR, FS, LP, RT
AR, FS, JS, RT
AR, BD, LP, JS
AR, BD, LP, RT
AR, BD, JS, RT
AR, LP, JS, RT
FS, BD, LP, JS
FS, BD, LP, RT
FS, BD, JS, RT
FS, LP, JS, RT
BD, LP, JS, RT

b) Write the initials of the representatives on six
separate pieces of paper, place the six slips of paper into a box, and then, while blindfolded, pick
four of the slips of paper.

c) $\frac{1}{15}$, $\frac{1}{15}$

CHAPTER 7

Exercises 7.1

____ 7.1 Dentists form a high income group whose
incomes are not representative of the incomes of people in general.

Exercises 7.2

____ 7.7
a) $\mu = \$68$ thous

b)

Sample	$\bar{x}$
76, 58	67.0
76, 64	70.0
76, 82	79.0
76, 60	68.0
58, 64	61.0
58, 82	70.0
58, 60	59.0
64, 82	73.0
64, 60	62.0
82, 60	71.0

c) See Figure A.10.

d)

Sample mean $\bar{x}$	Probability $P(\bar{x})$
59.0	0.1
61.0	0.1
62.0	0.1
67.0	0.1
68.0	0.1
70.0	0.2
71.0	0.1
73.0	0.1
79.0	0.1

e) 0.1

f) 0.5. If we take a random sample of two salaries, there is a 50% chance that the mean of the sample selected will be within four (that is, $4,000) of the population mean.

__ 7.9

b)

Sample	$\bar{x}$
76, 58, 64, 82	70.0
76, 58, 64, 60	64.5
76, 58, 82, 60	69.0
76, 64, 82, 60	70.5
58, 64, 82, 60	66.0

c) See Figure A.11.

d)

Sample mean $\bar{x}$	Probability $P(\bar{x})$
64.5	0.2
66.0	0.2
69.0	0.2
70.0	0.2
70.5	0.2

FIGURE A.10

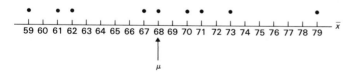

FIGURE A.11

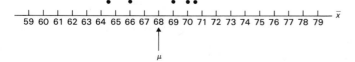

e) 0

f) 1. If we take a random sample of four salaries, there is a 100% chance (that is, it is certain) that the mean of the sample will be within four (that is, $4,000) of the population mean.

___ 7.11

a) $\mu = 15$ cm

b)

Sample	$\overline{x}$
19, 14	16.5
19, 15	17.0
19, 9	14.0
19, 16	17.5
19, 17	18.0
14, 15	14.5
14, 9	11.5
14, 16	15.0
14, 17	15.5
15, 9	12.0
15, 16	15.5
15, 17	16.0
9, 16	12.5
9, 17	13.0
16, 17	16.5

c) See Figure A.12 on the next page.

d)

Sample mean $\overline{x}$	Probability $P(\overline{x})$
11.5	1/15
12.0	1/15
12.5	1/15
13.0	1/15
14.0	1/15
14.5	1/15
15.0	1/15
15.5	2/15
16.0	1/15
16.5	2/15
17.0	1/15
17.5	1/15
18.0	1/15

e) $\frac{1}{15} = 0.067$

f) $\frac{6}{15} = 0.4$. If we take a random sample of two bullfrogs, there is a 40% chance that their mean length will be within 1 cm of the population mean length.

___ 7.13

b)

Sample	$\overline{x}$
19, 14, 15, 9	14.25
19, 14, 15, 16	16.00
19, 14, 15, 17	16.25
19, 14, 9, 16	14.50
19, 14, 9, 17	14.75
19, 14, 16, 17	16.50
19, 15, 9, 16	14.75
19, 15, 9, 17	15.00
19, 15, 16, 17	16.75
19, 9, 16, 17	15.25
14, 15, 9, 16	13.50
14, 15, 9, 17	13.75
14, 15, 16, 17	15.50
14, 9, 16, 17	14.00
15, 9, 16, 17	14.25

c) See Figure A.13 on the next page.

d)

Sample mean $\overline{x}$	Probability $P(\overline{x})$
13.50	1/15
13.75	1/15
14.00	1/15
14.25	2/15
14.50	1/15
14.75	2/15
15.00	1/15
15.25	1/15
15.50	1/15
16.00	1/15
16.25	1/15
16.50	1/15
16.75	1/15

e) $\frac{1}{15} = 0.067$

f) $\frac{10}{15} = 0.667$. If we take a random sample of four bullfrogs, there is a 66.7% chance that their mean length will be within 1 cm of the population mean length.

___ 7.15

b)

Sample	$\overline{x}$
19, 14, 15, 9, 16, 17	15.0

c) See Figure A.14 on the next page.

FIGURE A.12

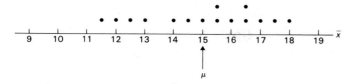

FIGURE A.13

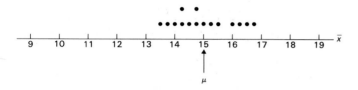

FIGURE A.14

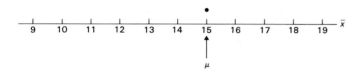

d)

Sample mean $\bar{x}$	Probability $P(\bar{x})$
15.0	1

e) 1

f) 1. If we take a random sample of six bullfrogs, there is a 100% chance (that is, it is certain) that their mean length will be within 1 cm of the population mean length.

The only possible sample here is identical to the population.

___ **7.17**   The larger the sample size, the smaller the sampling error tends to be in estimating a population mean, μ, by a sample mean, $\bar{x}$.

Exercises 7.3

___ **7.21**

a) $\mu = 68$

b)

$\bar{x}$	$P(\bar{x})$	$\bar{x}P(\bar{x})$
59.0	0.1	5.9
61.0	0.1	6.1
62.0	0.1	6.2
67.0	0.1	6.7
68.0	0.1	6.8
70.0	0.2	14.0
71.0	0.1	7.1
73.0	0.1	7.3
79.0	0.1	7.9
		68.0

$\mu_{\bar{x}} = \Sigma \bar{x}P(\bar{x}) = 68$

c) $\mu_{\bar{x}} = \mu = 68$

__ **7.23**

b)

$\bar{x}$	$P(\bar{x})$	$\bar{x}P(\bar{x})$
64.5	0.2	12.9
66.0	0.2	13.2
69.0	0.2	13.8
70.0	0.2	14.0
70.5	0.2	14.1
		68.0

$\mu_{\bar{x}} = \Sigma \bar{x}P(\bar{x}) = 68$

c) $\mu_{\bar{x}} = \mu = 68$

__ **7.25**

a) $\mu_{\bar{x}} = 6.9, \sigma_{\bar{x}} = 0.50$
b) $\mu_{\bar{x}} = 6.9, \sigma_{\bar{x}} = 0.19$

__ **7.27**

a) $\mu_{\bar{x}} = 7.4, \sigma_{\bar{x}} = 0.37$
b) $\mu_{\bar{x}} = 7.4, \sigma_{\bar{x}} = 0.18$

__ **7.29**   Increase the size of the sample.

Exercises 7.4

__ **7.39**

a) Approximately normally distributed with mean, $\mu_{\bar{x}} = 100$, and standard deviation, $\sigma_{\bar{x}} = 4$.
b) None.
c) No, since the distribution of the population is not specified, we need a sample size of at least 30 to apply Key Fact 7.6.

__ **7.41**

a) Normal with parameters μ and $\sigma/\sqrt{n}$.
b) No, since the population being sampled is normally distributed.
c) $\mu_{\bar{x}} = \mu$ and $\sigma_{\bar{x}} = \sigma/\sqrt{n}$

__ **7.43**

a)

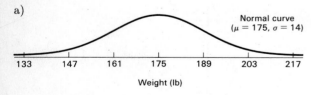

Weight (lb)

b) Normal with $\mu_{\bar{x}} = 175$ and $\sigma_{\bar{x}} = 9.90$.

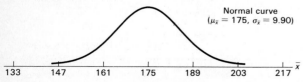

c) Normal with $\mu_{\bar{x}} = 175$ and $\sigma_{\bar{x}} = 4.67$.

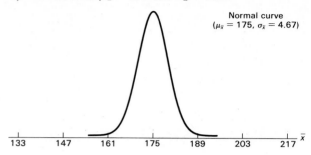

__ **7.45**

a) 0.7154
b) There is a 71.54% chance that the mean weight, $\bar{x}$, of the nine males obtained will be within five pounds of the population mean weight of 175 lb.
c) 71.54%   d) 0.95

__ **7.47**

a) 0.4972
b) No, because the sample size is at least 30.
c)

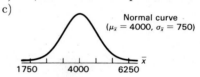

d) 0.9652
e) Because the population standard deviation is so large.

__ **7.49**   0.8064

__ **7.51**

a) $z = (\bar{x} - 100)/5.33$
b) standard normal distribution
c) Yes, because the sample size is less than 30.
d) 0.8990

__ **7.53**

a) 0.2514   b) 0.0174
c) No, because there is about a 25% chance that a randomly selected bag of water-softener salt will weigh 39 lb or less, if the company's claim is correct.
d) Yes, because if the company's claim is correct, there is less than a 2% chance that 10 randomly selected bags of water-softener salt will have a mean weight of 39 lb or less.

REVIEW TEST FOR CHAPTER 7

1. No, because parents of students at Yale tend to have higher incomes than parents of college students in general.

2. a) not random   b) random

3. a) The error resulting from using the mean income tax, $\bar{x}$, of the 171,700 tax returns sampled as an estimate of the mean income tax, μ, of all 1980 tax returns.
   b) $88
   c) No, not necessarily. However, increasing the sample size from 171,700 to 250,000 would increase the likelihood for small sampling error.
   d) Increase the sample size.

4. a) $\mu_{\bar{x}} = 506$, $\sigma_{\bar{x}} = 47.4$
   b) $\mu_{\bar{x}} = 506$, $\sigma_{\bar{x}} = 16.76$
   c) Smaller, because $\sigma_{\bar{x}} = \sigma/\sqrt{n}$, and so the larger the sample size, the smaller the value of $\sigma_{\bar{x}}$.

5. a) False. By the central limit theorem, the random variable $\bar{x}$ is approximately normally distributed. Furthermore, $\mu_{\bar{x}} = \mu = 40$ and $\sigma_{\bar{x}} = \sigma/\sqrt{n} = 10/\sqrt{100} = 1$. Thus, $P(30 \le \bar{x} \le 50)$ equals the area under the normal curve with parameters $\mu_{\bar{x}} = 40$ and $\sigma_{\bar{x}} = 1$ that lies between 30 and 50. Applying the usual techniques, we find that area to be 1.0000 to four decimal places. Hence, there is almost a 100% chance that the mean of the sample will be between 30 and 50.
   b) Not possible to tell, since we do not know the distribution of the population.
   c) True. Referring to part (a), we see that $P(39 \le \bar{x} \le 41)$ equals the area under the normal curve with parameters $\mu_{\bar{x}} = 40$ and $\sigma_{\bar{x}} = 1$ that lies between 39 and 41. Applying the usual techniques, we find that area to be 0.6826. Hence, there is about a 68.26% chance that the mean of the sample will be between 39 and 41.

6. a) False. Since the population is normally distributed, so is the random variable $\bar{x}$. Furthermore, $\mu_{\bar{x}} = \mu = 40$ and $\sigma_{\bar{x}} = \sigma/\sqrt{n} = 10/\sqrt{100} = 1$. Hence, as in Problem 5(a), we find that there is almost a 100% chance that the mean of the sample will be between 30 and 50.

b) True. Since the population is normally distributed, percentages for the population are equal to areas under the normal curve with parameters $\mu = 40$ and $\sigma = 10$. Applying the usual techniques, we find that the area under that normal curve between 30 and 50 is 0.6826.
   c) True. From part (a), we see that the random variable $\bar{x}$ is normally distributed with $\mu_{\bar{x}} = 40$ and $\sigma_{\bar{x}} = 1$. Hence, as in Problem 5(c), we find that there is about a 68.26% chance that the mean of the sample will be between 39 and 41.

7. a)

b) Normal with $\mu_{\bar{x}} = 285$ and $\sigma_{\bar{x}} = 25.98$.

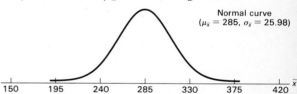

c) Normal with $\mu_{\bar{x}} = 285$ and $\sigma_{\bar{x}} = 15$.

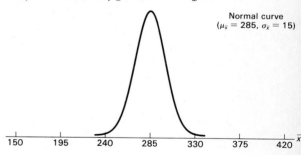

8. a) 0.2960
   b) There is a 29.6% chance that the mean monthly rent, $\bar{x}$, of the three studio apartments obtained will be within $10 of the population mean monthly rent of $285.
   c) 29.6%   d) 0.9452

9. a) For a normally distributed population, the random variable $\bar{x}$ is normally distributed, regardless of the sample size. Also, we know

that $\mu_{\bar{x}} = \mu$. Consequently, since the normal curve for a normally distributed population or random variable is centered at its μ-parameter, all three curves are centered at the same place.

b) Curve B. Since $\sigma_{\bar{x}} = \sigma/\sqrt{n}$, the larger the sample size, the smaller the value of $\sigma_{\bar{x}}$ and, hence, the smaller the spread of the normal curve for $\bar{x}$. Thus, Curve B, which has the smaller spread, corresponds to the larger sample size.

c) Because $\sigma_{\bar{x}} = \sigma/\sqrt{n}$ and the spread of a normal curve is determined by $\sigma_{\bar{x}}$. Thus, different sample sizes result in normal curves with different spreads.

d) Curve B. The smaller the value of $\sigma_{\bar{x}}$, the smaller the sampling error tends to be.

10. a) 0.6212
b) No, because the sample size is large and, therefore, $\bar{x}$ is approximately normally distributed, regardless of the distribution of the population of life-insurance amounts. Yes.
c) 0.9946

11. a) $z = (\bar{x} - 513)/8.32$
b) approximately standard normal
c) 0.0198
d) No, because we do not know the distribution of the population of monthly benefits to retired workers.

12. a) No. Assuming the manufacturer's claim is correct, the probability is 0.1587 that the paint will last 4.5 years or less on a (randomly selected) house painted with the paint. In other words, there is about a 16% chance that the paint would last 4.5 years or less.
b) Yes. Assuming the manufacturer's claim is correct, the probability is 0.0008 that the paint will last an average of 4.5 years or less for 10 (randomly selected) houses painted with the paint. In other words, there is less than a 0.1% chance that that would occur, if the manufacturer's claim is correct.
c) No. Assuming the manufacturer's claim is correct, the probability is 0.2643 that the paint will last an average of 4.9 years or less for 10 (randomly selected) houses painted with the paint. In other words, there is about a 26% chance that that would occur, if the manufacturer's claim is correct.

CHAPTER 8

Exercises 8.1

__ 8.1
a) 6.87 lb
b) No, because it is unlikely that a sample mean, $\bar{x}$, will be exactly equal to a population mean, μ; some sampling error is to be anticipated.

__ 8.3 $70.7

__ 8.5
a) A *point estimate* for a parameter is the value of a statistic used to estimate the parameter.
b) A *confidence-interval estimate* of a parameter consists of an interval of numbers obtained from a point estimate of the parameter together with a percentage that specifies how confident we are that the parameter lies in the interval.
c) The *confidence level* of a confidence-interval estimate is the percentage that specifies how confident we are that the parameter lies in the confidence interval.

__ 8.7
a) 6.22 to 7.51 lb
b) We can be 95.44% confident that the mean weight, μ, of all newborns is somewhere between 6.22 and 7.51 lb.
c) It may or may not, but we can be 95.44% confident that it does.

__ 8.9
a) $65.9 to $75.4
b) We can be 95.44% confident that the mean price, μ, of all science books is somewhere between $65.9 and $75.4.

Exercises 8.2

__ 8.15
a) Confidence level = 0.90; $\alpha = 0.10$.
b) Confidence level = 0.99; $\alpha = 0.01$.

__ 8.17
a) 617.3 to 668.7 square feet.
b) We can be 90% confident that the mean size, μ, of household vegetable gardens in the U.S. is somewhere between 617.3 and 668.7 square feet.

___ **8.19**

a) $10.85 to $13.17

b) We can be 95% confident that the mean hourly earnings, μ, of all persons employed in the aircraft industry is somewhere between $10.85 and $13.17.

___ **8.21**

a) 133.6 to 140.2 lb

b) We can be 90% confident that the mean weight, μ, of all American women 5 feet 4 inches tall and in the age group 18–24 years is somewhere between 133.6 and 140.2 lb.

___ **8.23** $8301.7 to $9172.3

___ **8.25**

a) 131.7 to 142.1 lb

b) The confidence level here is greater than that in Exercise 8.21.

c)

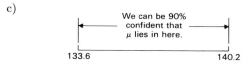

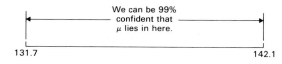

___ **8.27**

a) 3.1 to 3.3 persons

b) We can be 95% confident that the mean size, μ, of all American families is somewhere between 3.1 and 3.3 persons.

___ **8.29**

a) `MTB > ZINTERVAL 95%, sigma=3.25, data in 'EARNINGS'`

Note: The entire command should be typed on one line. It wasn't done that way here simply because of space restrictions.

___ **8.31**

a) `MTB > STDEV 'WEIGHTS'`
    `ST.DEV. =    12.769` ← *Minitab output*
  `MTB > ZINTERVAL 90%, sigma=12.769, data in 'WEIGHTS'`

___ **8.33**

a) 20.8 lb   b) 21.47 lb   c) 80   d) 625.16 lb
e) 621.33 to 628.99 lb

Exercises 8.3

___ **8.35**

a) 25.7 square feet.

b) We can be 90% confident that the maximum error made in using $\overline{x}$ to estimate μ is 25.7 square feet.

c) 355

___ **8.37**

a) $1.16

b) We can be 95% confident that the maximum error made in using $\overline{x}$ to estimate μ is $1.16.

c) 163   d) $12.37 to $13.37

___ **8.39**

a) 3.3 lb   b) 271   c) Because σ is unknown.
d) 132.2 to 136.2 lb

Exercises 8.4

___ **8.43**   a) 1.440   b) 2.447   c) 3.143

___ **8.45**

a) 1.323   b) 2.518   c) −2.080   d) ±1.721

___ **8.47**

a) $1614.98 to $1836.24

b) Yes, because the confidence interval does not contain, and is to the right of, the 1984 figure of $852.31.

___ **8.49**

a) 58.9 to 64.1 bushels

b) The one-acre yields using the new fertilizer are normally distributed.

c) Yes, because the confidence interval does not contain, and is to the right of, the national average of 58.4 bushels per acre.

___ **8.51**

a) 43.3 to 46.4 inches

b) We can be 95% confident that the mean height, μ, of all six-year-old girls is somewhere between 43.3 and 46.4 inches.

___ **8.53**

a) `MTB > TINTERVAL 95%, data in 'ELECBILL'`

___ **8.55**

a) 10.985 mg   b) 0.604 mg   c) 25
d) 10.736 to 11.235 mg

Exercises 8.5

__ **8.65**   A population proportion, p, is a parameter since it is a descriptive measure for a population. A sample proportion, $\hat{p}$, is a statistic since it is a descriptive measure for a sample.

__ **8.67**

a) 0.626 to 0.674

b) We can be 95% confident that the percentage of Americans who drink beer, wine, or hard liquor, at least occasionally, is somewhere between 62.6% and 67.4%.

__ **8.69**

a) 0.784 to 0.836

b) We can be 99% confident that the percentage of adult Americans who are in favor of "right to die" laws is somewhere between 78.4% and 83.6%.

__ **8.71**

a) 0.732 to 0.828

b) We can be 90% confident that the percentage of all real-estate experts who believe next year will be a good time to buy real estate is somewhere between 73.2% and 82.8%.

__ **8.73**   Not very well! I applied Procedure 8.3 without checking the two assumptions for its use; namely, that the number of successes, x, and the number of failures, $n - x$, are both at least 5. Since the number of successes here is only $0.008 \cdot 500 = 4$, I should not have used Procedure 8.3.

REVIEW TEST FOR CHAPTER 8

1. a) Approximately a normal distribution with mean, $\mu_{\bar{x}} = \mu$, and standard deviation, $\sigma_{\bar{x}} = \sigma/\sqrt{n}$.

   b) Approximately the standard normal distribution.

   c) Probabilities for the random variable in part (b) are approximately equal to areas under the standard normal curve. Thus, employ Table II.

2. a) 1.53   b) 0.25   c) 2.455

3. 54.2 to 62.9 yr

4. Only (c) provides a correct interpretation.

5. a) `MTB > STDEV 'AGES'`
   `      ST.DEV. =    13.362` ← *Minitab output*

```
MTB > ZINTERVAL 95%, sigma=13.362, data
         in 'AGES'
```

6. a) 13.4 yr   b) 13.36 yr   c) 36
   d) 58.53 yr   e) 54.16 to 62.90 yr

7. a) 58.5 to 61.7 hr

   b) We can be 99% confident that the mean battery life, μ, is somewhere between 58.5 and 61.7 hr.

8. a) 1.6 hr

   b) We can be 99% confident that the maximum error made in using $\bar{x}$ to estimate μ is 1.6 hr.

   c) 491   d) 59.3 to 60.3 hr

9. The one that looks more like the standard normal curve because, as the number of degrees of freedom gets larger, t-curves look increasingly like the standard normal curve.

10. a) A normal distribution with mean, $\mu_{\bar{x}} = \mu$, and standard deviation, $\sigma_{\bar{x}} = \sigma/\sqrt{n}$.

    b) t-distribution with $n - 1$ degrees of freedom.

    c) Probabilities for the random variable in part (b) are equal to areas under the t-curve with $df = n - 1$. Thus, employ Table III.

11. a) 2.101   b) 1.734   c) -1.330
    d) ± 2.878

12. a) 6.76 to 8.56 hr

    b) We can be 95% confident that the mean daily viewing time, μ, of all American households is somewhere between 6.76 and 8.56 hours.

    c) No, because the 1988 average of seven hours and six minutes does not lie completely to the left of the confidence interval in part (a).

13. a) `MTB > TINTERVAL 95%, data in 'VIEWTIME'`

14. a) 7.660 hr   b) 1.913 hr   c) 0.428 hr
    d) 20   e) 6.764 to 8.556 hr

15. a) 0.491 to 0.591

    b) We can be 95% confident that the percentage of all registered voters watching the debate who thought Mondale did a better job is somewhere between 49.1% and 59.1%.

    c) No, because the confidence interval does not lie completely to the right of 0.5.

CHAPTER 9

Exercises 9.1

__ **9.1**   A *hypothesis* is a statement that something is true.

___ 9.3 Let μ denote last year's mean amount spent for maintenance and repairs per residential property owner.
a) H_0: $\mu = \$280$ b) H_a: $\mu > \$280$
c) right-tailed test

___ 9.5 Let μ denote the mean daily iron intake of all adult females under the age of 51.
a) H_0: $\mu = 18$ mg   b) H_a: $\mu < 18$ mg
c) left-tailed test

___ 9.7 Let μ denote the mean weight of all "50-lb" bags of this dog food.
a) H_0: $\mu = 50$ lb   b) H_a: $\mu \neq 50$ lb
c) two-tailed test

___ 9.9 Let μ denote the mean gas mileage of all Orions.
a) H_0: $\mu = 26$ mpg   b) H_a: $\mu < 26$ mpg
c) left-tailed test

Exercises 9.2

___ 9.15
a) $z \geq 1.645$ b) $z < 1.645$ c) $z = 1.645$
d) $\alpha = 0.05$
e)

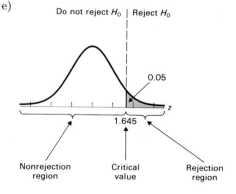

f) right-tailed test

___ 9.17
a) $z \leq -2.33$ b) $z > -2.33$ c) $z = -2.33$
d) $\alpha = 0.01$

e)

Reject H_0 | Do not reject H_0

0.01

−2.33

Rejection region   Critical value   Nonrejection region

f) left-tailed test

___ 9.19
a) $z \leq -1.645$ or $z \geq 1.645$
b) $-1.645 < z < 1.645$
c) $z = \pm 1.645$
d) $\alpha = 0.10$
e)

Reject H_0 | Do not reject H_0 | Reject H_0

0.05   0.05

−1.645   1.645

Nonrejection region

Critical values

Rejection region

f) two-tailed test

___ 9.21
a) A Type I error would occur if, in fact, $\mu = \$280$ but the results of the sampling lead to the conclusion that $\mu > \$280$.
b) A Type II error would occur if, in fact, $\mu > \$280$ but the results of the sampling fail to lead to that conclusion.
c) A correct decision would occur if, in fact, $\mu = \$280$ and the results of the sampling do not lead

to the rejection of that fact; or if, in fact, $\mu >$ \$280 and the results of the sampling lead to that conclusion.

d) correct decision   e) Type II error

__ 9.23

a) A Type I error would occur if, in fact, $\mu = 18$ mg but the results of the sampling lead to the conclusion that $\mu < 18$ mg.
b) A Type II error would occur if, in fact, $\mu < 18$ mg but the results of the sampling fail to lead to that conclusion.
c) A correct decision would occur if, in fact, $\mu = 18$ mg and the results of the sampling do not lead to the rejection of that fact; or if, in fact, $\mu < 18$ mg and the results of the sampling lead to that conclusion.
d) Type I error   e) correct decision

__ 9.25

a) A Type I error would occur if, in fact, $\mu = 50$ lb but the results of the sampling lead to the conclusion that $\mu \neq 50$ lb.
b) A Type II error would occur if, in fact, $\mu \neq 50$ lb but the results of the sampling fail to lead to that conclusion.
c) A correct decision would occur if, in fact, $\mu = 50$ lb and the results of the sampling do not lead to the rejection of that fact; or if, in fact, $\mu \neq 50$ lb and the results of the sampling lead to that conclusion.
d) correct decision   e) Type II error

__ 9.27

a) A Type I error would occur if, in fact, $\mu = 26$ mpg but the results of the sampling lead to the conclusion that $\mu < 26$ mpg.
b) A Type II error would occur if, in fact, $\mu < 26$ mpg but the results of the sampling fail to lead to that conclusion.
c) A correct decision would occur if, in fact, $\mu = 26$ mpg and the results of the sampling do not lead to the rejection of that fact; or if, in fact, $\mu < 26$ mpg and the results of the sampling lead to that conclusion.
d) correct decision   e) Type I error

__ 9.29   True, because the significance level, α, is the probability of rejecting a true null hypothesis. Thus, if α is small, then it is unlikely that a true null hypothesis will be rejected.

Exercises 9.3

__ 9.35   Critical value: $z_{0.01} = 2.33$.

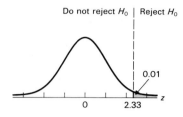

__ 9.37   Critical values: $\pm z_{0.05} = \pm 1.645$.

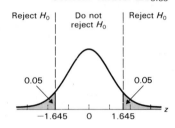

__ 9.39   Critical value: $-z_{0.05} = -1.645$.

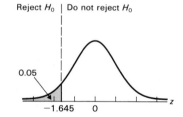

__ 9.41

STEP 1 $H_0: \mu = \$280$, $H_a: \mu > \$280$
STEP 2 $\alpha = 0.05$
STEP 3   Critical value $= 1.645$
STEP 4 $z = 0.26$
STEP 5   Do not reject H_0
STEP 6   The data do not provide sufficient evidence to conclude that last year's mean amount, μ, spent for maintenance and repairs has increased over the 1983 mean of \$280.

__ 9.43

STEP 1 $H_0: \mu = 18$ mg, $H_a: \mu < 18$ mg
STEP 2 $\alpha = 0.01$
STEP 3   Critical value $= -2.33$
STEP 4 $z = -7.22$
STEP 5   Reject H_0
STEP 6   The data provide sufficient evidence to conclude that adult females under the age

of 51 are, on the average, getting less than the RDA of 18 mg of iron.

___ 9.45

a) STEP 1 H_0: $\mu = 50$ lb, H_a: $\mu \neq 50$ lb
   STEP 2 $\alpha = 0.05$
   STEP 3   Critical values $= \pm 1.96$
   STEP 4 $z = 1.13$
   STEP 5   Do not reject H_0
   STEP 6   The data do not provide sufficient evidence to conclude that the mean weight, μ, of all "50-lb" bags of this dog food differs from the advertized weight of 50 lb.

b) STEPS 1–3 are the same as in part (a).
   STEP 4 $z = 2.17$
   STEP 5   Reject H_0
   STEP 6   The data provide sufficient evidence to conclude that the mean weight, μ, of all "50-lb" bags of this dog food differs from the advertized weight of 50 lb.

___ 9.47

STEP 1 H_0: $\mu = 26$ mpg, H_a: $\mu < 26$ mpg
STEP 2 $\alpha = 0.05$
STEP 3   Critical value $= -1.645$
STEP 4 $z = -2.65$
STEP 5   Reject H_0
STEP 6   The data support the consumer group's conjecture.

___ 9.49   For large samples, the sample standard deviation, s, is likely to be a good approximation of the population standard deviation, σ.

Exercises 9.4

___ 9.53   a) 0.0212   b) 0.6217

___ 9.55   a) 0.2296   b) 0.8770

___ 9.57   a) 0.0970   c) 0.6030

___ 9.59

STEP 1 H_0: $\mu = \$280$, H_a: $\mu > \$280$
STEP 2 $\alpha = 0.05$
STEP 3 $z = 0.26$
STEP 4 $P = 0.3974$
STEP 5   Do not reject H_0
STEP 6   The data do not provide sufficient evidence to conclude that last year's mean amount, μ, spent for maintenance and repairs has increased over the 1983 mean of $280.

___ 9.61

STEP 1 H_0: $\mu = 18$ mg, H_a: $\mu < 18$ mg
STEP 2 $\alpha = 0.01$
STEP 3 $z = -7.22$
STEP 4 $P = 0.0000$ (to four decimal places)
STEP 5   Reject H_0
STEP 6   The data provide sufficient evidence to conclude that adult females under the age of 51 are, on the average, getting less than the RDA of 18 mg of iron.

___ 9.63

a) STEP 1 H_0: $\mu = 50$ lb, H_a: $\mu \neq 50$ lb
   STEP 2 $\alpha = 0.05$
   STEP 3 $z = 1.13$
   STEP 4 $P = 0.2584$
   STEP 5   Do not reject H_0
   STEP 6   The data do not provide sufficient evidence to conclude that the mean weight, μ, of all "50-lb" bags of this dog food differs from the advertized weight of 50 lb.

b) STEPS 1 and 2 are the same as in part (a).
   STEP 3 $z = 2.17$
   STEP 4 $P = 0.0300$
   STEP 5   Reject H_0
   STEP 6   The data provide sufficient evidence to conclude that the mean weight, μ, of all "50-lb" bags of this dog food differs from the advertized weight of 50 lb.

___ 9.65

STEP 1 H_0: $\mu = 26$ mpg, H_a: $\mu < 26$ mpg
STEP 2 $\alpha = 0.05$
STEP 3 $z = -2.65$
STEP 4 $P = 0.0040$
STEP 5   Reject H_0
STEP 6   The data support the consumer group's conjecture.

___ 9.67

a)
```
MTB > ZTEST of mu=280, sigma=265, data in
        'EXPENDIT';
SUBC> ALTERNATIVE=1.
```

___ 9.69

a)
```
MTB > STDEV 'IRON'
     ST.DEV. =      3.0832   ←—— Minitab output
MTB > ZTEST of mu=18, sigma=3.0832, data
        in 'IRON';
SUBC> ALTERNATIVE=-1.
```

__ 9.71

a) H_0: $\mu = 7.1$ days, H_a: $\mu < 7.1$ days
b) `SUBC> ALTERNATIVE=-1.`
c) 7.70 days   d) 7.011 days   e) 40
f) 6.850 days   g) -0.21 h) 0.42   i) 0.42
j) The data do not provide sufficient evidence to conclude that the mean hospital stay, μ, for this year will be less than the 1982 mean of 7.1 days.

Exercises 9.5

__ 9.77

STEP 1 H_0: $\mu = 120$ min, H_a: $\mu > 120$ min
STEP 2 $\alpha = 0.05$
STEP 3   Critical value $= 1.729$
STEP 4 $t = 1.386$
STEP 5   Do not reject H_0
STEP 6   The data do not provide sufficient evidence to conclude that the mean drying time, μ, is actually greater than the manufacturer's claim of 120 minutes.

__ 9.79

a) STEP 1 H_0: $\mu = 36$ mo, H_a: $\mu < 36$ mo
    STEP 2 $\alpha = 0.01$
    STEP 3   Critical value $= -2.821$
    STEP 4 $t = -4.471$
    STEP 5   Reject H_0
    STEP 6   The data provide sufficient evidence to conclude that the mean life, μ, of the supplier's batteries is less than the claimed 36 months.
b) That battery life is approximately normally distributed.

__ 9.81

STEP 1 H_0: $\mu = 38.5¢/\text{lb}$, H_a: $\mu \neq 38.5¢/\text{lb}$
STEP 2 $\alpha = 0.05$
STEP 3   Critical values $= \pm 2.145$
STEP 4 $t = 2.622$
STEP 5   Reject H_0
STEP 6   The data provide sufficient evidence to conclude that the mean retail price, μ, for oranges now is different from the 1983 mean of 38.5 cents per pound.

__ 9.83

a) `MTB > TTEST of mu=120, data in 'DRYTIMES';`
    `SUBC> ALTERNATIVE=1.`

__ 9.85

a) H_0: $\mu = \$550$, H_a: $\mu < \$550$
b) `SUBC> ALTERNATIVE=-1.`

c) \$68.624   d) 15   e) \$496.667
f) -3.01 g) 0.0047   h) 0.0047
i) The data provide sufficient evidence to conclude that, on the average, the new bumper will reduce repair costs resulting from front-end collisions at low speeds.

Exercises 9.6

__ 9.95

STEP 1 H_0: $p = 0.638$, H_a: $p \neq 0.638$
STEP 2 $\alpha = 0.05$
STEP 3   Critical values $= \pm 1.96$
STEP 4 $z = -2.15$
STEP 5   Reject H_0
STEP 6   The data provide sufficient evidence to conclude that the proportion, p, of voters who think that repair of city streets is an important priority has changed from the 1988 proportion of 0.638.

__ 9.97

STEP 1 H_0: $p = 0.122$, H_a: $p > 0.122$
STEP 2 $\alpha = 0.10$
STEP 3   Critical value $= 1.28$
STEP 4 $z = 1.02$
STEP 5   Do not reject H_0
STEP 6   The data do not provide sufficient evidence to conclude that the proportion, p, of female physicians is higher now than in 1981.

__ 9.99

STEP 1 H_0: $p = 0.065$, H_a: $p > 0.065$
STEP 2 $\alpha = 0.01$
STEP 3   Critical value $= 2.33$
STEP 4 $z = 2.58$
STEP 5   Reject H_0
STEP 6   Evidently, the actual response rate, p, will exceed 6.5%. Hence, the decision will be for additional advertising and promotion.

__ 9.101

STEP 1 H_0: $p = 0.123$, H_a: $p < 0.123$
STEP 2 $\alpha = 0.05$
STEP 3   Critical value $= -1.645$
STEP 4 $z = -0.53$
STEP 5   Do not reject H_0
STEP 6   The study does not provide sufficient evidence to conclude that, in 1983, the proportion, p, of northeastern families below the poverty level was less than the national proportion of 0.123.

REVIEW TEST FOR CHAPTER 9

1. Let μ denote last year's mean cheese consumption by Americans.
   a) H_0: $\mu = 20.6$ lb   b) H_a: $\mu > 20.6$ lb
   c) right-tailed test

2. a) $z \geq 1.28$ b) $z < 1.28$ c) $z = 1.28$
   d) $\alpha = 0.10$
   e)

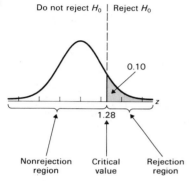

Do not reject H_0 | Reject H_0

0.10

1.28

Nonrejection region   Critical value   Rejection region

   f) right-tailed test

3. a) A Type I error would occur if, in fact, $\mu = 20.6$ lb but the results of the sampling lead to the conclusion that $\mu > 20.6$ lb.
   b) A Type II error would occur if, in fact, $\mu > 20.6$ lb but the results of the sampling fail to lead to that conclusion.
   c) A correct decision would occur if, in fact, $\mu = 20.6$ lb and the results of the sampling do not lead to the rejection of that fact; or if, in fact, $\mu > 20.6$ lb and the results of the sampling lead to that conclusion.
   d) Type I error   e) correct decision

4. a) STEP 1 H_0: $\mu = 20.6$ lb, H_a: $\mu > 20.6$ lb
      STEP 2 $\alpha = 0.10$
      STEP 3   Critical value = 1.28
      STEP 4 $z = 1.03$
      STEP 5   Do not reject H_0
      STEP 6   The data do not provide sufficient evidence to conclude that last year's mean cheese consumption, μ, for all Americans has increased over the 1983 mean of 20.6 lb.
   b) A Type II error because, given that the null hypothesis was not rejected, the only error that could be made is the error of not rejecting a false null hypothesis.

5. a) `MTB > ZTEST of mu=20.6, sigma=6.9, data in 'CHEESE';`
      `SUBC> ALTERNATIVE=1.`

6. a) H_0: $\mu = 20.6$ lb, H_a: $\mu > 20.6$ lb
   b) `SUBC> ALTERNATIVE=1.`
   c) 6.90 lb   d) 6.480 lb   e) 35
   f) 21.800 lb   g) 1.03   h) 0.15   i) 0.15
   j) The data do not provide sufficient evidence to conclude that last year's mean cheese consumption, μ, for all Americans has increased over the 1983 mean of 20.6 lb.

7. STEP 1 H_0: $p = 0.5$, H_a: $p > 0.5$
   STEP 2 $\alpha = 0.01$
   STEP 3   Critical value = 2.33
   STEP 4 $z = 3.89$
   STEP 5   Reject H_0
   STEP 6   The data provide sufficient evidence to conclude that a majority of all Americans were satisfied with President Reagan's performance.

8. a) STEP 1 H_0: $\mu = 29$ mpg, H_a: $\mu \neq 29$ mpg
      STEP 2 $\alpha = 0.05$
      STEP 3   Critical values = ± 2.145
      STEP 4 $t = -0.599$
      STEP 5   Do not reject H_0
      STEP 6   The data do not provide sufficient evidence to conclude that the company's report was incorrect.
   b) Yes, because the assumption for performing a t-test, Procedure 9.4 on page 458, is that the population under consideration is (approximately) normally distributed.

9. a) `MTB > TTEST of mu=29, data in 'MILEAGE'`

10. a) H_0: $\mu = 29$ mpg, H_a: $\mu \neq 29$ mpg
    b) No subcommand was used.
    c) 1.595 mpg   d) 15   e) 28.753 mpg
    f) -0.60 g) 0.56   h) 0.56
    i) The data do not provide sufficient evidence to conclude that the mean gas mileage, μ, for all cars of the model under consideration differs from the figure of 29 mpg reported by the company.

11. STEP 1 H_0: $\mu = \$178$, H_a: $\mu < \$178$
    STEP 2 $\alpha = 0.05$
    STEP 3   Critical value = -1.645
    STEP 4 $z = -1.43$
    STEP 5   Do not reject H_0
    STEP 6   The data do not provide sufficient evidence to conclude that the mean value

lost, μ, due to purse snatching has decreased from the 1983 mean of $178.

12. a) $P = 0.0764$
   b) STEP 1 H_0: $\mu = \$178$, H_a: $\mu < \$178$
      STEP 2 $\alpha = 0.05$
      STEP 3 $z = -1.43$
      STEP 4 $P = 0.0764$
      STEP 5   Do not reject H_0
      STEP 6   The data do not provide sufficient
               evidence to conclude that the mean
               value lost, μ, due to purse snatching
               has decreased from the 1983 mean
               of $178.

CHAPTER 10

Exercises 10.1

___ 10.1

a) μ_1, σ_1, μ_2, and σ_2 are parameters; $\bar{x}_1$, s_1, $\bar{x}_2$, and s_2 are statistics.
b) μ_1, σ_1, μ_2, and σ_2 are fixed numbers; $\bar{x}_1$, s_1, $\bar{x}_2$, and s_2 are random variables.

___ 10.3

STEP 1 H_0: $\mu_1 = \mu_2$, H_a: $\mu_1 > \mu_2$
STEP 2 $\alpha = 0.10$
STEP 3   Critical value = 1.28
STEP 4 $z = 1.19$
STEP 5   Do not reject H_0
STEP 6   The data do not provide sufficient evidence
         to conclude that, on the average, males stay
         in the hospital longer than females.

___ 10.5

STEP 1 H_0: $\mu_1 = \mu_2$, H_a: $\mu_1 \neq \mu_2$
STEP 2 $\alpha = 0.05$
STEP 3   Critical values = ± 1.96
STEP 4 $z = 6.32$
STEP 5   Reject H_0
STEP 6   Evidently, last year's mean annual fuel ex-
         penditure for households using natural gas
         is different from that for households using
         only electricity.

___ 10.7

a) STEP 1 H_0: $\mu_1 = \mu_2$, H_a: $\mu_1 < \mu_2$
   STEP 2 $\alpha = 0.05$
   STEP 3   Critical value = -1.645
   STEP 4 $z = -1.52$

STEP 5   Do not reject H_0
STEP 6   At the 5% significance level, the data
         do not provide sufficient evidence to
         conclude that the training program in-
         creases sales.

b) STEP 1 H_0: $\mu_1 = \mu_2$, H_a: $\mu_1 < \mu_2$
   STEP 2 $\alpha = 0.10$
   STEP 3   Critical value = -1.28
   STEP 4 $z = -1.52$
   STEP 5   Reject H_0
   STEP 6   At the 10% significance level, the data
            do provide sufficient evidence to con-
            clude that the training program in-
            creases sales.

___ 10.9

a) -0.2 to 4.1 days
b) We can be 80% confident that the difference,
   $\mu_1 - \mu_2$, between the mean lengths of stay in
   short-term hospitals by males and females is
   somewhere between -0.2 and 4.1 days.

___ 10.11

a) $175.20 to $332.72
b) We can be 95% confident that the difference,
   $\mu_1 - \mu_2$, between last year's mean fuel expenditures for households using natural gas and
   those using only electricity is somewhere between $175.20 and $332.72.

___ 10.13

a) $-\$66.73$ to $\$2.73$ b) $-\$59.03$ to $-\$4.97$

Exercises 10.2

___ 10.17

STEP 1 H_0: $\mu_1 = \mu_2$, H_a: $\mu_1 \neq \mu_2$
STEP 2 $\alpha = 0.05$
STEP 3   Critical values = ± 2.101
STEP 4 $t = -3.181$
STEP 5   Reject H_0
STEP 6   There appears to be a difference in mean
         lasting times for the two brands of paint.

___ 10.19

STEP 1 H_0: $\mu_1 = \mu_2$, H_a: $\mu_1 < \mu_2$
STEP 2 $\alpha = 0.05$
STEP 3   Critical value = -1.699
STEP 4 $t = -0.587$
STEP 5   Do not reject H_0

STEP 6   The data do not provide sufficient evidence to conclude that mine workers earn less, on the average, than construction workers.

___ 10.21

STEP 1 $H_0: \mu_1 = \mu_2$, $H_a: \mu_1 > \mu_2$
STEP 2 $\alpha = 0.05$
STEP 3   Critical value = 1.714
STEP 4 $t = 1.186$
STEP 5   Do not reject H_0
STEP 6   The data do not provide sufficient evidence to conclude that males in the age group 25–34 years are, on the average, taller than those who were in that same age group 20 years ago.

___ 10.23

a) −2.92 to −0.60 months
b) We can be 95% confident that the difference, $\mu_1 - \mu_2$, between the mean lasting times of Brand A and Brand B paints is somewhere between −2.92 and −0.60 months.

___ 10.25

a) −$1.91 to $0.93
b) We can be 90% confident that the difference, $\mu_1 - \mu_2$, between the mean hourly earnings of nonsupervisory mine and construction workers is somewhere between −$1.91 and $0.93.

___ 10.27

a) −0.72 to 3.94 inches
b) We can be 90% confident that the difference, $\mu_1 - \mu_2$, between the mean height of males in the age group 25–34 years and the mean height of males in the age group 45–54 years is somewhere between −0.72 and 3.94 inches.

___ 10.29

a) MTB > TWOSAMPLE T, 95% confidence, for
      'BRAND A' vs 'BRAND B';
  SUBC> POOLED.

___ 10.31

a) $H_0: \mu_1 = \mu_2$, $H_a: \mu_1 < \mu_2$
b) MTB > TWOSAMPLE T, 99% confidence, for
      'CURRENT' vs 'NEW';
  SUBC> POOLED;
  SUBC> ALTERNATIVE=-1.
c) $s_1 = 55.3$ hr, $s_2 = 44.6$ hr
d) $n_1 = 20$, $n_2 = 10$
e) $\bar{x}_1 = 1023.2$ hr, $\bar{x}_2 = 1093.0$ hr
f) −3.46   g) 0.0009   h) 0.0009

i) The data provide sufficient evidence to conclude that, on the average, the new bulb will outlast the bulb currently produced.
j) −126 to −14 hr
k) The lifetimes of both the current and new bulb are normally distributed with equal standard deviations. Also, the samples must be independent.
l) 52.1 hr

Exercises 10.3

___ 10.37

STEP 1 $H_0: \mu_1 = \mu_2$, $H_a: \mu_1 \neq \mu_2$
STEP 2 $\alpha = 0.05$
STEP 3   Critical values = ±2.052
STEP 4 $t = -0.955$
STEP 5   Do not reject H_0
STEP 6   The data do not provide sufficient evidence to conclude that there is a difference between the mean GPAs of sophomores and juniors at the university.

___ 10.39

a) STEP 1 $H_0: \mu_1 = \mu_2$, $H_a: \mu_1 < \mu_2$
   STEP 2 $\alpha = 0.01$
   STEP 3   Critical value = −2.518
   STEP 4 $t = -4.104$
   STEP 5   Reject H_0
   STEP 6   The data provide sufficient evidence to conclude that Mirror-sheen has a longer effectiveness time, on the average, than Sureglow.
b) Yes. Since the hypothesis test was performed at the 0.01 level of significance, there is only a 1% chance that we would conclude (as we did) that Mirror-sheen outlasts Sureglow, on the average, when in fact it does not.

___ 10.41

STEP 1 $H_0: \mu_1 = \mu_2$, $H_a: \mu_1 < \mu_2$
STEP 2 $\alpha = 0.05$
STEP 3   Critical value = −1.645 (approximately)
STEP 4 $t = -3.951$
STEP 5   Reject H_0
STEP 6   The data provide sufficient evidence to conclude that Idaho has a smaller mean potato yield than Nevada.

___ 10.43

a) −0.44 to 0.16
b) We can be 95% confident that the difference, $\mu_1 - \mu_2$, between the mean GPAs of sophomores

and juniors at the university is somewhere between -0.44 and 0.16.

___ 10.45

a) -3.8 to -0.9 days
b) We can be 98% confident that the difference, $\mu_1 - \mu_2$, between the mean effectiveness times of Sureglow and Mirror-sheen is somewhere between -3.8 and -0.9 days.

___ 10.47

a) -47.9 to -19.7 cwt
b) We can be 90% confident that the difference, $\mu_1 - \mu_2$, between the mean yields per acre of potatoes for Idaho and Nevada is somewhere between -47.9 and -19.7 cwt.

___ 10.49

a) MTB > TWOSAMPLE T, 98% confidence,
        for 'SUREGLOW' vs 'MIRROR';
   SUBC> ALTERNATIVE=-1.

___ 10.51

a) $H_0: \mu_1 = \mu_2$, $H_a: \mu_1 > \mu_2$
b) MTB > TWOSAMPLE T, 90% confidence,
        for 'MALES' vs 'FEMALES';
   SUBC> ALTERNATIVE=1.
c) $s_1 = 4.62$ yr, $s_2 = 7.25$ yr
d) $n_1 = 11$, $n_2 = 12$
e) $\bar{x}_1 = 35.69$ yr, $\bar{x}_2 = 32.81$ yr
f) 1.14 g) 0.13 h) 0.13
i) The data do not provide sufficient evidence to conclude that the mean age at the time of first divorce for males is greater than that for females.
j) -1.5 to 7.2 yr
k) The ages at the time of first divorce for both males and females are normally distributed. Also, the samples must be independent.

Exercises 10.4

___ 10.55

STEP 1 $H_0: \mu_1 = \mu_2$, $H_a: \mu_1 > \mu_2$
STEP 2 $\alpha = 0.01$
STEP 3   Critical value $= 2.624$
STEP 4   Paired differences, $d = x_1 - x_2$:

1	-3	4	0	7
-1	2	6	3	3
0	2	4	9	-2

STEP 5 $t = 2.696$

STEP 6   Reject H_0
STEP 7   The data provide sufficient evidence to conclude that the running program will, on the average, reduce heart rates.

___ 10.57

STEP 1 $H_0: \mu_1 = \mu_2$, $H_a: \mu_1 < \mu_2$
STEP 2 $\alpha = 0.05$
STEP 3   Critical value $= -1.761$
STEP 4   Paired differences, $d = x_1 - x_2$, are given.
STEP 5 $t = -1.420$
STEP 6   Do not reject H_0
STEP 7   The data do not provide sufficient evidence to conclude that nonsupervisory mine workers earn a smaller average hourly wage than nonsupervisory construction workers.

___ 10.59

STEP 1 $H_0: \mu_1 = \mu_2$, $H_a: \mu_1 > \mu_2$
STEP 2 $\alpha = 0.05$
STEP 3   Critical value $= 1.833$
STEP 4   Paired differences, $d = x_1 - x_2$:

1	-1	6	1	13
-1	4	-2	4	8

STEP 5 $t = 2.213$
STEP 6   Reject H_0
STEP 7   The data provide sufficient evidence to conclude that the mean age of married men is greater than the mean age of married women.

___ 10.61

a) 0.1 to 4.6
b) We can be 98% confident that the difference, $\mu_1 - \mu_2$, between the mean heart rates of all people before and after the running program is somewhere between 0.1 and 4.6.

___ 10.63

a) $-\$1.02$ to $\$0.11$
b) We can be 90% confident that the difference, $\mu_1 - \mu_2$, between the mean hourly earnings of nonsupervisory mine and construction workers is somewhere between $-\$1.02$ and $\$0.11$.

___ 10.65

a) 0.6 to 6.0 yr
b) We can be 90% confident that the difference, $\mu_1 - \mu_2$, between the mean age of married men and the mean age of married women is somewhere between 0.6 and 6.0 years.

Exercises 10.6

___ **10.71**

a) p_1 and p_2 are parameters and the other quantities are statistics.

b) p_1 and p_2 are fixed numbers and the other quantities are random variables.

___ **10.73**

STEP 1 H_0: $p_1 = p_2$, H_a: $p_1 \neq p_2$
STEP 2 $\alpha = 0.01$
STEP 3   Critical values $= \pm 2.575$
STEP 4 $z = -2.62$
STEP 5   Reject H_0
STEP 6   The data provide sufficient evidence to conclude that there is a difference between the percentages of American and French households that own washing machines.

___ **10.75**

STEP 1 H_0: $p_1 = p_2$, H_a: $p_1 > p_2$
STEP 2 $\alpha = 0.05$
STEP 3   Critical value $= 1.645$
STEP 4 $z = 4.62$
STEP 5   Reject H_0
STEP 6   We can conclude that the percentage of employed workers who have registered to vote exceeds the percentage of unemployed workers who have registered to vote.

___ **10.77**

STEP 1 H_0: $p_1 = p_2$, H_a: $p_1 < p_2$
STEP 2 $\alpha = 0.10$
STEP 3   Critical value $= -1.28$
STEP 4 $z = -0.46$
STEP 5   Do not reject H_0
STEP 6   The data do not provide sufficient evidence to conclude that the percentage of American dentists practicing in the Northeast is smaller than the percentage of American physicians practicing in the Northeast.

___ **10.79**

a) -0.140 to -0.002

b) We can be 99% confident that the difference, $p_1 - p_2$, between the proportions of American and French households that own washing machines is somewhere between -0.140 and -0.002.

___ **10.81**

a) 0.102 to 0.212

b) We can be 90% confident that the difference, $p_1 - p_2$, between the proportions of employed and unemployed workers who have registered to vote is somewhere between 0.102 and 0.212.

___ **10.83**

a) -0.063 to 0.029

b) We can be 80% confident that the difference, $p_1 - p_2$, between the proportion of American dentists who practice in the Northeast and the proportion of American physicians who practice in the Northeast is somewhere between -0.063 and 0.029.

REVIEW TEST FOR CHAPTER 10

1. STEP 1 H_0: $\mu_1 = \mu_2$, H_a: $\mu_1 > \mu_2$
   STEP 2 $\alpha = 0.01$
   STEP 3   Critical value $= 2.33$
   STEP 4 $z = 7.65$
   STEP 5   Reject H_0
   STEP 6   Evidently, the mean resale price for homes in Phoenix, Arizona, exceeds that for homes in Flint, Michigan.

2. \$23,586 to \$44,258. We can be 98% confident that the difference, $\mu_1 - \mu_2$, between the mean resale prices of homes in Phoenix, Arizona, and Flint, Michigan, is somewhere between \$23,586 and \$44,258.

3. a) STEP 1 H_0: $\mu_1 = \mu_2$, H_a: $\mu_1 \neq \mu_2$
      STEP 2 $\alpha = 0.10$
      STEP 3   Critical values $= \pm 1.833$
      STEP 4   Paired differences, $d = x_1 - x_2$:

82	−95	−49	0	−36
−152	49	−38	−43	−118

   STEP 5 $t = -1.766$
   STEP 6   Do not reject H_0
   STEP 7   The data do not provide sufficient evidence to conclude that there is a difference in mean results for the two speed reading programs.

   b) The population of all possible paired differences is normally distributed.

4. -81.5 to 1.5 words per minute. We can be 90% confident that the difference, $\mu_1 - \mu_2$, between the mean reading speed of people using Program 1 and the mean reading speed of people using Program 2 is somewhere between -81.5 and 1.5 words per minute.

5. STEP 1 $H_0: p_1 = p_2$, $H_a: p_1 < p_2$
   STEP 2 $\alpha = 0.05$
   STEP 3   Critical value $= -1.645$
   STEP 4 $z = -1.82$
   STEP 5   Reject H_0
   STEP 6   It appears that the percentage of unemployed male doctoral scientists is lower than the percentage of unemployed female doctoral scientists.

6. -0.094 to -0.003. We can be 90% confident that the difference, $p_1 - p_2$, between the proportions of unemployed male and female doctoral scientists is somewhere between -0.094 and -0.003.

7. STEP 1 $H_0: \mu_1 = \mu_2$, $H_a: \mu_1 \neq \mu_2$
   STEP 2 $\alpha = 0.05$
   STEP 3   Critical values $= \pm 1.96$ (approximately)
   STEP 4 $t = 0.266$
   STEP 5   Do not reject H_0
   STEP 6   The data do not provide sufficient evidence to conclude that there is a difference between the mean IQs of male and female students at the university.

8. -5.4 to 7.1. We can be 95% confident that the difference, $\mu_1 - \mu_2$, between the mean IQs of female and male students at the university is somewhere between -5.4 and 7.1.

9. a) MTB > TWOSAMPLE T, 95% confidence, for 'FEMALE' vs 'MALE';
      SUBC> POOLED.
   b) *Note:* The output resulting from the command and subcommand in part (a) will indicate that the 95% confidence interval is from -5.6 to 7.3. This differs from the 95% confidence interval obtained in Problem 8 because Minitab uses the exact value for $t_{0.025}$ instead of the approximate value, 1.96, used in Problem 8.

10. a) $H_0: \mu_1 = \mu_2$, $H_a: \mu_1 \neq \mu_2$
    b) MTB > TWOSAMPLE T, 95% confidence, for 'FEMALE' vs 'MALE';
       SUBC> POOLED.
    c) $s_1 = 8.80$, $s_2 = 11.2$
    d) $\bar{x}_1 = 120.40$, $\bar{x}_2 = 119.6$
    e) 0.27   f) 0.79   g) 0.79
    h) The data do not provide sufficient evidence to conclude that there is a difference between the mean IQs of male and female students at the university.

i) -5.6 to 7.3. (See note in the answer to Problem 9(b).)

j) The IQs of both female and male students at the university are (approximately) normally distributed with equal standard deviations. Also, the samples must be independent.

k) 10.1

11. STEP 1 $H_0: \mu_1 = \mu_2$, $H_a: \mu_1 < \mu_2$
    STEP 2 $\alpha = 0.05$
    STEP 3   Critical value $= -1.725$
    STEP 4 $t = -2.210$
    STEP 5   Reject H_0
    STEP 6   The data provide sufficient evidence to conclude that Germans consume less fish than Russians, on the average.

12. -8.5 to -1.0 kg. We can be 90% confident that the difference, $\mu_1 - \mu_2$, between last year's mean fish consumption by Germans and last year's mean fish consumption by Russians is somewhere between -8.5 and -1.0 kg.

13. a) MTB > TWOSAMPLE T, 90% confidence, for 'GERMANS' vs 'RUSSIANS';
       SUBC> ALTERNATIVE=-1.

14. a) $H_0: \mu_1 = \mu_2$, $H_a: \mu_1 < \mu_2$
    b) MTB > TWOSAMPLE T, 90% confidence, for 'GERMANS' vs 'RUSSIANS';
       SUBC> ALTERNATIVE=-1.
    c) $s_1 = 5.06$ kg, $s_2 = 5.61$ kg
    d) $n_1 = 10$, $n_2 = 15$
    e) $\bar{x}_1 = 11.30$ kg, $\bar{x}_2 = 16.07$ kg
    f) -2.21 g) 0.020   h) 0.020
    i) The data provide sufficient evidence to conclude that Germans consume less fish than Russians, on the average.
    j) -8.5 kg to -1.0 kg
    k) The fish consumptions are (approximately) normally distributed in each country. Also, the samples must be independent.

CHAPTER 11

Exercises 11.1

___ **11.1**   a) 32.852   b) 10.117
___ **11.3**   a) 18.307   b) 3.247
___ **11.5**   a) 1.646   b) 15.507
___ **11.7**   a) 0.831, 12.832   b) 9.591, 34.170

Exercises 11.2

___ 11.9 Because the hypothesis test is carried out by determining how well the observed frequencies fit the expected frequencies.

___ 11.11

STEP 1 H_0: The current distribution of the U.S. resident population by region is the same as the 1983 distribution.

         H_a: The current distribution of the U.S resident population by region is different from the 1983 distribution.

STEP 2   Expected frequencies:

Region	p	E
Northeast	0.212	106
Midwest	0.252	126
South	0.340	170
West	0.196	98

STEP 3   Assumptions 1 and 2 are satisfied since all expected frequencies are at least 5.

STEP 4 $\alpha = 0.05$

STEP 5   Critical value = 7.815

STEP 6 $\chi^2 = 1.827$

STEP 7   Do not reject H_0

STEP 8   The data do not provide sufficient evidence to conclude that the current distribution of the U.S. resident population by region is different from the 1983 distribution.

___ 11.13

STEP 1 H_0: The distribution of the number of years of school completed by Tennessee residents is the same as the national distribution.

         H_a: The distribution of the number of years of school completed by Tennessee residents is different from the national distribution.

STEP 2   Expected frequencies:

Years of school	p	E
8 or less	0.183	54.9
9–11	0.153	45.9
12	0.346	103.8
13–15	0.157	47.1
16 or more	0.161	48.3

STEP 3   Assumptions 1 and 2 are satisfied since all expected frequencies are at least 5.

STEP 4 $\alpha = 0.01$

STEP 5   Critical value = 13.277

STEP 6 $\chi^2 = 20.037$

STEP 7   Reject H_0

STEP 8   We conclude that the distribution of the number of years of school completed by Tennessee residents is different from the national distribution.

___ 11.15

STEP 1 H_0: The die is not loaded.

         H_a: The die is loaded.

STEP 2   Expected frequencies:

Number	p	E
1	1/6	25
2	1/6	25
3	1/6	25
4	1/6	25
5	1/6	25
6	1/6	25

STEP 3   Assumptions 1 and 2 are satisfied since all expected frequencies are at least 5.

STEP 4 $\alpha = 0.05$

STEP 5   Critical value = 11.070

STEP 6 $\chi^2 = 2.480$

STEP 7   Do not reject H_0

STEP 8   The data do not provide sufficient evidence to conclude that the die is loaded.

___ 11.17

STEP 1 H_0: The distribution of types of violent crime in New Jersey is the same as that for the nation as a whole.

         H_a: The distribution of types of violent crime in New Jersey is different from that for the nation as a whole.

STEP 2   Expected frequencies:

Violent crime	p	E
Murder	0.016	9.6
Forcible rape	0.064	38.4
Robbery	0.403	241.8
Aggravated assault	0.517	310.2

STEP 3   Assumptions 1 and 2 are satisfied since all expected frequencies are at least 5.

STEP 4 $\alpha = 0.01$
STEP 5 Critical value = 11.345
STEP 6 $\chi^2 = 18.003$
STEP 7 Reject H_0
STEP 8 The data provide sufficient evidence to conclude that the distribution of types of violent crime in New Jersey is different from that for the nation as a whole.

Exercises 11.3

__ 11.23
STEP 1 H_0: Annual income and educational level are statistically independent.

   H_a: Annual income and educational level are statistically dependent.

STEP 2 Observed and expected frequencies:

Years of schooling

	0–8	9–12	Over 12	Total
Under $10,000	34 15.8	36 37.7	10 26.5	80
$10,000–$24,999	41 29.5	72 70.2	36 49.3	149
$25,000–$39,999	10 28.9	78 68.8	58 48.3	146
$40,000 and over	4 14.8	26 35.3	45 24.8	75
Total	89	212	149	450

Annual income

STEP 3 Assumptions 1 and 2 are satisfied since all expected frequencies are at least 5.
STEP 4 $\alpha = 0.01$
STEP 5 Critical value = 16.812
STEP 6 $\chi^2 = 81.645$
STEP 7 Reject H_0
STEP 8 The data provide sufficient evidence to conclude that annual income and educational level are statistically dependent.

__ 11.25
STEP 1 H_0: Sex and occupation type for employed workers are nonassociated.

   H_a: Sex and occupation type for employed workers are associated.

STEP 2 Observed and expected frequencies:

Sex

Occupation type	Male	Female	Total
Managerial/ Professional	14 13.6	10 10.4	24
Technical sales Administrative	9 14.2	16 10.8	25
Service	4 5.7	6 4.3	10
Other	20 13.6	4 10.4	24
Total	47	36	83

STEP 3 Assumptions 1 and 2 are satisfied since all expected frequencies are at least 1, and only 12.5% (1/8) of the expected frequencies are less than 5.
STEP 4 $\alpha = 0.01$
STEP 5 Critical value = 11.345
STEP 6 $\chi^2 = 12.454$
STEP 7 Reject H_0
STEP 8 The data provide sufficient evidence to conclude that there is an association between the characteristics, sex and occupation type, for employed workers.

__ 11.27
a) See the table at the top of the next page.
b) STEP 1 H_0: Type of violent crime and age of the person arrested are nonassociated.

   H_a: Type of violent crime and age of the person arrested are associated.

STEP 2 Observed and expected frequencies: See the table at the top of the next page.
STEP 3 Assumptions 1 and 2 are satisfied since all expected frequencies are at least 1, and only 16.7% (2/12) of the expected frequencies are less than 5.
STEP 4 $\alpha = 0.01$
STEP 5 Critical value = 16.812
STEP 6 $\chi^2 = 31.241$
STEP 7 Reject H_0
STEP 8 We conclude that an association exists between the characteristics, type of violent crime committed and age of the person arrested.

Age

Type of violent crime	18–24	25–44	45+	Total
Murder	11 / 13.3	16 / 15.2	4 / 2.5	31
Forcible rape	21 / 21.9	26 / 25.0	4 / 4.1	51
Robbery	128 / 97.0	92 / 110.9	6 / 18.1	226
Aggravated assault	162 / 189.8	234 / 216.9	46 / 35.4	442
Total	322	368	60	750

__ 11.29

STEP 1 H_0: Net worth and marital status for top wealthholders are statistically independent.

H_a: Net worth and marital status for top wealthholders are statistically dependent.

STEP 2 Observed and expected frequencies:

Marital status

Net worth	Married	Single/ Divorced	Widowed	Total
$100,000– $249,999	227 / 223.9	54 / 53.0	63 / 67.1	344
$250,000– $499,999	60 / 63.1	15 / 14.9	22 / 18.9	97
$500,000– $999,999	20 / 20.2	4 / 4.8	7 / 6.0	31
$1,000,000 or more	10 / 9.8	2 / 2.3	3 / 2.9	15
Total	317	75	95	487

STEP 3 Assumption 2 is violated since 25% (3/12) of the expected frequencies are less than 5.

__ 11.31

a) MTB > CHISQUARE on '0-8' '9-12' 'OVER 12'

__ 11.33

a) 34   b) 164   c) 9   d) 13.21

e) 1.343   f) 6

g) H_0: Grade and study time for intermediate algebra students at Arizona State University are nonassociated.

H_a: Grade and study time for intermediate algebra students at Arizona State University are associated.

h) 10.574

i) Critical value = 12.592 (from Table IV) and $\chi^2 = 10.574$ (from part (h)). Do not reject H_0. The data do not provide sufficient evidence to conclude that an association exists between the characteristics, grade and study time, for intermediate algebra students at ASU.

Exercises 11.4

__ 11.41

STEP 1 H_0: $\sigma = 100$, H_a: $\sigma \neq 100$
STEP 2 $\alpha = 0.05$
STEP 3 Critical values = 12.401, 39.364
STEP 4 $\chi^2 = 29.561$
STEP 5 Do not reject H_0
STEP 6 The data do not provide sufficient evidence to conclude that the standard deviation, σ, of this year's verbal scores is different from the 1941 standard deviation of 100.

__ 11.43

STEP 1 H_0: $\sigma = 1$ second, H_a: $\sigma > 1$ second
STEP 2 $\alpha = 0.01$
STEP 3 Critical value = 36.191
STEP 4 $\chi^2 = 39.222$
STEP 5 Reject H_0
STEP 6 Evidently, the standard deviation, σ, of the weekly errors exceeds the one-second claim made by the manufacturer.

__ 11.45

STEP 1 H_0: $\sigma = 0.2$ fl oz, H_a: $\sigma < 0.2$ fl oz
STEP 2 $\alpha = 0.05$
STEP 3 Critical value = 6.571
STEP 4 $\chi^2 = 5.933$
STEP 5 Reject H_0
STEP 6 The data provide sufficient evidence to conclude that the standard deviation, σ, of the amounts of coffee being dispensed is less than 0.2 fl oz.

__ 11.47   86.7 to 154.4. We can be 95% confident that the standard deviation of this year's verbal SAT scores is somewhere between 86.7 and 154.4.

__ 11.49   1.04 to 2.27 seconds. We can be 98% confident that the standard deviation, σ, of the weekly errors of the watches manufactured is somewhere between 1.04 and 2.27 seconds.

__ 11.51   0.10 to 0.19 fl oz. We can be 90% confident that the standard deviation, σ, of the amounts of coffee being dispensed is somewhere between 0.10 and 0.19 fl oz.

__ 11.53   So as to insure that there will not be a large variation in the amount of coffee dispensed.

REVIEW TEST FOR CHAPTER 11

1. a) 6.408   b) 33.409   c) 27.587
   d) 8.672   e) 7.564, 30.191

2. STEP 1 H_0: The actual preference distribution for the number of bedrooms is the same as the distribution for the number of bedrooms in homes completed in 1987.

   H_a: The actual preference distribution for the number of bedrooms is different from the distribution for the number of bedrooms in homes completed in 1987.

   STEP 2   Expected frequencies:

No. bedrooms	p	E
2 or less	0.19	28.5
3	0.58	87.0
4 or more	0.23	34.5

   STEP 3   Assumptions 1 and 2 are satisfied since all expected frequencies are at least 5.
   STEP 4 $\alpha = 0.05$
   STEP 5   Critical value = 5.991
   STEP 6 $\chi^2 = 29.821$
   STEP 7   Reject H_0
   STEP 8   The data provide sufficient evidence to conclude that the actual preference distribution for the number of bedrooms is different from the distribution for the number of bedrooms in homes completed in 1987.

3. STEP 1 H_0: Response and educational level are statistically independent.

   H_a: Response and educational level are statistically dependent.

STEP 2   Observed and expected frequencies:

Response

Educational level	Favor	Oppose	No opinion	Total
College grad	264 229.1	17 38.5	6 19.3	287
Some college	205 190.0	26 31.9	7 16.0	238
High school grad	461 459.9	81 77.3	34 38.8	576
Non high school grad	290 340.9	81 57.3	56 28.8	427
Total	1220	205	103	1528

STEP 3   Assumptions 1 and 2 are satisfied since all expected frequencies are at least 5.
STEP 4 $\alpha = 0.01$
STEP 5   Critical value = 16.812
STEP 6 $\chi^2 = 77.837$
STEP 7   Reject H_0
STEP 8   The data provide sufficient evidence to conclude that response and educational level are statistically dependent.

4. a) `MTB > CHISQUARE on 'FAVOR' 'OPPOSE'`
                             `'NOOPIN'`

5. a) 205   b) 238   c) 26   d) 31.93
   e) 1.102   f) 6   g) 77.837
   h) Critical value = 16.812 (from Table IV) and $\chi^2 = 77.837$ (from part (g)). Reject H_0. The data provide sufficient evidence to conclude that response and educational level are statistically dependent.

6. a) STEP 1 H_0: $\sigma = 16$, H_a: $\sigma \neq 16$
        STEP 2 $\alpha = 0.10$
        STEP 3   Critical values = 13.848, 36.415
        STEP 4 $\chi^2 = 21.110$
        STEP 5   Do not reject H_0
        STEP 6   The data do not provide sufficient evidence to conclude that the standard deviation of all IQs is not equal to 16.
   b) That IQs are normally distributed.

7. 12.2 to 19.8. We can be 90% confident that the standard deviation of all IQ scores is somewhere between 12.2 and 19.8.

CHAPTER 12

Exercises 12.1

___ **12.1**

a) $y = 41.88 + 0.33x$ b) $b_0 = 41.88, b_1 = 0.33$

c)

Miles x	Cost ($) y
50	58.38
100	74.88
250	124.38

d)

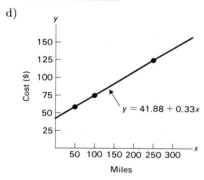

e) About $90; exact cost is $91.38.

___ **12.3**

a) $b_0 = 32, b_1 = 1.8$ b) $-40, 32, 68, 212$

c)

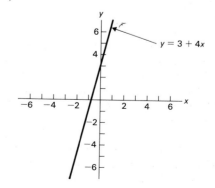

d) About 80°F; exact temperature is 82.4°F.

___ **12.5**

a) $b_0 = 41.88, b_1 = 0.33$

b) The y-intercept, $b_0 = 41.88$, gives the y-value at which the straight line, $y = 41.88 + 0.33x$, intersects the y-axis. The slope, $b_1 = 0.33$, indicates that the y-value increases by 0.33 units for every increase in x of one unit.

c) The y-intercept, $b_0 = 41.88$, is the cost (in dollars) for driving the car zero miles. The slope, $b_1 = 0.33$, represents the fact that the cost per mile is $0.33; it is the amount the total cost increases for each additional mile driven.

___ **12.7**

a) $b_0 = 32, b_1 = 1.8$

b) The y-intercept, $b_0 = 32$, gives the y-value at which the straight line, $y = 32 + 1.8x$, intersects the y-axis. The slope, $b_1 = 1.8$, indicates that the y-value increases by 1.8 units for every increase in x of one unit.

c) The y-intercept, $b_0 = 32$, is the Fahrenheit temperature corresponding to 0°C. The slope, $b_1 = 1.8$, represents the fact that the Fahrenheit temperature increases by 1.8° for every increase of the Celsius temperature of 1°.

___ **12.9**

a) $b_0 = 3, b_1 = 4$ b) slopes upward

c)

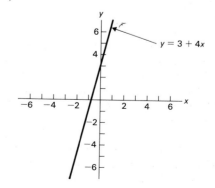

__ 12.11

a) $b_0 = 6$, $b_1 = -7$ b) slopes downward

c)

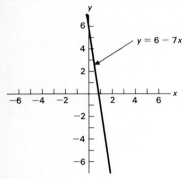

__ 12.17

a) $b_0 = 0$, $b_1 = 1.5$ b) slopes upward

c)

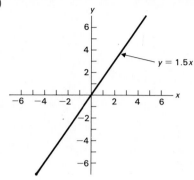

__ 12.13

a) $b_0 = -2$, $b_1 = 0.5$ b) slopes upward

c)

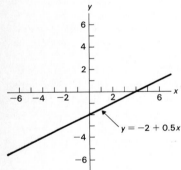

__ 12.19

a) slopes upward   b) $y = 5 + 2x$

c)

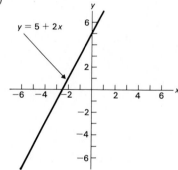

__ 12.15

a) $b_0 = 2$, $b_1 = 0$ b) horizontal

c)

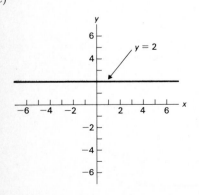

__ 12.21

a) slopes downward   b) $y = -2 - 3x$

c)

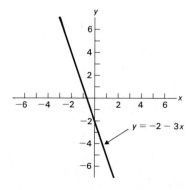

__ **12.23**

a) slopes downward   b) $y = -0.5x$

c)

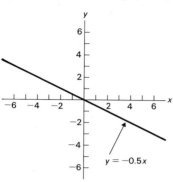

__ **12.29**

a)

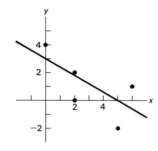

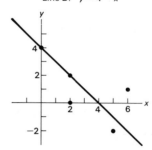

b) Line A: $y = 3 - 0.6x$

x	y	$\hat{y}$	e	e^2
0	4	3.0	1.0	1.00
2	2	1.8	0.2	0.04
2	0	1.8	−1.8	3.24
5	−2	0.0	−2.0	4.00
6	1	−0.6	1.6	2.56
				10.84

__ **12.25**

a) horizontal   b) $y = 3$

c)

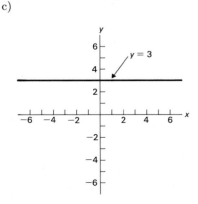

Line B: $y = 4 - x$

x	y	$\hat{y}$	e	e^2
0	4	4	0	0
2	2	2	0	0
2	0	2	−2	4
5	−2	−1	−1	1
6	1	−2	3	9
				14

c) Line A

___ **12.31**

a) $\hat{y} = 2.875 - 0.625x$

b)

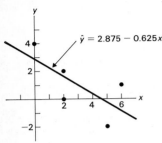

___ **12.33**

a) $\hat{y} = -174.49 + 4.84x$

b)

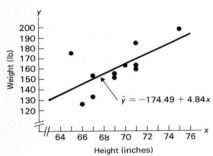

c) Weight tends to increase as height increases.

d) The weights of 18–24-year-old males increase an estimated 4.84 lb for each increase in height of one inch.

e) 149.54 lb; 178.56 lb (See the note at the beginning of Appendix B on page A-19.)

___ **12.35**

a) $\hat{y} = 222.27 - 1.14x$

b)

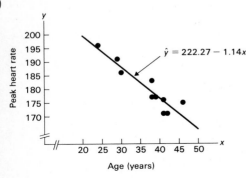

c) Peak heart rate tends to decrease as age increases.

d) The peak heart rate an individual can reach during intensive exercise decreases by an estimated 1.14 for each increase in age of one year.

e) 190.34 (See the note at the beginning of Appendix B on page A-19.)

___ **12.37**

a) $\hat{y} = 12.86 + 1.21x$

b)

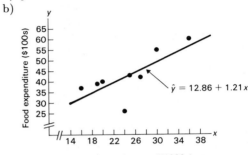

c) Annual food expenditure tends to increase as disposable income increases.

d) Annual food expenditures increase an estimated $121 (1.21 hundred dollars) for each increase in disposable income of $1000.

e) $4321 (See the note at the beginning of Appendix B on page A-19.)

___ **12.39**   Only the second one.

___ **12.41**

a) It is acceptable to use the regression equation to predict the weight of an 18–24-year-old male who is 68 inches tall since that height lies within the range of the heights in the sample data. It is not acceptable (and would be extrapolation) to use the regression equation to predict the weight of an 18–24-year-old male who is 60 inches tall since that height lies outside the range of the heights in the sample data.

b) Heights between 65 and 75 inches, inclusive.

___ **12.43**

a) MTB > PLOT 'FOODEX' versus 'INCOME'

b) MTB > REGRESS 'FOODEX' on 1 predictor
          'INCOME'

___ **12.45**   a) $\hat{y} = 2152 + 60.8x$ b) 3185.6 gm

Exercises 12.3

— 12.53

a) The coefficient of determination, r^2.
b) It equals the percentage reduction obtained in the total squared error by using the regression equation, instead of the sample mean, $\overline{y}$, to predict the observed y-values. It also equals the percentage of the total variation in the observed y-values that is explained by the regression.

— 12.55

a) $SST = 20$, $SSR = 9.375$, $SSE = 10.625$
b) $20 = 9.375 + 10.625$　　c) $r^2 = 0.469$
d) 46.9%　　e) 46.9%　　f) moderately useful

— 12.57

a) $SST = 4352.91$, $SSR = 1909.46$, $SSE = 2443.45$
b) 0.439
c) 43.9%. In words, 43.9% of the variation in the weight data is explained by height.
d) moderately useful

— 12.59

a) $SST = 642.10$, $SSR = 553.60$, $SSE = 88.50$
b) 0.862
c) 86.2%. In words, 86.2% of the variation in the peak-heart-rate data is explained by age.
d) extremely useful

— 12.61

a) $SST = 783.50$, $SSR = 429.95$, $SSE = 353.55$
b) 0.549
c) 54.9%. In words, 54.9% of the variation in the food-expenditure data is explained by disposable income.
d) moderately useful

— 12.63

a) MTB > REGRESS 'FOODEX' on 1 predictor
　　　'INCOME'

— 12.65

a) 0.372
b) $SSR = 2{,}505{,}745$, $SSE = 4{,}234{,}255$,
　　$SST = 6{,}740{,}000$
c) 37.2%

Exercises 12.4

— 12.71

a) $r = -0.685$

b) Suggest a moderately strong negative linear correlation.
c) Data points are clustered moderately closely about the regression line.
d) $r^2 = 0.469$

— 12.73

a) $r = 0.662$
b) Suggests a moderately strong positive linear correlation.
c) Data points are clustered moderately closely about the regression line.
d) $r^2 = 0.438$ (*Note:* In part (b) of Exercise 12.57, we found that $r^2 = 0.439$. The discrepancy between these two values of r^2 is due to the error resulting from rounding r to three decimal places before squaring.)

— 12.75

a) $r = -0.929$
b) Suggest an extremely strong negative linear correlation.
c) Data points are clustered extremely closely about the regression line.
d) $r^2 = 0.863$ (*Note:* In part (b) of Exercise 12.59, we found that $r^2 = 0.862$. The discrepancy between these two values of r^2 is due to the error resulting from rounding r to three decimal places before squaring.)

— 12.77

a) $r = 0.741$
b) Suggests a moderately strong positive linear correlation.
c) Data points are clustered moderately closely about the regression line.
d) $r^2 = 0.549$

— 12.79

a) $r = 0$
b) No, only that there is no *linear* relationship between the variables.
c)

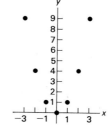

d) No, because the data points are not scattered about a straight line.

e) For each data point (x, y), we have $y = x^2$.

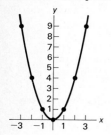

___ 12.81

a) `MTB > CORRELATION of 'INCOME' and 'FOODEX'`

Exercises 12.5

___ 12.89

a) $\hat{y} = 266 + 6.73x_1 + 3.26x_2 + 4.51x_3$

b) $367.4 million

c) $R^2 = 0.911$; 91.1% of the variation in the sales data is explained by television, magazine, and radio advertising expenditures.

___ 12.91

a) $\hat{y} = 0.648 + 0.562x_1 + 0.000267x_2 + 0.000223x_3$

b) 2.73

c) $R^2 = 0.349$

d) Moderately useful, at best.

___ 12.93

a) `MTB > REGRESS 'SALES' on 3 predictors`
`'TV', 'MAG', and 'RADIO'`

___

REVIEW TEST FOR CHAPTER 12

1. a) $y = 72 - 12x$ b) $b_0 = 72$, $b_1 = -12$
   c) The line slopes downward since $b_1 < 0$.
   d) $4800; $1200

e)

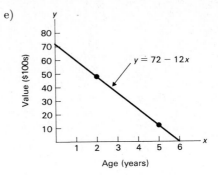

f) About $2500; exact value is $2400.

2. a)

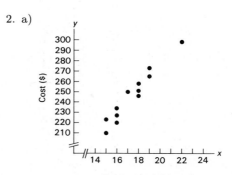

b) Yes, because the data points appear to be scattered about a straight line.

c) $\hat{y} = 35.80 + 12.08x$

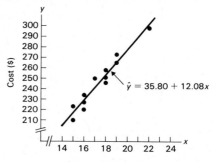

d) Cost of grading tends to increase as the number of papers graded increases.

e) Paper-grading costs increase an estimated $12.08 for each additional 100 papers graded.

f) $229.13 (See the note at the beginning of Appendix B on page A-19.)

3. a) $SST = 6984.25$, $SSR = 6558.31$,
      $SSE = 425.94$
   b) $r^2 = 0.939$ c) 93.9%   d) 93.9%
   e) extremely useful

4. a) MTB > PLOT 'COST' versus 'PAPERS'
   b) MTB > REGRESS 'COST' on 1 predictor
           'PAPERS'
   c) The same command as in part (b).

5. a) $\hat{y} = 35.8 + 12.1x$ b) $r^2 = 0.939$
   c) $SSR = 6558.3$, $SSE = 425.9$, $SST = 6984.3$

6. a) $r = 0.969$
   b) Suggest an extremely strong positive linear
      correlation.
   c) Data points are clustered extremely closely
      about the regression line.
   d) $r^2 = (0.969)^2 = 0.939$

7. a) MTB > CORRELATION of 'PAPERS' and
           'COST'

8. a) $\hat{y} = -40.9 + 0.772x_1 + 3.11x_2$
   b) \$33,564
   c) $R^2 = 0.766$. Thus, 76.6% of the variation in
      the annual-income data is explained by age
      and number of years of school completed.

9. MTB > REGRESS 'INCOME' on 2 predictors
        'AGE' and 'EDUC'

CHAPTER 13

Exercises 13.1

__ 13.1

a) 1. *Population regression line:* There is a straight
      line, $y = \beta_0 + \beta_1 x$, such that for each x-value,
      the mean of the corresponding population of
      y-values lies on that straight line.
   2. *Equal standard deviations:* The standard de-
      viation, σ, of the population of y-values corre-
      sponding to a particular x-value is the same,
      regardless of the x-value.
   3. *Normality:* For each x-value, the correspond-
      ing population of y-values is normally dis-
      tributed.

__ 13.3   It would mean there are constants, β_0,
β_1, and σ, such that for each height, x, the weights
of 18–24-year-old males of that height are normally
distributed with mean $\beta_0 + \beta_1 x$ and standard deviation σ.

__ 13.5   It would mean there are constants, β_0,
β_1, and σ, such that for each age, x, the peak heart
rates that can be reached during intensive exercise
by individuals of that age are normally distributed
with mean $\beta_0 + \beta_1 x$ and standard deviation σ.

__ 13.7   It would mean there are constants, β_0,
β_1, and σ, such that for each disposable-income
level, x, the annual food expenditures made by
middle-income families (father, mother, two children) at that level are normally distributed with
mean $\beta_0 + \beta_1 x$ and standard deviation σ.

__ 13.9

a) $s_e = 16.48$ lb
b) Presuming that the variables, height (x) and
   weight (y), for 18–24-year-old males satisfy Assumptions 1–3 for regression inferences, the standard error of the estimate, $s_e = 16.48$ lb, provides
   an estimate for the common population standard
   deviation, σ, of weights for all 18–24-year-old
   males of any particular height.

__ 13.11

a) $s_e = 3.33$
b) Presuming that the variables, age (x) and peak
   heart rate (y), for individuals satisfy Assumptions 1–3 for regression inferences, the standard
   error of the estimate, $s_e = 3.33$, provides an estimate for the common population standard deviation, σ, of peak heart rates for all individuals of
   any particular age.

__ 13.13

a) $s_e = \$7.68$ hundred
b) Presuming that the variables, disposable income
   (x) and annual food expenditure (y), for middle-
   income families with a father, mother, and two
   children satisfy Assumptions 1–3 for regression
   inferences, the standard error of the estimate,
   $s_e = 7.68$ (\$768), provides an estimate for the
   common population standard deviation, σ, of
   annual food expenditures for all middle-income
   families (father, mother, two children) with any
   particular disposable income.

__ 13.15 $s_e = 1.88$

__ 13.17

a) MTB > REGRESS 'FOODEX' on 1 predictor
        'INCOME'

__ 13.19 $s_e = 382.1$ gm

Exercises 13.2

___ **13.21**

STEP 1 $H_0: \beta_1 = 0$, $H_a: \beta_1 \neq 0$
STEP 2 $\alpha = 0.10$
STEP 3 Critical values $= \pm 1.833$
STEP 4 $t = 2.652$
STEP 5 Reject H_0
STEP 6 The data provide sufficient evidence to conclude that the slope of the population regression line is not zero and, hence, that height is useful as a predictor of weight for 18–24-year-old males.

___ **13.23**

STEP 1 $H_0: \beta_1 = 0$, $H_a: \beta_1 \neq 0$
STEP 2 $\alpha = 0.05$
STEP 3 Critical values $= \pm 2.306$
STEP 4 $t = -7.074$
STEP 5 Reject H_0
STEP 6 Evidently, age is useful as a predictor of peak heart rate.

___ **13.25**

STEP 1 $H_0: \beta_1 = 0$, $H_a: \beta_1 \neq 0$
STEP 2 $\alpha = 0.01$
STEP 3 Critical values $= \pm 3.707$
STEP 4 $t = 2.701$
STEP 5 Do not reject H_0
STEP 6 The data do not provide sufficient evidence to conclude that disposable income is useful as a predictor of annual food expenditure for middle-income families with a father, mother, and two children.

___ **13.27**

a) 1.49 to 8.18 lb
b) We can be 90% confident that, for 18–24-year-old males, the increase in mean weight per one inch increase in height is somewhere between 1.49 and 8.18 lb.

___ **13.29**

a) -1.51 to -0.77
b) We can be 95% confident that the drop in mean peak heart rate per one year increase in age is somewhere between 0.77 and 1.51.

___ **13.31**

a) -0.45 to 2.88
b) We can be 99% confident that, for middle-income families with a father, mother, and two children, the change in mean annual food expenditure per

$1000 increase in family disposable income is somewhere between $-$45 and $288.

___ **13.33**

a) `MTB > REGRESS 'FOODEX' on 1 predictor`
      `'INCOME'`

___ **13.35**

a) 60.82   b) 14.68   c) 4.14
d) 0.000 (to three decimal places)
e) The data provide sufficient evidence to conclude that estriol level is useful for predicting birth weights.
f) 30.80 to 90.84. We can be 95% confident that the increase in mean birth weight per increase of estriol level by 1 mg/24 hr is somewhere between 30.80 and 90.84 gm.

Exercises 13.3

___ **13.39**

a) 164.05 lb
b) 154.54 to 173.56 lb. We can be 90% confident that the mean weight of all 18–24-year-old males who are 70 inches tall is somewhere between 154.54 and 173.56 lb.
c) 164.05 lb
d) 132.38 to 195.71 lb. We can be 90% certain that the weight of a randomly selected 18–24-year-old male who is 70 inches tall will be somewhere between 132.38 and 195.71 lb.
e) See Figure A.15 at the top of the next page.
f) The error in the estimate of the mean weight of all 18–24-year-old males who are 70 inches tall is due only to the fact that the population regression line is being estimated by a sample regression line; whereas, the error in the prediction of the weight of a randomly selected 18–24-year-old male who is 70 inches tall is due to that fact plus the variation in weights of such males.

___ **13.41**

a) 176.65
b) 173.95 to 179.35. We can be 95% confident that the mean peak heart rate of all 40-year-olds is somewhere between 173.95 and 179.35.
c) 176.65
d) 168.52 to 184.78. We can be 95% certain that the peak heart rate of a randomly selected 40-year-old will be somewhere between 168.52 and 184.78.

FIGURE A.15

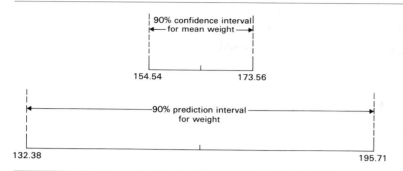

154.54   173.56

132.38   195.71

___ 13.43

a) 33.13 to 53.29. We can be 99% confident that the mean annual food expenditure of all middle-income families consisting of a father, mother, and two children that have a disposable income of $25,000 is somewhere between $3313 and $5329.

b) 13.02 to 73.39. We can be 99% certain that the annual food expenditure of a randomly selected middle-income family consisting of a father, mother, and two children that has a disposable income of $25,000 will be somewhere between $1302 and $7339.

___ 13.45

a) MTB > REGRESS 'FOODEX' on 1 predictor
        'INCOME';
   SUBC> PREDICT 25.

b) The same command and subcommand as those in part (a).

d) The REGRESS; PREDICT command provides only 95% confidence and prediction intervals. To obtain 99% confidence and prediction intervals using Minitab, we must proceed somewhat differently. See the *Minitab Supplement* for details.

___ 13.47

a) SUBC> PREDICT 15.

b) 3064.6 gm

c) 2909.1 to 3220.1 gm

d) 2267.6 to 3861.6 gm

Exercises 13.4

___ 13.49

STEP 1 $H_0: \rho = 0$, $H_a: \rho > 0$

STEP 2 $\alpha = 0.05$

STEP 3   Critical value = 1.833

STEP 4 $t = 2.652$ (See the note at the beginning of Appendix B on page A-19.)

STEP 5   Reject H_0

STEP 6   The data provide sufficient evidence to conclude that the variables, height and weight, are positively linearly correlated for 18–24-year-old males.

___ 13.51

STEP 1 $H_0: \rho = 0$, $H_a: \rho < 0$

STEP 2 $\alpha = 0.025$

STEP 3   Critical value = -2.306

STEP 4 $t = -7.074$ (See the note at the beginning of Appendix B on page A-19.)

STEP 5   Reject H_0

STEP 6   The data provide sufficient evidence to conclude that age and peak heart rate are negatively linearly correlated.

___ 13.53

STEP 1 $H_0: \rho = 0$, $H_a: \rho \neq 0$

STEP 2 $\alpha = 0.01$

STEP 3   Critical values = ± 3.707

STEP 4 $t = 2.701$ (See the note at the beginning of Appendix B on page A-19.)

STEP 5   Do not reject H_0

STEP 6   The data do not provide sufficient evidence to conclude that family disposable income and annual food expenditure are linearly correlated for middle-income families with a father, mother, and two children.

___ 13.55   No, ρ is a fixed number (constant) that measures the linear correlation between the population of all data points. Yes, r is a random variable—it measures the linear correlation between a sample of data points and, so, its value depends on chance; namely, on which data points are obtained from sampling.

Exercises 13.5

__ **13.57** It would mean there are constants, β_0, β_1, β_2, β_3, and σ, such that for each television advertising expenditure, x_1, magazine advertising expenditure, x_2, and radio advertising expenditure, x_3, the total sales is normally distributed with mean $\beta_0 + \beta_1 x_1 + \beta_2 x_2 + \beta_3 x_3$ and standard deviation σ.

__ **13.59**

a) $\hat{y} = 266 + 6.73x_1 + 3.26x_2 + 4.51x_3$, where x_1, x_2, and x_3 denote, respectively, television, magazine, and radio advertising expenditures (in millions of dollars) and $\hat{y}$ denotes predicted sales (in millions of dollars). Based on the sample data, this equation is the best estimate of the unknown population regression equation, $y = \beta_0 + \beta_1 x_1 + \beta_2 x_2 + \beta_3 x_3$.

b) $s_e = 4.418$. Based on the sample data, the best estimate for the common population standard deviation, σ, of sales, for any particular expenditures on television, magazine, and radio advertising, is $4.418 million.

c) Yes, because $p = 0.002$ in the line labeled **Regression,** $\alpha = 0.05$, and $0.002 \le 0.05$.

d) Yes, because $p = 0.002$ in the line labeled **TV,** $\alpha = 0.05$, and $0.002 \le 0.05$. Thus, the data provide sufficient evidence to conclude that, in conjunction with magazine and radio advertising expenditures, television advertising expenditure is useful as a predictor of sales.

e) No, because $p = 0.269$ in the line labeled **RADIO,** $\alpha = 0.05$, and $0.269 > 0.05$. Thus, the data do not provide sufficient evidence to conclude that, in conjunction with television and magazine advertising expenditures, radio advertising expenditure is useful as a predictor of sales.

f) $367.58 million

g) $361.24 to $373.92 million

h) $367.58 million

i) $355.05 to $380.12 million

__ **13.61**

a) Yes, because $p = 0.000$ in the line labeled **Regression,** $\alpha = 0.01$, and $0.000 \le 0.01$.

b) Yes, because $p = 0.001$ in the line labeled **HSGPA,** $\alpha = 0.01$, and $0.001 \le 0.01$. Thus, the

data provide sufficient evidence to conclude that, in conjunction with SAT math and verbal scores, high-school GPA is useful as a predictor of sophomore GPA for Arizona State University students.

c) No. For SAT math score, we see from the line labeled **SATMATH** that $p = 0.785$; and for SAT verbal score, we see from the line labeled **SATVERBL** that $p = 0.824$. These values of p show that the null hypothesis of "non-usefulness as a predictor of sophomore GPA" could not be rejected for either variable (SAT math score or SAT verbal score), even at quite high significance levels.

d) In view of parts (b) and (c), it may be worthwhile to perform a regression analysis with high-school GPA as the only predictor variable.

e) 2.4916 to 2.9682

f) 1.7209 to 3.7389

g) The Assumptions 1–3 for multiple regression inferences, given in Key Fact 13.9 on page 681.

__ **13.63**

a) MTB > REGRESS 'SALES' on 3 predictors
'TV', 'MAG', and 'RADIO';
SUBC> PREDICT 9.5 4.3 5.2.

REVIEW TEST FOR CHAPTER 13

1. It would mean there are constants, β_0, β_1, and σ, such that for each number of papers graded, x, the cost for grading, y, is normally distributed with mean $\beta_0 + \beta_1 x$ and standard deviation σ.

2. a) $\hat{y} = 35.80 + 12.08x$
   b) $s_e = \$6.53$
   c) The standard error of the estimate, $s_e = \$6.53$, provides an estimate for the common population standard deviation, σ, of all costs for grading any particular number of papers.

3. a) STEP 1 $H_0: \beta_1 = 0$, $H_a: \beta_1 \ne 0$
   STEP 2 $\alpha = 0.05$
   STEP 3 Critical values $= \pm 2.228$
   STEP 4 $t = 12.409$
   STEP 5 Reject H_0
   STEP 6 The data provide sufficient evidence to conclude that the number of papers graded is useful as a predictor of cost.

   b) 9.91 to 14.25. We can be 95% confident that the increase in mean cost per increase in the number of papers graded by 100 is somewhere between $9.91 and $14.25.

4. a) $229.13 (See the note at the beginning of Appendix B on page A-19.)
   b) $223.93 to $234.33. We can be 95% confident that the mean cost of grading 1600 papers is somewhere between $223.93 and $234.33.
   c) $229.13
   d) $213.69 to $244.58. We can be 95% certain that the cost of grading 1600 papers will be somewhere between $213.69 and $244.58.
   e) The error in the estimate of the mean cost for grading 1600 papers is due only to the fact that the population regression line is being estimated by a sample regression line; whereas, the error in the prediction of the cost for grading 1600 papers is due to that fact plus the variation in cost for grading 1600 papers.

5. a) MTB > REGRESS 'COST' on 1 predictor
          'PAPERS'
   b) The same command as in part (a).
   c) MTB > REGRESS 'COST' on 1 predictor
          'PAPERS';
      SUBC> PREDICT 16.

6. a) $\hat{y} = 35.8 + 12.1x$ b) 6.526
   c) 12.0835   d) 0.9738   e) 12.41
   f) 0.000 (to three decimal places)
   g) The data provide sufficient evidence to conclude that the number of papers graded is useful for predicting cost.
   h) 9.91 to 14.25. We can be 95% confident that the increase in mean cost per increase in the number of papers graded by 100 is somewhere between $9.91 and $14.25.
   i) SUBC> PREDICT 16.
   j) $229.13   k) $223.93 to $234.34
   l) $229.13   m) $213.68 to $244.58

7. STEP 1 $H_0: \rho = 0, H_a: \rho > 0$
   STEP 2 $\alpha = 0.025$
   STEP 3   Critical value = 2.228
   STEP 4 $t = 12.409$ (See the note at the beginning of Appendix B on page A-19.)
   STEP 5   Reject H_0
   STEP 6   The data provide sufficient evidence to conclude that the variables, number of papers graded and cost, are positively linearly correlated.

8. a) $\hat{y} = -40.9 + 0.772x_1 + 3.11x_2$, where x_1 and x_2 denote, respectively, age and number of years of school completed, and $\hat{y}$ denotes predicted annual income (in $thous). Based on the sample data, this equation is the best estimate of the unknown population regression equation, $y = \beta_0 + \beta_1 x_1 + \beta_2 x_2$.
   b) $s_e = 7.886$. Based on the sample data, the best estimate for the common population standard deviation, σ, of annual incomes, for any particular age and number of years of school completed, is $7,886.
   c) Yes, because $p = 0.000$ in the line labeled Regression, $\alpha = 0.05$, and $0.000 \le 0.05$.
   d) Yes, because $p = 0.000$ in the line labeled AGE, $\alpha = 0.05$, and $0.000 \le 0.05$. Thus, the data provide sufficient evidence to conclude that, in conjunction with the number of years of school completed, age is useful as a predictor of annual income for males (between the ages of 25 and 50 with at least a ninth-grade education).
   e) Yes, because $p = 0.000$ in the line labeled EDUC, $\alpha = 0.05$, and $0.000 \le 0.05$. Thus, the data provide sufficient evidence to conclude that, in conjunction with age, the number of years of school completed is useful as a predictor of annual income for males (between the ages of 25 and 50 with at least a ninth-grade education).
   f) The Assumptions 1–3 for multiple regression inferences, given in Key Fact 13.9 on page 681.

9. a) $33,528   b) $31,342 to $35,714
   c) $33,528   d) $17,653 to $49,403
   e) The Assumptions 1–3 for multiple regression inferences, given in Key Fact 13.9.

10. a) MTB > REGRESS 'INCOME' on 2 predictors
           'AGE' and 'EDUC'
    b) MTB > REGRESS 'INCOME' on 2 predictors
           'AGE' and 'EDUC';
       SUBC> PREDICT 32 16.

CHAPTER 14

Exercises 14.1

___ 14.1   a) 12   b) 7
___ 14.3   a) 1.79   b) 2.29

—— 14.5   a) 2.88   b) 2.10

—— 14.7   a) 2.25   b) 3.21

Exercises 14.2

—— 14.13   The pooled-t procedure, Procedure 10.3 on page 489.

—— 14.15   The common population standard deviation, σ, of the populations under consideration. (This presumes, of course, that the equal-standard-deviations assumption holds; that is, the third assumption in Key Fact 14.2 on page 705.)

—— 14.17

a) Error sum of squares:

$$SSE = (n_1 - 1)s_1^2 + (n_2 - 1)s_2^2 + \cdots + (n_k - 1)s_k^2$$

This represents the variation within the samples.

b) Treatment sum of squares:

$$SSTR = n_1(\bar{x}_1 - \bar{x})^2 + n_2(\bar{x}_2 - \bar{x})^2 + \cdots + n_k(\bar{x}_k - \bar{x})^2$$

This represents the variation among the sample means.

c) Total sum of squares:

$$SST = \Sigma(x - \bar{x})^2$$

This represents the total variation among all the sample data.

—— 14.19 $SST = SSTR + SSE$. The one-way ANOVA identity shows that the total variation among all the sample data can be partitioned into a component representing variation among the sample means and a component representing variation within the samples.

—— 14.21

a) $SSTR = 2.8$, $SSE = 10.8$, $SST = 13.6$

b) $13.6 = 2.8 + 10.8$

c) $MSTR = 1.4$, $MSE = 0.9$

d)

Source	df	SS	MS	F
Treatment	2	2.8	1.4	1.56
Error	12	10.8	0.9	
Total	14	13.6		

—— 14.23

a) $SSTR = 132{,}508.2$, $SSE = 95{,}877.6$, $SST = 228{,}385.8$

b) $228{,}385.8 = 132{,}508.2 + 95{,}877.6$

c) $MSTR = 44{,}169.40$, $MSE = 5{,}992.35$

d)

Source	df	SS	MS	F
Treatment	3	132,508.2	44,169.40	7.37
Error	16	95,877.6	5,992.35	
Total	19	228,385.8		

—— 14.25

Source	df	SS	MS	F
Treatment	2	43.304	21.652	3.08
Error	12	84.400	7.033	
Total	14	127.704		

—— 14.27

STEP 1 H_0: $\mu_1 = \mu_2 = \mu_3 = \mu_4$
         H_a: Not all the means are equal.

STEP 2 $\alpha = 0.05$

STEP 3   Critical value = 3.24

STEP 4 $SST = 530.95$, $SSTR = 70.95$, $SSE = 460.00$

STEP 5   One-way ANOVA table:

Source	df	SS	MS	F
Treatment	3	70.95	23.65	0.82
Error	16	460.00	28.75	
Total	19	530.95		

STEP 6 $F = 0.82$; do not reject H_0

STEP 7   The data do not provide sufficient evidence to conclude that there is a difference in mean lifetimes among the four brands of batteries.

—— 14.29

STEP 1 H_0: $\mu_1 = \mu_2 = \mu_3$
         H_a: Not all the means are equal.

STEP 2 $\alpha = 0.05$

STEP 3   Critical value = 4.26

STEP 4 $SST = 60.916$, $SSTR = 5.828$, $SSE = 55.088$

STEP 5 One-way ANOVA table:

Source	df	SS	MS	F
Treatment	2	5.828	2.914	0.48
Error	9	55.088	6.121	
Total	11	60.916		

STEP 6 $F = 0.48$; do not reject H_0
STEP 7 The data do not provide sufficient evidence to conclude that there is a difference in mean hourly earnings for nonsupervisory workers in the three industries.

___ 14.31
STEP 1 H_0: $\mu_1 = \mu_2 = \mu_3 = \mu_4 = \mu_5$
H_a: Not all the means are equal.
STEP 2 $\alpha = 0.01$
STEP 3 Critical value $= 3.65$
STEP 4 $SST = 1384.015$, $SSTR = 406.370$,
$SSE = 977.645$
STEP 5 One-way ANOVA table:

Source	df	SS	MS	F
Treatment	4	406.370	101.593	6.23
Error	60	977.645	16.294	
Total	64	1384.015		

STEP 6 $F = 6.23$; reject H_0
STEP 7 The data provide sufficient evidence to conclude that a difference exists in the mean times served by prisoners in the five offense groups.

___ 14.33 No, it means only that not all the means are equal. For instance, the null hypothesis would be false if $\mu_1 \neq \mu_2$ and $\mu_2 = \mu_3 = \cdots = \mu_k$.

___ 14.35
a) MTB > AOVONEWAY for 'WHTRADE' 'FIRE'
        'SERVICES'

___ 14.37
a) $SSTR = 6545.7$, $SSE = 7236.0$, $SST = 13,781.7$
b) $MSTR = 3272.9$, $MSE = 47.0$
c) $F = 69.65$
d) Let μ_1, μ_2, and μ_3 denote the mean annual incomes of women whose educational attainments are elementary school, secondary school, and college, respectively. Then the null and alternative hypotheses are

$$H_0: \mu_1 = \mu_2 = \mu_3$$
$$H_a: \text{Not all the means are equal.}$$

e) $P = 0.000$ (to three decimal places)
f) Reject H_0 and conclude that a difference exists among the mean annual incomes of women in the three categories of educational attainment.
g) $n_1 = 40$, $\bar{x}_1 = \$11,099$, $s_1 = \$6,720$
   $n_2 = 62$, $\bar{x}_2 = \$16,587$, $s_2 = \$7,324$
   $n_3 = 55$, $\bar{x}_3 = \$27,190$, $s_3 = \$6,388$
h) Referring to the second graph, we see that the required confidence interval is from approximately $15,000 to $18,600.

Exercises 14.3

___ 14.47 Two-way analysis of variance; one-way analysis of variance.

___ 14.49 To isolate and remove an extraneous source of variation so that it is easier to detect differences among the population means when such differences exist.

___ 14.51 $SST = SSTR + SSB + SSE$. The two-way ANOVA identity shows that the total variation among all the sample data can be partitioned into a component representing variation among the sample means for the treatments, a component representing variation among the sample means for the blocks, and a component representing variation within the samples for the treatments with the variation due to the blocks removed.

___ 14.53

Source	df	SS	MS	F
Treatment	2	43.334	21.667	5.91
Block	4	55.060	13.765	
Error	8	29.339	3.667	
Total	14	127.733		

14.55

Source	df	SS	MS	F
Treatment	6	250.488	41.748	9.80
Block	4	15.364	3.841	
Error	24	102.240	4.260	
Total	34	368.092		

14.57

a) The three fertilizers.
b) The five farms.

c)

Source	df	SS	MS	F
Treatment	2	188.933	94.467	7.33
Block	4	4858.933	1214.733	
Error	8	103.067	12.883	
Total	14	5150.933		

14.59

STEP 1 H_0: $\mu_1 = \mu_2 = \mu_3$
H_a: Not all the means are equal.
STEP 2 $\alpha = 0.05$
STEP 3 Critical value = 4.46
STEP 4 $SST = 5150.933$, $SSTR = 188.933$,
$SSB = 4858.933$, $SSE = 103.067$
STEP 5 See the answer to part (c) of Exercise 14.57.
STEP 6 $F = 7.33$; reject H_0
STEP 7 The data provide sufficient evidence to conclude that a difference exists among the mean wheat yields of the three fertilizers.

14.61

STEP 1 H_0: $\mu_1 = \mu_2 = \mu_3 = \mu_4$
H_a: Not all the means are equal.
STEP 2 $\alpha = 0.01$
STEP 3 Critical value = 6.99
STEP 4 $SST = 10{,}249.938$, $SSTR = 1020.688$,
$SSB = 8443.188$, $SSE = 786.063$

STEP 5 Two-way ANOVA table:

Source	df	SS	MS	F
Treatment	3	1,020.688	340.229	3.90
Block	3	8,443.188	2814.396	
Error	9	786.063	87.340	
Total	15	10,249.938		

STEP 6 $F = 3.90$; do not reject H_0
STEP 7 The data do not provide sufficient evidence to conclude that there is a difference in mean weekly food costs among the four U.S. regions.

14.63

a) STEP 1 H_0: $\mu_1 = \mu_2 = \mu_3 = \mu_4$
H_a: Not all the means are equal.
STEP 2 $\alpha = 0.05$
STEP 3 Critical value = 3.49
STEP 4 $SST = 959.2$, $SSTR = 94.0$,
$SSB = 815.2$, $SSE = 50.0$
STEP 5 Two-way ANOVA table:

Source	df	SS	MS	F
Treatment	3	94.0	31.333	7.52
Block	4	815.2	203.800	
Error	12	50.0	4.167	
Total	19	959.2		

STEP 6 $F = 7.52$; reject H_0
STEP 7 The data provide sufficient evidence to conclude that there is a difference in mean lifetimes among the four brands of batteries.

b) By blocking, the extraneous source of variation in battery life due to differences in flashlights was isolated and removed. This made it easier to detect differences among the mean lifetimes of the four brands of batteries.

14.65

a) MTB > TWOWAY on 'FOODCOST' 'TREATMNT'
        'BLOCK'

__ **14.67**

a) $SST = 3{,}210{,}339$, $SSTR = 596{,}452$, $SSB = 2{,}488{,}982$, $SSE = 124{,}905$

b) $MSTR = 198{,}817$, $MSB = 1{,}244{,}491$, $MSE = 20{,}818$

c) Let μ_1, μ_2, μ_3, and μ_4 denote the mean starting-monthly-salary offers for civil, chemical, electrical, and mechanical engineers, respectively. Then the null and alternative hypotheses are

$$H_0: \mu_1 = \mu_2 = \mu_3 = \mu_4$$

$$H_a: \text{Not all the means are equal.}$$

d) $F = 9.55$ e) Critical value $= 4.76$

f) Reject H_0 and conclude that a difference exists among the mean starting-monthly-salary offers in the four engineering fields.

REVIEW TEST FOR CHAPTER 14

1. a) 24   b) 5   c) 4.53   d) 9.47
   e) 4.53

2. $MSTR$ (or $SSTR$)

3. MSE (or SSE)

4. a) $\bar{x}_1 = 413.2$, $s_1 = 62.9$
      $\bar{x}_2 = 395.0$, $s_2 = 76.0$
      $\bar{x}_3 = 479.0$, $s_3 = 72.7$

   b) $MSTR = 11{,}583.6$, $MSE = 5076.2$

   c) The variation among the sample means.

   d) The variation within the samples.

5. a) The three assumptions for one-way ANOVA, given in Key Fact 14.2 on page 705: independent samples, normal populations, and equal standard deviations. Assumption 1 on independent samples is absolutely essential to the one-way ANOVA procedure. Assumption 2 on normality is not too critical as long as the populations are not too far from being normally distributed. Assumption 3 on equal standard deviations is also not that important provided that the sample sizes are roughly the same.

   b) STEP 1 $H_0: \mu_1 = \mu_2 = \mu_3$
         $H_a:$ Not all the means are equal.
      STEP 2 $\alpha = 0.05$
      STEP 3 Critical value $= 3.74$
      STEP 4 $SST = 94{,}234.0$, $SSTR = 23{,}167.2$,
         $SSE = 71{,}066.8$

STEP 5 One-way ANOVA table:

Source	df	SS	MS	F
Treatment	2	23,167.2	11,583.6	2.28
Error	14	71,066.8	5,076.2	
Total	16	94,234.0		

STEP 6 $F = 2.28$; do not reject H_0

STEP 7 The data do not provide sufficient evidence to conclude that there is a difference in mean losses among the three types of robberies.

6. a) The total sum of squares, SST, represents the total variation among all the sample data; the treatment sum of squares, $SSTR$, represents the variation among the sample means; and the error sum of squares, SSE, represents the variation within the samples.

   b) $SST = SSTR + SSE$. The one-way ANOVA identity shows that the total variation among all the sample data can be partitioned into a component representing variation among the sample means and a component representing variation within the samples.

7. STEP 1 $H_0: \mu_1 = \mu_2 = \mu_3 = \mu_4$
      $H_a:$ Not all the means are equal.
   STEP 2 $\alpha = 0.01$
   STEP 3 Critical value $= 5.29$
   STEP 4 $SST = 2936.55$, $SSTR = 228.15$,
      $SSE = 2708.40$
   STEP 5 One-way ANOVA table:

Source	df	SS	MS	F
Treatment	3	228.15	76.050	0.45
Error	16	2708.40	169.275	
Total	19	2936.55		

STEP 6 $F = 0.45$; do not reject H_0

STEP 7 We cannot conclude that a difference exists in mean sales among the four designs.

8. a) MTB > AOVONEWAY for 'DESIGN A'
         'DESIGN B' 'DESIGN C' 'DESIGN D'

9. a) $SSTR = 228$, $SSE = 2708$, $SST = 2937$
   b) $MSTR = 76$, $MSE = 169$
   c) $F = 0.45$ d) $P = 0.721$
   e) Do not reject H_0. We cannot conclude that a difference exists in mean sales among the four designs.
   f) $\bar{x}_1 = 47.40$, $s_1 = 5.03$, $n_1 = 5$
      $\bar{x}_2 = 54.80$, $s_2 = 12.66$, $n_2 = 5$
      $\bar{x}_3 = 53.60$, $s_3 = 14.54$, $n_3 = 5$
      $\bar{x}_4 = 47.60$, $s_4 = 16.74$, $n_4 = 5$

10. By employing a randomized block design, an extraneous source of variation can be isolated and removed, thus making it easier to detect differences among the population means when such differences exist.

11. a) The total sum of squares, SST, represents the total variation among all the sample data; the treatment sum of squares, $SSTR$, represents the variation among the sample means for the treatments; the block sum of squares, SSB, represents the variation among the sample means for the blocks; and the error sum of squares, SSE, represents the variation within the samples for the treatments with the variation due to the blocks removed.
    b) $SST = SSTR + SSB + SSE$. The two-way ANOVA identity shows that the total variation among all the sample data can be partitioned into a component representing variation among the sample means for the treatments, a component representing variation among the sample means for the blocks, and a component representing variation within the samples for the treatments with the variation due to the blocks removed.

12. a) The three assumptions for two-way ANOVA, given in Procedure 14.2 on page 732: blocked samples, normal populations, and equal standard deviations. (See also the footnote on page 732.)
    b) STEP 1　H_0: $\mu_1 = \mu_2 = \mu_3$
       H_a: Not all the means are equal.
       STEP 2　$\alpha = 0.05$
       STEP 3　Critical value $= 5.14$
       STEP 4　$SST = 1,001,012.917$,
       $SSTR = 189,228.167$,
       $SSB = 731,340.917$,
       $SSE = 80,443.833$

STEP 5　Two-way ANOVA table:

Source	df	SS	MS	F
Treatment	2	189,228.167	94,614.083	7.06
Block	3	731,340.917	243,780.306	
Error	6	80,443.833	13,407.306	
Total	11	1,001,012.917		

STEP 6　$F = 7.06$; reject H_0
STEP 7　The data provide sufficient evidence to conclude that a difference exists among last year's mean energy expenditures for households using the three types of fuel.

13. a) MTB > TWOWAY on 'ENERGY$' 'TREATMNT'
       'BLOCK'

14. a) $SST = 1,001,013$, $SSTR = 189,228$,
       $SSB = 731,341$, $SSE = 80,444$
    b) $MSTR = 94,614$, $MSB = 243,780$,
       $MSE = 13,407$
    c) $F = 7.06$ d) Critical value $= 5.14$
    e) Reject H_0. The data provide sufficient evidence to conclude that a difference exists among last year's mean energy expenditures for households using the three types of fuel.

CHAPTER 15

Exercises 15.1

___ 15.1　*Distribution-free* refers to inferential methods that do not depend on the distribution of the population(s) being sampled.

Exercises 15.2

___ 15.3　The binomial distribution.

___ 15.5
STEP 1　H_0: $\eta = 32.8$ yr, H_a: $\eta > 32.8$ yr
STEP 2　Rejection region: $x \geq 9$; $\alpha = 0.073$
STEP 3　Sign table:

37 (+)	28 (−)	36 (+)	33 (+)
37 (+)	43 (+)	41 (+)	28 (−)
24 (−)	44 (+)	27 (−)	24 (−)

STEP 4 $x = 7$
STEP 5 Do not reject H_0
STEP 6 The data do not provide sufficient evidence
to conclude that the median age of recipients of science and engineering doctoral degrees has increased since 1983.

___ 15.7
STEP 1 H_0: $\eta = \$112.8$ thous, H_a: $\eta < \$112.8$ thous
STEP 2 Rejection region: $x \leq 6$; $\alpha = 0.058$
STEP 3 Sign table:

84.4 (−)	88.9 (−)	109.7 (−)	128.9 (+)
97.2 (−)	112.5 (−)	96.1 (−)	98.6 (−)
71.7 (−)	96.9 (−)	91.2 (−)	119.5 (+)
98.5 (−)	105.0 (−)	110.3 (−)	104.0 (−)
84.1 (−)	97.1 (−)	84.5 (−)	91.3 (−)

STEP 4 $x = 2$
STEP 5 Reject H_0
STEP 6 Evidently, the 1986 median net earnings of office-based internists was less than that of all office-based MDs.

___ 15.9
STEP 1 H_0: $\eta = 38.5$¢/lb, H_a: $\eta \neq 38.5$¢/lb
STEP 2 Rejection region: $x \leq 3$ or $x \geq 12$; $\alpha = 0.034$
STEP 3 Sign table:

43.0 (+)	40.0 (+)	42.6 (+)	40.2 (+)	37.5 (−)
44.1 (+)	45.2 (+)	41.8 (+)	35.6 (−)	34.6 (−)
37.9 (−)	44.2 (+)	44.5 (+)	38.2 (−)	42.4 (+)

STEP 4 $x = 10$
STEP 5 Do not reject H_0
STEP 6 The data do not provide sufficient evidence to conclude that the median retail price for oranges now is different from the 1983 median of 38.5 cents per pound.

___ 15.11
a) MTB > STEST of eta=112.8, data in
        'EARNINGS';
   SUBC> ALTERNATIVE=-1.

___ 15.13
a) H_0: $\eta = 6.8$ yr, H_a: $\eta < 6.8$ yr
b) SUBC> ALTERNATIVE=-1.
c) 50   d) 27, 0, 23   e) 6.555 yr
f) 0.3359   g) 0.3359

h) Do not reject H_0. The data do not provide sufficient evidence to conclude that last year's median marriage duration is less than the 1985 median of 6.8 years.

Exercises 15.3

___ 15.19 Because the D-value for such a data value equals zero and, consequently, we cannot attach a sign to the rank of $|D|$.

___ 15.21
a) Wilcoxon signed-rank test   b) t-test
c) sign test

___ 15.23 Yes, provided that the population mean, μ, and the population median, η, are equal. This will be the case, for example, if the population has a symmetric distribution.

___ 15.25
STEP 1 H_0: $\eta = 32.1$ yr, H_a: $\eta > 32.1$ yr
STEP 2 $\alpha = 0.053$
STEP 3 Critical value = 44
STEP 4 Work table:

Age x	Difference $D = x - 32.1$	$|D|$	Rank of $|D|$	Signed rank R
38	5.9	5.9	3	3
41	8.9	8.9	4	4
58	25.9	25.9	10	10
45	12.9	12.9	6	6
10	−22.1	22.1	8	−8
35	2.9	2.9	2	2
53	20.9	20.9	7	7
7	−25.1	25.1	9	−9
32	−0.1	0.1	1	−1
22	−10.1	10.1	5	−5

STEP 5 $W = 32$
STEP 6 Do not reject H_0
STEP 7 The data do not provide sufficient evidence to conclude that the median age of today's United States residents has increased over the 1987 median age of 32.1 years.

___ 15.27
STEP 1 H_0: $\eta = 12$ min, H_a: $\eta < 12$ min
STEP 2 $\alpha = 0.052$
STEP 3 Critical value = 26

STEP 4 Work table:

Time x	Difference $D = x - 12$	$|D|$	Rank of $|D|$	Signed rank R
10.9	−1.1	1.1	2	−2
15.0	3.0	3.0	9	9
14.2	2.2	2.2	4.5	4.5
11.4	−0.6	0.6	1	−1
7.1	−4.9	4.9	12	−12
9.2	−2.8	2.8	7	−7
10.1	−1.9	1.9	3	−3
9.2	−2.8	2.8	7	−7
8.8	−3.2	3.2	10	−10
9.8	−2.2	2.2	4.5	−4.5
6.6	−5.4	5.4	13	−13
4.4	−7.6	7.6	14	−14
14.8	2.8	2.8	7	7
8.0	−4.0	4.0	11	−11

STEP 5 $W = 20.5$
STEP 6 Reject H_0
STEP 7 Evidently, the new antacid works faster.

____ 15.29

a) STEP 1 H_0: $\mu = 38.5 ¢/\text{lb}$, H_a: $\mu \neq 38.5¢/\text{lb}$
   STEP 2 $\alpha = 0.048$
   STEP 3 Critical values = 25, 95
   STEP 4 Work table:

Price x	Difference $D = x - 38.5$	$|D|$	Rank of $|D|$	Signed rank R
43.0	4.5	4.5	11	11
44.1	5.6	5.6	12	12
37.9	−0.6	0.6	2	−2
40.0	1.5	1.5	4	4
45.2	6.7	6.7	15	15
44.2	5.7	5.7	13	13
42.6	4.1	4.1	10	10
41.8	3.3	3.3	7	7
44.5	6.0	6.0	14	14
40.2	1.7	1.7	5	5
35.6	−2.9	2.9	6	−6
38.2	−0.3	0.3	1	−1
37.5	−1.0	1.0	3	−3
34.6	−3.9	3.9	8.5	−8.5
42.4	3.9	3.9	8.5	8.5

STEP 5 $W = 99.5$
STEP 6 Reject H_0

STEP 7 The data provide sufficient evidence to conclude that the mean retail price for oranges now is different from the 1983 mean of 38.5 cents per pound.

b) Because a normally distributed population is symmetric.

____ 15.31

a) STEP 1 H_0: $\mu = 310$ ml, H_a: $\mu < 310$ ml
   STEP 2 $\alpha = 0.05$
   STEP 3 Critical value = −1.753
   STEP 4 $t = -1.845$
   STEP 5 Reject H_0
   STEP 6 The data provide sufficient evidence to conclude that the mean content, μ, is less than the advertised content of 310 ml.

b) STEP 1 H_0: $\mu = 310$ ml, H_a: $\mu < 310$ ml
   STEP 2 $\alpha = 0.052$
   STEP 3 Critical value = 36
   STEP 4 Work table:

Content x	Difference $D = x - 310$	$|D|$	Rank of $|D|$	Signed rank R
297	−13	13	14	−14
311	1	1	2	2
322	12	12	12.5	12.5
315	5	5	7	7
318	8	8	9	9
303	−7	7	8	−8
307	−3	3	5	−5
296	−14	14	15	−15
306	−4	4	6	−6
291	−19	19	16	−16
312	2	2	4	4
309	−1	1	2	−2
300	−10	10	10.5	−10.5
298	−12	12	12.5	−12.5
300	−10	10	10.5	−10.5
311	1	1	2	2

STEP 5 $W = 36.5$
STEP 6 Do not reject H_0
STEP 7 The data do not provide sufficient evidence to conclude that the mean content, μ, is less than the advertised content of 310 ml.

c) Since the population is normally distributed, the t-test is more powerful than the Wilcoxon signed-rank test. That is, the t-test is more likely to detect a false null hypothesis.

___ 15.33

a) `MTB > WTEST of eta=32.1, data in 'AGES';`
   `SUBC> ALTERNATIVE=1.`

___ 15.35

a) H_0: $\eta = 6.8$ yr, H_a: $\eta < 6.8$ yr
b) `SUBC> ALTERNATIVE=-1.`
c) 50   d) 0   e) 7.231 yr
f) 0.716   g) 0.716
h) Do not reject H_0. The data do not provide sufficient evidence to conclude that last year's median marriage duration is less than the 1985 median of 6.8 years.
i) Neither hypothesis test led to rejection of the null hypothesis.
j) The sign test is probably more appropriate since, most likely, marriage durations do not have a symmetric distribution.

Exercises 15.4

___ 15.41

STEP 1 H_0: $\eta_1 = \eta_2$, H_a: $\eta_1 < \eta_2$
STEP 2 $\alpha = 0.05$
STEP 3   Critical value $= 33$
STEP 4   Work table:

Fewer than two years of HS algebra	Overall rank	Two or more years of HS algebra	Overall rank
58	3	84	14
81	11.5	67	7
74	8.5	65	6
61	4	75	10
64	5	74	8.5
43	1	92	15
		83	13
		52	2
		81	11.5

STEP 5 $M = 33$
STEP 6   Reject H_0
STEP 7   Evidently, students with fewer than two years of high-school algebra have a lower median semester average in this teacher's chemistry courses than do students with two or more years of high-school algebra.

___ 15.43

STEP 1 H_0: $\eta_1 = \eta_2$, H_a: $\eta_1 > \eta_2$
STEP 2 $\alpha = 0.05$
STEP 3   Critical value $= 54$
STEP 4   Work table:

Private	Overall rank	Public	Overall rank
289	9	79	6
953	13	41	4
23	2	516	11
64	5	15	1
117	7	24	3
650	12	411	10
		265	8

STEP 5 $M = 48$
STEP 6   Do not reject H_0
STEP 7   The data do not provide sufficient evidence to conclude that the median number of volumes held by public colleges and universities is less than that held by private colleges and universities.

___ 15.45

a) STEP 1 H_0: $\mu_1 = \mu_2$, H_a: $\mu_1 \neq \mu_2$
   STEP 2 $\alpha = 0.05$
   STEP 3   Critical values $= 79, 131$
   STEP 4   Work table:

Brand A	Overall rank	Brand B	Overall rank
35.6	3	37.2	12.5
37.0	11	39.7	19
34.9	1.5	37.2	12.5
36.0	5	38.8	17.5
36.6	9.5	37.7	15
36.1	6	36.4	7
35.8	4	37.5	14
34.9	1.5	40.5	20
38.8	17.5	38.2	16
36.5	8	36.6	9.5

STEP 5 $M = 67$
STEP 6   Reject H_0
STEP 7   There appears to be a difference in mean lasting times between the two paints.
b) Because two normal distributions with equal standard deviations have the same shape.

___ **15.47**

a) MTB > MANNWHITNEY for 'PRIVATE'
         vs 'PUBLIC';
   SUBC> ALTERNATIVE=1.

Note: In some versions of Minitab, the ALTER-
NATIVE subcommand does not work for the
Mann-Whitney test. For such versions, place the
ALTERNATIVE directly after MANNWHITNEY.
For example, in part (a), use:

   MTB > MANNWHITNEY, ALTERNATIVE=1, for
         'PRIVATE' vs 'PUBLIC'

___ **15.49**

a) H_0: $\eta_1 = \eta_2$, H_a: $\eta_1 < \eta_2$
b) MTB > MANNWHITNEY for 'CURRENT' vs 'NEW';
   SUBC> ALTERNATIVE=-1.
   (See the note in the answer to Exercise 15.47.)
c) 1033.5 hr, 1105.0 hr   d) 20, 10
e) 242.5 (Recall that Minitab uses W instead of M.)
f) 0.0016   g) 0.0016
h) Reject H_0. The data provide sufficient evidence
to conclude that the new bulb will outlast the
current bulb, on the average.
i) -109.0 to -25.0 hr
j) The lifetime distributions of the current bulb and
new bulb have the same shape. Also, the samples
must be independent.

Exercises 15.5

___ **15.57**   False. It suggests only that the variable,
y, tends to decrease as the variable, x, increases, not
necessarily in a linear manner.

___ **15.59**   negative; positive; negative

___ **15.61**

a) $r_s = -0.250$ b) $r_s = -0.250$
c) The variables, IQ and number of drinks per
month, seem to be uncorrelated for people who
drink alcoholic beverages.

___ **15.63** $r_s = 0.786$. There seems to be a moderately strong positive correlation between the wine
tastes of the two connoisseurs.

___ **15.65**

STEP 1 H_0: IQ and number of drinks per month
           are uncorrelated.
         H_a: IQ and number of drinks per month
           are correlated.
STEP 2 $\alpha = 0.05$

STEP 3   Critical values = ± 0.683
STEP 4 $r_s = -0.250$
STEP 5   Do not reject H_0
STEP 6   The data do not provide sufficient evidence
          to conclude that there is a correlation be-
          tween the variables, IQ and number of
          drinks per month, for people who drink al-
          coholic beverages.

___ **15.67**

STEP 1 H_0: The wine tastes of the two connois-
          seurs are uncorrelated.
         H_a: The wine tastes of the two connois-
          seurs are positively correlated.
STEP 2 $\alpha = 0.05$
STEP 3   Critical value = 0.643
STEP 4 $r_s = 0.786$
STEP 5   Reject H_0
STEP 6   The data provide sufficient evidence to con-
          clude that there is a positive correlation
          between the wine tastes of the two connois-
          seurs.

___ **15.69**

STEP 1 H_0: Birth rates and literacy rates for
          nations of the world are uncorrelated.
         H_a: Birth rates and literacy rates for
          nations of the world are negatively
          correlated.
STEP 2 $\alpha = 0.01$
STEP 3   Critical value = -0.745
STEP 4 $r_s = -0.870$
STEP 5   Reject H_0
STEP 6   The data provide sufficient evidence to con-
          clude that the variables, birth rate and lit-
          eracy rate, are negatively correlated.

REVIEW TEST FOR CHAPTER 15

1. STEP 1 H_0: $\eta = \$365$, H_a: $\eta < \$365$
    STEP 2   Rejection region: $x \leq 2$; $\alpha = 0.055$
    STEP 3   Sign table:

346 $(-)$	199 $(-)$	488 $(+)$	292 $(-)$	368 $(+)$
287 $(-)$	191 $(-)$	220 $(-)$	383 $(+)$	278 $(-)$

    STEP 4 $x = 3$
    STEP 5   Do not reject H_0
    STEP 6   The data do not provide sufficient evi-
           dence to conclude that the 1980 median

monthly cost for housing to Indiana homeowners was less than the national median of $365.

2. a) `MTB > STEST of eta=365, data in`
   `     'HOMECOST';`
   `SUBC> ALTERNATIVE=-1.`

3. a) $H_0: \eta = \$365$, $H_a: \eta < \$365$
   b) `SUBC> ALTERNATIVE=-1.`
   c) 10   d) 7; 0; 3   e) $289.5
   f) 0.1719   g) 0.1719
   h) The data do not provide sufficient evidence to conclude that the 1980 median monthly cost for housing to Indiana homeowners was less than the national median of $365.

4. a) STEP 1 $H_0: \mu = 29$ mpg, $H_a: \mu \neq 29$ mpg
      STEP 2 $\alpha = 0.048$
      STEP 3   Critical values = 25, 95
      STEP 4   Work table:

Mileage x	Difference $D = x - 29$	$\lvert D \rvert$	Rank of $\lvert D \rvert$	Signed rank R
27.3	−1.7	1.7	10.5	−10.5
30.9	1.9	1.9	12	12
25.9	−3.1	3.1	15	−15
31.2	2.2	2.2	13	13
29.7	0.7	0.7	6	6
28.8	−0.2	0.2	2	−2
29.4	0.4	0.4	3.5	3.5
28.5	−0.5	0.5	5	−5
28.9	−0.1	0.1	1	−1
31.6	2.6	2.6	14	14
27.8	−1.2	1.2	7.5	−7.5
27.8	−1.2	1.2	7.5	−7.5
28.6	−0.4	0.4	3.5	−3.5
27.3	−1.7	1.7	10.5	−10.5
27.6	−1.4	1.4	9	−9

STEP 5 $W = 48.5$
STEP 6   Do not reject H_0
STEP 7   The data do not provide sufficient evidence to conclude that the company's report was incorrect.

b) The distribution is symmetric.
c) Because a normally distributed population is symmetric.

5. a) `MTB > WTEST of eta=29, data in`
   `         'MILEAGES'`

6. a) $H_0: \eta = 29$ mpg, $H_a: \eta \neq 29$ mpg
   b) No subcommand was used.

c) 15   d) 0   e) 0.532   f) 0.532
g) The data do not provide sufficient evidence to conclude that the company's report was incorrect.

7. a) Yes, because the mean and median are identical for a normally distributed population.
   b) The t-test, since that hypothesis-testing procedure is designed specifically for a normally distributed population.

8. a) STEP 1 $H_0: \eta_1 = \eta_2$, $H_a: \eta_1 > \eta_2$
      STEP 2 $\alpha = 0.05$
      STEP 3   Critical value = 111
      STEP 4   Work table:

Male	Overall rank	Female	Overall rank
73	19	33	10
25	6	34	12
31	8	35	13
47	16.5	10	1
33	10	13	2
39	14	63	18
20	4.5	43	15
47	16.5	33	10
19	3	20	4.5
		30	7

STEP 5 $M = 97.5$
STEP 6   Do not reject H_0
STEP 7   The data do not provide sufficient evidence to conclude that the median age of this year's naturalized alien males exceeds the median age of this year's naturalized alien females.

b) They have the same shape.
c) Yes, because the pooled-t test is designed specifically for normally distributed populations.

9. a) `MTB > MANNWHITNEY for 'MALE' vs`
   `              'FEMALE';`
   `SUBC> ALTERNATIVE=1.`
   (See the note in the answer to Ex. 15.47.)

10. a) $H_0: \eta_1 = \eta_2$, $H_a: \eta_1 > \eta_2$
    b) `SUBC> ALTERNATIVE=1.`
    (See the note in the answer to Ex. 15.47.)
    c) 33 yr, 33 yr   d) 9, 10
    e) 97.5 (Recall that Minitab uses W instead of M.)

f) 0.2833   g) 0.2833

h) Do not reject H_0. The data do not provide sufficient evidence to conclude that the median age of this year's naturalized alien males exceeds the median age of this year's naturalized alien females.

i) 5.00 yr   j) −10.99 to 19.00 yr

k) The age distributions of this year's naturalized alien males and females have the same shape. Also, the samples must be independent.

11. a) Pooled-t test (Procedure 10.3 on p. 489)
    b) Non-pooled-t test (Procedure 10.5 on p. 501)
    c) Mann-Whitney test (Procedure 15.3 on p. 777)
    d) Two-sample-z test (Procedure 10.1 on p. 480)

12. a) 0.685   b) 0.685

13. STEP 1 H_0: Murder rate and burglary rate are uncorrelated.

STEP 1 H_a: Murder rate and burglary rate are positively correlated.

STEP 2 $\alpha = 0.05$

STEP 3 Critical value = 0.564

STEP 4 $r_s = 0.685$

STEP 5 Reject H_0

STEP 6 The data provide sufficient evidence to conclude that the variables, murder rate and burglary rate, for U.S. cities are positively correlated.

CHAPTER 16

Exercises 16.2

__ 16.1 Answers will vary.

__ 16.3 Answers will vary.

__ 16.5

a) Number the suites from 1 to 48, use a table of random numbers to randomly select three of the 48 suites, and take as the sample the 24 dormitory residents living in the three suites obtained.

b) Probably not, since friends are likely to have similar opinions.

c)

Stratum	Frequency	Percent of total	No. to be selected
Freshman	128	33.3 $\left(\frac{1}{3}\right)$	8
Sophomore	112	29.2 $\left(\frac{7}{24}\right)$	7
Junior	96	25.0 $\left(\frac{1}{4}\right)$	6
Senior	48	12.5 $\left(\frac{1}{8}\right)$	3
	384		24

Referring to the table above, we see that a stratified sample of 24 dormitory residents can be obtained as follows: Number the freshman dormitory residents from 1 to 128 and use a table of random numbers to randomly select eight of the 128 freshman dormitory residents; number the sophomore dormitory residents from 1 to 112 and use a table of random numbers to randomly select seven of the 112 sophomore dormitory residents; and so forth.

__ 16.7

a) MTB > SAMPLE 10 from 'INTER500'
         put into 'SRS'

Exercises 16.3

__ 16.9

a) A *Type I error* is the mistake of rejecting a true null hypothesis.

b) A *Type II error* is the mistake of not rejecting a false null hypothesis.

__ 16.11

a) A Type I error would occur if the mean content of the bottles of soda is not less than 1000 ml, but the results of the sampling lead to the conclusion that it is.

b) A Type II error would occur if the mean content of the bottles of soda is less than 1000 ml, but the results of the sampling fail to lead to that conclusion.

__ 16.13

a) A Type I error would occur if this year's mean age of trucks in use does not exceed the 1983 mean of 8.1 years, but the results of the sampling lead to the conclusion that it does.

b) A Type II error would occur if this year's mean age of trucks in use does exceed the 1983 mean of

8.1 years, but the results of the sampling fail to lead to that conclusion.

16.15

a) A Type I error would occur if last year's mean beef consumption per person is the same as the 1983 mean of 106.5 lb, but the results of the sampling lead to the conclusion that it is not.

b) A Type II error would occur if last year's mean beef consumption per person differs from the 1983 mean of 106.5 lb, but the results of the sampling fail to lead to that conclusion.

16.17

a) If $\bar{x} \leq 990.0$ ml, reject H_0; otherwise, do not reject H_0.

b) 0.05

c)

True mean μ	P(Type II error) β
970	0.0005
975	0.0068
980	0.0505
985	0.2061
990	0.5000
992.5	0.6591
995	0.7939
997.5	0.8907

d)

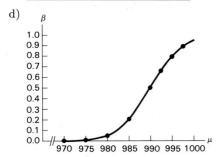

e)

True mean μ	Power $1 - \beta$
970	0.9995
975	0.9932
980	0.9495
985	0.7939
990	0.5000
992.5	0.3409
995	0.2061
997.5	0.1093

f) Power

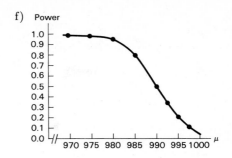

16.19

a) If $\bar{x} \geq 9.2$ years, reject H_0; otherwise, do not reject H_0.

b) 0.10

c)

True mean μ	P(Type II error) β
8.5	0.7939
9.0	0.5910
9.5	0.3632
10.0	0.1736
10.5	0.0643
11.0	0.0174

d)

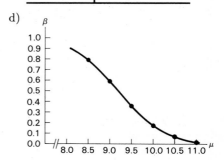

e)

True mean μ	Power $1 - \beta$
8.5	0.2061
9.0	0.4090
9.5	0.6368
10.0	0.8264
10.5	0.9357
11.0	0.9826

f)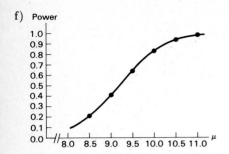

True mean μ	Power $1 - \beta$
96	0.9938
98	0.9505
100	0.7881
102	0.4841
104	0.1880
106	0.0563
108	0.1000
110	0.3195
112	0.6480
114	0.8907
116	0.9808
118	0.9982

f)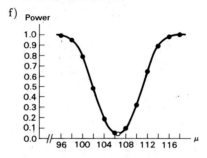

Power

___ 16.21

a) If $\bar{x} \le 101.9$ lb or $\bar{x} \ge 111.1$ lb, reject H_0; otherwise, do not reject H_0.

b) 0.05

c)

True mean μ	P(Type II error) β
96	0.0062
98	0.0495
100	0.2119
102	0.5159
104	0.8120
106	0.9437
108	0.9000
110	0.6805
112	0.3520
114	0.1093
116	0.0192
118	0.0018

___ 16.23

a) If $\bar{x} \le 994.5$ ml, reject H_0; otherwise, do not reject H_0.

b) 0.05

c)

True mean μ	P(Type II error) β
970	0.0000
975	0.0000
980	0.0000
985	0.0022
990	0.0885
992.5	0.2743
995	0.5596
997.5	0.8159

d)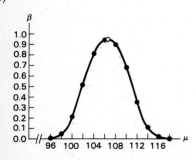

d)

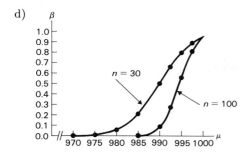

The principle being illustrated is that increasing the sample size for a hypothesis test, without changing the significance level, decreases the Type II error probabilities.

REVIEW TEST FOR CHAPTER 16

1. A literature search, because it is often possible to avoid the effort and expense of a study if someone else has already done that study and published the results.

2. a) Number the registered voters from 1 to 7246, use Table XI to obtain 50 random numbers between 1 and 7246, and take as the sample the 50 registered voters who are numbered with the numbers obtained.

   b)

6293	7075	4096	1946	5945
3331	6679	7132	3335	6562
6151	4106	0189	6417	5109
0940	6899	3094	0102	2930
1928	5236	0529	5669	0353
4124	2293	3909	4865	0538
0219	6984	1966	5203	3098
0341	1424	4204	2200	5306
0208	0315	5802	0788	1793
5470	4198	1065	6875	4830

3. a) `MTB > SAMPLE 50 from 'VOTERS'`
      `put into 'SRS'`
   b) *Note:* It is possible, depending on the Minitab system being used, that there may be insufficient storage available to carry out part (b). If that is the case, Minitab will print the following message:

   `* ERROR * Insufficient storage space`
   `for worksheet and scratch area * Erase`
   `unnecessary columns and matrices`

4.

Stratum	Freq.	Percent of total	No. to be selected
Full prof.	205	25	10
Associate prof.	328	40	16
Assistant prof.	246	30	12
Instructor	41	5	2
	820	100	40

a) The number of each rank that should be selected is displayed in the final column of the table above.

b) The procedure is as follows: Number the full professors from 1 to 205 and use Table XI to randomly select 10 of the 205 full professors; number the associate professors from 1 to 328 and use Table XI to randomly select 16 of the 328 associate professors; and so on.

5. a) A Type I error would occur if the mean daily intake of iron is not less than 18 mg, but the results of the sampling lead to the conclusion that it is. A Type II error would occur if the mean daily intake of iron is less than 18 mg, but the results of the sampling fail to lead to that conclusion.

   b) A Type I error would result in the researcher concluding that, on the average, adult females under the age of 51 get less than 18 mg of iron, when, in fact, they do not. A Type II error would result in the researcher failing to conclude that, on the average, adult females under the age of 51 get less than 18 mg of iron, when, in fact, they do.

6. a) If $\bar{x} \le 16.93$ mg, reject H_0; otherwise, do not reject H_0.

   b) 0.01

   c)

True mean μ	P(Type II error) β
15.75	0.0051
16.00	0.0212
16.25	0.0694
16.50	0.1736
16.75	0.3483
17.00	0.5596
17.25	0.7580
17.50	0.8925
17.75	0.9633

d)

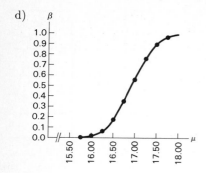

The operating characteristic curve portrays the probability of making a Type II error (that is, of not rejecting a false null hypothesis) for various values of μ. It provides a visual display of the overall effectiveness of the hypothesis test.

e)

True mean μ	Power $1 - \beta$
15.75	0.9949
16.00	0.9788
16.25	0.9306
16.50	0.8264
16.75	0.6517
17.00	0.4404
17.25	0.2420
17.50	0.1075
17.75	0.0367

f)

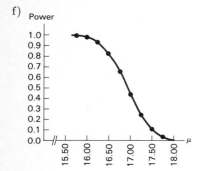

The power curve portrays the probability of not making a Type II error (that is, of rejecting a false null hypothesis) for various values of μ. It provides, as does the operating characteristic curve, a visual display of the overall effectiveness of the hypothesis test.

7. a) If $\bar{x} \le 17.20$ mg, reject H_0; otherwise, do not reject H_0.

b) 0.01

c)

True mean μ	P(Type II error) β
15.75	0.0000
16.00	0.0003
16.25	0.0029
16.50	0.0212
16.75	0.0951
17.00	0.2810
17.25	0.5596
17.50	0.8078
17.75	0.9452

d)

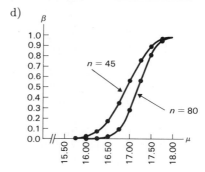

The principle being illustrated is that increasing the sample size for a hypothesis test, without changing the significance level, decreases the Type II error probabilities.

8. The general principle of experimental design is to ensure that all pertinent variables are taken into account. These include not only the variables of interest but also extraneous variables that influence the experimental results.

9. a) This experimental design does not take into account the extraneous source of variation in corn yield due to differences in farms (soil condition, climate, etc.).

b) A better method would be to use each fertilizer on all three farms. That is, on each farm, divide each field into three plots and randomly assign the three fertilizers to the three plots. This can be portrayed as shown in the figure in the first column on the following page.

Farm *A*	Farm *B*	Farm *C*
Fertilizer 3	Fertilizer 1	Fertilizer 2
Fertilizer 1	Fertilizer 2	Fertilizer 3
Fertilizer 2	Fertilizer 3	Fertilizer 1

Using this experimental design, which is a randomized block design, the fertilizers can be compared without any bias due to differences in the farms.

10. Answers will vary.

INDEX

TABLE IV
Values of χ^2_α

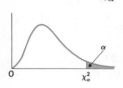

df	$\chi^2_{0.995}$	$\chi^2_{0.99}$	$\chi^2_{0.975}$	$\chi^2_{0.95}$	$\chi^2_{0.05}$	$\chi^2_{0.025}$	$\chi^2_{0.01}$	$\chi^2_{0.005}$	df
1	0.000	0.000	0.001	0.004	3.841	5.024	6.635	7.879	1
2	0.010	0.020	0.051	0.103	5.991	7.378	9.210	10.597	2
3	0.072	0.115	0.216	0.352	7.815	9.348	11.345	12.838	3
4	0.207	0.297	0.484	0.711	9.488	11.143	13.277	14.860	4
5	0.412	0.554	0.831	1.145	11.070	12.832	15.086	16.750	5
6	0.676	0.872	1.237	1.635	12.592	14.449	16.812	18.548	6
7	0.989	1.239	1.690	2.167	14.067	16.013	18.475	20.278	7
8	1.344	1.646	2.180	2.733	15.507	17.535	20.090	21.955	8
9	1.735	2.088	2.700	3.325	16.919	19.023	21.666	23.589	9
10	2.156	2.558	3.247	3.940	18.307	20.483	23.209	25.188	10
11	2.603	3.053	3.816	4.575	19.675	21.920	24.725	26.757	11
12	3.074	3.571	4.404	5.226	21.026	23.337	26.217	28.300	12
13	3.565	4.107	5.009	5.892	22.362	24.736	27.688	29.819	13
14	4.075	4.660	5.629	6.571	23.685	26.119	29.141	31.319	14
15	4.601	5.229	6.262	7.261	24.996	27.488	30.578	32.801	15
16	5.142	5.812	6.908	7.962	26.296	28.845	32.000	34.267	16
17	5.697	6.408	7.564	8.672	27.587	30.191	33.409	35.718	17
18	6.265	7.015	8.231	9.390	28.869	31.526	34.805	37.156	18
19	6.844	7.633	8.907	10.117	30.144	32.852	36.191	38.582	19
20	7.434	8.260	9.591	10.851	31.410	34.170	37.566	39.997	20
21	8.034	8.897	10.283	11.591	32.671	35.479	38.932	41.401	21
22	8.643	9.542	10.982	12.338	33.924	36.781	40.289	42.796	22
23	9.260	10.196	11.689	13.091	35.172	38.076	41.638	44.181	23
24	9.886	10.856	12.401	13.848	36.415	39.364	42.980	45.558	24
25	10.520	11.524	13.120	14.611	37.652	40.646	44.314	46.928	25
26	11.160	12.198	13.844	15.379	38.885	41.923	45.642	48.290	26
27	11.808	12.879	14.573	16.151	40.113	43.194	46.963	49.645	27
28	12.461	13.565	15.308	16.928	41.337	44.461	48.278	50.993	28
29	13.121	14.256	16.047	17.708	42.557	45.722	49.588	52.336	29
30	13.787	14.953	16.791	18.493	43.773	46.979	50.892	53.672	30

CRITICAL
VALUES